Grundlehren der mathematischen Wissenschaften 292

A Series of Comprehensive Studies in Mathematics

Editors

M. Artin S. S. Chern J. Coates J. M. Fröhlich
H. Hironaka F. Hirzebruch L. Hörmander S. MacLane
C. C. Moore J. K. Moser M. Nagata W. Schmidt
D. S. Scott Ya. G. Sinai J. Tits M. Waldschmidt
S. Watanabe

Managing Editors

M. Berger B. Eckmann S. R. S. Varadhan

Masaki Kashiwara   Pierre Schapira

Sheaves on Manifolds

With a Short History
«Les débuts de la théorie des faisceaux»
By Christian Houzel

Springer-Verlag
Berlin Heidelberg NewYork London
Paris Tokyo Hong Kong Barcelona

Masaki Kashiwara
Research Institute for Mathematical Sciences
Kyoto University
Kyoto 606, Japan

Pierre Schapira
Department of Mathematics
Paris-Nord University
F-93430 Villetaneuse, France

Mathematics Subject Classification (1980): 58G 32B 14F 18E 18F

Second Reprint 2002

ISBN 3-540-51861-4 Springer-Verlag Berlin Heidelberg NewYork
ISBN 0-387-51861-4 Springer-Verlag NewYork Berlin Heidelberg

Springer-Verlag Berlin Heidelberg New York
a member of BertelsmannSpringer Science+Business Media GmbH

Printed in Germany

Typesetting: ASCO Trade Typesetting Ltd., Hong Kong
41/3111 – 54321 – Printed on acid-free paper

Preface

For a long time after its introduction by Leray, sheaf theory was mainly applied to the theory of functions of several complex variables or to algebraic geometry, until it became a basic tool for almost all mathematicians, and cohomology a natural language for many people.

However, while there exists an extensive literature dealing with cohomology of sheaves (e.g. the famous book by Godement) or even with derived functors, there are in fact very few books developing sheaf theory within the beautiful framework of derived categories although its necessity is becoming more and more evident. Most of the constructions of the theory take on their full strength in this context, or even, do not make sense outside of it. This is particularly evident for the Poincaré-Verdier duality, which appeared in the sixties, as well as for the Sato microlocalization, introduced in 1969, which is only beginning to be fully understood.

Since the seventies, other fundamental ideas have emerged and sheaf theory (on manifolds) naturally includes the "microlocal" point of view. Our aim is to present here a self-contained work, starting from the beginning (derived categories and sheaves), dealing in detail with the main features of the theory, such as duality, Fourier transformation, specialization and microlocalization, microsupport and contact transformations, and also to give two main applications. The first of these deals with real analytic geometry, and includes the concepts of constructible sheaves, subanalytic cycles, Euler-Poincaré indices, Lefschetz formula, perverse sheaves, etc. The second one is the theory of linear partial differential equations, including D-modules, microfunctions, elliptic and microhyperbolic systems, and complex quantized contact transformations.

With this book we hope to illustrate the deep links that tie together branches of mathematics at first glance seemingly disconnected, such as for example here, algebraic topology and linear partial differential equations. At the same time, we want to emphasize the essentially geometrical nature of the problems encountered (most obvious in the involutivity theorem for sheaves), and to show how efficient the algebraic tools introduced by Grothendieck are in solving them, even for an analyst.

Of course, many important applications of the theory are just touched upon, such as for instance the theory of microdifferential systems (complete monographs on the topic are however available now), others are simply omitted, such as representation theory and equivariant sheaf theory.

Finally, we want to express our thanks to C. Houzel who agreed to write a short history of sheaf theory, to L. Illusie who helped us when preparing the "Historical Notes", to those who went through various parts of the book and made constructive comments, especially E. Andronikof, A. Arabia, J-M. Delort, E. Leichtnam and J-P. Schneiders, and also to Catherine Simon at Paris-Nord University and the secretarial staff of the RIMS at Kyoto, who had the patience to type the manuscripts.

May 1990 M. Kashiwara and P. Schapira

Table of contents

Introduction

The aim of this book is to give a self-contained exposition of sheaf theory.

Sheaves were created during the last world war by Jean Leray, while he was a prisoner of war in a German camp.

The purpose of sheaf theory is quite general: it is to obtain global information from local information, or else to define "obstructions" which characterize the fact that a local property does not hold globally any more: for example a manifold is not always orientable, or a differential equation can be locally solvable, but not globally. Hence, sheaf theory is a wide generalization of a part of algebraic topology (e.g. singular homology) which corresponds to *constant sheaves* or, more generally, to *locally constant sheaves*. There are many natural examples of sheaves, such as orientation sheaves, sheaves of differentiable or holomorphic functions, sheaves of solutions of systems of differential equations, constructible sheaves obtained as direct images, etc.

It was not clear at the beginning however whether such a general theory could have any application, until it was successfully applied to the theory of functions of several complex variables and one can imagine that the original work of Leray would have remained far from accessible without the substantial work developed in the fifties by Cartan, Serre, and later Grothendieck (cf. the Short History by Houzel, below).

Sheaf theory takes on its full strength when combined with the tools of homological algebra. In fact, Leray also introduced the notion of spectral sequences which, together with that of derived functors of Cartan-Eilenberg, leads naturally to the theory of derived categories, due to Grothendieck. After having been long reserved to some specialists of algebraic geometry, the theory of derived categories began to be fully recognized as a basic tool of mathematics. In particular, it would certainly not have been possible without it to give such a beautiful generalization of Poincaré's duality, as did Verdier in the sixties (after related work of Grothendieck in the framework of étale cohomology), or to treat systematically what Grothendieck calls "the six operations" on sheaves, that is, the functors Rf_*, f^{-1}, $Rf_!$, $f^!$, $\otimes^L$, and $R\mathcal{H}om$. As we shall see, this formalism leads to deep and powerful formulas which interpret, in a general context, classical results. We have already mentioned the Poincaré duality, but there are many other topics such as the Lefschetz fixed point formula or the Euler-Poincaré index. Of course these functors are an abstract version of classical operations on functions: direct image for integration, inverse image for composition, tensor product for product. (These three operations give six operations by "duality", an operation which has no counterpart for functions, but which we shall introduce here for constructible functions.)

At this stage, we have briefly explained what one could roughly call "the classical theory of sheaves". But a new and fundamental idea, due to Mikio Sato, appeared in 1969, which was to give a new perspective to sheaf theory, namely "the microlocal point of view", and indeed it is one of the aims of this book to develop sheaf theory within this new framework. Sato's main interest was the study of analytical singularities of solutions of systems of linear differential equations. Already in 1959 he had used local cohomology to define the sheaf of hyperfunctions and to interpret them as sums of boundary values of holomorphic functions. Ten years later he introduced the sheaf of microfunctions, to recognize "from which direction the boundary values come". To perform this, Sato introduced the functor ν_M of *specialization* (along a submanifold M of a manifold X), and its *Fourier transform* the functor μ_M of *microlocalization*. These functors send the derived categories of sheaves on X to the derived category of sheaves on the normal and conormal bundles to M in X respectively, and they allow us to analyze precisely a sheaf on a neighborhood of M, taking into account all normal (or conormal) directions to M.

When trying to apply Sato's theory to the study of microhyperbolic systems, the present authors gradually realized that the only information they were using was the characteristic variety of the system (in the cotangent bundle to a complex manifold X), and that it was possible to forget the complex structure of X and even the fact that the subject was partial differential equations. What they were doing was nothing more than non-characteristic deformations (of the complex of holomorphic solutions sheaves of the system), in the "non-forbidden" directions, that is, across non-characteristic real hypersurfaces. Note that these techniques of non-characteristic deformation had already appeared before, but here the authors were dealing with micro-differential systems, and really needed *microlocal geometry*. In particular, they introduced the γ-topology, a kind of microlocal cut-off. Later, in 1982, by abstracting their previous work on microhyperbolic systems, they introduced the notion of micro-support of a sheaf: roughly speaking, a point p of the cotangent bundle T^*X to a real manifold X does not belong to $SS(F)$, the micro-support of a sheaf F, if F has no cohomology supported by half-spaces whose conormals are close to p. This new definition allows us to study sheaves "microlocally" and, in particular, to make contact transformations (the natural transformations on cotangent bundles) operate on sheaves, similarly as quantized contact transformations operate on microfunctions in Sato-Kawai-Kashiwara [1] or on Fourier distributions in Hörmander [2].

The idea of micro-support is closely related to Morse theory. As is well-known, if ϕ is a real function on a compact real manifold X, the cohomology groups of the spaces $\{x \in X; \phi(x) < t\}$ are isomorphic as long as t does not meet a critical value of ϕ. If one considers the similar problem for the cohomology groups of a sheaf F on X, one obtains the corresponding result with the help of the micro-support of F. For complex manifolds, the micro-support of constructible sheaves may also be described using the *vanishing-cycle functor* of Grothendieck-Deligne. Notice that the vanishing-cycle functor appears as a particular case of Sato's microlocalization functor.

The micro-support has a deep geometrical meaning, and we shall prove that this set is involutive. This gives in particular a purely real and geometric proof of the classical corresponding result for the characteristic variety of systems of differential equations.

As we shall see all along in this book, the microlocal point of view on sheaves deepens the theory of sheaves and leads to many applications. Let us discuss here only a few examples.

(a) Many morphisms in sheaf theory become isomorphisms if some microlocal condition is satisfied. Consider for example a morphism of manifolds $f: Y \to X$, and let F be a sheaf (or better, a complex of sheaves) on X. Then there exists a canonical morphism

$$\omega_{Y/X} \otimes f^{-1}F \to f^!F \;, \tag{i.1}$$

($\omega_{Y/X}$ is the relative dualizing complex). It is well-known that this morphism is an isomorphism if f is smooth, but in fact there is a stronger result: this morphism is an isomorphism as soon as "f is non-characteristic for F"; (if f is a closed embedding, this means that the intersection of $SS(F)$ and the conormal bundle to Y in X is contained in the zero-section).

(b) Let X be a complex manifold, and $\mathcal{M}$ a system of linear differential equations on X, that is, a left coherent $\mathcal{D}_X$-module, where $\mathcal{D}_X$ denotes the sheaf of rings of holomorphic differential operators on X. Let

$$F = R\mathcal{H}om_{\mathcal{D}_X}(\mathcal{M}, \mathcal{O}_X) \tag{i.2}$$

be the complex of sheaves of holomorphic solutions of the system. The geometrical situation of linear partial differential equations with analytic coefficients is neatly expressed by the formula below, whose proof relies basically on the Cauchy-Kowalevski theorem:

$$SS(F) = \mathrm{char}(\mathcal{M}) \;. \tag{i.3}$$

Here $\mathrm{char}(\mathcal{M})$ denotes the characteristic variety of the system.

If one applies the result of (a) to the case where Y is a real analytic manifold and X a complexification of Y, then one finds that if $\mathcal{M}$ is elliptic, the complex of real analytic solutions and the complex of hyperfunction solutions of $\mathcal{M}$ are isomorphic. Hence, we get a purely sheaf-theoretical proof of a classical result of analysis.

(c) A complex of sheaves F on a real analytic manifold is weakly constructible if there exists a subanalytic stratification of X such that all cohomology groups are locally constant on the strata. This condition will be shown to be equivalent to a microlocal one, namely that the micro-support of F is a subanalytic Lagrangian subset of T^*X. Moreover if some finiteness condition is satisfied, one can associate to F a *Lagrangian cycle* supported by SS(F), and one obtains the global Euler-Poincaré index of F on X as the intersection number of this cycle with a cycle naturally associated to the zero-section. By this result, we have an efficient microlocal tool to calculate indices.

In this book, we hope to convince the reader that this point of view is crucially relevant. Starting from the beginning (derived categories), we shall present both the classical theory ("the six operations") and the microlocal theory (microlocalization, micro-support, contact transformations). Then we shall apply the machinery to the study of constructible sheaves on real manifolds, and finally deal briefly with its applications to linear partial differential equations.

In more detail, the contents of the book are as follows.

Chapter I contains the basic facts about homological algebra which are necessary for the rest of the book and in particular the theory of derived categories (with the exception of the notion of t-structures, postponed until Chapter X).

Chapters II and III contain the "classical" notions on sheaves in the language of derived categories, including the six operations, as well as the Fourier-Sato transformation, which interchanges sheaves (more precisely, objects of the derived category of sheaves) on a vector bundle, and sheaves on the dual vector bundle.

Chapter IV is devoted to microlocalization. After recalling the geometric construction of the normal deformation of a submanifold M in a manifold X, we define the specialization functor ν_M, which sends sheaves on X to sheaves on the normal bundle T_MX, and its Fourier-Sato transform, the microlocalization functor μ_M. We also define a natural generalization of μ_M, the functor μhom and we study the functorial properties of all these functors.

In **Chapter V** we introduce the micro-support of sheaves. After proving a global extension theorem for sheaves in terms of the geometry of their micro-support, we make use of the γ-topology to cut-off sheaves "microlocally". Then we study the behavior of the micro-support under the functorial operations in the non-characteristic case (for inverse images) or proper case (for direct images). As an application, *the Morse inequalities* for sheaves are obtained.

In **Chapter VI**, we use the micro-support to localize the derived category of sheaves $\mathbf{D}^b(X)$ with respect to a subset Ω of T^*X, (one obtains new triangulated categories, $\mathbf{D}^b(X;\Omega)$), and to define "microlocal" inverse or direct images. Next, we extend the results of Chapter V to the general case by studying the behavior of the micro-support with respect to the functorial operations. In particular, we prove that the micro-support of $\mu_M(F)$ is contained in the normal cone of SS(F) along T^*_MX. This inequality is a sheaf-theoretical version of a theorem on micro-hyperbolic systems, and will be used all along the book. This chapter also contains a crucial result, *the involutivity theorem* for micro-supports.

In **Chapter VII** we perform contact transformations for sheaves. If χ is a contact transformation between two open subsets Ω_X and Ω_Y of T^*X and T^*Y then, under suitable conditions, one can construct an isomorphism $\mathbf{D}^b(X;\Omega_X) \simeq$

$\mathbf{D}^b(Y;\Omega_Y)$, and this isomorphism is compatible with μhom. When calculating the image of the constant sheaf A_M on a submanifold M of X, one is led naturally to the notion of *pure sheaves* along a smooth Lagrangian manifold. A pure sheaf is "generically" (and microlocally) isomorphic to some sheaf $L_M\ [d]$ for some A-module L and some shift d, but the calculation of the shift requires the whole machinery of the inertia index of a triplet of Lagrangian planes. In this chapter we calculate in particular the shift of the microlocal composition of kernels.

In **Chapter VIII**, we make a detailed study of constructible sheaves on real manifolds. We introduce the (microlocal) notion of a μ-stratification, and then prove that a sheaf is weakly constructible if and only if its micro-support is subanalytic and isotropic (hence, Lagrangian). Then we can apply the preceding results to study the functorial operations on constructible sheaves. On a complex manifold, we prove that a sheaf is constructible if it is so on the real underlying manifold, and moreover if its micro-support is invariant by the action of $\mathbb{C}^\times$. From this, we deduce a theorem for non-proper direct images. Finally, we show that the nearby-cycle and the vanishing-cycle functors are particular case of the specialization and the microlocalization functors.

The notions of subanalytic chains and cycles are introduced in **Chapter IX**, with the help of the dualizing complex. Then, using the functor μhom, we associate to a constructible sheaf F its characteristic cycle and we show that the intersection number of this Lagrangian cycle with the zero-section of T^*X gives the global Euler-Poincaré index of F on X. We also calculate local Euler-Poincaré indices, and make the link between Lagrangian cycles and constructible functions on X, thus obtaining a new calculus on these functions. In this chapter, we also give a Lefschetz fixed point formula for constructible sheaves.

Chapter X develops the theory of perverse sheaves. After recalling the notion of t-structures, we define the perverse sheaves (i.e. perverse complexes) on a real manifold, and show that they form an abelian category. Then we study perverse sheaves on complex manifolds, give a microlocal characterization of perversity, and prove that perversity is preserved by various functorial operations.

In **Chapter XI**, we show briefly how to apply the theory of sheaves to the study of systems of linear partial differential equations. After a short review of the theory of $\mathcal{O}_X$ and $\mathcal{D}_X$-modules, we prove one inclusion in (i.3), and deduce that the complex of holomorphic solutions of a holonomic $\mathcal{D}_X$-module is perverse. We also introduce the sheaves of hyperfunctions and microfunctions, and deduce from (i.3) some basic results on the microfunction solutions of elliptic or hyperbolic systems. In the course of this chapter, we also make quantized contact transformations operate on the sheaf $\mathcal{O}_X$. This result has many important applications which shall not be discussed here.

We end this book with a short **Appendix** in which we collect all results (with some proofs) that we need on symplectic geometry, especially on the inertia index.

Each chapter opens with a short introduction, and includes exercises of varying difficulty. Some of these exercises (especially in Chapter I) are auxiliary results used in the course of the book. In general, the proof of such exercises is straightforward, otherwise a hint is given.

We have as far as possible avoided giving bibliographical references within the text. Instead, we have chosen to end each chapter with a few historical comments. The reason is that most of the time a theorem has a long and complicated history, and it would be tedious to quote each time everyone who contributed to a result. On the other hand, it seems improper to quote only the person who has initiated the subject or who has put it into final form.

The origin and the beginnings of sheaf theory are rather intricate, and this book benefits from the historical work of Christian Houzel who has agreed to contribute a detailed account of this part of the history.

A Short History: Les débuts de la théorie des faisceaux

by Christian Houzel

1. Le cours de Leray (1945)

Pendant qu'il était prisonnier de guerre à l'Oflag XVII en Autriche, Jean Leray a fait un cours de topologie algébrique à l'Université de captivité qu'il avait contribué à organiser. C'est un sujet qu'il avait déjà abordé en 1934 dans son article avec J. Schauder sur l'extension en dimension infinie de la notion de degré d'application et du théorème du point fixe de Brouwer [33]. Leray avait besoin d'un tel théorème dans des espaces fonctionnels pour obtenir l'existence de solutions des équations non linéaires rencontrées en hydrodynamique (pour lesquelles les solutions ne sont pas nécessairement régulières ni uniques).

Le cours de Leray a été publié à la fin de la guerre en 1945 dans le Journal de Liouville [29]. La topologie algébrique y est développée sur des bases nouvelles, évitant les hypothèses d'orientabilité ou de linéarité locale et les méthodes de subdivision ou d'approximation simpliciale. L'accent est mis sur la cohomologie, qui n'avait été clairement distinguée de l'homologie que juste avant la guerre [47], en particulier après les travaux de de Rham [37]; la cohomologie d'un espace à coefficients dans un anneau a toujours une structure multiplicative, et la structure multiplicative en homologie, avec laquelle on travaillait dans le cas des variétés compactes orientables, se déduit de celle de la cohomologie par dualité de Poincaré. Leray rebaptise "homologie" la cohomologie et parle de "groupes de Betti" quand il s'agit de l'homologie. Pour définir la cohomologie d'un espace topologique E à coefficients dans un anneau A, il s'inspire du procédé de Čech [9], mais il remplace la notion ensembliste de recouvrement par une notion mieux adaptée à la topologie algébrique: celle de couverture. Pour définir une couverture on se donne d'abord un "complexe abstrait", suite de groupes commutatifs libres de type fini correspondant aux diverses dimensions p et munis de bases $(X^{p\alpha})_\alpha$, avec la donnée d'un opérateur cobord $X^{p\alpha} \mapsto \dot{X}^{p\alpha}$ élément du groupe de dimension $p+1$ (on étend par linéarité aux autres éléments du groupe), soumis à l'axiome que le cobord d'un cobord est nul. On rend "concret" le complexe en associant à chaque $X^{p\alpha}$ un support $|X^{p\alpha}|$, partie non vide de E; on impose l'axiome que si $X^{q\beta}$ est adhérent à $X^{p\alpha}$ (c'est-à-dire lui est relié par une suite finie d'éléments de base dont chacun intervient dans le cobord du précédent), le support de $X^{q\beta}$ est contenu dans celui de $X^{p\alpha}$. Le complexe concret K est une couverture si les supports sont fermés, pour tout point x de E, le sous-complexe formé par les éléments dont le support contient x est un simplexe (sa cohomologie est triviale) et la somme K^0 des éléments de dimension 0 est un cocycle (dit (co-)cycle unité). Les classes de cohomologie de E à coefficients dans A sont celles des formes L^p de couvertures quelconques K de E (combinaisons linéaires à coefficients dans A des éléments de base de K en dimension p), en convenant d'identifier L^p avec "l'intersection" $L^p.K'^0$ chaque fois que K' est une autre couverture (on définit l'intersection à l'aide du complexe produit tensoriel; le support de $X^{p\alpha} \otimes X'^{q\beta}$ est l'intersection des supports de $X^{p\alpha}$ et de $X'^{q\beta}$ et on passe au quotient en annulant les éléments de support vide). Dans le cas d'un espace normal, on peut calculer la cohomologie en prenant seulement les couvertures d'une famille stable par intersection et contenant, pour tout recouvrement ouvert fini ρ de E, une couverture dont les supports sont "ρ-petits"; dans le cas d'un espace compact, on peut se contenter d'une seule couverture si ses supports sont "simples" (c'est-à-dire cohomologiquement triviaux). Leray étend les résultat de Hopf [23] sur certaines variétés orientables compactes au cas d'espaces topologiques compacts. Il développe la théorie de la dualité permettant de récupérer les groupes de Betti. La première partie de son cours se termine sur l'introduction du nombre de Lefschetz d'une application continue de E dans lui-même, dans le cas où E est compact connexe et admet un recouvrement fini "convexoïde" (c'est-à-dire par des fermés simples dont les intersections sont vides ou simples).

Dans la deuxième partie, Leray compare le nombre de Lefschetz de $\xi : E \to E$ à celui de la restriction de ξ à un fermé stable par ξ. Il introduit les "pseudo-cycles" (éléments de la limite projective des cohomologies de parties B compactes de E) pour étendre au cas non compact des résultats démontrés précédemment dans le cas compact. Il définit des couvertures ("dallages") à partir de décompositions cellulaires de variétés différentiables et il établit la dualité de Poincaré dans le cas orientable et sans bord. La troisième partie, qui est l'origine du travail de Leray en topologie algébrique, définit l'indice total $i(O)$ des solutions d'une équation $x = \xi(x)$ dans un ouvert O de E (où $\xi : F \to E$ est une application continue définie dans un fermé F qui contient $\bar{O}$); on suppose E "convexoïde" (compact connexe et admettant un recouvrement par des fermés simples dont les intersections finies sont vides ou simples et dont les intérieurs forment une base de la topologie) et on considère la cohomologie à coefficients entiers. L'indice total $i(O)$ est défini si l'équation considérée n'a aucune solution sur la frontière de O et il ne dépend que de la restriction de ξ à $\bar{O}$; il est invariant par homotopie (sur ξ) et est égal au nombre de Lefschetz dans le cas où $O = E$. Si toutes les solutions de l'équation dans O appartiennent à une réunion d'ouverts disjoints O_α contenus dans O, $i(O)$ est la somme des $i(O_\alpha)$. Dans le cas d'une solution isolée x, on définit son indice comme l'indice total $i(V)$ où V est un voisinage assez petit de x, et si O ne contient que des solutions isolées, $i(O)$ est la somme des indices de ces solutions. Un théorème d'unicité s'obtient dans la théorie des sillages avec grande résistance au courant en établissant que toute solution de l'équation est isolée et d'indice 1. Leray définit encore un indice pour des équations d'une forme un peu plus générale, et il applique sa théorie à l'équation de Fredholm. Il considère aussi le cas d'équations de la forme $x = \xi(x, x')$ où $\xi : F \to E$ est une application continue définie dans un fermé F de $E \times E'$ (E supposé convexoïde simple); l'indice est remplacé par une opération de projection sur E' des $Z^p.O$, Z^p pseudocycle de $E \times E'$.

2. La théorie des faisceaux et la suite spectrale

Les recherches de Leray en topologie algébrique pendant sa captivité l'avaient mené beaucoup plus loin. L'étude des relations entre l'homologie d'un espace fibré et celles de sa base et de sa fibre rendait nécessaire l'introduction de nouveaux outils: cohomologie à coefficients locaux, variant d'un point à un autre; calcul de la cohomologie par une suite d'approximations. Plus généralement, ces outils devaient servir, dans le cas d'une application continue $\xi : X \to Y$, à étudier la cohomologie de X à partir de celles de Y et des fibres de ξ; ce genre d'étude avait servi à Picard pour déterminer l'homologie des surfaces algébriques complexes [36], et Lefschetz avait étendu cette méthode aux dimensions supérieures [28]. L'idée des coefficients locaux était d'ailleurs venue indépendamment, sous la forme particulière des systèmes locaux de coefficients, à N. Steenrod [41].

Dans [31], qui reproduit ses cours au Collège de France des années 1947–48 et 1949–50, Leray appelle faisceau sur un espace topologique localement compact X la donnée, pour tout fermé F de X, d'un anneau $\mathscr{B}(F)$, et, pour toute inclusion $F_1 \subset F$ de fermés de X, d'un homomorphisme de "section" $b \mapsto F_1 b$ de $\mathscr{B}(F)$ dans $\mathscr{B}(F_1)$. Il impose que $\mathscr{B}(\emptyset) = 0$ et la transitivité de l'opération de section: $F_2(F_1 b) = F_2 b$ si $F_2 \subset F_1 \subset F$. Un faisceau est "continu" si la limite inductive des $\mathscr{B}(W)$, W voisinage fermé de ∞, est nulle et si, pour tout fermé F, $\mathscr{B}(F)$ est limite inductive des $\mathscr{B}(V)$ où V parcourt les voisinages fermés de $F \cup \infty$; il est dit propre s'il vérifie la première de ces deux conditions, la deuxième pour F non compact, et si $\mathscr{B}(K)$ est limite inductive des $\mathscr{B}(V)$, V voisinage fermé de K pour K compact. A côté des faisceaux, Leray introduit, comme dans son cours de 1945, les complexes et les couvertures; mais, selon une suggestion d'Henri Cartan [5], les complexes n'ont plus nécessairement de base, et ils sont munis d'une multiplication: ce sont des anneaux différentiels avec une loi associant à chaque élément k un fermé $S(k)$ de X appelé support de l'élément. Etant donnés un complexe $\mathscr{K}$ sur X et un fermé F de X, la section $F\mathscr{K}$ est le complexe sur F quotient de $\mathscr{K}$ par l'idéal des éléments dont le support ne rencontre pas F; on obtient ainsi un faisceau $\mathscr{B} : F \mapsto F\mathscr{K}$, dit associé à $\mathscr{K}$. Un complexe $\mathscr{K}$ est une couverture de X s'il est sans torsion, gradué en degrés $\geqslant 0$ avec un cobord de degré 1, et s'il possède un élément unité u de support X et tel que la cohomologie de $x\mathscr{K}$ se réduise aux multiples de xu quel que soit le point x de X. Pour définir la cohomologie de X à coefficients dans un faisceau propre $\mathscr{B}$, ou même l'hypercohomologie dans le cas où $\mathscr{B}$ est muni d'une dérivation qui en fait un faisceau différentiel, Leray utilise une couverture *fine* au lieu de se servir de toutes les couvertures comme dans le cours de 1945; un complexe $\mathscr{K}$ est dit fin si, pour tout recouvrement ouvert fini (V_ν) du compactifié $X \cup \infty$, on peut décomposer l'automorphisme identique de $\mathscr{K}$ en une somme d'applications linéaires λ_ν: $\mathscr{K} \to \mathscr{K}$ telles que $S(\lambda_\nu k) \subset \bar{V}_\nu \cap S(k)$ pour tout élément k de $\mathscr{K}$ et tout indice ν. L'existence de couvertures fines sur un espace localement compact X de dimension finie est assurée par la construction de Čech ou celle d'Alexander [1]; si X est de dimension n, on peut choisir la couverture fine nulle en degré $>n$. Si $\mathscr{X}$ est une couverture fine de X et $\mathscr{B}$ un faisceau différentiel propre sur X, on définit un anneau produit tensoriel $\mathscr{X} \otimes \mathscr{B}$, engendré par les $k \otimes b$ où $k \in \mathscr{K}$, $b \in \mathscr{B}(F)$ (F fermé de X) et $S(k) \subset F$; le support d'un élément $\sum k_\mu \otimes b_\mu$ est l'ensemble des points x tels que $x \sum k_\mu \otimes b_\mu$ soit non nul (si $x \in S(k)$, on convient que $x(k \otimes b) = xk \otimes xb$, sinon on le pose nul). Le complexe $\mathscr{X} \circ \mathscr{B}$ est le quotient de $\mathscr{X} \otimes \mathscr{B}$ par l'idéal des éléments à support vide (avec la même loi de supports). Leray montre que la cohomologie de $\mathscr{X} \circ \mathscr{B}$ ne dépend pas du choix de la couverture fine $\mathscr{X}$, et il note $\mathscr{H}(X \circ \mathscr{B})$ cette cohomologie.

Pour établir que la cohomologie d'un espace normal se calcule avec les couvertures d'une famille stable par intersection et dans laquelle il y a, pour tout recouvrement ouvert ρ, une couverture à supports ρ-petits, Leray se servait, dans son cours de 1945, d'un lemme fondamental d'après lequel $K^*.C'$ et C' ont même cohomologie si K^* est une couverture et C' un complexe tels que $K^*.e$ soit un simplexe pour tout support e de C'. En analysant la démonstration de ce lemme, il a été conduit à la notion d'*anneau spectral* et à l'introduction des anneaux filtrés (avec une filtration entière décroissante), qu'il appelait d'abord sous-valués (le terme "filtré" est dû à H. Cartan [3]). A un anneau différentiel filtré $\mathscr{A}$ (cf. Koszul [26]) est associée une suite $(\mathscr{H}_r\mathscr{A})_r$ d'anneaux différentiels gradués, où la différentielle de $\mathscr{H}_r\mathscr{A}$ est de degré r. Elle est définie de la manière suivante: on note $\mathscr{A}^{(p)}$ l'ensemble des éléments de $\mathscr{A}$ de filtration $\geqslant p$ et on pose $\mathscr{C}^p = \mathscr{C} \cap \mathscr{A}^{(p)}$ et $\mathscr{D}^p = \mathscr{D} \cap \mathscr{A}^{(p)}$, où $\mathscr{C}$ est l'ensemble des cocycles de $\mathscr{A}$ et $\mathscr{D}$ celui des cobords. On désigne ensuite par $\mathscr{C}_r^p$ l'ensemble des $a \in \mathscr{A}^{(p)}$ dont le cobord est de filtration $\geqslant p + r$ et par $\mathscr{D}_r^p$ l'ensemble des cobords δa pour $a \in \mathscr{C}_r^{p-r}$. Ces groupes apparaissent comme les approximations d'ordre r de $\mathscr{C}^p$ et $\mathscr{D}^p$ respectivement; on voit que $\mathscr{C}_r^p$ (resp. $\mathscr{D}_r^p$) est fonction décroissante (resp. croissante) de r et qu'il contient $\mathscr{C}^p$ (resp. est contenu dans $\mathscr{D}^p$). En degré p, $\mathscr{H}_r\mathscr{A}$ est défini comme le quotient $\mathscr{C}_r^p/(\mathscr{C}_{r-1}^{p+1} + \mathscr{D}_{r-1}^p)$, et le cobord $\delta_r h_r^{[p]}$ de la classe $h_r^{[p]}$ d'un $c_r^p \in \mathscr{C}_r^p$ dans ce quotient est, par définition, la classe de $\delta c_r^p \; mod. (\mathscr{C}_{r-1}^{p+1} + \mathscr{D}_{r-1}^p)$. On vérifie que la cohomologie de $\mathscr{H}_r\mathscr{A}$ s'identifie canoniquement à $\mathscr{H}_{r+1}\mathscr{A}$; ainsi, pour $s \geqslant r$, chaque $\mathscr{H}_s\mathscr{A}$ s'identifie à un sous-quotient de $\mathscr{H}_r\mathscr{A}$. Il en est de même de l'anneau gradué $\mathscr{H}_\infty\mathscr{A}$ valant $\mathscr{C}^p/(\mathscr{C}^{p+1} + \mathscr{D}^p)$ en degré p, qui est isomorphe au gradué associé à l'anneau de cohomologie $\mathscr{H}\mathscr{A}$ (muni de la filtration provenant de celle de $\mathscr{A}$). L'anneau spectral $(\mathscr{H}_r\mathscr{A})$ identifie ce gradué à un sous-anneau de la "limite inductive" des $\mathscr{H}_r\mathscr{A}$ et à cette limite elle-même lorsque la filtration de $\mathscr{A}$ est bornée supérieurement; si, de plus, la différentielle de $\mathscr{H}_r\mathscr{A}$ s'annule pour $r > l$, la suite $\mathscr{H}_r\mathscr{A}$ est stationnaire pour ces valeurs de r, et le gradué étudié s'identifie à $\mathscr{H}_r\mathscr{A}$. On voit ainsi que $\mathscr{H}\mathscr{A}$ est isomorphe à $\mathscr{H}_r\mathscr{A}$ pour $r > l$ si la filtration de $\mathscr{A}$ est bornée supérieurement et que $\mathscr{H}_{l+1}\mathscr{A}$ est concentré en degré 0. L'anneau spectral dépend fonctoriellement de $\mathscr{A}$; si un homomorphisme $\lambda: \mathscr{A}' \to \mathscr{A}$ d'anneaux filtrés à filtrations bornées supérieurement induit un isomorphisme de $\mathscr{H}_{l+1}\mathscr{A}'$

sur $\mathscr{H}_{l+1}\mathscr{A}$, il induit un isomorphisme respectant la filtration de $\mathscr{H}\mathscr{A}'$ sur $\mathscr{H}\mathscr{A}$.

L'anneau $\mathscr{A}$ étant toujours différentiel filtré, Leray considère encore un anneau différentiel gradué $\mathscr{K}$ (avec une différentielle de degré 1) et le produit tensoriel filtré $\mathscr{K}^l \otimes \mathscr{A}$ où les éléments de degré p de $\mathscr{K}$ sont considérés comme de filtration lp (l entier donné). Lorsque $\mathscr{K}$ est sans torsion $\mathscr{H}_{l+1}(\mathscr{K}^l \otimes \mathscr{A})$ est isomorphe (canoniquement) à la cohomologie de $\mathscr{K}^l \otimes \mathscr{H}_l\mathscr{A}$; on le voit en remarquant que le remplacement de la différentielle de $\mathscr{A}$ par 0 ne change pas $\mathscr{H}_l(\mathscr{K}^l \otimes \mathscr{A})$; ceci permet, en utilisant des résultats de Cartan, d'établir, sous des hypothèses convenables, une formule de Künneth.

Si $\mathscr{B}$ est un faisceau différentiel filtré propre sur un espace X, $\mathscr{X}$ une couverture fine de X et l un entier, l'anneau spectral $(\mathscr{H}_r(\mathscr{X}^l \circ \mathscr{B}))$ et la cohomologie $\mathscr{H}(\mathscr{X}^l \circ \mathscr{B})$ ne dépendent pas du choix de $\mathscr{X}$, et Leray les note respectivement $(\mathscr{H}_r(X^l \circ \mathscr{B}))$ et $\mathscr{H}(X^l \circ \mathscr{B})$; en particulier $\mathscr{H}_{l+1}(X^l \circ \mathscr{B})$ est isomorphe à la cohomologie de $X^l \circ \mathscr{F}_l\mathscr{B}$, où $(\mathscr{F}_r\mathscr{B})$ est le faisceau spectral associé à $\mathscr{B}$. Lorsque $\mathscr{B}$ est différentiel gradué propre, à degré borné inférieurement, avec une différentielle de degré > 0 et tel que, pour tout $x \in X$, $\mathscr{B}(x)$ soit concentré en degré 0, on en déduit, en prenant $l = 1$, que $\mathscr{H}(X \circ \mathscr{B}) = \mathscr{H}(X \circ \mathscr{F}\mathscr{B})$ où $\mathscr{F}\mathscr{B}$ est le faisceau de cohomologie de $\mathscr{B}$. Si (F_μ) est un recouvrement fermé fini, et si $\mathscr{K}^*$ est le complexe (libre) de Čech associé, l'hypothèse que $\mathscr{H}(F \circ \mathscr{B}) = \mathscr{H}\mathscr{B}(F)$ pour tout F intersection non vide des F_μ implique que $\mathscr{H}(\mathscr{K}^* \otimes \mathscr{B}) = \mathscr{H}(X \circ \mathscr{B})$.

A une application continue $\xi : X \to Y$ entre espaces localement compacts, Leray associe, pour tout faisceau différentiel filtré propre $\mathscr{B}$ sur X et tout couple d'entiers (l, m) avec $l < m$, un anneau spectral $\mathscr{H}_r(\xi^{-1} Y^m \circ X^l \circ \mathscr{B})$, défini au moyen de couvertures fines $\mathscr{X}$ de X et $\mathscr{Y}$ de Y; la couverture $\xi^{-1}\mathscr{Y}$ est le quotient de $\mathscr{Y}$ par l'idéal des éléments y tels que $\xi^{-1}(S(y))$ soit vide, et $\xi^{-1}\mathscr{Y}^m \circ \mathscr{X}^l$ est une couverture fine de X. Ainsi $\mathscr{H}(\xi^{-1}\mathscr{Y}^m \circ \mathscr{X}^l \circ \mathscr{B})$ est $\mathscr{H}(X \circ \mathscr{B})$ muni d'une certaine filtration indépendante du choix de $\mathscr{X}$ et de $\mathscr{Y}$, et son gradué est un sous-anneau de la limite inductive des $\mathscr{H}_r(\xi^{-1} Y^m \circ X^l \circ \mathscr{B})$ et coïncide avec cette limite sous des hypothèses convenables de dimension finie; on a $\mathscr{H}_{m+1}(\xi^{-1} Y^m \circ X^l \circ \mathscr{B}) = \mathscr{H}(Y^m \circ \xi\mathscr{F}_m(X^l \circ \mathscr{B}))$ où $\xi\mathscr{F}_r(X^l \circ \mathscr{B})$ est l'image par ξ du faisceau spectral associé à $\mathscr{H} \circ \mathscr{B}$ (l'image $\xi\mathscr{F}$ d'un faisceau $\mathscr{F}$ sur X par ξ est définie par $\xi\mathscr{F}(F) = \mathscr{F}(\xi^{-1}F)$ pour tout fermé F de Y).

La fin de l'article [30] est consacrée aux cas particuliers d'un faisceau constant ("faisceau identique à un anneau") et d'un système local de coefficients au sens de Steenrod [41] ("faisceau localement isomorphe à un anneau"). L'article [31], dont les résultats avaient été annoncés dans la note [29] de 1946 et le contenu exposé au Collège de France en 1950, applique la théorie générale au cas où ξ est une fibration de fibre F et où on considère la cohomologie de X à coefficients dans un anneau (constant) $\mathscr{A}$; le faisceau image $\mathscr{B} = \xi\mathscr{F}(X \circ \mathscr{A})$ sur Y est localement isomorphe à $\mathscr{H}(F \circ \mathscr{A})$. Une partie des résultats que Leray en déduit avait été obtenu indépendamment par G. Hirsch [22]. Lorsque F a même homologie qu'une sphère, on retrouve en particulier les résultats de Gysin [20] complétés par ceux de Chern et Spanier [10]; lorsque Y a même homologie qu'une sphère, on retrouve les résultats de Wang [44].

3. Le Séminaire Cartan et la démonstration par Weil des théorèmes de de Rham

Les idées de Leray ont inspiré A. Weil et H. Cartan pour des travaux de rénovation de la topologie algébrique. Le premier avait eu une conversation avec Leray en 1945 et en avait tiré l'idée d'une démonstration, maintenant classique, des théorèmes de de Rham; une première version de cette démonstration avait été communiquée à Cartan dans une lettre de 1947 [45], et la rédaction complète a été publiée dans les Commentarii Mathematici Helvetici en 1952 [46]. Weil met en dualité deux doubles complexes associés à un "recouvrement simple" $\mathscr{U} = (U_i)$ d'une variété différentiable paracompacte V; on impose à $\mathscr{U}$ d'être localement fini et à chaque intersection non vide U_J des U_i (J appartenant au nerf de $\mathscr{U}$) d'être différentiablement contractile. Le premier double complexe est constitué, en bidegré (m, p), des "coéléments" $\Omega = (\omega_H)$ où, pour tout $H = (i_0, \ldots, i_p)$ tel que $U_{|H|} = \bigcap_\nu U_{i_\nu} \neq \emptyset$, ω_H est une forme différentielle de degré m dans $U_{|H|}$; les differéntielles d et ∂, de degré 1, sont définies par la différentielle extérieure et la somme alternée des restrictions, et il y a deux opérateurs d'homotopie correspondants, définis respectivement à l'aide des contractions des $U_{|H|}$ en un point et d'une partition de l'unité subordonnée à $\mathscr{U}$. Ces opérateurs permettent d'établir un isomorphisme entre le groupe de de Rham de degré m (m-formes fermées sur V modulo les formes exactes) et le m-ième groupe de cohomologie du nerf de $\mathscr{U}$. Le deuxième complexe est formé, en bidegré (m, p), des "éléments" $T = (t_H)$, où, pour tout $H = (i_0, \ldots, i_p)$ comme ci-dessus, t_H est une chaîne singulière finie de dimension m, à support dans $U_{|H|}$; les deux différentielles (de degré -1) correspondent respectivement à celle de l'homologie singulière et à celle de l'homologie du nerf de $\mathscr{U}$, et il y a encore deux opérateurs d'homotopie qui donnent un isomorphisme entre l'homologie singulière de V et l'homologie du nerf. L'accouplement entre les deux complexes est donné par l'intégration des formes différentielles sur les chaînes singulières ((T, Ω) est défini seulement si T ou Ω est fini).

Le Séminaire Cartan à l'Ecole Normale Supérieure a été consacré à la topologie algébrique de 1948 à 1951. Par ailleurs, Cartan avait exposé certains résultats à un Colloque de topologie algébrique du CNRS en 1948. Le séminaire de 1948–49 contenait une première version de la théorie des faisceaux (exposés 12 à 17) qui a été retirée de la circulation pour être remplacée par une nouvelle présentation dans le séminaire de 1950–51. La forme adoptée alors pour la notion de faisceau est due à M. Lazard: un faisceau de K-modules F (K anneau commutatif) sur un espace topologique (régulier) $\mathscr{X}$ est un espace étalé (terme dû à Godement) $p: F \to \mathscr{X}$ dont chaque fibre $p^{-1}(x) = F_x$ a une structure de K-module, de manière que l'addition et la multiplication par les éléments de K (discret) soient continues pour la topologie de F. A chaque ouvert X de $\mathscr{X}$ on associe le module $\Gamma(F, X)$ des *sections* $s: X \to F$ de F dans X (caractérisées par $p \circ s = \mathrm{id}_x$), et chaque fibre F_x est la limite inductive des $\Gamma(F, X)$ pour X voisinage de x. On peut définir un faisceau F à partir de données associant à tout ouvert X un module F_X et à toute inclusion $X \subset Y$ d'ouverts un homomorphisme de restriction $f_{XY} : F_Y \to F_X$, avec une propriété de transitivité des restrictions; on prend comme fibre F_x la limite inductive des F_X pour X voisinage de x et on munit leur somme disjointe d'une topologie convenable. Notons que les morphismes canoniques $F_X \to \Gamma(F, X)$ ne sont en général ni injectifs ni surjectifs, bien qu'il en soit ainsi pour les divers faisceaux de fonctions que l'on définit naturellement de cette manière. Par exemple, le faisceau C^n des cochaînes d'Alexander-Spanier de degré n se définit ainsi à partir des données C_X^n = ensemble des applications de X^{n+1} dans K, et, pour $n \geqslant 1$, $C_X^n \to \Gamma(C^n, X)$ n'est pas injectif, mais il est surjectif si X est paracompact. Il en est de même pour le faisceau S^n des cochaînes singulières.

L'innovation la plus importante de Cartan consiste en l'introduction des familles de supports; on appelle ainsi une famille Φ de fermés paracompacts de $\mathscr{X}$ qui est héréditaire, stable pas réunion finie et telle que tout élément de Φ ait un voisinage appartenant à Φ. Leray n'avait considéré que le cas de la famille des compacts, dans le cas où $\mathscr{X}$ est localement compact. On note $\Gamma(F)$ le module des sections globales de F et $\Gamma_\Phi(F)$ le sous-module des sections s dont le support (ensemble des x tels que $s(x) \neq 0$; il est toujours fermé) appartient à Φ. Lorsque $f : F \to G$ est un homomorphisme surjectif de faisceaux, l'homomorphisme correspondant $\Gamma_\Phi(F) \to \Gamma_\Phi(G)$ n'est pas surjectif en général, mais il l'est si on suppose que le noyau F' de f est fin, c'est-à-dire que, pour tout recouvrement ouvert localement fini $(\mathscr{U}^i)$ de $\mathscr{X}$, l'automorphisme identique de F' est somme d'endomorphismes dont chacun est nul en dehors d'un fermé contenu dans l'ouvert de même indice. Le faisceau des cochaînes d'Alexander-Spanier, ou celui des cochaînes singulières sont des faisceaux fins; il en est de même du faisceau des formes différentielles extérieures sur une variété différentiable. Cartan définit la cohomologie de $\mathscr{X}$ à coefficients dans un faisceau et à supports dans une famille Φ d'une manière axiomatique, en imposant à l'anneau de base K d'être principal. Il s'agit d'une suite de foncteurs $F \mapsto H_\Phi^q(\mathscr{X}, F)$ à valeurs K-modules, nuls pour $q < 0$ et coïncidant avec $\Gamma_\Phi(F)$ pour $q = 0$; de plus, à chaque suite exacte (1) $0 \to F' \to F \to F'' \to 0$ de faisceaux est associé un homomorphisme $\delta_q : H_\Phi^q(\mathscr{X}, F'') \to H_\Phi^{q+1}(\mathscr{X}, F')$ fonctoriel par rapport à la suite. On impose à ces données les axiomes suivants: $H_\Phi^q(\mathscr{X}, F)$ est nul si F est fin et $q > 0$; la suite $\cdots \to H_\Phi^q(\mathscr{X}, F') \to H_\Phi^q(\mathscr{X}, F) \to H_\Phi^q(\mathscr{X}, F'') \xrightarrow{\delta_q} H_\Phi^{q+1}(\mathscr{X}, F') \to \cdots$ associée à (1) est exacte. Après avoir établi l'unicité d'une telle théorie, Cartan en démontre l'existence en prouvant que si $0 \to K \to C_0 \to C_1 \to \cdots \to C_n \to \cdots$ est une suite exacte où les C_n sont des faisceaux fins et sans torsion ("résolution fine du faisceau constant K"), on obtient une théorie de la cohomologie en posant $H_\Phi^q(\mathscr{X}, F) = H^q(\Gamma_\Phi(C \circ F))$ pour tout faisceau F, où le *faisceau fondamental* C est le complexe $C_0 \to C_1 \to \cdots$, et $\circ$ désigne le produit tensoriel des faisceaux sur $\mathscr{X}$. Or les cochaînes d'Alexander-Spanier donnent une résolution fine de K; lorsqu'il en est de même des cochaînes singulières, on dit que l'espace est *HLC*. Plus généralement, on peut calculer la cohomologie à coefficients dans F et à supports dans Φ à l'aide d'une Φ-résolution $0 \to F \to A_0 \to A_1 \to \cdots$ de F; il s'agit d'une suite exacte de faisceaux tels que $H_\Phi^q(\mathscr{X}, A_n) = 0$ pour $q \geqslant 1$ et $n \geqslant 0$. On a $H_\Phi^q(\mathscr{X}, F) \simeq H^q(\Gamma_\Phi(A))$. On obtient une Φ-résolution de F en tensorisant par F une résolution

$0 \to K \to C_0 \to C_1 \to \cdots$ où les C_n sont sans torsion et Φ-injectifs, c'est-à-dire tels que, pour tout épimorphisme $F' \to F''$, $\Gamma_\Phi(C \circ F') \to \Gamma_\Phi(C \circ F'')$ soit surjectif; on dit alors que le complexe C est un faisceau Φ-fondamental.

Aux couvertures de Leray correspondent les carapaces dans la présentation de Cartan. Une carapace sur $\mathscr{X}$ est un K-module A muni, pour chaque point x de $\mathscr{X}$, d'un module quotient $\varphi_x : A \to A_x$; le support $\sigma(\alpha)$ d'un élément α de A est l'ensemble des x tels que $\varphi(\alpha) \neq 0$, et on impose qu'il soit fermé et non vide sauf si $\alpha = 0$. On peut aussi définir une carapace par la donnée des supports, à la manière de Leray; dans ce cas, A_x est le quotient de A par le sous-module des α tels que $x \notin \sigma(\alpha)$. La somme disjointe $F = \mathscr{F}(A)$ des A_x ($x \in \mathscr{X}$) devient un faisceau si on la munit de la topologie dont une base d'ouverts est constituée des $\{\varphi_x(\alpha) | x \in X\}$ où $\alpha \in A$ et X est un ouvert de $\mathscr{X}$; le morphisme naturel de A dans $\Gamma(\mathscr{F}(A))$ est injectif, mais pas surjectif en général.

En vue d'obtenir directement la suite spectrale des espaces fibrés en homologie d'un manière proche de la présentation initiale de Leray [30], S. Eilenberg (exposé 8) place la théorie dans le cadre axiomatique de l'homologie des ensembles ordonnés. Il considère un ensemble (partiellement) ordonné d'éléments $A, B, \ldots$ avec un plus petit élément 0 et un plus grand élément 1 et associe à tout couple (A, B) avec $B < A$ un groupe commutatif $H(A, B)$ fonctoriel en (A, B) (morphismes donnés par l'ordre) et à tout triplet (A, B, C) un homomorphisme $d : H(A, B) \to H(B, C)$ fonctoriel en (A, B, C), de manière que la suite périodique $\cdots \to H(B, C) \to H(A, C) \to H(A, B) \to H(B, C) \to \cdots$ soit exacte. On écrit $H(A)$ au lieu de $H(A, 0)$ et la suite exacte associée au triplet $(A, A, 0)$ montre que $H(A, A) = 0$. Si $(A_p)_{p \in \mathbb{Z}}$ est une suite strictement croissante d'éléments majorés par A, on définit, pour tout p, B_p et C_p comme les images de $H(A_p)$ dans $H(A)$ et $H(A_p, A_{p-1})$ respectivement, et D_p comme le noyau de $H(A_p, A_{p-1}) \to H(A, A_{p-1})$; pour $k \geqslant 1$, on désigne encore par C_p^k l'image de $H(A_p, A_{p-k})$ dans $H(A_p, A_{p-1})$ et par D_p^k le noyau de $H(A_p, A_{p-1}) \to H(A_{p+k-1}, A_{p-1})$. On a $D_p^1 = 0$, $C_p^1 = H(A_p, A_{p-1})$; les C_p^k (resp. D_p^k) décroissent (resp. croissent) avec k et contiennent $C_p^\infty = C_p$ (resp. sont contenus dans $D_p^\infty = D_p$). On pose enfin $E_p = C_p/D_p$ et $E_p^k = C_p^k/D_p^k$ (ainsi $E_p^1 = H(A_p, A_{p-1})$); on démontre que $E_p \simeq B_p/B_{p-1}$. A l'aide des morphismes d des triplets $(A_p, A_{p-1}, A_{p-k-1})$ et $(A_p, A_{p-k}, A_{p-k-1})$ on construit une différentielle $d_p^k : E_p^k \to E_{p-k}^k$ faisant de $(E_p^k)_p$ un complexe, et on démontre que l'homologie de ce complexe s'identifie à E^{k+1}. Il y a aussi une théorie contravariante, où le sens des flèches est renversé (cohomologie). On retrouve les suites spectrales au sens algébrique de Koszul en prenant comme ensemble ordonné celui des sous-modules différentiels d'un Λ-module différentiel fixé A (Λ anneau de base); si $A' \supset B'$ sont deux tels sous-modules, on leur associe $H(A', B') = H(A'/B')$ et le d d'un triplet $A' \supset B' \supset C'$ est défini par la suite exacte $0 \to B'/C' \to A'/C' \to A'/B' \to 0$. La suite spectrale d'homologie d'un espace fibré $\tau : X \to B$, de fibre F et dont la base B est un complexe simplicial fini, s'obtient en filtrant X par les sous-espaces $X_p = \tau^{-1}(B^p)$ où B^p est le p-squelette de B; l'ensemble ordonné est celui des sous-espaces de X et on pose $H(X', Y') = H(X', Y'; G)$ (homologie relative à coefficients dans un groupe G) si $X' \supset Y'$ sont deux tels sous-espaces. On a donc une suite spectrale pour laquelle $E_{p+q,p}^1 =$

$H_{p+q}(X_p, X_{p-1}; G)$ (avec l'opérateur bord du triplet (X_p, X_{p-1}, X_{p-2})) et $E^2_{p+q,p} \simeq H_p(B; H_q(F; G))$ homologie de la base à coefficients dans le système local $H_q(F; G)$. Comme autre application de la suite spectrale, Cartan étudie (exposés 11 et 12) l'homologie et la cohomologie des espaces avec groupe d'opérateurs.

En théorie des faisceaux, Cartan définit les deux suites spectrales d'hypercohomologie définies par un complexe de faisceaux F sur $\mathscr{X}$. Si A est une carapace graduée avec cobord de degré 1 et dont les supports appartiennent à une famille Φ, on définit, pour tout p, un homomorphisme (fonctoriel) $H^{p+k}(A) \to H^p_\Phi(\mathscr{X}, H^k(\mathscr{F}(A)))$ si les conditions suivantes sont vérifiées: a) $\mathscr{X}$ est de Φ-dimension finie (c'est-à-dire qu'il existe un entier n tel que $H^q_\Phi(\mathscr{X}, F) = 0$ pour $q > n$ et pour tout faisceau F) ou bien le degré de A est borné inférieurement; b) $H^q(\mathscr{F}(A)) = 0$ pour $q > k$. C'est même un isomorphisme si on suppose que $H^q(\mathscr{F}(A)) = 0$ pour $q \neq k$, que $A \simeq \Gamma_\Phi(\mathscr{F}(A))$ et que A est homotopiquement Φ-fine. La dualité de Poincaré pour une variété différentiable $\mathscr{X}$ s'en déduit en considérant le faisceau S des chaînes singulières, noté S^{-p} en dimension p (de manière à avoir une différentielle de degré 1); ce faisceau est homotopiquement fin et $H^p(S)$ est nul sauf pour $p = -n$, où n est la dimension de $\mathscr{X}$. Le faisceau $T = H^{-n}(S)$ est localement isomorphe à K, et défini par l'orientation locale de $\mathscr{X}$; on a $H^{p-n}(\Gamma_\Phi(S \circ F)) \simeq H^p_\Phi(\mathscr{X}, T \circ F)$, où le premier membre, encore noté $H^\Phi_{n-p}(\mathscr{X}, F)$, est la Φ-homologie singulière de $\mathscr{X}$.

La suite spectrale d'une application continue $f: E \to B$ est exposée dans le cadre de la Φ-cohomologie par J-P. Serre (exposé 21); on se donne des familles de supports Ψ et Φ sur E et B respectivement, et on suppose qu'elles sont "adaptées" en un sens convenable. Si G est un faisceau sur E, on calcule sa cohomologie comme celle de la carapace $A^0 = \Gamma_\Psi(C \circ G)$ où C est un faisceau fondamental sur E; on obtient une carapace A sur B en associant à tout $x \in A^0$ le support $\overline{f(\sigma(x))}$ (où $\sigma(x)$ est le support de x dans E). Il y a alors une suite spectrale telle que $E^2_{p+q,p} = H^p_\Phi(B, H^q(\mathscr{F}(A)))$ et que E soit le gradué associé à $H_\Psi(E, G)$ convenablement filtré; les fibres du faisceau $H^q(\mathscr{F}(A))$ sur B sont les $H^q_\Psi(F_b, G)$ $(b \in B)$.

Des articles de K. Oka [34] et [35] de 1950 et 1951 introduisaient une notion très proche de celle de faisceau dans la théorie des fonctions analytiques de plusieurs variables: celle d'idéal de domaine indéterminé (on dirait maintenant faisceau d'idéaux). La réflexion sur ces travaux d'Oka a sans doute éclairé Cartan dans sa formulation de la théorie des faisceaux, et elle lui a permis de reformuler les résultats d'Oka comme des théorèmes de cohomologie des faisceaux (les fameux théorèmes A et B); le Séminaire Cartan de 1951–52 est consacré à ces questions, dans le cadre des faisceaux analytiques cohérents. La finitude de la cohomologie d'un espace analytique compact à coefficients dans un faisceau cohérent est établie par Cartan et Serre en 1953 [8][1]; la dualité pour les faisceaux analytiques localement libres sur une variété complexe compacte est due à J-P. Serre en 1953 [39]. Sur le modèle de la théorie des espaces analytiques, Serre [40]

[1] Ce théorème a été étendu au cas relatif par H. Grauert en 1960, voir "Ein Theorem der analytischen Garbentheorie und die Modulräume komplexer Strukturen", Publ. Math. IHES N°5, 1960, pp. 5–64.

a repris en 1954 les bases de la géométrie algébrique à l'aide de la cohomologie des faisceaux algébriques cohérents.

4. La période de maturité: 1955–58

Cette période est marquée par l'élaboration systématique de l'algèbre homologique et son application à la cohomologie des faisceaux. Le livre *Homological Algebra* [7] de Cartan et Eilenberg, publié en 1956, est resté classique depuis; il utilise le langage des foncteurs introduit en 1942 par Eilenberg et MacLane [12] et il rattache les théories cohomologiques aux notions de foncteur satellite et de foncteur dérivé. Les premiers satellites $S_1 T$ et $S^1 T$ d'un foncteur additif covariant T défini sur une catégorie de modules sont construits de la manière suivante: si A est un module, on considère des suites exactes $0 \to M \to P \to A \to 0$ et $0 \to A \to Q \to N \to 0$ avec P projectif et Q injectif, et on pose $S_1 T(A) = \operatorname{Ker}(T(M) \to T(P))$, $S^1 T(A) = \operatorname{Coker}(T(Q) \to T(N))$ (on montre que cela ne dépend pas, à isomorphisme unique près, du choix des suites exactes). On pose ensuite $S_{n+1} T = S_1(S_n T)$ et $S^{n+1} T = S^1(S^n T)$, et on note aussi $S^{-n}T$ pour $S_n T$ et $S_0 T = S^0 T = T$. Pour toute suite exacte $0 \to A' \to A \to A'' \to 0$ de modules, on a des homomorphismes de connexion $\theta_1 : S_1 T(A'') \to T(A')$ et $\theta^1 : T(A'') \to S^1 T(A')$; dans la suite infinie $\cdots \to S^n T(A') \to S^n T(A) \to S^n T(A'') \xrightarrow{\theta} S^{n+1} T(A') \to \cdots$, le composé de deux morphismes successifs est nul, et la suite tout entière est exacte si T est semi-exact. Il y a une théorie duale pour les foncteurs additifs contravariants. Les foncteurs dérivés droits d'un foncteur additif covariant T sont définis tous en même temps (et non plus par récurrence) en posant $R^n T(A) = H^n(T(X))$ où $0 \to A \to X^0 \to X^1 \to \cdots$ est une résolution injective du module A; pour un foncteur contravariant, on prendrait une résolution projective $\cdots \to Y^1 \to Y^0 \to A \to 0$. En prenant au contraire une résolution projective dans le cas où T est covariant et une résolution injective s'il est contravariant, on définit les foncteurs dérivés gauches $L_n T$. Lorsque T est covariant (resp. contravariant), toute suite exacte de modules $0 \to A' \to A \to A'' \to 0$ définit des homomorphismes de connexion $R^n T(A'') \to R^{n+1} T(A')$ et $L_n T(A'') \to L_{n-1} T(A')$ (resp. $R^n T(A') \to R^{n+1} T(A'')$ et $L_n T(A') \to L_{n-1} T(A'')$), et les suites longues de foncteurs dérivés sont exactes. On définit de la même manière les foncteurs dérivés de plusieurs arguments, en prenant des résolutions convenables de chaque argument. On a des morphismes fonctoriels $\tau^0 : T \to R^0 T$ et $\sigma_0 : L_0 T \to T$; pour que le premier (resp. le deuxième) soit un isomorphisme, il faut et il suffit que T soit exact à gauche (resp. à droite). Lorsque cette condition est vérifiée, pour que T soit exact, il faut et il suffit que $R^1 T = 0$ (resp. $L_1 T = 0$). A partir des morphismes σ_0 et τ^0, on définit des morphismes $\sigma_n : L_n T \to S_n T$ et $\tau^n : S^n T \to R^n T$; le premier est un isomorphisme si T est exact à droite et le second si T est exact à gauche. La théorie des foncteurs dérivés sert à définir les foncteurs Tor (dérivés du produit tensoriel) et Ext (dérivés du foncteur Hom); les foncteurs Tor permettent d'exprimer d'une manière satisfaisante les relations obtenues par Künneth [27] entre l'homologie d'un produit de deux espaces topologiques et celles des facteurs, et c'est pour comprendre ces relations que Cartan et Eilenberg avaient entrepris leur élaboration de l'algèbre homologique. La théorie ainsi construite permet aussi d'unifier diverses théories cohomologiques: cohomologie des algèbres (Hochschild [24]), des groupes finis (Tate), des algèbres de Lie (Chevalley-Eilenberg [11]). La théorie des suites spectrales est exposée au chapitre XV du livre, et ses applications au chapitre suivant; il s'agit essentiellement des suites spectrales de foncteurs composés, dont le terme E_2^{pq} est composé des foncteurs dérivés tandis que l'aboutissement est le gradué associé aux dérivés du foncteur composé (pour une filtration convenable). Le dernier chapitre du livre est consacré à l'hyperhomologie: si T est un foncteur additif de modules et A un complexe de modules, on montre que la cohomologie du double complexe $T(X)$, où X est une "résolution de Cartan-Eilenberg" de A (c'est un double complexe), ne dépend pas du choix de X. Cette cohomologie a une filtration telle que le gradué associé soit l'aboutissement de deux suites spectrales dont les termes E_2^{pq} sont respectivement $H^p(R^qT(A))$ et $(R^qT)(H^p(A))$.

D. Buchsbaum [2] a proposé un cadre abstrait plus général pour l'algèbre homologique: celui des catégories exactes, que Grothendieck a introduites de son côté sous le nom de catégories abéliennes. Les axiomes des catégories exactes sont donnés en appendice au livre de Cartan et Eilenberg. Le travail de Grothendieck [14] avait pour but de trouver un cadre commun à la cohomologie d'un espace topologique à coefficients dans un faisceau et à la théorie des foncteurs dérivés de foncteurs de modules, dont l'analogie est évidente; il avait été exposé au printemps 1955 à l'Université de Kansas. Grothendieck définit les catégories abéliennes comme des catégories additives dont tout morphisme a un noyau et un conoyau, le morphisme naturel de la coïmage dans l'image étant un isomorphisme. Si C est une catégorie abélienne où les sommes directes infinies existent, tout objet se plonge dans un object injectif si l'on impose l'existence d'un "générateur" (objet U tel que, tout monomorphisme $i: B \to A$ soit un isomorphisme dès que $\mathrm{Hom}(U, i)$ est bijectif; il revient au même de dire que tout objet est quotient d'un $U^{(I)}$) et la condition (AB5): si (A_i) est une famille filtrante croissante de sous-objets d'un objet A, pour tout $B \subset A$ on a $(\sum A_i) \cap B = \sum (A_i \cap B)$; on dit que C a suffisamment d'injectifs.

Grothendieck appelle ∂-foncteur sur C une suite (T^i) de foncteurs additifs, par exemple covariants, avec la donnée, pour toute suite exacte $0 \to A' \to A \to A'' \to 0$ dans C, d'une suite d'homomorphismes de connexion $\partial: T^i(A'') \to T^{i+1}(A')$, fonctoriels en la suite exacte, telle que le composé de deux morphismes consécutifs de la suite longue $\cdots \to T^i(A') \to T^i(A) \to T^i(A'') \to T^{i+1}(A') \to \cdots$ soit nul. Il définit la suite des satellites droits S^iF d'un foncteur covariant F comme un ∂-foncteur universel au sens que tout morphisme $f^0: F \to T^0$ où (T^i) est un ∂-foncteur s'étend en un unique morphisme $(S^iF) \to (T^i)$ de ∂-foncteurs; ceci reprend une description axiomatique des satellites donnée par Cartan et Eilenberg. Les satellites gauches se définissent de même, en considérant des ∂^*-foncteurs. Si tout objet se plonge dans un objet injectif, les satellites droits S^iF ($i \geqslant 0$) existent pour tout foncteur additif covariant F, et ils forment un ∂-foncteur exact par exemple si F est exact à gauche ou à droite. Les foncteurs dérivés sont définis comme dans le livre de Cartan-Eilenberg au moyen de résolutions injectives ou projectives, dont on doit supposer l'existence. Grothendieck étend la théorie des suites spectrales et de l'hypercohomologie au cas des catégories abéliennes. En particulier, il donne dans ce cadre la suite spectrale des foncteurs composés, pour des foncteurs additifs covariants $F : C \to C'$ et $G : C' \to C''$; on suppose que C et C' ont suffisamment d'injectifs, que G est exact à gauche et que $R^qG(F(I)) = 0$ pour tout $q > 0$ si I est un objet injectif de C, et on obtient un foncteur spectral aboutissant à $R(GF)$ (filtré) et tel que $E_2^{pq}(A) = R^pG(R^qF(A))$ pour tout objet A de C.

Grothendieck appelle *préfaisceau* d'ensembles sur un espace topologique X un système inductif d'ensembles $(F(U))$ défini sur les ouverts U de X (ordonnés par inclusion); les faisceaux sont pour lui des préfaisceaux particuliers, soumis à la condition suivante: si (U_i) est un recouvrement d'un ouvert U par des ouverts non vides et si (f_i) est une famille d'éléments $f_i \in F(U_i)$ telle que les restrictions de f_i et de f_j à $U_i \cap U_j$ coïncident dans $F(U_i \cap U_j)$ quel que soit le couple d'indices (i, j), alors il existe un unique $f \in F(U)$ dont la restriction à U_i soit f_i pour tout indice i. Cette condition exprime que les applications naturelles $F(U) \to \Gamma(\tilde{F}, U)$ sont bijectives, en désignant par $\tilde{F}$ le faisceau au sens de Cartan (espace étalé) associé au préfaisceau F. On définit de même les faisceaux à valeurs dans une catégorie, par exemple les faisceaux de groupes, les faisceaux d'anneaux, etc. Si O est un faisceau d'anneaux sur X, la catégorie des O-Modules (c'est-à-dire des faisceaux de O-modules à gauche) est abélienne et les sommes directes infinies existent dedans; elle admet un générateur (somme directe des faisceaux O_U coïncidant avec O dans U et avec 0 dans $X - U$, où U est un ouvert variable de X) et elle vérifie (AB5), donc elle a suffisamment d'injectifs.

La cohomologie à coefficients dans un faisceau est développée par Grothendieck sans faire aucune hypothèse restrictive sur l'espace, ce qui permet d'appliquer cette théorie aux espaces non séparés que l'on rencontre en géométrie algébrique (topologie de Zariski; Serre [40] se servait, pour les faisceaux cohérents, de la cohomologie de Čech d'un recouvrement par des ouverts affines après avoir établi que ceux-ci sont cohomologiquement triviaux pour les coefficients cohérents). Les familles de supports considérées par Grothendieck sont plus générales que celles de Cartan: ce sont des familles Φ non vides de parties fermées héréditaires et filtrantes croissantes. Le foncteur $F \to \Gamma_\Phi(F)$ (sections globales à supports dans Φ pour un faisceau abélien F, c'est-à-dire un faisceau de groupes commutatifs) est exact à gauche, et la cohomologie $H_\Phi^q(X, F)$ est définie comme la suite des foncteurs dérivés droits de ce foncteur; on peut la calculer au moyen d'une résolution injective de F. La suite spectrale d'une application continue $f : Y \to X$ apparaît comme un cas particulier de la suite spectrale des foncteurs composés; si Φ et Ψ sont des familles de supports sur X et Y respectivement, avec Φ paracompactifiante (c'est-à-dire vérifiant les conditions de Cartan) ou bien Ψ formée de tous les fermés de Y, il existe un foncteur spectral de terme $E_2^{pq}(F) = H^p(X, R^qf_\Psi(F))$ et d'aboutissement $H_{\Psi'}^n(Y, F)$ (F faisceau abélien sur Y), où f_Ψ est le foncteur image directe "à supports dans Ψ" et R^qf_Ψ est son q-ième dérivé droit, et Ψ' une modification convenable de Ψ (si Ψ comprend tous les fermés de Y, c'est l'ensemble des fermés A de Y tels que $\overline{f(A)} \in \Phi$). Le faisceau image directe supérieure $R^q f_\Psi(F)$ est associé au préfaisceau $U \mapsto H^q_{\Psi(U)}(f^{-1}(U), F)$ ($\Psi(U)$ est une localisation de Ψ au-dessus de U ouvert de X). De même la suite spectrale reliant la cohomologie de Čech d'un recouvrement ouvert $\mathscr{U}$ à la vraie cohomologie est une application de la suite spectrale des foncteurs composés; son terme E_2^{pq} vaut $H^p(\mathscr{U}, H^q(F))$ où $H^q(F)$ est le faisceau associé au préfaisceau $V \mapsto H^q(V, F)$ (c'est le q-ième dérivé droit du foncteur d'inclusion de la catégorie des faisceaux dans celle des préfaisceaux).

La fin de l'article contient des théories entièrement nouvelles sur les Ext de faisceaux de modules, avec la suite spectrale reliant les Ext locaux aux Ext globaux et sur la cohomologie des espaces à groupes d'opérateurs à coefficients dans des faisceaux où le groupe opère de façon équivariante. La théorie des Ext de faisceaux a servi à Grothendieck pour formuler et démontrer le théorème de dualité de Serre sur des variétés algébriques projectives dans le cas où il peut y avoir des singularités (voir [15] et [16]).

Le premier livre exposant d'une manière systématique la théorie des faisceaux a été publié par R. Godement en 1958 [13]. Le point de vue adopté est voisin de celui de Grothendieck; la principale innovation consiste en l'introduction de certaines classes de faisceaux acycliques extrêmement utiles: les faisceaux flasques et les faisceaux mous qui sont plus faciles à utiliser que les faisceaux injectifs et que les faisceaux fins pour construire des résolutions. On dit qu'un faisceau $\mathscr{F}$ est *flasque* si toute section de $\mathscr{F}$ dans un ouvert se prolonge à l'espace entier; un faisceau injectif est flasque. Lorsque le noyau d'un épimorphisme $\mathscr{L} \to \mathscr{L}''$ de faisceaux abéliens est flasque, l'homomorphisme $\mathscr{L}(U) \to \mathscr{L}''(U)$ est surjectif pour tout ouvert U. Tout faisceau $\mathscr{F}$ se plonge canoniquement dans un faisceau flaque $\mathscr{C}^0(\mathscr{F}): U \mapsto \prod_{x \in U} \mathscr{F}(x)$, et on en déduit une résolution flasque canonique $0 \to \mathscr{F} \to \mathscr{C}^0 \to \mathscr{C}^1 \to \cdots$ qui permet de calculer la cohomologie à coefficients dans $\mathscr{F}$ car les faisceaux flasques sont Φ-acycliques pour toute famille de supports Φ. On dit qu'un faisceau $\mathscr{F}$ sur l'espace X est *mou* si toute section de $\mathscr{F}$ au-dessus d'un fermé de X se prolonge à X entier; on dit qu'il est Φ-mou (Φ famille paracompactifiante) si, pour tout $S \in \Phi$, la restriction $\mathscr{F}|S$ est un faisceau mou. Si X est paracompact, tout faisceau flasque est mou. Si le noyau de l'épimorphisme $\mathscr{L} \to \mathscr{L}''$ est Φ-mou, l'homomorphisme $\Gamma_\Phi(\mathscr{L}) \to \Gamma_\Phi(\mathscr{L}'')$ est surjectif; il en résulte que les faisceaux Φ-mous sont Φ-acycliques et que l'on peut calculer la Φ-cohomologie à l'aide de résolutions Φ-molles. Si un faisceau d'anneau $\mathscr{A}$ est Φ-mou, il en est de même de tout $\mathscr{A}$-Module; les faisceaux abéliens Φ-fins sont caractérisés par le fait que leurs faisceaux d'endomorphismes locaux (sur $\mathbb{Z}$) sont Φ-mous.

5. Catégories dérivées et opérations sur les faisceaux

Pour les besoins de la géométrie algébrique, Grothendieck a considérablement développé l'algèbre homologique et la théorie des faisceaux dans les années 1957–1965, en introduisant le concept de catégorie dérivée et le formalisme des opérations sur les faisceaux dans le cadre des catégories dérivées. Par ailleurs, pour définir une bonne cohomologie des schémas à coefficients constants ("discrets"), il a été conduit à élargir la notion d'espace topologique en la remplaçant par celles de site et de topos. Le Séminaire de Géométrie algébrique de l'IHES, animé de 1960 à 1968 par Grothendieck porte la marque de ces rénovations de la théorie. L'année 1961–62 traite de la théorie de la dualité locale pour les faisceaux algébriques cohérents; on y trouve la cohomologie à supports dans un sous-espace localement fermé, la notion de module dualisant et l'étude des Ext de faisceaux quasi-cohérents sur des schémas. Pour obtenir le théorème de dualité sous une forme satisfaisante, il fallait disposer du langage des catégories dérivées, que J-L. Verdier a mis en forme dans sa thèse en 1963, d'après les idées de Grothendieck (voir [42]). Dans ce langage, on travaille non pas sur les invariants cohomologiques, mais directement sur des complexes, en enrichissant suffisamment l'ensemble des morphismes pour qu'un morphisme de complexes qui induit un isomorphisme en cohomologie devienne un isomorphisme; ceci évite l'opération dangereuse de passer à un sous-quotient. Les foncteurs dérivés totaux au sens des catégories dérivées se composent, sous des hypothèses convenables, comme les foncteurs que l'on dérive, et les suites spectrales n'apparaissent qu'au moment où on veut calculer la cohomologie. R. Hartshorne a consacré un séminaire à Harvard en 1963–64 à la théorie de la dualité des faisceaux algébriques cohérents d'après des "prénotes" de Grothendieck; il a publié ce séminaire en 1966 ([21]). Le théorème de dualité se formule dans une situation relative, avec $f : X \to Y$ morphisme de (pré)schémas noethériens; à côté de l'image inverse usuelle f^* de faisceaux de $\mathcal{O}_Y$-modules et de son dérivé total gauche Lf^* (au sens des catégories dérivées), on définit, sous des hypothèses convenables, un foncteur image inverse $f^!$ (dans les catégories dérivées), muni d'un morphism trace $\mathrm{Tr}_f : Rf_* f^! \to \mathrm{id}$, de manière à avoir un isomorphisme de dualité $R\mathrm{Hom}_{\mathcal{O}_X}(F, f^! G) \simeq R\mathrm{Hom}_{\mathcal{O}_Y}(Rf_* F, G)$ pour F dans la catégorie dérivée des $\mathcal{O}_X$-modules et G dans la catégorie dérivée des $\mathcal{O}_Y$-modules. Lorsque f est lisse de dimension relative n, $f^!(G) = f^*(G) \otimes \omega[n]$, où $\omega = \Omega^n_{X/Y}$ est le faisceau des formes différentielles relatives de degré n; lorsque f est fini, $f^!(G) = \mathcal{H}om_{\mathcal{O}_Y}(f_* \mathcal{O}_X, G)$. La construction générale est extrêmement compliquée. Une construction analogue a été faite en cohomologie étale par Grothendieck dans son séminaire de 1964–65, où il donne une théorie de la dualité pour la cohomologie l-adique; c'est l'analogue algébrique et relatif de la dualité de Poincaré. La théorie correspondante en topologie ordinaire a été traitée par Verdier [43] en 1965, donnant ainsi une généralisation relative de la dualité de Poincaré, dans le cas d'une application continue $f : X \to Y$ d'espaces topologiques séparés, avec Y localement paracompact et f "localement euclidienne" au sens suivant: localement sur X, elle est composée d'une immersion fermée suivie d'une projection $V \times \mathbb{R}^n \to V$ (ouvert de Y); avec une hypothèse supplémentaire de dimension cohomologique finie pour f et de dimension topologique finie pour X, on a un isomorphisme de dualité $R\mathrm{Hom}(f_!(F^\cdot), G^\cdot) \simeq R\mathrm{Hom}(F^\cdot, f^!(G^\cdot))$ où $f_!$ est le foncteur image direct à supports propres, et le foncteur adjoint à droite $f^!$ est obtenu en résolvant un foncteur $pf^! : G^\cdot \mapsto$ {faisceau associé au préfaisceau $(U \mapsto \mathrm{Hom}^\cdot(f_*(\mathcal{I}_U), G^\cdot))$}, en notant $\mathcal{I}$ une résolution injective du faisceau constant $\mathbb{Z}$ sur X.

Les méthodes de Grothendieck ont été adaptées par M. Sato et M. Kashiwara ([38] et [25]) à l'étude des systèmes d'équations aux dérivées partielles et à l'analyse microlocale. Il en est résulté un important développement et, par un retour des choses, vers la fin des années 70, les techniques de $\mathcal{D}$-Modules issues de ces travaux se sont révélées fécondes en géométrie algébrique.

Bibliographie

1 Alexander, J.W.: On the connectivity ring of an abstract space. Ann. Math. **37**, 698–708 (1936)
2 Buchsbaum, D.A.: Exact categories and Duality. Trans. A.M.S. **80**, 1–34 (1955)
3 Cartan, H.: Sur la cohomologie des espaces où opère un groupe. C.R. Acad. Sc. Paris **226**, 148–150, 303–305 (1948)
4 Cartan, H.: Séminaire «Topologie algébrique», 1e année (1948–49)
5 Cartan, H.: Sur la notion de carapace en topologie algébrique. Colloque de topologie algébrique, CNRS 1–2 (1947)
6 Cartan, H.: Séminaire «Cohomologie des groupes, suite spectrale, faisceaux». 3e année (1950–51)
7 Cartan, H. et Eilenberg, S.: Homological Algebra. Princeton Univ. Press (1956)
8 Cartan, H. et Serre, J-P.: Un théorème de finitude concernant les variétés analytiques compactes. C.R. Acad. Sc. Paris **237**, 128–130 (1953)
9 Čech, E.: Multiplications on a complex. Ann. Math. **37**, 681–697 (1936)
10 Chern, S.S. et Spanier, E.: The homology structure of fibre bundles. Proc. Nat. Acad. Sc. USA **36**, 248–255 (1950)
11 Chevalley, C. et Eilenberg, S.: Cohomology theory of the Lie groups and Lie algebras. Trans. A.M.S. **63**, 85–124 (1948)
12 Eilenberg, S. et Maclane, S.: Group extensions and homology. Ann. Math. **43**, 758–831 (1942)
13 Godement, R.: Théorie des faisceaux. Hermann, Paris (1958)
14 Grothendieck, A.: Sur quelques points d'algèbre homologique. Tôhoku Math. J. **9**, 119–221 (1957)
15 Grothendieck, A.: Théorèmes de dualité pour les faisceaux algébriques cohérents. Sém. Bourbaki **149** (1957)
16 Grothendieck, A.: The cohomology of abstract algebraic varieties. Intern. Congress of Math. at Edinburgh 1958, Cambridge 103–118 (1960)
17 Grothendieck, A.: Cohomologie locale des faisceaux cohérents et théorèmes de Lefschetz locaux et globaux. Sém. de Géométrie algébrique 1962 (SGA2), North-Holland, Amsterdam (1968)
18 Grothendieck, A.: Cohomologie l-adique et fonctions L. Sém. de Géométrie algébrique 1964–65 (SGA5). Lect. Notes Math. **589**. Springer, Berlin Heidelberg New York (1977)
19 Grothendieck, A.: Récoltes et semailles. Prépublication de l'Université de Montpellier
20 Gysin, W.: Zur Homologietheorie der Abbildungen und Faserungen der Mannigfaltigkeiten. Comment. Math. Helv. **14**, 61–122 (1941)
21. Hartshorne, R.: Residues and Duality. Lect. Notes Math. **20**. Springer, Berlin Heidelberg New York (1966)
22 Hirsch, G.: Sur les groupes d'homologie des espaces fibrés. Bull Soc. Math. Belgique 23–33 (1947–48)
23 Hopf, H.: Quelques problèmes de la théorie des représentations continues. Ens. Math. **35**, 334–347 (1936)
24 Hochschild, G.: On the cohomology groups of an associative algebra, Ann. Math. **46**, 58–67 (1945)
25 Kashiwara, M.: Algebraic study of systems of partial differential equations. Thesis, Univ. of Tokyo (1970)
26 Koszul, J-L.: Homologie et cohomologie des algèbres de Lie. Bull. Soc. Math. France **78**, 65–127 (1950)

27 Künneth, H.: Über die Torsionzahlen von Produktmannigfaltigkeiten. Math. Ann. **91**, 65–85 (1923)
28 Lefschetz, S.: On certain numerical invariants of algebraic varieties with applications to abelian varieties. Trans. A.M.S. **22**, 327–382 (1921)
29 Leray, J.: Sur la forme des espaces topologiques et sur les points fixes des représentations. J. Math. Pures et Appl., 9ᵉ série **24**, 95–167 (1945); Sur la position d'un ensemble fermé de points d'un espace topologique, ibid. 169–199; Sur les équations et les transformations, ibid. 201–248
30 Leray, J.: Propriétés de l'anneau d'homologie de la projection d'un espace fibré sur sa base. C.R. Acad. Sc. Paris **223**, 395–397 (1946)
31 Leray, J.: L'anneau spectral et l'anneau filtré d'homologie d'un espace localement compact et d'une application continue. J. Math. Pures et Appl., 9ᵉ série **29**, 1–139 (1950)
32 Leray, J.: L'homologie d'un espace fibré dont la fibre est connexe, ibid. 169–213
33 Leray, J. et Schauder, J.: Topologie et équations fonctionnelles. Ann. ENS **51**, 45–78 (1934)
34 Oka, K.: Sur quelques notions arithmétiques. Bull. Soc. Math. France **78**, 1–27 (1950)
35 Oka, K.: Lemme fondamental. Journ. Math. Soc. Japan **3**, 204–214 et 259–278 (1951)
36 Picard, E.: Mémoire sur les fonctions algébriques de deux variables indépendantes. J. Math. Pures et Appl. **5**, 135–319 (1889)
37 de Rham, G.: Sur l'Analysis Situs des variétés à *n* dimensions. J. Math. Pures et Appl., 9ᵉ série **10**, 115–200 (1931)
38 Sato, M.: Hyperfunctions and partial differential equations. Proc. Intern. Conference on Functional Analysis and related Topics Tokyo 1969, 91–94. Univ. Tokyo Press (1969)
39 Serre, J-P.: Un théorème de dualité. Comm. Math. Helv. **29**, 9–26 (1955)
40 Serre, J-P.: Faisceaux algébriques cohérents. Ann. Math. **61**, 197–278 (1955)
41 Steenrod, N.: Homology with local coefficients. Ann. Math. **44**, 610–627 (1943)
42 Verdier, J-L.: Catégories dérivées (Etat 0). Lect. Notes Math. **569**, 262–312. Springer, Berlin Heidelberg New York (1977)
43 Verdier, J-L.: Dualité dans la cohomologie des espaces localement compacts. Sém. Bourbaki **300** (1965–66)
44 Wang, H.C.: The homology groups of the fiber bundles over a sphere. Duke Math. J. **16**, 33–38 (1949)
45 Weil, A.: Lettre à H. Cartan. Oeuvres **2**, 44. Berlin (1985)
46 Weil, A.: Sur les théorèmes de de Rham. Comm. Math. Helv. **26** (1952)
47 Whitney, H.: On products in a complex. Ann. Math. **39**, 397–432 (1938)
48 Dieudonné, J.: A History of Algebraic and Differential Topology 1900–1960. Birkhäuser, Boston (1989)

Chapter I. Homological algebra

Summary

This chapter contains the bases of homological algebra which are necessary for the understanding of the rest of this book: categories and functors, triangulated categories, localization, derived categories, ind-objects and pro-objects, Mittag-Leffler condition.

Since it is not possible to present in a single chapter the whole theory with all details, we have left out as an exercise some auxiliary results and we have postponed to Chapter X the theory of t-structures, which is not used until there.

Of course, the reader will also consult with great benefit the (classical) books and papers on this subject, such as Bourbaki [1], Cartan-Eilenberg [1], Freyd [1], Gabriel-Zisman [1], Godement [1], Grothendieck [1], Hilton-Stammbach [1], Iversen [1], MacLane [1], Mitchell [1], Northcott [1], and expecially Deligne [1], Gelfand-Manin [1], Hartshorne [1] and Verdier [2] concerning derived categories.

Note. 1.0. As is well-known, to manipulate categories of sets may be dangerous. One way to avoid this danger is to remain in a given universe, which is what we shall assume. This has no implications for our purpose, and we shall not discuss any further this point.

1.1. Categories and functors

Definition 1.1.1. *A category $\mathscr{C}$ consists of the following data.*

(i) *A family* $\mathrm{Ob}(\mathscr{C})$, *whose members are called the objects of* $\mathscr{C}$,

(ii) *for all pairs* (X, Y) *of* $\mathrm{Ob}(\mathscr{C})$, *a set* $\mathrm{Hom}_{\mathscr{C}}(X, Y)$, *whose elements are called morphisms from* X *to* Y,

(iii) *for any triple* (X, Y, Z) *of* $\mathrm{Ob}(\mathscr{C})$, *a map from* $\mathrm{Hom}_{\mathscr{C}}(X, Y) \times \mathrm{Hom}_{\mathscr{C}}(Y, Z)$ *to* $\mathrm{Hom}_{\mathscr{C}}(X, Z)$, *called the composition map, and denoted* $(f, g) \mapsto g \circ f$.

These data satisfying:

(1.1.1) *the composition of morphisms is associative,*

(1.1.2) *for any* $X \in \mathrm{Ob}(\mathscr{C})$ *there exists* $\mathrm{id}_X \in \mathrm{Hom}_{\mathscr{C}}(X, X)$ *such that* $f \circ \mathrm{id}_X = f$ *and* $\mathrm{id}_X \circ g = g$ *for any* $f \in \mathrm{Hom}_{\mathscr{C}}(X, Y)$ *and any* $g \in \mathrm{Hom}_{\mathscr{C}}(Y, X)$.

Note that for each $X \in \mathrm{Ob}(\mathscr{C})$, id_X is unique.

We write for short $f : X \to Y$ to denote a morphism $f \in \mathrm{Hom}_{\mathscr{C}}(X, Y)$. A morphism $f : X \to Y$ is called an **isomorphism** if there exists $g : Y \to X$ such that $f \circ g = \mathrm{id}_Y$ and $g \circ f = \mathrm{id}_X$. If f is an isomorphism we write $f : X \xrightarrow{\sim} Y$ or else $X \simeq Y$.

Example 1.1.2. We denote by $\mathfrak{Set}$ the category of sets and maps of sets.

A **subcategory** $\mathscr{C}'$ of $\mathscr{C}$ is a category $\mathscr{C}'$ such that $\mathrm{Ob}(\mathscr{C}') \subset \mathrm{Ob}(\mathscr{C})$ and for any pair (X, Y) of $\mathrm{Ob}(\mathscr{C}')$, $\mathrm{Hom}_{\mathscr{C}'}(X, Y) \subset \mathrm{Hom}_{\mathscr{C}}(X, Y)$, with the induced composition law, and $\mathrm{id}_X \in \mathrm{Hom}_{\mathscr{C}'}(X, X)$.

If moreover $\mathrm{Hom}_{\mathscr{C}'}(X, Y) = \mathrm{Hom}_{\mathscr{C}}(X, Y)$, then $\mathscr{C}'$ is called a **full subcategory** of $\mathscr{C}$.

Example 1.1.3. We denote by $\mathfrak{Top}$ the category of topological spaces and continuous maps. Then $\mathfrak{Top}$ is a subcategory of $\mathfrak{Set}$, but not a full subcategory.

Let $\mathscr{C}$ be a category. The **opposite category**, denoted $\mathscr{C}^\circ$, is defined by:

$$(1.1.3) \qquad \begin{cases} \mathrm{Ob}(\mathscr{C}^\circ) = \mathrm{Ob}(\mathscr{C}) , \\ \mathrm{Hom}_{\mathscr{C}^\circ}(X, Y) = \mathrm{Hom}_{\mathscr{C}}(Y, X) \quad \text{for any pair } (X, Y) \text{ of } \mathscr{C} , \\ \text{with the obvious composition law.} \end{cases}$$

Let $f : X \to Y$ be a morphism in $\mathscr{C}$. One says that f is a **monomorphism** if for any $W \in \mathrm{Ob}(\mathscr{C})$ and any pair (g, g') of $\mathrm{Hom}_{\mathscr{C}}(W, X)$ such that $f \circ g = f \circ g'$, one has $g = g'$. One says f is an **epimorphism** if f is a monomorphism in $\mathscr{C}^\circ$, that is, for any $Z \in \mathrm{Ob}(\mathscr{C})$ and any pair (h, h') of $\mathrm{Hom}_{\mathscr{C}}(Y, Z)$ such that $h \circ f = h' \circ f$, one has $h = h'$.

In a category $\mathscr{C}$ an object P is called **initial** if $\mathrm{Hom}_{\mathscr{C}}(P, Y)$ has exactly one element for any $Y \in \mathrm{Ob}(\mathscr{C})$. Similarly an object Q is called **final** if $\mathrm{Hom}_{\mathscr{C}}(X, Q)$ has only one element, that is, if Q is initial in $\mathscr{C}^\circ$. Note that two initial (resp. final) objects are naturally isomorphic.

Definition 1.1.4. *Let* $\mathscr{C}$ *and* $\mathscr{C}'$ *be two categories. A functor* F *from* $\mathscr{C}$ *to* $\mathscr{C}'$ *consists of the following data and rules.*

(i) *A map* $F : \mathrm{Ob}(\mathscr{C}) \to \mathrm{Ob}(\mathscr{C}')$,
(ii) *for any pair* (X, Y) *of* $\mathrm{Ob}(\mathscr{C})$, *a map* $F : \mathrm{Hom}_{\mathscr{C}}(X, Y) \to \mathrm{Hom}_{\mathscr{C}'}(F(X), F(Y))$.

These data satisfying:

$$(1.1.4) \qquad F(\mathrm{id}_X) = \mathrm{id}_{F(X)} ,$$

$$(1.1.5) \qquad F(f \circ g) = F(f) \circ F(g) .$$

One sometimes says that F is a **covariant functor** from $\mathscr{C}$ to $\mathscr{C}'$, and a functor from $\mathscr{C}^\circ$ to $\mathscr{C}'$ is called a **contravariant functor** from $\mathscr{C}$ to $\mathscr{C}'$.

For example let $X \in \mathrm{Ob}(\mathscr{C})$. Then $\mathrm{Hom}_{\mathscr{C}}(X, \cdot): Z \mapsto \mathrm{Hom}_{\mathscr{C}}(X, Z)$ is a (covariant) functor from $\mathscr{C}$ to $\mathfrak{Set}$ and $\mathrm{Hom}_{\mathscr{C}}(\cdot, X): Z \mapsto \mathrm{Hom}_{\mathscr{C}}(Z, X)$ is a contravariant functor from $\mathscr{C}$ to $\mathfrak{Set}$.

Definition 1.1.5. *Let F_1 and F_2 be two functors from $\mathscr{C}$ to $\mathscr{C}'$. A morphism θ from F_1 to F_2 consists of the following data.*

$$\text{For any } X \in \mathrm{Ob}(\mathscr{C}) \text{ , an element } \theta(X) \in \mathrm{Hom}_{\mathscr{C}'}(F_1(X), F_2(X)) \ .$$

These data satisfying:

(1.1.6) *for any $f \in \mathrm{Hom}_{\mathscr{C}}(X, Y)$ the following diagram commutes:*

$$\begin{array}{ccc} F_1(X) & \xrightarrow{\theta(X)} & F_2(X) \\ \Big\downarrow{\scriptstyle F_1(f)} & & \Big\downarrow{\scriptstyle F_2(f)} \\ F_1(Y) & \xrightarrow{\theta(Y)} & F_2(Y) \ . \end{array}$$

Note that if $\mathscr{C}$ and $\mathscr{C}'$ are two categories, one gets a new category whose objects are functors from $\mathscr{C}$ to $\mathscr{C}'$ and morphisms are morphisms of such functors.

Definition 1.1.6. *A functor F from $\mathscr{C}$ to $\mathfrak{Set}$ is called representable if there exists $X \in \mathrm{Ob}(\mathscr{C})$ such that F is isomorphic to the functor* $\mathrm{Hom}_{\mathscr{C}}(X, \cdot)$.

In this case X is unique up to isomorphism, and is called a **representative** of F.

There is an analogous definition for contravariant functors.

Definition 1.1.7. *Let $F: \mathscr{C} \to \mathscr{C}'$ be a functor.*

(i) *One says F is fully faithful if:*

(1.1.7) *for any pair (X, Y) of* $\mathrm{Ob}(\mathscr{C})$ *the map* $\mathrm{Hom}_{\mathscr{C}}(X, Y) \to \mathrm{Hom}_{\mathscr{C}'}(F(X), F(Y))$ *is bijective.*

(ii) *One says F is an equivalence of categories if* (1.1.7) *is satisfied and moreover:*

(1.1.8) *for any $X' \in \mathrm{Ob}(\mathscr{C}')$ there exist $X \in \mathrm{Ob}(\mathscr{C})$ and an isomorphism $X' \to F(X)$.*

Note that F is an equivalence of categories iff there exists a functor $F': \mathscr{C}' \to \mathscr{C}$ and isomorphisms of functors $F \circ F' \simeq \mathrm{id}_{\mathscr{C}'}$ and $F' \circ F \simeq \mathrm{id}_{\mathscr{C}}$. In this case one says F' is a **quasi-inverse** of F.

We can embed a category $\mathscr{C}$ into the category $\mathscr{C}^\vee$ of all contravariant functors from $\mathscr{C}$ to $\mathfrak{Set}$.

Let $h: \mathscr{C} \to \mathscr{C}^\vee$ be the functor $X \mapsto \mathrm{Hom}_{\mathscr{C}}(\cdot, X)$.

Proposition 1.1.8. (i) *For any* $X \in \mathrm{Ob}(\mathscr{C})$ *and* $F \in \mathrm{Ob}(\mathscr{C}^\vee)$, we have:

$$\mathrm{Hom}_{\mathscr{C}^\vee}(h(X), F) \simeq F(X) \ .$$

(ii) *The functor h is fully faithful.*

Proof. (i) To $f \in \mathrm{Hom}_{\mathscr{C}^\vee}(h(X), F)$ we associate $\phi(f) \in F(X)$ as follows. Since f is a morphism of functors from $h(X)$ to F, $f(X)$ gives a map from $h(X)(X) = \mathrm{Hom}_{\mathscr{C}}(X, X)$ to $F(X)$. Then define $\phi(f)$ as the image of id_X. Conversely, to $s \in F(X)$ we can associate $\psi(s) \in \mathrm{Hom}_{\mathscr{C}^\vee}(h(X), F)$ as follows. For $Y \in \mathrm{Ob}(\mathscr{C})$, consider the maps:

$$\begin{aligned} h(X)(Y) &= \mathrm{Hom}_{\mathscr{C}}(Y, X) \\ &\to \mathrm{Hom}_{\mathfrak{Set}}(F(X), F(Y)) \\ &\to F(Y) \ , \end{aligned}$$

where the last arrow is defined by s. It is easily checked that ϕ and ψ are inverse to each other.

(ii) Applying (i) to $h(Y)$ we obtain:

$$\begin{aligned} \mathrm{Hom}_{\mathscr{C}^\vee}(h(X), h(Y)) &\simeq h(Y)(X) \\ &= \mathrm{Hom}_{\mathscr{C}}(X, Y) \ . \quad \square \end{aligned}$$

Note that a contravariant functor $F: \mathscr{C} \to \mathfrak{Set}$ is representable iff it is isomorphic to $h(X)$ for some $X \in \mathrm{Ob}(\mathscr{C})$.

Similarly one can consider the opposite category $\mathscr{C}^\wedge$ of $\mathscr{C}^\vee$, that is, the category of (covariant) functors from $\mathscr{C}$ to $\mathfrak{Set}$. There is a natural equivalence of categories:

$$\mathscr{C}^\wedge \simeq \mathscr{C}^{\circ\vee\circ}$$

and the functor $h': \mathscr{C} \to \mathscr{C}^\wedge$, $X \mapsto \mathrm{Hom}_{\mathscr{C}}(X, \cdot)$ is fully faithful.

1.2. Abelian categories

Definition 1.2.1. *An additive category $\mathscr{C}$ is a category $\mathscr{C}$ such that:*

(i) *for any pair* (X, Y) *of* $\mathrm{Ob}(\mathscr{C})$, $\mathrm{Hom}_{\mathscr{C}}(X, Y)$ *has a structure of additive* (i.e. *abelian*) *group, and the composition law is bilinear,*

(ii) *there exists an object* 0 *such that* $\mathrm{Hom}_{\mathscr{C}}(0, 0) = 0$,

(iii) *for any pair* (X, Y) *of* $\mathrm{Ob}(\mathscr{C})$ *the functor*

$$W \mapsto \mathrm{Hom}_{\mathscr{C}}(X, W) \times \mathrm{Hom}_{\mathscr{C}}(Y, W) \text{ is representable },$$

(iv) *for any pair* (X, Y) *of* $\mathrm{Ob}(\mathscr{C})$, *the functor*

$$W \mapsto \mathrm{Hom}_{\mathscr{C}}(W, X) \times \mathrm{Hom}_{\mathscr{C}}(W, Y) \text{ is representable }.$$

Note that (iii) and (iv) are equivalent under the conditions (i) and (ii). Moreover the representatives given in (iii) and (iv) are isomorphic (cf. Exercise I.3). We shall denote such a representative by $X \oplus Y$ and call it the **direct sum** of X and Y. The isomorphism:

$$\mathrm{Hom}_{\mathscr{C}}(X \oplus Y, X \oplus Y) \simeq \mathrm{Hom}_{\mathscr{C}}(X, X \oplus Y) \oplus \mathrm{Hom}_{\mathscr{C}}(Y, X \oplus Y)$$

defines two morphisms, associated to $\mathrm{id}_{X \oplus Y}$, $i_1 : X \to X \oplus Y$ and $i_2 : Y \to X \oplus Y$. These morphisms satisfy the following universal property. For any pair of morphisms, $f : X \to W$ and $g : Y \to W$, there exists $h : X \oplus Y \to W$ such that the diagram below commutes:

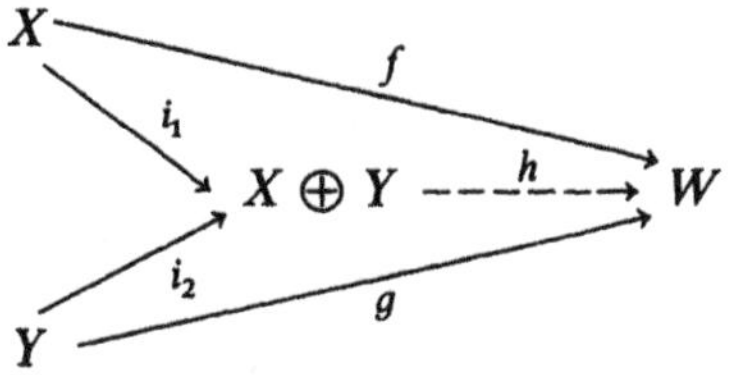

There is a similar remark by reversing the arrows.

If F is a functor from $\mathscr{C}$ to $\mathscr{C}'$, two additive categories, one says that F is **additive** if for any pair (X, Y) of $\mathrm{Ob}(\mathscr{C})$, the map F from $\mathrm{Hom}_{\mathscr{C}}(X, Y)$ to $\mathrm{Hom}_{\mathscr{C}'}(F(X), F(Y))$ is a group homomorphism.

Now we assume $\mathscr{C}$ is additive. Let $Z \in \mathrm{Ob}(\mathscr{C})$. The functor $\mathrm{Hom}_{\mathscr{C}}(Z, \cdot)$ associates to a morphism $f : X \to Y$ a group homomorphism:

$$\mathrm{Hom}_{\mathscr{C}}(Z, f) : \mathrm{Hom}_{\mathscr{C}}(Z, X) \to \mathrm{Hom}_{\mathscr{C}}(Z, Y) .$$

One defines similarly:

$$\mathrm{Hom}_{\mathscr{C}}(f, Z) : \mathrm{Hom}_{\mathscr{C}}(Y, Z) \to \mathrm{Hom}_{\mathscr{C}}(X, Z) .$$

Definition 1.2.2. *Let* $f \in \mathrm{Hom}_{\mathscr{C}}(X, Y)$.

(i) *If the functor*:

$$\mathrm{Ker}(\mathrm{Hom}_{\mathscr{C}}(\cdot, f)) : Z \mapsto \mathrm{Ker}(\mathrm{Hom}_{\mathscr{C}}(Z, f)) = \{u \in \mathrm{Hom}_{\mathscr{C}}(Z, X); f \circ u = 0\}$$

is representable, its representative is called the kernel of f, *and denoted* $\mathrm{Ker}\, f$.

(ii) *Similarly, if the functor*:

$$\operatorname{Ker}(\operatorname{Hom}_{\mathscr{C}}(f,\cdot)): Z \mapsto \operatorname{Ker}(\operatorname{Hom}_{\mathscr{C}}(f,Z)) = \{u \in \operatorname{Hom}_{\mathscr{C}}(Y,Z); u \circ f = 0\}$$

is representable, its representative is called the cokernel of f and denoted Coker f.

Note that the equality $\operatorname{Ker} f = 0$ (resp. $\operatorname{Coker} f = 0$) is equivalent to saying that f is a monomorphism (resp. an epimorphism).

Assume that f has a kernel. There exists a morphism of functors:

$$\beta : \operatorname{Hom}_{\mathscr{C}}(\cdot, \operatorname{Ker} f) \to \operatorname{Hom}_{\mathscr{C}}(\cdot, X)$$

and $\beta(\mathrm{id}_{\operatorname{Ker} f})$ defines a morphism:

$$\alpha : \operatorname{Ker} f \to X \ ,$$

so that $\beta = \operatorname{Hom}_{\mathscr{C}}(\cdot, \alpha)$.

Then α satisfies the following universal property: for any morphism $e : W \to X$ such that $f \circ e = 0$, e factors through α, that is, the dotted arrow below exists uniquely (making the following diagram commutative):

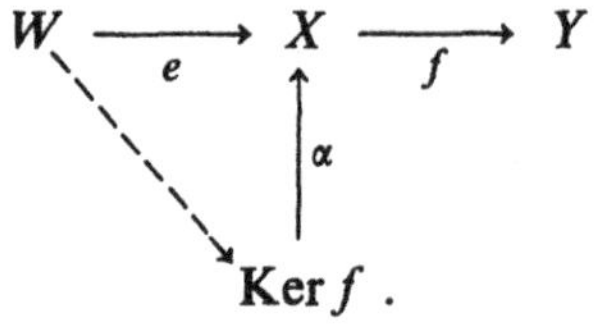

Similarly, if f has a cokernel, there exists a morphism of functors:

$$\delta : \operatorname{Hom}_{\mathscr{C}}(\operatorname{Coker} f, \cdot) \to \operatorname{Hom}_{\mathscr{C}}(Y, \cdot) \ ,$$

which defines:

$$\gamma : Y \to \operatorname{Coker} f$$

and γ is a solution of the universal problem represented by the following diagram (where $g \circ f = 0$):

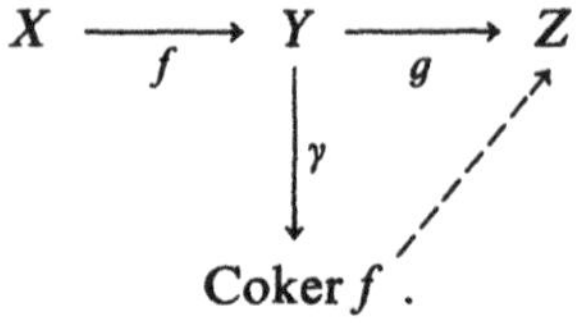

Assume that $\alpha : \operatorname{Ker} f \to X$ has a cokernel. In this case one denotes it by $\operatorname{Coim} f$ and calls it the **coimage** of f. Similarly assume $\gamma : Y \to \operatorname{Coker} f$ has a kernel: one denotes it by $\operatorname{Im} f$ and calls it the **image** of f. It follows from the universal properties of the kernel and cokernel that if $\operatorname{Coim} f$ and $\operatorname{Im} f$ exist, then there exists a natural morphism:

$$\operatorname{Coim} f \to \operatorname{Im} f \ . \tag{1.2.1}$$

Definition 1.2.3. *An additive category $\mathscr{C}$ is called an abelian category if its satisfies the two following conditions.*

(i) *For any morphism $f : X \to Y$,* $\operatorname{Ker} f$ *and* $\operatorname{Coker} f$ *exist.*
(ii) *The canonical morphism* $\operatorname{Coim} f \to \operatorname{Im} f$ *is an isomorphism.*

Examples 1.2.4. (a) Let $\mathfrak{Ban}(\mathbb{C})$ denote the category of Banach vector spaces over $\mathbb{C}$ and linear continuous maps. This is an additive category. Moreover if $f : X \to Y$ is a morphism, f admits a kernel and a cokernel, (take $Y/\overline{\operatorname{Im} f}$ as a cokernel, where $\overline{\operatorname{Im} f}$ denotes the closure of $\operatorname{Im} f$ in Y). Note that it may happen that $\operatorname{Ker} f = \operatorname{Coker} f = 0$, but f is not an isomorphism. Hence the category $\mathfrak{Ban}(\mathbb{C})$ is not abelian.

(b) Let A be a ring with unit. Then $\mathfrak{Mod}(A)$, the category of left A-modules and A-linear maps is an abelian category. Denoting by A^{op} the opposite ring of A, the abelian category $\mathfrak{Mod}(A^{op})$ is equal to the category of right A-modules.

In fact it has been proved that any abelian category is equivalent to a full subcategory of $\mathfrak{Mod}(A)$, for some ring A, (cf. Mitchell [1]).

(c) An interesting example of an additive category which is not abelian, is the category of filtered left modules over a filtered ring (cf. Chapter XI).

From now on, we assume $\mathscr{C}$ is an abelian category.

Definition 1.2.5. *A sequence of morphisms:*

$$X \xrightarrow[f]{} Y \xrightarrow[g]{} Z$$

is called an exact sequence if:

(i) $g \circ f = 0$,
(ii) *the natural morphism* $\operatorname{Im} f \to \operatorname{Ker} g$ *is an isomorphism.*

More generally a sequence of morphisms is called exact if any successive pair of arrows is exact.

Hence, if $f : X \to Y$ is a morphism, we get exact sequences:

$$0 \to \operatorname{Ker} f \to X \to \operatorname{Im} f \to 0 \ ,$$

$$0 \to \operatorname{Im} f \to Y \to \operatorname{Coker} f \to 0 \ .$$

Note that the sequence $0 \to X \xrightarrow[f]{} Y$ (resp. $X \xrightarrow[f]{} Y \to 0$) is exact iff f is a monomorphism (resp. an epimorphism).

Definition 1.2.6. *Let $\mathscr{C}$ and $\mathscr{C}'$ be two abelian categories. An additive functor F from $\mathscr{C}$ to $\mathscr{C}'$ is called left (resp. right) exact if for any exact sequence in $\mathscr{C}$:*

$$0 \to X' \to X \to X''$$

(resp.: $X' \to X \to X'' \to 0$) the sequence:

$$0 \to F(X') \to F(X) \to F(X'')$$

(resp.: $F(X') \to F(X) \to F(X'') \to 0$) is exact.

If F is both left and right exact, F is called exact.

A contravariant functor F from $\mathscr{C}$ to $\mathscr{C}'$ is called left exact (resp. right exact, resp. exact), if so is F regarded as a functor from $\mathscr{C}^\circ$ to $\mathscr{C}'$.

Example 1.2.7. Let $X \in \mathrm{Ob}(\mathscr{C})$. Then $\mathrm{Hom}_{\mathscr{C}}(X, \cdot)$ and $\mathrm{Hom}_{\mathscr{C}}(\cdot, X)$ are both left exact functors from $\mathscr{C}$ to $\mathfrak{Mod}(\mathbb{Z})$.

Definition 1.2.8. *Let $X \in \mathrm{Ob}(\mathscr{C})$. One says that X is injective (resp. projective) if the functor* $\mathrm{Hom}_{\mathscr{C}}(\cdot, X)$ *(resp.* $\mathrm{Hom}_{\mathscr{C}}(X, \cdot)$*) is exact.*

Note that $Z \in \mathrm{Ob}(\mathscr{C})$ is injective iff, for all diagrams as below, in which the row is exact, the dotted arrow exists (making the diagram commutative):

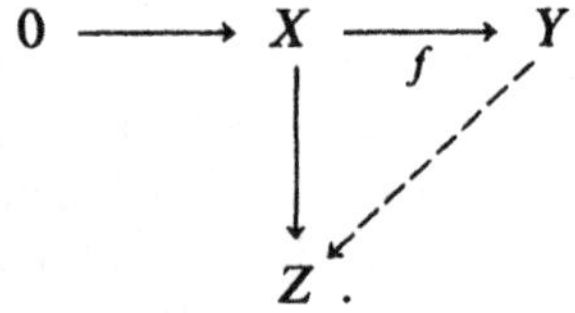

In fact Z is injective iff for all monomorphisms $f: X \to Y$, $\mathrm{Hom}_{\mathscr{C}}(f, Z)$ is surjective.

From this remark we obtain that if $0 \to X \to Y \to Z \to 0$ is an exact sequence and if X is injective, then the sequence splits (cf. Exercise I.5).

Moreover if $0 \to X' \to X \to X'' \to 0$ is an exact sequence in $\mathscr{C}$ and X' is injective, then X is injective iff X'' is injective.

There are similar statements for projective objects, by reversing the arrows.

Terminology 1.2.9. In an abelian category $\mathscr{C}$, if $0 \to X \to Y \to Z \to 0$ is an exact sequence, one sometimes says that X is a sub-object of Y and Z a quotient of Y. One sometimes even writes $Z = Y/X$.

Notation 1.2.10. We shall indifferently write $\mathfrak{Mod}(\mathbb{Z})$ or $\mathfrak{Ab}$ to denote the category of abelian groups.

If A is a ring and X is an A-module one writes $\mathrm{Hom}_A(X, \cdot)$ instead of $\mathrm{Hom}_{\mathfrak{Mod}(A)}(X, \cdot)$ and similarly for $\mathrm{Hom}_A(\cdot, X)$.

If A is commutative, these functors take their values in $\mathfrak{Mod}(A)$ and one keeps the same notations to denote the functors from $\mathfrak{Mod}(A)$ to $\mathfrak{Mod}(A)$ so obtained. One proceeds similarly with the functors $X \otimes_A \cdot$ or $\cdot \otimes_A X$.

1.3. Categories of complexes

Let $\mathscr{C}$ be an additive category.

Definition 1.3.1. *A complex X in $\mathscr{C}$ consists of the data $\{X^n, d_X^n\}_{n \in \mathbb{Z}}$, such that for any $n \in \mathbb{Z}$:*

$$(1.3.1) \qquad X^n \in \mathrm{Ob}(\mathscr{C}) \ , \quad \mathrm{d}_X^n \in \mathrm{Hom}_{\mathscr{C}}(X^n, X^{n+1}) \quad \textit{and} \quad \mathrm{d}_X^{n+1} \circ \mathrm{d}_X^n = 0 \ .$$

A morphism f from a complex X to a complex Y is a sequence $\{f^n\}_{n \in \mathbf{Z}}$ *of morphisms* $f^n : X^n \to Y^n$, *such that for any n:*

$$(1.3.2) \qquad \mathrm{d}_Y^n \circ f^n = f^{n+1} \circ \mathrm{d}_X^n \ .$$

We denote by $\mathbf{C}(\mathscr{C})$ the category of complexes of $\mathscr{C}$ thus obtained. This is also an additive category, and if $\mathscr{C}$ is abelian, then $\mathbf{C}(\mathscr{C})$ is abelian.

One often writes a complex as a sequence:

$$\cdots \longrightarrow X^{n-1} \xrightarrow[\mathrm{d}_X^{n-1}]{} X^n \xrightarrow[\mathrm{d}_X^n]{} X^{n+1} \longrightarrow \cdots \ .$$

The family $\mathrm{d}_X = \{\mathrm{d}_X^n\}_n$ is called the **differential** of the complex X. A complex X is said to be **bounded** (resp. **bounded below**, resp. **bounded above**) if $X^n = 0$ for $|n| \gg 0$ (resp. $n \ll 0$, resp. $n \gg 0$). The full subcategory of $\mathbf{C}(\mathscr{C})$ consisting of bounded complexes (resp. complexes bounded below, resp. complexes bounded above), is denoted $\mathbf{C}^b(\mathscr{C})$ (resp. $\mathbf{C}^+(\mathscr{C})$, resp. $\mathbf{C}^-(\mathscr{C})$).

We identify $\mathscr{C}$ with the full subcategory of $\mathbf{C}(\mathscr{C})$ consisting of complexes X such that $X^n = 0$ for $n \neq 0$.

Definition 1.3.2. *Let k be an integer, and let* $X \in \mathrm{Ob}(\mathbf{C}(\mathscr{C}))$. *One defines a new complex* $X[k]$ *by setting:*

$$(1.3.3) \qquad \begin{cases} X[k]^n = X^{n+k} \ , \\ \mathrm{d}_{X[k]}^n = (-1)^k \mathrm{d}_X^{n+k} \ . \end{cases}$$

For a morphism $f : X \to Y$ *in* $\mathbf{C}(\mathscr{C})$, *one defines* $f[k] : X[k] \to Y[k]$ *by setting:*

$$(1.3.4) \qquad f[k]^n = f^{n+k} \ .$$

The functor $[k]$ from $\mathbf{C}(\mathscr{C})$ to $\mathbf{C}(\mathscr{C})$ is called the **shift functor** of degree k.

Definition 1.3.3. *A morphism* $f : X \to Y$ *in* $\mathbf{C}(\mathscr{C})$ *is called homotopic to zero if there exist morphisms* $s^n : X^n \to Y^{n-1}$ *in* $\mathscr{C}$ *such that for any n:*

$$(1.3.5) \qquad f^n = s^{n+1} \circ \mathrm{d}_X^n + \mathrm{d}_Y^{n-1} \circ s^n \ .$$

One says f is **homotopic** to g if $f - g$ is homotopic to zero. We denote by $\mathrm{Ht}(X, Y)$ the subgroup of $\mathrm{Hom}_{\mathbf{C}(\mathscr{C})}(X, Y)$ consisting of morphisms homotopic to zero. One sees easily that the composition map $\mathrm{Hom}_{\mathbf{C}(\mathscr{C})}(X, Y) \times \mathrm{Hom}_{\mathbf{C}(\mathscr{C})}(Y, Z) \to \mathrm{Hom}_{\mathbf{C}(\mathscr{C})}(X, Z)$ sends $\mathrm{Ht}(X, Y) \times \mathrm{Hom}_{\mathbf{C}(\mathscr{C})}(Y, Z)$ and $\mathrm{Hom}_{\mathbf{C}(\mathscr{C})}(X, Y) \times \mathrm{Ht}(Y, Z)$ into $\mathrm{Ht}(X, Z)$. This permits to define a new category $\mathbf{K}(\mathscr{C})$ as follows.

Definition 1.3.4. *The category* $\mathbf{K}(\mathscr{C})$ *is defined by*

$$(1.3.6) \qquad \begin{cases} \mathrm{Ob}(\mathbf{K}(\mathscr{C})) = \mathrm{Ob}(\mathbf{C}(\mathscr{C})) \ , \\ \mathrm{Hom}_{\mathbf{K}(\mathscr{C})}(X, Y) = \mathrm{Hom}_{\mathbf{C}(\mathscr{C})}(X, Y)/\mathrm{Ht}(X, Y) \ . \end{cases}$$

One defines similarly the categories $\mathbf{K}^b(\mathscr{C})$, $\mathbf{K}^+(\mathscr{C})$ and $\mathbf{K}^-(\mathscr{C})$. They are full subcategories of $\mathbf{K}(\mathscr{C})$.

From now on, and until the end of this section, we assume that $\mathscr{C}$ is abelian.

Definition 1.3.5. *For $X \in \mathrm{Ob}(\mathbf{C}(\mathscr{C}))$, one sets:*

$$Z^k(X) = \operatorname{Ker} d_X^k \ , \qquad B^k(X) = \operatorname{Im} d_X^{k-1} \ ,$$

$$H^k(X) = \operatorname{Coker}(B^k(X) \to Z^k(X)) \ .$$

One calls $H^k(X)$ the k-th cohomology of the complex X.

In other words:

$$H^k(X) = \operatorname{Ker} d_X^k / \operatorname{Im} d_X^{k-1} \ . \tag{1.3.7}$$

Note that $H^k(\cdot)$ is an additive functor from $\mathbf{C}(\mathscr{C})$ to $\mathscr{C}$, and:

$$H^k(X) = H^0(X[k]) \ . \tag{1.3.8}$$

If $f : X \to Y$ is homotopic to zero, then $H^k(f): H^k(X) \to H^k(Y)$ is the zero morphism. Hence $H^k(\cdot)$ is a well-defined functor from $\mathbf{K}(\mathscr{C})$ to $\mathscr{C}$.

There are exact sequences:

$$X^{k-1} \longrightarrow Z^k(X) \longrightarrow H^k(X) \longrightarrow 0 \ ,$$

$$0 \longrightarrow H^k(X) \longrightarrow \operatorname{Coker}(d_X^{k-1}) \longrightarrow X^{k+1} \ ,$$

$$0 \longrightarrow Z^{k-1}(X) \longrightarrow X^{k-1} \longrightarrow B^k(X) \longrightarrow 0 \ ,$$

$$0 \longrightarrow B^k(X) \longrightarrow X^k \longrightarrow \operatorname{Coker}(d_X^{k-1}) \longrightarrow 0 \ ,$$

$$0 \to H^k(X) \to \operatorname{Coker} d_X^{k-1} \xrightarrow[d_X^k]{} Z^{k+1}(X) \longrightarrow H^{k+1}(X) \longrightarrow 0 \ . \tag{1.3.9}$$

Proposition 1.3.6. *Let $0 \to X \to Y \to Z \to 0$ be an exact sequence in $\mathbf{C}(\mathscr{C})$. Then there exists a canonical long exact sequence in $\mathscr{C}$:*

$$\cdots \longrightarrow H^n(X) \longrightarrow H^n(Y) \longrightarrow H^n(Z) \xrightarrow[\delta]{} H^{n+1}(X) \longrightarrow \cdots \ ,$$

more precisely, if

$$\begin{array}{ccccccccc} 0 & \longrightarrow & X & \longrightarrow & Y & \longrightarrow & Z & \longrightarrow & 0 \\ & & \downarrow & & \downarrow & & \downarrow & & \\ 0 & \longrightarrow & X' & \longrightarrow & Y' & \longrightarrow & Z' & \longrightarrow & 0 \end{array}$$

is a commutative diagram of exact sequences in $\mathbf{C}(\mathscr{C})$, then all the diagrams:

$$\begin{array}{ccc} H^n(Z) & \longrightarrow & H^{n+1}(X) \\ \downarrow & & \downarrow \\ H^n(Z') & \longrightarrow & H^{n+1}(X') \end{array}$$

commute.

Proof. Consider the commutative diagram with exact rows:

$$\begin{array}{ccccccccc} & & \mathrm{Coker}(\mathrm{d}_X^{n-1}) & \longrightarrow & \mathrm{Coker}(\mathrm{d}_Y^{n-1}) & \longrightarrow & \mathrm{Coker}(\mathrm{d}_Z^{n-1}) & \longrightarrow & 0 \\ & & \downarrow{\scriptstyle \mathrm{d}_X^n} & & \downarrow{\scriptstyle \mathrm{d}_Y^n} & & \downarrow{\scriptstyle \mathrm{d}_Z^n} & & \\ 0 & \longrightarrow & Z^{n+1}(X) & \longrightarrow & Z^{n+1}(Y) & \longrightarrow & Z^{n+1}(Z) . & & \end{array}$$

The result then follows from (1.3.9) and Exercise I.9. The functoriality of the construction is left to the reader. □

Let $X \in \mathrm{Ob}(\mathbf{C}(\mathscr{C}))$. We define the **truncated complexes** $\tau^{\leqslant n}(X)$ and $\tau^{\geqslant n}(X)$ by:

(1.3.10) $$\tau^{\leqslant n}(X): \cdots \to X^{n-2} \to X^{n-1} \to \mathrm{Ker}\,\mathrm{d}_X^n \to 0 \to \cdots ,$$

(1.3.11) $$\tau^{\geqslant n}(X): \cdots \to 0 \to \mathrm{Coker}\,\mathrm{d}_X^{n-1} \to X^{n+1} \to X^{n+2} \to \cdots .$$

Then we have morphisms in $\mathbf{C}(\mathscr{C})$:

(1.3.12) $$\tau^{\leqslant n}(X) \to X , \qquad X \to \tau^{\geqslant n}(X)$$

and for $n' \leqslant n$:

(1.3.13) $$\tau^{\leqslant n'}(X) \to \tau^{\leqslant n}(X) , \qquad \tau^{\geqslant n'}(X) \to \tau^{\geqslant n}(X) .$$

Moreover:

Proposition 1.3.7. (i) *The natural morphism $H^k(\tau^{\leqslant n}(X)) \to H^k(X)$ is an isomorphism for $k \leqslant n$ and $H^k(\tau^{\leqslant n}(X)) = 0$ for $k > n$.*

(ii) *The natural morphism $H^k(X) \to H^k(\tau^{\geqslant n}(X))$ is an isomorphism for $k \geqslant n$ and $H^k(\tau^{\geqslant n}(X)) = 0$ for $k < n$.*

The proof is straightforward.

Remark 1.3.8. Let X and Y be two objects of $\mathbf{C}(\mathscr{C})$. One sometimes says that X and Y are homotopically equivalent if they are isomorphic in $\mathbf{K}(\mathscr{C})$, that is, if there exists $f \in \mathrm{Hom}_{\mathbf{C}(\mathscr{C})}(X, Y)$ which is an isomorphism in $\mathbf{K}(\mathscr{C})$. Such an f is called a homotopy equivalence.

Notations 1.3.9. (i) Consider a sequence $\{X_n, \mathrm{d}_n^X\}_{n \in \mathbf{Z}}$ where $X_n \in \mathrm{Ob}(\mathscr{C})$, $\mathrm{d}_n^X \in \mathrm{Hom}_{\mathscr{C}}(X_n, X_{n-1})$ and $\mathrm{d}_n^X \circ \mathrm{d}_{n+1}^X = 0$. Then we shall still say that this sequence is a complex in $\mathscr{C}$. In fact setting $X^n = X_{-n}$, $\mathrm{d}^n_X = \mathrm{d}^X_{-n}$, the sequence $\{X^n, \mathrm{d}^n_X\}$ is a complex in our previous sense.

(ii) We sometimes denote by $X^\cdot$ (resp. $X_.$) a complex $\{X^n, \mathrm{d}^n_X\}$ (resp. $\{X_n, \mathrm{d}^X_n\}$). The object $\operatorname{Ker} \mathrm{d}^X_{n-1}/\operatorname{Im} \mathrm{d}^X_n$ is called the n-th holomology group of $X_.$ and denoted by $H_n(X_.)$.

Notation 1.3.10. In the sequel we shall set: $\tau^{<n}(X) = \tau^{\leqslant(n-1)}(X)$ and $\tau^{>n}(X) = \tau^{\geqslant(n+1)}(X)$.

1.4. Mapping cones

Let $\mathscr{C}$ be an additive category, and let $f : X \to Y$ be a morphism in $\mathbf{C}(\mathscr{C})$.

Definition 1.4.1. *The mapping cone of f, denoted by $M(f)$, is the object of $\mathbf{C}(\mathscr{C})$ defined as follows:*

$$(1.4.1) \qquad \begin{cases} M(f)^n = X^{n+1} \oplus Y^n , \\ \mathrm{d}^n_{M(f)} = \begin{pmatrix} \mathrm{d}^n_{X[1]} & 0 \\ f^{n+1} & \mathrm{d}^n_Y \end{pmatrix} . \end{cases}$$

Recall that $\mathrm{d}^n_{X[1]} = -\mathrm{d}^{n+1}_X$.

We define the morphisms $\alpha(f) : Y \to M(f)$ and $\beta(f) : M(f) \to X[1]$ by:

$$(1.4.2) \qquad \alpha(f)^n = \begin{pmatrix} 0 \\ \mathrm{id}_{Y^n} \end{pmatrix} ,$$

$$(1.4.3) \qquad \beta(f)^n = (\mathrm{id}_{X^{n+1}}, 0) .$$

Lemma 1.4.2. *For any $f : X \to Y$ in $\mathbf{C}(\mathscr{C})$, there exists $\phi : X[1] \to M(\alpha(f))$ such that:*

(1.4.4) *ϕ is an isomorphism in $\mathbf{K}(\mathscr{C})$,*

(1.4.5) *The diagram below commutes in $\mathbf{K}(\mathscr{C})$:*

$$\begin{array}{ccccccc} Y & \xrightarrow{\alpha(f)} & M(f) & \xrightarrow{\beta(f)} & X[1] & \xrightarrow{-f[1]} & Y[1] \\ \downarrow{\scriptstyle \mathrm{id}_Y} & & \downarrow{\scriptstyle \mathrm{id}_{M(f)}} & & \downarrow{\scriptstyle \phi} & & \downarrow{\scriptstyle \mathrm{id}_{Y[1]}} \\ Y & \xrightarrow[\alpha(f)]{} & M(f) & \xrightarrow[\alpha(\alpha(f))]{} & M(\alpha(f)) & \xrightarrow[\beta(\alpha(f))]{} & Y[1] . \end{array}$$

Note that such a result would not hold in $\mathbf{C}(\mathscr{C})$. Note further that ϕ is not unique even in $\mathbf{K}(\mathscr{C})$. This is a source of many problems which are not yet all well clarified.

Proof. We have:

$$M(\alpha(f))^n = Y^{n+1} \oplus M(f)^n = Y^{n+1} \oplus X^{n+1} \oplus Y^n \ .$$

We define $\phi^n : X[1]^n \to M(\alpha(f))^n$ and $\psi^n : M(\alpha(f))^n \to X[1]^n$ by:

$$\phi^n = \begin{pmatrix} -f^{n+1} \\ \mathrm{id}_{X^{n+1}} \\ 0 \end{pmatrix}, \qquad \psi^n = (0, \mathrm{id}_{X^{n+1}}, 0) \ .$$

Then the lemma follows from the following observations.

(a) $\phi = (\phi^n)_n$ and $\psi = (\psi^n)_n$ are morphisms of complexes,
(b) $\psi \circ \phi = \mathrm{id}_{X[1]}$,
(c) $\phi \circ \psi$ is homotopic to $\mathrm{id}_{M(\alpha(f))}$,
(d) $\psi \circ \alpha(\alpha(f)) = \beta(f)$,
(e) $\beta(\alpha(f)) \circ \phi = -f[1]$.

All these properties, except (c), can be checked directly. To get (c) we define $s^n : M(\alpha(f))^n \to M(\alpha(f))^{n-1}$ by:

$$s^n = \begin{pmatrix} 0 & 0 & \mathrm{id}_{Y^n} \\ 0 & 0 & 0 \\ 0 & 0 & 0 \end{pmatrix} .$$

Then one verifies that:

$$\mathrm{id}_{M(\alpha(f))^n} - \phi^n \circ \psi^n = s^{n+1} \circ d^n_{M(\alpha(f))} + d^{n-1}_{M(\alpha(f))} \circ s^n \ . \quad \square$$

One defines a **triangle** in $\mathbf{K}(\mathscr{C})$ as being a sequence of morphisms $X \to Y \to Z \to X[1]$ and a morphism of triangles as being a commutative diagram in $\mathbf{K}(\mathscr{C})$:

$$\begin{array}{ccccccc} X & \longrightarrow & Y & \longrightarrow & Z & \longrightarrow & X[1] \\ \downarrow{\scriptstyle \phi} & & \downarrow & & \downarrow & & \downarrow{\scriptstyle \phi[1]} \\ X' & \longrightarrow & Y' & \longrightarrow & Z' & \longrightarrow & X'[1] \ . \end{array}$$

Definition 1.4.3. *A triangle $X \to Y \to Z \to X[1]$ in $\mathbf{K}(\mathscr{C})$ is called a distinguished triangle, if it is isomorphic to a triangle $X' \xrightarrow{f} Y' \xrightarrow{\alpha(f)} M(f) \xrightarrow{\beta(f)} X'[1]$, for some f in $\mathbf{C}(\mathscr{C})$.*

Proposition 1.4.4. *The collection of distinguished triangles in $\mathbf{K}(\mathscr{C})$ satisfies the following properties,* (TR 0)–(TR 5).

(TR 0) *A triangle isomorphic to a distinguished triangle is distinguished.*
(TR 1) *For any $X \in \mathrm{Ob}(\mathbf{K}(\mathscr{C}))$, $X \xrightarrow{\mathrm{id}_X} X \longrightarrow 0 \longrightarrow X[1]$ is a distinguished triangle.*

(TR 2) *Any* $f : X \to Y$ *in* $\mathbf{K}(\mathscr{C})$ *can be embedded in a distinguished triangle* $X \xrightarrow{f} Y \to Z \to X[1]$.

(TR 3) $X \xrightarrow{f} Y \xrightarrow{g} Z \xrightarrow{h} X[1]$ *is a distinguished triangle if and only if* $Y \xrightarrow{g} Z \xrightarrow{h} X[1] \xrightarrow{-f[1]} Y[1]$ *is a distinguished triangle.*

(TR 4) *Given two distinguished triangles* $X \xrightarrow{f} Y \to Z \to X[1]$ *and* $X' \xrightarrow{f'} Y' \to Z' \to X'[1]$, *a commutative diagram*

$$\begin{array}{ccc} X & \xrightarrow{f} & Y \\ {\scriptstyle u}\downarrow & & \downarrow{\scriptstyle v} \\ X' & \xrightarrow{f'} & Y' \end{array}$$

can be embedded in a morphism of triangles (not necessarily unique).

(TR 5) (octahedral axiom). *Suppose given distinguished triangles:*

$$\begin{array}{l} X \xrightarrow{f} Y \longrightarrow Z' \longrightarrow X[1]\ , \\ Y \xrightarrow{g} Z \longrightarrow X' \longrightarrow Y[1]\ , \\ X \xrightarrow{g \circ f} Z \longrightarrow Y' \longrightarrow X[1]\ , \end{array}$$

then there exists a distinguished triangle

$$Z' \to Y' \to X' \to Z'[1]$$

such that the following diagram is commutative:

$$\begin{array}{ccccccc} X & \xrightarrow{f} & Y & \longrightarrow & Z' & \longrightarrow & X[1] \\ {\scriptstyle \mathrm{id}_X}\downarrow & & \downarrow{\scriptstyle g} & & \downarrow & & \downarrow{\scriptstyle \mathrm{id}_{X[1]}} \\ X & \xrightarrow{g \circ f} & Z & \longrightarrow & Y' & \longrightarrow & X[1] \\ {\scriptstyle f}\downarrow & & \downarrow{\scriptstyle \mathrm{id}_Z} & & \downarrow & & \downarrow{\scriptstyle f[1]} \\ Y & \xrightarrow{g} & Z & \longrightarrow & X' & \longrightarrow & Y[1] \\ \downarrow & & \downarrow & & \downarrow{\scriptstyle \mathrm{id}_{X'}} & & \downarrow \\ Z' & \longrightarrow & Y' & \longrightarrow & X' & \longrightarrow & Z'[1]\ . \end{array}$$

Proof. The properties (TR 0) and (TR 2) are obvious, and (TR 3) follows from Lemma 1.4.2.

Since the mapping cone of $f : 0 \to X$ is X, the triangle $0 \longrightarrow X \xrightarrow{\mathrm{id}_X} X \longrightarrow 0[1]$ is distinguished. Applying (TR 3) we get (TR 1). Let us prove (TR 4). We may assume that $X \xrightarrow{f} Y \to Z \to X[1]$ and $X' \xrightarrow{f'} Y' \to Z' \to X'[1]$ are $X \xrightarrow{f} Y \xrightarrow{\alpha(f)}$

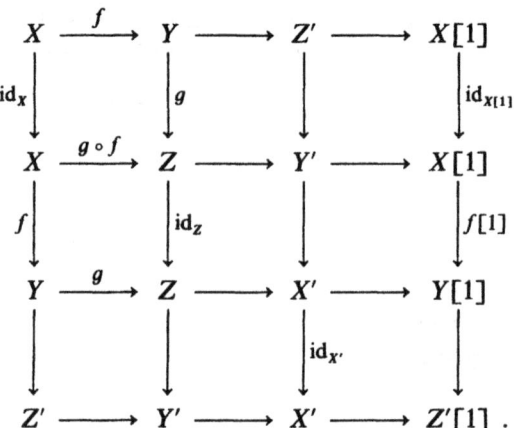

$M(f) \xrightarrow{\beta(f)} X[1]$ and $X' \xrightarrow{f'} Y' \xrightarrow{\alpha(f')} M(f') \xrightarrow{\beta(f')} X'[1]$, respectively. We shall construct a morphism $w: M(f) \to M(f')$ such that:

$$(1.4.4) \qquad \begin{cases} w \circ \alpha(f) = \alpha(f') \circ v , \\ u[1] \circ \beta(f) = \beta(f') \circ w . \end{cases}$$

By the definition of $\mathbf{K}(\mathscr{C})$, there exists $s^n: X^n \to Y'^{n-1}$ such that $v^n \circ f^n - f'^n \circ u^n = s^{n+1} \circ d_X^n + d_{Y'}^{n-1} \circ s^n$. We define $w^n: M(f)^n = X^{n+1} \oplus Y^n \to M(f')^n = X'^{n+1} \oplus Y'^n$ by:

$$w^n = \begin{pmatrix} u^{n+1} & 0 \\ s^{n+1} & v^n \end{pmatrix} .$$

Then a direct calculation shows that w is a morphism of complexes and satisfies (1.4.4).

Let us prove (TR 5). We may assume $Z' = M(f)$, $X' = M(g)$ and $Y' = M(g \circ f)$. Let us define $u: Z' \to Y'$ and $v: Y' \to X'$ by:

$$u^n: X^{n+1} \oplus Y^n \to X^{n+1} \oplus Z^n , \qquad u = \begin{pmatrix} \mathrm{id}_{X^{n+1}} & 0 \\ 0 & g^n \end{pmatrix} ,$$

$$v^n: X^{n+1} \oplus Z^n \to Y^{n+1} \oplus Z^n , \qquad v^n = \begin{pmatrix} f^{n+1} & 0 \\ 0 & \mathrm{id}_{Z^n} \end{pmatrix} .$$

We define $w: X' \to Z'[1]$ as the composite $X' \to Y[1] \to Z'[1]$. Then the diagram in (TR 5) is commutative, and it is enough to show that $Z' \xrightarrow{u} Y' \xrightarrow{v} X' \xrightarrow{w} Z'[1]$ is a distinguished triangle. For that purpose we shall construct an isomorphism $\phi: M(u) \to X'$ and its inverse $\psi: X' \to M(u)$ such that $\phi \circ \alpha(u) = v$ and $\beta(u) \circ \psi = w$. We have:

$$M(u)^n = M(f)^{n+1} \oplus M(g \circ f)^n = X^{n+2} \oplus Y^{n+1} \oplus X^{n+1} \oplus Z^n$$

and $X'^n = M(g)^n = Y^{n+1} \oplus Z^n$. We define ϕ and ψ by:

$$\phi^n = \begin{pmatrix} 0 & \mathrm{id}_{Y^{n+1}} & f^{n+1} & 0 \\ 0 & 0 & 0 & \mathrm{id}_{Z^n} \end{pmatrix} , \qquad \psi^n = \begin{pmatrix} 0 & 0 \\ \mathrm{id}_{Y^{n+1}} & 0 \\ 0 & 0 \\ 0 & \mathrm{id}_{X^{n+1}} \end{pmatrix} .$$

Then one checks easily that ϕ and ψ are morphisms of complexes and $\phi \circ \alpha(u) = v$, $\beta(u) \circ \psi = w$.

We have $\phi \circ \psi = \mathrm{id}_{X'}$. If we define:

$$s^n: M(u)^n \to M(u)^{n-1} , \qquad s^n = \begin{pmatrix} 0 & 0 & \mathrm{id}_{X^{n+1}} & 0 \\ 0 & 0 & 0 & 0 \\ 0 & 0 & 0 & 0 \\ 0 & 0 & 0 & 0 \end{pmatrix}$$

then:

$$(\mathrm{id}_{M(u)} - \psi \circ \phi)^n = s^{n+1} \circ \mathrm{d}^n_{M(u)} + \mathrm{d}^{n-1}_{M(u)} \circ s^n \ .$$

Hence $\psi \circ \phi$ equals $\mathrm{id}_{M(u)}$ in $\mathbf{K}(\mathscr{C})$. □

Remark 1.4.5. Property (TR 5) may be visualized by the following octahedral diagram:

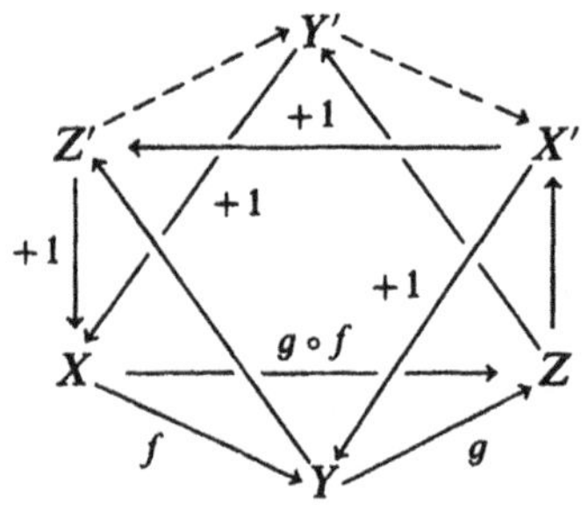

Diagram 1.4.1

1.5. Triangulated categories

We obtain the notion of triangulated category by abstracting the properties of $\mathbf{K}(\mathscr{C})$.

Let $\mathscr{C}$ be an additive category, together with an automorphism $T : \mathscr{C} \to \mathscr{C}$. We write sometimes $[1]$ for T and $[k]$ for T^k, (i.e. $X[1]$ for $T(X)$, or $f[1]$ for $T(f)$).

A triangle in $\mathscr{C}$ is a sequence of morphisms

$$X \to Y \to Z \to T(X) \ .$$

Definition 1.5.1. *A triangulated category $\mathscr{C}$ consists of the following data and rules.*

(1.5.1) *An additive category $\mathscr{C}$ together with an automorphism $T : \mathscr{C} \to \mathscr{C}$,*

(1.5.2) *a family of triangles, called distinguished triangles.*

These data satisfy the axioms (TR 0)–(TR 5) *of Proposition* 1.4.4 *when setting* $X[1] = T(X)$.

Let $(\mathscr{C}, T)$ and $(\mathscr{C}', T')$ be two triangulated categories. We say that an additive functor F from $\mathscr{C}$ to $\mathscr{C}'$ is a functor of triangulated categories if $F \circ T \simeq T' \circ F$, and F sends distinguished triangles of $\mathscr{C}$ into distinguished triangles of $\mathscr{C}'$.

Clearly, for an additive category $\mathscr{C}$, $\mathbf{K}(\mathscr{C})$ is a triangulated category. Now let $\mathscr{C}$ be a triangulated category and let $\mathscr{A}$ be an abelian category.

Definition 1.5.2. *An additive functor $F : \mathscr{C} \to \mathscr{A}$ is called a cohomological functor if for any distinguished triangle $X \to Y \to Z \to T(X)$, the sequence $F(X) \to F(Y) \to F(Z)$ is exact.*

For a cohomological functor F, we write F^k for $F \circ T^k$. Then for any distinguished triangle $X \to Y \to Z \to T(X)$ we obtain a long exact sequence:

$$\cdots \to F^{k-1}(Z) \to F^k(X) \to F^k(Y) \to F^k(Z) \to F^{k+1}(X) \to \cdots . \tag{1.5.3}$$

Proposition 1.5.3. (i) *If $X \xrightarrow{f} Y \xrightarrow{g} Z \to T(X)$ is a distinguished triangle, then $g \circ f = 0$.*

(ii) *For any $W \in \mathrm{Ob}(\mathscr{C})$, $\mathrm{Hom}_{\mathscr{C}}(W, \cdot)$ and $\mathrm{Hom}_{\mathscr{C}}(\cdot, W)$ are cohomological functors.*

Proof. (i) By (TR 1), $X \xrightarrow{\mathrm{id}_X} X \longrightarrow 0 \longrightarrow T(X)$ is a distinguished triangle. Therefore by (TR 4) there is a morphism $\phi : 0 \to Z$ which makes the following diagram commutative:

$$\begin{array}{ccccccc}
X & \longrightarrow & X & \longrightarrow & 0 & \longrightarrow & T(X) \\
{\scriptstyle \mathrm{id}_X}\downarrow & & {\scriptstyle f}\downarrow & & \vdots{\scriptstyle \phi} & & \downarrow \\
X & \xrightarrow{f} & Y & \xrightarrow{g} & Z & \longrightarrow & T(X) .
\end{array}$$

Hence $g \circ f = \phi \circ 0 = 0$.

(ii) Let $X \xrightarrow{f} Y \xrightarrow{g} Z \to T(X)$ be a distinguished triangle. In order to show that $\mathrm{Hom}_{\mathscr{C}}(W, \cdot)$ is a cohomological functor, it is enough to show that, for any $\phi \in \mathrm{Hom}_{\mathscr{C}}(W, Y)$ with $g \circ \phi = 0$, we can find $\psi \in \mathrm{Hom}_{\mathscr{C}}(W, X)$, with $\phi = f \circ \psi$. This follows from (TR 1), (TR 3) and (TR 4) which imply that the dotted arrow below can be completed:

$$\begin{array}{ccccccc}
W & \xrightarrow{\mathrm{id}_W} & W & \longrightarrow & 0 & \longrightarrow & T(W) \\
\vdots{\scriptstyle \psi} & & \downarrow{\scriptstyle \phi} & & \downarrow & & \\
X & \xrightarrow{f} & Y & \xrightarrow{g} & Z & \longrightarrow & T(X) .
\end{array}$$

The proof that $\mathrm{Hom}_{\mathscr{C}}(\cdot, W)$ is a cohomological functor is similar. □

Remark 1.5.4. Let $\mathscr{C}$ be an additive category, $f : X \to Y$ a morphism in $\mathbf{C}(\mathscr{C})$. In general the composite $X \xrightarrow{f} Y \xrightarrow{\alpha(f)} M(f)$ is not zero in $\mathbf{C}(\mathscr{C})$, but only in $\mathbf{K}(\mathscr{C})$.

Corollary 1.5.5. *Let*

$$\begin{array}{ccccccc} X & \longrightarrow & Y & \longrightarrow & Z & \longrightarrow & T(X) \\ \downarrow{\scriptstyle\phi} & & \downarrow{\scriptstyle\psi} & & \downarrow{\scriptstyle\theta} & & \downarrow{\scriptstyle T(\phi)} \\ X' & \longrightarrow & Y' & \longrightarrow & Z' & \longrightarrow & T(X') \end{array}$$

be a morphism of distinguished triangles. If ϕ and ψ are isomorphisms, then so is θ.

Proof. For any $W \in \mathrm{Ob}(\mathscr{C})$, let us apply the functor $\mathrm{Hom}_{\mathscr{C}}(W, \cdot)$ to the above diagram. We obtain a commutative diagram whose rows are exact. Since $\mathrm{Hom}_{\mathscr{C}}(W, \phi)$ and $\mathrm{Hom}_{\mathscr{C}}(W, \psi)$ are isomorphisms, as well as $\mathrm{Hom}_{\mathscr{C}}(W, T(\phi))$ and $\mathrm{Hom}_{\mathscr{C}}(W, T(\psi))$, we obtain that $\mathrm{Hom}_{\mathscr{C}}(W, \theta)$ is an isomorphism by Exercise I.8. This completes the proof, in view of Proposition 1.1.8. □

Proposition 1.5.6. *Let $\mathscr{C}$ be an abelian category. Then the functor $H^0(\cdot) : \mathbf{K}(\mathscr{C}) \to \mathscr{C}$ is a cohomological functor.*

Proof. It is enough to show that if $f : X \to Y$ is a morphism in $\mathbf{C}(\mathscr{C})$, then the sequence

$$H^0(Y) \to H^0(M(f)) \to H^0(X[1])$$

is exact.

Since $0 \to Y \to M(f) \to X[1] \to 0$ is an exact sequence in $\mathbf{C}(\mathscr{C})$, the result follows from Proposition 1.3.6. □

Definition 1.5.7. *Let $\mathscr{C}$ be an abelian category and let $f : X \to Y$ be a morphism in $\mathbf{K}(\mathscr{C})$. One says that f is a quasi-isomorphism (qis for short) if $H^n(f)$ is an isomorphism for each n.*

Hence f is a *qis* if and only if $H^n(M(f)) = 0$ for each n. If f is a *qis*, one writes $X \underset{qis}{\longrightarrow} Y$, for short.

Notations 1.5.8. Let $\mathscr{C}$ be a triangulated category. In the subsequent sections we shall often write $X \longrightarrow Y \longrightarrow Z \underset{+1}{\longrightarrow}$ instead of $X \to Y \to Z \to T(X)$, to denote a distinguished triangle.

Definition 1.5.9. *Let $\mathscr{C}$ be a triangulated category. A triangle $X \xrightarrow{\alpha} Y \xrightarrow{\beta} Z \xrightarrow{\gamma} T(X)$ in $\mathscr{C}$ is called antidistinguished if the triangle $X \xrightarrow{\alpha} Y \xrightarrow{\beta} Z \xrightarrow{-\gamma} T(X)$ is distinguished.*

Note that the category $\mathscr{C}$, endowed with the family of antidistinguished triangles, is a triangulated category. We denote it by $\mathscr{C}^a$.

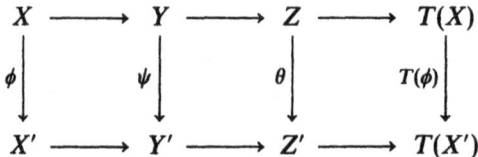

1.6. Localization of categories

Let $\mathscr{C}$ be a category, and let S be a family of morphisms in $\mathscr{C}$.

Definition 1.6.1. *One says that* S *is a multiplicative system if it satisfies* (S 1)–(S 4) *below.*

(S 1) *For any* $X \in \mathrm{Ob}(\mathscr{C})$, $\mathrm{id}_X \in S$.
(S2) *For any pair* (f, g) *of* S *such that the composition* $g \circ f$ *exists,* $g \circ f \in S$.
(S 3) *Any diagram:*

$$\begin{array}{ccc} & & Z \\ & & \downarrow{\scriptstyle g} \\ X & \xrightarrow{f} & Y \end{array}$$

with $g \in S$, *may be completed to a commutative diagram:*

$$\begin{array}{ccc} W & \longrightarrow & Z \\ {\scriptstyle h}\downarrow & & \downarrow{\scriptstyle g} \\ X & \xrightarrow{f} & Y \end{array}$$

with $h \in S$. *Ditto with all the arrows reversed.*
(S4) *If* f *and* g *belong to* $\mathrm{Hom}_{\mathscr{C}}(X, Y)$, *the following conditions are equivalent:*
  (i) *there exists* $t : Y \to Y'$, $t \in S$, *such that* $t \circ f = t \circ g$,
  (ii) *there exists* $s : X' \to X$, $s \in S$, *such that* $f \circ s = g \circ s$.

Definition 1.6.2. *Let* $\mathscr{C}$ *be a category,* S *a multiplicative system. The category* $\mathscr{C}_S$, *called the localization of* $\mathscr{C}$ *by* S, *is defined by:*

$$\mathrm{Ob}(\mathscr{C}_S) = \mathrm{Ob}(\mathscr{C}) \ , \tag{1.6.1}$$

(1.6.2) *for any pair* (X, Y) *of* $\mathrm{Ob}(\mathscr{C})$,

$$\mathrm{Hom}_{\mathscr{C}_S}(X, Y) = \{(X', s, f); X' \in \mathrm{Ob}(\mathscr{C}), s : X' \to X, f : X' \to Y, s \in S\}/\mathscr{R}$$

where $\mathscr{R}$ *is the following equivalence relation:*

$$(X', s, f)\mathscr{R}(X'', t, g)$$

iff there exists a commutative diagram

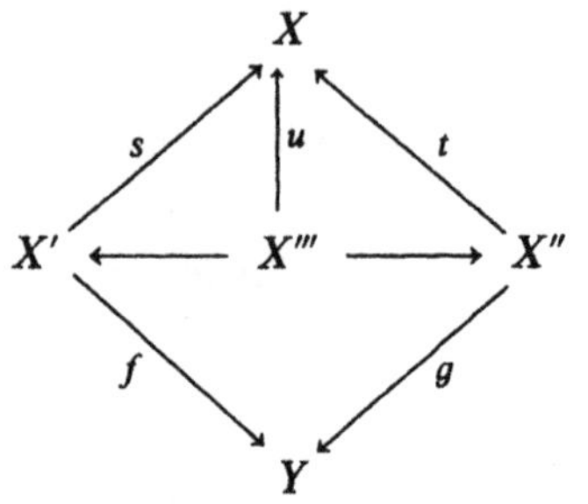

with $u \in S$.

The composition of $(X', s, f) \in \mathrm{Hom}_{\mathscr{C}_S}(X, Y)$ *and* $(Y', t, g) \in \mathrm{Hom}_{\mathscr{C}_S}(Y, Z)$ *is defined as follows. We use* (S 3) *to find a commutative diagram*:

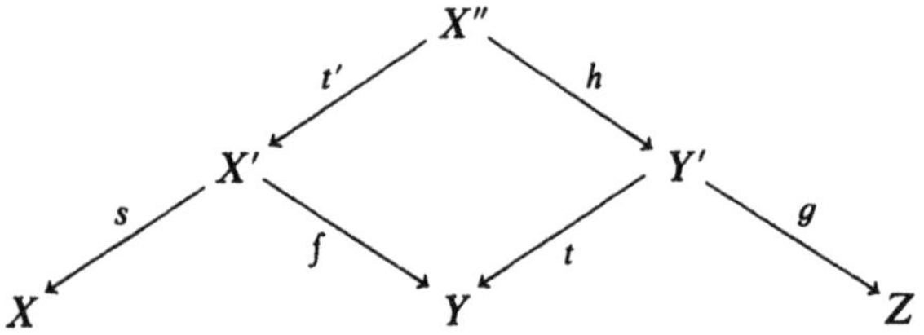

with $t' \in S$, *and we set*:

$$(Y', t, g) \circ (X', s, f) = (X'', s \circ t', g \circ h) \ .$$

One sees easily, using the axioms (S 1)–(S 4), that $\mathscr{C}_S$ is a category.

We shall denote by Q the functor:

$$Q : \mathscr{C} \to \mathscr{C}_S$$

defined by $Q(X) = X$ for $X \in \mathrm{Ob}(\mathscr{C})$, and $Q(f) = (X, \mathrm{id}_X, f)$ for $f \in \mathrm{Hom}_{\mathscr{C}}(X, Y)$.

Proposition 1.6.3. (i) *For* $s \in S$, $Q(s)$ *is an isomorphism in* $\mathscr{C}_S$.

(ii) *Let* $\mathscr{C}'$ *be another category,* $F : \mathscr{C} \to \mathscr{C}'$ *be a functor such that* $F(s)$ *is an isomorphism for all* $s \in S$. *Then* F *factors uniquely through* Q.

The proof is straightforward.

Remark 1.6.4. It follows from Proposition 1.6.3. that:

$$(\mathscr{C}^\circ)_S \simeq (\mathscr{C}_S)^\circ \ . \tag{1.6.3}$$

Hence we get an equivalent category to $\mathscr{C}_S$ by replacing condition (1.6.2) by:

(1.6.2)′ $\mathrm{Hom}_{\mathscr{C}_S}(X, Y) = \{(Y', t, g);\ Y' \in \mathrm{Ob}(\mathscr{C}), t : Y \to Y', g : X \to Y', t \in S\}/\mathscr{R}'$

where $\mathscr{R}'$ is defined similarly to $\mathscr{R}$.

Proposition 1.6.5. *Let $\mathscr{C}$ be a category, $\mathscr{C}'$ a full subcategory. Let S be a multiplicative system in $\mathscr{C}$, and let S' be the family of morphisms of $\mathscr{C}'$ which belong to S. Assume S' is a multiplicative system in $\mathscr{C}'$, and assume moreover that one of the following conditions holds:*

(i) *whenever $f : X \to Y$ is a morphism in S, with $Y \in \mathrm{Ob}(\mathscr{C}')$, there exists $g : W \to X$, with $W \in \mathrm{Ob}(\mathscr{C}')$ and $f \circ g \in S$,*
(ii) *the same as* (i) *with the arrows reversed.*

Then the localisation $\mathscr{C}'_{S'}$ is a full subcategory of $\mathscr{C}_S$.

The proof is straightforward.

When $\mathscr{C}$ is a triangulated category, it is possible to localize $\mathscr{C}$ with respect to a family of objects of $\mathscr{C}$.

Definition 1.6.6. *Let $\mathscr{C}$ be a triangulated category, and let $\mathscr{N}$ be a subfamily of* $\mathrm{Ob}(\mathscr{C})$. *One says that $\mathscr{N}$ is a null system if it satisfies* (N 1)–(N3) below.

(N 1) $0 \in \mathscr{N}$,
(N 2) $X \in \mathscr{N}$ *if and only if* $X[1] \in \mathscr{N}$,
(N 3) *If $X \to Y \to Z \to X[1]$ is a distinguished triangle, and $X \in \mathscr{N}$, $Y \in \mathscr{N}$, then $Z \in \mathscr{N}$.*

Now we set:

$$\left\{\begin{array}{l} S(\mathscr{N}) = \{f : X \to Y;\ f \text{ is embedded into a distinguished} \\ \text{triangle } X \xrightarrow{f} Y \to Z \to X[1], \text{ with } Z \in \mathscr{N}\}\ . \end{array}\right. \tag{1.6.4}$$

Proposition 1.6.7. *Assume $\mathscr{N}$ is a null system. Then $S(\mathscr{N})$ is a multiplicative system.*

Proof. The property (S 1) is deduced from (N 1) and (TR 1). Let us prove (S 2). Let $X \xrightarrow{f} Y \to Z' \to X[1]$ and $Y \xrightarrow{g} Z \to X' \to Y[1]$ be two distinguished triangles, with $X' \in \mathscr{N}$, $Z' \in \mathscr{N}$. By (TR 2) there exists a distinguished triangle $X \xrightarrow{g \circ f} Z \longrightarrow Y' \longrightarrow X[1]$, and by (TR 5) there exists a distinguished triangle $Z' \to Y' \to X' \to Z'[1]$. By (N 2), (N 3) and (TR 3), we have: $Y' \in \mathscr{N}$. Hence $g \circ f \in S(\mathscr{N})$.

To prove (S 3), consider a distinguished triangle $Z \xrightarrow{g} Y \xrightarrow{k} X' \to Z[1]$, with $X' \in \mathscr{N}$, and let $f : X \to Y$. There exists a distinguished triangle

$$W \longrightarrow X \xrightarrow{k \circ f} X' \longrightarrow W[1]\ .$$

Then by (TR 4) and (TR 3), we have a morphism of distinguished triangles:

$$\begin{array}{ccccccc} W & \xrightarrow{h} & X & \longrightarrow & X' & \longrightarrow & W[1] \\ \downarrow & & \downarrow & & \downarrow{\scriptstyle \mathrm{id}_{X'}} & & \downarrow \\ Z & \longrightarrow & Y & \longrightarrow & X' & \longrightarrow & Z[1]\ . \end{array}$$

Since $X' \in \mathscr{N}$, h belongs to $S(\mathscr{N})$.

A similar proof holds by reversing the arrows.

Finally we prove (S 4). Let $f: X \to Y$ and $t: Y \to Y'$, with $t \in S(\mathcal{N})$ and $t \circ f = 0$. We shall show that there exists $s: X' \to X, s \in S(\mathcal{N})$, such that $f \circ s = 0$. Let $Z \xrightarrow{g} Y \xrightarrow{t} Y' \to Z[1]$ be a distinguished triangle, with $Z \in \mathcal{N}$. By (TR 1), (TR 3), (TR 4), there exists $h: X \to Z$ such that $f = g \circ h$. If we embed h into a distinguished triangle $X' \xrightarrow{s} X \xrightarrow{h} Z \to X'[1]$ then s will satisfy the desired properties. The proof of the converse implication is similar. □

Notation 1.6.8. Let $\mathscr{C}$ be a triangulated category and $\mathcal{N}$ a null system in $\mathscr{C}$. We write $\mathscr{C}/\mathcal{N}$ instead of $\mathscr{C}_{S(\mathcal{N})}$.

Proposition 1.6.9. *Let $\mathscr{C}$ be a triangulated category and $\mathcal{N}$ a null system.*

(i) *$\mathscr{C}/\mathcal{N}$ becomes a triangulated category by taking for distinguished triangles those isomorphic to the image of a distinguished triangle in $\mathscr{C}$.*
(ii) *Denote by Q the natural functor $\mathscr{C} \to \mathscr{C}/\mathcal{N}$.* We have $Q(X) \simeq 0$ *for $X \in \mathcal{N}$.*
(iii) *Any functor $F: \mathscr{C} \to \mathscr{C}'$ of triangulated categories such that $F(X) \simeq 0$ for all $X \in \mathcal{N}$, factors uniquely through Q.*

The proof is obvious by Proposition 1.6.3.

Proposition 1.6.10. *Let $\mathscr{C}$ be a triangulated category, $\mathcal{N}$ a null system in $\mathscr{C}$, $\mathscr{C}'$ a full triangulated subcategory of $\mathscr{C}$ such that any distinguished triangle $X \to Y \to Z \to X[1]$ in $\mathscr{C}$, with $X \in \mathrm{Ob}(\mathscr{C}')$, $Y \in \mathrm{Ob}(\mathscr{C}')$, is a distinguished triangle in $\mathscr{C}'$. Let $\mathcal{N}' = \mathcal{N} \cap \mathrm{Ob}(\mathscr{C}')$. Then:*

(i) *$\mathcal{N}'$ is a null system in $\mathscr{C}'$.*
(ii) *Assume moreover that any morphism $Y \to Z$ in $\mathscr{C}$ with $Y \in \mathrm{Ob}(\mathscr{C}')$, $Z \in \mathcal{N}$ factorizes through an object of $\mathcal{N} \cap \mathrm{Ob}(\mathscr{C}')$. Then $\mathscr{C}'/\mathcal{N}'$ is a full subcategory of $\mathscr{C}/\mathcal{N}$.*

Proof. (i) is clear.

(ii) We shall verify condition (i) of Proposition 1.6.5.

Let $X \xrightarrow[f]{} Y \to Z \to X[1]$ be a distinguished triangle with $Y \in \mathrm{Ob}(\mathscr{C}')$, $Z \in \mathcal{N}$. By the hypothesis, the morphism $Y \to Z$ factorizes through $Y \to Z' \to Z$ with $Z' \in \mathcal{N} \cap \mathrm{Ob}(\mathscr{C}')$. Applying Axiom TR 5 to the morphisms $Y \to Z'$ and $Z' \to Z$ we find a distinguished triangle $Y \to Z' \to W \to Y[1]$ such that $W[-1] \to Y$ factorizes to $W[-1] \to X \xrightarrow{f} Y$. This completes the proof. □

Remark 1.6.11. Let $\mathcal{N}$ be a null system in $\mathscr{C}$, Q the functor $\mathscr{C} \to \mathscr{C}/\mathcal{N}$. Then $X \in \mathrm{Ob}(\mathscr{C})$ satisfies $Q(X) \simeq 0$ if and only if there exists $Y \in \mathrm{Ob}(\mathscr{C})$ such that $X \oplus Y \in \mathrm{Ob}(\mathcal{N})$. This is again equivalent to $X \oplus X[1] \in \mathrm{Ob}(\mathcal{N})$. The proof is straightforward.

1.7. Derived categories

In this section, $\mathscr{C}$ will denote an abelian category.

We shall apply the preceding construction to the triangulated category $\mathbf{K}(\mathscr{C})$. It is clear that:

$$\mathscr{N} = \{X \in \mathrm{Ob}(\mathbf{K}(\mathscr{C}));\ H^n(X) = 0 \text{ for any } n\} \tag{1.7.1}$$

is a null system. Note that, in view of Proposition 1.5.6, $S(\mathscr{N})$ consists of quasi-isomorphisms of $\mathbf{K}(\mathscr{C})$.

Definition 1.7.1. *We set* $\mathbf{D}(\mathscr{C}) = \mathbf{K}(\mathscr{C})/\mathscr{N}$ *and call* $\mathbf{D}(\mathscr{C})$ *the derived category of* $\mathscr{C}$.

By replacing $\mathbf{K}(\mathscr{C})$ with $\mathbf{K}^b(\mathscr{C})$ (resp. $\mathbf{K}^+(\mathscr{C})$, resp. $\mathbf{K}^-(\mathscr{C})$), we define similarly the derived categories $\mathbf{D}^b(\mathscr{C})$ (resp. $\mathbf{D}^+(\mathscr{C})$, resp. $\mathbf{D}^-(\mathscr{C})$). By Proposition 1.6.3 the functor $H^n(\cdot): \mathbf{K}(\mathscr{C}) \to \mathscr{C}$, factors through $\mathbf{D}(\mathscr{C})$. We still denote by $H^n(\cdot)$ the functor from $\mathbf{D}(\mathscr{C})$ to $\mathscr{C}$ so obtained.

Proposition 1.7.2. (i) $\mathbf{D}^b(\mathscr{C})$ *(resp.* $\mathbf{D}^+(\mathscr{C})$, *resp.* $\mathbf{D}^-(\mathscr{C})$*) is equivalent to the full subcategory of* $\mathbf{D}(\mathscr{C})$ *consisting of objects* X *such that* $H^n(X) = 0$ *for* $|n| \gg 0$ *(resp.* $n \ll 0$, *resp.* $n \gg 0$*).*

(ii) *By the composition of the functors* $\mathscr{C} \to \mathbf{K}(\mathscr{C}) \to \mathbf{D}(\mathscr{C})$, $\mathscr{C}$ *is equivalent to the full subcategory of* $\mathbf{D}(\mathscr{C})$ *consisting of objects* X *such that* $H^n(X) = 0$ *for* $n \neq 0$.

Proof. The Proposition follows immediately from Proposition 1.6.10 and the fact that, for $X \in \mathrm{Ob}(\mathbf{K}(\mathscr{C}))$, with $H^j(X) = 0$ for $j < n$ (resp. $H^j(X) = 0$ for $j > n$), $X \to \tau^{\geqslant n}(X)$, (resp. $\tau^{\leqslant n}(X) \to X$), is a quasi-isomorphism. □

Remark 1.7.3. Let $X \in \mathrm{Ob}(\mathbf{K}(\mathscr{C}))$, $Q(X)$ its image in $\mathbf{D}(\mathscr{C})$. Then $Q(X) = 0$ iff X is quasi-isomorphic to zero in $\mathbf{K}(\mathscr{C})$: this follows immediately from Remark 1.6.11 and from the additivity of the functors $H^n(\cdot)$. Now let $f: X \to Y$ be a morphism in $\mathbf{C}(\mathscr{C})$. By definition, f is 0 in $\mathbf{D}(\mathscr{C})$ iff there exists a quasi-isomorphism $g: X' \to X$, such that $f \circ g$ is homotopic to 0, or else, iff there exists a quasi-isomorphism $h: Y \to Y'$ such that $h \circ f$ is homotopic to 0. Note that in general, there exists no quasi-isomorphism $g: X' \to X$ (resp. $h: Y \to Y'$) such that $f \circ g = 0$ (resp. $h \circ f = 0$) in $\mathbf{C}(\mathscr{C})$.

Example 1.7.4. Let $\mathscr{C} = \mathfrak{Mod}(\mathbb{Z})$, and let $f: X \to Y$ be given by:

$$\begin{array}{ccccccccccc}
X & \cdots & \longrightarrow & 0 & \longrightarrow & 0 & \longrightarrow & \mathbb{Z} & \longrightarrow & 0 & \longrightarrow \cdots \\
\Big\downarrow f: & & & \Big\downarrow & & \Big\downarrow & & \Big\downarrow 1 & & \Big\downarrow & \\
Y & \cdots & \longrightarrow & 0 & \longrightarrow & \mathbb{Z} & \xrightarrow{\ 1\ } & \mathbb{Z} & \longrightarrow & 0 & \longrightarrow \cdots
\end{array}$$

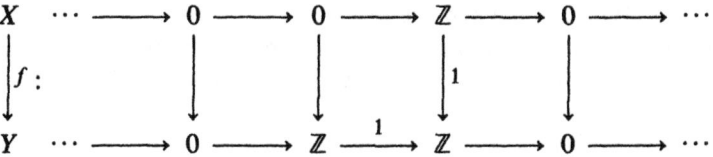

Then f is homotopic to 0, but there is no $g: X' \to X$ such that g is a quasi-isomorphism and $f \circ g = 0$. (In fact such a g would have to be 0 since the f^n are injective, which is absurd.)

Proposition 1.7.5. *Let $\mathscr{C}$ be an abelian category and let $0 \to X \xrightarrow{f} Y \xrightarrow{g} Z \to 0$ be an exact sequence in $\mathbf{C}(\mathscr{C})$. Let $M(f)$ be the mapping cone of f and let $\phi^n: M(f)^n = X^{n+1} \oplus Y^n \to Z^n$ be the morphism $(0, g^n)$. Then $\{\phi^n\}_n: M(f) \to Z$ is a morphism of complexes, $\phi \circ \alpha(f) = g$, and ϕ is a quasi-isomorphism.*

Proof. It is straightforward to see that ϕ is a morphism of complexes. Moreover we have an exact sequence:

$$0 \to M(\mathrm{id}_X) \xrightarrow{\gamma} M(f) \to Z \to 0$$

where γ is associated to the morphism $\mathrm{id}_X \to f$. This last morphism is described by the commutative diagram:

$$\begin{array}{ccc} X & \xrightarrow{\mathrm{id}} & X \\ \Big\downarrow{\scriptstyle \mathrm{id}} & & \Big\downarrow{\scriptstyle f} \\ X & \xrightarrow[f]{} & Y \end{array} .$$

By Proposition 1.3.6 it is enough to check that $H^n(M(\mathrm{id}_X)) = 0$ for all $n \in \mathbb{Z}$. Since $M(\mathrm{id}_X)$ is zero in $\mathbf{K}(\mathscr{C})$, this is evident. □

In the situation of Proposition 1.7.5 the distinguished triangle $X \to Y \to Z \xrightarrow{h} X[1]$ is called the distinguished triangle associated to the exact sequence $0 \to X \to Y \to Z \to 0$. Here $h = \beta(f) \circ \phi^{-1}$.

Note that the above distinguished triangle gives rise to a long exact sequence:

$$\cdots \longrightarrow H^n(X) \longrightarrow H^n(Y) \longrightarrow H^n(Z) \xrightarrow[H^n(h)]{} H^{n+1}(X) \longrightarrow \cdots$$

and $H^n(h) = -\delta$, δ being defined in Proposition 1.3.6.

Also note that if X, Y, Z are concentrated in degree zero (i.e.: are objects of $\mathscr{C}$) the morphism $h: Z \to X[1]$ is zero in $\mathbf{D}^+(\mathscr{C})$ if and only if the exact sequence splits. However one always has $H^n(h) = 0$, for all $n \in \mathbb{Z}$.

Remark 1.7.6. We have defined the functors $\tau^{\geqslant n}(\cdot)$ and $\tau^{\leqslant n}(\cdot)$ on $\mathbf{C}(\mathscr{C})$ in §3. It is straightforward that they transform a morphism homotopic to zero into a morphism homotopic to zero, and, by Proposition 1.3.7, they transform a quasi-isomorphism into a quasi-isomorphism. Thus one obtains the functors $\tau^{\geqslant n}: \mathbf{D}(\mathscr{C}) \to \mathbf{D}^+(\mathscr{C})$ and $\tau^{\leqslant n}: \mathbf{D}(\mathscr{C}) \to \mathbf{D}^-(\mathscr{C})$.

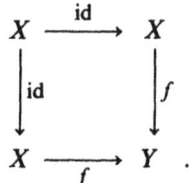

Applying Proposition 1.7.5, we get the distinguished triangles in $\mathbf{D}(\mathscr{C})$:

(1.7.2) $$\tau^{\leqslant n}(X) \longrightarrow X \longrightarrow \tau^{\geqslant n+1}(X) \xrightarrow[+1]{} ,$$

(1.7.3) $$\tau^{\leqslant n-1}(X) \longrightarrow \tau^{\leqslant n}(X) \longrightarrow H^n(X)[-n] \xrightarrow[+1]{} ,$$

(1.7.4) $$H^n(X)[-n] \longrightarrow \tau^{\geqslant n}(X) \longrightarrow \tau^{\geqslant n+1}(X) \xrightarrow[+1]{} .$$

In fact, for $X \in \mathrm{Ob}(\mathbf{C}(\mathscr{C}))$, $\tau^{\geqslant n+1}(X)$ is quasi-isomorphic to $\mathrm{Coker}(\tau^{\leqslant n}(X) \to X)$, $\tau^{\leqslant n-1}(X)$ is quasi-isomorphic to $\mathrm{Ker}(\tau^{\leqslant n}(X) \to H^n(X)[-n])$ and $\tau^{\geqslant n+1}(X)$ is quasi-isomorphic to $\mathrm{Coker}(H^n(X)[-n] \to \tau^{\geqslant n}(X))$.

Proposition 1.7.7. *Let $\mathscr{I}$ be a full additive subcategory of $\mathscr{C}$ such that:*

(1.7.5) *for any $X \in \mathrm{Ob}(\mathscr{C})$, there exists $X' \in \mathrm{Ob}(\mathscr{I})$ and an exact sequence $0 \to X \to X'$.*

Then:

(i) *for any $X \in \mathrm{Ob}(\mathbf{K}^+(\mathscr{C}))$, there exists $X' \in \mathrm{Ob}(\mathbf{K}^+(\mathscr{I}))$ and a quasi-isomorphism $f: X \to X'$,*

(ii) *let $\mathscr{N}$ be given by* (1.7.1) *and let $\mathscr{N}' = \mathscr{N} \cap \mathrm{Ob}(\mathbf{K}^+(\mathscr{I}))$. Then the canonical functor:*

$$\mathbf{K}^+(\mathscr{I})/\mathscr{N}' \to \mathbf{D}^+(\mathscr{C})$$

is an equivalence of categories.

Proof. By Proposition 1.6.5, (ii) follows from (i). Now let $X \in \mathrm{Ob}(\mathbf{K}^+(\mathscr{C}))$. We shall construct by induction a complex $X'_{\leqslant p}: \cdots \to X'^{p-1} \to X'^p \to 0 \to \cdots$ and a morphism of complexes $X \to X'_{\leqslant p}$ such that the X'^j's belong to $\mathscr{I}$ for all j, $H^j(X) \simeq H^j(X'_{\leqslant p})$ for $j < p$ and $H^p(X) \to \mathrm{Coker}\, d^{p-1}_{X'_{\leqslant p}}$ is a monomorphism.

This is possible for $p \ll 0$. Assume we have constructed $X'_{\leqslant p}$. Set $Z'^{p+1} = \mathrm{Coker}\, d^{p-1}_{X'_{\leqslant p}} \oplus_{\mathrm{Coker}\, d^{p-1}_X} X^{p+1}$ (cf. Exercise I.6) and choose $X'^{p+1} \in \mathrm{Ob}(\mathscr{I})$ such that there exists a monomorphism $Z'^{p+1} \to X'^{p+1}$.

Let us apply the results of Exercise I.6 to the diagram

$$\begin{array}{ccc} \mathrm{Coker}\, d^{p-1}_X & \longrightarrow & X^{p+1} \\ \downarrow & & \downarrow \\ \mathrm{Coker}\, d^{p-1}_{X'_{\leqslant p}} & \longrightarrow & X'^{p+1} \end{array}$$

(the morphisms $X'^p \to X'^{p+1}$ and $X^{p+1} \to X'^{p+1}$ are defined in the obvious way). Then $H^p(X) \simeq H^p(X'_{\leqslant p+1})$ follows from the fact that $H^p(X) \to \mathrm{Coker}\, d^{p-1}_{X'_{\leqslant p}}$ is a monomorphism, and moreover $H^{p+1}(X) \to \mathrm{Coker}\, d^p_{X'_{\leqslant p+1}}$ is a monomorphism.

□

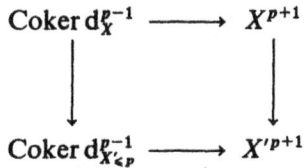

Corollary 1.7.8. *In the situation of Proposition* 1.7.7 *assume* (1.7.5) *and also*:

(1.7.6) *there exists an integer* $d \geqslant 0$ *such that, for any exact sequence in* $\mathscr{C}$, $X^0 \to X^1 \to \cdots \to X^d \to 0$ *with* $X^j \in \mathrm{Ob}(\mathscr{I})$ *for* $j < d$, *we have* $X^d \in \mathrm{Ob}(\mathscr{I})$.

Then for any $X \in \mathrm{Ob}(\mathbf{K}^b(\mathscr{C}))$, *there exists* $X' \in \mathrm{Ob}(\mathbf{K}^b(\mathscr{I}))$, *and a quasi-isomorphism* $X \to X'$.

Proof. By the preceding proposition, we can find $X' \in \mathrm{Ob}(\mathbf{K}^+(\mathscr{I}))$, and a quasi-isomorphism $X \to X'$.

If we assume $H^j(X) = 0$ for $j > n_0$, then we get $H^j(X') = 0$ for $j > n_0$. Therefore $\tau^{\leqslant n_0+d}(X') \to X'$ is a quasi-isomorphism. Now $(\tau^{\leqslant n_0+d}(X'))^k$ belongs to $\mathrm{Ob}(\mathscr{I})$ for $k < n_0 + d$, and is 0 for $k > n_0 + d$. Since $H^k(\tau^{\leqslant n_0+d}(X')) = 0$ for $k > n_0$, the condition (1.7.6) implies $(\tau^{\leqslant n_0+d}(X'))^{n_0+d} \in \mathrm{Ob}(\mathscr{I})$. Finally X being bounded, the quasi-isomorphism $X \to X'$ defines the quasi-isomorphism $X \to \tau^{\leqslant k}X'$, for $k \gg 0$. □

The last proposition is especially important in the following situation.

Definition 1.7.9. *One says* $\mathscr{C}$ *has enough injectives if for any* $X \in \mathrm{Ob}(\mathscr{C})$ *there exists an injective object* X' *in* $\mathscr{C}$ *and a monomorphism* $X \to X'$.

In other words $\mathscr{C}$ has enough injectives if the subcategory of injective objects satisfies (1.7.5).

Proposition 1.7.10. *Assume* $\mathscr{C}$ *has enough injectives, and let* $\mathscr{I}$ *denote the full subcategory of* $\mathscr{C}$ *of injective objects. Then the natural functor from* $\mathbf{K}^+(\mathscr{I})$ *to* $\mathbf{D}^+(\mathscr{C})$ *is an equivalence of categories.*

Proof. By Proposition 1.7.7, it is enough to show that:

$$\mathscr{N} \cap \mathrm{Ob}(\mathbf{K}^+(\mathscr{I})) = 0 \ . \tag{1.7.7}$$

That is, any $X \in \mathrm{Ob}(\mathbf{C}^+(\mathscr{I}))$ such that $H^n(X) = 0$ for any n, is homotopic to zero. Set $Z^n = \mathrm{Ker}\, \mathrm{d}_X^n$. Then we have exact sequences:

$$0 \longrightarrow Z^n \xrightarrow{i^n} X^n \xrightarrow{j^n} Z^{n+1} \longrightarrow 0 \ . \tag{1.7.8}$$

By induction on n, we get that all Z^n are injective. Therefore the sequences (1.7.8) split, and there are morphisms:

$$k^n : X^n \to Z^n \ , \qquad t^n : Z^{n+1} \to X^n$$

such that $k^n \circ i^n = \mathrm{id}_{Z^n}$, $j^n \circ t^n = \mathrm{id}_{Z^{n+1}}$, $k^n \circ t^n = 0$, and $\mathrm{id}_{X^n} = i^n \circ k^n + t^n \circ j^n$. Then $s^n = t^{n-1} \circ k^n : X^n \to X^{n-1}$ gives the homotopy, i.e.: $\mathrm{id}_{X^n} = \mathrm{d}_X^{n-1} \circ s^n + s^{n+1} \circ \mathrm{d}_X^n$. □

Now let $\mathscr{C}$ be an abelian category, $\mathscr{C}'$ a full abelian subcategory. Denote by $\mathbf{D}^+_{\mathscr{C}'}(\mathscr{C})$ the full triangulated subcategory of $\mathbf{D}^+(\mathscr{C})$ consisting of complexes whose cohomology objects belong to $\mathscr{C}'$. There is a natural functor:

$$\mathbf{D}^+(\mathscr{C}') \xrightarrow[\delta]{} \mathbf{D}^+_{\mathscr{C}'}(\mathscr{C}) \ . \tag{1.7.9}$$

We shall give a useful criterion which ensures that δ is an equivalence. First let us say that $\mathscr{C}'$ is a **thick subcategory** of $\mathscr{C}$ if for any exact sequence $Y \to Y' \to X \to Z \to Z'$ in $\mathscr{C}$ with Y, Y', Z, Z' in $\mathscr{C}'$, X belongs to $\mathscr{C}'$.

Proposition 1.7.11. *Let $\mathscr{C}$ be an abelian category, $\mathscr{C}'$ a thick full abelian subcategory. Assume that for any monomorphism $f: X' \to X$ with $X' \in \mathrm{Ob}(\mathscr{C}')$, there exists a morphism $g: X \to Y$, with $Y \in \mathrm{Ob}(\mathscr{C}')$ such that $g \circ f$ is a monomorphism. Then the functor δ in* (1.7.9) *is an equivalence of categories.*

Proof. By Proposition 1.6.5, it is enough to show:

$$\begin{cases} \text{for any } X \in \mathrm{Ob}(\mathbf{D}^+_{\mathscr{C}'}(\mathscr{C})) \text{ there exists } X' \in \mathrm{Ob}(\mathbf{K}^+(\mathscr{C}')) \\ \text{and a quasi-isomorphism } X \xrightarrow{\sim} X' \ . \end{cases} \tag{1.7.10}$$

The construction of X' will be similar to that of the proof of Proposition 1.7.7. Having defined a complex $X'_{\leqslant p}: \cdots \to X'^{p-1} \to X'^p \to 0 \to \cdots$ and a morphism $X \to X'_{\leqslant p}$ such that the X'^j's belong to $\mathscr{C}'$, $H^j(X) \simeq H^j(X'_{\leqslant p})$ for $j < p$ and $H^p(X) \to \mathrm{Coker}\, d^{p-1}_{X'_{\leqslant p}}$ is a monomorphism, we construct X'^{p+1} as follows.

Let $M = \mathrm{Coker}\, d^{p-1}_{X'_{\leqslant p}} \oplus_{\mathrm{Coker}\, d^{p-1}_X} \mathrm{Ker}\, d^{p+1}_X$ and $N = \mathrm{Coker}\, d^{p-1}_{X'_{\leqslant p}} \oplus_{\mathrm{Coker}\, d^{p-1}_X} X^{p+1}$. We have an exact sequence (cf. Exercise I.6 and (1.3.9)):

$$0 \to H^p(X) \to \mathrm{Coker}\, d^{p-1}_{X'_{\leqslant p}} \to M \to H^{p+1}(X) \to 0 \ .$$

Hence M belongs to $\mathscr{C}'$. Applying the hypothesis to the monomorphism $i: M \to N$, we find a morphism $g: N \to X'^{p+1}$ with $X'^{p+1} \in \mathrm{Ob}(\mathscr{C}')$ and $g \circ i$ is a monomorphism. One defines the morphisms $X'^p \to X'^{p+1}$ and $X^{p+1} \to X'^{p+1}$ in the natural way, and one checks as in the proof of Proposition 1.7.7 that the complex $X'_{\leqslant p+1}$ has the required properties. □

Remark 1.7.12. With the same hypotheses as in Proposition 1.7.11, δ induces an equivalence:

$$\mathbf{D}^b(\mathscr{C}') \simeq \mathbf{D}^b_{\mathscr{C}'}(\mathscr{C}) \ .$$

(Use the functor $\tau^{\leqslant n}$ for $n \gg 0$.)

Comments 1.7.13. Let us summarize the construction of $\mathbf{D}(\mathscr{C})$. We start with an abelian category $\mathscr{C}$, and consider the category $\mathbf{C}(\mathscr{C})$ of complexes of $\mathscr{C}$. Then we decide that a morphism in $\mathbf{C}(\mathscr{C})$ homotopic to zero is the zero morphism: we get the category $\mathbf{K}(\mathscr{C})$. The advantage of $\mathbf{K}(\mathscr{C})$ on $\mathbf{C}(\mathscr{C})$ is that many diagrams which do not commute in $\mathbf{C}(\mathscr{C})$, actually do so in $\mathbf{K}(\mathscr{C})$, making $\mathbf{K}(\mathscr{C})$ a triangulated category. Then we want a morphism in $\mathsf{K}(\mathscr{C})$ which induces an isomorphism on the cohomology to be invertible. For that purpose we "localize" $\mathsf{K}(\mathscr{C})$, and get $\mathsf{D}(\mathscr{C})$.

Notation 1.7.14. Let A be a ring. If there is no risk of confusion we shall write $\mathsf{D}(A)$ instead of $\mathsf{D}(\mathfrak{Mod}(A))$.

1.8. Derived functors

In this section $\mathscr{C}$ and $\mathscr{C}'$ will denote two abelian categories, and $F: \mathscr{C} \to \mathscr{C}'$ an additive functor.

We shall denote by Q the natural functor $\mathsf{K}^+(\mathscr{C}) \to \mathsf{D}^+(\mathscr{C})$ or $\mathsf{K}^+(\mathscr{C}') \to \mathsf{D}^+(\mathscr{C}')$.

Definition 1.8.1. *Let $T: \mathsf{D}^+(\mathscr{C}) \to \mathsf{D}^+(\mathscr{C}')$ be a functor of triangulated categories, and let s be a morphism of functors:*

$$s: Q \circ \mathsf{K}^+(F) \to T \circ Q \ ,$$

where $\mathsf{K}^+(F): \mathsf{K}^+(\mathscr{C}) \to \mathsf{K}^+(\mathscr{C}')$ is the functor naturally associated to F. Assume that for any functor of triangulated categories $G: \mathsf{D}^+(\mathscr{C}) \to \mathsf{D}^+(\mathscr{C}')$, the morphism:

$$\mathrm{Hom}(T, G) \underset{s}{\longrightarrow} \mathrm{Hom}(Q \circ \mathsf{K}^+(F), G \circ Q)$$

is an isomorphism.

Then (T, s), which is unique up to isomorphism, is called the right derived functor of F, and denoted RF. The functor $H^n \circ RF$, also denoted R^nF, is called the n-th derived functor of F.

Let us give a useful criterium which ensures the existence of RF. From now on and until Proposition 1.8.7, we assume F is left exact.

Definition 1.8.2. *A full additive subcategory $\mathscr{I}$ of $\mathscr{C}$ is called injective with respect to F (or F-injective, for short), if:*

(i) *condition* (1.7.5) *is satisfied,*
(ii) *if $0 \to X' \to X \to X'' \to 0$ is an exact sequence in $\mathscr{C}$, and if X' and X are in $\mathrm{Ob}(\mathscr{I})$, then X'' is also in $\mathrm{Ob}(\mathscr{I})$,*
(iii) *if $0 \to X' \to X \to X'' \to 0$ is an exact sequence in $\mathscr{C}$, and if X', X, X'', are in $\mathrm{Ob}(\mathscr{I})$, then the sequence $0 \to F(X') \to F(X) \to F(X'') \to 0$ is exact.*

Note that under conditions (i) and (ii), the condition (iii) is equivalent to the similar condition in which one only assumes $X' \in \mathrm{Ob}(\mathscr{I})$, because of the assumption that F is left exact.

Let $\mathscr{I}$ be F-injective. Then one can check easily that F transforms objects of $\mathbf{K}^+(\mathscr{I})$ quasi-isomorphic to zero into objects of $\mathbf{K}^+(\mathscr{C}')$ satisfying the same property. Therefore the composition of functors

$$\mathbf{K}^+(\mathscr{I}) \xrightarrow{\mathbf{K}^+(F)} \mathbf{K}^+(\mathscr{C}') \longrightarrow \mathbf{D}^+(\mathscr{C}')$$

factors through $\mathbf{K}^+(\mathscr{I})/\mathscr{N} \cap \mathrm{Ob}(\mathbf{K}^+(\mathscr{I}))$ where $\mathscr{N}$ is given by (1.7.1). Since $\mathbf{K}^+(\mathscr{I})/\mathscr{N} \cap \mathrm{Ob}(\mathbf{K}^+(\mathscr{I}))$ is equivalent to $\mathbf{D}^+(\mathscr{C})$ by Proposition 1.7.7, we obtain:

Proposition 1.8.3. *Assume there exists an F-injective subcategory $\mathscr{I}$ of $\mathscr{C}$. Then the functor from $\mathbf{K}^+(\mathscr{I})/\mathscr{N} \cap \mathrm{Ob}(\mathbf{K}^+(\mathscr{I}))$ to $\mathbf{D}^+(\mathscr{C}')$ constructed above is the right derived functor of F.*

Remark 1.8.4. It follows from the universal property of RF that the preceding construction does not depend on $\mathscr{I}$.

Remark 1.8.5. Let $\mathscr{I}$ be the full subcategory of injective objects of $\mathscr{C}$ and assume $\mathscr{C}$ has enough injectives, (i.e: (1.7.5) is satisfied). Then $\mathscr{I}$ is F-injective with respect to any left exact functor F, since any sequence in $\mathscr{I}$ splits, (cf. Exercise I.5). In particular RF always exists in this case.

Remark 1.8.6. Assume there exists an F-injective subcategory $\mathscr{I}$ of $\mathscr{C}$, let n be an integer and let $X \in \mathrm{Ob}(\mathbf{K}^+(\mathscr{C}))$ be such that $H^k(X) = 0$ for $k < n$. Then $R^kF(X) = 0$ for $k < n$ and $R^nF(X) = F(H^n(X))$. In fact, for such an X, one can find $X' \in \mathrm{Ob}(\mathbf{K}^+(\mathscr{I}))$ and a quasi-isomorphism $X \to X'$ such that $X'^k = 0$ for $k < n$. Of course, if $X \in \mathrm{Ob}(\mathbf{K}^+(\mathscr{I}))$, then $R^kF(X) = F(H^k(X))$ for all k. In particular if $X \in \mathrm{Ob}(\mathscr{I})$, $R^kF(X) = 0$ for $k \neq 0$. (An object X of $\mathscr{C}$ such that $R^kF(X) = 0$ for $k \neq 0$ is called **F-acyclic**, cf. Exercise I.19.)

Proposition 1.8.7. *Let $\mathscr{C}$, $\mathscr{C}'$, $\mathscr{C}''$ be three abelian categories and let $F : \mathscr{C} \to \mathscr{C}'$, $F' : \mathscr{C}' \to \mathscr{C}''$ be two left exact functors. Assume there exists a full additive subcategory $\mathscr{I}$ of $\mathscr{C}$ (resp. $\mathscr{I}'$ of $\mathscr{C}'$) which is F-injective (resp. F'-injective), and such that $F(\mathrm{Ob}(\mathscr{I})) \subset \mathrm{Ob}(\mathscr{I}')$. Then $\mathscr{I}$ is $(F' \circ F)$-injective, and we have:*

$$R(F' \circ F) = RF' \circ RF \ . \tag{1.8.1}$$

The proof is straightforward.

Note that if we only assume that RF, RF', $R(F' \circ F)$ exist, we find a canonical morphism of functors:

$$R(F' \circ F) \to RF' \circ RF \ . \tag{1.8.2}$$

In fact we have:

$$\mathrm{Hom}(R(F' \circ F), RF' \circ RF) = \mathrm{Hom}(Q \circ \mathbf{K}^+(F') \circ \mathbf{K}^+(F), RF' \circ RF \circ Q)$$

and the morphisms:

$$Q \circ \mathbf{K}^+(F) \xrightarrow{\alpha} RF \circ Q$$

$$Q \circ \mathbf{K}^+(F') \xrightarrow{\beta} RF' \circ Q$$

give the morphisms:

$$Q \circ \mathbf{K}^+(F') \circ \mathbf{K}^+(F) \xrightarrow{\beta \circ \mathbf{K}^+(F)} RF' \circ Q \circ \mathbf{K}^+(F) \xrightarrow{RF' \circ \alpha} RF' \circ RF \circ Q \quad .$$

Proposition 1.8.8. *Let $\mathscr{C}$ and $\mathscr{C}'$ be two abelian categories, let F', F, F'' be three left exact functors from $\mathscr{C}$ to $\mathscr{C}'$ and let $\lambda : F' \to F$ and $\mu : F \to F''$ be morphisms of functors. Assume there exists a full additive subcategory $\mathscr{I}$ of $\mathscr{C}$ which is injective with respect to F', F and F''. Assume further:*

(1.8.3) *for any $X \in \mathrm{Ob}(\mathscr{I})$, the sequence*

$$0 \to F'(X) \to F(X) \to F''(X) \to 0 \text{ is exact} \quad .$$

Then there exists naturally a morphism of functors $\nu : RF'' \to RF'[1]$ such that for any $X \in \mathrm{Ob}(\mathbf{D}^+(\mathscr{C}))$, the sequence:

$$RF'(X) \xrightarrow{R\lambda(X)} RF(X) \xrightarrow{R\mu(X)} RF''(X) \xrightarrow{\nu(X)} RF'(X)[1]$$

is a distinguished triangle in $\mathbf{D}^+(\mathscr{C}')$.

Proof. For any $X \in \mathrm{Ob}(\mathbf{K}^+(\mathscr{I}))$, we have an exact sequence:

$$0 \longrightarrow F'(X^n) \xrightarrow{\lambda(X^n)} F(X^n) \xrightarrow{\mu(X^n)} F''(X^n) \longrightarrow 0$$

and hence, by Remark 1.7.5, $F''(X)$ is isomorphic to the mapping cone $M(\lambda(X))$ of $\lambda(X)$ in $\mathbf{D}^+(\mathscr{C}')$.

The morphism $\alpha(\lambda(X)) : M(\lambda(X)) \to F'(X)[1]$ gives a morphism in $\mathbf{D}^+(\mathscr{C}')$ from $F''(X)$ to $F'(X)[1]$. Passing through the quotient, we obtain:

$$\nu : RF'' \to RF'[1] \quad .$$

The rest of the statement is straightforward. □

To conclude this section, let us consider the case of a right exact functor F. Then, by reversing the arrows, one defines the notion of an F-projective subcategory of $\mathscr{C}$, and the left derived functor of F, denoted LF.

To be more precise, a full additive subcategory $\mathscr{P}$ of $\mathscr{C}$ is called F-projective (F being right exact) if:

(i) for any $X \in \mathrm{Ob}(\mathscr{C})$, there exists $X' \in \mathrm{Ob}(\mathscr{P})$ and an exact sequence $X' \to X \to 0$,

(ii) if $0 \to X' \to X \to X'' \to 0$ is an exact sequence in $\mathscr{C}$, and if X'' and X are in $\mathrm{Ob}(\mathscr{P})$, then X' is also in $\mathrm{Ob}(\mathscr{P})$,

(iii) if $0 \to X' \to X \to X'' \to 0$ is an exact sequence in $\mathscr{C}$ and if X', X, X'' are in $\mathrm{Ob}(\mathscr{P})$, then the sequence $0 \to F(X') \to F(X) \to F(X'') \to 0$ is exact.

Then the construction of the left derived functor:

$$LF : \mathbf{D}^-(\mathscr{C}) \to \mathbf{D}^-(\mathscr{C}')$$

is similar to that of the right derived functors.

Example 1.8.9. Let A be a ring. Then the category $\mathfrak{Mod}(A)$ has enough injectives and enough projectives, (cf. Cartan-Eilenberg [1]). Moreover let M be a right A-module. Then the category of (left) flat A-modules is projective with respect to the functor $M \otimes_A \cdot$.

We shall meet numerous derived functors in the forthcoming chapters.

Remark 1.8.10. The construction of derived functors can be extended to functors which are only defined on $\mathbf{K}^+(\mathscr{C})$. More precisely, let F be a functor of triangulated categories from $\mathbf{K}^+(\mathscr{C})$ to $\mathbf{K}^+(\mathscr{C}')$. Assume to be given a full triangulated subcategory $\mathscr{I}$ of $\mathbf{K}^+(\mathscr{C})$ such that (1.8.4) and (1.8.5) below are satisfied.

(1.8.4) *For any* $X \in \mathrm{Ob}(\mathbf{K}^+(\mathscr{C}))$, *there is a quasi-isomorphism* $X \to X'$ *with* $X' \in \mathrm{Ob}(\mathscr{I})$.

(1.8.5) *If* $X \in \mathrm{Ob}(\mathscr{I})$ *is quasi-isomorphic to* 0, *then* $F(X)$ *is quasi-isomorphic to* 0.

Then one can define $RF : \mathbf{D}^+(\mathscr{C}) \to \mathbf{D}^+(\mathscr{C}')$ similarly as we did in Proposition 1.8.3, and RF will satisfy the universal property of Definition 1.8.1.

Remark 1.8.11. Let $F : \mathscr{C} \to \mathscr{C}'$ be a contravariant functor. Then it defines a contravariant functor $\mathbf{K}(F) : \mathbf{K}(\mathscr{C}) \to \mathbf{K}(\mathscr{C}')$ as follows. If $X = (X^n)_{n \in \mathbb{Z}}$ belongs to $\mathbf{C}(\mathscr{C})$ then:

$$\begin{cases} (K(F)(X))^n = F(X^{-n}) \\ d^n_{\mathbf{K}(F)(X)} = (-1)^{n+1} F(d_X^{-n-1}) \end{cases} \quad (1.8.6)$$

Taking for F the canonical contravariant functor $\mathscr{C}^\circ \to \mathscr{C}$, we have $\mathbf{K}(\mathscr{C}^\circ) \simeq (\mathbf{K}(\mathscr{C}))^\circ$ and if $\mathscr{C}$ is abelian, $\mathbf{D}(\mathscr{C}^\circ) \simeq (\mathbf{D}(\mathscr{C}))^\circ$. We also have $\mathbf{K}^\pm(\mathscr{C}^\circ) \simeq (\mathbf{K}^\mp(\mathscr{C}))^\circ$ and $\mathbf{D}^\pm(\mathscr{C}^\circ) \simeq (\mathbf{D}^\mp(\mathscr{C}))^\circ$.

1.9. Double complexes

Let $\mathscr{C}$ be an additive category.

Definition 1.9.1. *A double complex* (X, d_X) *in* $\mathscr{C}$ *consists of the data* $\{X^{n,m}, \mathrm{d}_X'^{n,m}, \mathrm{d}_X''^{n,m}\}_{n,m\in\mathbb{Z}}$ *where* $X^{n,m} \in \mathrm{Ob}(\mathscr{C})$, $\mathrm{d}_X'^{n,m} : X^{n,m} \to X^{n+1,m}$, $\mathrm{d}_X''^{n,m} : X^{n,m} \to X^{n,m+1}$ *for any pair* (n,m) *and*:

$$\text{(1.9.1)} \qquad \mathrm{d}_X'^2 = 0 \ , \qquad \mathrm{d}_X''^2 = 0 \ , \qquad \mathrm{d}_X' \circ \mathrm{d}_X'' = \mathrm{d}_X'' \circ \mathrm{d}_X' \ .$$

The meaning of (1.9.1) is the following: $\mathrm{d}_X'^{n+1,m} \circ \mathrm{d}_X'^{n,m} = 0$, and similarly for the other relations.

We shall sometimes call "simple complex" a complex in the sense of Definition 1.3.1.

Let X and Y be two double complexes. A morphism f from X to Y is defined in the obvious way. Then we get the category $\mathbf{C}^2(\mathscr{C})$ of double complexes on $\mathscr{C}$.

Let X be a double complex. For a given $n \in \mathbb{Z}$, let X_I^n denote the simple complex:

$$X_I^n = \{X^{n,m}, \mathrm{d}_X''^{n,m}\}_{m\in\mathbb{Z}} \ .$$

The family of morphisms $\{\mathrm{d}_X'^{n,m}\}_{m\in\mathbb{Z}}$ defines a morphism:

$$\mathrm{d}_I^n : X_I^n \to X_I^{n+1}$$

and clearly $\mathrm{d}_I^{n+1} \circ \mathrm{d}_I^n = 0$. Therefore we have constructed a functor:

$$F_I : \mathbf{C}^2(\mathscr{C}) \to \mathbf{C}(\mathbf{C}(\mathscr{C})) \ ,$$

$$X \mapsto \{X_I^n, \mathrm{d}_I^n\} \ .$$

This functor is clearly an equivalence of categories.

By reversing the first and the second index (i.e., d′ and d″), we get another equivalence:

$$F_{II} : \mathbf{C}^2(\mathscr{C}) \to \mathbf{C}(\mathbf{C}(\mathscr{C})) \ ,$$

$$X \mapsto \{X_{II}^m, \mathrm{d}_{II}^m\} \ ,$$

where $X_{II}^m = \{X^{n,m}, \mathrm{d}_X'^{n,m}\}_{n\in\mathbb{Z}}$, and $\mathrm{d}_{II}^m = \{\mathrm{d}_X''^{n,m}\}_{n\in\mathbb{Z}}$.

Assume X satisfies the following finiteness property:

(1.9.2) for any $k \in \mathbb{Z}$, the set $\{(n,m) \in \mathbb{Z} \times \mathbb{Z} : n + m = k, X^{n,m} \neq 0\}$ is finite.

Then it is possible to associate to X a simple complex $\mathrm{s}(X)$. One sets:

$$\mathrm{s}(X)^k = \bigoplus_{k=n+m} X^{n,m} \ .$$

Let $i_{n,m} : X^{n,m} \to \bigoplus_{k=n'+m'} X^{n',m'}$ and $p_{n,m} : \bigoplus_{k=n'+m'} X^{n',m'} \to X^{n,m}$ be the natural morphisms from $X^{n,m}$ to $s(X)^k$ and from $s(X)^k$ to $X^{n,m}$, respectively. One defines:

$$d^k_{s(X)} : s(X)^k \to s(X)^{k+1}$$

by:

$$p_{n',m'} \circ d^k_{s(X)} \circ i_{n,m} = \begin{cases} d'^{n,m}_X & \text{if } m = m' \\ (-1)^n d''^{n,m}_X & \text{if } n = n' \\ 0 & \text{otherwise} \end{cases} \tag{1.9.3}$$

for $n + m = k$, $n' + m' = k + 1$.

Proposition 1.9.2. *The data* $\{s(X)^k, d^k_{s(X)}\}_{k \in \mathbb{Z}}$, *define a complex in* $\mathscr{C}$, *that is,* $d^{k+1}_{s(X)} \circ d^k_{s(X)} = 0$.

The proof is straightforward.

We call $s(X)$ the **simple complex associated to** X.

One may consider the full additive subcategory $\mathbf{C}^2_f(\mathscr{C})$ of objects of $\mathbf{C}^2(\mathscr{C})$ satisfying (1.9.2), and $s(\cdot)$ becomes an additive functor:

$$s(\cdot) : \mathbf{C}^2_f(\mathscr{C}) \to \mathbf{C}(\mathscr{C}) \ . \tag{1.9.4}$$

Now we assume $\mathscr{C}$ is an abelian category.

We define the functors $\tau_I^{\leq n}$, $\tau_I^{\geq n}$, $\tau_{II}^{\leq m}$, $\tau_{II}^{\geq m}$ by using F_I or F_{II}. For example:

$$\tau_I^{\leq n} = (F_I)^{-1} \circ \tau^{\leq n} \circ F_I \ .$$

In other words, $\tau_I^{\leq n}(X)$ is the double complex:

$$\cdots \longrightarrow X^{n-1,\cdot} \xrightarrow[d'^{n-1,\cdot}_X]{} X^{n,\cdot} \longrightarrow \operatorname{Ker} d'^{n,\cdot}_X \longrightarrow 0 \longrightarrow \cdots$$

or equivalently the complex in $\mathbf{C}(\mathscr{C})$:

$$\cdots \longrightarrow X_I^{n-1} \xrightarrow[d_I^{n-1}]{} X_I^n \longrightarrow \operatorname{Ker} d_I^n \longrightarrow 0 \longrightarrow \cdots \ .$$

We also introduce the simple complex:

$$H_I^p(X) = H^p(F_I(X)) \ , \tag{1.9.5}$$

and the double complex:

$$H_I(X) : \cdots \longrightarrow H_I^p(X) \xrightarrow[0]{} H_I^{p+1}(X) \longrightarrow \cdots \tag{1.9.6}$$

and similarly for $H^q_{II}(X)$ and $H_{II}(X)$.

Theorem 1.9.3. *Let* $f : X \to Y$ *be a morphism of double complexes, where* X *and* Y *both satisfy* (1.9.2). *Assume* f *induces an isomorphism:*

$$f : H_I H_{II}(X) \simeq H_I H_{II}(Y) \ .$$

Then $\mathrm{s}(f) : \mathrm{s}(X) \to \mathrm{s}(Y)$ *is a quasi-isomorphism.*

Proof. Let $q \in \mathbb{Z}$. We have a commutative diagram of distinguished triangles:

$$\begin{array}{ccccccc}
\mathrm{s}\tau_{II}^{\leqslant q-1}(X) & \longrightarrow & \mathrm{s}\tau_{II}^{\leqslant q}(X) & \longrightarrow & H_{II}^q(X)[-q] & \xrightarrow[+1]{} \\
\Big\downarrow{\scriptstyle \mathrm{s}\tau_{II}^{\leqslant q-1}(f)} & & \Big\downarrow{\scriptstyle \mathrm{s}\tau_{II}^{\leqslant q}(f)} & & \Big\downarrow{\scriptstyle H_{II}^q(f)[-q]} & \\
\mathrm{s}\tau_{II}^{\leqslant q-1}(Y) & \longrightarrow & \mathrm{s}\tau_{II}^{\leqslant q}(Y) & \longrightarrow & H_{II}^q(Y)[-q] & \xrightarrow[+1]{}
\end{array}$$

(cf. Exercise I.25).

Assume for a while $X_{II}^q = 0$ for $q \ll 0$.

Then since $\mathrm{s}\tau_{II}^{\leqslant q}(f)$ is quasi-isomorphic to 0 for $q \ll 0$, and $H_{II}^q(f)$ is a quasi-isomorphism for all q by the hypothesis, we obtain that $\mathrm{s}\tau_{II}^{\leqslant q}(f)$ is a quasi-isomorphism for all q. Since $H^k(\mathrm{s}\tau_{II}^{\leqslant q}(f)) = H^k(\mathrm{s}(f))$ for $q \gg 0$, k being fixed (cf. Exercise I.25 again), we get the conclusion in this case. To treat the general case, we apply the preceding result to $\tau_{II}^{\geqslant q}(f)$. This gives the result since for a fixed k, $H^k(\mathrm{s}(f)) = H^k(\mathrm{s}\tau_{II}^{\geqslant q}(f))$ for $q \ll 0$. □

Finally, we note that s transforms a morphism homotopic to 0 into a morphism homotopic to 0. More precisely, a morphism $f : X \to Y$ in $\mathbf{C}_f^2(\mathscr{C})$ is called homotopic to 0 if there are morphisms $t_1^{n,m} : X^{n,m} \to Y^{n-1,m}$ and $t_2^{n,m} : X^{n,m} \to Y^{n,m-1}$ such that:

$$d''^{n-1,m} \circ t_1^{n,m} = t_1^{n+1,m} \circ d''^{n,m} \ ,$$

$$d'^{n,m-1} \circ t_2^{n,m} = t_2^{n,m+1} \circ d'^{n,m} \ ,$$

and:

$$f^{n,m} = d'^{n-1,m} \circ t_1^{n,m} + t_1^{n+1,m} \circ d'^{n,m} + d''^{n,m-1} \circ t_2^{n,m} + t_2^{n,m+1} \circ d''^{n,m} \ .$$

Then $\mathrm{s}(f) : \mathrm{s}(X) \to \mathrm{s}(Y)$ is homotopic to zero. In particular, if $F_I(f) : F_I(X) \to F_I(Y)$ is homotopic to zero, then $\mathrm{s}(f)$ is homotopic to zero, and similarly for $F_{II}(f)$.

1.10. Bifunctors

Let $\mathscr{C}$, $\mathscr{C}'$ and $\mathscr{C}''$ be three categories.

Definition 1.10.1. *A bifunctor F from* $\mathscr{C} \times \mathscr{C}'$ *to* $\mathscr{C}''$ *consists of the following data:*

(i) *a map* $F : \mathrm{Ob}(\mathscr{C}) \times \mathrm{Ob}(\mathscr{C}') \to \mathrm{Ob}(\mathscr{C}'')$,

(ii) *for any pair* $(X, Y) \in \mathrm{Ob}(\mathscr{C})$ *and any pair* $(X', Y') \in \mathrm{Ob}(\mathscr{C}')$, *a map* $F : \mathrm{Hom}_{\mathscr{C}}(X, Y) \times \mathrm{Hom}_{\mathscr{C}'}(X', Y') \to \mathrm{Hom}_{\mathscr{C}''}(F(X, X'), F(Y, Y'))$ *such that for any* $X \in \mathrm{Ob}(\mathscr{C})$ *(resp.* $X' \in \mathrm{Ob}(\mathscr{C}')$*),* $F(X, \cdot)$ *(resp.* $F(\cdot, X')$*) is a functor from* $\mathscr{C}$ *to* $\mathscr{C}''$ *(resp. from* $\mathscr{C}'$ *to* $\mathscr{C}''$*), and if* $f \in \mathrm{Hom}_{\mathscr{C}}(X, Y)$, $g \in \mathrm{Hom}_{\mathscr{C}'}(X', Y')$ *then* $F(f, Y') \circ F(X, g) = F(Y, g) \circ F(f, X')$.

A bifunctor is called additive, left exact, right exact, exact, a bifunctor of triangulated categories, a cohomological bifunctor, etc., if it is so with respect to each variable.

One defines in a natural way, morphisms of bifunctors.

Example 1.10.2. $\mathrm{Hom}_{\mathscr{C}}(\cdot, \cdot)$ is a bifunctor for $\mathscr{C}^{\circ} \times \mathscr{C}$ to $\mathfrak{Set}$.

Example 1.10.3. Let A be a ring. Then $\cdot \otimes_A \cdot$ is a bifunctor from $\mathfrak{Mod}(A^{op}) \times \mathfrak{Mod}(A)$ to $\mathfrak{Mod}(\mathbb{Z})$.

Now we assume that $\mathscr{C}$, $\mathscr{C}'$ and $\mathscr{C}''$ are three abelian categories, and F is an additive left exact bifunctor from $\mathscr{C} \times \mathscr{C}'$ to $\mathscr{C}''$.

For a complex X in $\mathrm{Ob}(\mathbf{C}^+(\mathscr{C}))$ and a complex X' in $\mathrm{Ob}(\mathbf{C}^+(\mathscr{C}'))$, $F(X,X')$ is a double complex in $\mathscr{C}''$. By associating to $F(X,X')$ the simple complex $s(F(X,X'))$, one defines a bifunctor:

$$\mathbf{C}^+(F) : \mathbf{C}^+(\mathscr{C}) \times \mathbf{C}^+(\mathscr{C}') \to \mathbf{C}^+(\mathscr{C}'')$$

and passing to the quotient, a bifunctor:

$$\mathbf{K}^+(F) : \mathbf{K}^+(\mathscr{C}) \times \mathbf{K}^+(\mathscr{C}') \to \mathbf{K}^+(\mathscr{C}'') \ .$$

(This is possible due to the final remark in §9.)

In order to define the derived functor of a bifunctor, it will be convenient to start with a bifunctor of triangulated categories. Therefore let $F : \mathbf{K}^+(\mathscr{C}) \times \mathbf{K}^+(\mathscr{C}') \to \mathbf{K}^+(\mathscr{C}'')$ be a bifunctor of triangulated categories. Let $\mathscr{I}$ (resp. $\mathscr{I}'$) be a full triangulated subcategory of $\mathbf{K}^+(\mathscr{C})$ (resp. $\mathbf{K}^+(\mathscr{C}')$).

Consider the two following conditions.

$$(1.10.1) \qquad \begin{cases} \textit{For any } X \in \mathrm{Ob}(\mathbf{K}^+(\mathscr{C}))\ (\textit{resp.}\ \mathrm{Ob}(\mathbf{K}^+(\mathscr{C}')))\ \textit{there is a} \\ \textit{quasi-isomorphism from } X \textit{ to an object of } \mathscr{I}\ (\textit{resp.}\ \mathscr{I}') \ . \end{cases}$$

$$(1.10.2) \qquad \begin{cases} \textit{For any } X \in \mathrm{Ob}(\mathscr{I}) \textit{ and any } X' \in \mathrm{Ob}(\mathscr{I}'), \textit{the complex } F(X,X') \\ \textit{is quasi-isomorphic to zero if either } X \textit{ or } X' \textit{ is quasi-isomorphic} \\ \textit{to zero} \ . \end{cases}$$

Proposition 1.10.4. *Under the assumptions* (1.10.1)–(1.10.2), *there exists a bifunctor of triangulated categories* $RF : \mathbf{D}^+(\mathscr{C}) \times \mathbf{D}^+(\mathscr{C}') \to \mathbf{D}^+(\mathscr{C}'')$ *making the diagram below commutative*:

$$\begin{array}{ccc} \mathscr{I} \times \mathscr{I}' & \xrightarrow{\ F\ } & \mathbf{K}^+(\mathscr{C}'') \\ \Big\downarrow{\scriptstyle Q \times Q'} & & \Big\downarrow{\scriptstyle Q''} \\ \mathbf{D}^+(\mathscr{C}) \times \mathbf{D}^+(\mathscr{C}') & \xrightarrow{\ RF\ } & \mathbf{D}^+(\mathscr{C}'') \ , \end{array}$$

where Q, Q', Q'' *are the localization functors from* $\mathscr{I}$, $\mathscr{I}'$, $\mathscr{C}''$ *to* $\mathbf{D}^+(\mathscr{C})$, $\mathbf{D}^+(\mathscr{C}')$, $\mathbf{D}^+(\mathscr{C}'')$ *respectively.*

Moreover RF satisfies the following universal property: for every bifunctor of triangulated categories $G : \mathbf{D}^+(\mathscr{C}) \times \mathbf{D}^+(\mathscr{C}') \to \mathbf{D}^+(\mathscr{C}'')$, *the canonical homomorphism*:

$$\mathrm{Hom}(RF, G) \to \mathrm{Hom}(Q'' \circ F, G \circ (Q \times Q'))$$

is an isomorphism. Here $Q'' \circ F$ *and* $G \circ (Q \times Q')$ *are bifunctors from* $\mathbf{K}^+(\mathscr{C}) \times \mathbf{K}^+(\mathscr{C}')$ *to* $\mathbf{D}^+(\mathscr{C}'')$.

The proof is straightforward.

Corollary 1.10.5. *Let* $F : \mathbf{K}^+(\mathscr{C}) \times \mathbf{K}^+(\mathscr{C}') \to \mathbf{K}^+(\mathscr{C}'')$ *be a bifunctor of triangulated categories. Assume there exists a full subcategory* $\mathscr{I}$ *of* $\mathbf{K}^+(\mathscr{C})$ *such that*:

(1.10.3) *for any* $Y \in \mathrm{Ob}(\mathbf{K}^+(\mathscr{C}'))$, *conditions* (1.8.4) *and* (1.8.5) *are satisfied with respect to the functor* $F(\cdot, Y)$.

(1.10.4) *For any* $X \in \mathrm{Ob}(\mathscr{I})$, $Y \in \mathrm{Ob}(\mathbf{K}^+(\mathscr{C}'))$, $F(X, Y)$ *is quasi-isomorphic to zero if* Y *is quasi-isomorphic to zero.*

Then F *admits a derived functor, and for* $Y \in \mathrm{Ob}(\mathbf{K}^+(\mathscr{C}'))$, *the functor* $F(\cdot, Y) : \mathbf{K}^+(\mathscr{C}) \to \mathbf{K}^+(\mathscr{C}'')$ *admits a derived functor, denoted* $R_I F(\cdot, Y)$. *Moreover for* $X \in \mathrm{Ob}(\mathbf{K}^+(\mathscr{C}))$ *one has a natural isomorphism*:

$$RF(X, Y) \simeq R_I F(X, Y) \ .$$

Proof. The existence of $R_I F(\cdot, Y)$ follows from Remark 1.8.10. The existence of RF follows from Proposition 1.10.4 by taking $\mathscr{I}' = \mathbf{K}^+(\mathscr{C}')$. The equality $RF(X, Y) = R_I F(X, Y)$ follows from the constructions. □

Let us come back to the situation where F is a left exact bifunctor from $\mathscr{C} \times \mathscr{C}'$ to $\mathscr{C}''$. Let $\mathscr{I}$ (resp. $\mathscr{I}'$) be a full additive subcategory of $\mathscr{C}$ (resp. $\mathscr{C}'$).

Definition 1.10.6. *We say that* $(\mathscr{I}, \mathscr{I}')$ *is F-injective if for any* $X \in \mathrm{Ob}(\mathscr{I})$ *and any* $X' \in \mathrm{Ob}(\mathscr{I}')$, $\mathscr{I}$ *is* $F(\cdot, X')$*-injective and* $\mathscr{I}'$ *is* $F(X, \cdot)$*-injective.*

Proposition 1.10.7. *Assume* $(\mathscr{I}, \mathscr{I}')$ *is F-injective. Then* $(\mathbf{K}^+(\mathscr{I}), \mathbf{K}^+(\mathscr{I}'))$ *satisfies* (1.10.1)–(1.10.2).

Proof. One gets (1.10.1) by applying Proposition 1.7.7. Let us verify (1.10.2). First we shall prove that $F(X, Y)$ is quasi-isomorphic to zero if $X \in \mathrm{Ob}(\mathscr{I})$, $Y \in \mathrm{Ob}(\mathbf{K}^+(\mathscr{I}'))$ and Y is quasi-isomorphic to zero.

In this case,

$$H^q_{II}(F(X, Y)) = 0 \qquad \text{for all } q \ .$$

Applying Theorem 1.9.3 we obtain that $F(X, Y)$ is quasi-isomorphic to zero. There is a similar proof by reversing the roles of X and Y. □

Corollary 1.10.8. *Let $F : \mathscr{C} \times \mathscr{C}' \to \mathscr{C}''$ be a left exact bifunctor of abelian categories. Assume there exists a full additive subcategory $\mathscr{I}$ of $\mathscr{C}$ satisfying conditions* (i), (ii), (iii) *of Definition* 1.8.2 *with respect to the functor $F(\cdot, Y)$, for any $Y \in \mathrm{Ob}(\mathscr{C}')$. Assume moreover that the functor $F(X, \cdot)$ is exact for any $X \in \mathrm{Ob}(\mathscr{I})$. Then the category $\mathbf{K}^+(\mathscr{I})$ satisfies* (1.10.3)–(1.10.4).

Proof. Apply Proposition 1.10.7 with $\mathscr{I}' = \mathscr{C}'$. □

Proposition 1.10.9. *Let $F : \mathscr{C} \times \mathscr{C}' \to \mathscr{C}''$ be a left exact bifunctor of abelian categories, and let $G : \mathscr{C}'' \to \mathscr{C}'''$ be a left exact functor of abelian categories. Assume there exist full additive subcategories $\mathscr{I}$, $\mathscr{I}'$ and $\mathscr{I}''$ of $\mathscr{C}$, $\mathscr{C}'$ and $\mathscr{C}''$ respectively such that $(\mathscr{I}, \mathscr{I}')$ is F-injective, $\mathscr{I}''$ is G-injective, and:*

$$F(\mathrm{Ob}(\mathscr{I}), \mathrm{Ob}(\mathscr{I}')) \subset \mathrm{Ob}(\mathscr{I}'') \ . \tag{1.10.5}$$

Then the derived functor $R(G \circ F) : \mathbf{D}^+(\mathscr{C}) \times \mathbf{D}^+(\mathscr{C}') \to \mathbf{D}^+(\mathscr{C}''')$ exists, and one has:

$$R(G \circ F) \simeq RG \circ RF \ .$$

The proof is straightforward.

There are similar results for other compositions of functors.

Remark 1.10.10. In Corollary 1.10.5, if we interchange the roles of $\mathscr{C}$ and $\mathscr{C}'$, then one can define $R_{II}F(X, \cdot)$. Of course one gets:

$$R_I F(X, Y) \simeq R_{II} F(X, Y) \simeq RF(X, Y) \ . \tag{1.10.6}$$

Example 1.10.11. Let $\mathscr{C}$ be an abelian category. Then $\mathrm{Hom}_{\mathscr{C}}(\cdot, \cdot)$ is an additive bifunctor from $\mathscr{C}^\circ \times \mathscr{C}$ to $\mathfrak{Mod}(\mathbb{Z})$. If $\mathscr{C}$ admits enough injectives (or else, enough projectives), then $\mathrm{Hom}_{\mathscr{C}}(\cdot, \cdot)$ admits a right derived functor:

$$R\mathrm{Hom}_{\mathscr{C}}(\cdot, \cdot) : (\mathbf{D}^-(\mathscr{C}))^\circ \times \mathbf{D}^+(\mathscr{C}) \to \mathbf{D}^+(\mathfrak{Mod}(\mathbb{Z})) \ .$$

The functor $H^n(\cdot) \circ R\mathrm{Hom}_{\mathscr{C}}(\cdot, \cdot)$ is denoted $\mathrm{Ext}^n_{\mathscr{C}}(\cdot, \cdot)$.

Example 1.10.12. Let A be a ring. Then $\cdot \otimes_A \cdot$ is an additive bifunctor from $\mathfrak{Mod}(A^{op}) \times \mathfrak{Mod}(A)$ to $\mathfrak{Mod}(\mathbb{Z})$. It is a right exact functor with respect to both variables. One denotes by $\cdot \otimes^L_A \cdot$ the left derived functor from $\mathbf{D}^-(\mathfrak{Mod}(A^{op})) \times \mathbf{D}^-(\mathfrak{Mod}(A))$ to $\mathbf{D}^-(\mathfrak{Mod}(\mathbb{Z}))$, and one sets: $\mathrm{Tor}^A_n(N, M) = H^{-n}(N \otimes^L_A M)$.

Notation 1.10.13. Let A be a commutative ring. One still denotes by $R\mathrm{Hom}_A(\cdot, \cdot)$ or $\cdot \otimes^L_A \cdot$ the derived functors of $\mathrm{Hom}_A(\cdot, \cdot)$ or $\cdot \otimes_A \cdot$, with values in $\mathbf{D}(\mathfrak{Mod}(A))$.

Remark 1.10.14. Let $F : \mathscr{C} \times \mathscr{C}' \to \mathscr{C}''$ be a bifunctor of abelian categories and let $G : \mathscr{C}' \times \mathscr{C} \to \mathscr{C}''$ be the associated bifunctor given by $G(Y, X) = F(X, Y)$. We define an isomorphism $r : \mathbf{K}^+(F) \xrightarrow{\sim} \mathbf{K}^+(G)$, as follows. Let $X = (X^n)_{n \in \mathbb{Z}}$, $Y = (Y^n)_{n \in \mathbb{Z}}$ be two objects of $\mathbf{C}(\mathscr{C})$ and $\mathbf{C}(\mathscr{C}')$, respectively. Then $r(X, Y)$ is the direct sum of $(-1)^{nm}\mathrm{id}\colon F(X^n, Y^m) \to G(Y^m, X^n)$. One checks that $r(X, Y)$ is a morphism of simple complexes.

Remark 1.10.15. Let $F\colon \mathscr{C}' \times \mathscr{C}^\circ \times \mathscr{C} \to \mathscr{C}''$ be a trifunctor and $G\colon \mathscr{C}' \to \mathscr{C}''$ a functor. Assume to be given for all $Y \in \mathrm{Ob}(\mathscr{C}')$ and $X \in \mathrm{Ob}(\mathscr{C})$ a morphism

$$\alpha(Y, X)\colon F(Y, X, X) \to G(Y)$$

functorial with respect to Y and such that for any morphism $f\colon X \to X'$ in $\mathscr{C}$ the diagram:

$$\begin{array}{ccccc} & & F(Y, X', X') & & \\ & \nearrow & & \searrow & \\ F(Y, X', X) & & & & G(Y) \\ & \searrow & & \nearrow & \\ & & F(Y, X, X) & & \end{array}$$

commutes. Then for $Y \in \mathrm{Ob}(\mathbf{K}(\mathscr{C}'))$ and $X \in \mathrm{Ob}(\mathbf{K}(\mathscr{C}))$ one can define the morphism

$$\mathbf{K}(\alpha)(Y, X)\colon K(F)(Y, X, X) \to K(G)(Y)$$

as the direct sum of the morphisms

$$F(Y^m, X^n, X^n) \to G(Y^m) \ .$$

One easily checks that $\mathbf{K}(\alpha)(X, Y)$ is a morphism of complexes, using the convention introduced in Remark 1.8.11.

For example, if A is a commutative ring with finite global dimension (cf. Exercise I.28), we can define for X and Y in $\mathbf{D}^b(\mathfrak{Mod}(A))$:

$$\mathbf{D}(\alpha)(Y, X)\colon R\mathrm{Hom}_A(X, Y) \overset{L}{\underset{A}{\otimes}} X \to Y \ . \tag{1.10.7}$$

In fact take a quasi-isomorphism $Y \to I$ where I is bounded and injective and a quasi-isomorphism $P \to X$ where P is bounded and projective.

Then the morphism:

$$\mathbf{K}(\alpha)(I, P)\colon \mathrm{Hom}_A(P, I) \otimes_A P \longrightarrow I$$

is well-defined by the above remark, and defines $\mathbf{D}(\alpha)(Y, X)$.

Remark 1.10.16. Let F be as in Remark 1.10.14. Then, we identify $F(X[p], Y) \simeq F(X, Y)[p]$ by the identity:

$$(F(X[p], Y))^n \simeq \prod_k F(X^{k+p}, Y^{n-k}) \simeq (F(X, Y)[p])^n$$

and $F(X, Y[q]) \simeq F(X, Y)[q]$ by the identity:

$$(F(X, Y[q]))^n \simeq \prod_k F(X^k, Y^{n+q-k}) \xrightarrow{(-)^{k\cdot q}} \prod_k F(X^k, Y^{n+q-k}) \simeq (F(X, Y)[q])^n,$$

in order that this becomes a morphism of complexes.

Then

$$\begin{array}{ccc} F(X[p], Y[q]) & \longrightarrow & F(X, Y[q])[p] \\ \downarrow & & \downarrow \\ F(X[p], Y)[q] & \longrightarrow & F(X, Y)[p+q] \end{array}$$

commutes or anticommutes according to the parity of pq.

Remark 1.10.17. Questions of sign are always very delicate. They are treated with great care in Deligne [1] and the first pages of Berthelot-Breen-Messing [1], but the conventions adopted by the last authors are different from ours. In particular, in (1.4.3), they replace β by $-\beta$.

1.11. Ind-objects and pro-objects

Let $\mathscr{C}^\vee$ be the category of contravariant functors from $\mathscr{C}$ to $\mathfrak{Set}$. Let $h: \mathscr{C} \to \mathscr{C}^\vee$ be the functor: $X \mapsto \mathrm{Hom}_{\mathscr{C}}(\cdot, X)$. By this functor we look upon $\mathscr{C}$ as a full subcategory of $\mathscr{C}^\vee$ (cf. §1). Recall that a contravariant functor F from $\mathscr{C}$ to $\mathfrak{Set}$ is representable iff F is isomorphic to an object of $\mathscr{C}$ in $\mathscr{C}^\vee$.

Let $\mathscr{I}$ be a category. For another category $\mathscr{C}$, an inductive system in $\mathscr{C}$ indexed by $\mathscr{I}$ is by definition a functor from $\mathscr{I}$ to $\mathscr{C}$. Similarly a projective system is a functor from $\mathscr{I}^\circ$ to $\mathscr{C}$. Let $F_\circ$ be the projective system in $\mathfrak{Set}$ indexed by $\mathscr{I}$ given by $F_\circ(i) = \{\mathrm{pt}\}$ for any $i \in \mathrm{Ob}(\mathscr{I})$. Then for a projective system $F: \mathscr{I}^\circ \to \mathfrak{Set}$ indexed by $\mathscr{I}$, the set of morphisms from $F_\circ$ to F is called a projective limit of F and is denoted by $\varprojlim_{\mathscr{I}} F$ or else $\varprojlim_{i \in \mathscr{I}} F(i)$. Namely $\varprojlim_{\mathscr{I}} F$ consists of $\{x(i)\}_{i \in \mathrm{Ob}(\mathscr{I})}$ such that $x(i) \in F(i)$ and $F(h)x(j) = x(i)$ for any $h: i \to j$.

We sometimes write $i \in \mathscr{I}$ instead of $i \in \mathrm{Ob}(\mathscr{I})$ and $\varprojlim_i$ instead of $\varprojlim_{i \in \mathscr{I}}$.

Definition 1.11.1. *Let $\mathscr{I}$ and $\mathscr{C}$ be two categories.*

(a) *For an inductive system F in $\mathscr{C}$ indexed by $\mathscr{I}$, one denotes by $\varinjlim_{\mathscr{I}} F$ the functor from $\mathscr{C}$ to $\mathfrak{Set}$:*

$$X \mapsto \varprojlim_{i \in \mathscr{I}} \mathrm{Hom}_{\mathscr{C}}(F(i), X) .$$

(b) *For a projective system F in $\mathscr{C}$ indexed by $\mathscr{I}$ one denotes by $\varprojlim_{\mathscr{I}} F$ the functor from $\mathscr{C}^\circ$ to $\mathfrak{Set}$:*

$$X \mapsto \varprojlim_{i \in \mathcal{I}} \mathrm{Hom}_{\mathcal{C}}(X, F(i)) \ .$$

If these functors are representable, one keeps the same notations to denote their representatives in $\mathcal{C}$, and one calls them inductive limit and projective limit, respectively.

If F is a projective system in $\mathfrak{Set}$ then Definition 1.11.1(b) coincides with the former one.

Definition 1.11.2. *A category $\mathcal{I}$ is called filtrant if it is non-empty and satisfies the following conditions.*

$$\text{(1.11.1)} \qquad \begin{cases} \text{For } i, j \in \mathrm{Ob}(\mathcal{I}), \text{ there exist } k \in \mathrm{Ob}(\mathcal{I}) \text{ and morphisms} \\ i \to k \text{ and } j \to k \ . \end{cases}$$

$$\text{(1.11.2)} \qquad \begin{cases} \text{For two morphisms } f, g \in \mathrm{Hom}_{\mathcal{I}}(i, j), \text{ there exists a morphism} \\ h : j \to k \text{ such that } h \circ f = h \circ g \ . \end{cases}$$

The following result is easily proved.

Proposition 1.11.3. *Let F be an inductive system in $\mathfrak{Set}$ indexed by a filtrant category $\mathcal{I}$. Then $\varinjlim_{\mathcal{I}} F$ is represented by $(\bigsqcup_{i \in \mathrm{Ob}(\mathcal{I})} F(i))/\sim$, where for $x \in F(i)$ and $y \in F(j)$, $x \sim y$ if and only if there exist $f : i \to k$ and $g : j \to k$ such that $F(f)x = F(g)y$.*

Definition 1.11.4. *Let $\mathcal{I}$ and $\mathcal{C}$ be two categories, $\mathcal{I}$ being filtrant.*

(a) *For an inductive system F in $\mathcal{C}$ indexed by $\mathcal{I}$, one denotes by "$\varinjlim_{\mathcal{I}}$" F the functor from $\mathcal{C}^\circ$ to $\mathfrak{Set}$:*

$$X \mapsto \varinjlim_{i \in \mathcal{I}} \mathrm{Hom}_{\mathcal{C}}(X, F(i)) \ .$$

(b) *For a projective system F in $\mathcal{C}$ indexed by $\mathcal{I}$ one denotes by "$\varprojlim_{\mathcal{I}}$" F the functor from $\mathcal{C}$ to $\mathfrak{Set}$:*

$$X \mapsto \varinjlim_{i \in \mathcal{I}} \mathrm{Hom}_{\mathcal{C}}(F(i), X) \ .$$

If these functors are representable one keeps the same notations to denote their representatives in $\mathcal{C}$.

A functor from $\mathcal{C}^\circ$ (resp. $\mathcal{C}$) to $\mathfrak{Set}$ (i.e. an object of $\mathcal{C}^\vee$ (resp. $\mathcal{C}^\wedge$)) is called an ind-object (resp. a pro-object) if it is isomorphic to "$\varinjlim_{\mathcal{I}}$" F (resp. "$\varprojlim_{\mathcal{I}}$" F) for an inductive (resp. projective) system F in $\mathcal{C}$ indexed by a filtrant category $\mathcal{I}$.

Example 11.1.5. Let $(I, \leqslant)$ be an ordered set. We associate to I a category, denoted $\mathscr{I}$, as follows. We set:

$$(1.11.3)\qquad \begin{cases} \mathrm{Ob}(\mathscr{I}) = I \ , \\ \mathrm{Hom}_{\mathscr{I}}(i,j) \textit{ consists of a single element } s_{ji} \textit{ if } i \leqslant j \textit{ and} \\ \textit{is empty otherwise} \ . \end{cases}$$

Then for a category $\mathscr{C}$, an inductive system F in $\mathscr{C}$ indexed by I is by definition an inductive system in $\mathscr{C}$ indexed by $\mathscr{I}$.

One usually writes $\{X_i, \rho_{j,i}\}$ instead of F, where $X_i = F(i)$, $\rho_{i,j} = F(s_{ij})$. Similarly one defines a projective system indexed by $\mathscr{I}$.

If I is a directed ordered set (i.e. for any $i \in I, j \in I$ there exists $k \in I$ with $k \geqslant i$, $k \geqslant j$), then the category $\mathscr{I}$ is filtrant.

Let F be an inductive system in $\mathscr{C}$ indexed by a filtrant category $\mathscr{I}$. Assume $\varinjlim_j F(j)$ is representable, and denote by K its representative. We get:

$$\mathrm{Hom}_{\mathscr{C}}(K, K) \simeq \varprojlim_j \mathrm{Hom}_{\mathscr{C}}(F(j), K) \ .$$

This isomorphism defines a family of morphisms, $\rho_j : F(j) \to K$ satisfying:

$$(1.11.4)\qquad \rho_j \circ F(s) = \rho_i \ , \qquad \text{for any } s : i \to j$$

and $\{K, \rho_i\}_i$ satisfies the universal property represented by the diagram:

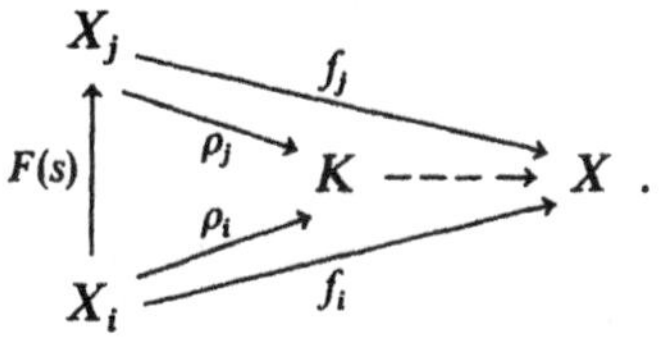

That is, for any family of morphisms $f_i : X_i \to X$ such that $f_j \circ F(s) = f_i$ for any $s : i \to j$, the dotted arrow exists uniquely, making the diagram commutative.

Now assume "$\varinjlim_j$" $F(j)$ is representable, and denote by L its representative. For any $i \in \mathrm{Ob}(\mathscr{I})$, the isomorphism:

$$\mathrm{Hom}_{\mathscr{C}}(F(i), L) \simeq \varinjlim_j \mathrm{Hom}_{\mathscr{C}}(F(i), F(j))$$

and $\mathrm{id}_{F(i)}$ define a morphism $\rho_i : F(i) \to L$, which satisfies (1.11.4). Moreover by the isomorphism:

$$\mathrm{Hom}_{\mathscr{C}}(L, L) \simeq \varinjlim_j \mathrm{Hom}(L, F(j)) \ ,$$

we find an $i_{\circ} \in \mathrm{Ob}(\mathscr{I})$ and a morphism $f : L \to F(i_{\circ})$ such that:

$$(1.11.5)\qquad \rho_{i_{\circ}} \circ f = \mathrm{id}_L \ .$$

For any $X \in \mathrm{Ob}(\mathscr{C})$ and any $i \in \mathrm{Ob}(\mathscr{I})$, the composition of the morphisms:

$$\mathrm{Hom}_{\mathscr{C}}(X, F(i)) \xrightarrow{\rho_i} \mathrm{Hom}_{\mathscr{C}}(X, L) \xrightarrow{f} \mathrm{Hom}_{\mathscr{C}}(X, F(i_\circ)) \longrightarrow \varinjlim_{j} \mathrm{Hom}_{\mathscr{C}}(X, F(j))$$

coincides with the canonical one. Therefore:

(1.11.6) *for any* $i \in \mathrm{Ob}(\mathscr{I})$ *there exist* $j \in \mathrm{Ob}(\mathscr{I})$, $s : i \to j$ *and* $t : i_\circ \to j$ *such that* $F(t) \circ f \circ \rho_i = F(s)$.

In fact these properties ensure the existence of "$\varinjlim$" in $\mathrm{Ob}(\mathscr{C})$.

Proposition 1.11.6. *Let* $L \in \mathrm{Ob}(\mathscr{C})$. *Then* L *represents* "$\varinjlim_{j}$" $F(j)$ *if and only if there exist* $\rho_i : F(i) \to L$ *for any* $i \in I$, *an* $i_\circ \in \mathrm{Ob}(\mathscr{I})$ *and* $f : L \to F(i_\circ)$ *such that* (1.11.4), (1.11.5) *and* (1.11.6) *are satisfied.*

Proof. We have already seen the necessity of these conditions. To prove the converse, note that for any $X \in \mathrm{Ob}(\mathscr{C})$, the two maps:

$$\varinjlim_{j} \mathrm{Hom}_{\mathscr{C}}(X, F(j)) \xrightarrow{\{\rho_j\}} \mathrm{Hom}_{\mathscr{C}}(X, L)$$

and

$$\mathrm{Hom}_{\mathscr{C}}(X, L) \xrightarrow{f} \mathrm{Hom}_{\mathscr{C}}(X, F(i_\circ)) \longrightarrow \varinjlim_{j} \mathrm{Hom}_{\mathscr{C}}(X, F(j))$$

are inverse to each other. □

Corollary 1.11.7. *If* "$\varinjlim_{j}$" $F(j)$ *is representable by* L, *then* $\varinjlim_{j} F(j)$ *is representable by* L.

Corollary 1.11.8. *Let* $\mathscr{C}'$ *be another category,* T *a functor from* $\mathscr{C}$ *to* $\mathscr{C}'$ *and let* F *be an inductive system in* $\mathscr{C}$. *If* "$\varinjlim_{j}$" $F(j)$ *exists in* $\mathscr{C}$, *then* "$\varinjlim_{j}$" $TF(j)$ *exists in* $\mathscr{C}'$, *and is represented by* $T($"$\varinjlim_{j}$" $F(j))$.

Proof. The existence of ρ_i, $i_\circ$, f in Proposition 1.11.6 satisfying (1.11.4)–(1.11.6) is preserved by T. □

Here we discussed inductive limits but the same results hold for projective limits, because a projective system in $\mathscr{C}$ is an inductive system in $\mathscr{C}^\circ$.

1.12. The Mittag-Leffler condition

In this section we shall mainly study projective systems of abelian groups, indexed by $\mathbb{N}$. We denote by $\mathfrak{Ab}$ the category of abelian groups. We consider the ordered set $(\mathbb{N}, \leqslant)$ as a category, and denote it by $\underline{\mathbb{N}}$ (cf. §11). Hence a projective system of abelian groups indexed by $\mathbb{N}$ is a functor from $\underline{\mathbb{N}}^\circ$ to $\mathfrak{Ab}$. Let X be such a functor. We also write $\{X_n, \rho_{n,p}\}$ instead of X. Hence $\rho_{n,p} : X_p \to X_n$ is a morphism of abelian groups defined for $p \geqslant n$, and these morphisms satisfy:

$$(1.12.1) \qquad \rho_{n,n} = \mathrm{id}_{X_n}, \qquad \rho_{n,p} \circ \rho_{p,q} = \rho_{n,q} \qquad \text{for } n \leqslant p \leqslant q .$$

Let $\mathrm{Hom}(\mathbb{N}^\circ, \mathfrak{Ab})$ be the category of contravariant functors from $\mathbb{N}$ to $\mathfrak{Ab}$ (that is, the category of projective systems of abelian groups indexed by $\mathbb{N}$). This is clearly an abelian category.

Since the category $\mathfrak{Ab}$ admits projective limits, (and also inductive limits, but we are not studying them for the moment), there exists a well defined functor:

$$\varprojlim : \mathrm{Hom}(\mathbb{N}^\circ, \mathfrak{Ab}) \to \mathfrak{Ab}$$

which associates to a projective system, its projective limit. If $X = \{X_n, \rho_{n,p}\}$ is a projective system, we shall write $\varprojlim X$ as well as $\varprojlim_n X_n$ to denote its projective limit.

Note that the functor $\varprojlim$ is left exact. We shall construct a full subcategory of $\mathrm{Hom}(\mathbb{N}^\circ, \mathfrak{Ab})$, injective with respect to this functor.

Definition 1.12.1. *Let $X = \{X_n, \rho_{n,p}\}$ be a projective system of abelian groups. One says that X satisfies the Mittag-Leffler condition (M-L for short) if for any $n \in \mathbb{N}$, the decreasing sequence $\{\rho_{n,p}(X_p)\}_{p \geqslant n}$ of subgroups of X_n is stationary.*

Proposition 1.12.2. *Let $0 \to X' \xrightarrow{f} X \xrightarrow{g} X'' \to 0$ be an exact sequence of projective systems of abelian groups.*

(i) *Assume X' and X'' satisfy the M-L condition. Then X satisfies the M-L condition.*
(ii) *Assume X satisfies the M-L condition. Then X'' satisfies the M-L condition.*

Proof. (i) For any n, N_0, N_1, p with $n \leqslant N_0 \leqslant N_1 \leqslant p$ we have a commutative diagram with exact rows:

$$\begin{array}{ccccccccc}
0 & \longrightarrow & X'_p & \xrightarrow{f} & X_p & \xrightarrow{g} & X''_p & \longrightarrow & 0 \\
 & & \downarrow & & \downarrow & & \downarrow & & \\
0 & \longrightarrow & X'_{N_1} & \xrightarrow{f} & X_{N_1} & \xrightarrow{g} & X''_{N_1} & \longrightarrow & 0 \\
 & & \downarrow & & \downarrow & & \downarrow & & \\
0 & \longrightarrow & X'_{N_0} & \xrightarrow{f} & X_{N_0} & \xrightarrow{g} & X''_{N_0} & \longrightarrow & 0 \\
 & & \downarrow & & \downarrow & & \downarrow & & \\
0 & \longrightarrow & X'_n & \xrightarrow{f} & X_n & \xrightarrow{g} & X''_n & \longrightarrow & 0 . \\
 & & \downarrow & & \downarrow & & \downarrow & & \\
 & & 0 & & 0 & & 0 & &
\end{array}$$

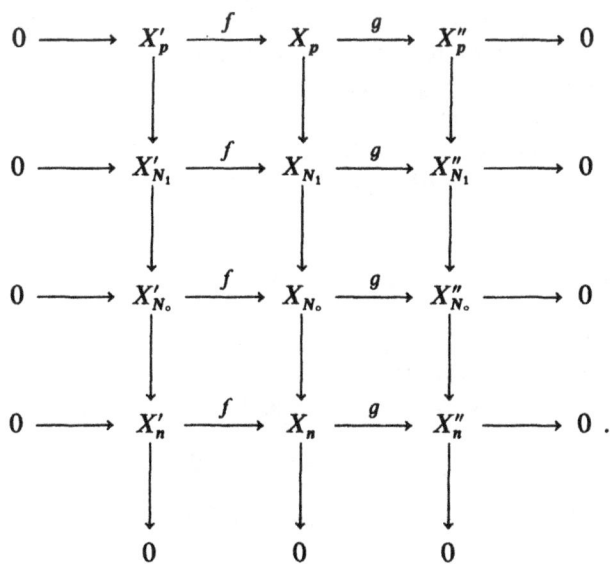

For n being fixed, we choose N_0 and then $N_1 \geqslant N_0$ such that

$$\operatorname{Im}(X'_p \to X'_n) = \operatorname{Im}(X'_{N_0} \to X'_n) \qquad \text{for any } p \geqslant N_0 \quad \text{and}$$

$$\operatorname{Im}(X''_p \to X''_{N_0}) = \operatorname{Im}(X''_{N_1} \to X''_{N_0}) \qquad \text{for any } p \geqslant N_1 \ .$$

Since $g(\operatorname{Im}(X_{N_1} \to X_{N_0})) = g(\operatorname{Im}(X_p \to X_{N_0}))$ we have:

$$\operatorname{Im}(X_{N_1} \to X_{N_0}) \subset \operatorname{Im}(X_p \to X_{N_0}) + f(X'_{N_0}) \ .$$

Hence:

$$\begin{aligned}\operatorname{Im}(X_{N_1} \to X_n) &\subset \operatorname{Im}(X_p \to X_n) + f(\operatorname{Im}(X'_{N_0} \to X'_n)) \\ &\subset \operatorname{Im}(X_p \to X_n) \ .\end{aligned}$$

(ii) is obvious. □

Proposition 1.12.3. *Let $0 \to X' \xrightarrow{f} X \xrightarrow{g} X'' \to 0$ be an exact sequence of projective systems. Assume X' satisfies the M-L condition. Then the sequence:*

$$0 \to \varprojlim X' \to \varprojlim X \to \varprojlim X'' \to 0$$

is exact.

Proof. Let $X'' = \{X''_n, \rho_{n,p}\}$, and let $x'' = \{x''_n\}_{n \in \mathbb{N}}$ be an element of $\varprojlim_n X''_n$. Let us choose an increasing sequence $\{v(n)\}_n$ such that $v(n) \geqslant n$ and $\operatorname{Im}(X'_p \to X'_n) = \operatorname{Im}(X'_{v(n)} \to X'_n)$ for any $p \geqslant v(n)$. First we note that if $x_{v(n)} \in X_{v(n)}$ satisfies $g(x_{v(n)}) = x''_{v(n)}$, then for each $p \geqslant v(n)$ there exists $x_p \in X_p$ which satisfies $g(x_p) = x''_p$ and $\rho_{n,p}(x_p) = \rho_{n,v(n)}(x_{v(n)})$. To check this point let $y_p \in X_p$ with $g(y_p) = x''_p$. Then $\rho_{v(n),p}(y_p) - x_{v(n)}$ belongs to $f(X'_{v(n)})$ thus $\rho_{n,p}(y_p) - \rho_{n,v(n)}(x_{v(n)}) \in \operatorname{Im}(f(X'_p) \to f(X'_n))$ and we may find $z_p \in f(X'_p)$ such that $y_p - z_p$ has the required properties. Now we construct a sequence $x = \{x_{v(n)}\}_n$ such that $g(x_{v(n)}) = x''_{v(n)}$ and $\rho_{n-1,v(n)}(x_{v(n)}) = \rho_{n-1,v(n-1)}(x_{v(n-1)})$.

Suppose we have already constructed $x_{v(i)} \in X_{v(i)}$ for $i < n_0$, then we can construct $x_{v(n_0)}$ by the preceding remark.

Then $x_n = \rho_{n,v(n)}(x_{v(n)})$ satisfies $g(x_n) = x''_n$ and $\rho_{n-1,n}(x_n) = x_{n-1}$. □

Now we consider projective systems of complexes of abelian groups, or equivalently, complexes of projective systems of abelian groups. Let $X^\cdot = \{X^k, \mathrm{d}^k\}$ be a complex of projective systems of abelian groups: for each k, $X^k = \{X^k_n, \rho^k_{n,p}\}$ is a projective system of abelian groups, and the morphisms d^k, $\rho^k_{n,p}$ satisfy the natural compatibility conditions.

To $X^\cdot$ we may associate a complex of abelian groups, denoted $X^\cdot_\infty$, by:

$$X^\cdot_\infty = \varprojlim X^\cdot = \{\varprojlim X^k, \mathrm{d}^k\} \ .$$

For each $n \in \mathbb{N}$, the natural morphisms $X^\cdot_\infty \to X^\cdot_n$ define for each $k \in \mathbb{Z}$, morphisms:

$$\phi_k : H^k(X_\infty^\cdot) \to \varprojlim_n H^k(X_n^\cdot) \ .$$

Proposition 1.12.4. *Assume that for each $k \in \mathbb{Z}$, the system X^k satisfies the M-L condition. Then:*

(a) *for each k, ϕ_k is surjective,*
(b) *if moreover, for a given i, the system $H^{i-1}(X^\cdot)$ (i.e. the projective system $\{H^{i-1}(X_n^\cdot), H^{i-1}(\rho_{n,p})\}$) satisfies the M-L condition, then ϕ_i is bijective.*

Proof. Set $Z_n^k = \operatorname{Ker}(d_n^k : X_n^k \to X_n^{k+1})$ and $B_n^k = \operatorname{Im}(d_{n-1}^k : X_{n-1}^k \to X_n^k)$. We have exact sequences:

$$0 \to Z_n^k \to X_n^k \to B_n^{k+1} \to 0 \ , \tag{1.12.2}$$

$$0 \to B_n^k \to Z_n^k \to H^k(X_n^\cdot) \to 0 \ , \tag{1.12.3}$$

$$0 \to \varprojlim_n B_n^k \to \varprojlim_n Z_n^k \to \varprojlim_n H^k(X_n^\cdot) \to 0 \ . \tag{1.12.4}$$

In fact (1.12.4) follows from Proposition 1.12.3 since the projective system $\{B_n^k\}_n$ satisfies the M-L condition by Proposition 1.12.2. Moreover, $\varprojlim(\cdot)$ being a left exact functor, we have:

$$\begin{aligned} \varprojlim_n Z_n^k &= \operatorname{Ker}\left(\varprojlim_n X_n^k \to \varprojlim_n X_n^{k+1}\right) \\ &= \operatorname{Ker}(X_\infty^k \to X_\infty^{k+1}) \ . \end{aligned} \tag{1.12.5}$$

Consider the diagram (1.12.6) whose rows are exact:

(1.12.6)

$$\begin{array}{ccccccccc} & & X_\infty^{k-1} & \longrightarrow & \operatorname{Ker}(X_\infty^k \to X_\infty^{k+1}) & \longrightarrow & H^k(X_\infty^\cdot) & \longrightarrow & 0 \\ & & \downarrow{\scriptstyle \psi_k} & & \downarrow{\scriptstyle \wr} & & \downarrow{\scriptstyle \phi_k} & & \\ 0 & \longrightarrow & \varprojlim_n B_n^k & \longrightarrow & \varprojlim_n Z_n^k & \longrightarrow & \varprojlim_n H^k(X_n^\cdot) & \longrightarrow & 0 \ . \end{array}$$

From this diagram we obtain that ϕ_k is surjective.

Then assume $\{H^{i-1}(X_n^\cdot)\}_n$ satisfies the M-L condition. By (1.12.3) and Proposition 1.12.2, we find that the projective system $\{Z_n^{i-1}\}_n$ satisfies the M-L condition. Applying Proposition 1.12.3 to (1.12.2) we get the exact sequence:

$$0 \to \varprojlim Z_n^{i-1} \to \varprojlim_n X_n^{i-1} \to \varprojlim_n B_n^i \to 0 \ .$$

Hence, in the diagram (1.12.6), ψ_{i-1} is surjective, and ϕ_i is injective. □

Remark 1.12.5. Let us denote by Hom($\mathscr{I}$, $\mathfrak{Ab}$) the category of functors from a category $\mathscr{I}$ to $\mathfrak{Ab}$, that is, the category of inductive systems of abelian groups indexed by $\mathscr{I}$. There is a well-defined functor:

$$\varinjlim : \mathrm{Hom}(\mathscr{I}, \mathfrak{Ab}) \to \mathfrak{Ab}$$

which associates to an inductive system, its inductive limit. Contrarily to the functor $\varprojlim$, the functor $\varinjlim$ is exact.

In particular, consider a complex of inductive systems of abelian groups, $X^{\cdot} = \{X^k, \mathrm{d}^k\}$, where each $X^k = \{X_i^k, \rho_{i,j}^k\}$ is an inductive system. Then we have for each $k \in \mathbb{Z}$:

$$H^k(\varinjlim X^{\cdot}) \simeq \varinjlim H^k(X^{\cdot}) \ . \tag{1.12.7}$$

To end this section, we shall recall a result (cf. Kashiwara [5]) on inductive and projective limits of sets indexed by $\mathbb{R}$.

Proposition 1.12.6. *Let $\{X_s, \rho_{s,t}\}$ be a projective system of sets indexed by $\mathbb{R}$. Assume that for each $s \in \mathbb{R}$, the canonical maps:*

$$\lambda_s : X_s \to \varprojlim_{r<s} X_r$$

and

$$\mu_s : \varinjlim_{t>s} X_t \to X_s$$

are both injective (resp. surjective). Then all the maps ρ_{s_0,s_1} $(s_0 \leqslant s_1)$ are injective (resp. surjective).

Proof. Let us first prove the injectivity. Let x and y belong to X_{s_1}, such that $\rho_{s_0,s_1}(x) = \rho_{s_0,s_1}(y)$ for some $s_0 < s_1$. Let:

$$I = \{s \in \mathbb{R}; s \leqslant s_1, \rho_{s,s_1}(x) = \rho_{s,s_1}(y)\} \ .$$

Then I contains s_0, and $s \in I$, $r < s$, imply $r \in I$. Let $s_2 = \sup I$. Since λ_{s_2} is injective, s_2 belongs to I. If $s_2 = s_1$, then $x = y$. Assume $s_2 < s_1$. Since μ_{s_2} is injective, there exists $s > s_2$ with $s \in I$. This is a contradiction.

Now we prove the surjectivity. Let $s_0 < s_1$, $x_0 \in X_{s_0}$. Let A be the set of pairs (s, x) with $s_0 \leqslant s \leqslant s_1$, $x \in X_s$, and $\rho_{s,s_0}(x) = x_0$. We can order A as follows:

$$(s, x) \leqslant (s', x') \quad \text{iff} \quad s \leqslant s' \quad \text{and} \quad \rho_{s,s'}(x') = x \ .$$

Let us show that A is inductively ordered. Let $B \subset A$, B being totally ordered, and let:

$$I = \{s \in \mathbb{R};\ s_0 \leqslant s \leqslant s_1, \text{there exists } x \in X_s, \text{with } (s, x) \in B\} \ .$$

Let $s_2 = \sup I$. If $s_2 \in I$, then B possesses a maximal element. If s_2 does not belong to I, there exists $(s_2, x_2) \in A$ greater than any element of B by the surjectivity of λ_{s_2}. This proves that A is inductively ordered.

Let (s, x) be a maximal element of A. If $s = s_1$, the proof is over. Otherwise there exists s', $s < s' \leqslant s_1$ and $x' \in X_{s'}$ such that $\rho_{s,s'}(x') = x$, since μ_s is surjective. This is a contradiction. □

Exercises to Chapter I

Exercise I.1. Let $\mathscr{C}$ be an additive category. Prove that there exists only one way to endow each $\mathrm{Hom}_{\mathscr{C}}(X, Y)$ of a structure of an additive group such that the composition law $\mathrm{Hom}_{\mathscr{C}}(X, Y) \times \mathrm{Hom}_{\mathscr{C}}(Y, Z) \to \mathrm{Hom}_{\mathscr{C}}(X, Z)$ is additive for all $X, Y, Z \in \mathrm{Ob}(\mathscr{C})$. (Hint: determine first $0 \in \mathrm{Hom}_{\mathscr{C}}(X, Y)$, then determine the addition of $\mathrm{Hom}_{\mathscr{C}}(X, Y)$ by using the existence of products.)

Exercise I.2. Let $\mathscr{C}$ and $\mathscr{C}'$ be two categories, and let $F: \mathscr{C} \to \mathscr{C}'$ and $G: \mathscr{C}' \to \mathscr{C}$ be functors.

(i) Prove that the following two conditions are equivalent.

(a) There exist morphisms of functors

$$\alpha: F \circ G \to \mathrm{id}_{\mathscr{C}'}, \qquad \beta: \mathrm{id}_{\mathscr{C}} \to G \circ F$$

such that the composition $G(Y) \xrightarrow{\beta(G(Y))} G \circ F \circ G(Y) \xrightarrow{G(\alpha(Y))} G(Y)$ is equal to $\mathrm{id}_{G(Y)}$ for any $Y \in \mathrm{Ob}(\mathscr{C}')$ and the composition $F(X) \xrightarrow{F(\beta(X))} F \circ G \circ F(X) \xrightarrow{\alpha(F(X))} F(X)$ is equal to $\mathrm{id}_{F(X)}$ for any $X \in \mathrm{Ob}(\mathscr{C})$.

(b) There exists an isomorphism of bifunctors from $\mathscr{C}^{\circ} \times \mathscr{C}'$ to $\mathfrak{Set}$: $\mathrm{Hom}_{\mathscr{C}'}(F(X), Y) \simeq \mathrm{Hom}_{\mathscr{C}}(X, G(Y))$.

In such a case, we say that G is a **right adjoint** functor to F and that F is a **left adjoint** functor to G.

(ii) Prove that, for a functor $F: \mathscr{C} \to \mathscr{C}'$ (resp. $G: \mathscr{C}' \to \mathscr{C}$), its right (resp. left) adjoint functor (if it exists) is unique up to an isomorphism.

(iii) Prove that a functor $F: \mathscr{C} \to \mathscr{C}'$ (resp. $G: \mathscr{C}' \to \mathscr{C}$) has a right (resp. left) adjoint functor if and only if for any $Y \in \mathrm{Ob}(\mathscr{C}')$, $X \mapsto \mathrm{Hom}_{\mathscr{C}'}(F(X), Y)$ is representable, (resp. for any $X \in \mathrm{Ob}(\mathscr{C})$, $Y \mapsto \mathrm{Hom}_{\mathscr{C}}(X, G(Y))$ is representable).

Exercise I.3. Under the conditions (i), (ii), (iii), of Definition 1.2.1, prove that Z is a representative of the functor $W \mapsto \mathrm{Hom}_{\mathscr{C}}(X, W) \oplus \mathrm{Hom}_{\mathscr{C}}(Y, W)$, if and only if there are morphisms $i_1: X \to Z$, $i_2: Y \to Z$, $p_1: Z \to X$, $p_2: Z \to Y$, such that $p_2 \circ i_1 = 0$, $p_1 \circ i_2 = 0$, $p_1 \circ i_1 = \mathrm{id}_X$, $p_2 \circ i_2 = \mathrm{id}_Y$ and $i_1 \circ p_1 + i_2 \circ p_2 = \mathrm{id}_Z$.

Exercise I.4. Let $X \xrightarrow{i_1} Z \xrightarrow{p_2} Y$ be a sequence of morphisms in an additive category $\mathscr{C}$, with $p_2 \circ i_1 = 0$. Prove that the following conditions are equivalent.

(i) For any $W \in \mathrm{Ob}(\mathscr{C})$, the sequence:

$$0 \to \mathrm{Hom}_{\mathscr{C}}(W, X) \to \mathrm{Hom}_{\mathscr{C}}(W, Z) \to \mathrm{Hom}_{\mathscr{C}}(W, Y) \to 0$$

is exact.

(ii) For any $W \in \mathrm{Ob}(\mathscr{C})$, the sequence:

$$0 \leftarrow \mathrm{Hom}_{\mathscr{C}}(X, W) \leftarrow \mathrm{Hom}_{\mathscr{C}}(Z, W) \leftarrow \mathrm{Hom}_{\mathscr{C}}(Y, W) \leftarrow 0$$

is exact.

(iii) One can find $i_2 : Y \to Z$ and $p_1 : Z \to X$ which satisfy the conditions of Exercise I.3.

If these conditions are satisfied, one says that the sequence $0 \to X \to Z \to Y \to 0$ splits. In such a case, Z is isomorphic to $X \oplus Y$, and one says that X is a **direct summand** of Z.

(iv) Prove that if $\mathscr{C}$ is an abelian category or a triangulated category, if there are $i_1 : X \to Z$ and $p_1 : Z \to X$ such that $p_1 \circ i_1 = \mathrm{id}_X$, then X is a direct summand of Z.

Exercise I.5. Let $0 \to X \to Z \to Y \to 0$ be an exact sequence in an abelian category. If either X is injective, or Y is projective, then the sequence splits.

Exercise I.6. Let $\mathscr{C}$ be an abelian category.

(i) Let $f : X \to Z$ and $g : Y \to Z$ be two morphisms in $\mathscr{C}$. Prove that $\mathrm{Ker}(X \oplus Y \to Z)$ represents the functor

$$W \mapsto \mathrm{Hom}_{\mathscr{C}}(W, X) \underset{\mathrm{Hom}_{\mathscr{C}}(W, Z)}{\times} \mathrm{Hom}_{\mathscr{C}}(W, Y) .$$

A representative of this functor is denoted $X \times_Z Y$.

Similarly, by reversing the arrows, one defines $X \oplus_Z Y$, for two morphisms $f : Z \to X$ and $g : Z \to Y$, as a representative of the functor $W \to \mathrm{Hom}_{\mathscr{C}}(X, W) \times_{\mathrm{Hom}_{\mathscr{C}}(Z, W)} \mathrm{Hom}_{\mathscr{C}}(Y, W)$ (i.e. $\mathrm{Coker}(Z \to X \oplus Y)$).

(ii) Show that, in the situation of (i), we have a commutative diagram:

$$\begin{array}{ccccccc}
 & & & & 0 & & 0 \\
 & & & & \downarrow & & \downarrow \\
 & & & & \mathrm{Ker}\, g' & \xrightarrow{\sim} & \mathrm{Ker}\, g \\
 & & & & \downarrow & & \downarrow \\
0 & \longrightarrow & \mathrm{Ker}\, f' & \longrightarrow & X \times_Z Y & \xrightarrow{f'} & Y \\
 & & \downarrow \wr & & \downarrow g' & & \downarrow g \\
0 & \longrightarrow & \mathrm{Ker}\, f & \longrightarrow & X & \xrightarrow{f} & Z ,
\end{array}$$

where f' and g' are defined in the obvious way.

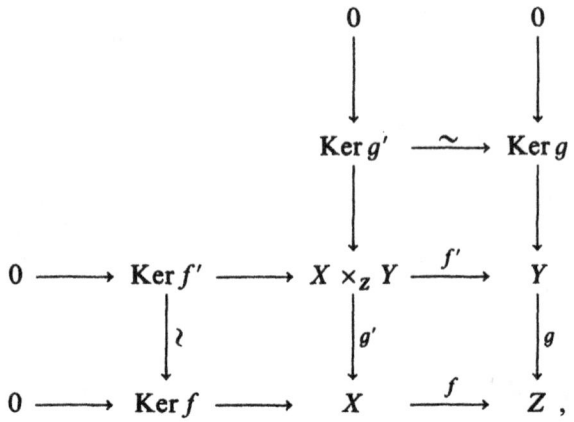

(iii) Let

$$\begin{array}{ccc} X' & \xrightarrow{f'} & Y' \\ {\scriptstyle g'}\downarrow & & \downarrow{\scriptstyle g} \\ X & \xrightarrow{f} & Y \end{array}$$

be a commutative diagram. Then the following conditions are equivalent:
(a) $X' \to X \times_Y Y'$ is an epimorphism,
(b) $X \oplus_{X'} Y' \to Y$ is a monomorphism,
(c) in the diagram:

$$\begin{array}{ccccccccc}
 & 0 & & 0 & & 0 & & & \\
 & \downarrow & & \downarrow & & \downarrow & & & \\
\longrightarrow & \operatorname{Ker} f' \times_{X'} \operatorname{Ker} g' & \longrightarrow & \operatorname{Ker} g' & \longrightarrow & \operatorname{Ker} g & \longrightarrow & 0 & \\
 & \downarrow & & \downarrow & & \downarrow & & \downarrow & \\
\longrightarrow & \operatorname{Ker} f' & \longrightarrow & X' & \xrightarrow{f'} & Y' & \longrightarrow & \operatorname{Coker} f' & \longrightarrow \\
 & \downarrow & & \downarrow{\scriptstyle g'} & & \downarrow{\scriptstyle g} & & \downarrow & \\
\longrightarrow & \operatorname{Ker} f & \longrightarrow & X & \xrightarrow{f} & Y & \longrightarrow & \operatorname{Coker} f & \longrightarrow \\
 & \downarrow & & \downarrow & & \downarrow & & \downarrow & \\
 & 0 & \longrightarrow & \operatorname{Coker} g' & \longrightarrow & \operatorname{Coker} g & \longrightarrow & \operatorname{Coker} f \oplus_Y \operatorname{Coker} g & \longrightarrow \\
 & & & \downarrow & & \downarrow & & \downarrow & \\
 & & & 0 & & 0 & & 0 &
\end{array}$$

all sequences are exact.

(iv) Let $f: X \to Y$ be a morphism in $\mathbf{C}(\mathscr{C})$. Assume that for each n, the square

$$\begin{array}{ccc} \operatorname{Coker} d_X^{n-1} & \longrightarrow & X^n \\ \downarrow & & \downarrow \\ \operatorname{Coker} d_Y^{n-1} & \longrightarrow & Y^n \end{array}$$

satisfies the equivalent conditions in (iii). Then f is a quasi-isomorphism. (Hint: use the diagram:

$$\begin{array}{ccccccccccc} 0 & \longrightarrow & H^n(X) & \longrightarrow & \operatorname{Coker} d_X^{n-1} & \longrightarrow & X^{n+1} & \longrightarrow & \operatorname{Coker} d_X^n & \longrightarrow & 0 \\ & & \downarrow & & \downarrow & & \downarrow & & \downarrow & & \\ 0 & \longrightarrow & H^n(Y) & \longrightarrow & \operatorname{Coker} d_Y^{n-1} & \longrightarrow & Y^{n+1} & \longrightarrow & \operatorname{Coker} d_Y^n & \longrightarrow & 0.) \end{array}$$

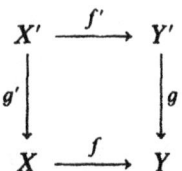

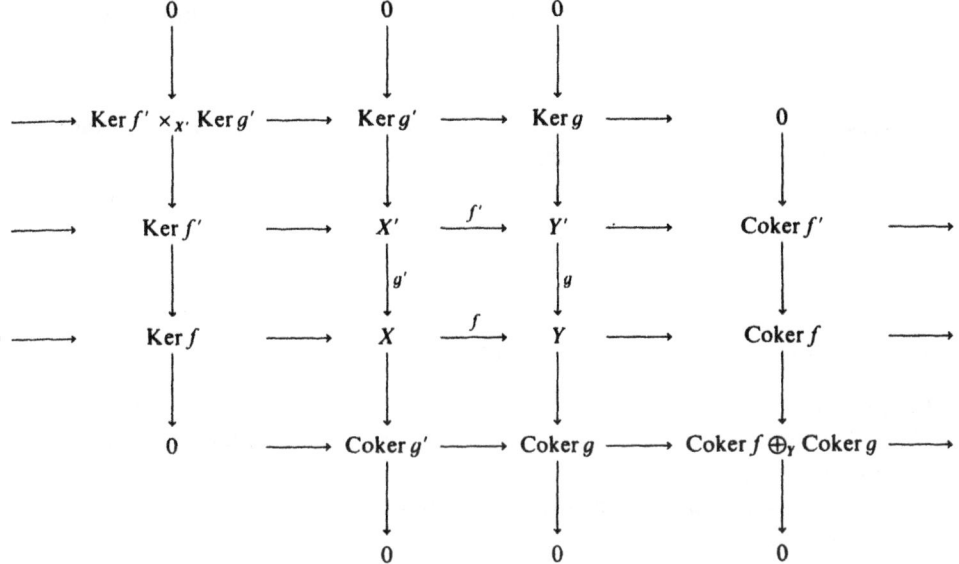

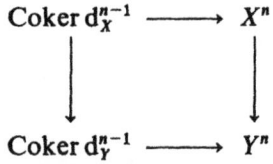

Exercise I.7. Let $\mathscr{C}$ be an abelian category.

(i) For an object Z of $\mathscr{C}$, let $\mathscr{P}(Z)$ be the category whose objects are the epimorphisms $f: Z' \to Z$, a morphism $(f: Z' \to Z) \to (f': Z'' \to Z)$ being defined by $h: Z' \to Z''$ with $f' \circ h = f$. Prove that $\mathscr{P}(Z)$ is cofiltrant, that is, $\mathscr{P}(Z)^\circ$ is filtrant.

(ii) For $X \in \mathrm{Ob}(\mathscr{C})$, set $\tilde{h}_Z(X) = \varinjlim_{\mathscr{P}(Z)} \mathrm{Hom}_{\mathscr{C}}(Z', X)$ where $\mathrm{Hom}_{\mathscr{C}}(Z', X)$ is the functor from $\mathscr{P}(Z)$ to $\mathfrak{Ab}$ which associates $\mathrm{Hom}_{\mathscr{C}}(Z', X)$ to $f: Z' \to Z$ in $\mathscr{P}(Z)$.

Prove that $\tilde{h}_Z$ is an exact functor from $\mathscr{C}$ to $\mathfrak{Ab}$ and prove that if f and f' in $\mathrm{Hom}_{\mathscr{C}}(X, X')$ satisfy $\tilde{h}_Z(f) = \tilde{h}_Z(f')$ for all $Z \in \mathrm{Ob}(\mathscr{C})$, then $f = f'$. Also prove that a sequence in $\mathscr{C}$ is exact if its image by $\tilde{h}_Z$ is exact for any $Z \in \mathrm{Ob}(\mathscr{C})$.

Note that this exercise will allow us to transform many problems on abelian categories to problems on $\mathfrak{Ab}$.

Exercise I.8 (The five lemma). Let $\mathscr{C}$ be an abelian category and consider a commutative diagram in $\mathscr{C}$ with exact rows:

$$\begin{array}{ccccccccc}
X^0 & \longrightarrow & X^1 & \longrightarrow & X^2 & \longrightarrow & X^3 & \longrightarrow & X^4 \\
\downarrow{\scriptstyle f_0} & & \downarrow{\scriptstyle f_1} & & \downarrow{\scriptstyle f_2} & & \downarrow{\scriptstyle f_3} & & \downarrow{\scriptstyle f_4} \\
Y^0 & \longrightarrow & Y^1 & \longrightarrow & Y^2 & \longrightarrow & Y^3 & \longrightarrow & Y^4 .
\end{array}$$

Prove that:

(i) if f_0 is an epimorphism and f_1 and f_3 are monomorphisms, then f_2 is a monomorphism,

(ii) if f_4 is a monomorphism and f_1 and f_3 are epimorphisms, then f_2 is an epimorphism.

Exercise I.9. Let $\mathscr{C}$ be an abelian category. Consider the commutative diagram with exact rows in $\mathscr{C}$:

$$\begin{array}{ccccccccc}
 & & X & \xrightarrow{f} & Y & \xrightarrow{g} & Z & \longrightarrow & 0 \\
 & & \downarrow{\scriptstyle \alpha} & & \downarrow{\scriptstyle \beta} & & \downarrow{\scriptstyle \gamma} & & \\
0 & \longrightarrow & X' & \xrightarrow{f'} & Y' & \xrightarrow{g'} & Z' & & .
\end{array}$$

Prove that there is a natural exact sequence:

$$\mathrm{Ker}\,\alpha \longrightarrow \mathrm{Ker}\,\beta \longrightarrow \mathrm{Ker}\,\gamma \xrightarrow{\varphi} \mathrm{Coker}\,\alpha \longrightarrow \mathrm{Coker}\,\beta \longrightarrow \mathrm{Coker}\,\gamma \ ,$$

so that the following diagram commutes:

$$\begin{array}{ccccc}
 & & Y & \xrightarrow{g} & Z \\
 & & \uparrow & & \uparrow \\
Y & \longleftarrow & \operatorname{Ker}\gamma\circ g & \longrightarrow & \operatorname{Ker}\gamma \\
\downarrow & & \downarrow & & \downarrow{\scriptstyle\varphi} \\
Y' & \xleftarrow{f'} & X' & \longrightarrow & \operatorname{Coker}\alpha\ .
\end{array}$$

Exercise I.10. Let $\mathscr{C}$ be an abelian category. Consider the diagram of exact sequences:

$$\begin{array}{ccccccccc}
0 & \longrightarrow & M & \longrightarrow & M_0 & \longrightarrow & M_1 & \longrightarrow & 0 \\
 & & \| & & & & & & \\
0 & \longrightarrow & M & \longrightarrow & M_0' & \longrightarrow & M_1' & \longrightarrow & 0
\end{array}$$

and assume M_0 and M_0' are injective. Construct an isomorphism $M_0 \oplus M_1' \simeq M_0' \oplus M_1$.

Exercise I.11. Let $\mathscr{C}$ be an abelian category and let $X \in \mathrm{Ob}(\mathbf{C}(\mathscr{C}))$ be such that for any $Y \in \mathrm{Ob}(\mathscr{C})$, the complex of abelian groups $\mathrm{Hom}_{\mathscr{C}}(Y, X)$ is exact. Prove that X is 0 in $\mathbf{K}(\mathscr{C})$, (cf. Exercise I.4).

Exercise I.12. Let $\mathscr{C}$ be a triangulated category, and consider a commutative diagram in $\mathscr{C}$:

$$\begin{array}{ccccccc}
X & \xrightarrow{f} & Y & \xrightarrow{g} & Z & \xrightarrow{h} & X[1] \\
\| & & \| & & \downarrow & & \| \\
X & \xrightarrow{f} & Y & \xrightarrow{g'} & Z' & \xrightarrow{h'} & X[1]\ ,
\end{array}$$

where the first row is a distinguished triangle and $f[1] \circ h' = 0$." Prove that the second row is also a distinguished triangle under one of the following hypotheses:

(i) for any $P \in \mathrm{Ob}(\mathscr{C})$, the sequence $\mathrm{Hom}(P, X) \to \mathrm{Hom}(P, Y) \to \mathrm{Hom}(P, Z') \to \mathrm{Hom}(P, X[1])$ is exact,

(ii) for any $Q \in \mathrm{Ob}(\mathscr{C})$, the sequence $\mathrm{Hom}(X[1], Q) \to \mathrm{Hom}(Z', Q) \to \mathrm{Hom}(Y, Q) \to \mathrm{Hom}(X, Q)$ is exact.

Exercise I.13. Let $X_i \longrightarrow Y_i \longrightarrow Z_i \xrightarrow[+1]{} (i = 1, 2)$ be two triangles in a triangulated category. Prove that they are distinguished triangles if and only if their direct sum $X_1 \oplus X_2 \longrightarrow Y_1 \oplus Y_2 \longrightarrow Z_1 \oplus Z_2 \xrightarrow[+1]{}$ is a distinguished triangle. (Hint:

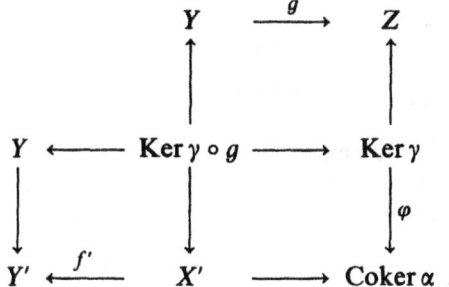

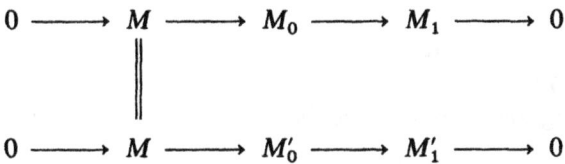

use Exercise I.12. To prove it is sufficient, consider a distinguished triangle $\delta : X_1 \longrightarrow Y_1 \longrightarrow U \underset{+1}{\longrightarrow}$ and prove that the composition $\delta_1 \to \delta_1 \oplus \delta_2 \to \delta$ is an isomorphism, where δ_1 and δ_2 are the given triangles.)

Exercise I.14. Let $\mathscr{C}$ be a category, S a multiplicative system. For $X \in \mathrm{Ob}(\mathscr{C})$ let S_X be the category whose objects are morphisms $s : X' \to X$ with $s \in S$ and morphisms are defined as follows. For $s : X' \to X$ and $s' : X'' \to X$, we set:

$$\mathrm{Hom}_{S_X}(s, s') = \{h \in \mathrm{Hom}_{\mathscr{C}}(X'', X'); s' = s \circ h\} \ .$$

(i) Show that the category $(S_X)^\circ$ is filtrant.

(ii) Prove that for $X, Y \in \mathrm{Ob}(\mathscr{C})$:

$$\mathrm{Hom}_{\mathscr{C}_S}(X, Y) = \varinjlim_{S_X^\circ} \mathrm{Hom}_{\mathscr{C}}(X', Y) \ .$$

Here $\mathrm{Hom}_{\mathscr{C}}(X', Y)$ is the functor from S_X° to $\mathfrak{Set}$ which associates $\mathrm{Hom}_{\mathscr{C}}(X', Y)$ to $s' : X' \to X$.

(iii) By reversing the arrows, define the category S_Y^a and prove:

$$\mathrm{Hom}_{\mathscr{C}_S}(X, Y) = \varinjlim_{S_Y^a} \mathrm{Hom}_{\mathscr{C}}(X, Y') \ .$$

Exercise I.15. (See Deligne [1].) Let $\mathscr{C}$ be a category. One defines the category $\mathrm{Ind}(\mathscr{C})$ as the full subcategory of $\mathscr{C}^\vee$ (cf. §1) consisting of objects isomorphic to "$\varinjlim_{\mathscr{I}}$" F for some inductive system F in $\mathscr{C}$ indexed by a filtrant category $\mathscr{I}$.

Now assume $\mathscr{C}$ is abelian and for $X \in \mathrm{Ob}(\mathbf{K}^+(\mathscr{C}))$ denote by S_X the category whose objects consist of quasi-isomorphisms $u : X \to X'$, a morphism $h : (u : X \to X') \to (u' : X \to X'')$ being defined by $v : X' \to X''$ with $v \circ u = u'$.

(i) Prove that the functor $\sigma : \mathbf{D}^+(\mathscr{C}) \to \mathrm{Ind}(\mathbf{K}^+(\mathscr{C}))$, $X \mapsto$ "$\varinjlim_{S_X}$" X' is well-defined and fully faithful. Here X' is the functor from S_X to $\mathbf{K}^+(\mathscr{C})$ which associates X' to $u : X \to X'$.

(ii) Let $F : \mathscr{C} \to \mathscr{C}'$ be a left exact functor of abelian categories. Define the functor $T : \mathbf{D}^+(\mathscr{C}) \to \mathrm{Ind}(\mathbf{K}^+(\mathscr{C}'))$ by $T(X) =$ "$\varinjlim_{S_X}$" $F(X')$. Let us say that F is derivable at $X \in \mathrm{Ob}(\mathbf{D}^+(\mathscr{C}))$ if there exists $Y \in \mathrm{Ob}(\mathbf{D}^+(\mathscr{C}'))$ such that $T(X) \simeq \sigma(Y)$. Prove that in such a case Y is unique, and that if F is derivable at each $X \in \mathrm{Ob}(\mathbf{D}^+(\mathscr{C}))$ then F admits a right derived functor and $\sigma \circ RF \simeq T$.

Exercise I.16. Let $\mathscr{C}$ be an additive category.

(i) For $X \in \mathbf{C}^-(\mathscr{C})$ and $Y \in \mathbf{C}^+(\mathscr{C})$ prove that:

$$Z^0(s(\mathrm{Hom}_{\mathscr{C}}(X, Y))) = \mathrm{Hom}_{\mathbf{C}(\mathscr{C})}(X, Y) \ ,$$

$$B^0(s(\mathrm{Hom}_{\mathscr{C}}(X, Y))) = \mathrm{Ht}(X, Y) \ ,$$

$$H^0(s(\mathrm{Hom}_{\mathscr{C}}(X, Y))) = \mathrm{Hom}_{\mathbf{K}(\mathscr{C})}(X, Y) \ .$$

(s is defined in (1.9.4).)

(ii) Assume moreover that $\mathscr{C}$ is an abelian category with enough injectives (or else, enough projectives). Prove that for $X \in \mathrm{Ob}(\mathbf{D}^-(\mathscr{C}))$ and $Y \in \mathrm{Ob}(\mathbf{D}^+(\mathscr{C}))$ one has:

$$H^0(R\mathrm{Hom}_{\mathscr{C}}(X, Y)) = \mathrm{Hom}_{\mathbf{D}(\mathscr{C})}(X, Y) .$$

Exercise I.17. Let $\mathscr{C}$ be an abelian category. One says $\mathscr{C}$ has homological dimension $\leqslant n$, where n is a non-negative integer, if $\mathrm{Ext}^j(X, Y) = 0$ for $j > n$ and for any $X, Y \in \mathrm{Ob}(\mathscr{C})$. Here $\mathrm{Ext}^j(X, Y) = \mathrm{Hom}_{\mathbf{D}(\mathscr{C})}(X, Y[j])$. Assume $\mathscr{C}$ has enough injectives. Prove that the following conditions are equivalent.

(i) $\mathscr{C}$ has homological dimension $\leqslant n$.

(ii) For any $X \in \mathrm{Ob}(\mathscr{C})$ there exists an injective resolution of X of length $\leqslant n$, i.e. an exact sequence: $0 \to X \to I^0 \to \cdots \to I^n \to 0$ with all I^j's injective.

The smallest $n \in \mathbb{N} \cup \{\infty\}$ such that these equivalent conditions are satisfied is called the **homological dimension** of $\mathscr{C}$ and denoted $\mathrm{hd}(\mathscr{C})$.

Exercise I.18. Let $\mathscr{C}$ be an abelian category with $\mathrm{hd}(\mathscr{C}) \leqslant 1$. Prove that for any $X \in \mathrm{Ob}(\mathbf{D}^b(\mathscr{C}))$, one has an isomorphism in $\mathbf{D}^b(\mathscr{C})$:

$$X \simeq \bigoplus_k H^k(X)[-k] .$$

(Example: $\mathscr{C} = \mathfrak{Mod}(A)$ where A is a principal ideal domain. In particular when A is a field, or else $A = \mathbb{Z}$.)

Exercise I.19. Let $\mathscr{C}$ and $\mathscr{C}'$ be two abelian categories, $F: \mathscr{C} \to \mathscr{C}'$ a left exact functor, and let $\mathscr{I}$ be an F-injective subcategory of $\mathscr{C}$. We call an object X of $\mathscr{C}$, **F-acyclic** if $R^kF(X) = 0$ for $k \neq 0$. Let $\mathscr{J}$ be the full subcategory of $\mathscr{C}$ consisting of F-acyclic objects.

(i) Prove that $\mathscr{J}$ is F-injective.

(ii) Prove that for an integer $n \geqslant 0$, the following conditions are equivalent:

(a) $R^kF(X) = 0$ for any $k > n$ and any $X \in \mathrm{Ob}(\mathscr{C})$,

(b) for any $X \in \mathrm{Ob}(\mathscr{C})$, there exists an exact sequence

$$0 \to X \to X^0 \to \cdots \to X^n \to 0$$

with $X^j \in \mathrm{Ob}(\mathscr{J})$ for $0 \leqslant j \leqslant n$,

(c) if $X^0 \to \cdots \to X^n \to 0$ is an exact sequence, and if $X^j \in \mathrm{Ob}(\mathscr{J})$ for $j < n$, then X^n belongs to $\mathscr{J}$.

In such a case, one says F has **cohomological dimension** $\leqslant n$.

Exercise I.20. In the situation of Proposition 1.8.7, assume F (resp. F') has cohomological dimension $\leqslant r$ (resp. $\leqslant r'$), (cf. Exercise I.19). Prove that $F' \circ F$ has cohomological dimension $\leqslant r + r'$.

Exercise I.21. Let $\mathscr{C}$ and $\mathscr{C}'$ be two abelian categories, $F: \mathscr{C} \to \mathscr{C}'$ a left exact functor. Assume that there exists an F-injective subcategory $\mathscr{I}$ of $\mathscr{C}$. Let

$X \in \mathrm{Ob}(\mathbf{D}^+(\mathscr{C}))$ be such that $R^iF(H^j(X)) = 0$ for all $i > 0$, all $j \leqslant j_\circ$. Prove the isomorphisms:

$$R^jF(X) \simeq F(H^j(X)) \quad \text{for all} \quad j \leqslant j_\circ \ .$$

Exercise I.22. In the situation of Proposition 1.8.7, let $X \in \mathrm{Ob}(\mathbf{D}^+(\mathscr{C}))$, and assume $R^jF(X) = 0$ for $j < n$. Prove that:

$$R^n(F' \circ F)(X) \cong F' \circ R^nF(X)$$

(Hint: apply Remark 1.8.6.)

Exercise I.23. Let $\mathscr{C}$ be an abelian category, $\mathscr{I}$ a full subcategory. Assume $\mathscr{I}$ satisfies condition (1.7.5), (1.7.6) and:

If $0 \to X' \to X \to X'' \to 0$ is an exact sequence in $\mathscr{C}$, and if X' belongs to $\mathscr{I}$, then X'' belongs to $\mathscr{I}$ if and only if X belongs to $\mathscr{I}$.

Let $* = \varnothing$ or b or $-$ or $+$.

(a) Prove that any $X \in \mathrm{Ob}(\mathbf{C}^*(\mathscr{C}))$ is quasi-isomorphic to an object Y of $\mathbf{C}^*(\mathscr{I})$.

(b) Let $\mathscr{C}'$ be another abelian category, $F: \mathscr{C} \to \mathscr{C}'$ a left exact functor. Assume the category $\mathscr{I}$ is F-injective. Prove that RF exists as a functor from $\mathbf{D}^*(\mathscr{C})$ to $\mathbf{D}^*(\mathscr{C}')$.

(c) Let $\mathscr{C}''$ be a third abelian category, $G: \mathscr{C} \times \mathscr{C}' \to \mathscr{C}''$ a left exact bifunctor. Assume that for each $X' \in \mathrm{Ob}(\mathscr{C}')$, the category $\mathscr{I}$ is $G(\cdot, X')$-injective. Prove that RG exists as a bifunctor from $\mathbf{D}^-(\mathscr{C}) \times \mathbf{D}^-(\mathscr{C}')$ to $\mathbf{D}^-(\mathscr{C}'')$ and $\mathbf{D}^*(\mathscr{C}) \times \mathbf{D}^b(\mathscr{C}')$ to $\mathbf{D}^*(\mathscr{C}'')$, (cf. Hartshorne [1, p. 42]).

Exercise I.24. (i) Let $F: \mathscr{C} \to \mathscr{C}'$ be a left exact functor of abelian categories. Let $X \in \mathrm{Ob}(\mathbf{D}^+(\mathscr{C}))$. Construct the natural morphisms: $H^j(RF(X)) \to F(H^j(X))$.

(ii) Let $\mathscr{C}, \mathscr{C}', \mathscr{C}''$ be abelian categories and let F be a bifunctor from $\mathscr{C} \times \mathscr{C}'$ to $\mathscr{C}''$. Let $X \in \mathrm{Ob}(\mathbf{D}^*(\mathscr{C}))$, $Y \in \mathrm{Ob}(\mathbf{D}^*(\mathscr{C}'))$, with $* = +$ or $-$.

(a) Assume F is left exact and $* = +$ (resp. F is right exact and $* = -$). Construct the natural morphisms, for $p, q \in \mathbb{Z}$:

$$H^{p+q}(RF(X, Y)) \to F(H^p(X), H^q(Y))$$

(resp. $F(H^p(X), H^q(Y)) \to H^{p+q}(LF(X, Y))$).

(b) Assume F is exact. Prove the isomorphisms for $n \in \mathbb{Z}$:

$$H^n(F(X, Y)) \simeq \bigoplus_{p+q=n} F(H^p(X), H^q(Y)) \ .$$

Exercise I.25. Let $\mathscr{C}$ be an abelian category and let X be a double complex in $\mathscr{C}$, satisfying condition (1.9.2).

(i) Prove that the following triangles are distinguished in $\mathbf{D}(\mathscr{C})$:

$$s\tau_{II}^{\leqslant n-1}(X) \longrightarrow s\tau_{II}^{\leqslant n}(X) \longrightarrow H_{II}^n(X)[-n] \underset{+1}{\longrightarrow} ,$$

$$H_{II}^n(X)[-n] \longrightarrow s\tau_{II}^{\geqslant n}(X) \longrightarrow s\tau_{II}^{\geqslant n+1}(X) \underset{+1}{\longrightarrow} .$$

(ii) Let $k \in \mathbb{Z}$ be fixed. Prove that the natural morphism:

$$H^k(s\tau_{II}^{\leqslant n}(X)) \to H^k(s(X))$$

(resp.: $H^k(s(X)) \to H^k(s\tau_{II}^{\geqslant n}(X))$), is an isomorphism for $n \gg 0$, (resp. for $n \ll 0$).

(iii) The integer $k \in \mathbb{Z}$ being fixed, prove that:

$$H^k(s\tau_{II}^{\leqslant n}(X)) = 0 \qquad \text{for} \quad n \ll 0$$

and

$$H^k(s\tau_{II}^{\geqslant n}(X)) = 0 \qquad \text{for} \quad n \gg 0 \ .$$

(Hint: $H^k(s(X))$ depends only on $\{X^{n,m}; k-1 \leqslant n+m \leqslant k+1\}$.)

Exercise I.26. (J.-P. Schneiders). In the situation of Exercise I.25, assume:

$$H_{II}^q(X) \underset{qis}{\simeq} 0 \qquad \text{except for} \quad q = q_0, q_1$$

where $q_0 < q_1$. Prove that we have a distinguished triangle:

$$H_{II}^{q_0}(X)[-q_0] \longrightarrow s(X) \longrightarrow H_{II}^{q_1}(X)[-q_1] \xrightarrow{+1}$$

(Hint: use Exercise I.25.)

Exercise I.27. Let $\mathscr{C}$ be an abelian (resp. a triangulated) category. One denotes by $\mathrm{K}(\mathscr{C})$ the abelian group obtained as the quotient of the free abelian group generated by the objects of $\mathscr{C}$ by the relation $X = X' + X''$ if there is an exact sequence $0 \to X' \to X \to X'' \to 0$ in $\mathscr{C}$ (resp. a distinguished triangle $X' \longrightarrow X \longrightarrow X'' \underset{+1}{\longrightarrow}$ in $\mathscr{C}$). One calls $\mathrm{K}(\mathscr{C})$ the **Grothendieck group** of $\mathscr{C}$.

Now let $\mathscr{C}$ be an abelian category. Prove that the functor $i: \mathscr{C} \to \mathbf{D}^b(\mathscr{C})$, $X \mapsto X$ induces a group isomorphism $\mathrm{K}(\mathscr{C}) \simeq \mathrm{K}(\mathbf{D}^b(\mathscr{C}))$ and its inverse is given by $X \mapsto \sum_j (-1)^j [H^j(X)]$, where $[Z]$ is the class of Z in $\mathrm{K}(\mathscr{C})$.

Exercise I.28. Let A be a ring. Prove that the following conditions (i)–(iii) are equivalent.

(i) $\mathfrak{Mod}(A)$ has homological dimension $\leqslant n$.
(ii) Any left module M has an injective resolution of length $\leqslant n$.
(iii) Any left module M has a projective resolution of length $\leqslant n$,
(i.e. there exists an exact sequence $0 \to P_n \to \cdots \to P_0 \to M \to 0$, with all P_j's projective).

One sets $\mathrm{gld}(A) = \sup(\mathrm{hd}(\mathfrak{Mod}(A)), \mathrm{hd}(\mathfrak{Mod}(A^{op})))$ and calls $\mathrm{gld}(A)$ the **global homological dimension** of A.

Exercise I.29. Let A be a ring.

(i) Prove that free modules are projective.
(ii) Prove that projective modules are direct summands of free modules.
(iii) An A-module M is called **flat** if the functor $\cdot \otimes_A M$ is exact. Prove that projective modules are flat.

(iv) Let n be a non-negative integer. Prove that the following conditions are equivalent.

(a) $\mathrm{Tor}_j^A(N, M) = 0$ for any $j > n$ and for any right module N and left module M.

(b) Any left module M has a flat resolution of length $\leqslant n$, (i.e. there exists an exact sequence $0 \to P^n \to \cdots \to P^0 \to M \to 0$ with all P^j's flat).

(b)^{op}: the same as (b) with left replaced by right.

One defines the **weak global dimension** of A, $\mathrm{wgld}(A)$, as the smallest $n \in \mathbb{N} \cup \{+\infty\}$ such that these conditions are satisfied.

(v) Prove that $\mathrm{wgld}(A) \leqslant \mathrm{gld}(A)$.

Exercise I.30. Let A be a commutative ring. An object X of $\mathbf{D}^b(\mathfrak{Mod}(A))$ is called **perfect** if it is isomorphic to a bounded complex of finitely generated projective A-modules.

(i) Prove that if $X \longrightarrow Y \longrightarrow Z \xrightarrow[+1]{}$ is a distinguished triangle in $\mathbf{D}^b(\mathfrak{Mod}(A))$ and X, Y are perfect, then Z is perfect.

(ii) Prove that a direct summand of a perfect object is perfect. (Hint: for a bounded complex of projective modules P' and a quasi-isomorphism $P' \to X' \oplus Y'$ with $P^j = X^j = Y^j = 0$ for $j > 0$, let $\tilde{P}$, $\tilde{X}$, $\tilde{Y}$ denote the complexes obtained by replacing P^0, P^{-1}, X^{-1}, Y^{-1} with 0, $P^{-1} \oplus P^0$, $X^{-1} \oplus P^0$, $Y^{-1} \oplus P^0$ respectively. Then construct a quasi-isomorphism $\tilde{P} \to \tilde{X} \oplus \tilde{Y}$ by using a morphism $P^0 \xrightarrow{h} \tilde{X}^{-1} \oplus \tilde{Y}^{-1}$ such that $P^0 \to \tilde{X}^{-1} \oplus \tilde{Y}^{-1} \xrightarrow{\phi} X^0 \oplus Y^0 \oplus P^0$ is $(0, 0, id_{P^0})$. Here $\phi|_{P^0 \oplus P^0 \to P^0} = (\mathrm{id}_{P^0}, -\mathrm{id}_{P^0})$.)

(iii) Let $M \in \mathrm{Ob}(\mathbf{D}^b(\mathfrak{Mod}(A)))$ and assume M is perfect. Let $M^* = R\mathrm{Hom}(M, A)$. Prove that M^* is perfect and the canonical morphism $M \to M^{**}$ is an isomorphism.

Now assume A is Noetherian and $\mathrm{gld}(A) < \infty$.

(iv) Let $\mathfrak{Mod}^f(A)$ denote the abelian category of finitely generated A-modules. Prove that any object of $\mathbf{D}^b(\mathfrak{Mod}^f(A))$ is perfect.

(v) Denote by $\mathbf{D}^b_f(\mathfrak{Mod}(A))$ the full subcategory of $\mathbf{D}^b(\mathfrak{Mod}(A))$ consisting of objects whose cohomology groups belong to $\mathfrak{Mod}^f(A)$. Prove that the natural functor $\mathbf{D}^b(\mathfrak{Mod}^f(A)) \to \mathbf{D}^b_f(\mathfrak{Mod}(A))$ is an equivalence. (Hint: use Proposition 1.7.11) (cf. [SGA6] Exposé I.)

Exercise I.31. (i) Let $M \in \mathrm{Ob}(\mathbf{D}^b(\mathfrak{Mod}(\mathbb{Z})))$ and let $M^* = R\mathrm{Hom}(M, \mathbb{Z})$. Prove that $M^* = 0$ implies $M = 0$. (Hint: one can assume $H^k(M) = 0$ for $k > 0$. Using the distinguished triangle $\tau^{\leqslant -1}(M) \longrightarrow M \longrightarrow H^0(M) \xrightarrow[+1]{}$, the proof reduces to the following:

"*let M be a $\mathbb{Z}$-module such that* $\mathrm{Hom}(M, \mathbb{Z}) = \mathrm{Ext}^1(M, \mathbb{Z}) = 0$. *Then* $M = 0$" .

To prove this result, prove that such an M is torsion free, then that M is divisible (multiplication by $n \in \mathbb{Z} \setminus \{0\}$ is surjective), then that $M \simeq \mathbb{Q} \otimes_{\mathbb{Z}} M$. If M is not zero, one can write $M = \mathbb{Q} \oplus L$, and this would imply $\mathrm{Ext}^1(\mathbb{Q}, \mathbb{Z}) = 0$, which is not true.)

(ii) Prove that if $M^* \in \mathrm{Ob}(\mathbf{D}^b(\mathfrak{Mod}^f(\mathbb{Z})))$ then $M \in \mathrm{Ob}(\mathbf{D}^b(\mathfrak{Mod}^f(\mathbb{Z})))$.

Exercise I.32. Let k be a commutative field, $X \in \mathrm{Ob}(\mathbf{D}^b(\mathfrak{Mod}(k)))$. One sets $X^* = R\mathrm{Hom}(X, k)$. (Remember that the differential is given by Remark 1.8.11.)

(i) Assuming X belongs to $\mathbf{D}_f^b(\mathfrak{Mod}(k))$, prove the natural isomorphisms:

$$X \xrightarrow{\sim} X^{**} , \qquad X^* \otimes X \xrightarrow{\sim} R\mathrm{Hom}(X, X) ,$$

and construct the morphism $X^* \otimes X \to k$ as the direct product of the morphisms $(X^n)^* \otimes X^n \to k$.

(ii) Let $X \in \mathrm{Ob}(\mathbf{D}_f^b(\mathfrak{Mod}(k)))$ and let $v \in \mathrm{Hom}(X, X)$. One sets:

$$\mathrm{tr}(v) = \sum_j (-1)^j \,\mathrm{tr}(H^j(v)) ,$$

where $\mathrm{tr}(H^j(v))$ is the trace of the endomorphism $H^j(v): H^j(X) \to H^j(X)$. Now let $Y \in \mathrm{Ob}(\mathbf{K}^b(\mathfrak{Mod}^f(k)))$ and let $v \in \mathrm{Hom}(Y, Y)$. Prove that:

$$\mathrm{tr}(v) = \sum_j (-1)^j \,\mathrm{tr}(v^j) .$$

(iii) Consider an endomorphism of distinguished triangles in $\mathbf{D}_f^b(\mathfrak{Mod}(k))$:

$$\begin{array}{ccccccc}
X' & \longrightarrow & X & \longrightarrow & X'' & \xrightarrow[+1]{} & \\
\downarrow{\scriptstyle v'} & & \downarrow{\scriptstyle v} & & \downarrow{\scriptstyle v''} & & \\
X' & \longrightarrow & X & \longrightarrow & X'' & \xrightarrow[+1]{} & .
\end{array}$$

Prove that $\mathrm{tr}(v) = \mathrm{tr}(v') + \mathrm{tr}(v'')$.

(iv) In the situation of (ii), prove that $\mathrm{tr}(v)$ is the image of v by the morphism:

$$H^0(R\mathrm{Hom}(X, X)) \simeq H^0(X^* \otimes X) \to k .$$

Usually one sets, for $X \in \mathrm{Ob}(\mathbf{D}_f^b(\mathfrak{Mod}(k)))$:

$$\chi(X) = \sum_j (-1)^j \dim H^j(X) .$$

Of course, $\chi(X) = \mathrm{tr}(\mathrm{id}_X)$ in k (cf. [SGA 6] Exposé I).

Exercise I.33. Let k be a commutative field, V a k-vector space, $u: V \to V$ an endomorphism. We say that u is in the trace class if $\dim(u^n(V)) < \infty$ for some $n > 0$, and in this case we set:

$$\mathrm{tr}(u) = \mathrm{tr}(u|_{u^n(V)}) .$$

(i) Prove that the definition of $\mathrm{tr}(u)$ is independent of the choice of n.

(ii) Let $V \xrightarrow{u} W \xrightarrow{v} V$ be morphisms of k-vector spaces. Prove that $u \circ v$ is in the trace class if and only if so is $v \circ u$, and that in this case $\mathrm{tr}(u \circ v) = \mathrm{tr}(v \circ u)$.

(iii) Consider the commutative diagram below in which the rows are exact:

$$\begin{array}{ccccccccc}
0 & \longrightarrow & V' & \xrightarrow{v'} & V & \xrightarrow{v''} & V'' & \longrightarrow & 0 \\
 & & \downarrow{\scriptstyle u'} & & \downarrow{\scriptstyle u} & & \downarrow{\scriptstyle u''} & & \\
0 & \longrightarrow & V' & \xrightarrow{v'} & V & \xrightarrow{v''} & V'' & \longrightarrow & 0 .
\end{array}$$

Prove that u is in the trace class if and only if u' and u'' are in the trace class, and prove in this case that $\mathrm{tr}(u) = \mathrm{tr}(u') + \mathrm{tr}(u'')$.

Exercise I.34. Let k be a commutative field, $X \in \mathrm{Ob}(\mathbf{D}^b_f(\mathfrak{Mod}(k)))$. One sets:

$$\mathrm{b}_i(X) = \dim H^i(X) \ , \qquad \mathrm{b}_i^*(X) = (-1)^i \sum_{j \leqslant i} (-1)^j \mathrm{b}_j(X) \ .$$

Let $X' \longrightarrow X \longrightarrow X'' \xrightarrow[+1]{}$ be a distinguished triangle in $\mathbf{D}^b_f(\mathfrak{Mod}(k))$. Prove:

$$\chi(X) = \chi(X') + \chi(X'') \ ,$$

$$\mathrm{b}_i^*(X) \leqslant \mathrm{b}_i^*(X') + \mathrm{b}_i^*(X'') \ .$$

($\chi(X)$ is defined in Exercise I.32).

Exercise I.35. (i) Let I be a directed set and $\{X_i\}$ an inductive system indexed by I in a category $\mathscr{C}$. Prove that "$\varinjlim$" X_i is an inductive limit in $\mathscr{C}^\vee$. More precisely prove that for any $F \in \mathrm{Ob}(\mathscr{C}^\vee)$

$$\mathrm{Hom}_{\mathscr{C}^\vee}(\text{“}\varinjlim\text{”}\, X_i, F) \simeq \varprojlim F(X_i) \ .$$

(ii) If $\{Y_j\}$ is an inductive system indexed by a directed set J in $\mathscr{C}$, then prove that

$$\mathrm{Hom}_{\mathscr{C}^\vee}\left(\text{“}\varinjlim_i\text{”}\, X_i, \varinjlim_j Y_j\right) \simeq \varprojlim_i \varinjlim_j \mathrm{Hom}_{\mathscr{C}}(X_i, Y_j) \ .$$

Exercise I.36. Let A be a Noetherian ring, and let $\mathfrak{Mod}^f(A)$ be the category of finitely generated A-modules. Let $\{X_i, \rho_{i,j}\}$ be an inductive system in this category, indexed by a directed ordered set I. Prove that if $\varinjlim_j X_j$ exists in $\mathfrak{Mod}^f(A)$, then it represents "$\varinjlim_j$" X_j.

Exercise I.37. Let $\mathscr{C}$ be an additive category. We denote by $\mathrm{End}(\mathscr{C})$ the set of endomorphisms of the functor $\mathrm{id}_{\mathscr{C}} : \mathscr{C} \to \mathscr{C}$.

(i) Prove that $\mathrm{End}(\mathscr{C})$ is a commutative ring.

(ii) Prove that for a ring A, $\mathrm{End}(\mathfrak{Mod}(A))$ is isomorphic to the center of A (i.e. $\{a \in A; ax = xa$ for any $x \in A\}$).

(iii) If A is a commutative ring and if a ring homomorphism $A \to \mathrm{End}(\mathscr{C})$ is given, we call $\mathscr{C}$ an additive category over A. Prove that $\mathrm{Hom}_{\mathscr{C}}(X, Y)$ has a structure of an A-module and prove that the composition of morphisms is A-bilinear.

(iv) Assume $\mathscr{C}$ is an abelian category over a commutative Noetherian ring A.

(a) Prove that, for $M \in \mathrm{Ob}(\mathfrak{Mod}^f(A))$ and $X \in \mathrm{Ob}(\mathscr{C})$, the functor $Y \mapsto \mathrm{Hom}_A(M, \mathrm{Hom}_{\mathscr{C}}(X, Y))$ is representable. We denote its representative by $X \otimes_A M$.

(b) Prove that $\otimes_A$ is a right exact bifunctor from $\mathscr{C} \times \mathfrak{Mod}^f(A)$ to $\mathscr{C}$.

(c) Prove that the bifunctor above has a left derived functor $\otimes_A^L : \mathbf{D}^-(\mathscr{C}) \times \mathbf{D}^-(\mathfrak{Mod}^f(A)) \to \mathbf{D}^-(\mathscr{C})$.

(d) Discuss similarly the functor $\mathrm{Hom}_A(\cdot,\cdot):\mathfrak{Mod}^f(A)^\circ \times \mathscr{C} \to \mathscr{C}$ (i.e. the functor $\otimes_A$ in $\mathscr{C}^\circ$).

Exercise I.38. Let $\mathscr{I}$ and $\mathscr{I}'$ be filtrant categories and $\phi:\mathscr{I}\to\mathscr{I}'$ a functor. We say that $\mathscr{I}$ and $\mathscr{I}'$ are **cofinal** (by ϕ) if we have:
(a) For any $i' \in \mathrm{Ob}(\mathscr{I}')$, there exist $i \in \mathrm{Ob}(\mathscr{I})$ and a morphism $i' \to \phi(i)$.
(b) For $i \in \mathrm{Ob}(\mathscr{I})$, $i' \in \mathrm{Ob}(\mathscr{I}')$ and a morphism $f:\phi(i)\to i'$, there exists a morphism $g: i \to i_1$ in $\mathscr{I}$ such that $\phi(g):\phi(i)\to\phi(i_1)$ factors through f.
Let $\mathscr{C}$ be a category and let F be an inductive (resp. projective) system indexed by $\mathscr{I}'$. Prove that:

$$\varinjlim_{\mathscr{I}} F\circ\phi \to \varinjlim_{\mathscr{I}'} F \qquad \text{and} \qquad \text{“}\varinjlim_{\mathscr{I}}\text{”}\, F\circ\phi \to \text{“}\varinjlim_{\mathscr{I}'}\text{”}\, F$$

(resp. $\varprojlim_{\mathscr{I}'} F \to \varprojlim_{\mathscr{I}} F\circ\phi$ and $\text{“}\varprojlim_{\mathscr{I}'}\text{”}\, F \to \text{“}\varprojlim_{\mathscr{I}}\text{”}\, F\circ\phi$) are isomorphisms.

Exercise I.39. Let $\mathscr{C}$ be an abelian category. For X, $Y \in \mathrm{Ob}(\mathbf{D}^b(\mathscr{C}))$, we set $\mathrm{Ext}^j(X,Y) = \mathrm{Hom}_{\mathbf{D}(\mathscr{C})}(X, Y[j])$.
(i) Let X, $Y \in \mathrm{Ob}(\mathscr{C})$ and $n \geqslant 1$. For an exact sequence

$$E: 0 \to Y \to Z_n \to Z_{n-1} \to \cdots \to Z_1 \to X \to 0$$

define an element $\mathrm{C}(E) \in \mathrm{Ext}^n(X,Y)$. One calls such an exact sequence an n-extension of X by Y.
(ii) Prove that any element of $\mathrm{Ext}^n(X,Y)$ can be written as $\mathrm{C}(E)$ for some extension E of X by Y.
(iii) Let $E': 0 \to Y \to Z'_n \to \cdots \to Z'_1 \to X \to 0$ be another extension. Prove that $\mathrm{C}(E) = \mathrm{C}(E')$ if and only if there exist an extension $E'': 0 \to Y \to Z''_n \to \cdots \to Z''_1 \to X \to 0$ and a commutative diagram:

$$\begin{array}{ccccccccc}
Y & \longrightarrow & Z_n & \longrightarrow & \cdots & \longrightarrow & Z_1 & \longrightarrow & X \\
\| & & \uparrow & & & & \uparrow & & \| \\
Y & \longrightarrow & Z''_n & \longrightarrow & \cdots & \longrightarrow & Z''_1 & \longrightarrow & X \\
\| & & \downarrow & & & & \downarrow & & \| \\
Y & \longrightarrow & Z'_n & \longrightarrow & \cdots & \longrightarrow & Z'_1 & \longrightarrow & X
\end{array}$$

(usually $\mathrm{Ext}^n(X,Y)$ is called the Yoneda extension).

Notes

We refer to the “Short History” by C. Houzel, at the beginning of this book, for a detailed history of cohomology theories.

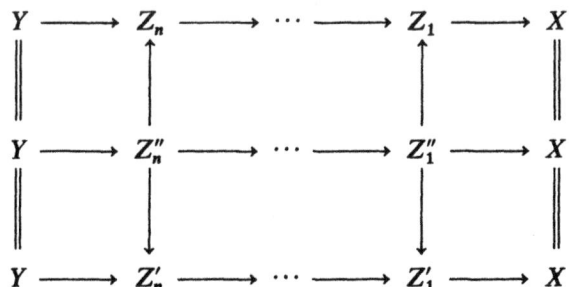

Let us only mention here that it is generally admitted that the idea of derived categories is due to Grothendieck [4] and has been developed by Verdier in his thesis. Unfortunately this thesis has never been published, with the exception of the short paper Verdier [2].

Before this sophisticated theory appeared, people used "derived functors" (cf. Cartan-Eilenberg [1] and the appendix by Buchsbaum), and the fundamental techniques of "spectral sequences" introduced by Leray [1, 2] in 1945. The famous paper of Grothendieck [1] at Tohoku had an enormous impact in its time, by clarifying and unifying the theory.

In this book no use is made of spectral sequences. In fact, we do not need here this calculus in its full generality, and Theorem 1.9.3 is sufficient for our purpose.

Finally, let us recall that ind-objects and pro-objects have been introduced in Grothendieck [2], and the Mittag-Leffler condition in Grothendieck [3].

Chapter II. Sheaves

Summary

In this chapter we construct the abelian category of sheaves on a topological space, and the usual associated functors, such as the inverse image f^{-1}, the direct image f_*, the proper direct image $f_!$, the tensor product $\otimes$ and the inner hom $\mathcal{H}om$. Making use of the results of the first chapter, one then defines the derived category $\mathbf{D}^b(X)$ of sheaves, and the derived functors of the preceding ones. In the course of the chapter, we also introduce the notions of injective sheaves, flat sheaves, flabby sheaves, c-soft sheaves, and give the tools of sheaf theory that we shall use later: non-characteristic deformation lemma and homotopy invariance of cohomology. Although we do not really need it, we (briefly) present Čech cohomology. We end this chapter by recalling some natural sheaves on real or complex manifolds.

Most of the results we explain here are classical, and we refer to Bredon [1], Godement [1], Iversen [1] for further developments.

2.1. Presheaves

Let X be a topological space. We denote by $\mathrm{OP}(X)$ the set of open subsets of X, ordered by the inclusion relation, and we associate to it (cf. I §1) a category $\mathfrak{O}\mathfrak{P}(X)$ by setting:

$$\begin{cases} \mathrm{Ob}(\mathfrak{O}\mathfrak{P}(X)) = \mathrm{OP}(X) & \\ \mathrm{Hom}_{\mathfrak{O}\mathfrak{P}(X)}(V, U) = \{\mathrm{pt}\} & \text{if } V \subset U\ , \\ \qquad\qquad = \emptyset & \text{otherwise}\ . \end{cases} \tag{2.1.1}$$

(Here V and U are open subsets of X and $\{\mathrm{pt}\}$ is the set with one element.)

Definition 2.1.1. *Let $\mathscr{C}$ be a category. A presheaf F on X with values in $\mathscr{C}$ is a contravariant functor from $\mathfrak{O}\mathfrak{P}(X)$ to $\mathscr{C}$, and a morphism of presheaves is a morphism of such functors.*

In other words, a presheaf on X is a projective system indexed by $\mathfrak{O}\mathfrak{P}(X)$ (cf. I §11).

In this book we shall only consider presheaves (and sheaves) of abelian groups. Therefore, unless otherwise specified, we take $\mathscr{C} = \mathfrak{Ab}$ in the preceding definition and a presheaf means a presheaf of abelian groups.

Thus a presheaf assigns to each open set $U \subset X$ an abelian group $F(U)$ and to each open inclusion $V \subset U$ a group homomorphism $\rho_{V,U}: F(U) \to F(V)$, called the restriction morphism, with the conditions:

$$\begin{cases} \rho_{U,U} = \mathrm{id}_{F(U)}, \rho_{W,U} = \rho_{W,V} \circ \rho_{V,U} \text{ for} \\ \text{any triplet } W \subset V \subset U \text{ of open subsets} \end{cases} . \tag{2.1.2}$$

A morphism of presheaves $\phi: F \to G$ is a family of group homomorphisms $\phi_U: F(U) \to G(U)$, compatible with the restriction morphisms. Namely, if one denotes by the same letters $\rho_{V,U}$ the restriction morphism for F and for G, the diagram below commutes, for all pair U, V of open sets with $V \subset U$:

$$\begin{array}{ccc} F(U) & \xrightarrow{\phi_U} & G(U) \\ {\scriptstyle \rho_{V,U}}\downarrow & & \downarrow{\scriptstyle \rho_{V,U}} \\ F(V) & \xrightarrow[\phi_V]{} & G(V) \end{array} .$$

One defines the presheaf 0 by $U \mapsto 0$, and one defines the direct sum $F \oplus G$ of two presheaves F and G by $U \mapsto F(U) \oplus G(U)$. Thus the category of presheaves (of abelian groups) and morphisms of presheaves on X is an additive category. We denote it by $\mathfrak{PSh}(X)$.

Let $\phi: F \to G$ be a morphism of presheaves. The correspondence: $U \mapsto \operatorname{Ker} \phi_U$ (resp. $U \mapsto \operatorname{Coker} \phi_U$) defines a presheaf on X and this presheaf is a kernel (resp. a cokernel) of ϕ in $\mathfrak{PSh}(X)$. One denotes it by $\operatorname{Ker} \phi$ (resp. $\operatorname{Coker} \phi$). As the property (ii) of Definition 1.2.3 is clearly satisfied we find that $\mathfrak{PSh}(X)$ is an abelian category.

Terminology 2.1.2. Let F be a presheaf on X, U an open subset of X. An element $s \in F(U)$ is called a **section** of F on U. If V is an open subset of U, one often writes $s|_V$ instead of $\rho_{V,U}(s)$, and call it the **restriction** of s to V.

One sets for $x \in X$:

$$F_x = \varinjlim_{U} F(U) , \tag{2.1.3}$$

where U runs through the family of open neighborhoods of x. The group F_x is called the **stalk** of F at x and the image of $s \in F(U)$ (where $x \in U$) in F_x is called the **germ** of s at x and denoted by s_x.

One denotes by $F|_U$ the presheaf on U defined by $U \supset V \mapsto F(V)$, and calls it the restriction of F to U.

2.2. Sheaves

Let X be a topological space.

Definition 2.2.1. *A presheaf (of abelian groups) F on X is called a sheaf if it satisfies conditions* (S 1)–(S 2) *below.*

(S 1) *For any open set $U \subset X$, any open covering $U = \bigcup_{i \in I} U_i$, any section $s \in F(U)$, $s|_{U_i} = 0$ for all i implies $s = 0$.*

(S 2) *For any open set $U \subset X$, any open covering $U = \bigcup_{i \in I} U_i$, any family $s_i \in F(U_i)$ satisfying $s_i|_{U_i \cap U_j} = s_j|_{U_i \cap U_j}$ for all pairs (i, j), there exists $s \in F(U)$ such that $s|_{U_i} = s_i$ for all i.*

Note that we could have defined with slight modifications, the notion of sheaves of sets, but we shall not use it in this book.

Note moreover that (S 1) and (S 2) are equivalent to saying that for any open subset $U \subset X$ and any open covering $U = \bigcup_{i \in I} U_i$, stable by finite intersections, the morphism $F(U) \to \varprojlim_i F(U_i)$ is an isomorphism.

If F is a sheaf, $F(\emptyset) = 0$.

If F is a sheaf on X, and U is open in X, then $F|_U$ is a sheaf on U.

One defines the **support of a sheaf** F, denoted supp(F), as the complementary of the union of open sets $U \subset X$ such that $F|_U = 0$. Similarly one defines the **support of a section** s of F on U as the complementary in U of the union of the open sets $V \subset U$ such that $s|_V = 0$. We denote it by supp(s). We have supp$(s) = \{x \in U; s_x \neq 0\}$.

One defines a morphism of sheaves as a morphism of the underlying presheaves. Then the category of sheaves, that we denote by $\mathfrak{Sh}(X)$, is clearly a full additive subcategory of $\mathfrak{PSh}(X)$. One denotes by $\Gamma(U; \cdot)$ the functor $F \mapsto F(U)$ from $\mathfrak{Sh}(X)$ to $\mathfrak{Ab}$. Hence $\Gamma(U; F) = F(U)$.

Proposition 2.2.2. *Let $\phi : F \to G$ be a morphism of sheaves. Then ϕ is an isomorphism if and only if for any $x \in X$, the induced morphism $\phi_x : F_x \to G_x$ is an isomorphism.*

Proof. The necessity is clear.

Conversely assume ϕ_x is an isomorphism for all $x \in X$, and let U be an open subset of X. Let us prove that $\phi_U : F(U) \to G(U)$ is injective. Let $s \in F(U)$ with $\phi_U(s) = 0$. Then $(\phi_U(s))_x = \phi_x(s_x) = 0$, and $s_x = 0$ for all $x \in U$.

By the axiom (S 1) this implies $s = 0$. Let us prove the surjectivity of $\phi_U : F(U) \to G(U)$. Let $t \in G(U)$. Then by the assumption, there is an open covering $\{U_i\}$ of U and $s_i \in F(U_i)$ such that $\phi(s_i) = t|_{U_i}$. Since $\phi(s_i)|_{U_i \cap U_j} = \phi(s_j)|_{U_i \cap U_j}$, the injectivity of ϕ implies $s_i|_{U_i \cap U_j} = s_j|_{U_i \cap U_j}$ and hence there exists $s \in F(U)$ such that $s|_{U_i} = s_i$. Then one can check easily that $\phi(s) = t$. □

Proposition 2.2.3. *Given a presheaf F on X, there exists a sheaf F^+ and a morphism $\theta : F \to F^+$ such that for any sheaf G the homomorphism given by θ:*

$$\mathrm{Hom}_{\mathfrak{Sh}(X)}(F^+, G) \to \mathrm{Hom}_{\mathfrak{PSh}(X)}(F, G)$$

is an isomorphism. In other word, $F \mapsto F^+$ is the left adjoint functor of the inclusion functor $\mathfrak{Sh}(X) \to \mathfrak{PSh}(X)$ (cf. Exercise I.2).

Moreover (F^+, θ) is unique up to isomorphism, and for any $x \in X$, $\theta_x : F_x \to F_x^+$ is an isomorphism.

Proof. For any open set $U \subset X$, let $F^+(U)$ be the set of functions s from U to $\bigsqcup_{x \in U} F_x$ such that for any $x \in U$, $s(x) \in F_x$ and there is an open neighborhood V of x, $V \subset U$, and $t \in F(V)$ such that $t_y = s(y)$ for all $y \in V$. Then F^+ clearly satisfies (S 1) and (S 2).

The morphism $\theta : F \to F^+$ is defined as follows: to $s \in F(U)$ one assigns the map from U to $\bigsqcup_{x \in U} F_x$ given by $x \mapsto s_x$. Then it is easy to see that, for any $x \in X$, $\theta_x : F_x \to (F^+)_x$ is an isomorphism. In particular if G is a sheaf, then $\theta : G \to G^+$ is an isomorphism by the preceding proposition. Hence for a presheaf F and a sheaf G, we can construct a homomorphism $\mathrm{Hom}_{\mathfrak{PSh}(X)}(F, G) \to \mathrm{Hom}_{\mathfrak{Sh}(X)}(F^+, G)$ as the composition of $\mathrm{Hom}_{\mathfrak{PSh}(X)}(F, G) \to \mathrm{Hom}_{\mathfrak{Sh}(X)}(F^+, G^+) \xleftarrow{\sim} \mathrm{Hom}_{\mathfrak{Sh}(X)}(F^+, G)$. It is easily checked that the obtained homomorphism is the inverse of $\mathrm{Hom}_{\mathfrak{Sh}(X)}(F^+, G) \to \mathrm{Hom}_{\mathfrak{PSh}(X)}(F, G)$. □

We call F^+ the **associated sheaf** to the presheaf F.

Let $\phi : F \to G$ be a morphism of sheaves. The presheaf $U \mapsto \mathrm{Ker}\, \phi_U$ is a sheaf, and is a kernel of ϕ in the category $\mathfrak{Sh}(X)$. On the contrary, $U \mapsto \mathrm{Coker}\, \phi_U$ is not always a sheaf, but the associated sheaf is a cokernel of ϕ in $\mathfrak{Sh}(X)$. Hence we shall denote by $\mathrm{Ker}\, \phi$ the sheaf $U \mapsto \mathrm{Ker}\, \phi_U$, and by $\mathrm{Coker}\, \phi$ the sheaf associated to the presheaf $U \mapsto \mathrm{Coker}\, \phi_U$. One shall beware that $\mathrm{Coker}\, \phi$ has not the same meaning in $\mathfrak{Sh}(X)$ and in $\mathfrak{PSh}(X)$. As we shall always work with sheaves, there will be no risk of confusion.

Note that for $x \in X$:

(2.2.1) $$(\mathrm{Ker}\, \phi)_x = \mathrm{Ker}\, \phi_x ,$$

(2.2.2) $$(\mathrm{Coker}\, \phi)_x = \mathrm{Coker}\, \phi_x ,$$

where $\phi_x : F_x \to G_x$ is the morphism associated to the family ϕ_U, $x \in U$.

The direct sum of two sheaves F and G is defined as the direct sum of the underlying presheaves. This is clearly a sheaf, and a direct sum in $\mathfrak{Sh}(X)$. We denote it by $F \oplus G$.

Proposition 2.2.4. *The category $\mathfrak{Sh}(X)$ of sheaves of abelian groups on X is an abelian category.*

Proof. Let $\phi : F \to G$ be a morphism of sheaves. Let $K = \mathrm{Coim}\, \phi$, $L = \mathrm{Im}\, \phi$ and let $\psi : K \to L$ be the natural morphism. For each $x \in X$, $\psi_x : K_x \to L_x$ is an isomorphism, by (2.2.1) and (2.2.2). Then the result follows from Proposition 2.2.2. □

Remark 2.2.5. By applying Proposition 2.2.2 we find that a complex of sheaves $F' \to F \to F''$ is exact if and only if for each $x \in X$ the sequence of groups $F'_x \to F_x \to F''_x$ is exact. In particular the functor $F \mapsto F_x$ from $\mathfrak{Sh}(X)$ to $\mathfrak{Ab}$ is exact. On the other hand, the functor $\Gamma(X; \cdot)$ from $\mathfrak{Sh}(X)$ to $\mathfrak{Ab}$ is only left exact.

Most of the sheaves one naturally encounters have a richer structure than merely a structure of abelian groups.

Let us denote by $\mathfrak{Ring}$ the category of unitary rings and morphisms of unitary rings. A presheaf with values in $\mathfrak{Ring}$ is called a presheaf of rings. If such a presheaf is a sheaf (with values in $\mathfrak{Ab}$), it is called a sheaf of rings.

Definition 2.2.6. *Let $\mathscr{R}$ be a sheaf of rings on X. An $\mathscr{R}$-module M, (or a sheaf of modules over $\mathscr{R}$), is a sheaf M such that for each open set $U \subset X$, $M(U)$ is a left $\mathscr{R}(U)$-module, and for any open inclusion $V \subset U$, the restriction morphism is compatible with the structure of module, that is, $\rho_{V,U}(sm) = \rho_{V,U}(s) \cdot \rho_{V,U}(m)$ for any $s \in \mathscr{R}(U)$, $m \in M(U)$.*

One defines in an obvious way the notion of morphisms of $\mathscr{R}$-modules.

One defines similarly the notion of sheaves of right $\mathscr{R}$-modules.

One denotes by $\mathfrak{Mod}(\mathscr{R})$ the category of (left) $\mathscr{R}$-modules. Denoting by $\mathscr{R}^{op}$ the sheaf of rings opposite to $\mathscr{R}$ (i.e. $\mathscr{R}^{op}(U) = \mathscr{R}(U)^{op}$), $\mathfrak{Mod}(\mathscr{R}^{op})$ is equivalent to the category of right $\mathscr{R}$-modules.

For F, $G \in \mathrm{Ob}(\mathfrak{Mod}(\mathscr{R}))$, we shall write $\mathrm{Hom}_{\mathscr{R}}(F, G)$ instead of $\mathrm{Hom}_{\mathfrak{Mod}(\mathscr{R})}(F, G)$.

Therefore $\mathrm{Hom}_{\mathscr{R}}(\cdot, \cdot)$ is a bifunctor from $\mathfrak{Mod}(\mathscr{R})^\circ \times \mathfrak{Mod}(\mathscr{R})$ to $\mathfrak{Ab}$.

One checks immediately that $\mathfrak{Mod}(\mathscr{R})$ is an abelian category. Moreover a sequence in $\mathfrak{Mod}(\mathscr{R})$ is exact if and only if it is exact considered as a sequence in $\mathfrak{Sh}(X)$.

Let us denote by $\mathbb{Z}_X$ the sheaf on X associated to the presheaf $U \mapsto \mathbb{Z}$ (cf. Definition 2.2.11 below for a generalization of this construction). Then $\mathbb{Z}_X$ is a sheaf of rings, and one has:

$$\mathfrak{Sh}(X) = \mathfrak{Mod}(\mathbb{Z}_X) \ . \tag{2.2.3}$$

We shall use indifferently one or the other notation. Moreover for F, $G \in \mathrm{Ob}(\mathfrak{Sh}(X))$, we shall write $\mathrm{Hom}(F, G)$ instead of $\mathrm{Hom}_{\mathbb{Z}_X}(F, G) = \mathrm{Hom}_{\mathfrak{Sh}(X)}(F, G)$.

Now let $\mathscr{R}$ be a sheaf of rings on X and let F and G be two $\mathscr{R}$-modules. Consider the presheaf:

$$U \mapsto \mathrm{Hom}_{\mathscr{R}|_U}(F|_U, G|_U) \ . \tag{2.2.4}$$

This presheaf is evidently a sheaf of abelian groups (even a sheaf of $\mathscr{R}$-modules, in case $\mathscr{R}$ is commutative).

Definition 2.2.7. *One denotes by $\mathscr{H}om_{\mathscr{R}}(F, G)$ the sheaf defined in* (2.2.4), *and calls it the sheaf of solutions of F in G (over $\mathscr{R}$).*

Note that in general, the natural morphism:

$$(2.2.5) \qquad (\mathcal{H}om_{\mathcal{R}}(F, G))_x \to \mathrm{Hom}_{\mathcal{R}_x}(F_x, G_x)$$

is neither injective, nor surjective. By the construction, we have:

$$(2.2.6) \qquad \Gamma(X; \mathcal{H}om_{\mathcal{R}}(F, G)) = \mathrm{Hom}_{\mathcal{R}}(F, G) \ .$$

The bifunctor $\mathcal{H}om_{\mathcal{R}}(\cdot, \cdot)$ is left exact with respect to each of its arguments.

Now let F be a right $\mathcal{R}$-module and G a left $\mathcal{R}$-module.

Definition 2.2.8. *One denotes by $F \otimes_{\mathcal{R}} G$ the sheaf associated to the presheaf $U \mapsto F(U) \otimes_{\mathcal{R}(U)} G(U)$, and calls it the tensor product of F and G (over $\mathcal{R}$).*

When $\mathcal{R}$ is commutative, $F \otimes_{\mathcal{R}} G$ is a sheaf of $\mathcal{R}$-modules. When $\mathcal{R} = \mathbb{Z}_X$, one shall simply write $F \otimes G$ instead of $F \otimes_{\mathbb{Z}_X} G$. By the construction of the tensor product, one has for $x \in X$:

$$(2.2.7) \qquad \left(F \underset{\mathcal{R}}{\otimes} G\right)_x = F_x \underset{\mathcal{R}_x}{\otimes} G_x \ .$$

Hence the functor $\cdot \otimes_{\mathcal{R}} \cdot$ is right exact in each of its arguments. Let $\mathcal{S}$ be another sheaf of rings and let H be a bi-$(\mathcal{R}, \mathcal{S})$-module (i.e. H is endowed with both a left $\mathcal{R}$-module structure and a right $\mathcal{S}$-module structure such that the two actions commute). Then one has a natural isomorphism:

$$(2.2.8) \qquad F \underset{\mathcal{R}}{\otimes} \left(H \underset{\mathcal{S}}{\otimes} G\right) \simeq \left(F \underset{\mathcal{R}}{\otimes} H\right) \underset{\mathcal{S}}{\otimes} G \ .$$

In such a situation, we simply write $F \otimes_{\mathcal{R}} H \otimes_{\mathcal{S}} G$.

Proposition 2.2.9. *Let $\mathcal{R}$ be a sheaf of rings, $\mathcal{S}$ a sheaf of commutative rings, and $\mathcal{S} \to \mathcal{R}$ a morphism of sheaves of rings such that its image is contained in the center of $\mathcal{R}$. Let F and G be two $\mathcal{R}$-modules and H an $\mathcal{S}$-module. Then one has canonical isomorphisms:*

$$(2.2.9) \qquad \begin{aligned} \mathcal{H}om_{\mathcal{R}}\left(H \underset{\mathcal{S}}{\otimes} F, G\right) &\simeq \mathcal{H}om_{\mathcal{R}}(F, \mathcal{H}om_{\mathcal{S}}(H, G)) \\ &\simeq \mathcal{H}om_{\mathcal{S}}(H, \mathcal{H}om_{\mathcal{R}}(F, G)) \ . \end{aligned}$$

Proof. For each open set U of X one has natural isomorphisms:

$$\begin{aligned} \mathrm{Hom}_{\mathcal{R}(U)}\left(H(U) \underset{\mathcal{S}(U)}{\otimes} F(U), G(U)\right) &\simeq \mathrm{Hom}_{\mathcal{R}(U)}(F(U), \mathrm{Hom}_{\mathcal{S}(U)}(H(U), G(U))) \\ &\simeq \mathrm{Hom}_{\mathcal{S}(U)}(H(U), \mathrm{Hom}_{\mathcal{R}(U)}(F(U), G(U))) \ . \end{aligned}$$

Let us denote by $H \otimes_{\mathcal{S}}^{\vee} F$ the presheaf $U \mapsto H(U) \otimes_{\mathcal{S}(U)} F(U)$ and let $\mathrm{Hom}_{\mathcal{R}^{\vee}}(\cdot, \cdot)$

denote the group of morphisms in the category of presheaves of $\mathscr{R}$-modules. Then:

$$\mathrm{Hom}_{\mathscr{R}^\vee}\left(H \overset{\vee}{\underset{\mathscr{S}}{\otimes}} F, G\right) \simeq \mathrm{Hom}_{\mathscr{R}}(F, \mathrm{Hom}_{\mathscr{S}}(H, G))$$

$$\simeq \mathrm{Hom}_{\mathscr{S}}(H, \mathscr{H}om_{\mathscr{R}}(F, G))$$

and it remains to apply Proposition 2.2.3. □

Corollary 2.2.10. *In the situation of Proposition 2.2.9, one has canonical morphisms:*

$$\mathscr{H}om_{\mathscr{R}}(F, G) \underset{\mathscr{S}}{\otimes} F \to G \qquad \text{in } \mathfrak{Mod}(\mathscr{R}) , \tag{2.2.10}$$

$$\mathscr{H}om_{\mathscr{R}}(F, G) \underset{\mathscr{S}}{\otimes} H \to \mathscr{H}om_{\mathscr{R}}\left(F, G \underset{\mathscr{S}}{\otimes} H\right) \qquad \text{in } \mathfrak{Mod}(\mathscr{S}) , \tag{2.2.11}$$

$$\mathscr{H}om_{\mathscr{R}}(F, G) \to \mathscr{H}om_{\mathscr{R}^{op}}(\mathscr{H}om_{\mathscr{R}}(G, \mathscr{R}), \mathscr{H}om_{\mathscr{R}}(F, \mathscr{R})) \qquad \text{in } \mathfrak{Mod}(\mathscr{S}) . \tag{2.2.12}$$

Proof. (i) By (2.2.9) we have:

$$\mathscr{H}om_{\mathscr{R}}\left(\mathscr{H}om_{\mathscr{R}}(F, G) \underset{\mathscr{S}}{\otimes} F, G\right) \simeq \mathscr{H}om_{\mathscr{S}}(\mathscr{H}om_{\mathscr{R}}(F, G), \mathscr{H}om_{\mathscr{R}}(F, G)) .$$

The identity of $\mathscr{H}om_{\mathscr{R}}(F, G)$ defines (2.2.10).

(ii) Tensoring by H in (2.2.10), we have a morphism

$$\mathscr{H}om_{\mathscr{R}}(F, G) \underset{\mathscr{S}}{\otimes} H \underset{\mathscr{S}}{\otimes} F \to G \underset{\mathscr{S}}{\otimes} H .$$

This defines (2.2.11) in view of (2.2.9).

(iii) Set $\mathrm{D}_{\mathscr{R}} F = \mathscr{H}om_{\mathscr{R}}(F, \mathscr{R})$ for short. Then $\mathrm{D}_{\mathscr{R}}$ is a functor from $\mathfrak{Mod}(\mathscr{R})$ to $\mathfrak{Mod}(\mathscr{R}^{op})$. We have the morphisms:

$$\mathscr{H}om_{\mathscr{R}^{op}}(\mathrm{D}_{\mathscr{R}} F, \mathrm{D}_{\mathscr{R}} F) \to \mathscr{H}om_{\mathscr{R}^{op}}\left(\mathscr{H}om_{\mathscr{R}}\left(F, G \underset{\mathscr{S}}{\otimes} \mathrm{D}_{\mathscr{R}} G\right), \mathrm{D}_{\mathscr{R}} F\right)$$

$$\to \mathscr{H}om_{\mathscr{R}^{op}}\left(\mathscr{H}om_{\mathscr{R}}(F, G) \underset{\mathscr{S}}{\otimes} \mathrm{D}_{\mathscr{R}} G, \mathrm{D}_{\mathscr{R}} F\right)$$

$$\simeq \mathscr{H}om_{\mathscr{S}}(\mathscr{H}om_{\mathscr{R}}(F, G), \mathscr{H}om_{\mathscr{R}^{op}}(\mathrm{D}_{\mathscr{R}} G, \mathrm{D}_{\mathscr{R}} F)) .$$

The identity of $\mathrm{D}_{\mathscr{R}} F$ defines (2.2.12). □

Now we shall construct projective and inductive limits in the category $\mathfrak{Mod}(\mathscr{R})$.

Let $\{F_i\}_{i \in I}$ be a projective system of $\mathscr{R}$-modules indexed by an ordered set I. The presheaf $U \mapsto \varprojlim_{i \in I} F_i(U)$ is clearly a sheaf, and it is a projective limit of $\{F_i\}$

in $\mathfrak{Mod}(\mathcal{R})$. Therefore we denote it by $\varprojlim_i F_i$. In other words, the projective limit in the category $\mathfrak{Mod}(\mathcal{R})$ exists and $\Gamma(U;\cdot)$ commutes with $\varprojlim$, i.e.

$$\Gamma(U; \varprojlim F_i) \simeq \varprojlim \Gamma(U; F_i) \ . \tag{2.2.13}$$

Note that if $x \in X$, the natural morphism $(\varprojlim_i F_i)_x \to \varprojlim_i (F_i)_x$ is not necessarily an isomorphism.

Similarly if $\{F_j\}_{j \in J}$ is an inductive system of $\mathcal{R}$-modules indexed by an ordered set J, we denote by $\varinjlim_j F_j$ the sheaf associated to the presheaf $U \mapsto \varinjlim_{j \in J} F_j(U)$. Note that for $x \in X$, one has an isomorphism:

$$\left(\varinjlim_j F_j\right)_x \simeq \varinjlim_j (F_j)_x \ . \tag{2.2.14}$$

However, the canonical homomorphism $\varinjlim \Gamma(U; F_j) \to \Gamma(U; \varinjlim F_j)$ is not necessarily an isomorphism.

As a particular case of the preceding construction (i.e: when the order on I or J is the trivial one), we obtain the notions of the product $\prod_{i \in I} F_i$ or the direct sum $\bigoplus_{j \in J} F_j$ of families of sheaves.

Before ending this section, let us define the "simplest" sheaves one can encounter on X.

Definition 2.2.11. (i) *Let M be an abelian group. One denotes by M_X the sheaf associated to the presheaf $U \mapsto M$, (U open in X), and says that M_X is the constant sheaf on X with stalk M.*

(ii) *Let F be a sheaf on X. One says F is locally constant on X if there exists an open covering $X = \bigcup_i U_i$ such that for each i, $F|_{U_i}$ is a constant sheaf.*

Note that if M_X is a constant sheaf with stalk M, then $(M_X)_x = M$ for all $x \in X$. Note also that, for an open set U, $\Gamma(U; M_X)$ is isomorphic to the set of continuous functions from U to the topological space M endowed with the discrete topology. We shall give examples of locally constant sheaves in §9.

Remark 2.2.12. Let $\{\mathrm{pt}\}$ denote the set with a single element and let A be a ring. Then A defines a sheaf of rings on the space $\{\mathrm{pt}\}$, and it will be sometimes useful to identify $\mathfrak{Mod}(A)$ to $\mathfrak{Mod}(A_{\{\mathrm{pt}\}})$, i.e. to consider an A-module as a sheaf of $A_{\{\mathrm{pt}\}}$-modules.

2.3. Operations on sheaves

Let X and Y be two topological spaces, $f : Y \to X$ a continuous map.

Definition 2.3.1. (i) *Let G be a sheaf on Y. The direct image of G by f, denoted f_*G, is the sheaf on X defined by:*

$$U \mapsto f_* G(U) = G(f^{-1}(U)) \ , \qquad U \textit{ open in } X \ .$$

(ii) *Let F be a sheaf on X. The inverse image of F by f, denoted $f^{-1}F$, is the sheaf on Y associated to the presheaf:*

$$V \mapsto \varinjlim_U F(U), \ V \textit{ open in } Y, \textit{ where } U \textit{ ranges through the family of open neighboroods of } f(V) \textit{ in } X \ .$$

One defines in an obvious way the direct (resp. the inverse) image of a morphism of sheaves on Y (resp. on X). Hence we get two functors:

$$f_* : \mathfrak{Sh}(Y) \to \mathfrak{Sh}(X) \ ,$$

$$f^{-1} : \mathfrak{Sh}(X) \to \mathfrak{Sh}(Y) \ .$$

If $\mathscr{R}$ (resp. $\mathscr{S}$) is a sheaf of rings on X (resp. Y), then $f^{-1}\mathscr{R}$ (resp. $f_*\mathscr{S}$) is a sheaf of rings on Y (resp. on X), and f_* and f^{-1} induce functors, still denoted f_* and f^{-1}:

$$f_* : \mathfrak{Mod}(\mathscr{S}) \to \mathfrak{Mod}(f_*\mathscr{S}) \ ,$$

$$f^{-1} : \mathfrak{Mod}(\mathscr{R}) \to \mathfrak{Mod}(f^{-1}\mathscr{R}) \ .$$

Example 2.3.2. Let $a_X : X \to \{\mathrm{pt}\}$. (Recall that $\{\mathrm{pt}\}$ is the set with one element.) Let M be an abelian group. Then:

$$M_X = a_X^{-1} M_{\{\mathrm{pt}\}} \ . \tag{2.3.1}$$

Let F be a sheaf on X. Then:

$$\Gamma(X; F) = \Gamma(\{\mathrm{pt}\}; a_{X*}F) \simeq a_{X*}F \ . \tag{2.3.2}$$

Let $y \in Y$ and let F be a sheaf on X. Then:

$$(f^{-1}F)_y = F_{f(y)} \ . \tag{2.3.3}$$

By this formula, we see that f^{-1} is an exact functor. On the other hand, the functor f_* is only left exact, as follows from the left exactness of the functor $\Gamma(U; \cdot)$.

There are natural morphisms of functors.

$$f^{-1} \circ f_* \to \mathrm{id} \qquad \text{in } \mathfrak{Mod}(f^{-1}\mathscr{R}) \ , \tag{2.3.4}$$

$$\mathrm{id} \to f_* \circ f^{-1} \qquad \text{in } \mathfrak{Mod}(\mathscr{R}) \ . \tag{2.3.5}$$

Proposition 2.3.3. *Let $\mathscr{R}$ be a sheaf of rings on X. Let $F \in \mathrm{Ob}(\mathfrak{Mod}(\mathscr{R}))$ and $G \in \mathrm{Ob}(\mathfrak{Mod}(f^{-1}\mathscr{R}))$. Then:*

(2.3.6) $$\mathrm{Hom}_{\mathscr{R}}(F, f_* G) \simeq \mathrm{Hom}_{f^{-1}\mathscr{R}}(f^{-1}F, G) \ .$$

In other words, f^{-1} is a left adjoint to f_* and f_* a right adjoint to f^{-1}.

Proof. We have homomorphisms:

$$\mathrm{Hom}_{\mathscr{R}}(F, f_* G) \underset{\alpha}{\to} \mathrm{Hom}_{f^{-1}\mathscr{R}}(f^{-1}F, f^{-1}f_* G) \underset{\beta}{\to} \mathrm{Hom}_{f^{-1}\mathscr{R}}(f^{-1}F, G)$$

and

$$\mathrm{Hom}_{f^{-1}\mathscr{R}}(f^{-1}F, G) \underset{\gamma}{\to} \mathrm{Hom}_{f_* f^{-1}\mathscr{R}}(f_* f^{-1}F, f_* G) \underset{\delta}{\to} \mathrm{Hom}_{\mathscr{R}}(F, f_* G) \ .$$

The homomorphisms α and γ are defined in the obvious way, and β and δ are deduced from (2.3.4) and (2.3.5) respectively. Then one checks easily that $\beta \circ \alpha$ and $\delta \circ \gamma$ are inverse one to each other. □

Corollary 2.3.4. *In the situation of Proposition* 2.3.3, *one has*:

(2.3.7) $$\mathscr{H}om_{\mathscr{R}}(F, f_* G) \simeq f_* \mathscr{H}om_{f^{-1}\mathscr{R}}(f^{-1}F, G) \ .$$

Proof. Let U be an open subset of X. Then:

$$\begin{aligned} \Gamma(U; f_* \mathscr{H}om_{f^{-1}\mathscr{R}}(f^{-1}F, G)) &= \mathrm{Hom}_{f^{-1}\mathscr{R}|_{f^{-1}(U)}}(f^{-1}F|_{f^{-1}(U)}, G|_{f^{-1}(U)}) \\ &= \mathrm{Hom}_{\mathscr{R}|_U}(F|_U, f_* G|_U) \\ &= \Gamma(U; \mathscr{H}om_{\mathscr{R}}(F, f_* G)) \ . \end{aligned}$$ □

Let Z be a another topological space, $g: Z \to Y$ a continuous map. By the construction of the direct or the inverse image, we immediately see that:

(2.3.8) $$(f \circ g)_* = f_* \circ g_* \ ,$$

(2.3.9) $$(f \circ g)^{-1} = g^{-1} \circ f^{-1} \ .$$

Inverse image commutes to tensor product. More precisely:

Proposition 2.3.5. *Let F_1 (resp. F_2) be a right (resp. left) $\mathscr{R}$-module. Then there is a canonical isomorphism*:

(2.3.10) $$f^{-1}F_1 \underset{f^{-1}\mathscr{R}}{\otimes} f^{-1}F_2 \simeq f^{-1}\left(F_1 \underset{\mathscr{R}}{\otimes} F_2\right) .$$

Proof. The morphism (2.3.10) is induced by:

$$F_1(U) \underset{\mathscr{R}(U)}{\otimes} F_2(U) \to \left(F_1 \underset{\mathscr{R}}{\otimes} F_2\right)(U) \ , \qquad U \text{ open in } X \ .$$

To prove this is an isomorphism, take $y \in Y$ and set $x = f(y)$. Then:

$$\left(f^{-1}F_1 \underset{f^{-1}\mathscr{R}}{\otimes} f^{-1}F_2\right)_y \simeq (f^{-1}F_1)_y \underset{(f^{-1}\mathscr{R})_y}{\otimes} (f^{-1}F_2)_y$$

$$\simeq (F_1)_x \underset{\mathscr{R}_x}{\otimes} (F_2)_x$$

$$\simeq \left(F_1 \underset{\mathscr{R}}{\otimes} F_2\right)_x . \quad \square$$

We shall now construct functors associated to a subset Z of X. We endow Z with the induced topology, and denote by $j: Z \hookrightarrow X$ the inclusion. We set for $F \in \mathrm{Ob}(\mathfrak{Sh}(X))$:

(2.3.11) $$F|_Z = j^{-1}F ,$$

(2.3.12) $$\Gamma(Z; F) = \Gamma(Z; j^{-1}F) .$$

Note that (2.3.12) agrees with the notation previously introduced for Z open.

There is a natural morphism $\Gamma(X; F) \to \Gamma(Z; F|_Z)$. If $s \in \Gamma(X; F)$ we denote by $s|_Z$ its image in $\Gamma(Z; F|_Z)$, and call it the restriction of s to Z.

Assume for a while that Z is closed in X. In this case we set:

$$F_Z = j_* j^{-1} F .$$

Thus we have a natural morphism $F \to F_Z$. Moreover:

(2.3.13) $$\begin{cases} F_Z|_Z = F|_Z , \\ F_Z|_{X\setminus Z} = 0 . \end{cases}$$

In particular, $(F_Z)_x = F_x$ if $x \in Z$, $(F_Z)_x = 0$ if $x \notin Z$. When Z is a locally closed subset of X, it is still possible to construct a sheaf F_Z on X satisfying (2.3.13). If Z is open in X, one sets:

$$F_Z = \mathrm{Ker}(F \to F_{X\setminus Z}) .$$

In the general case, let us write $Z = U \cap A$, where U is open in X and A is closed in X. We set:

$$F_Z = (F_U)_A .$$

This definition does not depend on the choice of U and A (see (i) in the following proposition).

Let us list the main properties of the functor $(\cdot)_Z : F \mapsto F_Z$.

Proposition 2.3.6. *Let Z be a locally closed subset of X and let F be a sheaf on X.*

(i) *The sheaf* F_Z *satisfies* (2.3.13). *Moreover any sheaf on* X *satisfying* (2.3.13) *is isomorphic to* F_Z.
(ii) *The functor* $(\cdot)_Z : F \mapsto F_Z$ *is exact.*
(iii) *Let* Z' *be another locally closed subset of* X. *Then*:

$$(F_Z)_{Z'} = F_{Z \cap Z'} \ .$$

(iv) *Assume* Z *is closed in* X, *and let* $j : Z \hookrightarrow X$ *be the inclusion of* Z *in* X. *Then*:

$$F_Z = j_* j^{-1} F \ .$$

(v) *Let* Z' *be a closed subset of* Z, (Z *being locally closed*). *Then the sequence below is exact*:

$$0 \to F_{Z \setminus Z'} \to F_Z \to F_{Z'} \to 0 \ .$$

(vi) *Let* Z_1 *and* Z_2 *be two closed subsets of* X. *Then the sequence*:

$$0 \longrightarrow F_{Z_1 \cup Z_2} \xrightarrow[\alpha]{} F_{Z_1} \oplus F_{Z_2} \xrightarrow[\beta]{} F_{Z_1 \cap Z_2} \longrightarrow 0$$

is exact. Here $\alpha = (\alpha_1, \alpha_2)$ *and* $\beta = (\beta_1, -\beta_2)$ *are induced by the natural morphisms* $F_{Z_1 \cup Z_2} \to F_{Z_i}$ *and* $F_{Z_i} \to F_{Z_1 \cap Z_2}$ ($i = 1, 2$), *respectively.*
(vii) *Let* U_1 *and* U_2 *be two open subsets of* X. *Then the sequence*:

$$0 \longrightarrow F_{U_1 \cap U_2} \xrightarrow[\gamma]{} F_{U_1} \oplus F_{U_2} \xrightarrow[\delta]{} F_{U_1 \cup U_2} \longrightarrow 0$$

is exact. Here $\gamma = (\gamma_1, \gamma_2)$ *and* $\delta = (\delta_1, -\delta_2)$ *are induced by the natural morphisms* $F_{U_1 \cap U_2} \to F_{U_i}$ *and* $F_{U_i} \to F_{U_1 \cap U_2}$ ($i = 1,2$), *respectively.*

Proof. One verifies easily (i) and the other assertions follow. □

Corollary 2.3.7. *Let* Z_1 *and* Z_2 *be two closed subsets of* X. *Then the sequence*:

$$0 \to \Gamma(Z_1 \cup Z_2; F) \to \Gamma(Z_1; F) \oplus \Gamma(Z_2; F) \to \Gamma(Z_1 \cap Z_2; F)$$

is exact.

Proof. Apply the left exact functor $\Gamma(X; \cdot)$ to the exact sequence of Proposition 2.3.6 (vi), and note that $\Gamma(X; F_Z) \simeq \Gamma(Z; F)$, for Z a closed subset of X. □

There is another sheaf which can be functiorially associated to a locally closed subset Z of X. Let U be an open subset of X, Z a closed subset of U. One sets:

$$\Gamma_Z(U; F) = \mathrm{Ker}(F(U) \to F(U \setminus Z)) \ . \tag{2.3.14}$$

Hence $\Gamma_Z(U; F)$ is the subgroup of $\Gamma(U; F)$ consisting of sections whose support is contained in Z.

Let V be an open subset of U containing Z. The canonical morphism $\Gamma_Z(U; F) \to \Gamma_Z(V; F)$ is an isomorphism. Thus for a locally closed subset Z of X

we may define $\Gamma_Z(X;F)$ as $\Gamma_Z(U;F)$, where U is any open subset of X containing Z as a closed subset. Note that the presheaf $U \mapsto \Gamma_{Z\cap U}(U;F)$ is a sheaf.

Definition 2.3.8. *One denotes by $\Gamma_Z(F)$ the sheaf $U \mapsto \Gamma_{Z\cap U}(U;F)$, and calls it the sheaf of sections of F supported by Z.*

Proposition 2.3.9. *Let Z be a locally closed subset of X, and let F be a sheaf on X.*

(i) *The functors $\Gamma_Z(X;\cdot): F \mapsto \Gamma_Z(X;F)$ from $\mathfrak{Sh}(X)$ to $\mathfrak{Ab}$ and $\Gamma_Z(\cdot): F \mapsto \Gamma_Z(F)$ from $\mathfrak{Sh}(X)$ to $\mathfrak{Sh}(X)$ are left exact. Moreover:*

$$\Gamma_Z(X;\cdot) = \Gamma(X;\cdot) \circ \Gamma_Z(\cdot) \ .$$

(ii) *Let Z' be another locally closed subset of X. Then:*

$$\Gamma_{Z'}(\cdot) \circ \Gamma_Z(\cdot) = \Gamma_{Z\cap Z'}(\cdot) \ .$$

(iii) *Assume Z is open in X, and let $i: Z \hookrightarrow X$ be the injection of Z in X. Then:*

$$\Gamma_Z(\cdot) = i_* \circ i^{-1} \ .$$

(iv) *Let Z' be a closed subset of the locally closed subset Z of X. Then the sequence below is exact:*

$$0 \to \Gamma_{Z'}(F) \to \Gamma_Z(F) \to \Gamma_{Z\setminus Z'}(F) \ .$$

(v) *Let U_1 and U_2 be two open subsets of X. Then the sequence:*

$$0 \longrightarrow \Gamma_{U_1\cup U_2}(F) \xrightarrow[\alpha]{} \Gamma_{U_1}(F) \oplus \Gamma_{U_2}(F) \xrightarrow[\beta]{} \Gamma_{U_1\cap U_2}(F)$$

is exact. Here $\alpha = (\alpha_1, \alpha_2)$ and $\beta = (\beta_1, -\beta_2)$ are induced by the morphisms $\Gamma_{U_1\cup U_2}(F) \xrightarrow[\alpha_i]{} \Gamma_{U_i}(F)$ and $\Gamma_{U_i}(F) \xrightarrow[\beta_i]{} \Gamma_{U_1\cap U_2}(F)$ $(i = 1,2)$, respectively.

(vi) *Let Z_1 and Z_2 be two closed subsets of X. Then the sequence:*

$$0 \longrightarrow \Gamma_{Z_1\cap Z_2}(F) \xrightarrow[\gamma]{} \Gamma_{Z_1}(F) \oplus \Gamma_{Z_2}(F) \xrightarrow[\delta]{} \Gamma_{Z_1\cup Z_2}(F)$$

is exact. Here $\gamma = (\gamma_1, \gamma_2)$ and $\delta = (\delta_1, -\delta_2)$ are induced by the morphisms $\Gamma_{Z_1\cap Z_2}(F) \xrightarrow[\gamma_i]{} \Gamma_{Z_i}(F)$ and $\Gamma_{Z_i}(F) \xrightarrow[\delta_i]{} \Gamma_{Z_1\cup Z_2}(F)$ $(i = 1,2)$, respectively.

The proof is straightforward.

We have defined several functors on the category $\mathfrak{Sh}(X)$: the functors $\mathcal{H}om(\cdot,\cdot)$, $\cdot \otimes \cdot$, f_*, f^{-1}, $(\cdot)_Z$, $\Gamma_Z(\cdot)$, $\Gamma(X;\cdot)$. We shall study now some of their relations.

Proposition 2.3.10. *Let $\mathcal{R}$ be a sheaf of rings on X and let $F \in \mathrm{Ob}(\mathfrak{Mod}(\mathcal{R}))$. Let Z be a locally closed subset of X. Then we have natural isomorphisms:*

$$\mathcal{R}_Z \underset{\mathcal{R}}{\otimes} F \simeq F_Z \ , \tag{2.3.15}$$

$$\mathcal{H}om_{\mathcal{R}}(\mathcal{R}_Z, F) \simeq \Gamma_Z(F) \ . \tag{2.3.16}$$

Proof. The first statement (2.3.15) follows from

$$\left(\mathcal{R}_Z \underset{\mathcal{R}}{\otimes} F\right)\bigg|_Z \simeq (\mathcal{R}_Z|_Z) \underset{\mathcal{R}|_Z}{\otimes} (F|_Z) \simeq (\mathcal{R}|_Z) \underset{\mathcal{R}|_Z}{\otimes} (F|_Z) \simeq F|_Z$$

and $(\mathcal{R}_Z \otimes_{\mathcal{R}} F)|_{X \setminus Z} = 0$, combined with Proposition 2.3.6 (i).

To prove (2.3.16) assume first Z is open in X. For any open subset U of X, the support of an element of $\Gamma(U; \mathcal{R}_Z)$ is closed in $U \cap Z$. From this remark we deduce that the natural morphism:

$$\mathrm{Hom}_{\mathcal{R}}(\mathcal{R}_Z, F) \to \mathrm{Hom}_{\mathcal{R}|_Z}(\mathcal{R}_Z|_Z, F|_Z)$$

is an isomorphism. The term on the right-hand side is $\Gamma(Z; F)$. Then by replacing X by U, for U open in X, in this isomorphism, we get the result for Z open. The case where Z is closed follows by applying the functor $\mathcal{H}om_{\mathcal{R}}(\cdot, F)$ to the exact sequence $0 \to \mathcal{R}_{X \setminus Z} \to \mathcal{R} \to \mathcal{R}_Z \to 0$ and comparing with the exact sequence $0 \to \Gamma_Z(F) \to F \to \Gamma_{X \setminus Z}(F)$.

Finally assume $Z = A \cap U$, where A is closed and U is open in X. We get:

$$\begin{aligned}
\mathcal{H}om_{\mathcal{R}}(\mathcal{R}_Z, F) &= \mathcal{H}om_{\mathcal{R}}(\mathcal{R}_{A \cap U}, F) \\
&= \mathcal{H}om_{\mathcal{R}}(\mathcal{R}_A, \Gamma_U(F)) \\
&= \Gamma_A(\Gamma_U(F)) \\
&= \Gamma_{U \cap A}(F) \ . \quad \square
\end{aligned}$$

Remark 2.3.11. By using Propositions 2.2.9, 2.3.3, 2.3.4, 2.3.5, 2.3.9, 2.3.10 we may obtain many other morphisms or isomorphisms. For example, let $f : Y \to X$ be a continuous map, $\mathcal{R}$ a sheaf of rings on X, Z a locally closed subset of X, and let F, F_1, F_2 (resp. G, G_1, G_2) be sheaves of $\mathcal{R}$-modules (resp. $f^{-1}\mathcal{R}$-modules). (When considering the tensor product, F_1, G and G_1 will be right modules.) Then there are natural morphisms or isomorphisms:

$$\left(F_1 \underset{\mathcal{R}}{\otimes} F_2\right)_Z \simeq F_1 \underset{\mathcal{R}}{\otimes} (F_2)_Z \simeq (F_1)_Z \underset{\mathcal{R}}{\otimes} F_2 \ , \tag{2.3.17}$$

$$\mathcal{H}om_{\mathcal{R}}((F_1)_Z, F_2) \simeq \mathcal{H}om_{\mathcal{R}}(F_1, \Gamma_Z(F_2)) \simeq \Gamma_Z \mathcal{H}om_{\mathcal{R}}(F_1, F_2) \ , \tag{2.3.18}$$

$$f^{-1}F_Z \simeq (f^{-1}F)_{f^{-1}(Z)} \ , \tag{2.3.19}$$

$$\Gamma_Z f_* G \simeq f_* \Gamma_{f^{-1}(Z)}(G) \ , \tag{2.3.20}$$

$$f_* G \underset{\mathcal{R}}{\otimes} F \to f_*\left(G \underset{f^{-1}\mathcal{R}}{\otimes} f^{-1}F\right) \ , \tag{2.3.21}$$

(2.3.22) $$f_*G_1 \underset{\mathscr{R}}{\otimes} f_*G_2 \to f_*\left(G_1 \underset{f^{-1}\mathscr{R}}{\otimes} G_2\right) ,$$

(2.3.23) $$f_*\mathscr{H}om_{f^{-1}\mathscr{R}}(G_1, G_2) \to \mathscr{H}om_{\mathscr{R}}(f_*G_1, f_*G_2) ,$$

(2.3.24) $$f^{-1}\mathscr{H}om_{\mathscr{R}}(F_1, F_2) \to \mathscr{H}om_{f^{-1}\mathscr{R}}(f^{-1}F_1, f^{-1}F_2) .$$

For example, let us construct the morphism (2.3.21). Consider the chain of morphisms (we shall not write $\mathscr{R}$, for short):

$$\begin{aligned}\mathrm{Hom}(G \otimes f^{-1}F, G \otimes f^{-1}F) &\to \mathrm{Hom}(f^{-1}f_*G \otimes f^{-1}F, G \otimes f^{-1}F)\\ &\simeq \mathrm{Hom}(f^{-1}(f_*G \otimes F), G \otimes f^{-1}F)\\ &\simeq \mathrm{Hom}(f_*G \otimes F, f_*(G \otimes f^{-1}F)) .\end{aligned}$$

The image of the identity morphism of $G \otimes f^{-1}F$ gives the desired morphism.

Notation 2.3.12. Let $p_X : X \to S$ and $p_Y : Y \to S$ be two continuous maps, and let $X \times_S Y = \{(x, y) \in X \times Y;\ p_X(x) = p_Y(y)\}$ be the fiber product of X and Y over S. We denote by q_1 and q_2 the projections from $X \times_S Y$ to X and Y respectively, and by p the projection $X \times_S Y \to S$ (cf. Diagram 2.3.25).

(2.3.25)

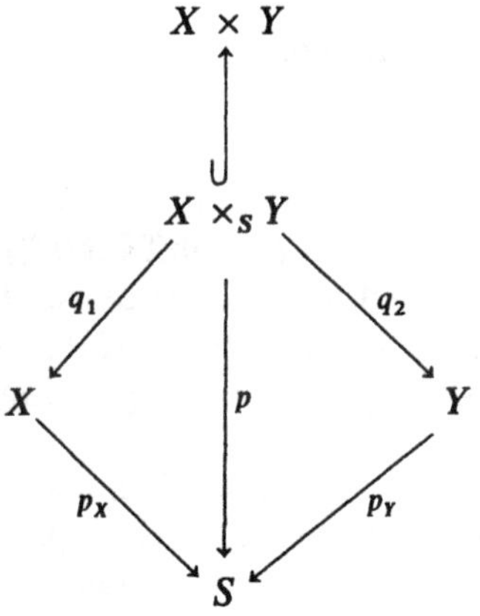

Let $\mathscr{R}$ be a sheaf of rings on S, let F (resp. G) be a sheaf of $p_X^{-1}(\mathscr{R}^{op})$-modules (resp. $p_Y^{-1}\mathscr{R}$-modules). One sets:

(2.3.26) $$F \underset{S}{\boxtimes_{\mathscr{R}}} G = q_1^{-1}F \underset{p^{-1}\mathscr{R}}{\otimes} q_2^{-1}G .$$

If there is no risk of confusion, we simply write $F \boxtimes_S G$, and if $S = \{\mathrm{pt}\}$, we do not write S in these formulas. Note that if $S = \{\mathrm{pt}\}$, $X = Y$ and if one denotes by δ the diagonal embedding, then:

(2.3.27) $$F \otimes G \simeq \delta^{-1}(F \boxtimes G) .$$

The sheaf $F \boxtimes_S G$ is called the **external tensor product** of F and G (over S).

2.4. Injective, flabby and flat sheaves

Let X be a topological space, $\mathscr{R}$ a sheaf of rings on X, F an object of $\mathfrak{Mod}(\mathscr{R})$. We shall say that F is $\mathscr{R}$-injective if F is injective in the category $\mathfrak{Mod}(\mathscr{R})$.

Proposition 2.4.1. (i) *Assume F is $\mathscr{R}$-injective. Let U be an open subset of X. Then $F|_U$ is $\mathscr{R}|_U$-injective.*

(ii) *Let $f : Y \to X$ be a continuous map, and let G be an $f^{-1}\mathscr{R}$-injective sheaf on Y. Then f_*G is $\mathscr{R}$-injective.*

Proof. (i) Let $i : U \hookrightarrow X$ be the inclusion map. Then for an $\mathscr{R}|_U$-module G we have by (2.3.18):

$$\begin{aligned}\mathrm{Hom}_{\mathscr{R}|U}(G, F|_U) &= \mathrm{Hom}_{\mathscr{R}|U}((i_*G)|_U, F|_U)\\ &= \mathrm{Hom}_{\mathscr{R}}((i_*G)_U, F) \ .\end{aligned}$$

Since the functor $G \mapsto (i_*G)_U$ is exact, the functor $G \mapsto \mathrm{Hom}_{\mathscr{R}|U}(G, F|_U)$ is exact.

(ii) Apply Proposition 2.3.3. □

Corollary 2.4.2. *Assume F is $\mathscr{R}$-injective. Then the functor $\mathscr{H}om_{\mathscr{R}}(\cdot, F)$ is exact.*

Proposition 2.4.3. *Let $\mathscr{R}$ be a sheaf of rings on X. Then the category $\mathfrak{Mod}(\mathscr{R})$ has enough injectives.*

Proof. Let $\hat{X}$ be the space X endowed with the discrete topology, and let $f : \hat{X} \to X$ be the natural map. Let $F \in \mathrm{Ob}(\mathfrak{Mod}(\mathscr{R}))$ and assume we have found an $f^{-1}\mathscr{R}$-injective sheaf I and a monomorphism $f^{-1}F \to I$. Applying the left exact functor f_* and using the morphism (2.3.5) we get the exact sequence $0 \to F \to f_*I$. Since f_*I is $\mathscr{R}$-injective by Proposition 2.4.1, it is enough to prove the result on $\hat{X}$.

Let $F \in \mathrm{Ob}(\mathfrak{Mod}(f^{-1}\mathscr{R}))$, and for each $x \in X$, let $0 \to F_x \to I_x$ be an exact sequence in $\mathfrak{Mod}(\mathscr{R}_x)$, with I_x injective (cf. Example 1.8.9). Then $\prod_{x \in X} I_x$ defines an injective sheaf I on $\hat{X}$, and $0 \to F \to I$ is exact. □

Remark 2.4.4. In general, the category $\mathfrak{Mod}(\mathscr{R})$ does not have enough projectives (cf. Exercise II.23).

Definition 2.4.5. *A sheaf F on X is flabby if for any open subset U of X the restriction morphism $\Gamma(X; F) \to \Gamma(U; F)$ is surjective.*

Proposition 2.4.6. *Let F be a flabby sheaf on X.*

(i) *For any open subset $U \subset X$, the sheaf $F|_U$ on U is flabby.*
(ii) *Let $f : X \to Y$ be a continuous map. Then the sheaf f_*F is flabby.*
(iii) *Let Z be a locally closed subset of X. Then the sheaf $\Gamma_Z(F)$ is flabby.*

(iv) *Let Z be a locally closed subset of X, Z' a closed subset of Z. Then the sequence below is exact:*

$$0 \to \Gamma_{Z'}(F) \to \Gamma_Z(F) \to \Gamma_{Z \setminus Z'}(F) \to 0 \ .$$

(v) *Let U_1 and U_2 be two open subsets of X. Then the sequence:*

$$0 \longrightarrow \Gamma_{U_1 \cup U_2}(F) \underset{\alpha}{\longrightarrow} \Gamma_{U_1}(F) \oplus \Gamma_{U_2}(F) \underset{\beta}{\longrightarrow} \Gamma_{U_1 \cap U_2}(F) \longrightarrow 0$$

is exact, (α and β are defined in Proposition 2.3.9).

(vi) *Let Z_1 and Z_2 be two closed subsets of X. Then the sequence:*

$$0 \longrightarrow \Gamma_{Z_1 \cap Z_2}(F) \underset{\gamma}{\longrightarrow} \Gamma_{Z_1}(F) \oplus \Gamma_{Z_2}(F) \underset{\delta}{\longrightarrow} \Gamma_{Z_1 \cup Z_2}(F) \longrightarrow 0$$

is exact, (γ and δ are defined in Proposition 2.3.9).

(vii) *Let $\mathscr{R}$ be a sheaf of rings on X, G an $\mathscr{R}$-module, H an $\mathscr{R}$-injective module. Then $\mathscr{H}om_{\mathscr{R}}(G, H)$ is flabby. In particular $\mathscr{R}$-injective modules are flabby.*

Proof. (i) and (ii) are obvious.

(iii) Replacing X by U, where U is an open subset of X containing Z as a closed subset, we may assume from the beginning that Z is closed. Let U be an open subset of X. We have to prove that the restriction morphism: $\Gamma_Z(X; F) \to \Gamma_{Z \cap U}(U; F)$ is surjective. Let $s \in \Gamma_{Z \cap U}(U; F)$. We first extend s by 0 on $X \setminus Z$. Let s' be this extension. Then we extend s' to X using the flabbiness of F.

(iv) For any open set U, $\Gamma(U; \Gamma_Z(F)) \to \Gamma(U; \Gamma_{Z \setminus Z'}(F)) \simeq \Gamma(U \setminus Z'; \Gamma_Z(F))$ is surjective by (iii).

(v) By Proposition 2.3.9, it remains to prove that β is an epimorphism, but the restriction morphism $\Gamma_{U_1}(F) \to \Gamma_{U_1 \cap U_2}(F)$ is itself an epimorphism.

(vi) Let $s \in \Gamma_{Z_1 \cup Z_2}(X; F)$. We may find $s_i \in \Gamma_{Z_i}(X \setminus Z_1 \cap Z_2; F)$ $(i = 1, 2)$ such that $s = s_1 - s_2$ on $X \setminus Z_1 \cap Z_2$. We extend s_1 and s_2 all over X. Let s_1' and s_2' be these extensions. Then $s_1' - s_2' = s + s'$, $s' \in \Gamma_{Z_1 \cap Z_2}(X; F)$, and $(s_1' - s') - s_2' = s$.

(vii) Let U be an open subset of X. Applying the exact functor $\mathrm{Hom}_{\mathscr{R}}(\cdot, H)$ to the exact sequence $0 \to G_U \to G \to G_{X \setminus U} \to 0$, we get the result in view of Corollary 2.4.2. □

Proposition 2.4.7. *Let $0 \to F' \to F \to F'' \to 0$ be an exact sequence in $\mathfrak{Sh}(X)$. Assume F' is flabby. Then the sequence $0 \to \Gamma(X; F') \to \Gamma(X; F) \to \Gamma(X; F'') \to 0$ is exact.*

Proof. Let $s'' \in \Gamma(X; F'')$, and let $\mathfrak{S}$ be the set of pairs (U, s) such that U is open in X, $s \in \Gamma(U; F)$, and s is sent to $s''|_U$. We order $\mathfrak{S}$ by setting $(U, s) \leqslant (V, t)$ if $U \subset V$ and $t|_U = s$. Then $\mathfrak{S}$ is clearly inductively ordered. Let (U, s) be a maximal element and suppose $U \neq X$. Let $x \in X \setminus U$. There exists an open neighborhood V of x and a section $t \in \Gamma(V; F)$ such that t is sent to $s''|_V$. On $U \cap V$, $s - t$ belongs to $\Gamma(U \cap V; F')$. Let $r \in \Gamma(X; F')$ which extends $s - t$ on $U \cap V$. Replacing t by

$t - r$, we may assume $t = s$ on $U \cap V$. Hence s can be extended to $U \cup V$, which is a contradiction. □

Corollary 2.4.8. *In the situation of Proposition 2.4.7, let Z be a locally closed subset of X. Then the sequences:*

$$0 \to \Gamma_Z(X; F') \to \Gamma_Z(X; F) \to \Gamma_Z(X; F'') \to 0$$

and

$$0 \to \Gamma_Z(F') \to \Gamma_Z(F) \to \Gamma_Z(F'') \to 0$$

are exact.

Proof. For an open set U such that $U \cap Z$ is closed in U we have a commutative diagram:

$$\begin{array}{ccccccccc}
 & & 0 & & 0 & & 0 & & \\
 & & \downarrow & & \downarrow & & \downarrow & & \\
0 & \longrightarrow & \Gamma_{Z \cap U}(U; F') & \longrightarrow & \Gamma_{Z \cap U}(U; F) & \longrightarrow & \Gamma_{Z \cap U}(U; F'') & \longrightarrow & 0 \\
 & & \downarrow & & \downarrow & & \downarrow & & \\
0 & \longrightarrow & \Gamma(U; F') & \longrightarrow & \Gamma(U; F) & \longrightarrow & \Gamma(U; F'') & \longrightarrow & 0 \\
 & & \downarrow & & \downarrow & & \downarrow & & \\
0 & \longrightarrow & \Gamma(U \backslash Z; F') & \longrightarrow & \Gamma(U \backslash Z; F) & \longrightarrow & \Gamma(U \backslash Z; F'') & \longrightarrow & 0 \\
 & & \downarrow & & \downarrow & & \downarrow & & \\
 & & 0 & & 0 & & 0 & &
\end{array}$$

Here the second and third rows, as well as all the columns are exact, by Proposition 2.4.7. Hence the top row is exact, (cf. Exercise I.8). □

Corollary 2.4.9. *Let $0 \to F' \to F \to F'' \to 0$ be an exact sequence in $\mathfrak{Sh}(X)$. Assume F' and F are flabby. Then F'' is flabby.*

Proof. Let U be an open subset of X. In the diagram below, α and β are surjective:

$$\begin{array}{ccc}
\Gamma(X; F) & \longrightarrow & \Gamma(X; F'') \\
\downarrow{\scriptstyle \alpha} & & \downarrow{\scriptstyle \gamma} \\
\Gamma(U; F) & \xrightarrow{\beta} & \Gamma(U; F'') .
\end{array}$$

Hence γ is surjective. □

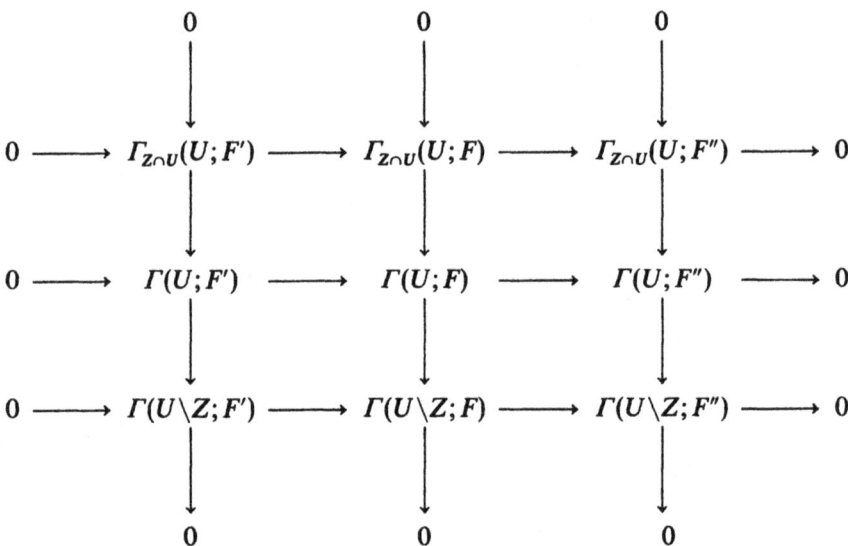

In view of Definition 1.8.2, we find that the full subcategory of $\mathfrak{Sh}(X)$ consisting of flabby sheaves is injective with respect to the functors $\Gamma(X;\cdot)$, $\Gamma_Z(\cdot)$, f_*.

The properties of being injective or flabby are local properties.

Proposition 2.4.10. *Let $X = \bigcup_{i\in I} U_i$ be an open covering.*

(i) *Let $F \in \mathrm{Ob}(\mathfrak{Sh}(X))$. If $F|_{U_i}$ is flabby for all i, then F is flabby.*
(ii) *Let $\mathscr{R}$ be a sheaf of rings on X and let $F \in \mathrm{Ob}(\mathfrak{Mod}(\mathscr{R}))$. If $F|_{U_i}$ is $\mathscr{R}|_{U_i}$-injective for all i, F is $\mathscr{R}$-injective.*

Proof. In order to prove (i), it is enough to show

(2.4.1) *A sheaf F is flabby if and only if, for any open set U, $F \to \Gamma_U(F)$ is an epimorphism.*

Since the flabbiness implies that $F \mapsto \Gamma_U(F)$ is an epimorphism, we shall show the converse.

Let s_0 be a section of F over an open subset V_0 of X. In order to prove that s_0 extends to a global section, let $\mathfrak{S}$ be the family of pairs (s, V) of open subsets V and $s \in \Gamma(V;F)$. We order $\mathfrak{S}$ by setting $(s, V) \leqslant (s', V')$ if $V \subset V'$ and $s'|_V = s$. Then $\mathfrak{S}$ is inductively ordered. Therefore there exists a maximal element (s, V) such that $(s, V) \geqslant (s_0, V_0)$. We shall show $V = X$. Otherwise there would be $x \in X \setminus V$. Since $F_x \to \Gamma_V(F)_x$ is surjective, there would exist an open neighborhood W of x and $t \in \Gamma(W;F)$ such that $t|_{V\cap W} = s|_{V\cap W}$. Then there would exist $s' \in \Gamma(W \cup V;F)$ such that $s'|_W = t$ and $s'|_V = s$. This contradicts the maximality of (s, V). Hence $V = X$ and we obtain (2.4.1).

(ii) The functor $\mathscr{H}om_{\mathscr{R}}(\cdot, F)$ is exact, and for any sheaf of $\mathscr{R}$-modules G, the sheaf $\mathscr{H}om_{\mathscr{R}}(G, F)$ is flabby. Thus the functor $\mathrm{Hom}_{\mathscr{R}}(\cdot, F) = \Gamma(X;\cdot) \circ \mathscr{H}om_{\mathscr{R}}(\cdot, F)$ is exact. □

Definition 2.4.11. *Let $\mathscr{R}$ be a sheaf of rings on X, $F \in \mathrm{Ob}(\mathfrak{Mod}(\mathscr{R}))$. One says that F is $\mathscr{R}$-flat if the functor $\cdot \otimes_{\mathscr{R}} F$, from the category of right $\mathscr{R}$-modules to $\mathfrak{Sh}(X)$, is exact.*

When there is no risk of confusion, we shall say "flat" instead of "$\mathscr{R}$-flat".

By (2.2.7), F is $\mathscr{R}$-flat if and only if F_x is an $\mathscr{R}_x$-flat module for all $x \in X$.

Proposition 2.4.12. *Let $F \in \mathrm{Ob}(\mathfrak{Mod}(\mathscr{R}))$. Then there exists an epimorphism $P \to F$, where P is $\mathscr{R}$-flat.*

Proof. Let $\mathfrak{S}$ denote the family of pairs (U, s) where U is open in X and $s \in \Gamma(U;F)$. For $(U,s) \in \mathfrak{S}$, we denote by $\mathscr{R}(U,s)$ the sheaf $\mathscr{R}_U$ indexed by (U,s). Then we set:

(2.4.2) $$P = \bigoplus_{(U,s)\in\mathfrak{S}} \mathscr{R}(U,s)\ .$$

The chain of morphisms $\mathscr{R}_U \to F_U \to F$, where the section $1 \in \Gamma(U;\mathscr{R}_U)$ is sent to

$s \in \Gamma(U; F_U)$, defines the epimorphism $P \to F$, and P is flat since for each $x \in X$, P_x is a free $\mathscr{R}_x$-module. □

Proposition 2.4.13. *Let $0 \to F' \to F \to F'' \to 0$ be an exact sequence in $\mathfrak{Mod}(\mathscr{R})$. If F and F'' are $\mathscr{R}$-flat, then so is F'.*

Proof. This follows immediately from the corresponding property for $\mathscr{R}_x$-modules, for all $x \in X$. □

Note that, by the two preceding results, the category $\mathfrak{Mod}(\mathscr{R})$ has enough projective objects with respect to the functor $G \otimes_{\mathscr{R}} \cdot$, for any right $\mathscr{R}$-module G.

2.5. Sheaves on locally compact spaces

When the topological space X has some finiteness properties, such as for example locally compact spaces, new classes of sheaves, and new functors are of interest. In this section, unless otherwise stated (such as in Proposition 2.5.1 and Remark 2.5.3), all spaces are supposed to be locally compact, and in particular Hausdorff.

Proposition 2.5.1. *Let X be a (not necessarily locally compact) topological space, Z a subspace, F a sheaf on X. Consider the canonical morphism:*

$$\psi : \varinjlim_{U} \Gamma(U; F) \to \Gamma(Z; F) ,$$

where U ranges through the family of open neighborhoods of Z in X.

(i) *The morphism ψ is injective.*
(ii) *Assume that X is Hausdorff and Z is compact. Then ψ is an isomorphism.*
(iii) *Assume that X is paracompact and Z is closed. Then ψ is an isomorphism.*

Before entering into the proof, recall that a paracompact space X is a Hausdorff space such that for each open covering $(U_i)_{i \in I}$ of X, there exists an open covering $(V_j)_{j \in J}$ of X, finer than $(U_i)_{i \in I}$ (i.e. for each $j \in J$ there exists $i \in I$ such that $V_j \subset U_i$) and locally finite. Also recall that if X is paracompact and $(U_i)_{i \in I}$ is an open locally finite covering, there exists an open covering $(V_i)_{i \in I}$ such that $\bar{V}_i \subset U_i$ for all i. A closed subspace of a paracompact space is paracompact, and locally compact spaces countable at infinity, as well as metric spaces are paracompact.

Proof of Proposition 2.5.1. (i) If $s \in \Gamma(U; F)$ is zero in $\Gamma(Z; F)$, then $s_x = 0$ for all $x \in Z$. Therefore s is zero on an open neighborhood of Z.

(ii) and (iii) Let $s \in \Gamma(Z; F)$. Then there exist a family of open sets $\{U_i\}_{i \in I}$ and $s_i \in \Gamma(U_i; F)$ such that $s|_{U_i \cap Z} = s_i|_{U_i \cap Z}$ and $\bigcup U_i \supset Z$. In the case (ii) we may assume I finite and in the case (iii) we may assume $\{U_i\}$ is a locally finite covering of X. In either case we can find a family of open subsets $\{V_i\}_{i\in I}$ such that $\bar{V}_i \subset U_i$, $\{\bar{V}_i\}$ is locally finite and $\bigcup V_i \supset Z$. For $x \in X$ set $I(x) = \{i \in I; x \in \bar{V}_i\}$ and define

$$W = \{x \in \bigcup V_i; s_{ix} = s_{jx} \text{ for any } i, j \in I(x)\} .$$

Then $I(x)$ is a finite set and each x has a neighborhood W_x such that $I(y) \subset I(x)$ for any $y \in W_x$. Therefore W is open and W contains Z by its construction. Since $s_i|_{W\cap V_i\cap V_j} = s_j|_{W\cap V_i\cap V_j}$, there exists $\tilde{s} \in \Gamma(W; F)$ such that $\tilde{s}|_{W\cap V_i} = s_i|_{W\cap V_i}$. Then $\tilde{s}$ satisfies $\psi(\tilde{s}) = s$. $\square$

Let $f: Y \to X$ be a continuous map (X and Y being not necessarily locally compact). Recall that one says that f is proper if f is closed (i.e: the image of any closed subset of Y is closed in X) and its fibers are relatively Hausdorff (two distinct points in the fiber have disjoint neighborhoods in Y) and compact. If X and Y are locally compact, f is proper if and only if the inverse image of any compact subset of X is compact. Let G be a sheaf on Y. One defines the subsheaf $f_!G$ of f_*G by setting for U open in X:

$$\Gamma(U; f_!G) = \{s \in \Gamma(f^{-1}(U); G); f: \operatorname{supp}(s) \to U \text{ is proper}\} . \quad (2.5.1)$$

Since the property of being proper is local on X, it is clear that the presheaf defined by (2.5.1) is a subsheaf of $f_*(G)$. This sheaf is called the **direct image with proper supports** of G. One denotes by $f_!$ the functor $G \mapsto f_!G$ from $\mathfrak{Sh}(Y)$ to $\mathfrak{Sh}(X)$. If $\mathscr{R}$ is a sheaf of rings on X, this functor induces a functor, still denoted by $f_!$, from $\mathfrak{Mod}(f^{-1}\mathscr{R})$ to $\mathfrak{Mod}(\mathscr{R})$. The functor $f_!$ is clearly left exact. One also sets:

$$\Gamma_c(X; F) = \{s \in \Gamma(X; F); \operatorname{supp}(s) \text{ is compact and Hausdorff}\} . \quad (2.5.2)$$

If a_X is the map: $X \to \{\text{pt}\}$, we find $\Gamma_c(X; F) \simeq a_{X!}F$.
Let $g: Z \to Y$ be another continuous map. We have:

$$f_! \circ g_! = (f \circ g)_! . \quad (2.5.3)$$

In particular, we have:

$$\Gamma_c(X; f_!G) \simeq \Gamma_c(Y; G) . \quad (2.5.4)$$

Proposition 2.5.2. *Let X and Y be locally compact spaces (in particular, they are Hausdorff spaces), $f: Y \to X$ a continuous map and G a sheaf on Y. Then for $x \in X$, the canonical morphism:*

$$\alpha: (f_!G)_x \to \Gamma_c(f^{-1}(x); G|_{f^{-1}(x)})$$

is an isomorphism.

Proof. Let us prove first that α is injective. Let V be an open neighborhood of $x \in X$ and let $t \in \Gamma(V; f_!G)$. Then t is defined by a section $s \in \Gamma(f^{-1}(V); G)$

such that $\operatorname{supp}(s) \to V$ is proper. If $\alpha(t) = 0$, then $\operatorname{supp}(s) \cap f^{-1}(x) = \emptyset$ and $x \notin f(\operatorname{supp}(s))$. This last set being closed, there exists an open neighborhood of x on which $t = 0$.

Next we prove that α is surjective. Let $s \in \Gamma_c(f^{-1}(x); G|_{f^{-1}(x)})$, and put $K = \operatorname{supp}(s)$. By Proposition 2.5.1 there exists an open neighborhood U of K in X and $t \in \Gamma(U; G)$ such that $t|_K = s|_K$. By shrinking U, we may assume $t|_{U \cap f^{-1}(x)} = s|_{U \cap f^{-1}(x)}$. Let V be a relatively compact open neighborhood of K, with $\bar{V} \subset U$. Since x does not belong to $f(\bar{V} \cap \operatorname{supp}(t) \setminus V)$, there exists an open neighborhood W of x such that $f^{-1}(W) \cap \bar{V} \cap \operatorname{supp}(t) \subset V$. One defines $\tilde{s} \in \Gamma(f^{-1}(W); G)$ by setting:

$$\tilde{s}|_{f^{-1}(W) \setminus (\operatorname{supp}(t) \cap \bar{V})} = 0 \ ,$$

$$\tilde{s}|_{f^{-1}(W) \cap V} = t|_{f^{-1}(W) \cap V} \ .$$

Since $\operatorname{supp}(\tilde{s})$ is contained in $f^{-1}(W) \cap \operatorname{supp}(t) \cap \bar{V}$, f is proper on this set. Moreover we have $\tilde{s}|_{f^{-1}(x)} = s$. □

Remark 2.5.3. Let $f\colon Y \to X$ be a continuous map, Y and X not being necessarily locally compact, and let G be a sheaf on Y. Assume f is proper on $\operatorname{supp}(G)$. By a similar argument as in the proof of Proposition 2.5.2 we find that for $x \in X$, the natural morphism:

$$(f_* G)_x \to \Gamma(f^{-1}(x); G|_{f^{-1}(x)})$$

is an isomorphism.

In Proposition 2.5.4 below, we do not assume X is locally compact.

Proposition 2.5.4. *Let Z be a locally closed subset of X, $i\colon Z \hookrightarrow X$ the inclusion map.*

(i) *The functor $i_!$ is exact.*
(ii) *Let $F \in \operatorname{Ob}(\mathfrak{Sh}(X))$. Then:*

$$F_Z \simeq i_! \circ i^{-1}(F) \ .$$

Proof. (i) $(i_! F)_x \simeq F_x$ or 0 according to $x \in Z$ or $x \notin Z$.

(ii) We have $(i_! \circ i^{-1}F)|_{X \setminus Z} = 0$ and $(i_! \circ i^{-1}F)|_Z = i^{-1}F$. Then apply Proposition 2.3.6 (i). □

Until the end of this section, all spaces will again be assumed to be locally compact spaces.

Definition 2.5.5. *Let $F \in \operatorname{Ob}(\mathfrak{Sh}(X))$. One says F is c-soft if for any compact subset K of X, the restriction morphism $\Gamma(X; F) \to \Gamma(K; F)$ is surjective.*

In view of Proposition 2.5.1, flabby sheaves, and in particular injective sheaves, are c-soft. In §9 we shall give many examples of c-soft sheaves.

Proposition 2.5.6. *Let $F \in \mathrm{Ob}(\mathfrak{Sh}(X))$. Then F is c-soft if and only if for any closed subset Z of X, the restriction morphism $\Gamma_c(X;F) \to \Gamma_c(Z;F|_Z)$ is surjective.*

Proof. If K is compact, $\Gamma(K;F) = \Gamma_c(K;F|_K)$. This proves the sufficiency of the condition.

Conversely assume F is c-soft and let $s \in \Gamma(Z;F|_Z)$ with compact support K. Let U be a relatively compact open neighborhood of K in X. Define $\tilde{s} \in \Gamma(\partial U \cup (Z \cap \bar{U}); F)$ by setting $\tilde{s}|_{Z \cap \bar{U}} = s$, $\tilde{s}|_{\partial U} = 0$, and extend $\tilde{s}$ to a section $t \in \Gamma(X;F)$. Since $t = 0$ on a neighborhood of ∂U, we may assume t is supported by $\bar{U}$. □

Proposition 2.5.7. *Let F be a c-soft sheaf on X.*

(i) *Let Z be a locally closed subset of X. Then $F|_Z$ is c-soft.*
(ii) *Let $f: X \to Y$ be a continuous map. Then $f_!F$ is c-soft.*
(iii) *For Z as in* (i), *F_Z is c-soft.*

Proof. (i) If Z is open, the result is clear. If Z is closed, apply Proposition 2.5.6.

(ii) If K is a compact subset of Y, then $\Gamma(K;f_!F) = \Gamma_c(f^{-1}(K);F)$. Since $\Gamma_c(Y;f_!F) = \Gamma_c(X;F)$, the result follows from Proposition 2.5.6.

(iii) follows from (i) and (ii) since $F_Z = f_!(F|_Z)$, f denoting the inclusion map $Z \hookrightarrow X$. □

Proposition 2.5.8. *Let $0 \to F' \to F \to F'' \to 0$ be an exact sequence of sheaves on X, and assume F' is c-soft. Let $f: X \to Y$ be a continuous map. Then the sequence $0 \to f_!F' \to f_!F \to f_!F'' \to 0$ is exact. In particular the sequence $0 \to \Gamma_c(X;F') \to \Gamma_c(X;F) \to \Gamma_c(X;F'') \to 0$ is exact.*

Proof. For all $y \in Y$, the sheaf $F'|_{f^{-1}(y)}$ is c-soft on $f^{-1}(y)$. Hence by Proposition 2.5.2 it is enough to prove the result in the particular case $f: X \to \{\mathrm{pt}\}$.

Let $s'' \in \Gamma_c(X;F'')$ and let U be a relatively compact open neighborhood of $\mathrm{supp}(s'')$. We shall show that s'' is in the image of $\Gamma_c(X;F) \to \Gamma_c(X;F'')$. By replacing F', F and F'' with F'_U, F_U, F''_U, and then X with $\bar{U}$, we may assume that X is compact. For $s'' \in \Gamma(X;F'')$, let $\{K_i\}_{i=1,\dots,n}$ be a finite covering of X by compact subsets such that there exists $s_i \in \Gamma(K_i;F)$ whose image is $s''|_{K_i}$. Let us argue by induction on n. For $n \geqslant 2$, on $K_1 \cap K_2$, $s_1 - s_2$ defines an element of $\Gamma(K_1 \cap K_2;F')$ hence extends to $s' \in \Gamma(X;F')$. Replacing s_2 by $s_2 + s'$, we may assume $s_1|_{K_1 \cap K_2} = s_2|_{K_1 \cap K_2}$. Therefore there exists $t \in \Gamma(K_1 \cup K_2;F)$ such that $t|_{K_i} = s_i$ $(i = 1, 2)$. Thus the induction proceeds. □

Corollary 2.5.9. *Let $0 \to F' \to F \to F'' \to 0$ be an exact sequence in $\mathfrak{Sh}(X)$. Assume F' and F are c-soft. Then F'' is c-soft.*

Proof. Similar to that of Corollary 2.4.9. □

By the preceding results, we find that the category of c-soft sheaves on X is injective with respect to the functors $\Gamma_c(X;\cdot)$, $f_!$ and $\Gamma(K;\cdot)$, (K compact). When X is countable at infinity, we have a better result.

Proposition 2.5.10. *Assume X is locally compact and countable at infinity. Then the category of c-soft sheaves is injective with respect to the functor $\Gamma(X;\cdot)$.*

Proof. Let $0 \to F' \to F \to F'' \to 0$ be an exact sequence in $\mathfrak{Sh}(X)$, with F' c-soft. Let $\{K_n\}_{n\in\mathbb{N}}$ be an increasing sequence of compact subsets of X, such that $X = \bigcup_n K_n$, and $K_n \subset \mathrm{Int}(K_{n+1})$ for all n. All the sequences $0 \to \Gamma(K_n; F') \to \Gamma(K_n; F) \to \Gamma(K_n; F'') \to 0$ are exact by Propositions 2.5.7 and 2.5.8. Since $\Gamma(K_{n+1}; F') \to \Gamma(K_n; F')$ is surjective for all n, we may apply Proposition 1.12.3 and the sequence:

$$0 \to \varprojlim_n \Gamma(K_n; F') \to \varprojlim_n \Gamma(K_n; F) \to \varprojlim_n \Gamma(K_n; F'') \to 0$$

is exact. Since for any sheaf G on X, $\Gamma(X; G) \simeq \varprojlim_n \Gamma(K_n; G)$, the proof is complete. □

Let us study some relations between the functor $f_!$ and other functors previously introduced. Consider first a Cartesian square of locally compact spaces:

$$\begin{array}{ccc} Y' & \xrightarrow{f'} & X' \\ {\scriptstyle g'}\downarrow & \square & \downarrow{\scriptstyle g} \\ Y & \xrightarrow[f]{} & X \end{array} \tag{2.5.5}$$

Recall that this means that the diagram is commutative and Y' is isomorphic to the fiber product $Y \times_X X' = \{(y, x') \in Y \times X';\ f(y) = g(x')\}$ as a topological space. (The square inside the diagram is here to emphasize that the diagram is Cartesian.)

Proposition 2.5.11. *One has a canonical isomorphism of functors:*

$$g^{-1} \circ f_! \simeq f'_! \circ g'^{-1} . \tag{2.5.6}$$

Proof. First we shall construct the canonical morphism:

$$f_! \circ g'_* \to g_* \circ f'_! . \tag{2.5.7}$$

Let $G \in \mathrm{Ob}(\mathfrak{Sh}(Y'))$, and let V be an open subset of X. A section $t \in \Gamma(V; f_! \circ g'_* G)$ is defined by a section $s \in \Gamma((f \circ g')^{-1}(V); G)$ such that $\mathrm{supp}(s) \subset g'^{-1}(Z)$ for a subset Z of $f^{-1}(V)$ proper over V. Then $g'^{-1}(Z) \to g^{-1}(V)$ is proper, and s defines a section of $g_* f'_! G$. This defines (2.5.7).

To prove (2.5.6), consider $G \in \mathrm{Ob}(\mathfrak{Sh}(Y))$. By Proposition 2.3.3 we have:

$$\mathrm{Hom}(g^{-1} \circ f_! G, f'_! \circ g'^{-1} G) = \mathrm{Hom}(f_! G, g_* \circ f'_! \circ g'^{-1} G) .$$

The morphism $f_! \to f_! \circ g'_* \circ g'^{-1} \to g_* \circ f'_! \circ g'^{-1}$ induces the morphism (2.5.6).

To prove this is an isomorphism, let us take $x' \in X'$. Then:

$$(g^{-1} \circ f_! G)_{x'} = (f_! G)_{g(x')}$$
$$= \Gamma_c(f^{-1}(g(x')); G) .$$

The map g' induces a homeomorphism $f'^{-1}(x') \simeq f^{-1}(g(x'))$, and an isomorphism $\Gamma_c(f^{-1}(g(x')); G) \simeq \Gamma_c(f'^{-1}(x'); g'^{-1}G) \simeq (f'_! \circ g'^{-1}G)_{x'}$. □

Finally, we shall study the relation of $f_!$ with tensor products.

Lemma 2.5.12. *Let A be a ring, and let M be a flat A-module. Let F be a sheaf of right A_X-modules. Then there is a natural isomorphism:*

$$\Gamma_c(X; F) \underset{A}{\otimes} M \xrightarrow{\sim} \Gamma_c\left(X; F \underset{A_X}{\otimes} M_X\right) .$$

In particular, if F is c-soft, then $F \otimes_{A_X} M_X$ is c-soft.

Proof. We may assume X is compact. If $X = \bigcup_j K_j$ is a finite covering of X by compact subsets, we have an exact sequence:

$$0 \longrightarrow \Gamma(X; F) \xrightarrow{\lambda} \bigoplus_j \Gamma(K_j; F) \xrightarrow{\mu} \bigoplus_{j,k} \Gamma(K_j \cap K_k; F) . \tag{2.5.8}$$

Since M is flat, this sequence remains exact after applying the functor $\cdot \otimes_A M$. Therefore we have a commutative diagram with exact rows:

$$\begin{array}{ccccccc}
0 \longrightarrow & \Gamma(X; F) \underset{A}{\otimes} M & \xrightarrow{\lambda} & \bigoplus_j \Gamma(K_j; F) \underset{A}{\otimes} M & \xrightarrow{\mu} & \bigoplus_{j,k} \Gamma(K_j \cap K_k; F) \underset{A}{\otimes} M \\
& \downarrow{\scriptstyle \alpha} & & \downarrow{\scriptstyle \beta} & & \downarrow{\scriptstyle \gamma} \\
0 \longrightarrow & \Gamma\left(X; F \underset{A_X}{\otimes} M_X\right) & \xrightarrow{\lambda'} & \bigoplus_j \Gamma\left(K_j; F \underset{A_X}{\otimes} M_X\right) & \xrightarrow{\mu'} & \bigoplus_{j,k} \Gamma\left(K_j \cap K_k; F \underset{A_X}{\otimes} M_X\right) .
\end{array}$$

Let us first show α is injective. We have an isomorphism for $x \in X$:

$$\varinjlim_U \left(\Gamma(U; F) \underset{A}{\otimes} M\right) \xrightarrow{\sim} \varinjlim_U \Gamma\left(U; F \underset{A_X}{\otimes} M_X\right) , \tag{2.5.9}$$

where U ranges through the family of open neighborhoods of x: in fact both sides of (2.5.9) are isomorphic to $F_x \otimes_A M$. Hence, if $s \in \Gamma(X; F) \otimes_A M$ satisfies $\alpha(s) = 0$, we can find a finite covering $X = \bigcup_j K_j$ such that $\lambda(s) = 0$. Then $s = 0$. If we apply this result to K_j and $K_j \cap K_k$ instead of X, we find that β and γ are injective.

To prove that α is surjective, let us take $t \in \Gamma(X; F \otimes_{A_X} M_X)$. By (2.5.9) there exists a finite covering $X = \bigcup_j K_j$, such that $\lambda'(t)$ is in the image of β. Then the injectivity of γ implies that t is in the image of α. □

Proposition 2.5.13. *Let $f: Y \to X$ be a continuous map, let $\mathscr{R}$ be a sheaf of rings on X, let $F \in \mathrm{Ob}(\mathfrak{Mod}(\mathscr{R}))$ and let $G \in \mathrm{Ob}(\mathfrak{Mod}(f^{-1}\mathscr{R}^{op}))$.*

(i) *There is a natural morphism:*

$$f_! G \underset{\mathscr{R}}{\otimes} F \to f_!\left(G \underset{f^{-1}\mathscr{R}}{\otimes} f^{-1}F\right). \tag{2.5.10}$$

(ii) *Assume F is a flat $\mathscr{R}$-module. Then (2.5.10) is an isomorphism.*

Proof. Note first that the morphism (2.5.10) is induced by the morphism (2.3.21). Now assume F is flat. To prove (2.5.10) is an isomorphism, let us choose $x \in X$, and apply Proposition 2.5.2 and Lemma 2.5.12. We get:

$$\begin{aligned}\left(f_!\left(G \underset{f^{-1}\mathscr{R}}{\otimes} f^{-1}F\right)\right)_x &\simeq \Gamma_c\left(f^{-1}(x); G \underset{f^{-1}\mathscr{R}}{\otimes} f^{-1}F\right)\\ &\simeq \Gamma_c\left(f^{-1}(x); G \underset{(\mathscr{R}_x)_Y}{\otimes} (F_x)_Y\right)\\ &\simeq \Gamma_c(f^{-1}(x); G) \underset{\mathscr{R}_x}{\otimes} F_x\\ &\simeq (f_!G)_x \underset{\mathscr{R}_x}{\otimes} F_x\\ &\simeq \left(f_!G \underset{\mathscr{R}}{\otimes} F\right)_x . \qquad \square\end{aligned}$$

Of course, using Proposition 2.5.13, we can get new natural morphisms. For example if G_1 and G_2 denote two $f^{-1}\mathscr{R}$-modules, we have the morphisms (when considering the tensor product, G_1 will be a right module):

$$f_!G_1 \underset{\mathscr{R}}{\otimes} f_*G_2 \longrightarrow f_!\left(G_1 \underset{f^{-1}\mathscr{R}}{\otimes} G_2\right), \tag{2.5.11}$$

$$\begin{array}{ccc} f_!\mathscr{H}om_{f^{-1}\mathscr{R}}(G_1, G_2) & \longrightarrow & \mathscr{H}om_{\mathscr{R}}(f_*G_1, f_!G_2)\\ \downarrow & & \downarrow\\ f_*\mathscr{H}om_{f^{-1}\mathscr{R}}(G_1, G_2) & \longrightarrow & \mathscr{H}om_{\mathscr{R}}(f_!G_1, f_!G_2) . \end{array} \tag{2.5.12}$$

Formula (2.5.11) follows immediately from (2.5.10) and (2.3.4). Let us explain the top arrow in (2.5.12). Set $H = \mathscr{H}om_{f^{-1}\mathscr{R}}(G_1, G_2)$. Then there are natural morphisms:

$$f_!H \otimes f_*G_1 \to f_!(H \otimes G_1) \to f_!G_2$$

and Proposition 2.2.9 gives the top arrow. The other arrows in (2.5.12) are similarly defined. Note that if $\mathscr{R}$ is commutative, these morphisms are $\mathscr{R}$-linear.

2.6. Cohomology of sheaves

We shall apply the results of Chapter I to the category of sheaves. Let X be a topological space, $\mathscr{R}$ a sheaf of rings on X. The category $\mathfrak{Mod}(\mathscr{R})$ of sheaves of (left) $\mathscr{R}$-modules on X is an abelian category. Hence we may consider the derived category $\mathbf{D}(\mathfrak{Mod}(\mathscr{R}))$ and its full triangulated subcategories $\mathbf{D}^*(\mathfrak{Mod}(\mathscr{R}))$, where $* = +, -, b$. We shall write for short:

$$\mathbf{D}^*(\mathscr{R}) = \mathbf{D}^*(\mathfrak{Mod}(\mathscr{R})) , \qquad * = \emptyset, +, -, b . \tag{2.6.1}$$

In particular, if A is a ring, the category $\mathbf{D}(A_X)$ is the derived category of the category of sheaves of A-modules on X. For example $\mathbf{D}(\mathbb{Z}_X) = \mathbf{D}(\mathfrak{Sh}(X))$.

In order to define the derived functors of the functors precedingly introduced, we consider the following situation: Z is a locally closed subset of X, $f: Y \to X$ is a continuous map, $g: W \to Y$ is a continuous map, $\mathscr{S} \to \mathscr{R}$ is a morphism of sheaves of rings on X whose image is contained in the center of $\mathscr{R}$ and $\mathscr{S}$ is commutative. When considering the functor $f_!$ (or $g_!$, etc.) we shall always implicitly assume all spaces are locally compact.

Since the category $\mathfrak{Mod}(\mathscr{R})$ has enough injectives, we may derive all left exact functors. We get the functors:

$$\begin{aligned}
&R\Gamma_Z(X;\cdot) &&: \mathbf{D}^+(\mathscr{R}) \to \mathbf{D}^+(\mathfrak{Ab}) ,\\
&R\Gamma_Z(\cdot) &&: \mathbf{D}^+(\mathscr{R}) \to \mathbf{D}^+(\mathscr{R}) ,\\
&R\Gamma(Z;\cdot) &&: \mathbf{D}^+(\mathscr{R}) \to \mathbf{D}^+(\mathfrak{Ab}) ,\\
&Rf_* &&: \mathbf{D}^+(f^{-1}\mathscr{R}) \to \mathbf{D}^+(\mathscr{R}) ,\\
&R\mathscr{H}om_{\mathscr{R}}(\cdot,\cdot) &&: \mathbf{D}^-(\mathscr{R})^\circ \times \mathbf{D}^+(\mathscr{R}) \to \mathbf{D}^+(\mathscr{S}) ,\\
&R\mathscr{H}om_{\mathscr{S}}(\cdot,\cdot) &&: \mathbf{D}^-(\mathscr{S})^\circ \times \mathbf{D}^+(\mathscr{R}) \to \mathbf{D}^+(\mathscr{R}) ,\\
&R\Gamma_c(X;\cdot) &&: \mathbf{D}^+(\mathscr{R}) \to \mathbf{D}^+(\mathfrak{Ab}) ,\\
&Rf_! &&: \mathbf{D}^+(f^{-1}\mathscr{R}) \to \mathbf{D}^+(\mathscr{R}) .
\end{aligned}$$

Moreover, the functors $(\cdot)_Z : F \to F_Z$ and $f^{-1} : F \to f^{-1}F$, being exact, they extend naturally to the derived categories. We get the functors, for $* = \emptyset, +, -, b$:

$$\begin{aligned}
&(\cdot)_Z : \mathbf{D}^*(\mathscr{R}) \to \mathbf{D}^*(\mathscr{R}) ,\\
&f^{-1} : \mathbf{D}^*(\mathscr{R}) \to \mathbf{D}^*(f^{-1}\mathscr{R}) .
\end{aligned}$$

Recall that to calculate these derived functors, one replaces a complex of sheaves F by a complex of injective sheaves I quasi-isomorphic to F, and one applies these functors to I. Moreover, for a given functor, it is enough to choose the I^j's in a subcategory injective with respect to this functor.

Example 2.6.1. Let $F \in \mathrm{Ob}(\mathfrak{Sh}(X))$. Consider an exact sequence $0 \to F \to F^0 \to \cdots$ where the F^j's are flabby. Identifying F to the complex $\cdots \to 0 \to F \to 0 \to \cdots$, where F stands in degree zero, F is quasi-isomorphic to the complex:

$$F^{\cdot} : \cdots \to 0 \to F^0 \to F^1 \to \cdots$$

(where F^0 stands in degree zero). Then $R\Gamma_Z(X; F)$ is represented by the complex $\Gamma_Z(X; F^{\cdot})$.

In order to define the left derived functor $\cdot \otimes^L_{\mathscr{R}} \cdot$, we need an assumption on $\mathscr{R}$.

Definition 2.6.2. *Let $\mathscr{R}$ be a sheaf of rings on X. One sets (cf. Exercise I.29):*

$$\mathrm{wgld}(\mathscr{R}) = \sup_{x \in X} (\mathrm{wgld}(\mathscr{R}_x)) \ .$$

One calls $\mathrm{wgld}(\mathscr{R})$ *the weak global dimension of* $\mathscr{R}$.

When considering tensor products over a sheaf of rings $\mathscr{R}$, we shall always assume $\mathscr{R}$ has finite weak global dimension:

$$\mathrm{wgld}(\mathscr{R}) < \infty \ . \tag{2.6.2}$$

Applying Proposition 2.4.12, Corollary 1.7.8 and Exercise I.23, we find that if $F \in \mathrm{Ob}(\mathbf{D}^b(\mathscr{R}))$ (resp. $\mathrm{Ob}(\mathbf{D}^+(\mathscr{R}))$), then F is quasi-isomorphic to a bounded complex (resp. a complex bounded from below) of flat $\mathscr{R}$-modules. Therefore we may define the left derived functors:

$$\cdot \overset{L}{\underset{\mathscr{R}}{\otimes}} \cdot : \mathbf{D}^*(\mathscr{R}^{op}) \times \mathbf{D}^*(\mathscr{R}) \to \mathbf{D}^*(\mathscr{S}) \ ,$$

$$\cdot \overset{L}{\underset{\mathscr{S}}{\otimes}} \cdot : \mathbf{D}^*(\mathscr{R}) \times \mathbf{D}^*(\mathscr{S}) \to \mathbf{D}^*(\mathscr{R})$$

with $* = -, +, b$.

In particular consider the situation of Notation 2.3.12. We get the functors for $* = -, +, b$:

$$\cdot \overset{L}{\underset{S}{\boxtimes}}{}_{\mathscr{R}} \cdot : \mathbf{D}^*(p_X^{-1}(\mathscr{R}^{op})) \times \mathbf{D}^*(p_Y^{-1}\mathscr{R}) \to \mathbf{D}^*(p^{-1}\mathscr{S}) \ .$$

(Here $\mathscr{R}$ and $\mathscr{S}$ are sheaves of rings on S.)

Now we shall briefly describe some of the numerous relations existing between all these functors, making a systematic use of Propositions 1.8.7 and 1.10.9.

Let $F \in \mathrm{Ob}(\mathbf{D}^+(\mathscr{R}))$. Then:

$$R\Gamma_Z(X;F) \simeq R\Gamma(X;R\Gamma_Z(F)) \ . \tag{2.6.3}$$

In fact, if F is injective, $\Gamma_Z(F)$ is injective.

Let $F \in \mathrm{Ob}(\mathbf{D}^b(\mathscr{R}))$ and let $G \in \mathrm{Ob}(\mathbf{D}^+(\mathscr{R}))$. Then:

$$R\mathrm{Hom}_{\mathscr{R}}(F,G) \simeq R\Gamma(X;R\mathscr{H}om_{\mathscr{R}}(F,G)) \ . \tag{2.6.4}$$

In fact the category of flabby sheaves is injective with respect to the functor $\Gamma(X;\cdot)$, and if F and G are sheaves of $\mathscr{R}$-modules with F injective, then $\mathscr{H}om_{\mathscr{R}}(G,F)$ is flabby.

Let $F \in \mathrm{Ob}(\mathbf{D}^+(g^{-1}f^{-1}\mathscr{R}))$. Then:

$$R(f\circ g)_*F \simeq Rf_*Rg_*F \ . \tag{2.6.5}$$

In fact, if F is flabby, g_*F is flabby, and flabby sheaves form an injective category with respect to direct images. Similarly, replacing "flabby" by "c-soft", one gets:

$$R(f\circ g)_!F \simeq Rf_!Rg_!F \ . \tag{2.6.6}$$

Proposition 2.6.3. *Let $\phi : \mathbf{D}^+(\mathscr{R}) \to \mathbf{D}^+(\mathscr{S})$ be the functor obtained from the forgetful functor $\mathfrak{Mod}(\mathscr{R}) \to \mathfrak{Mod}(\mathscr{S})$. Let $F \in \mathrm{Ob}(\mathbf{D}^-(\mathscr{R}))$, $G \in \mathrm{Ob}(\mathbf{D}^+(\mathscr{R}))$ and $H \in \mathrm{Ob}(\mathbf{D}^-(\mathscr{S}))$, (with $\mathrm{wgld}(\mathscr{S}) < \infty$). Then:*

(i) $\phi(R\mathscr{H}om_{\mathscr{S}}(H,G)) \simeq R\mathscr{H}om_{\mathscr{S}}(H,\phi(G))$.

(ii) *We have:*

$$\begin{aligned} R\mathscr{H}om_{\mathscr{R}}\left(F \overset{L}{\underset{\mathscr{S}}{\otimes}} H, G\right) &\simeq R\mathscr{H}om_{\mathscr{R}}(F, R\mathscr{H}om_{\mathscr{S}}(H,G)) \\ &\simeq R\mathscr{H}om_{\mathscr{S}}(H, R\mathscr{H}om_{\mathscr{R}}(F,G)) \qquad \text{in } \mathbf{D}^+(\mathscr{S}) \ . \end{aligned} \tag{2.6.7}$$

Proof. By Proposition 2.2.9, if H is flat over $\mathscr{S}$ and G is injective, then $\mathscr{H}om_{\mathscr{S}}(H,G)$ is a complex of injective $\mathscr{R}$-modules. Hence $R\mathscr{H}om_{\mathscr{R}}(F \otimes^L_{\mathscr{S}} H, G) = \mathscr{H}om_{\mathscr{R}}(F \otimes_{\mathscr{S}} H, G)$ and $R\mathscr{H}om_{\mathscr{R}}(F, R\mathscr{H}om_{\mathscr{S}}(H,G)) \simeq \mathscr{H}om_{\mathscr{R}}(F, \mathscr{H}om_{\mathscr{S}}(H,G))$. Then the first isomorphism in (ii) follows from Proposition 2.2.9. In order to prove (i) and the second isomorphism in (ii), we shall replace H with a (bounded above) complex H such that H^n is a direct sum of $\mathscr{S}_U$'s by the construction in Proposition 2.4.12. Then for any flabby $\mathscr{S}$-module K, $\mathrm{Ext}^j_{\mathscr{S}}(H^n,K) = 0$ for $j \neq 0$. Hence, for a complex of flabby $\mathscr{S}$-modules K, $R\mathrm{Hom}_{\mathscr{S}}(H,K) \simeq \mathrm{Hom}_{\mathscr{S}}(H,K)$. Now assume G injective. Then (i) follows from the fact that $\phi(G)$ is a complex of flabby $\mathscr{S}$-modules. Similarly, since $\mathscr{H}om_{\mathscr{R}}(F,G)$ is a complex of a flabby $\mathscr{S}$-modules by Proposition 2.4.6 (vii), $R\mathscr{H}om_{\mathscr{S}}(H, R\mathscr{H}om_{\mathscr{R}}(F,G)) \simeq$

$\mathscr{H}om_{\mathscr{S}}(H, \mathscr{H}om_{\mathscr{R}}(F, G))$. Since H is flat over $\mathscr{S}$, $R\mathscr{H}om_{\mathscr{R}}(F \otimes^L_{\mathscr{S}} H, G) \simeq \mathscr{H}om_{\mathscr{R}}(F \otimes_{\mathscr{S}} H, G)$. Then we can apply Proposition 2.2.9. □

By applying $R\Gamma(X; \cdot)$ and taking the 0-th cohomology, we get:

$$\text{(2.6.8)} \qquad \operatorname{Hom}_{\mathbf{D}^+(\mathscr{R})}\left(F \overset{L}{\underset{\mathscr{S}}{\otimes}} H, G\right) \simeq \operatorname{Hom}_{\mathbf{D}^+(\mathscr{S})}(H, R\mathscr{H}om_{\mathscr{R}}(F, G))$$
$$\simeq \operatorname{Hom}_{\mathbf{D}^+(\mathscr{R})}(F, R\mathscr{H}om_{\mathscr{S}}(H, G)) \ .$$

Choosing $H = \mathscr{S}_Z$ in (2.6.7), we get:

$$\text{(2.6.9)} \qquad R\mathscr{H}om_{\mathscr{R}}(F_Z, G) \simeq R\Gamma_Z R\mathscr{H}om_{\mathscr{R}}(F, G)$$
$$\simeq R\mathscr{H}om_{\mathscr{R}}(F, R\Gamma_Z(G)) \qquad \text{in } \mathbf{D}^+(\mathscr{S}) \ .$$

Moreover, arguing as in Corollary 2.2.10, one gets the natural morphisms:

$$\text{(2.6.10)} \qquad R\mathscr{H}om_{\mathscr{R}}(F, G) \overset{L}{\underset{\mathscr{S}}{\otimes}} F \to G \qquad \text{in } \mathbf{D}^+(\mathscr{R}) \ ,$$

$$\text{(2.6.11)} \quad R\mathscr{H}om_{\mathscr{R}}(F, G) \overset{L}{\underset{\mathscr{S}}{\otimes}} H \to R\mathscr{H}om_{\mathscr{R}}\left(F, G \overset{L}{\underset{\mathscr{S}}{\otimes}} H\right) \qquad \text{in } \mathbf{D}^+(\mathscr{S}) \ ,$$

$$\text{(2.6.12)} \quad R\mathscr{H}om_{\mathscr{R}}(F, G) \to R\mathscr{H}om_{\mathscr{R}}(R\mathscr{H}om_{\mathscr{R}}(G, \mathscr{R}), R\mathscr{H}om_{\mathscr{R}}(F, \mathscr{R})) \quad \text{in } \mathbf{D}^+(\mathscr{R}) \ .$$

In (2.6.12), we assume that $\mathscr{R}$ is commutative, $\operatorname{wgld}(\mathscr{R}) < \infty$ and F, G, $R\mathscr{H}om_{\mathscr{R}}(G, \mathscr{R})$ are bounded. The morphism (2.6.12) is defined as follows.

By (2.6.11) we have:

$$R\mathscr{H}om_{\mathscr{R}}(F, G) \overset{L}{\underset{\mathscr{R}}{\otimes}} R\mathscr{H}om_{\mathscr{R}}(G, \mathscr{R}) \to R\mathscr{H}om_{\mathscr{R}}\left(F, G \overset{L}{\underset{\mathscr{R}}{\otimes}} R\mathscr{H}om_{\mathscr{R}}(G, \mathscr{R})\right) .$$

On the other hand, (2.6.10) gives $G \otimes^L_{\mathscr{R}} R\mathrm{Hom}_{\mathscr{R}}(G, \mathscr{R}) \to \mathscr{R}$. Combining them, we obtain:

$$R\mathscr{H}om_{\mathscr{R}}(F, G) \overset{L}{\underset{\mathscr{R}}{\otimes}} R\mathscr{H}om_{\mathscr{R}}(G, \mathscr{R}) \to R\mathscr{H}om_{\mathscr{R}}(F, \mathscr{R}) \ .$$

Then by (2.6.8) this gives (2.6.12).

Proposition 2.6.4. (i) *Let* $G \in \operatorname{Ob}(\mathbf{D}^+(f^{-1}\mathscr{R}))$ *and* $F \in \operatorname{Ob}(\mathbf{D}^-(\mathscr{R}))$. *Then*:

$$\text{(2.6.13)} \qquad R\mathrm{Hom}_{\mathscr{R}}(F, Rf_*G) \simeq R\mathrm{Hom}_{f^{-1}\mathscr{R}}(f^{-1}F, G) \ .$$

(ii) *The functors* $f^{-1} : \mathbf{D}^+(\mathscr{R}) \to \mathbf{D}^+(f^{-1}\mathscr{R})$ *and* $Rf_* : \mathbf{D}^+(f^{-1}\mathscr{R}) \to \mathbf{D}^+(\mathscr{R})$ *are adjoint to each other, i.e.*:

$$\text{(2.6.14)} \qquad \operatorname{Hom}_{\mathbf{D}^+(\mathscr{R})}(F, Rf_*G) \simeq \operatorname{Hom}_{\mathbf{D}^+(f^{-1}\mathscr{R})}(f^{-1}F, G) \ .$$

Proof. (i) If G is injective, f_*G is injective. Hence $R\mathrm{Hom}_{\mathscr{R}}(F, Rf_*(\cdot))$ is the derived functor of $\mathrm{Hom}_{\mathscr{R}}(F, f_*(\cdot))$. Since $R\mathrm{Hom}_{f^{-1}\mathscr{R}}(f^{-1}F, \cdot)$ is the derived functor of $\mathrm{Hom}_{f^{-1}\mathscr{R}}(f^{-1}F, \cdot)$, the result follows from Proposition 2.3.3.

(ii) The proof is similar. □

The same argument gives:

$$R\mathscr{H}om_{\mathscr{R}}(F, Rf_*G) \simeq Rf_*R\mathscr{H}om_{f^{-1}\mathscr{R}}(f^{-1}F, G) \quad \text{in } \mathbf{D}^+(\mathscr{S}) \ . \tag{2.6.15}$$

By (2.6.14), we get natural morphisms:

$$\mathrm{id} \to Rf_* \circ f^{-1} \quad \text{in } \mathbf{D}^+(\mathscr{R}) \ , \tag{2.6.16}$$

$$f^{-1} \circ Rf_* \to \mathrm{id} \quad \text{in } \mathbf{D}^+(f^{-1}\mathscr{R}) \ . \tag{2.6.17}$$

Proposition 2.6.5. *Let* $F_1 \in \mathrm{Ob}(\mathbf{D}^+(\mathscr{R}^{op}))$ *and* $F_2 \in \mathrm{Ob}(\mathbf{D}^+(\mathscr{R}))$ *(with* $\mathrm{wgld}(\mathscr{R}) < \infty$*). Then:*

$$f^{-1}F_1 \overset{L}{\underset{f^{-1}\mathscr{R}}{\otimes}} f^{-1}F_2 \simeq f^{-1}\left(F_1 \overset{L}{\underset{\mathscr{R}}{\otimes}} F_2\right) \quad \text{in} \quad \mathbf{D}^+(f^{-1}\mathscr{S}) \ . \tag{2.6.18}$$

Proof. If F is flat over $\mathscr{R}$, $f^{-1}F$ is flat over $f^{-1}\mathscr{R}$. Hence $f^{-1}(\cdot) \otimes^L_{f^{-1}\mathscr{R}} f^{-1}(\cdot)$ is the derived functor of $f^{-1}(\cdot) \otimes_{f^{-1}\mathscr{R}} f^{-1}(\cdot)$, and it remains to apply Proposition 2.3.5. □

Proposition 2.6.6. *Let* $G \in \mathrm{Ob}(\mathbf{D}^+(f^{-1}\mathscr{R}^{op}))$ *and let* $F \in \mathrm{Ob}(\mathbf{D}^+(\mathscr{R}))$ *with* $\mathrm{wgld}(\mathscr{R}) < \infty$. *Then:*

$$Rf_!G \overset{L}{\underset{\mathscr{R}}{\otimes}} F \simeq Rf_!\left(G \overset{L}{\underset{f^{-1}\mathscr{R}}{\otimes}} f^{-1}F\right) \quad \text{in } \mathbf{D}^+(\mathscr{S}) \ . \tag{2.6.19}$$

(According to the conventions at the beginning of this section, X and Y are locally compact spaces.)

Proof. First assume F is a flat sheaf. By Lemma 2.5.12, one sees that the functor $\cdot \otimes_{f^{-1}\mathscr{R}} f^{-1}F$ sends c-soft sheaves to $f_!$-injective sheaves. Hence $Rf_!(\cdot \otimes_{f^{-1}\mathscr{R}} f^{-1}F)$ is the derived functor of $f_!(\cdot \otimes_{f^{-1}\mathscr{R}} f^{-1}F)$. Since $Rf_!(\cdot) \otimes_{\mathscr{R}} F$ is the derived functor of $f_!(\cdot) \otimes_{\mathscr{R}} F$, the result follows from Proposition 2.5.13 in this case. To treat the general case we note that if $F \in \mathrm{Ob}(\mathbf{D}^+(\mathscr{R}))$, then F is quasi-isomorphic to a complex bounded from below of flat sheaves. □

Proposition 2.6.7. *Consider the Cartersian square* (2.5.5) *and let* $G \in \mathrm{Ob}(\mathbf{D}^+(f^{-1}\mathscr{R}))$. *Then:*

$$g^{-1} \circ Rf_!G \simeq Rf'_! \circ g'^{-1}G \quad \text{in } \mathbf{D}^+(g^{-1}\mathscr{R}) \ . \tag{2.6.20}$$

Proof. Since $g^{-1} \circ Rf_!$ is the derived functor of $g^{-1} \circ f_!$, it is enough to prove that $Rf_!' \circ g'^{-1}$ is the derived functor of $f_!' \circ g'^{-1}$, by Proposition 2.5.11.

Denote by I_Y the subcategory of $\mathfrak{Mod}(f^{-1}\mathscr{R})$ consisting of sheaves G such that for all $x \in X$, $G|_{f^{-1}(x)}$ is c-soft, and define similarly $\mathrm{I}_{Y'}$ in $\mathfrak{Mod}(g'^{-1}f^{-1}\mathscr{R})$. Then I_Y is injective with respect to g'^{-1}, g'^{-1} sends I_Y to $\mathrm{I}_{Y'}$ and $\mathrm{I}_{Y'}$ is injective with respect to $f_!'$. This completes the proof. □

There are many other natural morphisms or isomorphisms which can be obtained by combining the formulas above. Let us mention few of them, leaving the proofs to the reader. In these formulas, when $\otimes_{\mathscr{R}}$ appears, we assume $\mathrm{wgld}(\mathscr{R}) < \infty$. Let $F \in \mathrm{Ob}(\mathbf{D}^+(\mathscr{R}))$, $G \in \mathrm{Ob}(\mathbf{D}^+(f^{-1}\mathscr{R}^{op}))$. We get:

(2.6.21) $$Rf_*G \overset{L}{\underset{\mathscr{R}}{\otimes}} F \to Rf_*\left(G \overset{L}{\underset{f^{-1}\mathscr{R}}{\otimes}} f^{-1}F\right) \quad \text{in } \mathbf{D}^+(\mathscr{S}) .$$

Let $G_1 \in \mathrm{Ob}(\mathbf{D}^+(f^{-1}\mathscr{R}^{op}))$, $G_2 \in \mathrm{Ob}(\mathbf{D}^+(f^{-1}\mathscr{R}))$. We get:

(2.6.22) $$Rf_*G_1 \overset{L}{\underset{\mathscr{R}}{\otimes}} Rf_*G_2 \to Rf_*\left(G_1 \overset{L}{\underset{f^{-1}\mathscr{R}}{\otimes}} G_2\right) \quad \text{in } \mathbf{D}^+(\mathscr{S}) ,$$

(2.6.23) $$Rf_!G_1 \overset{L}{\underset{\mathscr{R}}{\otimes}} Rf_*G_2 \to Rf_!\left(G_1 \overset{L}{\underset{f^{-1}\mathscr{R}}{\otimes}} G_2\right) \quad \text{in } \mathbf{D}^+(\mathscr{S}) .$$

Let $G_1 \in \mathrm{Ob}(\mathbf{D}^b(f^{-1}\mathscr{R}))$, $G_2 \in \mathrm{Ob}(\mathbf{D}^+(f^{-1}\mathscr{R}))$ and assume f_* and $f_!$ have finite cohomological dimension (cf. Exercise I.19). We get:

(2.6.24) $$Rf_*R\mathscr{H}om_{f^{-1}\mathscr{R}}(G_1, G_2) \to R\mathscr{H}om_{\mathscr{R}}(Rf_*G_1, Rf_*G_2) \quad \text{in } \mathbf{D}^+(\mathscr{S}) ,$$

(2.6.25) $$Rf_*R\mathscr{H}om_{f^{-1}\mathscr{R}}(G_1, G_2) \to R\mathscr{H}om_{\mathscr{R}}(Rf_!G_1, Rf_!G_2) \quad \text{in } \mathbf{D}^+(\mathscr{S}) ,$$

(2.6.26) $$Rf_!R\mathscr{H}om_{f^{-1}\mathscr{R}}(G_1, G_2) \to R\mathscr{H}om_{\mathscr{R}}(Rf_*G_1, Rf_!G_2) \quad \text{in } \mathbf{D}^+(\mathscr{S}) .$$

Let $F_1 \in \mathrm{Ob}(\mathbf{D}^b(\mathscr{R}))$ and $F_2 \in \mathrm{Ob}(\mathbf{D}^+(\mathscr{R}))$. We get:

(2.6.27) $$f^{-1}R\mathscr{H}om_{\mathscr{R}}(F_1, F_2) \to R\mathscr{H}om_{f^{-1}\mathscr{R}}(f^{-1}F_1, f^{-1}F_2) \quad \text{in } \mathbf{D}^+(f^{-1}\mathscr{S}) .$$

Finally we shall apply Propositions 1.8.8, 2.3.6, 2.3.9 and 2.4.6. Let U_1 and U_2 (resp. Z_1 and Z_2) be two open (resp. closed) subsets of X, let Z be a locally closed subset of X, Z' a closed subset of Z.

For $F \in \mathrm{Ob}(\mathbf{D}^+(\mathscr{R}))$, we get the distinguished triangles:

(2.6.28) $$R\Gamma_{U_1 \cup U_2}(F) \longrightarrow R\Gamma_{U_1}(F) \oplus R\Gamma_{U_2}(F) \longrightarrow R\Gamma_{U_1 \cap U_2}(F) \underset{+1}{\longrightarrow} ,$$

(2.6.29) $$R\Gamma_{Z_1 \cap Z_2}(F) \longrightarrow R\Gamma_{Z_1}(F) \oplus R\Gamma_{Z_2}(F) \longrightarrow R\Gamma_{Z_1 \cup Z_2}(F) \underset{+1}{\longrightarrow} ,$$

(2.6.30) $$F_{U_1 \cap U_2} \longrightarrow F_{U_1} \oplus F_{U_2} \longrightarrow F_{U_1 \cup U_2} \underset{+1}{\longrightarrow} ,$$

(2.6.31) $$F_{Z_1 \cup Z_2} \longrightarrow F_{Z_1} \oplus F_{Z_2} \longrightarrow F_{Z_1 \cap Z_2} \underset{+1}{\longrightarrow} ,$$

(2.6.32) $$R\Gamma_{Z'}(F) \longrightarrow R\Gamma_Z(F) \longrightarrow R\Gamma_{Z\setminus Z'}(F) \underset{+1}{\longrightarrow} ,$$

(2.6.33) $$F_{Z\setminus Z'} \longrightarrow F_Z \longrightarrow F_{Z'} \underset{+1}{\longrightarrow} .$$

Notations 2.6.8. Let $\mathscr{R}$ be a sheaf of rings on X, Z a locally closed subset of X, $F \in \mathrm{Ob}(\mathbf{D}^+(\mathscr{R}))$, $K \in \mathrm{Ob}(\mathbf{D}^b(\mathscr{R}))$, $G \in \mathrm{Ob}(\mathbf{D}^+(\mathscr{R}^{op}))$. One sets:

$$H^j_Z(F) = H^j(R\Gamma_Z(F)) ,$$

$$H^j(X;F) = H^j(R\Gamma(X;F)) ,$$

$$H^j_Z(X;F) = H^j(R\Gamma_Z(X;F)) ,$$

$$H^j(Z;F) = H^j(R\Gamma(Z;F)) ,$$

$$H^j_c(X;F) = H^j(R\Gamma_c(X;F)) ,$$

$$\mathscr{E}xt^j_{\mathscr{R}}(K,F) = H^j(R\mathscr{H}om_{\mathscr{R}}(K,F)) ,$$

$$\mathrm{Ext}^j_{\mathscr{R}}(K,F) = H^j(R\mathrm{Hom}_{\mathscr{R}}(K,F)) ,$$

$$\mathscr{T}or^{\mathscr{R}}_j(G,F) = H^{-j}\left(G \overset{L}{\underset{\mathscr{R}}{\otimes}} F\right) .$$

Note that if X is a locally compact space, then $H^j_c(X;F) \xleftarrow{\sim} \varinjlim_K H^j_K(X;F)$, where K ranges through the family of compact subsets of X.

Remark 2.6.9. For a not necessarily locally closed subset Z of X, we have defined the functor $F \mapsto \Gamma(Z;F) = \Gamma(Z;F|_Z)$. It is thus possible to consider its derived functor $F \mapsto R\Gamma(Z;F)$. One shall beware that $R\Gamma(Z;F)$ may be different from the functor $R\Gamma(Z;\cdot)$ applied to $F|_Z$. However there is a natural isomorphism $R\Gamma(Z;F) \simeq R\Gamma(Z;F|_Z)$ in one of the following situations:

(i) Z is open,
(ii) X is Hausdorff and Z is compact,
(iii) Z is closed in a paracompact open subset U of X.

In all these cases, one also has:

$$H^j(Z;F) \xleftarrow{\sim} \varinjlim H^j(U;F)$$

where U ranges through the family of open neighborhoods of Z in X.

Remark 2.6.10. When applying the functor $R\Gamma(X;\cdot)$ to the distinguished triangles (2.6.28)–(2.6.32), we get new distinguished triangles that we do not write. When applying the functor $H^0(\cdot)$ to these triangles, we get long exact sequences. The long exact sequences deduced from (2.6.28) to (2.6.31) are called **Mayer-Vietoris sequences.** For example, one has the exact sequence:

$$\cdots \to H^{k-1}(U_1 \cap U_2; F) \to H^k(U_1 \cup U_2; F) \to H^k(U_1; F) \oplus H^k(U_2; F) \to \cdots$$

and the exact sequence, obtained by applying $R\Gamma_c(X; \cdot)$ to (2.6.33) with $Z = X$, $Z' = K$ compact:

$$\cdots \to H^{k-1}(K; F) \to H^k_c(X \setminus K; F) \to H^k_c(X; F) \to \cdots .$$

Notations 2.6.11. (i) Let $F \in \mathrm{Ob}(\mathbf{D}^+(\mathbb{Z}_X))$. Its support, denoted $\mathrm{supp}(F)$, is the closed subset of X:

$$\mathrm{supp}(F) = \overline{\bigcup_{j \in \mathbb{Z}} \mathrm{supp}\, H^j(F)} \ . \tag{2.6.34}$$

(ii) Let A be a ring and let $M \in \mathrm{Ob}(\mathbf{D}^+(\mathfrak{Mod}(A)))$. One sets:

$$M_X = a_X^{-1} M \ ,$$

where a_X is the map $X \to \{\mathrm{pt}\}$.

(iii) Let A be a commutative ring. When working in the category $\mathbf{D}^+(A_X)$, we shall call A "the base ring". If there is no risk of confusion we shall write Hom, $\otimes$, $\boxtimes$, instead of Hom_A, $\otimes_A$, $\boxtimes_A$. From Chapter III, we shall often write $\mathbf{D}^+(X)$ instead of $\mathbf{D}^+(A_X)$.

2.7. Some vanishing theorems

In this section, we collect some useful results on the vanishing of the cohomology groups of sheaves, or equivalently, results on isomorphisms of cohomology groups, in some special situations. These results will be obtained either by "non-characteristic deformations", as in Proposition 2.7.2, or by homotopy methods, with the help of the Mittag-Leffler procedure, that we state now for sheaves.

Proposition 2.7.1. *Let X be a topological space and let $F \in \mathrm{Ob}(\mathbf{D}^+(\mathbb{Z}_X))$. Let $\{U_n\}_{n \in \mathbb{N}}$ be an increasing sequence of open subsets of X, $\{Z_n\}_{n \in \mathbb{N}}$ a decreasing sequence of closed subsets of X. Set $U = \bigcup_n U_n$, $Z = \bigcap_n Z_n$.*

(i) *For any j, the natural map $\varphi_j : H^j_Z(U; F) \to \varprojlim_n H^j_{Z_n}(U_n; F)$ is surjective.*

(ii) *Assume that for a given j, the projective system $\{H^{j-1}_{Z_n}(U_n; F)\}_n$ satisfies the M-L condition. Then φ_j is bijective.*

(iii) *Let now $\{X_n\}_{n \in \mathbb{N}}$ be an increasing family of subsets of X satisfying $X = \bigcup_n X_n$ and $X_n \subset \mathrm{Int}(X_{n+1})$ for all n. Assume that for a given j, the projective system $\{H^{j-1}(X_n; F)\}_n$ satisfies the M-L condition. Then the natural map $H^j(X; F) \to \varprojlim_n H^j(X_n; F)$ is bijective.*

Proof. We may assume F is a complex of flabby sheaves.

To prove (i) and (ii) denote by $E^{\cdot}_n$ the simple complex associated with the double complex:

$$\begin{array}{ccccccc}\cdots & \longrightarrow & \Gamma(U_n; F^{i-1}) & \longrightarrow & \Gamma(U_n; F^i) & \longrightarrow & \cdots \\ & & \downarrow & & \downarrow & & \\ \cdots & \longrightarrow & \Gamma(U_n \backslash Z_n; F^{i-1}) & \longrightarrow & \Gamma(U_n \backslash Z_n; F^i) & \longrightarrow & \cdots .\end{array}$$

Then $H^j_{Z_n}(U_n; F) \simeq H^j(E^{\cdot}_n)$, and $H^j_Z(U; F) \simeq H^j\left(\varprojlim_n E^{\cdot}_n\right)$. Since the systems $\{E^i_n\}_n$ satisfy the M-L condition for all i, the result follows from Proposition 1.12.4.

(iii) follows from the fact that $\{H^{j-1}(\operatorname{Int} X_n; F)\}_n$ satisfies the M-L condition and $\varprojlim H^j(X_n; F) \simeq \varprojlim H^j(\operatorname{Int} X_n; F)$. □

Proposition 2.7.2 (the non-characteristic deformation lemma). *Let X be a Hausdorff space, $F \in \operatorname{Ob}(\mathbf{D}^+(\mathbb{Z}_X))$, and let $\{U_t\}_{t \in \mathbb{R}}$ be a family of open subsets of X. We assume the following conditions:*

(i) $U_t = \bigcup_{s<t} U_s$, *for all $t \in \mathbb{R}$.*

(ii) *For all pairs (s,t) with $s \leqslant t$, the set $\overline{U_t \backslash U_s} \cap \operatorname{supp}(F)$ is compact.*

(iii) *Setting $Z_s = \bigcap_{t>s} (\overline{U_t \backslash U_s})$, we have for all pairs (s,t) with $s \leqslant t$, and all $x \in Z_s \backslash U_t$:*

$$(R\Gamma_{(X \backslash U_t)}(F))_x = 0 \; .$$

Then we have for all $t \in \mathbb{R}$, the isomorphism:

$$R\Gamma\left(\bigcup_s U_s; F\right) \xrightarrow{\sim} R\Gamma(U_t; F) \; .$$

Proof. Consider the assertions:

$$(a)^s_k: \quad \varinjlim_{t>s} H^k(U_t; F) \xrightarrow{\sim} H^k(U_s; F) \; ,$$

$$(b)^t_k: \quad \varprojlim_{s<t} H^k(U_s; F) \xleftarrow{\sim} H^k(U_t; F) \; .$$

Assume $(a)^s_k$ is proved for all $s \in \mathbb{R}$ and all $k \in \mathbb{Z}$, and $(b)^t_k$ is proved for all $t \in \mathbb{R}$, all $k < k_o$. By Proposition 1.12.6 we get:

$$H^k(U_t; F) \xrightarrow{\sim} H^k(U_s; F) \tag{2.7.1}$$

for all $k < k_o$, all pairs (s,t) with $s \leqslant t$. Then for t being fixed, the sequence $\{H^{k_o - 1}(U_{t-1/n}; F)\}_n$ satisfies the M-L condition, and $(b)^t_{k_o}$ is satisfied by Proposition 1.12.4. Arguing by induction on k, we find that $(b)^t_k$ will be satisfied for all $t \in \mathbb{R}$ all $k \in \mathbb{Z}$. Applying again Proposition 2.7.1 to the systems $\{H^k(U_n; F)\}_{n \in \mathbb{N}}$, we obtain the result by induction on k.

Now we shall prove $(a)^s_k$. Replacing X by $\operatorname{supp}(F)$, we may assume from the beginning that $\overline{U_t \backslash U_s}$ is compact for all $t \geqslant s$. Consider the distinguished triangles:

$$R\Gamma_{(X\setminus U_t)}(F)|_{Z_s} \longrightarrow R\Gamma_{(X\setminus U_s)}(F)|_{Z_s} \longrightarrow R\Gamma_{(U_t\setminus U_s)}(F)|_{Z_s} \underset{+1}{\longrightarrow} .$$

Since the two first terms are zero by the hypothesis, we get $R\Gamma_{(U_t\setminus U_s)}(F)|_{Z_s} = 0$. Therefore, for all $k \in \mathbb{Z}$, all $t \geqslant s$ we have (cf. Notations 2.6.8):

$$\begin{aligned} 0 &= H^k(Z_s; R\Gamma_{(U_t\setminus U_s)}(F)) \\ &= \varinjlim_{U \supset Z_s} H^k(U \cap U_t; R\Gamma_{(X\setminus U_s)}(F)) , \end{aligned}$$

where U ranges through the family of open neighborhoods of Z_s. Since for any such U there exists t' with $t \geqslant t' > s$ such that $U \cap U_t \supset U_{t'}\setminus U_s$, we get:

$$0 = \varinjlim_{t>s} H^k(U_t; R\Gamma_{(X\setminus U_s)}(F)) .$$

This implies $(a)_k^s$. □

We shall study systematically sheaves on Euclidian spaces in Chapter III, but we need a special result now.

Lemma 2.7.3. *Let I be the interval $[0, 1]$ of $\mathbb{R}$. Let F be a sheaf on I. Then:*

(i) *$H^j(I; F) = 0$ for all $j > 1$.*
(ii) *Assume $F(I) \to F_t$ is surjective for all $t \in I$. Then $H^j(I; F) = 0$ for all $j \geqslant 1$.*
(iii) *If F is the constant sheaf M_I on I, then the morphisms $M \to R\Gamma(I; F) \to F_t$ are isomorphisms for any $t \in I$.*

Proof. Let $j \geqslant 1$, *and let* $s \in H^j(I; F)$. Let f_{t_1,t_2} $(t_1 \leqslant t_2)$ be the natural map:

$$f_{t_1,t_2} : H^j(I; F) \to H^j([t_1, t_2]; F) .$$

Set $J = \{t \in [0, 1]; f_{0,t}(s) = 0\}$. Then $0 \in J$, and $0 \leqslant t' \leqslant t$, $t \in J$, implies $t' \in J$. Moreover J is open since we have for all $t_\circ < 1$:

$$H^j([0, t_\circ]; F) = \varinjlim_{t>t_\circ} H^j([0, t]; F) ,$$

and $f_{0,t_\circ}(s) = 0$ implies $f_{0,t}(s) = 0$ for some $t > t_\circ$.

Consider the Mayer-Vietoris sequence associated to the decomposition $[0, t_\circ] = [0, t] \cup [t, t_\circ]$, where $0 \leqslant t \leqslant t_\circ \leqslant 1$:

$$\cdots \to H^j([0, t_\circ]; F) \to H^j([0, t]; F) \oplus H^j([t, t_\circ]; F) \to H^j(\{t\}; F) \to \cdots .$$

For $j > 1$, or for $j = 1$ under the assumption that $F(I) \to F_t$ is surjective, we get:

$$H^j([0, t_\circ]; F) \simeq H^j([0, t]; F) \oplus H^j([t, t_\circ]; F) . \tag{2.7.2}$$

Then choose $t_\circ = \sup J$. Since $\varinjlim_{t<t_\circ} H^j([t, t_\circ]; F) = 0$, there exists $t < t_\circ$ with

$f_{t,t_0}(s) = 0$. On the other hand $f_{0,t}(s) = 0$. Thus $f_{0,t_0}(s) = 0$ by (2.7.2), and $J = [0, 1]$.

(iii) Since $H^j(I; F) = 0$ for $j > 0$ by (ii), $R\Gamma(I; F) \simeq \Gamma(I; F)$. Moreover the composition $M \to \Gamma(I; F) \to F_t \simeq M$ is the identity. Hence it is enough to show that if $s \in \Gamma(I; F)$ and $s_t = 0$, then s vanishes. This follows from the fact that supp(s) is open and closed. □

Now consider three topological spaces S, X, Y and a commutative diagram of continuous maps:

$$\text{(2.7.3)} \qquad \begin{array}{ccc} Y & \xrightarrow{\ f\ } & X \\ & {\scriptstyle p_Y}\searrow \quad \swarrow{\scriptstyle p_X} & \\ & S & \end{array} \quad .$$

In this situation, one says that $f : Y \to X$ is a continuous map over S.

Let $\mathscr{R}$ denote a sheaf of rings over S.

Let $F \in \mathrm{Ob}(\mathbf{D}^+(\mathscr{R}))$. The morphism (2.3.5) induces a morphism:

$$\text{(2.7.4)} \qquad Rp_{X*} \circ p_X^{-1}F \to Rp_{X*} \circ Rf_* \circ f^{-1} \circ p_X^{-1}F \simeq Rp_{Y*} \circ p_Y^{-1}F \ ,$$

and if f is proper, a morphism of functors

$$\text{(2.7.5)} \qquad Rp_{X!} \circ p_X^{-1}F \to Rp_{X!} \circ Rf_* \circ f^{-1} \circ p_X^{-1}F \simeq Rp_{Y!} \circ p_Y^{-1}F \ .$$

We denote by $f^\#$ (resp. $f_c^\#$) the morphism:

$$Rp_{X*} \circ p_X^{-1} \to Rp_{Y*} \circ p_Y^{-1} \qquad (\text{resp. } Rp_{X!} \circ p_X^{-1} \to Rp_{Y!} \circ p_Y^{-1})$$

given by (2.7.4) (resp. 2.7.5).

Definition 2.7.4. (i) *Let $f_0 : Y \to X$ and $f_1 : Y \to X$ be two continuous maps over S. One says f_0 and f_1 are homotopic over S if there exists a continuous map over S, $h : Y \times I \to X$, ($I = [0, 1]$), such that, denoting by j_t ($t \in I$) the map $y \mapsto (y, t)$ from Y to $Y \times I$, one has $f_i = h \circ j_i$ ($i = 0, 1$). Moreover, if h is proper, one says f_0 and f_1 are properly homotopic.*

(ii) *Let $f : Y \to X$ be a continuous map over S. We say that f is a homotopical isomorphism over S if there exists a continuous map $g : X \to Y$ over S such that $f \circ g$ and $g \circ f$ are homotopic to id_X and id_Y, respectively.*

Proposition 2.7.5. (i) *Let $f_0 : Y \to X$ and $f_1 : Y \to X$ be two homotopic maps over S. Then $f_0^\# = f_1^\#$.*

(ii) *If f_0 and f_1 are properly homotopic, then $f_{0c}^\# = f_{1c}^\#$.*

Proof. (i) Let $h : Y \times I \to X$ be a homotopy between f_0 and f_1 above S. We have a commutative diagram:

(2.7.6)
$$\begin{array}{ccc}
Y & & \\
{\scriptstyle j_i}\downarrow & \searrow^{f_i} & \\
Y \times I & \xrightarrow{h} & X \\
{\scriptstyle p}\downarrow & & \downarrow{\scriptstyle p_X} \\
Y & \xrightarrow[p_Y]{} & S\,,
\end{array}$$

where $p : Y \times I \to Y$ denotes the projection.

By Remark 2.5.3 and Lemma 2.7.3 we have isomorphisms of functors from $\mathbf{D}^+(\mathscr{R})$ to $\mathbf{D}^+(p_Y^{-1}\mathscr{R})$:

(2.7.7)
$$p_Y^{-1} \xrightarrow{\sim} Rp_* \circ p^{-1} \circ p_Y^{-1} \xrightarrow{\sim} j_i^{-1} \circ p^{-1} \circ p_Y^{-1} \simeq p_Y^{-1}$$

and the composition of all these isomorphisms is the identity on p_Y^{-1}.

We get a commutative diagram of morphisms of functors from $\mathbf{D}^+(\mathscr{R})$ to $\mathbf{D}^+(\mathscr{R})$:

$$\begin{array}{ccc}
Rp_{X*} \circ p_X^{-1} & \longrightarrow & Rp_{X*} \circ Rh_* \circ h^{-1} \circ p_X^{-1} \\
\downarrow{\scriptstyle f_i^{\#}} & & \downarrow \wr \\
Rp_{Y*} \circ f_i^{-1} \circ p_X^{-1} & & Rp_{Y*} \circ Rp_* \circ p^{-1} \circ p_Y^{-1} \\
\Vert & \swarrow_{\alpha_i} & \\
Rp_{Y*} \circ p_Y^{-1} & &
\end{array}$$

Here the morphism α_i is induced by:

$$Rp_* \circ p^{-1} \circ p_Y^{-1} \to j_i^{-1} \circ p^{-1} \circ p_Y^{-1} = p_Y^{-1} \;.$$

Hence $\alpha_0 = \alpha_1$ by (2.7.7), and $f_0^{\#} = f_1^{\#}$.

(ii) The proof is similar. □

Remark 2.7.6. Let X' (resp. Y') be a closed subset of X (resp. Y), and let $f : Y \to X$ be a map over S such that $f^{-1}(X') \subset Y'$. Let $F \in \mathrm{Ob}(\mathbf{D}^+(\mathscr{R}))$. We have the natural morphisms:

$$\begin{aligned}
Rp_{X*} \circ R\Gamma_{X'} \circ p_X^{-1}F &\to Rp_{X*} \circ Rf_* \circ f^{-1} \circ R\Gamma_{X'} \circ p_X^{-1}F \\
&\to Rp_{X*} \circ Rf_* \circ R\Gamma_{f^{-1}(X')} \circ f^{-1} \circ p_X^{-1}F \\
&\to Rp_{Y*} \circ R\Gamma_{Y'} \circ p_Y^{-1}F \;.
\end{aligned}$$

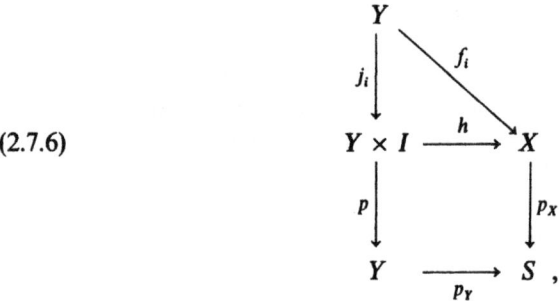

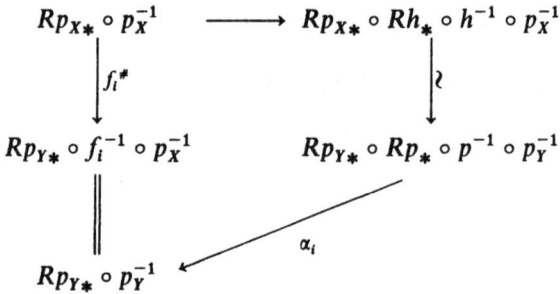

Let us denote by $f^{\#}_{Y',X'}$ this morphism of functors:

(2.7.8) $$f^{\#}_{Y',X'}: Rp_{X*} \circ R\Gamma_{X'} \circ p_X^{-1} \to Rp_{Y*} \circ R\Gamma_{Y'} \circ p_Y^{-1} \ .$$

If $Y' = f^{-1}(X')$, we write $f^{\#}_{X'}$ instead of $f^{\#}_{Y',X'}$. Now suppose that f_0, f_1, h in Definition 2.7.4 satisfy:

(2.7.9) $$\text{for each } \lambda \in I,\ h_\lambda^{-1}(X') \subset Y', \text{ where } h_\lambda(\cdot) = h(\cdot, \lambda) \ .$$

Then by modifying the proof of Proposition 2.7.5, we get:

(2.7.10) $$f^{\#}_{0Y',X'} = f^{\#}_{1Y',X'} \ .$$

Corollary 2.7.7. (i) *Let $f: Y \to X$ be a homotopical isomorphism over S. Then, for any $F \in \mathrm{Ob}(\mathbf{D}^+(\mathscr{R}))$, $f^{\#}: Rp_{X*}p_X^{-1}F \to Rp_{Y*}p_Y^{-1}F$ is an isomorphism.*

(ii) *Let $f: Y \to X$ be a continuous map and $s: X \to Y$ a continuous section of f. Assume that $s \circ f$ is homotopic to the identity over X. Then for any $F \in \mathrm{Ob}(\mathbf{D}^+(A_X))$, $F \to Rf_*f^{-1}F$ and the composition $Rf_*f^{-1}F \to Rf_*Rs_*s^{-1}f^{-1}F \simeq F$ are isomorphisms.*

(iii) *In particular, if X is a contractible topological space, then $M \xrightarrow{\sim} R\Gamma(X; M_X)$ for any $M \in \mathrm{Ob}(\mathbf{D}^+(\mathfrak{Mod}(A)))$.*

(iv) *Let $f: Y \to X$ be a continuous map. Assume f is proper with contractible fibers (and hence surjective). Then for $F \in \mathrm{Ob}(\mathbf{D}^+(A_X))$, $F \to Rf_*f^{-1}F$ is an isomorphism.*

Proof. (i) Let $g: X \to Y$ be a map as in Definition 2.7.4 (ii). Then, by Proposition 2.7.5, $f^{\#} \circ g^{\#} = \mathrm{id}$ and $g^{\#} \circ f^{\#} = \mathrm{id}$.

(ii) is a special case of (i).

(iii) is a special case of (ii).

(iv) Let $F \in \mathrm{Ob}(\mathbf{D}^+(A_X))$ and let $x \in X$. By Remark 2.5.3 we have:

$$\begin{aligned}(Rf_* \circ f^{-1}F)_x &\simeq R\Gamma(f^{-1}(x); f^{-1}F)\\ &\simeq R\Gamma(f^{-1}(x); (f|_{f^{-1}(x)})^{-1}F_x)\\ &\simeq F_x \ ,\end{aligned}$$

by (iii). □

Note that Corollary 2.7.7 (iv) is often referred to as the **Vietoris-Begle theorem**. We shall make it a little more precise, and extend it to non-proper situations.

Let $f: Y \to X$ be a continuous map. Let us denote by $\mathfrak{Mod}(A_Y/f)$ the full subcategory of $\mathfrak{Mod}(A_Y)$ consisting of sheaves G such that for each $x \in X$, the sheaf $G|_{f^{-1}(x)}$ is locally constant, and let us denote by $\mathbf{D}^+_f(A_Y)$ the full subcategory of $\mathbf{D}^+(A_Y)$ consisting of objects G such that $H^j(G) \in \mathfrak{Mod}(A_Y/f)$ for all j. Note that $\mathfrak{Mod}(A_Y/f)$ is a thick subcategory of $\mathfrak{Mod}(A_Y)$ (cf. I §7) and $\mathbf{D}^+_f(A_Y)$ is a triangulated category. The functors f_* and f^{-1} induce functors, that we still denote by f_* and f^{-1}:

$$\mathfrak{Mod}(A_X) \underset{f_*}{\overset{f^{-1}}{\rightleftarrows}} \mathfrak{Mod}(A_Y/f) \ . \tag{2.7.11}$$

The "inclusion" functor, $\mathfrak{Mod}(A_Y/f) \to \mathfrak{Mod}(A_Y)$ induces a functor: $\mathbf{D}^+(\mathfrak{Mod}(A_Y/f)) \to \mathbf{D}_f^+(A_Y)$. Hence we obtain functors (that we still denote by Rf_* and f^{-1}):

$$\mathbf{D}^+(A_X) \underset{Rf_*}{\overset{f^{-1}}{\rightleftarrows}} \mathbf{D}_f^+(A_Y) \ . \tag{2.7.12}$$

We shall give a sufficient condition in order that these functors are inverse one to each other. Assume to be given a family $\{Y_n\}_{n\in\mathbb{N}}$ of closed subsets of Y, satisfying:

$$\begin{cases} Y = \bigcup_n Y_n, \ Y_n \subset \mathrm{Int}(Y_{n+1}) & \text{for all } n \ , \\ f|_{Y_n} & \text{is proper with contractible fibers for all } n \ . \end{cases} \tag{2.7.13}$$

Proposition 2.7.8. *Under the hypothesis* (2.7.13), *the functors f^{-1} and f_* (resp. f^{-1} and Rf_*) in* (2.7.11) *(resp.* (2.7.12)*) are inverse one to each other.*

Proof. We shall only give the proof in case of formula (2.7.12), the other case being similar and simpler.

Note first that a contractible space being non empty, $f(Y_n) = X$ for all n.

Note moreover that a locally constant sheaf on a contractible space is a constant sheaf, (cf. Exercise II.4). Let us first prove the result when f is proper.

Let $G \in \mathrm{Ob}(\mathbf{D}_f^+(A_Y))$, and let $y \in Y$. Then $H^j(G)|_{f^{-1}f(y)}$ is a constant sheaf, and therefore $H^i(f^{-1}f(y); H^j(G)) = 0$ for all $j \in \mathbb{Z}$, all $i > 0$. We get:

$$\begin{aligned} H^j(f^{-1} \circ Rf_*G)_y &\simeq H^j(Rf_*G)_{f(y)} \\ &\simeq H^j(f^{-1}f(y); G) \\ &\simeq \Gamma(f^{-1}f(y); H^j(G)) \\ &\simeq H^j(G)_y \end{aligned}$$

Thus $f^{-1} \circ Rf_* \simeq \mathrm{id}_{\mathbf{D}_f^+(A_Y)}$, and we know by Corollary 2.7.7 that $Rf_* \circ f^{-1} \simeq \mathrm{id}_{\mathbf{D}^+(A_X)}$.

Now we treat the general case. Let $F \in \mathrm{Ob}(\mathbf{D}^+(A_X))$, and let V be an open subset of X. For each j, the family $\{H^j(f^{-1}(V) \cap Y_n; f^{-1}F)\}_n$ satisfies the M-L condition. Applying Lemma 2.7.1, we get:

$$\begin{aligned} H^j(V; F) &\simeq H^j(f^{-1}(V) \cap Y_n; f^{-1}F) \\ &\simeq H^j(f^{-1}(V); f^{-1}F) \ . \end{aligned}$$

Therefore $F \mapsto Rf_* \circ f^{-1}F$ is an isomorphism.

Finally let $G \in \mathrm{Ob}(\mathbf{D}_f^+(A_Y))$, let $y \in Y$ and let V be an open neighborhood of $f(y)$ in X. Since for all j, the family $\{H^j(f^{-1}(V) \cap Y_n; G)\}_n$ satisfies the M-L condition, we get:

$$H^j(f^{-1}(V); G) \simeq H^j(f^{-1}(V) \cap Y_n; G) .$$

Therefore:

$$\begin{aligned} H^j(f^{-1} \circ Rf_*G)_y &\simeq H^j(Rf_*G)_{f(y)} \\ &\simeq \varinjlim_{V \ni f(y)} H^j(f^{-1}(V); G) \\ &\simeq \varinjlim_{V \ni f(y)} H^j(f^{-1}(V) \cap Y_n; G) \\ &\simeq H^j(G)_y , \end{aligned}$$

and $G \mapsto f^{-1} \circ Rf_*G$ is an isomorphism. □

2.8. Cohomology of coverings

Let X be a topological space, and let $\mathscr{U} = \{U_j\}_{j \in J}$ be a family of open subsets of X.

For an integer $p \geqslant 0$, and $\alpha = (\alpha_0, \ldots, \alpha_p) \in J^{p+1}$, we set:

$$U_\alpha = \bigcap_{\nu=0}^{p} U_{\alpha_\nu} .$$

If σ is a permutation of $\{0, \ldots, p\}$, we denote by $\operatorname{sgn} \sigma$ its signature. If $\alpha = (\alpha_0, \ldots, \alpha_p)$ we set $\alpha^\sigma = (\alpha_{\sigma(0)}, \ldots, \alpha_{\sigma(p)})$.

Let F be a sheaf on X. For two open subsets U and V of X, with $U \subset V$, we denote by $\rho_{V,U}$ the canonical morphism $F_U \to F_V$ (cf. §3).

To $\mathscr{U}$ and F we associate a complex of sheaves on X (cf. Notations 1.3.9), denoted $\mathscr{C}_\cdot(\mathscr{U}; F)$, as follows.

For $p \in \mathbb{Z}$, $p < 0$, $\mathscr{C}_p(\mathscr{U}; F) = 0$.

For $p \geqslant 0$, $\mathscr{C}_p(\mathscr{U}; F)$ is the subsheaf of $\bigoplus_{\alpha \in J^{p+1}} F_{U_\alpha}$ consisting of alternating sections: $(s_\alpha)_{\alpha \in J^{p+1}}$ defines a section of $\mathscr{C}_p(\mathscr{U}; F)$ iff the two conditions below are satisfied.

$$\text{(2.8.1)} \qquad \begin{cases} \text{For any permutation } \sigma \text{ of } \{0, \ldots, p\} , \\ s_\alpha = (\operatorname{sgn} \sigma) s_{\alpha^\sigma} . \end{cases}$$

$$\text{(2.8.2)} \qquad \begin{cases} \text{If } \alpha = (\alpha_0, \ldots, \alpha_p) \text{ and } \alpha_\nu = \alpha_{\nu'}, \text{ for a pair of distinct } \nu \text{ and } \nu' \\ \text{in } \{0, \ldots, p\}, \text{ then } s_\alpha = 0 . \end{cases}$$

The differential $d_p : \mathscr{C}_p(\mathscr{U}; F) \to \mathscr{C}_{p-1}(\mathscr{U}; F)$ is given by:

$$(2.8.3)\qquad \begin{cases} \text{for } \beta \in J^p \text{ and } s = (s_\alpha)_\alpha \in \mathscr{C}_p(\mathscr{U};F)\ , \\ (\mathrm{d}_p(s))_\beta = \sum_{j \in J} \rho_{U_\beta, U_{(j,\beta)}}(s_{(j,\beta)})\ . \end{cases}$$

Here, $(j, \beta) = (j, \beta_0, \ldots, \beta_{p-1}) \in J^{p+1}$.

One checks immediately that:

$$(2.8.4)\qquad \mathrm{d}_{p-1} \circ \mathrm{d}_p = 0\ .$$

Example 2.8.1. (i) If $J = \{0, 1\}$, then $\mathscr{C}_.(\mathscr{U}; F)$ is the complex:

$$0 \to F_{U_1 \cap U_2} \to F_{U_1} \oplus F_{U_2} \to 0\ .$$

(ii) If $J = \{0, 1, 2\}$, then $\mathscr{C}_.(\mathscr{U}; F)$ is the complex:

$$0 \to F_{U_1 \cap U_2 \cap U_3} \to F_{U_2 \cap U_3} \oplus F_{U_3 \cap U_1} \oplus F_{U_1 \cap U_2} \to F_{U_1} \oplus F_{U_2} \oplus F_{U_3} \to 0\ .$$

Set $U = \bigcup_j U_j$. One defines the "augmentation map", $\delta : \mathscr{C}_.(\mathscr{U}; F) \to F_U$, by setting $\delta|_{F_{U_i}} = \rho_{U, U_i}$. Then one verifies immediately:

$$(2.8.5)\qquad \delta \circ \mathrm{d}_1 = 0\ .$$

By this relation, δ induces a morphism of complexes $\mathscr{C}_.(\mathscr{U}; F) \to F_U$, where F_U is identified to the complex $\cdots \to 0 \to F_U \to 0 \to \cdots$.

Lemma 2.8.2. *The morphism $\mathscr{C}_.(\mathscr{U}; F) \to F_U$ is a quasi-isomorphism.*

Proof. Let $\tilde{\mathscr{C}}_.(\mathscr{U}; F)$ denote the augmented complex $\mathscr{C}_.(\mathscr{U}; F) \to F$ (i.e: the complex $\cdots \xrightarrow[\mathrm{d}_1]{} \mathscr{C}_0(\mathscr{U}; F) \xrightarrow[\delta]{} F \longrightarrow 0 \longrightarrow \cdots$). Let $j_\circ \in J$ and let $J' = J \setminus \{j_\circ\}$, $\mathscr{U}' = \{U_j\}_{j \in J'}$. The morphism $\rho_{X, U_{j_\circ}}$ induces a morphism:

$$\alpha_{j_\circ} : \tilde{\mathscr{C}}_.(\mathscr{U}'; F_{U_{j_\circ}}) \to \tilde{\mathscr{C}}_.(\mathscr{U}'; F)\ .$$

Let us note first that $\tilde{\mathscr{C}}_.(\mathscr{U}; F)$ is the mapping cone of $\alpha_{j_\circ}$. In fact let us denote by $\bigoplus'_{\alpha \in J^{p+1}} F_{U_\alpha}$ the subsheaf of $\bigoplus_{\alpha \in J^{p+1}} F_{U_\alpha}$ consisting of sections satisfying (2.8.1) and (2.8.2). Then:

$$\begin{aligned} \tilde{\mathscr{C}}_p(\mathscr{U}; F) &= \bigoplus_{\alpha \in J^{p+1}}{}' F_{U_\alpha} \\ &= \Big(\bigoplus_{\alpha = (j_\circ, \beta), \beta \in J^p}{}' F_{U_{j_\circ} \cap U_\beta}\Big) \oplus \Big(\bigoplus_{\alpha \in J'^{p+1}}{}' F_{U_\alpha}\Big) \\ &= \tilde{\mathscr{C}}_{p-1}(\mathscr{U}'; F_{U_{j_\circ}}) \oplus \tilde{\mathscr{C}}_p(\mathscr{U}'; F) \end{aligned}$$

and one checks that the differential of $\tilde{\mathscr{C}}_p(\mathscr{U}; F)$ is that of the mapping cone of $\alpha_{j_\circ}$. Since $\alpha_{j_\circ}|_{U_{j_\circ}}$ is an isomorphism, the complex $\tilde{\mathscr{C}}_.(\mathscr{U}; F)|_{U_{j_\circ}}$ is homotopic to zero. Therefore $\mathscr{C}_.(\mathscr{U}; F) \to F_U$ is a quasi-isomorphism on U, and both terms being zero outside U, this completes the proof. $\square$

Remark 2.8.3. We have in fact proved that, locally on U, $\mathscr{C}_\cdot(\mathscr{U};F)$ is homotopic to F.

Now we assume $\mathscr{U}$ is an open covering of X. We set:

$$\mathscr{C}^\cdot(\mathscr{U};F) = \mathscr{H}om_{\mathbb{Z}_X}(\mathscr{C}_\cdot(\mathscr{U};\mathbb{Z}_X),F) \ , \tag{2.8.6}$$

$$C^\cdot(\mathscr{U};F) = \mathrm{Hom}_{\mathbb{Z}_X}(\mathscr{C}_\cdot(\mathscr{U};\mathbb{Z}_X),F) \ . \tag{2.8.7}$$

Then $\Gamma(X;\mathscr{C}^\cdot(\mathscr{U};F)) = C^\cdot(\mathscr{U};F)$. Moreover since $\mathrm{Hom}_{\mathbb{Z}_X}(\mathbb{Z}_U;F) = \Gamma(U;F)$, $C^p(\mathscr{U};F)$ is the submodule of $\prod_{\alpha\in J^{p+1}} F(U_\alpha)$ consisting of sections which satisfy (2.8.1)–(2.8.2).

Proposition 2.8.4. *The augmentation map δ induces a quasi-isomorphism of sheaves*: $F \xrightarrow{\sim} \mathscr{C}^\cdot(\mathscr{U};F)$.

Proof. Apply Remark 2.8.3 to the sheaf $\mathbb{Z}_X$. □

Proposition 2.8.5 (the Leray acyclic covering theorem). *Let $\mathscr{U} = \{U_j\}_{j\in J}$ be an open covering of X and let F be a sheaf on X. Assume that for all $\alpha \in J^p$ $(p \geqslant 0)$, all $k > 0$, $H^k(U_\alpha;F) = 0$. Then $R\Gamma(X;F)$ is quasi-isomorphic to $C^\cdot(\mathscr{U};F)$.*

Proof. Let $I^\cdot$ be a complex of flabby sheaves quasi-isomorphic to F. Then $R\Gamma(X;F) \simeq \Gamma(X;I^\cdot)$. It follows from Proposition 2.8.4 and Theorem 1.9.3 that $\Gamma(X;I^\cdot)$ is quasi-isomorphic to $s(\Gamma(X;\mathscr{C}^\cdot(\mathscr{U};I^\cdot)))$, where $s(\cdot)$ means the simple complex associated to the double complex. Since $\Gamma(X;\mathscr{C}^\cdot(\mathscr{U};I^\cdot)) \simeq C^\cdot(\mathscr{U};I^\cdot)$, it remains to note that $C^\cdot(\mathscr{U};F)$ is quasi-isomorphic to $s(C^\cdot(\mathscr{U};I^\cdot))$ by Theorem 1.9.3 and the hypothesis. □

Remark 2.8.6. The cohomology of coverings $H^n(C^\cdot(\mathscr{U};F))$ is often referred to as the **Čech cohomology**.

2.9. Examples of sheaves on real and complex manifolds

We shall explain here a few examples on the majority of which we shall come back at length in Chapter XI.

2.9.1. The sheaf $\mathscr{C}_X^0$. On a topological space X, the presheaf which associates to an open subset U of X the space $\mathscr{C}^0(U)$ of complex valued continuous functions, with the usual operation of restriction of functions, is clearly a sheaf. We denote it by $\mathscr{C}_X^0$. Note that the constant sheaf $\mathbb{Z}_X$ may be identified to the subsheaf of $\mathscr{C}_X^0$ of functions with values in $\mathbb{Z}$.

2.9.2. The sheaf $\mathscr{L}^1_{loc,dx}$. Let U be an open subset of the Euclidian space $\mathbb{R}^n$, and let $L^1(U;dx)$ denote the space of integrable functions on U, with respect to

dx, the Lebesgue measure on $\mathbb{R}^n$. The presheaf $U \mapsto L^1(U; dx)$ is not a sheaf. The associated sheaf on $\mathbb{R}^n$ is denoted $\mathscr{L}^1_{loc,dx}$.

2.9.3. Ringed spaces. A ringed space $(X, \mathscr{A}_X)$ is a topological space X endowed with a sheaf of rings $\mathscr{A}_X$. A morphism of ringed spaces, $f:(Y, \mathscr{A}_Y) \to (X, \mathscr{A}_X)$ is a continuous map $f: Y \to X$ together with a morphism of sheaves of rings $f^{-1}\mathscr{A}_X \to \mathscr{A}_Y$. If A is a ring and $\mathscr{A}_X$ is a sheaf of A-algebras (i.e. we have a morphism $A_X \to \mathscr{A}_X$), then $(X, \mathscr{A}_X)$ is called an A-ringed space.

2.9.4. C^α-manifolds. Let α be an integer $(0 \leqslant \alpha < \infty)$ or $\alpha = \infty$ or else $\alpha = \omega$. We denote by $\mathscr{C}^\alpha_{\mathbb{R}^n}$ the sheaf on $\mathbb{R}^n$ of complex valued functions of class C^α (C^ω means real analytic). A real manifold M of class C^α of dimension n is a locally compact space M countable at infinity endowed with a sheaf of rings $\mathscr{C}^\alpha_M$ such that $(M, \mathscr{C}^\alpha_M)$ is locally isomorphic to $(\mathbb{R}^n, \mathscr{C}^\alpha_{\mathbb{R}^n})$ as a $\mathbb{C}$-ringed space.

One denotes by $\dim X$ (or $\dim_{\mathbb{R}} X$), the dimension of a real manifold X. Note that in the literature one usually denotes by $\mathscr{A}_M$ the sheaf $\mathscr{C}^\omega_M$.

We refer to Guillemin – Pollack [1] for a basic course on differential geometry.

2.9.5. Orientation, forms and densities. On a C^0-manifold M, we shall also have to consider the orientation sheaf or_M. This sheaf is locally isomorphic to $\mathbb{Z}_M$, and the choice of an orientation on M (if it exists) is equivalent to the choice of an isomorphism $or_M \simeq \mathbb{Z}_M$. We shall study or_M carefully in the next chapter.

Now assume $\alpha = \infty$ or $\alpha = \omega$. Let p be an integer. We denote by $\mathscr{C}^{\alpha,(p)}_M$ the sheaf of differential forms of degree p with coefficients in $\mathscr{C}^\alpha_M$. We denote by $\mathrm{d}: \mathscr{C}^{\alpha,(p)}_M \to \mathscr{C}^{\alpha,(p+1)}_M$ the exterior derivative.

Recall that if $(x_1, \ldots, x_n)$ is a system of local coordinates on M, then a p-form f can be uniquely written:

$$f = \sum_{|I|=p} f_I \, \mathrm{d}x_I ,$$

where $I = \{i_1, \ldots, i_p\} \subset \{1, \ldots, n\}$, $(i_1 < i_2 < \cdots < i_p)$, $\mathrm{d}x_I = dx_{i_1} \wedge \cdots \wedge dx_{i_p}$, and f_I is a section of $\mathscr{C}^\alpha_M$. Then:

$$\mathrm{d}f = \sum_{i=1}^{n} \sum_{|I|=p} \frac{\partial f_I}{\partial x_i} dx_i \wedge dx_I .$$

We also introduce the sheaf:

$$\mathscr{V}^\alpha_M = \mathscr{C}^{\alpha,(n)}_M \otimes or_M$$

($\alpha = \infty$ or $\alpha = \omega$), and call it the sheaf of C^α-densities on M.

One can integrate C^∞-densities with compact support. We denote by $\int_M \cdot$ the integration map:

$$\int_M \cdot : \Gamma_c(M; \mathscr{V}^\infty_M) \to \mathbb{C} . \tag{2.9.1}$$

The sheaves $\mathscr{C}^{\alpha,(p)}_M$ and $\mathscr{V}^\alpha_M$ are sheaves of $\mathscr{C}^\alpha_M$-modules.

From the existence of "partitions of unity", one obtains that the sheaves $\mathscr{C}_M^\alpha$, $\mathscr{C}_M^{\alpha,(p)}$, $\mathscr{V}_M^\alpha$ are c-soft for $\alpha \neq \omega$. The sheaves $\mathscr{C}_M^\omega$, $\mathscr{C}_M^{\omega,(p)}$, $\mathscr{V}_M^\omega$ are acyclic with respect to the functor $\Gamma(M;\cdot)$, (i.e. $H^j(M;\mathscr{C}_M^\omega) = 0$ for $j > 0$. Cf. Grauert [1]).

2.9.6. Distributions and hyperfunctions. On a C^∞-manifold M, one also defines naturally the sheaf $\mathscr{D}\mathscr{b}_M$ of Schwartz distributions (cf. Schwartz [2], de Rham [1]). Recall that $\mathscr{D}\mathscr{b}_M$ is a c-soft sheaf, and $\Gamma_c(M;\mathscr{D}\mathscr{b}_M)$ is the topological dual of $\Gamma(M;\mathscr{V}_M^\infty)$, this last space being endowed with its natural topology of Fréchet space.

On a C^ω-manifold M, one can define similarly the sheaf $\mathscr{B}_M$ of Sato hyperfunctions (cf. Sato [1]). This is a flabby sheaf, and $\Gamma_c(M;\mathscr{B}_M)$ is the topological dual of $\Gamma(M;\mathscr{V}_M^\omega)$, this last space being endowed with its natural topology of DFS-space, (cf. Martineau [1] or Schapira [1] for a detailed exposition). However Sato's construction is purely cohomological, and we shall recall it in the subsection 2.9.13 below.

The integration map (2.9.1) defines a pairing:

$$(2.9.2)\qquad \begin{cases} \Gamma(M;\mathscr{C}_M^\infty) \times \Gamma_c(M;\mathscr{V}_M^\infty) \to \mathbb{C} \\ (f,g) \mapsto \int_M fg \end{cases} .$$

By this pairing, one gets a morphism of sheaves from $\mathscr{C}_M^\infty$ to $\mathscr{D}\mathscr{b}_M$, and one proves that this morphism is injective. Moreover on a real analytic manifold M, the injection $\Gamma(M;\mathscr{V}_M^\omega) \to \Gamma(M;\mathscr{V}_M^\infty)$ induces a morphism $\mathscr{D}\mathscr{b}_M \to \mathscr{B}_M$, which is also injective.

One may also introduce the sheaves $\mathscr{D}\mathscr{b}_M^{(p)} = \mathscr{C}_M^{\infty,(p)} \otimes_{\mathscr{C}_M^\infty} \mathscr{D}\mathscr{b}_M$ (resp. $\mathscr{B}_M^{(p)} = \mathscr{C}_M^{\omega,(p)} \otimes_{\mathscr{C}_M^\omega} \mathscr{B}_M$) of p-forms with distributions (resp. hyperfunctions) as coefficients. These are c-soft (resp. flabby) sheaves.

2.9.7. De Rham complexes. Let M be a C^∞-manifold. By the *Poincaré lemma*, the sequence:

$$(2.9.3)\qquad 0 \to \mathbb{C}_M \to \mathscr{C}_M^{\infty,(0)} \underset{d}{\to} \cdots \to \mathscr{C}_M^{\infty,(n)} \to 0$$

is exact. Hence the sheaf $\mathbb{C}_M$ is quasi-isomorphic to a complex of c-soft sheaves:

$$(2.9.4)\qquad \mathbb{C}_M \underset{qis}{\longrightarrow} (0 \longrightarrow \mathscr{C}_M^{\infty,(0)} \underset{d}{\longrightarrow} \cdots \longrightarrow \mathscr{C}_M^{\infty,(n)} \longrightarrow 0) .$$

This permits to calculate explicitly the cohomology groups $H^j(M;\mathbb{C}_M)$ or $H_c^j(M;\mathbb{C}_M)$. For example, by applying $R\Gamma(M;\cdot)$ to (2.9.4) one gets:

$$(2.9.5)\qquad R\Gamma(M;\mathbb{C}_M) \simeq (0 \longrightarrow \Gamma(M;\mathscr{C}_M^{\infty,(0)}) \underset{d}{\longrightarrow} \cdots \longrightarrow \Gamma(M;\mathscr{C}_M^{\infty,(n)}) \longrightarrow 0) .$$

The same results hold with $\mathscr{C}_M^\infty$ replaced by $\mathscr{D}\mathscr{b}_M$. If M is real analytic, the same results hold with $\mathscr{C}_M^\infty$ replaced by $\mathscr{C}_M^\omega$ or $\mathscr{B}_M$, but one shall notice that $\mathscr{C}_M^\omega$ is not any more c-soft but only $\Gamma(M;\cdot)$-acyclic. On the other hand $\mathscr{B}_M$ is not only c-soft but also flabby, which permits to calculate the cohomology group $H_Z^j(M;\mathbb{C}_M)$ for a locally closed set Z in M.

The complex (2.9.3) is referred to as the *de Rham complex* on M.

2.9.8. Complex manifolds. Let $\mathcal{O}_{\mathbb{C}^n}$ denote the sheaf of holomorphic functions on $\mathbb{C}^n$. A complex manifold X of dimension n is a $\mathbb{C}$-ringed space $(X, \mathcal{O}_X)$ locally isomorphic to $(\mathbb{C}^n, \mathcal{O}_{\mathbb{C}^n})$.

One denotes by $\dim_{\mathbb{C}} X$ the dimension of a complex manifold X. We refer to Wells [1] for basic notions on complex differential geometry, and we recommend the book by Banica-Stanasila [1] for further developments on analytic geometry.

We denote by $\mathcal{O}_X^{(p)}$ the sheaf of holomorphic p-forms on X, and by ∂ the holomorphic differential. One sometimes sets:

$$\Omega_X = \mathcal{O}_X^{(n)} \otimes \mathit{or}_X \ , \tag{2.9.6}$$

where or_X is the orientation sheaf on X (cf. Chapter III). The Poincaré lemma being valid with holomorphic coefficients, the sheaf $\mathbb{C}_X$ is quasi-isomorphic to the complex:

$$0 \longrightarrow \mathcal{O}_X^{(0)} \underset{\partial}{\longrightarrow} \cdots \longrightarrow \mathcal{O}_X^{(n)} \longrightarrow 0 \ . \tag{2.9.7}$$

2.9.9. Dolbeault complexes. Let $(X, \mathcal{O}_X)$ be a complex manifold. We denote by $(\bar{X}, \mathcal{O}_{\bar{X}})$ the topological space X endowed with the sheaf $\mathcal{O}_{\bar{X}}$ of anti-holomorphic functions on X. (Recall that $f: X \to \mathbb{C}$ is anti-holomorphic if its composite with the complex conjugation map on $\mathbb{C}$, is holomorphic.) Hence $(\bar{X}, \mathcal{O}_{\bar{X}})$ is another complex manifold.

We denote by $X^{\mathbb{R}}$ the underlying real analytic manifold. By identifying $X^{\mathbb{R}}$ to the diagonal of $X \times \bar{X}$, we find that $X \times \bar{X}$ is a complexification of $X^{\mathbb{R}}$. In fact:

$$\mathcal{O}_{X \times \bar{X}}|_{X^{\mathbb{R}}} \simeq \mathscr{C}^{\omega}_{X^{\mathbb{R}}} \ . \tag{2.9.8}$$

On $X \times \bar{X}$ we can consider the holomorphic differential ∂ of X and $\bar{\partial}$ of $\bar{X}$. Hence the differential d on $X \times \bar{X}$ is decomposed as $\mathrm{d} = \partial + \bar{\partial}$, which induces a splitting of the sheaves $\mathscr{C}^{\alpha,(r)}_{X^{\mathbb{R}}}$ ($\alpha = \infty$ or $\alpha = \omega$), as

$$\mathscr{C}^{\alpha,(r)}_{X^{\mathbb{R}}} = \bigoplus_{p+q=r} \mathscr{C}^{\alpha,(p,q)}_X \ ,$$

where $\mathscr{C}^{\alpha,(p,q)}_X$ is the sheaf of (p,q)-forms on X. In a system of local holomorphic coordinates $(z_1, \ldots, z_n)$ on X, a section f of $\mathscr{C}^{\alpha,(p,q)}_X$ is uniquely written as:

$$f = \sum_{|I|=p, |J|=q} f_{I,J}\, \mathrm{d}z_I \wedge \mathrm{d}\bar{z}_J \ ,$$

where $\mathrm{d}z_I = \mathrm{d}z_{i_1} \wedge \cdots \wedge \mathrm{d}z_{i_p}$, $\mathrm{d}\bar{z}_J = \mathrm{d}\bar{z}_{j_1} \wedge \cdots \wedge \mathrm{d}\bar{z}_{j_q}$, similarly to the notations of §9.5. In particular:

$$\partial f = \sum_I \sum_J \sum_{i=1}^n \frac{\partial f_{I,J}}{\partial z_i} \mathrm{d}z_i \wedge \mathrm{d}z_I \wedge \mathrm{d}\bar{z}_J \ .$$

The Dolbeault lemma asserts that the complex:

$$0 \longrightarrow \mathcal{O}_X^{(p)} \longrightarrow \mathcal{C}_X^{\infty,(p,0)} \xrightarrow[\bar\partial]{} \mathcal{C}_X^{\infty,(p,1)} \longrightarrow \cdots \longrightarrow \mathcal{C}_X^{\infty,(p,n)} \longrightarrow 0$$

is exact, and there is a similar result with $\mathcal{C}_X^{\infty,(p,q)}$ replaced by $\mathcal{C}_X^{\omega,(p,q)}$, or $\mathcal{D}b_X^{(p,q)}$ or else $\mathcal{B}_X^{(p,q)}$. In particular we find that $\mathcal{O}_X^{(p)}$ is quasi-isomorphic to the complex of flabby sheaves (cf. Komatsu [1] or Schapira [1]):

$$0 \longrightarrow \mathcal{B}_X^{(p,0)} \xrightarrow[\bar\partial]{} \cdots \longrightarrow \mathcal{B}_X^{(p,n)} \longrightarrow 0 \ . \tag{2.9.9}$$

As shown by Golovin [1], this is a complex of injective $\mathcal{O}_X$-modules.

2.9.10. Operations on $\mathcal{O}_X$. Let $f\colon Y \to X$ be a morphism of complex manifolds. By the definition of a morphism of ringed spaces, f induces a morphism $f^{-1}\mathcal{O}_X \to \mathcal{O}_Y$. There is another morphism which can be defined in the derived category $\mathbf{D}^+(\mathbb{C}_X)$:

$$Rf_!\Omega_Y[\dim_{\mathbb{C}} Y] \to \Omega_X[\dim_{\mathbb{C}} X] \ . \tag{2.9.10}$$

This morphism can be described as follows.

Let $n = \dim_{\mathbb{C}} X$, $m = \dim_{\mathbb{C}} Y$, $l = m - n$. The morphism: $f^{-1}\mathcal{C}_X^{\infty,(m-p,m-q)} \to \mathcal{C}_Y^{\infty,(m-p,m-q)}$ defines, by duality, a morphism:

$$f_!\mathcal{D}b_Y^{(p,q)} \otimes or_Y \to \mathcal{D}b_X^{(p-l,q-l)} \otimes or_X \ . \tag{2.9.11}$$

Then (2.9.10) is deduced from (2.9.11) and the Dolbeault resolutions of Ω_Y and Ω_X.

2.9.11. Cohomology of $\mathcal{O}_X$. We refer to Hörmander [1] for a detailed study of the cohomology of $\mathcal{O}_X$. Let us only recall that if Ω is open in $\mathbb{C}^n$, then one says that Ω is pseudo-convex if $H^j(\Omega; \mathcal{O}_X) = 0$ for all $j > 0$. For example, convex domains are pseudo-convex, and if $n = 1$, all domains are pseudo-convex. This last result generalizes as follows:

$$\begin{cases} \textit{if } \Omega \textit{ is open in } \mathbb{C}^n, & \textit{then} \\ H^j(\Omega; \mathcal{O}_{\mathbb{C}^n}) = 0 & \textit{for all } j \geqslant n \ , \end{cases} \tag{2.9.12}$$

(cf. Malgrange [1] who obtains (2.9.12) using Dolbeault's resolution and the fact that the equation $\left(\sum_{j=1}^{n} \frac{\partial}{\partial z_j}\frac{\partial}{\partial \bar z_j}\right) f = g$ is always solvable in $\Gamma(\Omega; \mathcal{C}_{\mathbb{R}^{2n}}^{\infty})$).

Let X be a complex manifold of dimension n, Z a locally closed subset of X, $x \in Z \setminus \operatorname{Int} Z$. Then:

$$H_Z^j(\mathcal{O}_X)_x = 0 \qquad \text{for } j \notin [1, n] \ . \tag{2.9.13}$$

In fact for $j = 0$ this is nothing but the "analytic continuation principle", and for $j > n$ this follows from (2.9.12) or else from (2.9.9) (i.e.: $\mathcal{O}_X$ has flabby dimension n, cf. Exercise II.9).

There is a useful criterium for the vanishing of $H^j_Z(\mathcal{O}_X)$, due to Martineau [2], and Kashiwara (cf. Sato-Kawai-Kashiwara [1]). Let $X = \mathbb{C}^n$, and let Z be a closed convex subset of X and $x \in Z$.

$$\text{(2.9.14)} \qquad \begin{cases} \textit{If there exist no affine complex space } L \textit{ of dimension} \\ d \textit{ through } x, \textit{ such that } L \cap Z \textit{ is a neighborhood} \\ \textit{of } x \textit{ in } L, \textit{ then } H^j_Z(\mathcal{O}_X)_x = 0 \textit{ for } j \leqslant n - d \ . \end{cases}$$

2.9.12. Boundary values of holomorphic functions. Let Ω be a strictly pseudo-convex open subset of $\mathbb{C}^n$, with C^2-boundary. Recall that, locally on $\partial\Omega$, there exists a holomorphic change of coordinates which interchanges Ω with a strictly convex open subset of $\mathbb{C}^n$.

Let j be the embedding $\Omega \hookrightarrow \bar{\Omega}$. We have a distinguished triangle on $\bar{\Omega}$:

$$\text{(2.9.15)} \qquad \mathcal{O}_X|_{\bar{\Omega}} \to Rj_*\mathcal{O}_\Omega \longrightarrow R\Gamma_{\partial\Omega}(\mathcal{O}_X|_{\bar{\Omega}}) \xrightarrow[+1]{} \ .$$

Since $H^0_{\partial\Omega}(\mathcal{O}_X|_{\bar{\Omega}}) = 0$, the presheaf $U \mapsto H^1_{U \cap \partial\Omega}(U \cap \bar{\Omega}; \mathcal{O}_X|_{\bar{\Omega}})$ is equal to the sheaf $H^1_{\partial\Omega}(\mathcal{O}_X|_{\bar{\Omega}})$, (cf. Exercise II.13). Moreover, $H^k_{\partial\Omega}(\mathcal{O}_X|_{\bar{\Omega}}) = 0$ for $k > 1$, since $R^k j_*\mathcal{O}_\Omega = 0$ for $k > 0$. We also have:

$$\text{(2.9.16)} \qquad \textit{the sheaf } H^1_{\partial\Omega}(\mathcal{O}_X|_{\bar{\Omega}}) \textit{ is flabby} \ .$$

To prove (2.9.16), we may assume by Proposition 2.4.10 that Ω is strictly convex. Let U be a convex open subset of $\mathbb{C}^n$. By applying the functor $R\Gamma(U; \cdot)$ to the triangle (2.9.15), we find:

$$\Gamma(U \cap \bar{\Omega}; H^1_{\partial\Omega}(\mathcal{O}_X|_{\bar{\Omega}})) \simeq \mathcal{O}_X(\Omega \cap U)/\mathcal{O}_X(\bar{\Omega} \cap U) \ .$$

In fact, $H^k(U \cap \bar{\Omega}; \mathcal{O}_X) = 0$ for $k > 0$, since $U \cap \bar{\Omega}$ admits a fundamental system of convex open neighborhoods in U.

Let ω be an open subset of $\partial\Omega$. There exists a convex open subset U of $\mathbb{C}^n$, such that $U \cap \bar{\Omega} = \omega$ and $U \cup \Omega$ is convex. Then the Mayer-Vietoris sequence:

$$0 \to \mathcal{O}_X(U \cup \Omega) \to \mathcal{O}_X(U) \oplus \mathcal{O}_X(\Omega) \to \mathcal{O}_X(U \cap \Omega) \to 0$$

is exact, and the map $\mathcal{O}_X(\Omega)/\mathcal{O}_X(\bar{\Omega}) \to \mathcal{O}_X(\Omega \cap U)/\mathcal{O}_X(\bar{\Omega} \cap U)$ is surjective, which proves (2.9.16).

2.9.13. Sato hyperfunctions. Let M be a real analytic manifold of dimension n, and let X be a complexification of M. (Recall that X is uniquely defined in a neighborhood of M.) The sheaf $\mathcal{B}_M$ of Sato's hyperfunctions is defined as:

$$\text{(2.9.17)} \qquad \mathcal{B}_M = H^n_M(\mathcal{O}_X) \otimes or_{M/X} \ ,$$

where $or_{M/X} = or_M \otimes or_X$ (cf. Chapter III).

Note that in view of (2.9.14), the complex $R\Gamma_M(\mathcal{O}_X)[n]$ is concentrated in degree 0, and we have:

$$\mathcal{B}_M \simeq R\Gamma_M(\mathcal{O}_X)[n] \otimes or_{M/X} \ .$$

Since the sheaves $H^j_M(\mathcal{O}_X)$ are zero for $j < n$, we find (Exercise II.13) that the presheaf $U \mapsto H^n_{U \cap M}(U; \mathcal{O}_X)$ is a sheaf, and is equal to $\mathscr{B}_M$. (We shall often identify the sheaf $\mathscr{B}_M$ on X and its restriction to M.) Moreover it follows from (2.9.12) that this sheaf is flabby. This sheaf coincides with the sheaf described in section 2.9.6 (for more details, cf. Chapter XI).

2.9.14. An example of a locally constant sheaf. Take $X = \mathbb{C}$, and let z be a holomorphic coordinate on X. Let α be a complex number, and let P be the holomorphic differential operator $z\frac{\partial}{\partial z} - \alpha$. Consider the complex of sheaves on X:

(2.9.18) $$F := 0 \longrightarrow \mathcal{O}_X \xrightarrow[P]{} \mathcal{O}_X \longrightarrow 0 \ .$$

The sheaf $H^0(F)|_{X \setminus \{0\}} \simeq \mathrm{Ker}(P)|_{X \setminus \{0\}}$ is a locally constant sheaf, since on each open connected and simply connected subset U of $X \setminus \{0\}$, $H^0(F)|_U$ is isomorphic to the constant sheaf $\mathbb{C}_U$ generated by (a branch of) z^α. However, if $\alpha \notin \mathbb{Z}$, $\Gamma(X \setminus \{0\}; H^0(F)) = 0$, since there exists no non-zero holomorphic function f on $X \setminus \{0\}$, solution of $Pf = 0$. One has for $\alpha \notin \mathbb{Z}$:

$$H^0(F)|_{X \setminus \{0\}} : \text{locally constant sheaf of rank one} \ .$$

$$H^0(F)|_{\{0\}} = 0 \ ,$$

$$H^1(F) = 0 \ .$$

When $\alpha = 0, 1, 2, \ldots$, we have:

$$H^0(F) \simeq \mathbb{C}_X \ ,$$

$$H^1(F) \simeq \mathbb{C}_{\{0\}} \ .$$

When $\alpha = -1, -2, \ldots$, we have:

$$H^0(F) \simeq \mathbb{C}_{X \setminus \{0\}} \qquad \text{and} \qquad H^1(F) = 0 \ .$$

The complex F is a simple example of what is called a "perverse sheaf". Such complexes will be studied in Chapters VIII and X.

Exercises to Chapter II

Exercise II.1. Let $\underset{\sim}{\mathbb{N}}$ be the set $\mathbb{N}$ endowed with the topology for which the open subsets are the sets $\{0, 1, \ldots, n\}$, $(n \geqslant -1)$ and $\mathbb{N}$. After identifying a presheaf on $\underset{\sim}{\mathbb{N}}$ and a projective system of groups on $\mathbb{N}$, prove that a presheaf F is a sheaf iff $\Gamma(\underset{\sim}{\mathbb{N}}; F) = \varprojlim F$. Also prove that

(a) $H^j(\underset{\sim}{\mathbb{N}}; F) = 0$ for $j \neq 0, 1$.

(b) $H^1(\mathbb{N}; F) = \{\{x_n\}_{n\in\mathbb{N}}; x_n \in F_n\}/\{\{x_n\}_{n\in\mathbb{N}};\ x_n \in F_n$ and there are $y_n \in F_n$ with $x_n = y_n - y_{n+1}\}$.

Exercise II.2. Let A and B be two closed subsets of X with $A \cup B = X$. For $F \in \mathrm{Ob}(\mathbf{D}^+(X))$, prove the natural isomorphism $(R\Gamma_B(F))_A \simeq R\Gamma_B(F_A)$.

Exercise II.3. (i) Let U be an open subset of X, $x \in \bar{U}\backslash U$. By considering the sheaf $F = Z_U$, prove that the formula $(\mathscr{H}om(F, G))_x \simeq \mathrm{Hom}(F_x, G_x)$ is false in general.

(ii) Give an example of a sheaf F on X, a closed subset Z of X, an open subset U of X, such that $Z \cap U = \emptyset$ but $R\Gamma_Z(F_U) \neq 0$. Observe that $\Gamma_Z(F_U) = 0$, and hence in this case the derived functor of the composition is not the composition of the derived functors.

Exercise II.4. Prove that a locally constant sheaf F on a contractible space X is a constant sheaf.

Exercise II.5. Let X be a paracompact space. One says that a sheaf F on X is **soft** if for all closed subset Z of X, the natural map $\Gamma(X; F) \to \Gamma(Z; F)$ is surjective. Prove that if F is soft, then $H^i(X; F) = 0$ for $i > 0$.

Exercise II.6. Let X be a locally compact space and let F be a sheaf on X.
(a) Prove that F is c-soft if and only if $H^i_c(U; F) = 0$ for all $i > 0$, all open subset U of X; (one writes $H^i_c(U; F)$ instead $H^i_c(U; F|_U)$).
(b) Assume moreover that X is countable at infinity. Prove that if F is c-soft, then F is soft.
(c) Prove that the property of being c-soft is local on X.

Exercise II.7. Let $\mathscr{R}$ be a c-soft sheaf of rings. Prove that any sheaf of $\mathscr{R}$-modules is c-soft.

Exercise II.8. Let X be a locally compact space countable at infinity. A sheaf F on X is said to be **supple** if for any pair of closed subsets Z_1 and Z_2 of an open subset U of X, $\Gamma_{Z_1}(U; F) \oplus \Gamma_{Z_2}(U; F) \to \Gamma_{Z_1 \cup Z_2}(U; F)$ is surjective (cf. Bengel-Schapira [1]).
(a) Prove that flabby sheaves are supple.
(b) Prove that if a sheaf F is supple then $\Gamma_Z(F)$ is c-soft for any closed subset Z.
(c) Prove that the property of being supple is a local property on X.

(Note: on a real analytic manifold M, the sheaf $\mathscr{D}\ell_M$ is c-soft but not supple. On the other hand the quotient sheaf $\mathscr{D}\ell_M/\mathscr{A}_M$ is supple but not flabby – the sheaves $\mathscr{D}\ell_M$ and $\mathscr{A}_M$ are those defined in §9.)

Exercise II.9. Let X be a topological space.
(a) For a sheaf F, and a non-negative integer n, prove that the following conditions are equivalent.

(i) There exists an exact sequence $0 \to F \to F^0 \to \cdots \to F^n \to 0$ where the F^j's are flabby.
(ii) If $0 \to F \to F^0 \to \cdots \to F^n \to 0$ is exact and if F^j is flabby for $0 \leqslant j < n$, then F^n is flabby.
(iii) For any closed set S, $H^k_S(X; F) = 0$ for $k > n$.
(iv) For any locally closed set S, $H^k_S(X;F) = 0$ for $k > n$.
(v) For any closed set S, $H^k_S(F) = 0$ for $k > n$.
(vi) For any locally closed set S, $H^k_S(F) = 0$ for $k > n$.

The smallest n which satisfies these equivalent conditions is called the **flabby dimension** of F. The maximum of the flabby dimensions of all the sheaves F on X is called the flabby dimension of X.

(b) For a locally compact space X, define the **c-soft dimension** of a sheaf F on X and the c-soft dimension of X and give the corresponding equivalent conditions (i)–(iv) in this case.
(c) Prove that, for a sheaf F and a locally compact space X the c-soft dimension of $F \leqslant$ the flabby dimension of $F \leqslant 1 +$ the c-soft dimension of F.

Exercise II.10. Let $\mathscr{R}$ be a sheaf of rings on X, and let $M \in \mathrm{Ob}(\mathfrak{Mod}(\mathscr{R}))$.
(a) Prove that M is injective if and only if for any sub-$\mathscr{R}$-module $\mathscr{I}$ of $\mathscr{R}$ (we say $\mathscr{I}$ is an ideal of $\mathscr{R}$), the natural homomorphism:

$$\Gamma(X; M) \simeq \mathrm{Hom}_{\mathscr{R}}(\mathscr{R}, M) \to \mathrm{Hom}_{\mathscr{R}}(\mathscr{I}, M)$$

is surjective.
(b) Let A be a field. Prove that any ideal of A_X is isomorphic to a sheaf A_U, where U is open in X. Deduce from this that an A_X-module M is injective iff the sheaf M is flabby.

Exercise II.11. Let $f: Y \to X$ be a continuous map of locally compact spaces. Let G be a sheaf on Y. Prove that the two conditions below are equivalent.
(a) For any $x \in X$, $G|_{f^{-1}(x)}$ is c-soft.
(b) For any open subset U of Y, any $j > 0$, $R^j f_! G_U = 0$.

Exercise II.12. Let X be a topological space.

(a) Let $(F_\lambda)_{\lambda \in \Lambda}$ be an inductive system of sheaves on X indexed by a directed ordered set Λ. Assuming X is compact and Hausdorff, prove the isomorphisms $\varinjlim_\lambda H^k(X; F_\lambda) \simeq H^k(X; \varinjlim_\lambda F_\lambda)$ for all $k \in \mathbb{N}$.

(b) Let $(F_n)_{n \in \mathbb{N}}$ be a projective system of sheaves on X and let Z be a locally closed subset of X. Assuming that $\{H^{k-1}_Z(X; F_n)\}_n$ satisfies the M-L condition, prove the isomorphism $H^k_Z(X; \varprojlim_n F_n) \xrightarrow{\sim} \varprojlim_n H^k_Z(X; F_n)$.

Exercise II.13. Let F be a sheaf on X, and let Z be a locally closed subset of X. Assume $R^j\Gamma_Z(F) = 0$ for $j < n$. Prove that the presheaf $U \mapsto H^n_Z(U; F)$ is a sheaf, and is equal to $R^n\Gamma_Z(F)$, (cf. Exercise I.22).

Exercise II.14. Let $X = \bigcup_{i \in I} U_i$ be an open covering of X.

For each $i \in I$, let F_i be a sheaf on U_i, and for each pair (i, j) let ϕ_{ij} be an isomorphism $F_j|_{U_i \cap U_j} \xrightarrow{\sim} F_i|_{U_i \cap U_j}$. Assume that $\phi_{ii} = \mathrm{id}_{F_i}$ and $\phi_{ij} \circ \phi_{jk} = \phi_{ik}$ on

$U_i \cap U_j \cap U_k$. Then prove that there exists a sheaf F on X and for each i, isomorphisms $\phi_i : F|_{U_i} \simeq F_i$, such that $\phi_{ij} = \phi_i \circ \phi_j^{-1}$ on $U_i \cap U_j$, for all pair (i, j). Prove that F is unique up to isomorphism.

Exercise II.15. (i) Let F be a bounded from below complex of sheaves on X. Construct the natural morphism:

$$H^j(\Gamma(X; F^\cdot)) \to H^j(R\Gamma(X; F^\cdot)) \ .$$

(ii) Let $\mathscr{U} = \{U_i\}_i$ be an open covering of X and let F be a sheaf on X. Construct the canonical morphism:

$$H^j(C^\cdot(\mathscr{U}; F)) \to H^j(X; F) \ .$$

Exercise II.16. Let A be a commutative ring, $A^\times$ the group of invertible elements of A. Let X be a topological space, $\mathscr{U} = \{U_i\}_{i \in I}$ an open covering of X, and let $c \in C^2(\mathscr{U}; A_X^\times)$, with $\delta c = 0$ (cf. §8). Let c′ be the class of c in $H^2(C^\cdot(\mathscr{U}; A_X^\times))$ and c″ the image of c′ in $H^2(X; A_X^\times)$, (cf. Exercise II.15). One defines a category $\mathfrak{Sh}(X; c)$ whose objects are the families $\{F_i, \rho_{ij}\}$, F_i is an A_{U_i}-module and $\rho_{ij} : F_j|_{U_i \cap U_j} \simeq F_i|_{U_i \cap U_j}$ an isomorphism, such that:

$$\rho_{ij}\rho_{jk}\rho_{ki} = c_{ijk} \cdot \mathrm{id}_{F_i|U_i \cap U_j \cap U_k} \qquad \text{for all } i, j, k \ ,$$

and the morphisms of this category are defined in the obvious way.

(i) Prove that $\mathfrak{Sh}(X; c)$ is an abelian category.

(ii) Let $\tilde{c}$ be another element of $C^2(\mathscr{U}; A_X^\times)$ and assume $\tilde{c}'' = c''$. Prove that there exists an equivalence of categories between $\mathfrak{Sh}(X; c)$ and $\mathfrak{Sh}(X; \tilde{c})$.

Exercise II.17. Let X be a locally compact space, $\mathscr{R}$ a sheaf of commutative rings on X with $\mathrm{wgld}(\mathscr{R}) < \infty$, and let Z_1 and Z_2 be two locally closed subsets of X.

(i) For F_1 and F_2 in $\mathbf{D}^+(\mathscr{R})$, construct the natural morphism:

$$R\Gamma_{Z_1}(F_1) \overset{L}{\underset{\mathscr{R}}{\otimes}} R\Gamma_{Z_2}(F_2) \to R\Gamma_{Z_1 \cap Z_2}\left(F_1 \overset{L}{\underset{\mathscr{R}}{\otimes}} F_2\right) .$$

(ii) Assume $\mathscr{R} = A_X$ (where A is a commutative ring). Construct the morphism:

$$R\Gamma_{Z_1}(X; F_1) \overset{L}{\underset{A}{\otimes}} R\Gamma_{Z_2}(X; F_2) \to R\Gamma_{Z_1 \cap Z_2}\left(X; F_1 \overset{L}{\underset{A}{\otimes}} F_2\right)$$

and the morphisms $(p, q \in \mathbb{Z})$:

$$H^p_{Z_1}(X; F_1) \underset{A}{\otimes} H^q_{Z_2}(X; F_2) \to H^{p+q}_{Z_1 \cap Z_2}\left(X; F_1 \overset{L}{\underset{A}{\otimes}} F_2\right) .$$

This last morphism is usually called the **cup product**. (Hint: use Exercise I.24.)

Exercise II.18. Let $f_i: Y_i \to X_i$, $i = 1, 2$ be continuous maps of locally compact spaces over a space S (cf. (2.7.3)). Let p_{Y_i} denote the map $Y_i \to S$ and set $f = f_1 \times_S f_2 : Y_1 \times_S Y_2 \to X_1 \times_S X_2$. Let $\mathscr{R}$ be a sheaf of commutative rings on S, with $\mathrm{wgld}(\mathscr{R}) < \infty$, and let $G_i \in \mathrm{Ob}(\mathbf{D}^+(p_{Y_i}^{-1}\mathscr{R}))$.

(i) Prove the isomorphism:

$$Rf_{1!}G_1 \underset{S}{\overset{L}{\boxtimes}}{}_{\mathscr{R}} Rf_{2!}G_2 \simeq Rf_!\left(G_1 \underset{S}{\overset{L}{\boxtimes}}{}_{\mathscr{R}} G_2\right).$$

This isomorphism is known as the **Künneth formula.** (Hint: reduce the problem to the case where $S = X_1 = X_2 = \{\mathrm{pt}\}$. Then use isomorphism (2.6.19), similarly to the proof of Proposition 3.1.15 below.)

(ii) Assume $S = X_1 = X_2 = \{\mathrm{pt}\}$ and $\mathscr{R}$ is a field. Prove:

$$H_c^n(Y_1 \times Y_2; G_1 \boxtimes G_2) \simeq \bigoplus_{p+q=n} (H_c^p(Y_1; G_1) \otimes H_c^q(Y_2; G_2)).$$

(Hint: use Exercise I.24.)

Exercise II.19. Assume X is locally compact and A is commutative with $\mathrm{wgld}(A) < \infty$. Let $F \in \mathrm{Ob}(\mathbf{D}^+(A_X))$, let Ω (resp. Z) be an open (resp. closed) subset of X and denote by a_X the map $X \to \{\mathrm{pt}\}$. Prove the isomorphisms:

$$R\Gamma(\Omega; F) \simeq Ra_{X*}R\mathscr{H}om(A_\Omega, F),$$

$$R\Gamma_c(\Omega; F) \simeq Ra_{X!}\left(A_\Omega \overset{L}{\otimes} F\right),$$

$$R\Gamma_Z(X; F) \simeq Ra_{X*}R\mathscr{H}om(A_Z, F),$$

$$R\Gamma_c(Z; F) \simeq Ra_{X!}\left(A_Z \overset{L}{\otimes} F\right).$$

Exercise II.20. (cf. Exercise IX.10). Let A be a commutative ring with $\mathrm{wgld}(A)$ finite and let E be a real finite-dimensional vector space. On $\mathrm{Ob}(\mathbf{D}^+(A_E))$ one defines the convolution operation by setting $F * G = Rs_!(F \boxtimes^L G)$, where s is the map $E \times E \to E$, $(x, y) \mapsto x + y$.

(a) Prove that for F, G, H in $\mathbf{D}^+(A_E)$ we have $F * G \simeq G * F$, $F * (G * H) \simeq (F * G) * H$, $A_{\{0\}} * F \simeq F$.

(b) Let Z_1 and Z_2 be two compact convex subsets of E. Prove that $A_{Z_1} * A_{Z_2} \simeq A_{Z_1 + Z_2}$.

(c) Prove that $A_\gamma * A_{\mathrm{Int}\,\gamma} = 0$, where γ is a proper closed convex cone.

(d) Assume $E = \mathbb{R}^n$. Let $Z_1 = [-1, 1]^n$, $Z_2 =]-1, 1[^n$. Prove that $A_{Z_1} * A_{Z_2} \simeq A_{\{0\}}[-n]$.

Exercise II.21. Let X be a topological space, $\{X_n\}_{n \in \mathbb{Z}}$ a decreasing sequence of closed subsets with $\bigcap_n X_n = \varnothing$ and $X_n = X$ for $n \ll 0$. Let $F \in \mathrm{Ob}(\mathbf{D}^+(X))$

satisfying: $H^k_{X_n \setminus X_{n+1}}(F) = 0$ for $k \neq n$. Using the distinguished triangle $R\Gamma_{X_{n+1}\setminus X_{n+2}}(F) \longrightarrow R\Gamma_{X_n\setminus X_{n+2}}(F) \longrightarrow R\Gamma_{X_n\setminus X_{n+1}}(F) \xrightarrow[+1]{}$, one defines d^n: $H^n_{X_n\setminus X_{n+1}}(F) \longrightarrow H^{n+1}_{X_{n+1}\setminus X_{n+2}}(F)$ and one sets $K^n = H^n_{X_n\setminus X_{n+1}}(F)$.

(i) Prove that $(K^\cdot, \mathrm{d}^\cdot)$ is a complex of sheaves on X.

(ii) Prove that $H^k_{X_n}(F) = 0$ for $k < n$ and prove the isomorphism $H^n_{X_{n-1}}(F) \xrightarrow{\sim} H^n(F)$.

(iii) Let $G^n = \Gamma_{X_n}(F^n) \cap (\mathrm{d}^n_F)^{-1}(\Gamma_{X_{n+1}}(F^{n+1}))$, where F is a complex of sheaves $\{F^n, \mathrm{d}^n_F\}_{n\in\mathbb{Z}}$. Construct the morphism $\mathrm{d}^n_G : G^n \to G^{n+1}$, prove that $G = (G^\cdot, \mathrm{d}^\cdot_G)$ is a complex, construct the morphisms $G \to K$ and $G \to F$ and prove that if all the F^n are flabby these are quasi-isomorphisms. Conclude that $F \simeq K$ in $\mathbf{D}^+(X)$. (See Hartshorne [1].)

Exercise II.22. Let $X = U_1 \cup U_2$ be a covering, either open or closed, of X and let $F_i \in \mathrm{Ob}(\mathbf{D}^+(U_i))$, $i = 1, 2$. Let ϕ be an isomorphism $F_1|_{U_1\cap U_2} \xrightarrow{\sim} F_2|_{U_1\cap U_2}$ in $\mathbf{D}^+(U_1 \cap U_2)$. Construct $F \in \mathrm{Ob}(\mathbf{D}^+(X))$ and isomorphisms $\psi_i : F|_{U_i} \simeq F_i$ such that $\psi_2|_{U_1\cap U_2} = \phi \circ \psi_1|_{U_1\cap U_2}$.

Exercise II.23. Let $X = \mathbb{R}^n$ with $n \geqslant 1$ and let k be a commutative field. Prove that if $P \in \mathrm{Ob}(\mathfrak{Mod}(k_X))$ is projective, then $P = 0$. (Hint: assuming $P \neq 0$, show that there exists a non-empty open subset U of X such that k_U is a direct sumand of P and deduce a contradiction.)

Exercise II.24. Let k be a commutative ring. In this exercise, by "a ring" we mean a ring A together with a morphism $k \to A$ such that the image of k is in the center of A. Let X be a topological space and let $\mathscr{A}, \mathscr{B}, \mathscr{C}, \mathscr{D}$ be sheaves of rings over X. We assume that $\mathscr{A}, \mathscr{B}, \mathscr{C}, \mathscr{D}$ are flat over k.

(i) Prove that a flat (resp. injective) $(\mathscr{A} \otimes_k \mathscr{B})$-module is flat (resp. injective) over $\mathscr{A}$.

(ii) Prove that the functor $\cdot \otimes_{\mathscr{B}} \cdot$:

$$\mathfrak{Mod}\left(\mathscr{A} \underset{k}{\otimes} \mathscr{B}^{op}\right) \times \mathfrak{Mod}\left(\mathscr{B} \underset{k}{\otimes} \mathscr{C}^{op}\right) \to \mathfrak{Mod}\left(\mathscr{A} \underset{k}{\otimes} \mathscr{C}^{op}\right)$$

has a left derived functor $\cdot \otimes^L_{\mathscr{B}} \cdot$:

$$\mathbf{D}^-\left(\mathscr{A} \underset{k}{\otimes} \mathscr{B}^{op}\right) \times \mathbf{D}^-\left(\mathscr{B} \underset{k}{\otimes} \mathscr{C}^{op}\right) \to \mathbf{D}^-\left(\mathscr{A} \underset{k}{\otimes} \mathscr{C}^{op}\right)$$

and prove the isomorphism in $\mathbf{D}^-(\mathscr{A} \otimes_k \mathscr{D}^{op})$:

$$\left(K \overset{L}{\underset{\mathscr{B}}{\otimes}} M\right) \overset{L}{\underset{\mathscr{C}}{\otimes}} N \simeq K \overset{L}{\underset{\mathscr{B}}{\otimes}} \left(M \overset{L}{\underset{\mathscr{C}}{\otimes}} N\right)$$

for $K \in \mathrm{Ob}(\mathbf{D}^-(\mathscr{A} \otimes_k \mathscr{B}^{op}))$, $M \in \mathrm{Ob}(\mathbf{D}^-(\mathscr{B} \otimes_k \mathscr{C}^{op}))$, $N \in \mathrm{Ob}(\mathbf{D}^-(\mathscr{C} \otimes_k \mathscr{D}^{op}))$.

(iii) Prove that the functor $\mathcal{H}om_{\mathcal{A}}(\cdot,\cdot)$:

$$\mathfrak{Mod}\left(\mathcal{A}\underset{k}{\otimes}\mathcal{B}^{op}\right)^{\circ}\times\mathfrak{Mod}\left(\mathcal{A}\underset{k}{\otimes}\mathcal{C}^{op}\right)\to\mathfrak{Mod}\left(\mathcal{B}\underset{k}{\otimes}\mathcal{C}^{op}\right)$$

has a right derived functor $R\mathcal{H}om_{\mathcal{A}}(\cdot,\cdot)$:

$$\mathbf{D}^{-}\left(\mathcal{A}\underset{k}{\otimes}\mathcal{B}^{op}\right)^{\circ}\times\mathbf{D}^{+}\left(\mathcal{A}\underset{k}{\otimes}\mathcal{C}^{op}\right)\to\mathbf{D}^{+}\left(\mathcal{B}\underset{k}{\otimes}\mathcal{C}^{op}\right),$$

and prove the isomorphism in $\mathbf{D}^{+}(\mathcal{C}\otimes_k\mathcal{D}^{op})$

$$R\,\mathcal{H}om_{\mathcal{A}}\left(M\overset{L}{\underset{\mathcal{B}}{\otimes}}N,K\right)\simeq R\mathcal{H}om_{\mathcal{B}}(N,R\mathcal{H}om_{\mathcal{A}}(M,K))$$

for $M\in\mathrm{Ob}(\mathbf{D}^{-}(\mathcal{A}\otimes_k\mathcal{B}^{op}))$, $N\in\mathrm{Ob}(\mathbf{D}^{-}(\mathcal{B}\otimes_k\mathcal{C}^{op}))$ and $K\in\mathrm{Ob}\cdot(\mathbf{D}^{+}(\mathcal{A}\otimes_k\mathcal{D}^{op}))$.

(iv) Let $f: Y\to X$ be a continuous map and $\mathcal{B}$ and $\mathcal{A}$ be sheaves of rings over Y and X, respectively, both flat over k. Assuming the existence of $n\in\mathbb{N}$ such that $\mathrm{wgld}(\mathcal{A}_x)\leqslant n$ for all $x\in X$, construct for $K\in\mathrm{Ob}(\mathbf{D}^{b}(\mathcal{B}\otimes_k f^{-1}\mathcal{A}^{op}))$ a morphism, functorial with respect to $M\in\mathrm{Ob}(\mathbf{D}^{-}(\mathcal{A}))$ and $N\in\mathrm{Ob}(\mathbf{D}^{+}(\mathcal{A}))$:

$$f^{-1}R\mathcal{H}om_{\mathcal{A}}(M,N)\to R\mathcal{H}om_{\mathcal{B}}\left(K\overset{L}{\underset{f^{-1}\mathcal{A}}{\otimes}}f^{-1}M,K\overset{L}{\underset{f^{-1}\mathcal{A}}{\otimes}}f^{-1}N\right)$$

in $\mathbf{D}^{+}(k_Y)$.

Exercise II.25. Let X be a topological space, S a closed subset, and let $\{S_k\}_{k\in Z}$ be a decreasing sequence of closed subsets of S such that $S_k=S$ for $k\ll 0$ and $S_k=\emptyset$ for $k\gg 0$. Let $F\in\mathrm{Ob}(\mathbf{D}^{+}(X))$ and let $r\in\mathbb{Z}$. Assume that $H^j_{S_k}(F)|_{S_k\setminus S_{k+1}}=0$ for $j<r$ and for all k. Prove that $H^j_S(F)=0$ for any $j<r$.

(Hint: arguing by induction on k, assume $H^j_S(F)|_{X\setminus S_{k-1}}=0$ for $j<r$ and consider the distinguished triangle:

$$R\Gamma_{(S_{k-1}\setminus S_k)}(F)|_{X\setminus S_k}\longrightarrow R\Gamma_S(F)|_{X\setminus S_k}\longrightarrow Ri_*i^{-1}R\Gamma_S(F)\xrightarrow[+1]{}$$

where i denotes the open embedding $X\setminus S_{k-1}\to X\setminus S_k$.)

Exercise II.26. Let $F\in\mathrm{Ob}(\mathbf{D}^{b}(\mathbb{Z}_X))$ and let $\mathcal{U}=\{U_j\}_{j=1,\dots,r}$ be a finite open covering of X. Construct the natural morphism:

$$H^0\left(\bigcap_j U_j;F\right)\to H^{r-1}(X;F).$$

(Hint: apply Exercise II.15.)

Notes

We refer to the "Short History" by C. Houzel, at the beginning of this book, for a detailed history of sheaf theory.

Let us only briefly recall that sheaves were introduced by Leray [1, 2] around 1945, as well as spectral sequences, the basic tool at that time for studying sheaf cohomology. The theory took its full importance under the impulse of Cartan [1], and perhaps can one date the modern formalism of cohomology of sheaves with the paper of Grothendieck [1] at Tohoku's. The book of Godement [1] also greatly contributed to the popularity of the theory, and the importance of the functorial operations on sheaves in the frame of derived categories was emphasized by Grothendieck, first in the coherent case (Grothendieck [4]), and then in the case of discrete coefficients ([SGA 4], [SGA 5]).

In the course of Chapter II, the reader will have met "Čech cohomology", the "Künneth formula", the "Mayer-Vietoris sequences", "the Vietoris-Begle theorem", "the Poincaré lemma", "de Rham complexes" etc. Among other important contributions to sheaf theory let us mention the introduction of flabby sheaves and soft sheaves by Godement, and the introduction of local cohomology (the functor $\Gamma_Z(\cdot)$ and its derived functor) by Grothendieck [1]. It should also be mentioned that the "non-characteristic deformation lemma" (Proposition 2.7.2), which will play a crucial role in Chapter V, is due to Kashiwara [3, 5]. It is a variation on Morse theory, based on the so-called Mittag-Leffler theorem of Grothendieck [3].

Chapter III. Poincaré-Verdier duality and Fourier-Sato transformation

Summary

In this chapter we introduce two fundamental tools for the study of sheaves on manifolds. First, following Verdier [1], we construct a right adjoint $f^!$ to the functor $Rf_!$. If $f : Y \to X$ is a continuous map of locally compact spaces satisfying suitable conditions, and if F (resp. G) belongs to $\mathbf{D}^+(A_X)$ (resp. $\mathbf{D}^+(A_Y)$), one gets the formula:

(3.0.1) $$\mathrm{Hom}(Rf_!G, F) = \mathrm{Hom}(G, f^!F) .$$

This generalizes the classical Poincaré duality. In fact if X is reduced to a single point, if $F = A$ and if Y is a topological manifold, one calculates $f^!A$ (called "the dualizing complex" on Y), and shows it is isomorphic to the orientation sheaf on Y, shifted by the dimension. With the functor $f^!$ in hands, it is then possible to derive many nice new formulas of sheaf theory.

Secondly, we make a detailed study of the Fourier-Sato transformation, an operation which interchanges conic sheaves (in the derived category) on a vector bundle and conic sheaves on the dual vector bundle.

In the course of this chapter we study sheaves on topological manifolds: orientation sheaf, flabby dimension, cohomologically constructible sheaves, and we also introduce the γ-topology, as a preparation to Chapter V.

For another approach to the contents of §1–§4 one may also consult Borel et al [1], Gelfand-Manin [1], Iversen [1] and [SHS].

Much of the material of this chapter may be considered as classical, though it had not received a systematic treatment beforehand.

Convention 3.0. In this, and all forthcoming chapters with the exception of Chapter XI, in order to make the presentation less intricate we shall work with sheaves of A_X-modules on a topological space X, where the base ring A is assumed to be commutative with finite global dimension. (Recall, cf. Exercise I.29, that $\mathrm{wgld}(A) \leqslant \mathrm{gld}(A)$.)

If there is no risk of confusion we write $\mathbf{D}^+(X)$ or $\mathbf{D}^b(X)$ instead of $\mathbf{D}^+(A_X)$ or $\mathbf{D}^b(A_X)$ respectively. We also write $F \otimes^L G$ and $R\mathcal{H}om(F, G)$ instead of $F \otimes^L_{A_X} G$ and $R\mathcal{H}om_{A_X}(F, G)$.

All manifolds are finite-dimensional and countable at infinity. A submanifold is always locally closed.

In this chapter, all topological spaces are assumed locally compact except those concerned with the γ-topology.

When composing two functors, such as for example $f^! \circ Rf_!$, we shall not always write the symbol $\circ$.

3.1. Poincaré-Verdier duality

Let $f: Y \to X$ be a continuous map of locally compact spaces. The aim of this section is to construct a right adjoint $f^!$ to the functor $Rf_!: \mathbf{D}^+(A_Y) \to \mathbf{D}^+(A_X)$. The functor $f^!$ will thus satisfy:

$$\text{(3.1.1)} \qquad \mathrm{Hom}_{\mathbf{D}^+(A_X)}(Rf_!G, F) = \mathrm{Hom}_{\mathbf{D}^+(A_Y)}(G, f^!F)$$

for $F \in \mathrm{Ob}(\mathbf{D}^+(A_X))$, $G \in \mathrm{Ob}(\mathbf{D}^+(A_Y))$.

If one considers the particular case where Y is an n-dimensional oriented manifold, $X = \{\mathrm{pt}\}$, $A = \mathbb{Q}$, $G = \mathbb{Q}_Y$, $F = \mathbb{Q}_{\{\mathrm{pt}\}}$, we shall see that $f^!\mathbb{Q}_{\{\mathrm{pt}\}} \simeq \mathbb{Q}_Y[n]$, hence:

$$\mathrm{Hom}(R\Gamma_c(Y; \mathbb{Q}_Y)[n], \mathbb{Q}) \simeq R\Gamma(Y; \mathbb{Q}_Y) .$$

Taking the j-th cohomology group, we get:

$$\text{(3.1.2)} \qquad (H_c^{n-j}(Y; \mathbb{Q}_Y))^* \simeq H^j(Y; \mathbb{Q}_Y) ,$$

where * means the dual vector space over $\mathbb{Q}$. This is the classical Poincaré duality.

Before entering into the construction of $f^!$, let us explain it heuristically. For $F \in \mathrm{Ob}(\mathbf{D}^+(A_X))$ and for V an open subset of Y, we have:

$$\begin{aligned} R\Gamma(V; f^!F) &= R\mathrm{Hom}(A_V, f^!F) \\ &= R\mathrm{Hom}(Rf_!A_V, F) \end{aligned}$$

and this last complex can be calculated by taking a c-soft resolution K of A_X. Then $Rf_!A_V = f_!K_V$, and if F is a complex of injective sheaves, we get:

$$R\Gamma(V; f^!F) = \mathrm{Hom}(f_!K_V, F) .$$

To perform this construction, we shall need some assumptions on f.

Definition 3.1.1. *A sheaf G on Y is called f-soft if for any $x \in X$, the sheaf $G|_{f^{-1}(x)}$ is c-soft.*

In view of Exercise II.6, G is f-soft if and only if for any open subset V of Y and any $j \neq 0$, $R^jf_!G_V = 0$.

To construct the functor $f^!$ we assume:

(3.1.3) *the functor $f_!: \mathfrak{Mod}(\mathbb{Z}_Y) \to \mathfrak{Mod}(\mathbb{Z}_X)$ has finite cohomological dimension.*

This means that there exists an integer $r \geqslant 0$ such that $R^j f_! = 0$ for $j > r$. This is also equivalent to one of the following conditions (cf. Exercise I.19):

$$\text{(3.1.4)} \qquad \begin{cases} \textit{for any } G \in \mathrm{Ob}(\mathfrak{Sh}(Y)), \textit{ there exists an exact} \\ \textit{sequence } 0 \to G \to G^0 \to \cdots \to G^r \to 0, \textit{ where the } G^j\textit{'s} \\ \textit{are } f\textit{-soft} \;, \end{cases}$$

$$\text{(3.1.4)}' \qquad \begin{cases} \textit{for any exact sequence in } \mathfrak{Sh}(Y): \\ G^0 \to \cdots \to G^r \to 0, \textit{ with } G^j \; f\textit{-soft for } j < r, \\ \textit{the sheaf } G^r \textit{ is } f\textit{-soft} \;. \end{cases}$$

Note that $f_!$ has cohomological dimension $\leqslant r$ if and only if, for any $x \in X$, $\Gamma_c(f^{-1}(x); \cdot)$ has cohomological dimension $\leqslant r$.

Let K be a $\mathbb{Z}_Y$-module and F an A_X-module. Let us define the presheaf $f_K^! F$ of A-modules by:

$$(f_K^! F)(V) = \mathrm{Hom}_{A_X}\left(f_!\left(A_Y \underset{\mathbb{Z}_Y}{\otimes} K_V\right), F\right) .$$

(Recall that for two A_X-modules F and G, we write $\mathrm{Hom}_{A_X}(G, F)$ instead of $\mathrm{Hom}_{\mathfrak{Mod}(A_X)}(G, F)$.)

For $V' \subset V$, $f_!(A_Y \otimes_{\mathbb{Z}_Y} K_{V'}) \to f_!(A_Y \otimes_{\mathbb{Z}_Y} K_V)$ defines the restriction map $(f_K^! F)(V) \to (f_K^! F)(V')$.

Lemma 3.1.2. *Let K be a flat and f-soft $\mathbb{Z}_Y$-module.*

(i) *For any sheaf G on Y, $G \otimes_{\mathbb{Z}_Y} K$ is f-soft.*
(ii) *$G \mapsto f_!(G \otimes_{\mathbb{Z}_Y} K)$ is an exact functor from $\mathfrak{Mod}(\mathbb{Z}_Y)$ to $\mathfrak{Mod}(\mathbb{Z}_X)$.*

Proof. (i) Any sheaf G on Y has a resolution:

$$\to G^{-r} \to \cdots \to G^0 \to G \to 0 \;,$$

where each G^j is a direct product of sheaves $\mathbb{Z}_V$, V open in Y, by the construction in the proof of Proposition 2.4.12. Hence $G^j \otimes_{\mathbb{Z}_Y} K$ is an f-soft sheaf. Since

$$\to G^{-r} \underset{\mathbb{Z}_Y}{\otimes} K \to \cdots \to G^0 \underset{\mathbb{Z}_Y}{\otimes} K \to G \underset{\mathbb{Z}_Y}{\otimes} K \to 0$$

is exact, (3.1.4)′ gives the desired result, taking r large enough.

(ii) follows immediately from (i). □

Lemma 3.1.3. *Let K be a flat and f-soft $\mathbb{Z}_Y$-module and let F be an injective A_X-module.*

(i) *The presheaf $f_K^! F$ is a sheaf and this sheaf is an injective A_Y-module.*
(ii) *We have a canonical isomorphism, functorial with respect to $G \in \mathrm{Ob}(\mathfrak{Mod}(A_Y))$:*

$$\mathrm{Hom}_{A_X}\left(f_!\left(G \underset{\mathbb{Z}_Y}{\otimes} K\right), F\right) \simeq \mathrm{Hom}_{A_Y}(G, f_K^! F) .$$

Proof. In the proof, we shall write $\cdot \otimes K$ instead of $\cdot \otimes_{\mathbb{Z}_Y} K$.

Let us first prove that $f_K^! F$ is a sheaf. Let $V = \bigcup V_j$ be an open covering of an open subset V of Y. Then there is an exact sequence:

$$\bigoplus_{j,k} A_{V_j \cap V_k} \to \bigoplus_j A_{V_j} \to A_V \to 0 \; .$$

Applying Lemma 3.1.2 (ii), we get the exact sequence:

$$f_!\left(\bigoplus_{j,k} A_{V_j \cap V_k} \otimes K\right) \to f_!\left(\bigoplus_j A_{V_j} \otimes K\right) \to f_!(A_V \otimes K) \to 0 \; .$$

Since F is injective, the sequence:

$$0 \to \mathrm{Hom}_{A_X}(f_!(A_V \otimes K), F) \to \mathrm{Hom}_{A_X}\left(f_!\left(\bigoplus_j A_{V_j} \otimes K\right), F\right)$$

$$\to \mathrm{Hom}_{A_X}\left(f_!\left(\bigoplus_{j,k} A_{V_j \cap V_k} \otimes K\right), F\right)$$

is exact, and this last sequence is isomorphic to the sequence:

$$0 \to (f_K^! F)(V) \to \prod_j (f_K^! F)(V_j) \to \prod_{j,k} (f_K^! F)(V_j \cap V_k) \; .$$

This shows that $f_K^! F$ is a sheaf.

Now let us define a homomorphism:

$$\alpha(G) : \mathrm{Hom}_{A_X}(f_!(G \otimes K), F) \to \mathrm{Hom}_{A_Y}(G, f_K^! F) \; .$$

Let $\phi \in \mathrm{Hom}_{A_X}(f_!(G \otimes K), F)$. Then for any open set V of Y, we have a chain of morphisms of A_X-modules:

$$\begin{aligned} G(V) \underset{A}{\otimes} f_!(A_Y \otimes K_V) &\longrightarrow f_!(G \otimes K_V) \\ &\longrightarrow f_!(G \otimes K) \\ &\xrightarrow{\phi} F \; . \end{aligned}$$

This gives a morphism from $G(V)$ to $(f_K^! F)(V) = \mathrm{Hom}(f_!(A_Y \otimes K_V), F)$, and this morphism being functorial with respect to V open in Y, we obtain an element $\alpha(G)(\phi) \in \mathrm{Hom}_{A_Y}(G, f_K^! F)$.

We shall prove that $\alpha(G)$ is an isomorphism in three steps (a), (b), (c).

(a) When $G = A_V$, V open in Y, $\alpha(G)$ is an isomorphism. In fact,

$$\begin{aligned} \mathrm{Hom}_{A_X}(f_!(G \otimes K), F) &= \mathrm{Hom}_{A_X}(f_!(A_V \otimes K), F) \\ &\simeq (f_K^! F)(V) \\ &\simeq \mathrm{Hom}_{A_Y}(G, f_K^! F) \; . \end{aligned}$$

(b) When $G = \bigoplus_j A_{V_j}$, for a family of open subsets V_j of Y, $\alpha(G)$ is an isomorphism. In fact this follows from (a) since $\alpha(G) = \prod_j \alpha(A_{V_j})$.
(c) Now let G be an arbitrary A_Y-module. Then by Proposition 2.4.12, there is an exact sequence $0 \to G'' \to G' \to G \to 0$, where $G' \simeq \bigoplus_j A_{V_j}$. Hence $\alpha(G')$ is an isomorphism. Consider the commutative diagram:

$$\begin{array}{ccccccc} 0 \longrightarrow & \mathrm{Hom}(f_!(G \otimes K), F) & \longrightarrow & \mathrm{Hom}(f_!(G' \otimes K), F) & \longrightarrow & \mathrm{Hom}(f_!(G'' \otimes K), F) & \\ & \downarrow{\scriptstyle \alpha(G)} & & \downarrow{\scriptstyle \alpha(G')} & & \downarrow{\scriptstyle \alpha(G'')} & \\ 0 \longrightarrow & \mathrm{Hom}(G, f_K^! F) & \longrightarrow & \mathrm{Hom}(G', f_K^! F) & \longrightarrow & \mathrm{Hom}(G'', f_K^! F) & . \end{array}$$

Both rows are exact by Lemma 3.1.2. Since $\alpha(G')$ is bijective, $\alpha(G)$ is injective. Applying this result to the sheaf G'', we find that $\alpha(G'')$ is injective. Hence $\alpha(G)$ is bijective.

As a consequence of this fact, we obtain that $\mathrm{Hom}_{A_Y}(\cdot, f_K^! F)$ is an exact functor on $\mathfrak{Mod}(A_Y)$. Therefore $f_K^! F$ is injective. □

Let $f_!$ have cohomological dimension $\leqslant r$.

Lemma 3.1.4. *The sheaf $\mathbb{Z}_Y$ admits a resolution $0 \to \mathbb{Z}_Y \to K^0 \to \cdots \to K^r \to 0$ such that all K^j's are flat and f-soft $\mathbb{Z}_Y$-modules.*

Proof. We shall construct a resolution of $\mathbb{Z}_Y$ by the same method as for the proof of Proposition 2.4.3. Let $p: \hat{Y} \to Y$ be the natural map from $\hat{Y}$, the set Y endowed with the discrete topology, to Y. Then we define: $K^0 = p_* p^{-1} \mathbb{Z}_Y$, $K^1 = p_* p^{-1}(K^0/\mathbb{Z}_Y)$, $\ldots$, $K^j = p_* p^{-1}(\mathrm{Coker}(K^{j-2} \to K^{j-1}))$, for $1 < j < r$ and $K^r = \mathrm{Coker}(K^{r-2} \to K^{r-1})$.

Then we have the exact sequence $0 \to \mathbb{Z}_Y \to K^0 \to \cdots \to K^r \to 0$, the K^j's are flabby for $0 \leqslant j < r$, hence f-soft, and it follows from (3.1.4)$'$ that K^r is also f-soft.

Thus it remains to show that the K^j's are flat $\mathbb{Z}_Y$-modules for $0 \leqslant j \leqslant r$. In order to prove it, we shall show that if G is flat then $p_* p^{-1} G$ and $p_* p^{-1} G/G$ are flat. Since for $y \in Y$,

$$(p_* p^{-1} G)_y = \varinjlim_{y \in U} \prod_{y' \in U} G_{y'} \ ,$$

and

$$(p_* p^{-1} G/G)_y = \varinjlim_{y \in U} \prod_{y' \in U \setminus \{y\}} G_{y'}$$

(U ranges through a neighborhood system of y), these $\mathbb{Z}$-modules are torsion-free and hence flat. □

Let $\mathscr{I}(X)$ denote the full subcategory of $\mathfrak{Mod}(A_X)$ consisting of injective objects. Then $\mathbf{K}^+(\mathscr{I}(X)) \to \mathbf{D}^+(A_X)$ is an equivalence of triangulated categories. Let K be a complex as in Lemma 3.1.4. For $F \in \mathbf{K}^+(\mathscr{I}(X))$, let $f_K^! F$ be the simple complex associated with the double complex $(f_{K^{-q}}^!(F^p))^{p,q}$. Then $f_K^! F$ is a complex of injective A_Y-modules. It is easy to check that $f_K^!$ sends a morphism homotopic to zero to a morphism homotopic to zero. Passing to the quotient, we obtain a functor of triangulated categories:

$$f_K^! : \mathbf{K}^+(\mathscr{I}(X)) \to \mathbf{K}^+(\mathscr{I}(Y)) .$$

Theorem 3.1.5. *Let $f: Y \to X$ be a continuous map of locally compact spaces such that $f_!$ has finite cohomological dimension. Then there exist a functor of triangulated categories $f^!: \mathbf{D}^+(A_X) \to \mathbf{D}^+(A_Y)$ and an isomorphism of bifunctors on $\mathbf{D}^+(A_Y)^\circ \times \mathbf{D}^+(A_X)$:*

$$\mathrm{Hom}_{\mathbf{D}^+(A_X)}(Rf_!(\cdot), \cdot) \simeq \mathrm{Hom}_{\mathbf{D}^+(A_Y)}(\cdot, f^!(\cdot)) .$$

In other words, $f^!$ is a right adjoint to $Rf_!$.

(Note that $f^!$ is unique, up to isomorphism; cf. Exercise I.2.)

Proof. We shall show that $f_K^!$ has the desired property.

For $F \in \mathrm{Ob}(\mathbf{K}^+(\mathscr{I}(X)))$ and $G \in \mathrm{Ob}(\mathbf{K}^+(\mathscr{I}(Y)))$, Lemma 3.1.3 implies:

$$\mathrm{Hom}_{\mathbf{K}^+(\mathfrak{Mod}(A_X))}\left(f_!\left(G \underset{\mathbb{Z}_Y}{\otimes} K\right), F\right) \simeq \mathrm{Hom}_{\mathbf{K}^+(\mathfrak{Mod}(A_Y))}(G, f_K^! F) .$$

Since $G \simeq G \otimes_{\mathbb{Z}_Y} \mathbb{Z}_Y \to G \otimes_{\mathbb{Z}_Y} K$ is a quasi-isomorphism, and $G \otimes_{\mathbb{Z}_Y} K$ is a complex of f-soft sheaves by Lemma 3.1.2, $Rf_! G \simeq f_!(G \otimes_{\mathbb{Z}_Y} K)$ in $\mathbf{D}^+(A_X)$. Hence $\mathrm{Hom}_{\mathbf{K}^+(\mathfrak{Mod}(A_X))}(f_!(G \otimes_{\mathbb{Z}_Y} K), F) \simeq \mathrm{Hom}_{\mathbf{D}^+(A_X)}(Rf_! G, F)$. On the other hand, since $f_K^! F$ is a complex of injective A_Y-modules,

$$\mathrm{Hom}_{\mathbf{K}^+(\mathfrak{Mod}(A_Y))}(G, f_K^! F) \simeq \mathrm{Hom}_{\mathbf{D}^+(A_Y)}(G, f_K^! F) .$$

This completes the proof. □

Remark 3.1.6. (i) Theorem 3.1.5 could be generalized as follows. Let $f: Y \to X$ be as above, let $\mathscr{R}$ (resp. $\mathscr{S}$) be a sheaf of rings on X (resp. Y), and assume to be given a morphism of sheaves of rings $f^{-1}\mathscr{R} \to \mathscr{S}$. Then there exists a functor of triangulated categories $f^!: \mathbf{D}^+(\mathscr{R}) \to \mathbf{D}^+(\mathscr{S})$ which is right adjoint to $Rf_!: \mathbf{D}^+(\mathscr{S}) \to \mathbf{D}^+(\mathscr{R})$. The proof is similar by defining $(f_K^! F)(V)$ as $\mathrm{Hom}_{\mathfrak{Mod}(\mathscr{R})}(f_!(\mathscr{S} \otimes_{\mathbb{Z}_Y} K_V), F)$.

(ii) The isomorphism in Theorem 3.1.5 is compatible with shifts. Namely, for $F \in \mathrm{Ob}(\mathbf{D}^+(A_X))$ and $G \in \mathrm{Ob}(\mathbf{D}^+(A_Y))$, the diagram below commutes:

$$\begin{array}{ccc}
\mathrm{Hom}_{\mathbf{D}^+(A_X)}(Rf_! G, F) & \longrightarrow & \mathrm{Hom}_{\mathbf{D}^+(A_Y)}(G, f^! F) \\
\downarrow \wr & & \downarrow \wr \\
\mathrm{Hom}_{\mathbf{D}^+(A_X)}((Rf_! G)[1], F[1]) & & \mathrm{Hom}_{\mathbf{D}^+(A_Y)}(G[1], (f^! F)[1]) \\
\wr | & & | \wr \\
\mathrm{Hom}_{\mathbf{D}^+(A_X)}(Rf_!(G[1]), F[1]) & \longrightarrow & \mathrm{Hom}_{\mathbf{D}^+(A_Y)}(G[1], f^!(F[1])) .
\end{array}$$

To stress the base ring A, we write for a while $f_A^!$ instead of $f^!$. Hence $f_A^!$ is a functor from $\mathbf{D}^+(A_X)$ to $\mathbf{D}^+(A_Y)$. Let $\phi_X : \mathbf{D}^+(A_X) \to \mathbf{D}^+(\mathbb{Z}_X)$ and $\phi_Y : \mathbf{D}^+(A_Y) \to \mathbf{D}^+(\mathbb{Z}_Y)$ be the forgetful functors.

Proposition 3.1.7. *One has:*

$$\phi_Y \circ f_A^! \simeq f_{\mathbb{Z}}^! \circ \phi_X \ .$$

Proof. Let $G \in \mathrm{Ob}(\mathbf{D}^+(\mathbb{Z}_Y))$, $F \in \mathrm{Ob}(\mathbf{D}^+(A_X))$. We have:

$$\begin{aligned}
\mathrm{Hom}_{\mathbf{D}^+(\mathbb{Z}_Y)}(G, f_{\mathbb{Z}}^! \circ \phi_X F) &\simeq \mathrm{Hom}_{\mathbf{D}^+(\mathbb{Z}_X)}(Rf_! G, \phi_X F) \\
&\simeq \mathrm{Hom}_{\mathbf{D}^+(A_X)}\left(A_X \overset{L}{\underset{\mathbb{Z}_X}{\otimes}} Rf_! G, F\right) \\
&\simeq \mathrm{Hom}_{\mathbf{D}^+(A_X)}\left(Rf_!\left(A_Y \overset{L}{\underset{\mathbb{Z}_Y}{\otimes}} G\right), F\right) \\
&\simeq \mathrm{Hom}_{\mathbf{D}^+(A_Y)}\left(A_Y \overset{L}{\underset{\mathbb{Z}_Y}{\otimes}} G, f_A^! F\right) \\
&\simeq \mathrm{Hom}_{\mathbf{D}^+(\mathbb{Z}_Y)}(G, \phi_Y \circ f_A^! F) \ .
\end{aligned}$$

Therefore $f_{\mathbb{Z}}^! \circ \phi_X F \simeq \phi_Y \circ f_A^! F$. □

Now let $g : Z \to Y$ be another continuous map of locally compact spaces.

Proposition 3.1.8. *Assume $f_!$ and $g_!$ have finite cohomological dimension. Then $(f \circ g)_!$ has finite cohomological dimension, and one has:*

$$(f \circ g)^! \simeq g^! \circ f^!$$

as functors from $\mathbf{D}^+(A_X)$ to $\mathbf{D}^+(A_Z)$.

Proof. It follows immediately from the formula: $(f \circ g)_! \simeq f_! \circ g_!$. □

Proposition 3.1.9. *Consider a Cartesian square of continuous maps and locally compact spaces:*

$$\begin{array}{ccc}
Y' & \xrightarrow{f'} & X' \\
{\scriptstyle g'}\downarrow & \square & \downarrow{\scriptstyle g} \\
Y & \xrightarrow{f} & X \ .
\end{array}$$

Assume $f_!$ has finite cohomological dimension. Then:

(i) *$f'_!$ has finite cohomological dimension.*
(ii) *One has a canonical isomorphism of functors from $\mathbf{D}^+(A_{X'})$ to $\mathbf{D}^+(A_Y)$:*

$$f^! \circ Rg_* \simeq Rg'_* \circ f'^! \ .$$

(iii) *We have a morphism of functors from* $\mathbf{D}^+(A_X)$ *to* $\mathbf{D}^+(A_{Y'})$:

$$g'^{-1}f^! \to f'^! g^{-1} .$$

Proof. (i) Let $x' \in X'$. Then $f'^{-1}(x')$ is homeomorphic to $f^{-1}(g(x'))$. Therefore the cohomological dimension of $f'_!$ is bounded by that of $f_!$.

(ii) We have, functorially with respect to $F \in \mathrm{Ob}(\mathbf{D}^+(A_{X'}))$, $G \in \mathrm{Ob}(\mathbf{D}^+(A_Y))$:

$$\begin{aligned} \mathrm{Hom}_{\mathbf{D}^+(A_Y)}(G, f^! Rg_* F) &\simeq \mathrm{Hom}_{\mathbf{D}^+(A_X)}(Rf_! G, Rg_* F) \\ &\simeq \mathrm{Hom}_{\mathbf{D}^+(A_{X'})}(g^{-1} Rf_! G, F) \\ &\simeq \mathrm{Hom}_{\mathbf{D}^+(A_{X'})}(Rf'_! \circ g'^{-1} G, F) \\ &\simeq \mathrm{Hom}_{\mathbf{D}^+(A_{Y'})}(g'^{-1} G, f'^! F) \\ &\simeq \mathrm{Hom}_{\mathbf{D}^+(A_Y)}(G, Rg'_* \circ f'^! F) . \end{aligned}$$

This implies the desired result.

(iii) By Proposition 2.5.11, for $F \in \mathrm{Ob}(\mathbf{D}^+(A_X))$, we have $Rf'_! g'^{-1} f^! F \simeq g^{-1} Rf_! f^! F \to g^{-1} F$. This gives the desired homomorphism $g'^{-1} f^! F \to f'^! g^{-1} F$. □

Proposition 3.1.10. *Assume* $f_!$ *has finite cohomological dimension. Then for* $F \in \mathrm{Ob}(\mathbf{D}^+(A_X))$, $G \in \mathrm{Ob}(\mathbf{D}^b(A_Y))$, *we have*:

$$R\mathrm{Hom}(Rf_! G, F) \simeq R\mathrm{Hom}(G, f^! F) ,$$

$$R\mathcal{H}om(Rf_! G, F) \simeq Rf_* R\mathcal{H}om(G, f^! F) .$$

Proof. The first isomorphism follows from the second one, by applying the functor $R\Gamma(X; \cdot)$.

We have the canonical morphism:

$$Rf_* R\mathcal{H}om(G, f^! F) \to R\mathcal{H}om(Rf_! G, Rf_! f^! F) .$$

By composition with the morphism $Rf_! \circ f^! F \to F$, we obtain the morphism:

$$Rf_* R\mathcal{H}om(G, f^! F) \to R\mathcal{H}om(Rf_! G, F) .$$

Let V be an open subset of X. Then:

$$\begin{aligned} H^j(R\Gamma(V; Rf_* R\mathcal{H}om(G, f^! F))) &\simeq \mathrm{Hom}_{\mathbf{D}^+(A_{f^{-1}(V)})}(G|_{f^{-1}(V)}, f^! F[j]|_{f^{-1}(V)}) \\ &\simeq \mathrm{Hom}_{\mathbf{D}^+(A_V)}((Rf_! G)|_V, F[j]|_V) \\ &\simeq H^j(R\Gamma(V; R\mathcal{H}om(Rf_! G, F))) . \end{aligned}$$

This completes the proof. □

Proposition 3.1.11. *Assume* $f_!$ *has finite cohomological dimension. Then there exists a natural morphism of functors from* $\mathbf{D}^+(A_X) \times \mathbf{D}^+(A_X)$ *to* $\mathbf{D}^+(A_Y)$:

$$f^!(\cdot) \overset{L}{\underset{A_Y}{\otimes}} f^{-1}(\cdot) \to f^!\left(\cdot \overset{L}{\underset{A_X}{\otimes}} \cdot\right) .$$

Proof. Let F, F_1 and F_2 belong to $\mathbf{D}^+(A_X)$. We have:

$$\begin{aligned}\mathrm{Hom}_{\mathbf{D}^+(A_Y)}\left(f^!F_1 \overset{L}{\underset{A_Y}{\otimes}} f^{-1}F_2, f^!F\right) &\simeq \mathrm{Hom}_{\mathbf{D}^+(A_X)}\left(Rf_!\left(f^!F_1 \overset{L}{\underset{A_Y}{\otimes}} f^{-1}F_2\right), F\right) \\ &\simeq \mathrm{Hom}_{\mathbf{D}^+(A_X)}\left(Rf_!f^!F_1 \overset{L}{\underset{A_X}{\otimes}} F_2, F\right) .\end{aligned}$$

Take $F = F_1 \otimes^L_{A_X} F_2$. Then we get the desired morphism as the image of $Rf_!f^!F_1 \otimes^L_{A_X} F_2 \to F_1 \otimes^L_{A_X} F_2$. □

We shall see in section 3 that if f is a *topological submersion*, then the morphism of Proposition 3.1.11 becomes an isomorphism.

There is a situation in which the functor $f^!$ is familiar.

Proposition 3.1.12. *Assume $f: Y \to X$ is a homeomorphism from Y onto a locally closed subset of X. Then:*

$$f^!(\cdot) \simeq f^{-1} \circ R\Gamma_{f(Y)}(\cdot) .$$

Proof. Set $Z = f(Y)$. Let $F \in \mathrm{Ob}(\mathbf{D}^+(A_X))$, $G \in \mathrm{Ob}(\mathbf{D}^+(A_Y))$. Then:

$$\begin{aligned}\mathrm{Hom}_{\mathbf{D}^+(A_X)}(f_!G, F) &\simeq \mathrm{Hom}_{\mathbf{D}^+(A_X)}(f_!G, R\Gamma_Z(F)) \\ &\simeq \mathrm{Hom}_{\mathbf{D}^+(A_Y)}(f^{-1} \circ f_!G, f^{-1}R\Gamma_Z(F)) \\ &\simeq \mathrm{Hom}_{\mathbf{D}^+(A_Y)}(G, f^{-1}R\Gamma_Z(F)) .\end{aligned}$$

Hence $f^!F \simeq f^{-1} \circ R\Gamma_Z(F)$. □

Proposition 3.1.13. *Assume $f_!$ has finite cohomological dimension. Then for $F_1 \in \mathrm{Ob}(\mathbf{D}^b(A_X))$, $F_2 \in \mathrm{Ob}(\mathbf{D}^+(A_X))$ we have:*

$$f^!R\mathcal{H}om(F_1, F_2) \simeq R\mathcal{H}om(f^{-1}F_1, f^!F_2) .$$

Proof. Let $G \in \mathrm{Ob}(\mathbf{D}^+(Y))$. Then:

$$\begin{aligned}\mathrm{Hom}_{\mathbf{D}^+(A_Y)}(G, f^!R\mathcal{H}om(F_1, F_2)) &\simeq \mathrm{Hom}_{\mathbf{D}^+(A_X)}(Rf_!G, R\mathcal{H}om(F_1, F_2)) \\ &\simeq \mathrm{Hom}_{\mathbf{D}^+(A_X)}\left(Rf_!G \overset{L}{\otimes} F_1, F_2\right) \\ &\simeq \mathrm{Hom}_{\mathbf{D}^+(A_X)}\left(Rf_!\left(G \overset{L}{\otimes} f^{-1}F_1\right), F_2\right) \\ &\simeq \mathrm{Hom}_{\mathbf{D}^+(A_X)}\left(G \overset{L}{\otimes} f^{-1}F_1, f^!F_2\right) \\ &\simeq \mathrm{Hom}_{\mathbf{D}^+(A_Y)}(G, R\mathcal{H}om(f^{-1}F_1, f^!F_2)) .\end{aligned}$$

Since this is true for any G, the proof follows. □

Proposition 3.1.14. *Assume that X has finite c-soft dimension (cf. Exercise II.9). Let $F \in \mathrm{Ob}(\mathbf{D}^+(A_X))$, $G \in \mathrm{Ob}(\mathbf{D}^b(A_X))$. Let q_1 and q_2 denote the first and second projections on $X \times X$, and Δ the diagonal. Then:*

$$R\mathcal{H}om(G,F) \simeq Rq_{1*}R\Gamma_\Delta R\mathcal{H}om(q_2^{-1}G, q_1^!F) .$$

Proof. Denote by δ the diagonal embedding. Then:

$$\begin{aligned} Rq_{1*}R\Gamma_\Delta R\mathcal{H}om(q_2^{-1}G, q_1^!F) &\simeq \delta^! R\mathcal{H}om(q_2^{-1}G, q_1^!F) \\ &\simeq R\mathcal{H}om(\delta^{-1}q_2^{-1}G, \delta^! q_1^! F) \\ &\simeq R\mathcal{H}om(G,F) . \quad \square \end{aligned}$$

Finally consider two locally compact spaces X and Y, and denote by q_1 and q_2 the projection from $X \times Y$ to X and Y respectively.

Proposition 3.1.15. *Assume that Y has finite c-soft dimension. Let $F \in \mathrm{Ob}(\mathbf{D}^+(A_X))$, $G \in \mathrm{Ob}(\mathbf{D}^b(A_Y))$. Then there is a canonical isomorphism:*

$$R\Gamma(X \times Y; R\mathcal{H}om(q_2^{-1}G, q_1^!F)) \simeq \mathrm{RHom}(R\Gamma_c(Y;G), R\Gamma(X;F)) .$$

Proof. Denote by a_X (resp. a_Y) the projection $X \to \{\mathrm{pt}\}$ (resp. $Y \to \{\mathrm{pt}\}$). Then:

$$\begin{aligned} R\Gamma(X \times Y; R\mathcal{H}om(q_2^{-1}G, q_1^!F)) &\simeq R\mathrm{a}_{X*}Rq_{1*}R\mathcal{H}om(q_2^{-1}G, q_1^!F)) \\ &\simeq R\mathrm{a}_{X*}R\mathcal{H}om(Rq_{1!}q_2^{-1}G, F) \\ &\simeq R\mathrm{a}_{X*}R\mathcal{H}om(\mathrm{a}_X^{-1}R\mathrm{a}_{Y!}G, F) \\ &\simeq \mathrm{RHom}(R\mathrm{a}_{Y!}G, R\mathrm{a}_{X*}F) . \quad \square \end{aligned}$$

Definition 3.1.16. (i) *Let $f: Y \to X$ be a continuous map, and assume $f_!$ has finite cohomological dimension. One sets:*

$$\omega_{Y/X} = f^! A_X$$

and calls $\omega_{Y/X}$ the relative dualizing complex. If $X = \{\mathrm{pt}\}$, one sets $\omega_Y = \omega_{Y/X}$ and calls ω_Y the dualizing complex on Y.

(ii) *Assume X has finite c-soft dimension, and let $F \in \mathrm{Ob}(\mathbf{D}^b(X))$. One sets:*

$$\mathrm{D}_X F = R\mathcal{H}om(F, \omega_X) , \qquad \mathrm{D}'_X F = R\mathcal{H}om(F, A_X) .$$

One calls $\mathrm{D}_X F$ the dual of F.

It there is no risk of confusion we write $\mathrm{D}F$ instead of $\mathrm{D}_X F$.

By Proposition 3.1.11, there is a natural morphism of functors:

$$f^{-1}(\cdot) \otimes \omega_{Y/X} \to f^!(\cdot) . \tag{3.1.6}$$

By Theorem 3.1.5 and Proposition 3.1.10, for $F \in \mathrm{Ob}(\mathbf{D}^b(A_X))$, we have:

$$\mathrm{Hom}_{\mathbf{D}^b(A_X)}(F, \omega_X) \simeq \mathrm{Hom}_{\mathbf{D}^b(\mathfrak{Mod}(A))}(R\Gamma_c(X;F), A) \quad \text{and} \tag{3.1.7}$$

$$R\mathrm{Hom}(F, \omega_X) \simeq R\mathrm{Hom}(R\Gamma_c(X;F), A) . \tag{3.1.8}$$

In particular, taking $F = A_Z$, we have $R\Gamma_Z(X;\omega_X) \simeq R\mathrm{Hom}(R\Gamma_c(Z;A_Z), A)$ for a locally closed subset Z.

3.2. Vanishing theorems on manifolds

Let V be a real vector space of dimension n.

Lemma 3.2.1. *For any sheaf F on V we have:*

$$H_c^j(V;F) = 0 \quad \textit{for} \quad j > n .$$

Proof. First assume $n = 1$, and set $\tilde{F} = i_! F$ where $i: V \to [0,1]$ is a homeomorphism from V onto the interval $]0,1[$. Then we have:

$$H_c^j(V;F) \simeq H^j([0,1];\tilde{F}) .$$

Hence the result follows from Lemma 2.7.3. In the general case, let V' be an $(n-1)$-dimensional vector space, and let f be a surjective linear map $V \to V'$. By the preceding result, $R^j f_! F = 0$ for $j \neq 0, 1$. Since $R^0 f_! F \simeq \tau^{\leqslant 0} Rf_! F$, and $\tau^{\geqslant 1} Rf_! F \simeq (R^1 f_! F)[-1]$, we get a distinguished triangle:

$$R^0 f_! F \longrightarrow Rf_! F \longrightarrow (R^1 f_! F)[-1] \xrightarrow[+1]{} .$$

Let us apply the functor $\Gamma_c(V'; \cdot)$ to this triangle. We obtain the long exact sequence:

$$\cdots \to H_c^j(V'; R^0 f_! F) \to H_c^j(V;F) \to H_c^{j-1}(V'; R^1 f_! F) \to \cdots$$

and the result follows by induction on n. □

Proposition 3.2.2. *Let X be an n-dimensional C^0-manifold and let F be a sheaf on X. Then:*

(i) *F admits a resolution of length at most n by c-soft sheaves,*
(ii) *F admits a resolution of length at most $n + 1$ by flabby sheaves,*
(iii) *$H_c^j(X;F) = 0$ for $j > n$,*
(iv) *$H^j(X;F) = 0$ for $j > n$.*
(v) *Let Z be a locally closed subset of X. Then $H_Z^j(X;F) = 0$ for $j > n + 1$.*

Recall that, by its definition, a C^0-manifold is countable at infinity.

Proof. Let $0 \to F \to F_0 \to F_1 \to \cdots$ be a flabby resolution of F, (cf. II §4). Consider the exact sequence:

$$0 \to F \to F_0 \to F_1 \to \cdots \to F_{n-1} \to G_n \to 0 \ ,$$

where $G_n = \mathrm{Im}(F_{n-1} \to F_n)$. We shall prove that G_n is c-soft. This property being local on X, (cf. Exercise II.6), we may assume X is open in some real vector space V of dimension n. Let U be an open subset of X. We have:

$$H_c^j(U;F) = H_c^j(V;F_U)$$

and this group is zero for $j > n$, by Lemma 3.2.1. Therefore $H_c^j(U;G_n) = 0$ for $j > 0$ and G_n is c-soft, (Exercise II.6).

Now (iii) follows from (i), as well as (iv), since $H^j(X;G) = 0$ for $j > 0$ if G is c-soft, (Proposition 2.5.10).

By the long exact sequence associated to $Z \hookrightarrow X$, we obtain (v) from (iv), and (ii) follows from (v). □

Proposition 3.2.3. *Let V be a real vector space of dimension n. Let $M \in \mathrm{Ob}(D^+(\mathfrak{Mod}(A)))$, and let $x \in V$.*

(i) *The natural morphism $R\Gamma(V;M_V) \to (M_V)_x \simeq M$ is an isomorphism.*
(ii) *The natural morphism $R\Gamma_{\{x\}}(V;M_V) \to R\Gamma_c(V;M_V)$ is an isomorphism.*
(iii) *There is an isomorphism:*

$$R\Gamma_c(\mathbb{R}^n;M_{\mathbb{R}^n}) \simeq M[-n] \ .$$

(iv) *Let ϕ be a linear automorphism of V and let $\phi_c^\#$ be the associated automorphism of $R\Gamma_c(V;M_V)$. Then $\phi_c^\# = \mathrm{sgn}(\phi)\cdot \mathrm{id}_{R\Gamma_c(V;M_V)}$, where $\mathrm{sgn}(\phi)$ is the sign of the determinant of ϕ.*
(v) *To each orientation of V is canonically associated an isomorphism $R\Gamma_c(V;M_V) \simeq M[-n]$ compatible with the isomorphism in* (iii).

Proof. (i) Apply Corollary 2.7.7.

(ii) We may assume $x = 0$. Set $F = M_V$. As in (i), $R\Gamma(\{y;|y| > 0\};F) \to R\Gamma(\{y;|y| > a\};F)$ is an isomorphism for $a > 0$. Hence $R\Gamma_{\{0\}}(V;F) \to R\Gamma_{\{y;|y| \leqslant a\}}(V;F)$ is an isomorphism. Since $H_c^j(V;F) = \varinjlim_a H^j_{\{y;|y|\leqslant a\}}(V;F)$ we obtain the desired result.

(iii), (iv) First assume $n = 1$. Denote by U_1 and U_2 the connected components of $V \setminus \{0\}$ and consider the distinguished triangle:

$$R\Gamma_{\{0\}}(V;M_V) \xrightarrow{\alpha} R\Gamma(V;M_V) \xrightarrow{\beta} R\Gamma(U_1;M_V) \oplus R\Gamma(U_2;M_V) \xrightarrow[+1]{\gamma} \ .$$

Applying (i), β admits a left inverse, and hence $\alpha = 0$ and we have an epimorphism:

$$R\Gamma(U_1;M_V) \oplus R\Gamma(U_2;M_V) \xrightarrow{\gamma} R\Gamma_{\{0\}}(V;M_V)[1] \ , \tag{3.2.1}$$

where $\beta = (\beta_1, \beta_2)$ and β_i is the isomorphism $\Gamma(V; M_V) \simeq \Gamma(U_i; M_V)$. Then (iii) follows in this case as well as (iv), since $\phi_c^{\#}[1] \circ \gamma = \pm 1$ according to whether ϕ interchanges U_1 and U_2 or not, in view of (3.2.1). To treat the general case we argue by induction on dim V and choose a linear decomposition $V \simeq L \times V'$, with $\dim L = 1$. By the Künneth formula (Exercise II.18), we get:

$$(3.2.2) \qquad R\Gamma_c(V; M_V) \simeq R\Gamma_c(L; M_L) \overset{L}{\otimes} R\Gamma_c(V'; A_{V'}) \; .$$

Then (iii) follows. To prove (iv) we note that $\phi_c^{\#}$ is locally constant with respect to $\phi \in \mathrm{GL}(V)$ (Proposition 2.7.5), and $\mathrm{GL}(V)$ has two connected components. Hence it is enough to find ϕ with $\phi_c^{\#} = -1$. We take $\phi = \psi \otimes \mathrm{id}_{V'}$, where ψ is the antipodal map on L, and conclude by (3.2.2).

(v) Follows from (iii) an (iv). □

In §5, we shall need the following result.

Proposition 3.2.4. *Let V be a real finite dimensional vector space, X an open subset of V, F a sheaf over X. Assume that for any convex compact subset K of X, the restriction map $\Gamma(X; F) \to \Gamma(K; F)$ is surjective. Then for any convex open subset U of X, $H^j(U; F) = 0$ for all $j > 0$.*

Proof. We may assume $X = U$ is convex. By considering an increasing sequence of compact convex subsets $\{K_n\}$ of X and applying Proposition 2.7.1, it is enough to show that $H^j(K; F) = 0$ for $j > 0$, for any compact convex subset K of X.

We proceed by induction on dim V, and note that if $\dim V = 1$, the result is a particular case of Lemma 2.7.3. Let $V = V' \oplus L$ be a linear decomposition of V, with $\dim L = 1$, and let f be the projection $V \to V'$. Let f_K be the restriction of f to K. If $y \in V'$, the sheaf $F|_{f^{-1}(y) \cap K}$ satisfies the hypothesis of the Proposition. Therefore $R^j f_{K*}(F|_K) = 0$ for $j > 0$. Moreover the sheaf $f_{K*}(F|_K)$ also satisfies the hypothesis of the Proposition. Thus by the induction hypothesis, we get:

$$H^j(K; F) = H^j(f(K); f_{K*}F) = 0$$

for $j > 0$. □

3.3. Orientation and duality

Let $f : Y \to X$ be a continuous map of locally compact spaces.

Definition 3.3.1. *We shall say that f is a topological submersion with fiber dimension l if for any $y \in Y$ there exists an open neighborhood V of y such that $U = f(V)$ is open in X, and there exists a commutative diagram:*

$$\begin{array}{ccc} V & \xrightarrow{f|_V} & U \\ \uparrow{\scriptstyle h} & \nearrow{\scriptstyle p} & \\ U \times \mathbb{R}^l & & \end{array} ,$$

where p is the projection, and h is a homeomorphism.

Note that if X and Y are C^1-manifolds and f is a C^1-submersion, then f is a topological submersion. If f is a topological submersion, then $f_!$ has finite cohomological dimension, by Proposition 3.2.2.

Proposition 3.3.2. *Assume f is a topological submersion with fiber dimension l. Then:*

(i) $H^k(f^!A_X) = 0$ *for* $k \neq -l$ *and* $H^{-l}(f^!A_X)$ *is locally isomorphic to* A_Y.
(ii) *The morphism of functors* $f^!A_X \otimes_{A_Y} f^{-1}(\cdot) \to f^!(\cdot)$ *is an isomorphism.*

Proof. We shall prove first (i) when $Y = \mathbb{R}^l$ and $X = \{\mathrm{pt}\}$. By (3.1.8), for any open set U of Y, $R\Gamma(U; f^!A_X) \simeq R\mathrm{Hom}(R\Gamma_c(U; A_Y), A)$. If furthermore U is homeomorphic to $\mathbb{R}^l$, $R\Gamma_c(U; A_Y) \simeq A[-l]$ by Proposition 3.2.3 and hence we obtain $H^j(U; f^!A_X) = 0$ for $j \neq -l$, and $\Gamma(U; H^{-l}(f^!A_X)) \simeq \mathrm{Hom}(H^l_c(U; A_Y), A)$.

Since $H^l_c(Y; A_Y) \leftarrow H^l_c(U; A_Y)$ is an isomorphism by Proposition 3.2.3 (ii), $A \simeq \Gamma(Y; H^{-l}(f^!A_X)) \to \Gamma(U; H^{-l}(f^!A_X))$ is an isomorphism. This shows that $H^{-l}(f^!A_X)$ is a constant sheaf isomorphic to A_Y. Now, let us prove the general case. The question being local, we may assume $Y = \mathbb{R}^l \times X$ and f is the projection. Let p be the projection from Y to $\mathbb{R}^l$ and let a_X, $a_{\mathbb{R}^l}$ be the projections from X and $\mathbb{R}^l$ to $\{\mathrm{pt}\}$, respectively. By Proposition 3.1.9, we have a morphism $p^{-1}\omega_{\mathbb{R}^l} \to f^!A_X$. Hence for any $F \in \mathrm{Ob}(\mathbf{D}^+(A_X))$, we have the chain of morphisms:

$$p^{-1}\omega_{\mathbb{R}^l} \otimes f^{-1}F \to f^!A_X \otimes f^{-1}F \to f^!F \ . \tag{3.3.1}$$

It is enough to show that the composition is an isomorphism. For an open set U of $\mathbb{R}^l$, and an open set V of X, if U is homeomorphic to $\mathbb{R}^l$, then we have a chain of isomorphisms:

$$\begin{aligned} R\Gamma(U \times V; f^!F) &\simeq R\mathrm{Hom}(A_{U\times V}, f^!F) \\ &\simeq R\mathrm{Hom}(Rf_!A_{U\times V}, F) \\ &\simeq R\mathrm{Hom}\left(R\Gamma_c(U; A_U) \overset{L}{\otimes} A_V, F\right) \\ &\leftarrow\!\!\!\simeq R\mathrm{Hom}(R\Gamma_c(U; A_U), A) \overset{L}{\otimes} R\mathrm{Hom}(A_V, F) \\ &\leftarrow\!\!\!\simeq R\Gamma(U; \omega_{\mathbb{R}^l}) \overset{L}{\otimes} R\Gamma(V; F) \ . \quad \square \end{aligned}$$

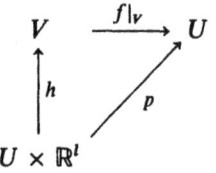

Definition 3.3.3. *Let $f: Y \to X$ be a topological submersion with fiber dimension l. One sets:*

$$\mathit{or}_{Y/X} = H^{-l}(\omega_{Y/X})$$

where $\omega_{Y/X}$ is the relative dualizing complex (Definition 3.1.16), and calls $\mathit{or}_{Y/X}$ the relative orientation sheaf. If $X = \{\mathrm{pt}\}$, (in which case Y is a topological manifold), one writes or_Y instead of $\mathit{or}_{Y/X}$, and calls or_Y the orientation sheaf on Y.

By Proposition 3.3.2 we have:

$$\omega_{Y/X} \simeq \mathit{or}_{Y/X}[l] \ . \tag{3.3.2}$$

Proposition 3.3.4. *Let $f: Y \to X$ be a topological submersion with fiber dimension l.*

(i) *Let $\omega^{\mathbb{Z}}_{Y/X}$ and $\mathit{or}^{\mathbb{Z}}_{Y/X}$ be the relative dualizing complex and the relative orientation sheaf with respect to the base ring $\mathbb{Z}$, respectively. Then we have:*

$$\omega_{Y/X} \simeq A_Y \underset{\mathbb{Z}_Y}{\otimes} \omega^{\mathbb{Z}}_{Y/X} \qquad \text{and} \qquad \mathit{or}_{Y/X} \simeq A_Y \underset{\mathbb{Z}_Y}{\otimes} \mathit{or}^{\mathbb{Z}}_{Y/X} \ .$$

(ii) *One has canonical isomorphisms:*

$$\mathit{or}_{Y/X} \otimes \mathit{or}_{Y/X} \simeq A_Y \ ,$$

$$\mathcal{H}om(\mathit{or}_{Y/X}, A_Y) \simeq \mathit{or}_{Y/X} \ .$$

(iii) *Let $g: Z \to Y$ be a continuous map, and assume $f \circ g$ is a topological submersion with fiber dimension m. Let $F \in \mathrm{Ob}(\mathbf{D}^+(A_X))$. Then:*

$$g^! \circ f^{-1}F \simeq (f \circ g)^{-1}F \otimes \mathit{or}_{Z/X} \otimes g^{-1}\mathit{or}_{Y/X}[m-l] \ .$$

Proof. (i) follows from Proposition 3.3.2.

(ii) follows from Exercise III.3.

(iii) By Proposition 3.3.2 and the preceding result, we have:

$$(f \circ g)^! F \simeq (f \circ g)^{-1}F \otimes \mathit{or}_{Z/X}[m] \ ,$$

$$f^! F \simeq f^{-1}F \otimes \mathit{or}_{Y/X}[l] \ .$$

Since $\mathit{or}_{Y/X}$ is locally isomorphic to A_Y, the result follows from:

$$g^! \circ f^{-1}F \otimes \mathit{or}_{Y/X}[l] \simeq g^! \circ f^! F \simeq (f \circ g)^{-1}F \otimes \mathit{or}_{Z/X}[m] \ . \quad \square$$

Remark 3.3.5. Let $\begin{array}{ccc} Y & \xrightarrow{f} & X \\ & {\scriptstyle p_Y}\searrow \ \swarrow{\scriptstyle p_X} & \\ & S & \end{array}$ be a commutative diagram of continuous maps of locally compact spaces. Assume p_X and p_Y are topological submersions with fiber dimension m and n respectively. Then $f^! A_X \simeq \mathit{or}_{Y/S} \otimes f^{-1}\mathit{or}_{X/S}[n-m]$, and it is natural to set:

(3.3.3) $$\mathit{or}_{Y/X} = \mathit{or}_{Y/S} \otimes f^{-1}\mathit{or}_{X/S} \ .$$

Note that if $g : Z \to Y$ is another continuous map such that $p_Y \circ g$ is a topological submersion then:

(3.3.4) $$\omega_{Z/X} \simeq \omega_{Z/Y} \otimes g^{-1}\omega_{Y/X} \quad \text{and} \quad \mathit{or}_{Z/X} \simeq \mathit{or}_{Z/Y} \otimes g^{-1}\mathit{or}_{Y/X} \ .$$

Proposition 3.3.6. *Let X be an n-dimensional C^0-manifold.*

(i) *or_X is the sheaf associated to the presheaf $U \mapsto \mathrm{Hom}(H^n_c(U; A_X), A)$.*
(ii) *For $x \in X$, there is a canonical isomorphism $\mathit{or}_{X,x} \simeq \mathrm{Hom}(H^n_{\{x\}}(X; A_X), A) \simeq H^n_{\{x\}}(X; A_X)$.*
(iii) *Assume X is differentiable and oriented. Then there is an isomorphism $\mathit{or}_X \simeq A_X$, and such an isomorphism changes sign when one reverses the orientation of X.*

Proof. (i) Let U be open subset of X homeomorphic to $\mathbb{R}^n$. Then:

$$\begin{aligned} R\Gamma(U; \omega_X) &\simeq R\mathrm{Hom}(A_U; \omega_X) \\ &\simeq R\mathrm{Hom}(R\Gamma_c(X; A_U), A) \\ &\simeq \mathrm{Hom}(H^n_c(U; A_U), A)[n] \ , \end{aligned}$$

in view of (3.1.7).

(ii) It follows from the preceding discussion and Propositions 3.2.3 and 3.3.4 (ii).

(iii) Let $X = \bigcup_i U_i$ be an open covering of X by open subsets diffeomorphic to $\mathbb{R}^n$, with compatible orientations. Then the isomorphisms $\phi_{U_i} : \mathit{or}_X|_{U_i} \simeq A_{U_i}$ will glue together by Lemma 3.3.7 below, and ϕ_{U_i} is replaced by $-\phi_{U_i}$ if one reverses the orientation, by Proposition 3.2.3. □

Lemma 3.3.7. *Let E be the Euclidian space $\mathbb{R}^n$, and fix an isomorphism $\mathit{or}_E \simeq A_E$. Let U and V be two open subsets of E and $f : U \to V$ a diffeomorphism. Assume the Jacobian of f is positive at each point of U. Then the following diagram is commutative:*

$$\begin{array}{ccccc} \mathit{or}_U & \xrightarrow{\sim} & \mathit{or}_{E/U} & \xrightarrow{\sim} & A_U \\ \uparrow{\scriptstyle f^{\#}_{\mathit{or}}} & & & & \uparrow{\scriptstyle f^{\#}_A} \\ f^{-1}(\mathit{or}_V) & \xrightarrow{\sim} & f^{-1}(\mathit{or}_{E/V}) & \xrightarrow{\sim} & f^{-1}(A_V) \ . \end{array}$$

(Here the morphisms $f^{\#}_{\mathit{or}}$ and $f^{\#}_A$ are defined as follows. Let a_U (resp. a_V) be the projection of U (resp. V) on $\{\mathrm{pt}\}$. Then $f^{\#}_A$ is the isomorphism $f^{-1} \circ \mathrm{a}_V^{-1} \xrightarrow{\sim} \mathrm{a}_U^{-1}$ and $f^{\#}_{\mathit{or}}$ the isomorphism $f^{-1} \circ \mathrm{a}_V^{!}[-n] \xrightarrow{\sim} f^{!} \circ \mathrm{a}_V^{!}[-n] \xrightarrow{\sim} \mathrm{a}_U^{!}[-n]$. Note that $f^{-1} \simeq f^{!}$.)

Proof. It is enough to prove that for each $x_\circ \in U$, the following diagram commutes:

$$\begin{array}{ccc} H^n_{\{f(x_\circ)\}}(E;A_E) & \xrightarrow[f^\#]{} & H^n_{\{x_\circ\}}(E;A_E) \; . \\ & \searrow \wr \quad \wr \swarrow & \\ & H^n_c(E;A_E) & \end{array}$$

We may assume $x_\circ = f(x_\circ) = 0$. Set $u = f'(0)$, $f_\lambda(x) = u(x) + \lambda(f(x) - u(x))$, $0 \leqslant \lambda \leqslant 1$. We may find open neighborhoods U' and V' of 0 such that $f_\lambda(U') \subset V'$ and $f_\lambda^{-1}(\{0\}) \cap U' \subset \{0\}$. Since $u^\#$ operates as the identity on $H^n_{\{0\}}(E;A_E)$, (Proposition 3.2.3), the same is true for $f^\#$ by Remark 2.7.6. □

One can recover the functor $f^!$ from well-known functors. In fact let $f: Y \to X$ be a map of C^0-manifolds. We may decompose f as the composite of a closed embedding and a submersion, $f = p \circ j$:

(3.3.5) $$f: Y \xhookrightarrow[j]{} Y \times X \xrightarrow[p]{} X \; ,$$

where p is the projection and j is the graph map, $j(y) = (y, f(y))$. Applying Propositions 3.1.8, 3.1.12 and 3.3.2, we obtain for $F \in \mathrm{Ob}(\mathbf{D}^+(A_X))$:

(3.3.6) $$f^!F \simeq j^{-1}R\Gamma_{j(Y)}(p^{-1}F) \otimes or_Y[\dim Y] \; .$$

Choosing $F = A_X$, we get:

(3.3.7) $$or_{Y/X} \simeq j^{-1}(H^{\dim X}_{j(Y)}(A_{Y\times X})) \otimes or_Y \; .$$

For $f = \mathrm{id}_X$, we obtain:

(3.3.8) $$or_X \simeq H^{\dim X}_X(A_{X\times X})|_X$$

(where X is identified with the diagonal of $X \times X$).

Notation 3.3.8. In this book, at the exception of X§3, we denote by $\dim X$ the dimension of a real manifold X. If $f: Y \to X$ is a morphism of C^0-manifolds we set:

(3.3.9) $$\dim Y/X = \dim Y - \dim X \; .$$

If Y is a submanifold we also write:

$$\mathrm{codim}_X Y = -\dim Y/X \; ,$$

If there is no risk of confusion we write $\mathrm{codim}\, Y$ instead of $\mathrm{codim}_X Y$. Note that:

(3.3.10) $$\omega_{Y/X} \simeq or_{Y/X}[\dim Y/X] \; .$$

It is then natural to set:

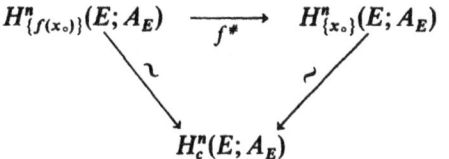

(3.3.11)
$$\omega_{Y/X}^{\otimes -1} = R\mathcal{H}om(\omega_{Y/X}, A_Y) \,,$$
$$\simeq or_{Y/X}[-\dim Y/X] \,.$$

Proposition 3.3.9. *Let $f: Y \to X$ be a continuous map of locally compact spaces. Assume:*

(i) *f is a topological submersion.*
(ii) *$Rf_! f^! \mathbb{Z}_X \to \mathbb{Z}_X$ is an isomorphism.*

Then for $F \in \mathrm{Ob}(\mathbf{D}^+(\mathbb{Z}_X))$, the morphism $F \mapsto Rf_* f^{-1} F$ is an isomorphism.

Proof. Let l be the fiber dimension of f. We have $f^!F \simeq f^{-1}F \otimes f^!\mathbb{Z}_X$, and $f^!\mathbb{Z}_X$ is locally isomorphic to $\mathbb{Z}_Y[l]$. Therefore:

$$f^{-1}F \simeq R\mathcal{H}om(f^!\mathbb{Z}_X, f^!F) \,,$$
$$Rf_* f^{-1}F \simeq Rf_* R\mathcal{H}om(f^!\mathbb{Z}_X, f^!F)$$
$$\simeq R\mathcal{H}om(Rf_! f^!\mathbb{Z}_X, F)$$
$$\xleftarrow{\sim} F \,. \quad \square$$

Remark 3.3.10. Let $f: Y \to X$ be a continuous map of locally compact spaces, and assume f is a topological submersion with fiber dimension l. Then condition (ii) of Proposition 3.3.9 will be satisfied if and only if, for any $x \in X$ we have the isomorphism:

(3.3.12)
$$R\Gamma_c(f^{-1}(x); \omega_{f^{-1}(x)}) \xrightarrow{\sim} \mathbb{Z} \,.$$

This isomorphism is equivalent to the isomorphism:

(3.3.13)
$$\mathbb{Z} \xrightarrow{\sim} R\Gamma(f^{-1}(x); \mathbb{Z}_{f^{-1}(x)}) \,.$$

In fact, set $M = R\Gamma_c(f^{-1}(x); \omega_{f^{-1}(x)})$ and $M^* = R\mathrm{Hom}(M; \mathbb{Z})$. Then $M^* \simeq R\Gamma(f^{-1}(x); \mathbb{Z}_{f^{-1}(x)})$, and the isomorphism $M \xrightarrow{\sim} \mathbb{Z}$ (in $\mathbf{D}^b(\mathfrak{Mod}(\mathbb{Z}))$) is equivalent to the isomorphism $\mathbb{Z} \xrightarrow{\sim} M^*$, in view of Exercise I.31.

Now let X be a C^0-manifold of dimension n and let a_X denote the map $X \to \{\mathrm{pt}\}$. The morphism $Ra_{X!} a_X^! A_{\{\mathrm{pt}\}} \to A_{\{\mathrm{pt}\}}$ defines the morphism:

(3.3.14)
$$Ra_{X!}\omega_X \to A \,.$$

Taking the 0-th cohomology, we get the "integration morphism", that we shall denote by $\int_X$:

(3.3.15)
$$\int_X : H_c^n(X; or_X) \to A \,.$$

On the other hand, when $A = \mathbb{C}$ and X is a C^∞-manifold, there is a well-known morphism $H^n_c(X; or_X) \to \mathbb{C}$, obtained as follows. The sheaf or_X is quasi-isomorphic to the de Rham complex (cf. II §9):

$$0 \longrightarrow \mathscr{C}_X^{\infty,(0)} \otimes or_X \xrightarrow[d]{} \cdots \longrightarrow \mathscr{C}_X^{\infty,(n)} \otimes or_X \longrightarrow 0 .$$

Since $\mathscr{C}_X^{\infty,(j)} \otimes or_X$ is c-soft, we have

$$H^n_c(X; or_X) \simeq \Gamma_c(X; \mathscr{C}^{\infty,(n)} \otimes or_X)/\mathrm{d}\Gamma_c(X; \mathscr{C}_X^{\infty,(n-1)} \otimes or_X) .$$

If ϕ is a compacty supported density, i.e. an element of $\Gamma_c(X; \mathscr{C}_X^{\infty,(n)} \otimes or_X)$, $\int_X \phi$ makes sense and is zero if $\phi = \mathrm{d}\psi$, for some $\psi \in \Gamma_c(X; \mathscr{C}_X^{\infty,(n-1)} \otimes or_X)$, by Stokes's theorem. Hence $\int_X$ defines a morphism:

$$\text{(3.3.16)} \qquad \int_X : \Gamma_c(X; \mathscr{C}_X^{\infty,(n)} \otimes or_X)/\mathrm{d}\Gamma_c(X; \mathscr{C}_X^{\infty,(n-1)} \otimes or_X) \to \mathbb{C} .$$

This morphism (3.3.16) coincides with (3.3.15), up to the sign.

We shall not prove it here and leave it as an exercise (cf. Exercise III.20). We only note that if X is connected, then $H^n_c(X; or_X) \simeq \mathrm{Hom}(H^0(X; \mathbb{C}_X); \mathbb{C}) \simeq \mathbb{C}$, which implies that (3.3.15) and (3.3.16) are equal up to a non-zero constant.

We shall come back on this integration morphism in Chapter IX. To end this section, we shall give a bound to $\mathrm{hd}(\mathfrak{Mod}(A_X))$.

Proposition 3.3.11. *Let X be a C^0-manifold of dimension n and let A be a ring. Then the homological dimension (cf. Exercise I.17) of the category $\mathfrak{Mod}(A_X)$ is bounded by $3n + \mathrm{gld}(A) + 1$.*

Proof. Let F and G belong to $\mathfrak{Mod}(A_X)$. We want to prove:

$$\text{(3.3.17)} \qquad \mathrm{Hom}_{\mathrm{D}(A_X)}(G, F[j]) = 0 \qquad \text{for} \qquad j > 3n + \mathrm{gld}(A) + 1 .$$

Let δ_X denote the diagonal embedding $X \hookrightarrow X \times X$. By Proposition 3.1.14 we have:

$$R\mathscr{H}om(G, F) \simeq \delta_X^! R\mathscr{H}om(q_2^{-1}G, q_1^! F) .$$

Hence:

$$\begin{aligned} \mathrm{Hom}_{\mathrm{D}(A_X)}(G, F[j]) &\simeq H^j(R\Gamma(X; R\mathscr{H}om(G, F))) \\ &\simeq H^j(R\Gamma_\Delta(X \times X; R\mathscr{H}om(q_2^{-1}G, q_1^! F))) . \end{aligned}$$

By Proposition 3.1.15, we have

$$R\Gamma(U \times V; R\mathscr{H}om(q_2^{-1}G, q_1^! F)) \simeq R\mathrm{Hom}(R\Gamma_c(U; G), R\Gamma(U; F)) .$$

Therefore $H^j(R\mathscr{H}om(q_2^{-1}G, q_1^! F)) = 0$ for $j > n + \mathrm{gld}(A)$, by Proposition 3.2.2 (iv), and the result follows by Proposition 3.2.2 (v). □

Note that the estimate above is far from the best.

Corollary 3.3.12. *Let X be a C^0-manifold. Then $R\mathcal{H}om(\cdot,\cdot)$ is a well-defined functor from $\mathbf{D}^b(A_X)^\circ \times \mathbf{D}^b(A_X)$ to $\mathbf{D}^b(A_X)$.*

(Recall that we have assumed $\mathrm{gld}(A) < \infty$.)

3.4. Cohomologically constructible sheaves

Let X be a locally compact space with finite c-soft dimension. Recall (cf. Exercise I.30) that an object M of $\mathbf{D}^b(\mathfrak{Mod}(A))$ is called "perfect" if it is quasi-isomorphic to a bounded complex of finitely generated projective A-modules.

Definition 3.4.1. *An object F of $\mathbf{D}^b(A_X)$ is called cohomologically constructible, if for all $x \in X$, the conditions below are satisfied.*

(i) *$\text{“}\varinjlim_{x\in U}\text{”}\, R\Gamma(U;F)$ and $\text{“}\varprojlim_{x\in U}\text{”}\, R\Gamma_c(U;F)$ are representable (U ranges through the family of open neighborhoods of x).*

(ii) *$\text{“}\varinjlim_{x\in U}\text{”}\, R\Gamma(U;F) \to F_x$ and $R\Gamma_{\{x\}}(X;F) \to \text{“}\varprojlim_{x\in U}\text{”}\, R\Gamma_c(U;F)$ are isomorphisms.*

(iii) *The complexes F_x and $R\Gamma_{\{x\}}(X;F)$ are perfect.*

For "$\varprojlim$" and "$\varinjlim$", see I §11.

Remark 3.4.2. Note that (ii) follows from (i). In fact the first isomorphism is obvious. To prove the second one, let us take a decreasing sequence $\{K_n\}_n$ of compact subsets which forms a neighborhood system of x. Then "$\varprojlim$" $R\Gamma_c(X;F) \simeq$ "$\varprojlim$" $R\Gamma_{K_n}(X;F)$. Therefore, for each $k \in \mathbb{Z}$, "$\varprojlim$" $H^k_{K_n}(X;F)$ is representable, and the projective system $\{H^k_{K_n}(X;F)\}_n$ satisfies the M-L condition. By Proposition 2.7.1 we get $\varprojlim H^k_{K_n}(X;F) \simeq H^k_{\{x\}}(X;F)$.

Proposition 3.4.3. *Assume F is cohomologically constructible. Then:*

(i) *$\mathrm{D}F$ is cohomologically constructible (cf. Definition 3.1.16),*

(ii) *$F \to \mathrm{DD}F$ is an isomorphism,*

(iii) *for any $x \in X$, $R\Gamma_{\{x\}}(X;\mathrm{D}F) \simeq R\mathrm{Hom}(F_x, A)$ and $(\mathrm{D}F)_x \simeq R\mathrm{Hom}(R\Gamma_{\{x\}}(X;F), A)$.*

Proof. (i) and (iii) By (3.1.8), we have:

$$R\Gamma(U;\mathrm{D}F) \simeq R\mathrm{Hom}(R\Gamma_c(U;F), A)\ , \tag{3.4.1}$$

Applying the functor "$\varinjlim_{x\in U}$" we get:

$$\text{(3.4.2)} \qquad \text{“}\varinjlim_{x\in U}\text{”}\, R\Gamma(U;\mathrm{D}F) \simeq R\mathrm{Hom}\left(\text{“}\varprojlim_{x\in U}\text{”}\, R\Gamma_c(U;F), A\right)$$

$$\simeq R\mathrm{Hom}(R\Gamma_{\{x\}}(X;F), A) \ ,$$

which shows the second isomorphisms in (iii).

Therefore “$\varinjlim_{x\in U}$” $R\Gamma(U;\mathrm{D}F)$ is representable and perfect. Let K be a compact neighborhood of x, $\mathring{K}$ its interior. We have:

$$\begin{aligned} R\Gamma_K(X;\mathrm{D}F) &\simeq R\mathrm{Hom}(A_K, \mathrm{D}F) \\ &\simeq R\mathrm{Hom}(F_K, \omega_X) \\ &\simeq R\mathrm{Hom}(R\Gamma(X;F_K), A) \ . \end{aligned}$$

Applying the functor “$\varprojlim$”, we get:

$$\begin{aligned} \text{“}\varprojlim_{x\in U}\text{”}\, R\Gamma_c(U;\mathrm{D}F) &\simeq \text{“}\varprojlim_{x\in \mathring{K}}\text{”}\, R\Gamma_K(X;\mathrm{D}F) \\ &\simeq R\mathrm{Hom}\left(\text{“}\varinjlim_{x\in \mathring{K}}\text{”}\, R\Gamma(X;F_K), A\right) \\ &\simeq R\mathrm{Hom}\left(\text{“}\varinjlim_{x\in U}\text{”}\, R\Gamma(U;F), A\right) \\ &\simeq R\mathrm{Hom}(F_x, A) \ , \end{aligned}$$

and this complex is perfect, which shows the first isomorphism in (iii).

(ii) By (i), $\mathrm{D}F$ and $\mathrm{DD}F$ are both constructible. For $x \in X$ we have, by (iii):

$$\begin{aligned} (\mathrm{DD}F)_x &\simeq R\mathrm{Hom}(R\Gamma_{\{x\}}(X;\mathrm{D}F), A) \\ &\simeq R\mathrm{Hom}(R\mathrm{Hom}(F_x, A), A) \simeq F_x \ . \end{aligned}$$

Hence $F \to \mathrm{DD}F$ is an isomorphism. □

Proposition 3.4.4. *Let X and Y be two locally compact spaces with finite c-soft dimension, and let q_1 and q_2 denote the first and second projection from $X \times Y$ to X and Y respectively. Let $F \in \mathrm{Ob}(\mathbf{D}^b(A_X))$ and $G \in \mathrm{Ob}(\mathbf{D}^+(A_Y))$, and assume F is cohomologically constructible. Then:*

$$\mathrm{D}F \overset{L}{\boxtimes} G \to R\mathcal{H}om(q_1^{-1}F, q_2^! G)$$

is an isomorphism. If X is a C^0-manifold, then

$$\mathrm{D}'F \overset{L}{\boxtimes} G \to R\mathcal{H}om(q_1^{-1}F, q_2^{-1} G)$$

is an isomorphism.

Proof. It is enough to prove the first isomorphism.

Let U and V be open subsets of X and Y respectively. By Proposition 3.1.15 we have:

$$R\Gamma(U \times V; R\mathcal{H}om(q_1^{-1}F, q_2^!G)) \simeq R\mathrm{Hom}(R\Gamma_c(U;F), R\Gamma(V;G)) .$$

Applying the functor "$\varinjlim_{x \in U}$" we get:

$$\begin{aligned}
\text{“}\varinjlim_{x \in U}\text{”}\, R\Gamma(U \times V; R\mathcal{H}om(q_1^{-1}F, q_2^!G)) &\simeq R\mathrm{Hom}\left(\text{“}\varprojlim_{x \in U}\text{”}\, R\Gamma_c(U;F), R\Gamma(V;G)\right) \\
&\simeq R\mathrm{Hom}(R\Gamma_{\{x\}}(X;F), R\Gamma(V;G)) \\
&\simeq R\mathrm{Hom}(R\Gamma_{\{x\}}(X;F), A) \overset{L}{\otimes} R\Gamma(V;G) \\
&\simeq (\mathrm{D}F)_x \overset{L}{\otimes} R\Gamma(V;G) .
\end{aligned}$$

Therefore:

$$\begin{aligned}
R\mathcal{H}om(q_1^{-1}F, q_2^!G) &\simeq q_1^{-1}\mathrm{D}F \overset{L}{\otimes} q_2^{-1}G \\
&\simeq q_1^{-1}\mathrm{D}'F \overset{L}{\otimes} q_2^!G . \quad \square
\end{aligned}$$

Examples 3.4.5. (i) Let X be a C^0-manifold, Y a closed submanifold of codimension p. Then A_Y and $R\Gamma_Y(A_X)$ are cohomologically constructible on X, and moreover:

$$\mathrm{D}_X(A_Y) \simeq R\Gamma_Y(\omega_X) \simeq \omega_Y . \tag{3.4.3}$$

(ii) Assume X is a real finite-dimensional vector space, and let Z be a closed (resp. open) convex subset of X. Then A_Z and $R\Gamma_Z(A_X)$ are cohomologically constructible on X (cf. Exercise III.4).

Proposition 3.4.6. *Let F and G be two cohomologically constructible objects of $\mathrm{D}^b(X)$. Then:*

$$\begin{aligned}
R\mathcal{H}om(G, F) &\simeq R\mathcal{H}om(\mathrm{D}F, \mathrm{D}G) \\
&\simeq \mathrm{D}\left(\mathrm{D}(F) \overset{L}{\otimes} G\right) .
\end{aligned}$$

Proof. To prove that the canonical morphism:

$$R\mathcal{H}om(G, F) \to R\mathcal{H}om(\mathrm{D}F, \mathrm{D}G)$$

is an isomorphism, it is enough to check (cf. Proposition 3.1.14):

$$q_2^{-1}\mathrm{D}G \overset{L}{\otimes} q_1^! F \simeq q_1^{-1}(\mathrm{DD}F) \overset{L}{\otimes} q_2^! \mathrm{D}G \ ,$$

which is obvious.

On the other hand we have:

$$\begin{aligned} R\mathscr{H}om\left(\mathrm{D}F \overset{L}{\otimes} G, \omega_X\right) &\simeq R\mathscr{H}om(G, R\mathscr{H}om(\mathrm{D}F, \omega_X)) \\ &\simeq R\mathscr{H}om(G, \mathrm{DD}F) \\ &\simeq R\mathscr{H}om(G, F) \ . \quad \square \end{aligned}$$

We shall meet constructible sheaves all along this book.

3.5. γ-topology

Let V be a real finite-dimensional vector space, and let γ be a closed convex cone (with vertex at 0) in V. One may define a new topology on V, associated to γ.

Definition 3.5.1. *The γ-topology on V is the topology for which the open sets Ω satisfy:*

(i) *Ω is open for the usual topology,*
(ii) *$\Omega + \gamma = \Omega$.*

Let X be a subset of V. One denotes by X_γ the set X endowed with the induced γ-topology, and by

$$\phi_\gamma : X \to X_\gamma$$

the natural continuous map from X (with the usual topology) to X_γ. One may also write ϕ_γ^X instead of ϕ_γ to emphasize the space X.

Note that the $\{0\}$-topology is the usual topology, and if $\gamma_1 \subset \gamma_2$ are two closed convex cones, the natural map $X_{\gamma_1} \to X_{\gamma_2}$ is continuous.

Example 3.5.2. Take $X = \mathbb{R}$, $\gamma = [0, +\infty[$. The γ-open sets are the intervals $]c, +\infty[$; $-\infty \leqslant c \leqslant \infty$.

In the sequel we shall denote by γ^a the opposite cone to γ, that is, $\gamma^a = -\gamma$.

If Ω is a subset of X, we shall say that Ω is γ-open (resp. γ-closed, resp. a γ-neighborhood of x) if Ω is open (resp. closed, resp. a neighborhood of x) in the γ-topology.

Proposition 3.5.3. *Let X be a γ-open subset of V, and let F be a sheaf over X_γ.*

(i) *Let U be a convex open subset of X. Then the natural morphism $R\Gamma(U + \gamma; F) \to R\Gamma(U; \phi_\gamma^{-1} F)$ is an isomorphism.*

(ii) *Let K be a convex compact subset of X. Then the natural morphism $R\Gamma(K + \gamma; \phi_\gamma^{-1}F) \to R\Gamma(K; \phi_\gamma^{-1}F)$ is an isomorphism.*
(iii) *The natural morphism $F \to R\phi_{\gamma *}\phi_\gamma^{-1}F$ is an isomorphism.*

Proof. We decompose the proof into several steps.

(a) Let U be a convex subset of X, and let ψ be the natural map: $\Gamma(U + \gamma; F) \to \Gamma(U; \phi_\gamma^{-1}F)$. Let us prove that ψ is injective. Let s be a section of F over $U + \gamma$, such that $s_x = 0$ in $(\phi_\gamma^{-1}F)_x$ for all $x \in U$. For such an x, there exists a γ-open set W which contains x such that $s_y = 0$ for all $y \in W$. Therefore $s_{x+v} = 0$ for all $x \in U$, $v \in \gamma$.

Let us prove now that ψ is surjective. A section s of $\phi_\gamma^{-1}F$ all over U is defined by an open covering $U = \bigcup_{i \in I} U_i$ and sections s_i of F over $U_i + \gamma$, with $s_x = s_{i,x}$ for any $x \in U_i$. We shall show that $x \in (U_i + \gamma) \cap (U_j + \gamma)$ implies $s_{i,x} = s_{j,x}$. Let $x_i \in U_i \cap (x + \gamma^a)$, $x_j \in U_j \cap (x + \gamma^a)$, $x_t = tx_i + (1 - t)x_j$, $(t \in [0, 1])$. Then $x_t \in U \cap (x + \gamma^a)$. Let:

$$A = \{t \in [0, 1]; \text{for any } k \in I \text{ such that } x_t \in U_k, \text{ we have } s_{i,x} = s_{k,x}\} .$$

The set A can also be defined as:

$$A = \{t \in [0, 1]; \text{there exists } k \in I \text{ such that if } x_t \in U_k \text{ we have } s_{i,x} = s_{k,x}\} .$$

Therefore A is open and closed, and contains 1. Thus $A = [0, 1]$. Hence $s_i|_{(U_i+\gamma)\cap(U_j+\gamma)} = s_j|_{(U_i+\gamma)\cap(U_j+\gamma)}$. Therefore there is $\tilde{s} \in \Gamma(U + \gamma; F)$ such that $\tilde{s}|_{U_i+\gamma} = s_i$. This shows the surjectivity of ψ. Hence we have obtained the isomorphism:

$$\Gamma(U + \gamma; F) \simeq \Gamma(U; \phi_\gamma^{-1}F) \tag{3.5.1}$$

(in particular, $\Gamma(U + \gamma; F) \simeq \Gamma(U + \gamma; \phi_\gamma^{-1}F)$).

(b) Let K be a convex compact subset of X. By (a),

$$\varinjlim_U \Gamma(U; \phi_\gamma^{-1}F) \to \Gamma(K; \phi_\gamma^{-1}F)$$

is an isomorphism, where U ranges through the family of γ-open neighborhoods of K. For any convex compact set L with $K \subset L \subset K + \gamma$, any γ-open neighborhood of K contains L, and therefore

$$\Gamma(L; \phi_\gamma^{-1}F) \xrightarrow{\sim} \Gamma(K; \phi_\gamma^{-1}F) .$$

Since $\Gamma(K + \gamma; \phi_\gamma^{-1}F) \simeq \varprojlim_L \Gamma(L; \phi_\gamma^{-1}F)$, we get the isomorphism:

$$\Gamma(K + \gamma; \phi_\gamma^{-1}F) \xrightarrow{\sim} \Gamma(K; \phi_\gamma^{-1}F) . \tag{3.5.2}$$

(c) The natural morphism $F \to \phi_{\gamma *}\phi_\gamma^{-1}F$ is an isomorphism. In fact if U is a convex γ-open subset of X, then:

$$\Gamma(U;\phi_{\gamma*}\phi_\gamma^{-1}F) = \Gamma(U;\phi_\gamma^{-1}F)$$
$$\simeq \Gamma(U;F) \ .$$

(d) Assume F is flabby, and let Ω be a convex open subset of X, K a convex compact subset of Ω. The restriction morphism $\Gamma(\Omega;\phi_\gamma^{-1}F) \to \Gamma(K;\phi_\gamma^{-1}F)$ is surjective by (3.5.1). Applying Proposition 3.2.4 we obtain:

(3.5.3) $$R^j\Gamma(\Omega;\phi_\gamma^{-1}F) = 0 \quad \text{for} \quad j \neq 0 \ .$$

This implies that $R^j\Gamma(K;\phi_\gamma^{-1}F)$ and $R^j\phi_{\gamma*}\phi_\gamma^{-1}F$ are zero for $j \neq 0$. Therefore we have proved the Proposition under the additional hypothesis that F is flabby.

(e) Let F be a sheaf on X_γ and let $F^\cdot$ be a complex bounded from below of flabby sheaves, $F^\cdot$ being quasi-isomorphic to F. By (3.5.3) we have $R\Gamma(U+\gamma;F) \simeq \Gamma(U+\gamma;F^\cdot) \simeq \Gamma(U;\phi_\gamma^{-1}F^\cdot) \simeq R\Gamma(U;\phi_\gamma^{-1}F)$. This proves (i), and the proofs of (ii) and (iii) are similar. □

We shall now describe another method to obtain $\phi_\gamma^{-1}R\phi_{\gamma*}F$. Assume $X = V$, and set:

(3.5.4) $$Z(\gamma) = \{(x,y) \in X \times X; y - x \in \gamma\} \ .$$

Let us denote by q_1 and q_2 the first and second projection from $X \times X$.

Proposition 3.5.4. *Let $F \in \mathrm{Ob}(\mathbf{D}^+(X))$. There is a natural isomorphism:*

$$Rq_{1*}((q_2^{-1}F)_{Z(\gamma)}) \simeq \phi_\gamma^{-1}R\phi_{\gamma*}F \ .$$

Proof. Let $\tilde{q}_j : Z(\gamma) \to X$ be the restriction of q_j ($j = 1, 2$). Then we have, for any γ-open subset Ω of X:

$$\tilde{q}_1^{-1}\Omega = \{(x,y) \in X \times X; x \in \Omega, y \in x + \gamma\} \subset \tilde{q}_2^{-1}\Omega \ .$$

Hence for any sheaf G on $Z(\gamma)$, we have a morphism $(\phi_\gamma \circ \tilde{q}_2)_* G \to (\phi_\gamma \circ \tilde{q}_1)_* G$, and we obtain a morphism of functors from $\mathbf{D}^+(Z(\gamma))$ to $\mathbf{D}^+(X_\gamma)$: $R(\phi_\gamma \circ \tilde{q}_2)_* \to R(\phi_\gamma \circ \tilde{q}_1)_*$.

Thus for any $F \in \mathrm{Ob}(\mathbf{D}^+(X))$, we obtain the morphisms:

$$R\phi_{\gamma*}F \to R\phi_{\gamma*}R\tilde{q}_{2*}\tilde{q}_2^{-1}F \to R\phi_{\gamma*}R\tilde{q}_{1*}\tilde{q}_2^{-1}F \ ,$$

and this gives:

(3.5.5) $$\phi_\gamma^{-1}R\phi_{\gamma*}F \to R\tilde{q}_{1*}\tilde{q}_2^{-1}F \ .$$

Let us prove that this is an isomorphism.

For any convex compact set K, the projection $\tilde{q}_1^{-1}K \to \tilde{q}_2\tilde{q}_1^{-1}K = K + \gamma$ has contractible and proper fibers. Hence by Corollary 2.7.7 (iv),

$$R\Gamma(\tilde{q}_1^{-1}K;\tilde{q}_2^{-1}F) \simeq R\Gamma(K+\gamma;F) \ .$$

Thus, for any $x \in X$ and an integer j,

$$\begin{aligned} H^j(R\tilde{q}_{1*}\tilde{q}_2^{-1}F)_x &\simeq \varinjlim_K H^j(R\Gamma(K;R\tilde{q}_{1*}\tilde{q}_2^{-1}F)) \\ &\simeq \varinjlim_K H^j(R\Gamma(\tilde{q}_1^{-1}K;\tilde{q}_2^{-1}F)) \\ &\simeq \varinjlim_K H^j(R\Gamma(K+\gamma;F)) \\ &\simeq H^j(\phi_\gamma^{-1}R\phi_{\gamma*}F)_x , \end{aligned}$$

where K ranges over the family of convex compact neighborhoods of x. This shows that (3.5.5) is an isomorphism. □

3.6. Kernels

Let X and Y be two locally compact spaces with finite c-soft dimension and let q_1 and q_2 denote the first and second projection on $X \times Y$, respectively. Let $K \in \mathrm{Ob}(\mathbf{D}^b(X \times Y))$.

Definition 3.6.1. *We define the functors* $\Phi_K : \mathbf{D}^+(Y) \to \mathbf{D}^+(X)$ *and* $\Psi_K : \mathbf{D}^+(X) \to \mathbf{D}^+(Y)$ *by*:

$$\Phi_K(G) = Rq_{1!}\left(K \overset{L}{\otimes} q_2^{-1}G\right) ,$$

$$\Psi_K(F) = Rq_{2*}R\mathscr{H}om(K, q_1^!F) .$$

Proposition 3.6.2. *There exists an isomorphism of bifunctors on* $\mathbf{D}^+(Y)^\circ \times \mathbf{D}^+(X)$:

$$\mathrm{Hom}_{\mathbf{D}^+(X)}(\Phi_K(\cdot), \cdot) \simeq \mathrm{Hom}_{\mathbf{D}^+(Y)}(\cdot, \Psi_K(\cdot)) .$$

In other words, Ψ_K *and* Φ_K *are adjoint functors.*

Proof. Let $F \in \mathrm{Ob}(\mathbf{D}^+(X))$, $G \in \mathrm{Ob}(\mathbf{D}^+(Y))$. Then:

$$\begin{aligned} \mathrm{Hom}_{\mathbf{D}^+(X)}(\Phi_K(G), F) &= \mathrm{Hom}_{\mathbf{D}^+(X)}\left(Rq_{1!}\left(K \overset{L}{\otimes} q_2^{-1}G\right), F\right) \\ &\simeq \mathrm{Hom}_{\mathbf{D}^+(X\times Y)}\left(K \overset{L}{\otimes} q_2^{-1}G, q_1^!F\right) \\ &\simeq \mathrm{Hom}_{\mathbf{D}^+(X\times Y)}(q_2^{-1}G, R\mathscr{H}om(K, q_1^!F)) \\ &\simeq \mathrm{Hom}_{\mathbf{D}^+(Y)}(G, Rq_{2*}R\mathscr{H}om(K, q_1^!F)) . \quad \square \end{aligned}$$

Proposition 3.6.3. *Assume* $X = Y$, *and* $K = A_\Delta$, *where* Δ *is the diagonal of* $X \times X$. *Then* Φ_K *and* Ψ_K *are isomorphic to the identity in* $\mathbf{D}^+(X)$.

Proof. By Proposition 3.1.14 we have:

$$\begin{aligned} F &\simeq Rq_{2*}R\Gamma_\Delta R\mathscr{H}om(q_2^{-1}A_X, q_1^! F) \\ &\simeq Rq_{2*}R\mathscr{H}om(A_\Delta, q_1^! F) \\ &\simeq \Psi_K(F) \ . \end{aligned}$$

Since $A_\Delta \otimes q_2^{-1}G \simeq (q_2^{-1}G)_\Delta$, we have similarly $\Phi_K(G) \simeq G$. □

Let Z be a third locally compact space with finite c-soft dimension. We denote by q_1' and q_2' (resp. q_1'' and q_2'') the first and second projection on $X \times Z$ (resp. on $Y \times Z$) and we introduce the notation q_{ij} to denote the (i, j)-projection on $X \times Y \times Z$. For example q_{13} is the projection onto $X \times Z$.

(3.6.1)

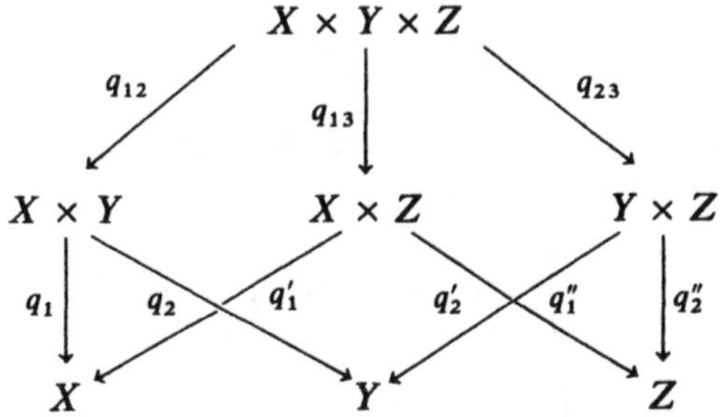

Proposition 3.6.4. *Let* $K_1 \in \mathrm{Ob}(\mathbf{D}^b(X \times Y))$, $K_2 \in \mathrm{Ob}(\mathbf{D}^b(Y \times Z))$. *Set:*

$$K = Rq_{13!}\left(q_{12}^{-1}K_1 \overset{L}{\otimes} q_{23}^{-1}K_2\right) .$$

Then $\Psi_{K_2} \circ \Psi_{K_1} \simeq \Psi_K$ *and* $\Phi_{K_1} \circ \Phi_{K_2} \simeq \Phi_K$.

Proof. Let $F \in \mathrm{Ob}(\mathbf{D}^+(A_X))$. Then:

$$\begin{aligned} \Psi_{K_2} \circ \Psi_{K_1}(F) &= Rq_{2*}''R\mathscr{H}om(K_2, q_1''^! Rq_{2*}R\mathscr{H}om(K_1, q_1^! F)) \\ &\simeq Rq_{2*}''R\mathscr{H}om(K_2, Rq_{23*}q_{12}^! R\mathscr{H}om(K_1, q_1^! F)) \\ &\simeq Rq_{2*}''Rq_{23*}R\mathscr{H}om(q_{23}^{-1}K_2, R\mathscr{H}om(q_{12}^{-1}K_1, q_{12}^! q_1^! F)) \\ &\simeq Rq_{2*}'Rq_{13*}R\mathscr{H}om\left(q_{12}^{-1}K_1 \overset{L}{\otimes} q_{23}^{-1}K_2, q_{13}^! q_1'^! F\right) \\ &\simeq Rq_{2*}'R\mathscr{H}om\left(Rq_{13!}\left(q_{12}^{-1}K_1 \overset{L}{\otimes} q_{23}^{-1}K_2\right), q_1'^! F\right) \\ &= \Psi_K(F) \ . \end{aligned}$$

The proof for Φ_K is similar. □

We set:

$$K_1 \circ K_2 = Rq_{13!}\left(q_{12}^{-1}K_1 \overset{L}{\otimes} q_{23}^{-1}K_2\right) . \tag{3.6.2}$$

Corollary 3.6.5. *Assume $Z = X$, $K_2 \circ K_1 \simeq A_{\Delta_X}[l\,]$, $K_1 \circ K_2 \simeq A_{\Delta_Y}[l']$ for some shifts l and l' (Δ_X and Δ_Y denote the diagonal of $X \times X$ and $Y \times Y$, respectively). Then Φ_{K_1}, Φ_{K_2}, Ψ_{K_1} and Ψ_{K_2} are equivalence of categories.*

Note that in this case, l and l' must be equal if $X \neq \emptyset$.

Example 3.6.6. Let $\tau : E \to X$ be a real vector bundle over a locally compact space X, with fiber dimension n, and let $\pi : E^* \to X$ be the dual vector bundle. Let $\dot{E}$ (resp. $\dot{E}^*$) be the space E (resp. E^*) with the zero-section removed, and put:

$$S = \dot{E}/\mathbb{R}^+ \ , \qquad S^* = \dot{E}^*/\mathbb{R}^+ \ . \tag{3.6.3}$$

The projections from S and S^* to X are topological submersions with fiber dimension $n - 1$. We introduce the sets:

$$\begin{cases} D = \left\{(x, y) \in S \underset{X}{\times} S^*; \langle x, y\rangle \geqslant 0\right\} \ , \\ I = \left\{(y, x) \in S^* \underset{X}{\times} S; \langle y, x\rangle > 0\right\} \ , \end{cases} \tag{3.6.4}$$

and the objects of $\mathbf{D}^b(S \times S^*)$ and $\mathbf{D}^b(S^* \times S)$:

$$\begin{cases} K_1 = A_D \otimes \omega_{S^*/X} \ , \\ K_2 = A_I \ . \end{cases} \tag{3.6.5}$$

Proposition 3.6.7. *One has:*

$$K_1 \circ K_2 \simeq A_{\Delta_S} \qquad \textit{in } \mathbf{D}^b(S \times S) \ ,$$

$$K_2 \circ K_1 \simeq A_{\Delta_{S^*}} \qquad \textit{in } \mathbf{D}^b(S^* \times S^*) \ .$$

Proof. We denote by p_{ij} the (i, j)-th projection on $S \times S^* \times S$. Consider the diagram:

$$\begin{array}{ccccc} & & S \times S^* \times S & & \\ & \swarrow^{p_{12}} & \downarrow{\scriptstyle p_{13}} & \searrow^{p_{23}} & \\ S \times S^* & & S \times S & & S^* \times S \ . \end{array}$$

Set $L = p_{12}^{-1}(D) \cap p_{23}^{-1}(I)$. Then $K_1 \circ K_2 \simeq Rp_{13!}(A_L \otimes \omega_{S^*/X})$.

Let $(x, x') = z \in S \times S$. If $z \notin S \times_X S$, then $p_{13}^{-1}(z) \cap L = \emptyset$. If $z \in S \times_X S$, but $z \notin S \times_S S$, then $p_{13}^{-1}(z) \cap L$ is homeomorphic to a closed half-space of $\mathbb{R}^{n-1}$ or else is empty. Therefore in this case $(K_1 \circ K_2)_z = 0$.

Let $U = p_{13}^{-1}(\Delta_S) \cap L$. Since $(Rp_{13!}A_L)_{\Delta_S} \simeq Rp_{13!}(A_U)$, and $U \to \Delta_S$ is isomorphic to the map $\pi : I \to S$, we obtain:

$$K_1 \circ K_2|_{\Delta_S} \simeq R\pi_!(A_I \otimes \omega_{S^*/X}) \simeq R\pi_!\pi^! A_S \ .$$

The map $\pi : I \to S$ is isomorphic to the projection $S \times \mathbb{R}^{n-1} \to S$ locally on S. Therefore $R\pi_!\pi^! A_S \to A_S$ is an isomorphism.

The proof for $K_2 \circ K_1$ is similar. □

By Corollary 3.6.5, we see that Φ_{K_1}, Φ_{K_2}, Ψ_{K_1} and Ψ_{K_2} are equivalences of categories.

3.7. Fourier-Sato transformation

Let us first define **conic sheaves.** Let $\mathbb{R}^+$ denote the multiplicative group of strictly positive numbers, and let X be a locally compact space, endowed with an action of $\mathbb{R}^+$. In other words we have a continuous map:

$$\mu : X \times \mathbb{R}^+ \to X \ ,$$

which satisfies for each $x \in X$, $t_1, t_2 \in \mathbb{R}^+$:

$$\text{(3.7.1)} \qquad \begin{cases} \mu(x, t_1 t_2) = \mu(\mu(x, t_1), t_2) \ , \\ \mu(x, 1) = x \ . \end{cases}$$

Definition 3.7.1. (i) *We denote by $\mathfrak{Mod}_{\mathbb{R}^+}(A_X)$ the full subcategory of $\mathfrak{Mod}(A_X)$ consisting of sheaves F such that for any orbit b of $\mathbb{R}^+$ in X, $F|_b$ is a locally constant sheaf.*

(ii) *We denote by $\mathbf{D}^+_{\mathbb{R}^+}(A_X)$ (or simply $\mathbf{D}^+_{\mathbb{R}^+}(X)$) the full subcategory of $\mathbf{D}^+(A_X)$ consisting of objects F such that for all $j \in \mathbb{Z}$, $H^j(F) \in \mathfrak{Mod}_{\mathbb{R}^+}(A_X)$.*

(iii) *We call an object of $\mathfrak{Mod}_{\mathbb{R}^+}(A_X)$ (resp. $\mathbf{D}^+_{\mathbb{R}^+}(X)$) a conic object.*

Consider the maps:

$$\text{(3.7.2)} \qquad X \xhookrightarrow[j]{} X \times \mathbb{R}^+ \overset{\mu}{\underset{p}{\rightrightarrows}} X \ ,$$

where $j(x) = (x; 1)$, and p is the projection. One has natural morphisms:

$$\text{(3.7.3)} \qquad \mu^{-1}F \xleftarrow[\alpha]{} p^{-1}Rp_*\mu^{-1}F \xrightarrow[\beta]{} p^{-1}F \ ,$$

where β is defined by the morphism of evaluation at 1:

$$Rp_*\mu^{-1}F \to Rp_*Rj_*j^{-1}\mu^{-1}F \simeq F \ .$$

Proposition 3.7.2. *Let $F \in \mathrm{Ob}(\mathbf{D}^+(X))$. Then the following conditions are equivalent.*

(i) $F \in \mathrm{Ob}(\mathbf{D}^+_{\mathbb{R}^+}(X))$.
(ii) *The morphisms α and β in (3.7.3) are isomorphisms.*
(iii) *For all $j \in \mathbb{Z}$, $H^j(\mu^{-1}F)$ is locally constant on the fibers of p.*
(iv) $\mu^{-1}F \simeq p^{-1}F$.
(v) $\mu^! F \simeq p^! F$.

Proof. (iv) $\Leftrightarrow$ (v) and (i) $\Leftrightarrow$ (iii) are clear, as well as (ii) $\Rightarrow$ (iv) and (iv) $\Rightarrow$ (iii). Finally, (iii) $\Rightarrow$ (ii) follows from Corollary 2.7.7. □

Corollary 3.7.3. *Let U be an open subset of X. Assume that for any orbit b of $\mathbb{R}^+$ in X, $b \cap U$ is contractible (in particular, non empty). Then for $F \in \mathrm{Ob}(\mathbf{D}^+_{\mathbb{R}^+}(X))$ the restriction morphism $R\Gamma(X;F) \to R\Gamma(U;F)$ is an isomorphism.*

Proof. Let $\mu' : U \times \mathbb{R}^+ \to X$ be the restriction of μ. Then μ' has contractible fibers.

Applying Proposition 3.3.9, we get:

$$F \simeq R\mu'_* \mu'^{-1} F \ ,$$

hence:

$$\begin{aligned} R\Gamma(X;F) &\simeq R\Gamma(U \times \mathbb{R}^+; \mu'^{-1}F) \\ &\simeq R\Gamma(U \times \mathbb{R}^+; p^{-1}F) \\ &\simeq R\Gamma(U;F) \ . \quad \square \end{aligned}$$

Let X and Y be two spaces endowed with an action of $\mathbb{R}^+$. Then the space $X \times Y$ is endowed with an action of $\mathbb{R}^+ \times \mathbb{R}^+$ and we leave to the reader the definition of the category $\mathbf{D}^+_{\mathbb{R}^+ \times \mathbb{R}^+}(X \times Y)$ and the notion of **biconic sheaf**.

If $F \in \mathrm{Ob}(\mathbf{D}^+_{\mathbb{R}^+}(X))$ and $G \in \mathrm{Ob}(\mathbf{D}^+_{\mathbb{R}^+}(Y))$ then $F \boxtimes^L G \in \mathrm{Ob}(\mathbf{D}^+_{\mathbb{R}^+ \times \mathbb{R}^+}(X \times Y))$. Note that a biconic sheaf on $X \times Y$ is in particular a conic sheaf, for the diagonal action of $\mathbb{R}^+$ on $X \times Y$ via the diagonal embedding $\mathbb{R}^+ \hookrightarrow \mathbb{R}^+ \times \mathbb{R}^+$.

Now let $f: Y \to X$ be a continuous map, and assume $f_!$ has finite cohomological dimension, and f commutes with the action of $\mathbb{R}^+$. We get a commutative diagram:

(3.7.4)
$$\begin{array}{ccccc} Y & \xleftarrow{\mu} & Y \times \mathbb{R}^+ & \xrightarrow{p_Y} & Y \\ {\scriptstyle f}\downarrow & & {\scriptstyle f}\downarrow & & \downarrow{\scriptstyle f} \\ X & \xleftarrow{\mu} & X \times \mathbb{R}^+ & \xrightarrow{p_X} & X \ . \end{array}$$

Proposition 3.7.4. (i) *Let $F \in \mathrm{Ob}(\mathbf{D}^+_{\mathbb{R}^+}(X))$. Then $f^{-1}F$ and $f^! F$ are conic.*

(ii) *Let $G \in \mathrm{Ob}(\mathbf{D}^+_{\mathbb{R}^+}(Y))$. Then $Rf_* G$ and $Rf_! G$ are conic.*

(iii) *Let* F_1 *and* F_2 *belong to* $\mathrm{Ob}(\mathbf{D}^+_{\mathbb{R}^+}(X))$. *Then* $F_1 \otimes^L F_2$ *is conic. If* $F_1 \in \mathrm{Ob}(\mathbf{D}^b_{\mathbb{R}^+}(X))$, *then* $R\mathcal{H}om(F_1, F_2)$ *is conic.*

Proof. We shall apply Proposition 3.7.2.

(i) We have the isomorphisms:

$$\mu^{-1}f^{-1}F \simeq \tilde{f}^{-1}\mu^{-1}F \simeq \tilde{f}^{-1}p_X^{-1}F \simeq p_Y^{-1}f^{-1}F \ .$$

Thus $f^{-1}F$ is conic. The proof for $f^!$ is similar.

(ii) We have the isomorphisms:

$$\mu^! Rf_* G \simeq R\tilde{f}_* \mu^! G \simeq R\tilde{f}_* p_Y^! G \simeq p_X^! Rf_* G \ .$$

Thus $Rf_* G$ is conic. The proof for $Rf_!$ is similar.

(iii) We have:

$$\begin{aligned}\mu^{-1}\left(F_1 \overset{L}{\otimes} F_2\right) &\simeq \mu^{-1}F_1 \overset{L}{\otimes} \mu^{-1}F_2 \\ &\simeq p^{-1}F_1 \overset{L}{\otimes} p^{-1}F_2 \\ &\simeq p^{-1}\left(F_1 \overset{L}{\otimes} F_2\right) ,\end{aligned}$$

$$\begin{aligned}\mu^! R\mathcal{H}om(F_1, F_2) &\simeq R\mathcal{H}om(\mu^{-1}F_1, \mu^! F_2) \\ &\simeq R\mathcal{H}om(p^{-1}F_1, p^! F_2) \\ &\simeq p^! R\mathcal{H}om(F_1, F_2) \ . \quad \square\end{aligned}$$

Let $\tau : E \to Z$ be a real vector bundle with fiber dimension n over a locally compact space Z.

We identify Z with the zero-section of E, and denote by $i : Z \hookrightarrow E$ this embedding. We set:

$$\dot{E} = E \backslash Z \ , \qquad \dot{\tau} = \tau|_{\dot{E}} \ . \tag{3.7.5}$$

We denote by "a" the **antipodal map** $x \mapsto -x$ on E, and if A is a subset of E, we denote by A^a its image by a. We say that a subset A of E is **convex** (resp. is **conic**, resp. is a **proper cone**) if for any $z \in Z$, the set $\tau^{-1}(z) \cap A$ is convex (resp. is conic, resp. is a proper cone). Recall that a cone is proper if it contains no lines.

Proposition 3.7.5. *Let* $F \in \mathrm{Ob}(\mathbf{D}^+_{\mathbb{R}^+}(E))$. *Then*:

(i) $R\tau_* F \simeq i^{-1}F$.
(ii) $R\tau_! F \simeq i^! F$.

Proof. (i) The morphism $\tau^{-1}R\tau_* F \to F$ defines the morphism $R\tau_* F \simeq i^{-1}\tau^{-1}R\tau_* F \to i^{-1}F$. To check this is an isomorphism, let us take $z \in Z$, and let

U be a convex open neighborhood of $i(z)$. Then by Corollary 3.7.3:

$$\begin{aligned} R\Gamma(U;F) &\simeq R\Gamma(\mathbb{R}^+U;F) \\ &\simeq R\Gamma(\tau(U);R\tau_*F) \ . \end{aligned}$$

Taking the cohomology of both sides, and the inductive limit when U ranges through the family of open convex neighborhoods of $i(z)$, we get the result.

(ii) The morphism $i_!i^!F \to F$ defines:

$$i^!F \simeq R\tau_!i_!i^!F \to R\tau_!F \ .$$

To prove this is an isomorphism, we argue as in (i). Let U be a convex open neighborhood of an open subset V of Z such that $\tau^{-1}V \cap \bar{U} \to V$ is proper. Then by Corollary 3.7.3:

$$R\Gamma_Z(\tau^{-1}V;F) \simeq R\Gamma_{\bar{U}}(\tau^{-1}V;F) \ .$$

Taking the cohomology of both sides, and the inductive limit when U ranges through the family of subsets of $\tau^{-1}V$ satisfying the property above, we get $R\Gamma(V;i^!F) \simeq R\Gamma(V;R\tau_!F)$, which shows $i^!F \simeq R\tau_!F$. □

Let $\pi: E^* \to Z$ be the dual vector bundle. We define similarly $i: Z \hookrightarrow E^*$, $\dot{E}^*$ and $\dot{\pi}$.

If A is a subset of E, the **polar set** A° is defined by:

(3.7.6) $A^\circ = \{y \in E^*; \pi(y) \in \tau(A) \text{ and } \langle x,y\rangle \geq 0 \text{ for all } x \in \tau^{-1}\pi(y) \cap A\}$.

We denote by p_1 and p_2 the first and second projections from $E \times_Z E^*$:

$$\begin{array}{ccccc} & & E \underset{Z}{\times} E^* & & \\ & \swarrow^{p_1} & & \searrow^{p_2} & \\ E & & & & E^* \\ & \searrow_{\tau} & & \swarrow_{\pi} & \\ & & Z & & \end{array} .$$

We introduce:

$$P = \left\{(x,y) \in E \underset{Z}{\times} E^*; \langle x,y\rangle \geq 0\right\} ,$$

$$P' = \left\{(x,y) \in E \underset{Z}{\times} E^*; \langle x,y\rangle \leq 0\right\}$$

and the functors (cf. III §6):

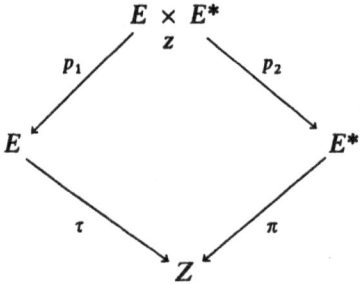

$$(3.7.7)\qquad \begin{cases} \tilde{\Psi}_{P'} = Rp_{1*} \circ R\Gamma_{P'} \circ p_2^! \ , \\ \tilde{\Phi}_{P'} = Rp_{2!} \circ (\cdot)_{P'} \circ p_1^{-1} \ , \\ \tilde{\Psi}_{P} = Rp_{2*} \circ R\Gamma_{P} \circ p_1^{-1} \ , \\ \tilde{\Phi}_{P} = Rp_{1!} \circ (\cdot)_{P} \circ p_2^! \ . \end{cases}$$

It follows from Proposition 3.7.4 that these functors are well-defined from $\mathbf{D}^+_{\mathbb{R}^+}(E)$ to $\mathbf{D}^+_{\mathbb{R}^+}(E^*)$ or from $\mathbf{D}^+_{\mathbb{R}^+}(E^*)$ to $\mathbf{D}^+_{\mathbb{R}^+}(E)$.

Let $\mathscr{or}_{E/Z}$ be the sheaf of relative orientation of E over Z. Since this sheaf is constant on the fibers of τ, we sometimes identify $\mathscr{or}_{E/Z}$ and its restriction to Z. Note that $\mathscr{or}_{E/Z}|_Z \simeq \mathscr{or}_{Z/E}$ (where $Z \to E$ is defined by the zero embedding). Moreover $\mathscr{or}_{Z/E}$ and $\mathscr{or}_{Z/E^*}$ are naturally isomorphic, since an orientation on a vector space defines an orientation on the dual space (cf. Remark 3.7.11 below).

In order to study the functors in (3.7.7) we need a lemma.

Lemma 3.7.6. *Let* $F \in \mathrm{Ob}(\mathbf{D}^+_{\mathbb{R}^+}(E))$. *Then* $\mathrm{supp}((R\Gamma_P(p_1^{-1}F))_{P'})$ *is contained in* $Z \times_Z E^* \subset E \times_Z E^*$.

Proof. Set $U = E \times_Z E^* \setminus Z \times_Z E^*$. Locally on U, the projection p_1 is isomorphic to the projection $\mathbb{R}^n \times E \to E$ and the sets P and P' are locally isomorphic to the sets $\{x_1 \geqslant 0\}$ and $\{x_1 \leqslant 0\}$, respectively, where x_1 denotes the first coordinate on $\mathbb{R}^n$. Hence in order to prove that $(R\Gamma_P(p_1^{-1}F))_{P'}$ is zero on U, it is enough to prove that if X is a locally compact space, if p denotes the projection $X \times \mathbb{R} \to X$ and if t is a coordinate on $\mathbb{R}$, then:

$$(3.7.8)\qquad (R\Gamma_{\{t \geqslant 0\}}(p^{-1}G))_{\{t=0\}} = 0 \ ,$$

for any $G \in \mathrm{Ob}(\mathbf{D}^+(X))$. Since $p^{-1}G$ is conic on the vector bundle $X \times \mathbb{R}$ over X, we have:

$$(R\Gamma_{\{t \geqslant 0\}}(p^{-1}G))|_{t=0} \simeq Rp_* R\Gamma_{\{t \geqslant 0\}}(p^{-1}G) \ .$$

In order to see that it vanishes, it is enough to show the same statement with $p^{-1}G$ replaced by $p^!G$. Then $Rp_* R\Gamma_{\{t \geqslant 0\}}(p^!G) \simeq R\mathscr{H}om(Rp_! A_{\{t \geqslant 0\}}, G)$ and $Rp_! A_{\{t \geqslant 0\}} = 0$. □

Theorem 3.7.7. *The two functors from* $\mathbf{D}^+_{\mathbb{R}^+}(E)$ *to* $\mathbf{D}^+_{\mathbb{R}^+}(E^*)$, $\tilde{\Phi}_{P'}$ *and* $\tilde{\Psi}_P$ *are naturally isomorphic.*

Proof. Let $F \in \mathrm{Ob}(\mathbf{D}^+_{\mathbb{R}^+}(E))$. We have the isomorphisms:

$$\begin{aligned} \tilde{\Phi}_{P'}(F) &= Rp_{2!}(p_1^{-1}F)_{P'} \\ &\simeq Rp_{2!} R\Gamma_P((p_1^{-1}F)_{P'}) \\ &\simeq Rp_{2!}((R\Gamma_P(p_1^{-1}F))_{P'}) \\ &\simeq Rp_{2*}((R\Gamma_P(p_1^{-1}F))_{P'}) \\ &\simeq Rp_{2*} R\Gamma_P(p_1^{-1}F) \ . \end{aligned}$$

In fact the first isomorphism follows from Proposition 3.7.5 (ii), the second one from Exercise II.2, the third one from Lemma 3.7.6 and the last one from Proposition 3.7.5 (i). □

This implies that the two functors from $\mathbf{D}^+_{\mathbb{R}^+}(E^*)$ to $\mathbf{D}^+_{\mathbb{R}^+}(E)$, $\tilde{\Phi}_P$ and $\tilde{\Psi}_{P'}$ are also isomorphic.

Definition 3.7.8. *Let $F \in \mathrm{Ob}(\mathbf{D}^+_{\mathbb{R}^+}(E))$. We set:*

$$F^\wedge = \tilde{\Phi}_{P'}(F)(= Rp_{2!}(p_1^{-1}F)_{P'})$$

$$(\simeq \tilde{\Psi}_P(F) = Rp_{2*}R\Gamma_P(p_1^{-1}F))$$

and call $F^\wedge$ the Fourier-Sato transform of F.

Let $G \in \mathrm{Ob}(\mathbf{D}^+_{\mathbb{R}^+}(E^*))$. We set:

$$G^\vee = \tilde{\Psi}_{P'}(G)(= Rp_{1*}R\Gamma_{P'}(p_2^!G))$$

$$(\simeq \tilde{\Phi}_P(G) = Rp_{1!}(p_2^!G)_P)$$

and call $G^\vee$ the inverse Fourier-Sato transform of G.

Of course one can use the same formulas to define $F^\vee$ and $G^\wedge$ by interchanging E and E^*. Note that:

$$F^\vee \simeq (F^\wedge)^a \otimes or_{E^*/Z}[n] \simeq (F^\wedge)^a \otimes \omega_{E^*/Z} \, , \tag{3.7.9}$$

where $(F^\wedge)^a$ is the inverse image of $F^\wedge$ by the antipodal map on E^*.

In fact (3.7.9) follows from the definition since $p_1^!F \simeq p_1^{-1}F \otimes \omega_{E\times_Z E^*/E}$.

Theorem 3.7.9. *The functors $^\wedge$ from $\mathbf{D}^+_{\mathbb{R}^+}(E)$ to $\mathbf{D}^+_{\mathbb{R}^+}(E^*)$ and $^\vee$ from $\mathbf{D}^+_{\mathbb{R}^+}(E^*)$ to $\mathbf{D}^+_{\mathbb{R}^+}(E)$, are equivalences of categories, inverse to each other. In particular if F and F' belong to $\mathbf{D}^+_{\mathbb{R}^+}(E)$, then:*

$$\mathrm{Hom}_{\mathbf{D}^+_{\mathbb{R}^+}(E)}(F', F) \simeq \mathrm{Hom}_{\mathbf{D}^+_{\mathbb{R}^+}(E^*)}(F'^\wedge, F^\wedge) \, .$$

Proof. By Proposition 3.6.2 and Theorem 3.7.7 we know that $\tilde{\Phi}_{P'}$ and $\tilde{\Psi}_{P'}$ are adjoint functors and $\tilde{\Phi}_P$ and $\tilde{\Psi}_P$ are adjoint functors, after having chosen an isomorphism $or_{Z/E} \simeq or_{Z/E^*}$. Therefore if $F \in \mathrm{Ob}(\mathbf{D}^+_{\mathbb{R}^+}(E))$, we get a morphism:

$$F \to F^{\wedge\vee} \, . \tag{3.7.10}$$

In order to prove that (3.7.10) is an isomorphism it is enough to prove that for each open convex subset U of E, it induces an isomorphism:

$$H^j(U; F) \simeq H^j(U; F^{\wedge\vee}) \, .$$

Applying Corollary 3.7.3, we may assume U is a convex open cone. In such a case we have:

$$\begin{aligned} H^j(U;F^{\wedge\vee}) &= \mathrm{Hom}_{\mathbf{D}^+_{\mathbb{R}^+}(E)}(A_U, \tilde{\Psi}_{P'}\tilde{\Phi}_{P'}(F)[j]) \\ &\simeq \mathrm{Hom}_{\mathbf{D}^+_{\mathbb{R}^+}(E^*)}(\tilde{\Phi}_{P'}(A_U), \tilde{\Phi}_{P'}(F)[j]) \\ &\simeq \mathrm{Hom}_{\mathbf{D}^+_{\mathbb{R}^+}(E^*)}(\tilde{\Phi}_{P'}(A_U), \tilde{\Psi}_{P}(F)[j]) \\ &\simeq \mathrm{Hom}_{\mathbf{D}^+_{\mathbb{R}^+}(E)}(\tilde{\Phi}_P\tilde{\Phi}_{P'}(A_U), F[j]) \\ &\overset{\alpha}{\leftarrow} \mathrm{Hom}_{\mathbf{D}^+_{\mathbb{R}^+}(E)}(A_U, F[j]) \ , \end{aligned}$$

and α comes from the morphisms $\tilde{\Phi}_P\tilde{\Phi}_{P'}(A_U) \simeq \tilde{\Phi}_P\tilde{\Psi}_P(A_U) \to A_U$. Hence it is enough to prove that $\tilde{\Phi}_P\tilde{\Psi}_P(A_U)$ is isomorphic to A_U. This is done in Lemma 3.7.10 below. Similarly we can prove that for $G \in \mathrm{Ob}(\mathbf{D}^+_{\mathbb{R}^+}(E^*))$, $G^{\vee\wedge} \to G$ is an isomorphism. □

Lemma 3.7.10. (i) *Let γ be a proper closed convex cone of E containing the zero-section. Then*:

$$(A_\gamma)^\wedge \simeq A_{\mathrm{Int}\,\gamma^\circ} \ .$$

(ii) *Let U be a convex open cone of E. Then*:

$$(A_U)^\wedge \simeq A_{U^{\circ a}} \otimes or_{E^*/Z}[-n] \ .$$

Proof. (i) Let $y \in E^*$. Then:

$$\begin{aligned} ((A_\gamma)^\wedge)_y &\simeq R\Gamma_c(p_2^{-1}(y); A_{(\gamma\times_Z E^*)\cap P'}) \\ &\simeq R\Gamma_c(p_1(p_2^{-1}(y)\cap P')\cap\gamma; A_E) \ . \end{aligned}$$

Set $\gamma_y = p_1(p_2^{-1}(y)\cap P')\cap\gamma$. If $y \notin \mathrm{Int}\,\gamma^\circ$, γ_y is a closed convex proper cone which contains a half-line. Then $\tau^{-1}(\pi(y))$ and $\tau^{-1}(\pi(y))\backslash\gamma_y$ being homeomorphic, we find $((A_\gamma)^\wedge)_y = 0$ in that case. If $y \in \mathrm{Int}\,\gamma^\circ$, then $\gamma_y = \{0\}$ and the map $((A_\gamma)^\wedge)_y \to ((A_Z)^\wedge)_y$ is an isomorphism. This shows that $(A_\gamma)^\wedge \simeq ((A_Z)^\wedge)_{\mathrm{Int}\,\gamma^\circ}$, and (i) follows since $(A_Z)^\wedge \simeq A_{E^*}$.

(ii) The proof is similar. We have:

$$\begin{aligned} ((A_U)^\wedge)_y &\simeq R\Gamma_c(p_2^{-1}(y); A_{(U\times E^*)\cap P'}) \\ &\simeq R\Gamma_c(p_1(p_2^{-1}(y)\cap P')\cap U; A_E) \ . \end{aligned}$$

Set $\gamma_y = p_1(p_2^{-1}(y)\cap P')\cap U$. If $y \notin U^{\circ a}$ then $\gamma_y = \emptyset$ and $((A_U)^\wedge)_y = 0$. If $y \in U^{\circ a}$, $((A_U)^\wedge)_y \to (Rp_{2!}A_{E\times_Z E^*})_y$ is an isomorphism.

Hence $(A_U)^\wedge \simeq (Rp_{2!}A_{E\times_Z E^*})_{U^{\circ a}}$. Then the result follows from the isomorphism $Rp_{2!}A_{E\times_Z E^*} \simeq or_{E^*/Z}[-n]$. □

Remark 3.7.11. Since $\tilde{\Phi}_{P'}$ and $\tilde{\Psi}_{P'}$ as well as $\tilde{\Phi}_P$ and $\tilde{\Psi}_P$ are adjoint to each other, we have morphisms:

$$\alpha'(F): F \to \tilde{\Psi}_{P'}\tilde{\Phi}_{P'}(F) \ ,$$

$$\beta'(G): \tilde{\Phi}_{P'}\tilde{\Psi}_{P'}(G) \to G \ ,$$

$$\alpha(G): G \to \tilde{\Psi}_{P}\tilde{\Phi}_{P}(G) \ ,$$

$$\beta(F): \tilde{\Phi}_{P}\tilde{\Psi}_{P}(F) \to F \ .$$

Here, in order to define $\alpha(G)$ and $\beta(F)$ we need to identify $R\tau_* \omega_{E/Z}$ and $R\pi_* \omega_{E^*/Z}$. We shall take the following identification (note that the problem is local).

Choose any negative definite symmetric form on E. The associated isomorphism $E \xrightarrow{\sim} E^*$ yields an isomorphism between $R\tau_* \omega_{E/Z}$ and $R\pi_* \omega_{E^*/Z}$.

Then, after having identified $\tilde{\Phi}_{P'}$ and $\tilde{\Psi}_P$ as well as $\tilde{\Phi}_P$ and $\tilde{\Psi}_{P'}$ (Theorem 3.7.7), we can show that $\alpha'(F)$ (resp. $\alpha(G)$) and $\beta(F)$ (resp. $\beta'(G)$) are inverse to each other. But we shall not give here the proofs.

Let us summarize some properties of the Fourier-Sato transformation.

Proposition 3.7.12. *Let* $F \in \mathrm{Ob}(\mathbf{D}^+_{\mathbb{R}^+}(E))$.

(i) $F^{\wedge\wedge} \simeq F^a \otimes or_{E/Z}[-n]$.

(ii) *Let U be a convex open subset of E^*. Then*:

$$R\Gamma(U; F^\wedge) \simeq R\Gamma_{U^\circ}(\tau^{-1}\pi(U); F) \simeq R\Gamma_{U^\circ}(E; F) \ .$$

(iii) *Let γ be a closed convex proper cone of E^* containing the zero-section. Then*:

$$R\Gamma_\gamma(E^*; F^\wedge) \simeq R\Gamma(\mathrm{Int}\, \gamma^{\circ a}; F) \otimes or_{E/Z}[-n] \ .$$

(iv) *We have*:

$$(\mathrm{D}'F)^\vee \simeq \mathrm{D}'(F^\wedge) \ , \qquad (\mathrm{D}F)^\vee \simeq \mathrm{D}(F^\wedge) \ .$$

Proof. (i), (ii), (iii) follow immediately from the preceding results.

(iv) We have:

$$\begin{aligned} R\mathcal{H}om(F^\wedge, A_{E^*}) &= R\mathcal{H}om(Rp_{2!}(p_1^{-1}F)_{P'}, A_{E^*}) \\ &\simeq Rp_{2*}R\mathcal{H}om((p_1^{-1}F)_{P'}, p_2^! A_{E^*}) \\ &\simeq Rp_{2*}R\mathcal{H}om((p_1^{-1}F)_{P'}, p_1^{-1}\omega_{E/Z}) \\ &\simeq Rp_{2*}R\Gamma_{P'}(p_1^!(\mathrm{D}'F)) \\ &\simeq (\mathrm{D}'F)^\vee \ . \end{aligned}$$

The proof for D is similar. □

Now we shall study some functorial properties of the Fourier-Sato transformation.

Let Z' be another locally compact space, $f: Z' \to Z$ a continuous map. We set $E' = Z' \times_Z E$ and denote by f_τ (resp. f_π) the map from E' to E (resp. E'^* to E^*) associated to f.

Proposition 3.7.13. (i) *Let* $F \in \mathrm{Ob}(\mathbf{D}^+_{\mathbb{R}^+}(E))$. *Then*:

$$(f_\tau^! F)^\wedge \simeq f_\pi^!(F^\wedge) ,$$

$$(f_\tau^{-1} F)^\wedge \simeq f_\pi^{-1}(F^\wedge) .$$

(ii) *Let* $G \in \mathrm{Ob}(\mathbf{D}^+_{\mathbb{R}^+}(E'))$. *Then*:

$$(Rf_{\tau *} G)^\wedge \simeq Rf_{\pi *}(G^\wedge) ,$$

$$(Rf_{\tau !} G)^\wedge \simeq Rf_{\pi !}(G^\wedge) .$$

Proof. Consider the diagram below where the squares are Cartesian:

$$\begin{array}{ccc} E'^* & \xrightarrow{f_\pi} & E^* \\ {\scriptstyle q_2'}\uparrow & \square & \uparrow{\scriptstyle q_2} \\ Z' \underset{Z}{\times} P & \xrightarrow{\tilde{f}} & P \\ {\scriptstyle q_1'}\downarrow & \square & \downarrow{\scriptstyle q_1} \\ E' & \xrightarrow[f_\tau]{} & E \end{array} \tag{3.7.11}$$

Then:

$$\begin{aligned} f_\pi^! F^\wedge \otimes \omega_{E^*/Z} &= f_\pi^! Rq_{2*} q_1^! F \\ &\simeq Rq'_{2*} \tilde{f}^! q_1^! F \\ &\simeq Rq'_{2*} q_1'^! f_\tau^! F \\ &\simeq (f_\tau^! F)^\wedge \otimes \omega_{E^*/Z} . \end{aligned}$$

This shows the formula $f_\pi^!(F^\wedge) \simeq (f_\tau^! F)^\wedge$. The proofs of the other formulas are similar. □

Now let E_1 and E_2 be two vector bundles over Z, $f: E_1 \to E_2$ a morphism of such bundles. We denote by ${}^t f: E_2^* \to E_1^*$ the dual morphism. Then $\omega_{E_1/E_2} \simeq f^! A_{E_2}$ and $\omega_{E_2^*/E_1^*} \simeq {}^t f^! A_{E_1^*}$.

Proposition 3.7.14. (i) *Let* $F \in \mathrm{Ob}(\mathbf{D}^+_{\mathbb{R}^+}(E_1))$. *Then*:

$${}^t f^{-1}(F^\wedge) \simeq (Rf_! F)^\wedge ,$$

$${}^t f^!(F^\vee) \simeq (Rf_* F)^\vee ,$$

$${}^t f^!(F^\wedge) \simeq (Rf_* F)^\wedge \otimes \omega_{E_2^*/E_1^*} ,$$

$${}^t f^{-1}(F^\vee) \simeq (Rf_! F)^\vee \otimes \omega_{E_2^*/E_1^*} .$$

(ii) *Let* $G \in \mathrm{Ob}(\mathbf{D}^+_{\mathbb{R}^+}(E_2))$. *Then*:

$$(f^{-1}G)^\vee \simeq R^t f_!(G^\vee) ,$$

$$(f^!G)^\wedge \simeq R^t f_*(G^\wedge) ,$$

$$(\omega_{E_1/E_2} \otimes f^!G)^\vee \simeq R^t f_*(G^\vee) ,$$

$$(\omega_{E_1/E_2} \otimes f^{-1}G)^\wedge \simeq R^t f_!(G^\wedge) .$$

Proof. Set $\tilde{P}' = E_1 \times_{E_2} P_2' = P_1' \times_{E_1^*} E_2^* = \{(x,y) \in E_1 \times_Z E_2^*; \langle x, {}^t f(y)\rangle = \langle f(x), y\rangle \leqslant 0\}$, and consider the diagram below where the two squares are Cartesian:

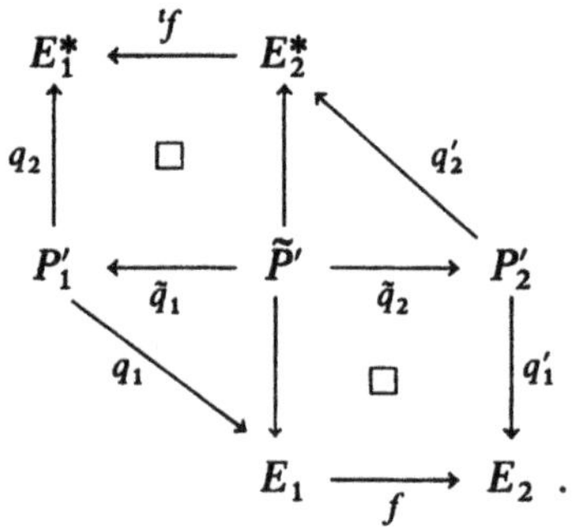

Then:

$$\begin{aligned} {}^t f^{-1} F^\wedge &\simeq {}^t f^{-1} Rq_{2!} q_1^{-1} F \\ &\simeq Rq'_{2!} R\tilde{q}_{2!} \tilde{q}_1^{-1} q_1^{-1} F \\ &\simeq Rq'_{2!} q_1'^{-1} Rf_! F \\ &\simeq (Rf_! F)^\wedge . \end{aligned}$$

Similarly:

$$\begin{aligned} {}^t f^! F^\vee &\simeq {}^t f^! Rq_{2*} q_1^! F \\ &\simeq Rq'_{2*} R\tilde{q}_{2*} \tilde{q}_1^! q_1^! F \\ &\simeq Rq'_{2*} q_1'^! Rf_* F \\ &\simeq (Rf_* F)^\vee . \end{aligned}$$

The other formulas follow by setting $F = G^\vee$ or $F = G^\wedge$. □

Finally we study the external tensor product. Let E_1 and E_2 be two vector bundles over Z. We denote by the same symbol $^\wedge$, the Fourier-Sato transformation on E_i $(i = 1, 2)$, or on $E_1 \times_Z E_2$.

Proposition 3.7.15. *Let $F_i \in \mathrm{Ob}(\mathbf{D}^+_{\mathbb{R}^+}(E_i))$, $i = 1, 2$. Then:*

$$F_1^\wedge \overset{L}{\underset{Z}{\boxtimes}} F_2^\wedge \simeq \left(F_1 \overset{L}{\underset{Z}{\boxtimes}} F_2\right)^\wedge .$$

Proof. For $i = 1, 2$ denote by $p_i^j (j = 1, 2)$ and p_i the i-th projection defined on $E_j \times_Z E_j^* \ (j = 1, 2)$ and $(E_1 \times_Z E_2) \times_Z (E_1^* \times_Z E_2^*)$, respectively. Let $P_j' \ (j = 1, 2)$ and P' denote the closed subsets $\{\langle x, y\rangle \leqslant 0\}$ of $E_j \times_Z E_j^* \ (j = 1, 2)$ and $(E_1 \times_Z E_2) \times_Z (E_1^* \times_Z E_2^*)$. We have (cf. Exercise II. 18):

$$F_1^\wedge \overset{L}{\underset{Z}{\boxtimes}} F_2^\wedge \simeq Rp_{2!}\left((p_1^1)^{-1}F_1 \overset{L}{\underset{Z}{\boxtimes}} (p_1^2)^{-1}F_2\right)_{P_1' \times_Z P_2'} ,$$

$$\left(F_1 \overset{L}{\underset{Z}{\boxtimes}} F_2\right)^\wedge \simeq Rp_{2!}\left((p_1^1)^{-1}F_1 \overset{L}{\underset{Z}{\boxtimes}} (p_1^2)^{-1}F_2\right)_{P'} .$$

We set $G = (p_1^1)^{-1}F_1 \boxtimes^L_Z (p_1^2)^{-1}F_2$. The map p_2 decomposes as follows:

$$\left(E_1 \underset{Z}{\times} E_2\right) \underset{Z}{\times} \left(E_1^* \underset{Z}{\times} E_2^*\right) \underset{\beta}{\longrightarrow} E_1^* \underset{Z}{\times} E_2^* \times \mathbb{R} \times \mathbb{R} \underset{\alpha}{\longrightarrow} E_1^* \underset{Z}{\times} E_2^* ,$$

where $\beta(x_1, x_2, y_1, y_2) = (y_1, y_2, \langle x_1, y_1\rangle, \langle x_2, y_2\rangle)$ and $\alpha(y_1, y_2, t_1, t_2) = (y_1, y_2)$. Since $R\beta_! G_{P'} \simeq (R\beta_! G)_{\{t_1 + t_2 \leqslant 0\}}$ and $R\beta_! G_{P_1' \times_Z P_2'} \simeq (R\beta_! G)_{\{t_1 \leqslant 0, t_2 \leqslant 0\}}$, it remains to show that

$$R\alpha_!((R\beta_! G)_{\{t_1 + t_2 \leqslant 0\}}) \to R\alpha_!((R\beta_! G)_{\{t_1 \leqslant 0, t_2 \leqslant 0\}})$$

is an isomorphism.

We shall prove it at each point of $E_1^* \times_Z E_2^*$. Hence we have to show that if H is a biconic object of $\mathbf{D}^+(\mathbb{R} \times \mathbb{R})$, then:

$$R\Gamma_c(\mathbb{R} \times \mathbb{R}; H_{\{t_1 + t_2 \leqslant 0\}}) \xrightarrow{\sim} R\Gamma_c(\mathbb{R} \times \mathbb{R}; H_{\{t_1 \leqslant 0, t_2 \leqslant 0\}}) , \tag{3.7.12}$$

or equivalently:

$$R\Gamma_c(\mathbb{R} \times \mathbb{R}; H_{\{0 < t_1 \leqslant t_2\} \cup \{0 < t_2 \leqslant -t_1\}}) = 0 . \tag{3.7.13}$$

Then this follows from the fact that the cohomology sheaves of H are constant on each connected component of the open set $\{t_1 t_2 \neq 0\}$. □

Remark 3.7.16. In the course of the book, we neglect the problem of signs. For example, the identification of $or_{Y/X}$ and $or_Y \otimes f^{-1} or_X$ (for a morphism $f: Y \to X$), or the identification $or_{Z/E} \simeq or_{Z/E^*}$ for a vector bundle $E \to Z$, (and, as a particular case, the identification $or_{T^*X} \simeq A_{T^*X}$) will not in general be described.

Exercises to Chapter III

Exercise III.1. We take $A = \mathbb{Q}$. Set $X = \mathbb{R}$, $B = \{1/n; n \in \mathbb{N}\setminus\{0\}\}$, $Z = B \cup \{0\}$.

(i) Let $F = \mathbb{Q}_B$. Calculate $D'F$ and $D'D'F$. Show that $F \neq D'D'F$.

(ii) Let $G = \mathbb{Q}_Z$. Prove that G is soft and $H^1_{\{0\}}(X; G)$ is infinite dimensional over $\mathbb{Q}$.

(Hint: show that $H^0(X; F)$ (resp. $H^0(X; G)$) is isomorphic to the space of all sequences (resp. all stationary sequences) with values in $\mathbb{Q}$.)

(iii) Prove that $H^2_{\{0\}}(X; \mathbb{Q}_{(X\setminus Z)}) \neq 0$.

Exercise III.2. Show that the soft dimension of $\mathbb{R}^n$ is n and the flabby dimension of $\mathbb{R}^n$ is $n+1$. (Hint: use Exercise III.1.)

Exercise III.3. Let X be a topological space, F a sheaf on X locally isomorphic to $\mathbb{Z}_X$. Prove that there are canonical isomorphisms:

$$F \otimes F \simeq \mathbb{Z}_X \ , \qquad D'F \simeq F \ .$$

Exercise III.4. Let X be a C^0-manifold. We shall say that Ω is locally cohomologically trivial in X (*l.c.t.* for short) if for any $x \in \bar{\Omega}\setminus\Omega$, one has:

$$(R\Gamma_{\bar{\Omega}}(A_X))_x = 0 \ , \qquad (R\Gamma_{\Omega}(A_X))_x \simeq A \ ,$$

(cf. Schapira [3]).

(i) Prove that Ω is *l.c.t.* in X iff $D'(A_\Omega) \simeq A_{\bar{\Omega}}$ and $D'(A_{\bar{\Omega}}) \simeq A_\Omega$.

(ii) Prove that if Ω is *l.c.t.* in X then $\Omega = \mathrm{Int}(\bar{\Omega})$.

(iii) Prove that if Ω is convex in $\mathbb{R}^n$ then Ω is *l.c.t.* in $\mathbb{R}^n$ and moreover A_Ω and $A_{\bar{\Omega}}$ are cohomologically constructible.

Exercise III.5. Let V be a real n-dimensional vector space, and let q be a quadratic form on V. For $a \in \mathbb{R}$, we set $Z_a = \{x \in V; q(x) \leqslant a\}$. Let ε^- be the number of non-positive eigenvalues of q. Prove:

(i) $R\Gamma_c(V; A_{Z_a})[\varepsilon^-] \simeq \begin{cases} A & a \geqslant 0 \ , \\ 0 & a < 0 \ . \end{cases}$

(ii) Using Proposition 3.1.10, deduce from (i) that:

$$R\Gamma_{Z_a}(V; A_V)[n - \varepsilon^-] \simeq \begin{cases} A & a \geqslant 0 \ , \\ 0 & a < 0 \ . \end{cases}$$

Exercise III.6. Let $f: Y \to X$ be a sphere bundle with fiber dimension $n \geqslant 1$. Prove the existence of a distinguished triangle:

$$R^0 f_* A_Y \longrightarrow Rf_* A_Y \longrightarrow R^n f_* A_Y[-n] \underset{+1}{\longrightarrow} \ ,$$

which leads to:

$$A_X \longrightarrow Rf_* A_Y \longrightarrow f_* or_{Y/X}[-n] \underset{+1}{\longrightarrow} \ .$$

(Hint: use Exercise I.26.)

Exercise III.7. Let $\tau: E \to Z$ be a vector bundle with fiber dimension n, $i: Z \hookrightarrow E$ the zero-section embedding. We assume E endowed with a relative orientation (i.e. we assume to be given an isomorphism $A_E \simeq \mathcal{or}_{E/Z}$).

(i) Construct the isomorphism $\Gamma(Z; A_Z) \simeq H_Z^n(E; A_E)$. (The image of 1 is called the **Thom class.**)

(ii) Construct the commutative diagram:

$$\begin{array}{ccc} R\tau_! A_E & \longrightarrow & R\tau_* A_E \\ \Big| \wr & & \Big| \wr \\ A_Z[-n] & \longrightarrow & A_Z \end{array}$$

and the morphism $\Gamma(Z; A_Z) \to H^n(Z; A_Z)$. (The image of 1 is called the **Euler class.**)

(iii) Prove that the Euler class is the image of the Thom class by the maps: $H_Z^n(E; A_E) \to H^n(E; A_E) \simeq H^n(Z; A_Z)$.

(iv) Prove that if the bundle admits a nowhere vanishing continuous section, then the Euler class is zero, (cf. Exercise V.11).

Exercise III.8. Let X be a topological manifold, Y a closed submanifold of codimension l and $j: Y \hookrightarrow X$ the inclusion. Assume to be given an isomorphism $\mathcal{or}_{Y/X} \simeq A_Y$. Then construct the long exact sequence:

$$H^k(Y; A_Y) \xrightarrow{\alpha} H^{k+l}(X; A_X) \longrightarrow H^{k+l}(X \setminus Y; A_X) \longrightarrow H^{k+1}(Y; A_Y) \; .$$

(The map α is called the **Gysin map**. The image of $1 \in H^0(Y; A_Y)$ is again called the Euler class.)

Exercise III.9. We keep the hypotheses of Proposition 3.1.9.

(i) Prove that there are natural morphisms of functors and commutative diagrams:

$$
\begin{array}{ccccc}
Rf_! \circ Rg'_! & & \longrightarrow & & Rf_* \circ Rg'_* \\
\downarrow & & & & \wr \\
Rf_! \circ Rg'_* & \longrightarrow & Rg_* \circ Rf'_! & \longrightarrow & Rg_* \circ Rf'_* \ ,
\end{array}
$$

$$
g'^{-1} \circ f^! \longrightarrow f'^! \circ g^{-1} \ ,
$$

$$
\begin{array}{ccc}
g^{-1} \circ Rf_! & \xrightarrow{\sim} & Rf'_! \circ g'^{-1} \\
\downarrow & & \downarrow \\
g^{-1} \circ Rf_* & \longrightarrow & Rf'_* \circ g'^{-1} \ ,
\end{array}
$$

$$
\begin{array}{ccc}
Rf'_! \circ g'^! & \longrightarrow & g^! \circ Rf_! \\
\downarrow & & \downarrow \\
Rf'_* \circ g'^! & \xrightarrow{\sim} & g^! \circ Rf_* \ .
\end{array}
$$

(ii) For $G \in \mathrm{Ob}(\mathbf{D}^+(A_Y))$, $F \in \mathrm{Ob}(\mathbf{D}^+(A_X))$, construct a canonical commutative diagram:

$$
\begin{array}{ccc}
Rf_! R\mathcal{H}om(G, f^! F) & \longrightarrow & R\mathcal{H}om(Rf_* G, F) \\
\downarrow & & \downarrow \\
Rf_* R\mathcal{H}om(G, f^! F) & \xrightarrow{\sim} & R\mathcal{H}om(Rf_! G, F) \ .
\end{array}
$$

(iii) For F_1 and F_2 in $\mathbf{D}^b(A_X)$, construct the commutative diagram:

$$
\begin{array}{ccccc}
 & & R\mathcal{H}om(f^{-1}F_1, f^{-1}F_2) & & \\
 & \nearrow & & \searrow & \\
f^{-1}R\mathcal{H}om(F_1, F_2) & & & & R\mathcal{H}om(f^{-1}F_1 \otimes \omega_{Y/X}, f^! F_2) \ . \\
 & \searrow & & \nearrow & \\
 & & R\mathcal{H}om(f^! F_1, f^! F_2) & &
\end{array}
$$

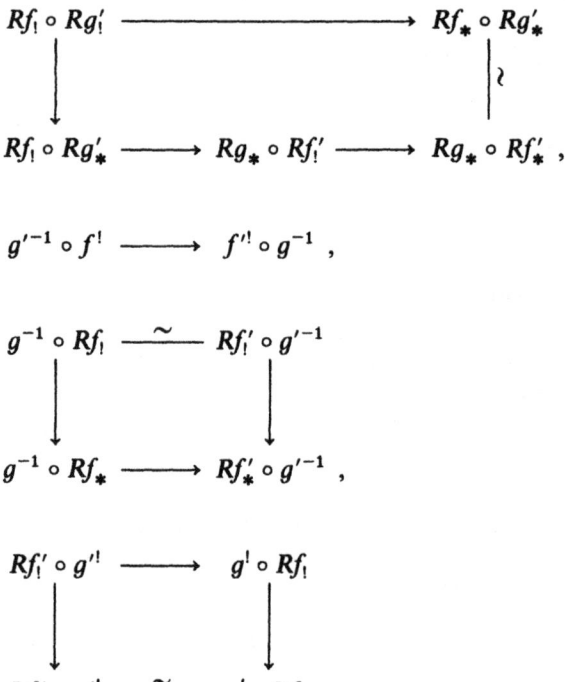

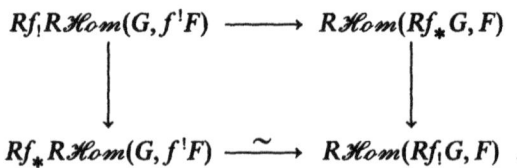

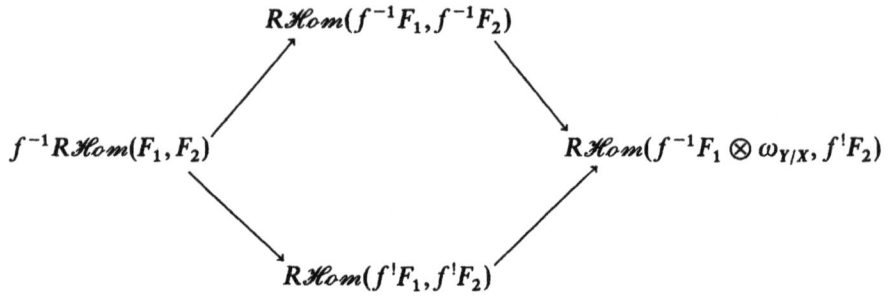

Exercise III.10. Prove that with the hypotheses of Proposition 3.3.9, for any $F \in \mathrm{Ob}(\mathbf{D}^+(A_X))$, the morphism $Rf_! f^! F \to F$ is an isomorphism.

Exercise III.11. Let k be a commutative field, X a compact n-dimensional C^0-manifold, F a cohomologically constructible object of $\mathbf{D}^b(k_X)$, $\mathrm{D}F$ its dual.

(i) Prove that $H^j(X;F)$ and $H^{-j}(X;\mathrm{D}F)$ are finite-dimensional k-vector spaces, dual one to each other. (Note that $H^{-j}(X;\mathrm{D}F) = H^{n-j}(X;\mathrm{D}'F \otimes or_X)$). (Hint: prove by induction on j that, for any compact subset K and any open subset $U \supset K$, $\mathrm{Im}(H^j(U;F) \to H^j(K;F))$ is finite-dimensional.)

(ii) Assume now $k = \mathbb{R}$, X is compact, oriented of class C^∞. Describe the duality between $H^j(X;\mathbb{R}_X)$ and $H^{n-j}(X;\mathbb{R}_X)$ using the de Rham complexes with C^∞-functions and with distributions as coefficients (cf. II §9).

Exercise III.12. Let E be a finite-dimensional real vector space, γ a closed convex cone with vertex at 0, $\gamma^a = -\gamma$.

(i) Prove the isomorphism:

$$A_{\gamma^a} \xrightarrow{\sim} \phi_\gamma^{-1} R\phi_{\gamma *}(A_{\gamma^a}) \ .$$

(ii) Let Ω be a γ-open subset of E. Prove the isomorphism:

$$A_\Omega \simeq \phi_\gamma^{-1} R\phi_{\gamma *} A_\Omega \ .$$

Exercise III.13. Let $X = \mathbb{R}^2$ with coordinates (x, y), $A = \{(x,y); y = 0\}$, $B = \left\{(x,y); x \neq 0, y = x \sin\frac{1}{x}\right\}$. Let $F = \mathbb{Q}_A$, $G = \mathbb{Q}_{\bar{B}}$. Prove that F and G are cohomologically constructible (over $\mathbb{Q}$) but $F \otimes G$ and $R\mathcal{H}om(F, G)$ are not.

Exercise III.14. Let $\mathbf{S}^n$ be the unit sphere of the Euclidian space $\mathbb{R}^n$, let $a \in \mathbb{R}$ with $-1 \leqslant a < 1$, and set:

$$\Omega = \{(x,y) \in \mathbf{S}^n \times \mathbf{S}^n; \langle x, y\rangle > a\} \ , \qquad K = A_\Omega \ .$$

Prove that the functors Ψ_K and Φ_K (cf. III §6) are equivalences of categories on $\mathbf{D}^+(A_{\mathbf{S}^n})$.

Exercise III.15. Let V be the vector space $\mathbb{R}^n$ endowed with its orientation, and let m be an integer, $1 \leqslant m \leqslant n$. Set:

$$X_m = \{\lambda; \lambda \text{ is an oriented } m\text{-dimensional linear subspace of } V\} \ .$$

$$\Omega = \{(\lambda, \mu) \in X_m \times X_{n-m}; \lambda \text{ and } \mu \text{ are transversal and the orientation of } \lambda \oplus \mu \text{ is that of } V\} \ .$$

$$K = A_\Omega \ .$$

Prove that Ψ_K and Φ_K define equivalences of categories between $\mathbf{D}^+(A_{X_m})$ and $\mathbf{D}^+(A_{X_{n-m}})$.

Exercise III.16. Let $(t, x) = (t, x_1, \ldots, x_n)$ be the coordinates on $\mathbb{R}^{1+n}$, and let γ_i $(i = 1, 2, 3)$ be the closed cone:

$$\gamma_1 = \left\{(t, x); t^2 \geqslant \sum_{i=1}^{n} x_i^2\right\} ,$$

$$\gamma_2 = \left\{(t, x); t^2 \leqslant \sum_{i=1}^{n} x_i^2\right\} ,$$

$$\gamma_3 = \left\{(t, x); t^2 = \sum_{i=1}^{n} x_i^2\right\} .$$

Set $\gamma_i^+ = \gamma_i \cap \{(x, t); t \geqslant 0\}$, and also: $G_i = A_{\gamma_i}$, $G_i^+ = A_{\gamma_i^+}$. Calculate the Fourier-Sato transform of all these sheaves.

Exercise III.17. Let $\tau: E \to Z$ be a vector bundle over a locally compact space Z. We use the same notations as in Example 3.6.6, and we denote by the same letter γ the projections $\dot{E} \to S$ and $\dot{E}^* \to S^*$.

Let $j: \dot{E} \hookrightarrow E$ be the inclusion map. Prove the commutativity of the diagrams, where $\tilde{\Psi}_D$ and $\tilde{\Phi}_D$ are defined similarly to (3.7.7):

$$\begin{array}{ccc}
\mathbf{D}^+(S) & \xrightarrow[\tilde{\Psi}_D]{} & \mathbf{D}^+(S^*) \\
\downarrow{\scriptstyle \gamma^{-1}} & & \uparrow{\scriptstyle R\gamma_*} \\
\mathbf{D}^+_{\mathbb{R}^+}(\dot{E}) & & \mathbf{D}^+_{\mathbb{R}^+}(\dot{E}^*) \\
\downarrow{\scriptstyle Rj_*} & & \uparrow{\scriptstyle j^{-1}} \\
\mathbf{D}^+_{\mathbb{R}^+}(E) & \xrightarrow[\wedge]{} & \mathbf{D}^+_{\mathbb{R}^+}(E^*)
\end{array}
\qquad
\begin{array}{ccc}
\mathbf{D}^+(S^*) & \xrightarrow[\tilde{\Phi}_D]{} & \mathbf{D}^+(S) \\
\downarrow{\scriptstyle \gamma^{-1}} & & \uparrow{\scriptstyle R\gamma_*} \\
\mathbf{D}^+_{\mathbb{R}^+}(\dot{E}^*) & & \mathbf{D}^+_{\mathbb{R}^+}(\dot{E}) \\
\downarrow{\scriptstyle Rj_!} & & \uparrow{\scriptstyle j^{-1}} \\
\mathbf{D}^+_{\mathbb{R}^+}(E^*) & \xrightarrow[\vee]{} & \mathbf{D}^+_{\mathbb{R}^+}(E) .
\end{array}$$

Exercise III.18. Let X be a C^0-manifold of dimension n, let $x \in X$ and let U be an open neighborhood of x. One defines the **residue morphism** at x, denoted $\mathrm{Res}(x; \cdot)$ as the composite:

$$H^{n-1}(U \backslash \{x\}; \mathit{or}_X) \to H^n_{\{x\}}(X; \mathit{or}_X) \xrightarrow{\sim} A .$$

(i) Assume X is compact, and let Z be a finite subset of X. Let $u \in H^{n-1}(X \backslash Z; \mathit{or}_X)$. Prove:

$$\sum_{x \in Z} \mathrm{Res}(x; u) = 0 .$$

(Hint: send $H^n_Z(X; \mathit{or}_X)$ into $H^n(X; \mathit{or}_X)$.)

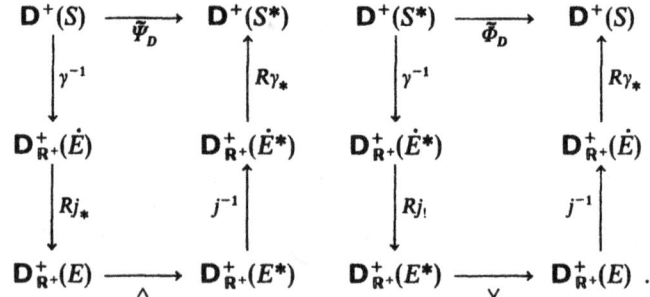

(ii) One does not assume X compact any more. On the contrary, one assumes $H^{n-1}(X; \mathit{or}_X) = H^n(X; \mathit{or}_X) = 0$. Let Y be a compact C^0-manifold of dimension $n - 1$ and let γ be a continuous map, $\gamma: Y \to X$. One assumes to be given an isomorphism $\gamma^{-1}\mathit{or}_X \simeq \mathit{or}_Y$. If Z is a closed subset of X which does not intersect $\gamma(Y)$, one denotes by γ^{-1} the map $H^{n-1}(X\setminus Z; or_X) \to H^{n-1}(Y; or_Y)$ deduced from $\gamma^{-1} or_X \xrightarrow{\sim} or_Y$.

Let $x \in X\setminus\gamma(Y)$. Since $H^{n-1}(X\setminus\{x\}; or_X) \simeq A$, there is only one $u \in H^{n-1}(X\setminus\{x\}; or_X)$ such that $\operatorname{Res}(x; u) = 1$. One sets:

$$\operatorname{Ind}(x;\gamma) = \int_Y \gamma^{-1}(u)$$

where $\int_Y$ denotes the map $H^{n-1}(Y; or_Y) \to A$.

Now let Z be a finite subset of $X\setminus\gamma(Y)$ and let $v \in H^{n-1}(X\setminus Z; or_X)$. Prove the **Cauchy residues formula**:

$$\int_Y \gamma^{-1}(v) = \sum_{x \in Z} \operatorname{Ind}(x;\gamma)\operatorname{Res}(x;v) \ .$$

(This exercise is extracted from Iversen [2].)

Exercise III.19. We keep the notations of Proposition 2.7.5 and diagram (2.7.5), and we assume all spaces are locally compact and the functors $h_!$, $p_{Y!}$, $p_{X!}$ have finite cohomological dimension. Moreover we assume h is proper. Then one defines for $j = 0, 1$, the morphisms of functors $f_{j\#} : Rp_{Y!} \circ p_Y^! \to Rp_{X!} \circ p_X^!$ by using $Rf_{j!} \circ f_j^! \to \mathrm{id}_X$. Prove that $f_{0\#} = f_{1\#}$.

Exercise III.20. In (i) and (ii) below we set $X = \mathbb{R}$. We choose $a, b \in X$ with $a < b$ and we set $I_b^- =]-\infty, b]$, $I_a^+ = [a, +\infty[$.

(i) Show that $\mathbb{Z}_X$ is quasi-isomorphic to the complex:

$$K : 0 \to \mathbb{Z}_{I_b^-} \oplus \mathbb{Z}_{I_a^+} \to \mathbb{Z}_{I_b^- \cap I_a^+} \to 0 \ .$$

Let $\mathscr{C}_X^{\cdot}$ denote the de Rham complex $0 \to \mathscr{C}_X^{\infty} \xrightarrow{d} \mathscr{C}_X^{\infty,(1)} \to 0$ (cf. II §9). Construct a morphism of complexes: $K \xrightarrow{\phi} \mathscr{C}_X^{\cdot}$ such that the composition

$$\mathbb{Z}_X \simeq H^0(K) \to H^0(\mathscr{C}_X^{\cdot}) \simeq \mathbb{C}_X$$

coincides with the morphism $\mathbb{Z}_X \to \mathbb{C}_X$ associated to $\mathbb{Z} \subset \mathbb{C}$.

(ii) Prove that $H_c^1(X; \mathbb{Z}_X) \simeq H^1(\Gamma_c(X; K))$, and prove that the diagram below commutes up to sign; (see (3.3.16) for the definition of $\int_X$).

$$\begin{array}{ccccc} H_c^1(X;\mathbb{Z}_X) & \xrightarrow{\sim} & H^1(\Gamma_c(X;K)) & \xrightarrow{\phi} & H^1(\Gamma_c(X;\mathscr{C}_X^{\cdot})) \\ \downarrow & & & & \downarrow{\scriptstyle\int_X} \\ \mathbb{Z} & & \longrightarrow & & \mathbb{C} \end{array} \ .$$

(iii) Prove that when $A = \mathbb{C}$ the homomorphisms (3.3.15) and (3.3.16) are equal up to sign.

Exercise III.21. Let $f : Y \to X$ be a topological submersion of locally compact spaces, and let Z_1, Z_2 be two closed subsets of Y such that $Z_1 \subset Z_2$. Assume that $H_c^j((Z_2 \setminus Z_1) \cap f^{-1}(x); \mathbb{Z}_Y) = 0$ for any j. Prove that for any $F \in \mathrm{Ob}(\mathbf{D}^b(X))$, $Rf_* R\Gamma_{Z_1}(f^{-1}F) \to Rf_* R\Gamma_{Z_2}(f^{-1}F)$ is an isomorphism.

Notes

We refer to the "Short History" by C. Houzel, at the beginning of this book for a detailed history of the duality of sheaves and the functor $f^!$.

Let us only recall that the functor $f^!$ was first introduced by Grothendieck [4] in 1963 in his theory of duality for coherent sheaves on locally noetherian schemes. This theory served as a model for duality theories in other contexts: locally compact spaces (Verdier [1]), étale cohomology ([SGA 4], [SGA 5]) and more recently $\mathscr{D}$-modules (cf. XI §2). The theory we present here (§1 and §4) is that of Verdier, and is detailed in the Heidelberg-Strasbourg Seminar [SHS]. Note that many results of §2 and §3 have been known for a long time (e.g. cf. Borel [1], Borel-Moore [1]).

The Fourier-Sato transformation plays an important role in several parts of Mathematics (cf. Kashiwara-Hotta [1], Malgrange [2], Brylinski [2]) and there exists a similar construction in the algebraic case due to Deligne (cf. Illusie [1], Katz-Laumon [1], Laumon [1]). This transformation was first introduced by Sato (cf. Sato-Kawai-Kashiwara [1]) for sphere bundles around 1970, and it is only after 1975 (especially after the paper of Kashiwara [4]), that people systematically worked on vector bundles (i.e. included the zero-section) in these questions. Apart from the introduction of the zero-section, which is performed in Brylinski-Malgrange-Verdier [1], all results of §7 are essentially contained in Sato-Kawai-Kashiwara [1] and Kashiwara-Kawai [2].

Chapter IV. Specialization and microlocalization

Summary

Let X be a manifold, M a closed submanifold. We first construct a new manifold $\tilde{X}_M$, the normal deformation of M in X. This manifold is of dimension one more than the dimension of X, and is endowed with a map $(p, t): \tilde{X}_M \to X \times \mathbb{R}$ such that $t^{-1}(c)$ is isomorphic to X for $c \neq 0$ and $t^{-1}(0)$ is isomorphic to $T_M X$, the normal bundle to M in X.

We use this manifold to associate to a sheaf F on X (or more generally to $F \in \mathrm{Ob}(\mathbf{D}^b(X))$) an object $\nu_M(F)$ of $\mathbf{D}^b(T_M X)$ called the specialization of F along M. Its Fourier-Sato transform $\mu_M(F)$ is the microlocalization of F along M.

Having defined the functors ν_M and μ_M and studied their functorial properties, we then proceed to study the functor μhom.

This new functor generalizes the microlocalization functor and will play a central role throughout the rest of the book. The results of §2 and §3 originated from Sato-Kawai-Kashiwara [1].

Convention 4.0. We keep convention 3.0. Moreover all manifolds and morphisms of manifolds are assumed to be either of class C^∞ or else real analytic.

From now on we shall mainly work in categories of bounded complexes of sheaves (cf. Corollary 3.3.12). Of course, many results still hold with weaker hypotheses.

4.1. Normal deformation and normal cones

Let X be an n-dimensional manifold. Let M be a closed submanifold of codimension l, $T_M X$ the normal bundle to M in X, that is, the vector bundle on M defined by the exact sequence:

$$0 \to TM \to M \underset{X}{\times} TX \to T_M X \to 0 . \tag{4.1.1}$$

We shall construct a new manifold $\tilde{X}_M$ and two maps:

$$\begin{cases} p: \tilde{X}_M \to X , \\ t: \tilde{X}_M \to \mathbb{R} , \end{cases} \tag{4.1.2}$$

such that

$$
(4.1.3)\qquad \begin{cases} p^{-1}(X\setminus M) & \text{is isomorphic to } (X\setminus M)\times(\mathbb{R}\setminus\{0\})\ , \\ t^{-1}(\mathbb{R}\setminus\{0\}) & \text{is isomorphic to } X\times(\mathbb{R}\setminus\{0\})\ , \\ t^{-1}(0) & \text{is isomorphic to } T_M X\ . \end{cases}
$$

The manifold $\tilde{X}_M$ will be called the **normal deformation** of M in X.

For that purpose consider an open covering $X = \bigcup_i U_i$, and open embeddings $\phi_i : U_i \hookrightarrow \mathbb{R}^n$, such that $U_i \cap M = \phi_i^{-1}(\{0\}^l \times \mathbb{R}^{n-l})$. Set $x = (x', x'') \in \mathbb{R}^l \times \mathbb{R}^{n-l}$, and define:

$$
V_i = \{(x,t) \in \mathbb{R}^n \times \mathbb{R}; (tx', x'') \in \phi_i(U_i)\}\ .
$$

We denote by $t_{V_i} : V_i \to \mathbb{R}$ the projection $(x,t) \mapsto t$, and by $p_{V_i} : V_i \to U_i$ the map: $(x,t) \mapsto \phi_i^{-1}(tx', x'')$. One defines the map

$$
\psi_{ji} : V_i \underset{U_i}{\times} (U_i \cap U_j) \to \mathbb{R}^n
$$

by setting $\psi_{ji}(x,t) = (\psi'_{ji}(x,t), \psi''_{ji}(x,t))$ with:

$$
(t\psi'_{ji}(x,t), \psi''_{ji}(x,t)) = \phi_j \phi_i^{-1}(tx', x'')\ .
$$

This is possible, since the first l components of $\phi_j\phi_i^{-1}(tx', x'')$ vanish when $t = 0$.

Let $\mathscr{R}$ be the equivalence relation which identifies $(x_i, t_i) \in V_i$ and $(x_j, t_j) \in V_j$ if $t_i = t_j$ and $x_j = \psi_{ji}(x_i, t_i)$. We set:

$$
\tilde{X}_M = \Big(\bigsqcup_i V_i\Big)\Big/\mathscr{R}\ .
$$

It is straightforward to check that $\tilde{X}_M$ is actually a manifold and that the maps $p : \tilde{X}_M \to X$ and $t : \tilde{X}_M \to \mathbb{R}$ defined by $p_{|V_i} = p_{V_i}$ and $t_{|V_i} = t_{V_i}$ respectively, are well-defined, and satisfy the first two statements in (4.1.3). We say that $\tilde{X}_M$ is the normal deformation of M in X.

Note that the multiplicative group $\mathbb{R}\setminus\{0\}$ acts on $\tilde{X}_M$ by:

$$
(c, (x', x'', t)) \mapsto (cx', x'', c^{-1}t)\ .
$$

Let us show that the hypersurface $t^{-1}(0)$ of $\tilde{X}_M$ may be canonically identified with the normal bundle $T_M X$. Set $\psi_{ji}(x, t) = (\psi'_{ji}, \psi''_{ji})$ as above, and also set $\phi_j \circ \phi_i^{-1} = \phi_{ji} = (\phi'_{ji}, \phi''_{ji})$. Then:

$$
(4.1.4)\qquad \begin{cases} \psi'_{ji}(x,0) = \displaystyle\sum_{k=1}^{l} x_k \frac{\partial}{\partial x_k} \phi'_{ji}(0, x'')\ , \\ \psi''_{ji}(x,0) = \phi''_{ji}(0, x'')\ . \end{cases}
$$

Therefore the family $\{V_i \cap t^{-1}(0), \psi_{ji}(x,0)\}$ defines a covering of $T_M X$ by local charts and $x = (x', x'') \in V_i \cap t^{-1}(0)$ is identified with the normal vector x' at $(0, x'') \in M$.

We denote by Ω the open subset of $\tilde{X}_M$ obtained as the inverse image of $\mathbb{R}^+$ by the map t, by j the embedding $\Omega \hookrightarrow \tilde{X}_M$, by $\tilde{p}$ the map $p \circ j$, by s the immersion $T_M X \hookrightarrow \tilde{X}_M$. We denote by τ the projection $T_M X \to M$, and by i the inclusion $M \hookrightarrow X$. When there is no fear of confusion, we sometimes write τ instead of $i \circ \tau$.

$$
\begin{array}{ccccc}
T_M X & \xrightarrow[s]{} & \tilde{X}_M & \xleftarrow{j} & \Omega \\
\downarrow{\scriptstyle\tau} & & \downarrow{\scriptstyle p} & \swarrow{\scriptstyle\tilde{p}} & \\
M & \xrightarrow{i} & X & &
\end{array}
\tag{4.1.5}
$$

Note that $\tilde{p}$ is smooth and Ω is isomorphic to $X \times \mathbb{R}^+$ by the map $(\tilde{p}, t)$. Note that $p^{-1}(M)$ is the union of $T_M X$ and $M \times \mathbb{R}$, and $T_M X \cap (M \times \mathbb{R}) = M \times \{0\}$ coincides with the zero-section of $T_M X$.

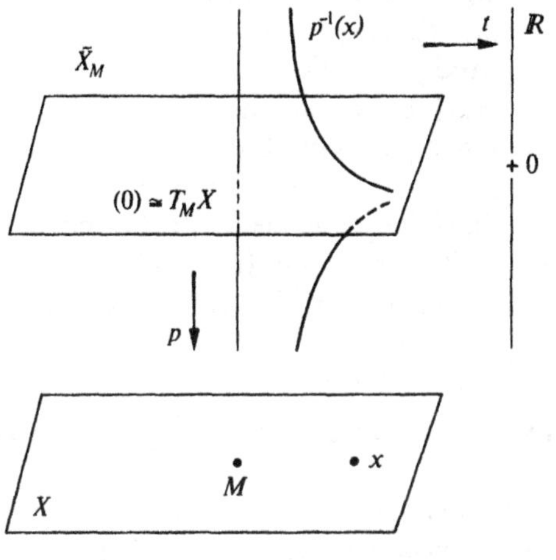

Fig 4.1.1

Definition 4.1.1. (i) *Let S be a subset of X. The "normal cone to S along M", denoted $C_M(S)$, is the set:*

$$C_M(S) = T_M X \cap \overline{\tilde{p}^{-1}(S)} \ .$$

(ii) *Let S_1 and S_2 be two subsets of X. The normal cone $C(S_1, S_2)$ is the set $C_{\Delta_X}(S_1 \times S_2)$ of TX, where Δ_X is the diagonal of $X \times X$ and TX is identified with $T_{\Delta_X}(X \times X)$ by the first projection.*

By its construction, $C_M(S)$ is a closed conic subset of $T_M X$ and its projection onto M is the set $M \cap \bar{S}$.

Similarly, $C(S_1, S_2)$ is a closed conic subset of TX. Notice that if M is a closed submanifold, then $C(S, M)$ is the inverse image of $C_M(S)$ by the projection $M \times_X TX \to T_M X$ as seen by the following proposition.

Proposition 4.1.2. (i) *Let (x) be a system of local coordinates on X, and let S_1 and S_2 be two subsets of X. Let $(x_\circ; v_\circ) \in TX$. Then:*

$$(x_\circ; v_\circ) \in C(S_1, S_2) \Leftrightarrow \text{there exists a sequence } \{(x_n, y_n, c_n)\} \text{ in } S_1 \times S_2 \times \mathbb{R}^+ \text{ such that:}$$

$$(4.1.7) \qquad x_n \xrightarrow[n]{} x_\circ\,, \qquad y_n \xrightarrow[n]{} x_\circ\,, \qquad c_n(x_n - y_n) \xrightarrow[n]{} v_\circ\,.$$

(ii) *Let $(x) = (x', x'')$ be a system of local coordinates on X, $M = \{x; x' = 0\}$, and let S be a subset of X. Let $x_\circ = (0, x''_\circ) \in M$, $(x_\circ; v_\circ) \in T_M X$. Then:*

$$(x_\circ; v_\circ) \in C_M(S) \Leftrightarrow \text{there exists a sequence } \{(x_n, c_n)\} \text{ in } S \times \mathbb{R}^+ \text{ such that, setting } x_n = (x'_n, x''_n) : x_n \xrightarrow[n]{} x_\circ\,,\ c_n x'_n \xrightarrow[n]{} v_\circ\,.$$

Proof. It is enough to prove (ii). Let (x', x'', t) be the coordinates on $\tilde{X}_M$, p: $\tilde{X}_M \to X$ being the map $(x', x'', t) \mapsto (tx', x'')$.

(a) Assume $(x_\circ, v_\circ)$ belongs to $C_M(S)$. Then there exists a sequence $\{(x'_n, x''_n, t_n)\}$ in $\tilde{p}^{-1}(S)$ such that $(x'_n, x''_n, t_n) \xrightarrow[n]{} (v_\circ, x''_\circ, 0)$. The sequence $\{((t_n x'_n, x''_n), t_n^{-1})\}$ in $S \times \mathbb{R}^+$ has the required properties.

(b) Conversely let $\{(x_n, c_n)\}$ be a sequence in $S \times \mathbb{R}^+$ such that $x_n \xrightarrow[n]{} (0, x''_\circ)$, $c_n x'_n \xrightarrow[n]{} v_\circ$. If the sequence $\{c_n\}$ is not bounded, we can assume by extracting a subsequence that $c_n \xrightarrow[n]{} +\infty$. Then the sequence $\{(c_n x'_n, x''_n, c_n^{-1})\}$ in $\tilde{p}^{-1}(S)$ converges to $(v_\circ, x''_\circ, 0)$. If the sequence $\{c_n\}$ is bounded, then $v_\circ = 0$. We may find a sequence of positive numbers $\{\varepsilon_n\}$ such that $\varepsilon_n \xrightarrow[n]{} 0$, $\varepsilon_n^{-1} x'_n \xrightarrow[n]{} 0$. Then $\{(\varepsilon_n^{-1} x'_n, x''_n, \varepsilon_n)\}$ is a sequence in $\tilde{p}^{-1}(S)$ which converges to $(0, x''_\circ, 0)$. $\square$

Proposition 4.1.3. *Let V be a conic open subset of $T_M X$.*

(i) *Let W be an open neighborhood of V in $\tilde{X}_M$, and let $U = \tilde{p}(W \cap \Omega)$. Then $V \cap C_M(X \setminus U) = \emptyset$.*

(ii) *Conversely, let U be an open subset of X such that $V \cap C_M(X \setminus U) = \emptyset$. Then $\tilde{p}^{-1}(U) \cup V$ is an open neighborhood of V in $\bar{\Omega} = \Omega \cup T_M X$.*

Proof. (i) since $W \cap \tilde{p}^{-1}(X \setminus U) = \emptyset$, we have $V \cap \tilde{p}^{-1}(X \setminus U) = \emptyset$.

(ii) By the definition of $C_M(X \setminus U)$, $C_M(X \setminus U) \cup \tilde{p}^{-1}(X \setminus U)$ is closed in $\bar{\Omega}$. Since $T_M X \setminus V$ is closed and $C_M(X \setminus U)$ is contained in $T_M X \setminus V$, $(T_M X \setminus V) \cup \tilde{p}^{-1}(X \setminus U)$ is closed in $\bar{\Omega}$. Its complementary set is $\tilde{p}^{-1}(U) \cup V$. $\square$

The next result will be crucial in the proof of Theorem 4.2.3 below.

Proposition 4.1.4. *Let V be a conic open subset of $T_M X$. Then the family of open neighborhoods W of V in $\tilde{X}_M$ such that all the fibers of the map $p : W \cap \Omega \to X$ are connected, forms a neighborhood system of V.*

Proof. Let W be an open neighborhood of V in $\tilde{X}_M$. Set:

$$X' = \tilde{X}_M\,, \qquad \mathring{X}' = \dot{T}_M X \cup p^{-1}(X \setminus M) = X' \setminus (M \times \mathbb{R})\,, \qquad S = \mathring{X}'/\mathbb{R}^+\,.$$

(Recall that $\dot{T}_M X = T_M X \setminus M$, and $p^{-1}(M) = T_M X \cup (M \times \mathbb{R})$.)

Then S is a manifold and $\alpha : X' \to S$ is an $\mathbb{R}^+$-bundle. Moreover S contains $\dot{T}_M X/\mathbb{R}^+$ as a hypersurface and $S \to X$ is a proper map. Shrinking X if necessary, let us choose a section σ of $\mathring{X}' \to S$ by extending a section of $\dot{T}_M X \to \dot{T}_M X/\mathbb{R}^+$. Set:

$$W' = \bigcup_{x \in \sigma^{-1}(W)} \{\text{the connected component of } \alpha^{-1}(x) \cap W \text{ containing } \sigma(x)\} \ .$$

By the construction, the fibers of the map $W' \to S$ are connected, and W' is an open neighborhood of $V \cap \dot{T}_M X$.

Then we define $W'' = W' \cup V \cup (W \cap t^{-1}\mathbb{R}^-)$. This is an open neighborhood of V, $W'' \subset W$, and the fibers of $p : W'' \cap \Omega = W' \to X$ are connected. $\square$

Let $f : Y \to X$ be a morphism of manifolds, let N be a closed submanifold of Y of codimension k and assume $f(N) \subset M$. We denote by $f_{|N}$ the map from N to M induced by f.

We denote by f' the map from TY to $Y \times_X TX$ associated to f and by f_τ the base change $Y \times_X TX \to TX$. Then the tangent map Tf is the composite map $f_\tau \circ f'$:

$$Tf : TY \xrightarrow[f']{} Y \underset{X}{\times} TX \xrightarrow[f_\tau]{} TX \ . \tag{4.1.8}$$

We denote by $T_N f$ the map from $T_N Y$ to $T_M X$ associated with Tf and we denote by f'_N the map from $T_N Y$ to $N \times_M T_M X$, and $f_{N\tau}$ the base change $N \times_M T_M X \to T_M X$:

$$T_N f : T_N Y \xrightarrow[f'_N]{} N \underset{M}{\times} T_M X \xrightarrow[f_{N\tau}]{} T_M X \ . \tag{4.1.9}$$

If there is no fear of confusion, we still write f_τ instead of $f_{N\tau}$, and f' instead of f'_N.

Let us denote by p_X, j_X, t_X, s_X, $\tilde{p}_X$ the maps associated with the normal deformation of M in X, and let $\Omega_X = t_X^{-1}(\mathbb{R}^+)$, and similarly for p_Y, j_Y, t_Y, s_Y, $\tilde{p}_Y$, Ω_Y. The map f defines a map $\tilde{f}'$ from $\tilde{Y}_N$ to $\tilde{X}_M$. If $y = (y', y'')$ (resp. $x = (x', x'')$) is a local coordinate system on Y (resp. on X) such that $N = \{y; y' = 0\}$ (resp. $M = \{x; x' = 0\}$), and $f = (f_1, f_2)$, then:

$$\begin{cases} \tilde{f}'(y', y'', t) = \left(\dfrac{1}{t} f_1(ty', y''), f_2(ty', y''), t\right) & \text{for} \quad t \neq 0 \ , \\ \tilde{f}'(y', y'', 0) = (\mathrm{d}_{y'} f_1(0, y'') \cdot y', f_2(0, y''), 0) \ . \end{cases} \tag{4.1.10}$$

Therefore we get the commutative diagram below, where all the squares are Cartesian.

$$\begin{array}{ccccccc} T_N Y & \xhookrightarrow[s_Y]{} & \tilde{Y}_N & \xhookleftarrow[j_Y]{} & \Omega_Y & \xrightarrow[\tilde{p}_Y]{} & Y \\ \downarrow{\scriptstyle T_N f} & \square & \downarrow{\scriptstyle \tilde{f}'} & \square & \downarrow{\scriptstyle \tilde{f}} & \square & \downarrow{\scriptstyle f} \\ T_M X & \xhookrightarrow[s_X]{} & \tilde{X}_M & \xhookleftarrow[j_X]{} & \Omega_X & \xrightarrow[\tilde{p}_X]{} & X \ . \end{array} \tag{4.1.11}$$

Note that $\tilde{p}_X$ and $\tilde{p}_Y$ are smooth and $t_Y = t_X \circ \tilde{f}'$. However the diagram:

$$
\text{(4.1.12)} \qquad \begin{array}{ccc} \tilde{Y}_N & \xrightarrow{p_Y} & Y \\ {\scriptstyle \tilde{f}'}\downarrow & & \downarrow{\scriptstyle f} \\ \tilde{X}_M & \xrightarrow[p_X]{} & X \end{array}
$$

is not Cartesian in general, and even if f is proper, $\tilde{f}'$ may be not proper. Such a pathological situation will not appear if f is "clean" or better "transversal" with respect to M. In order to recall these definitions, it is more convenient to work with cotangent and conormal bundles (cf. §3 below).

Let $f : Y \to X$ be a morphism of manifolds, M a closed submanifold of X.

Definition 4.1.5. (i) *One says f is clean with respect to M if $f^{-1}(M)$ is a submanifold N of Y, and the map ${}^t f'_N : N \times_M T^*_M X \to T^*_N Y$ is surjective.*

(ii) *One says f is transversal to M if the map ${}^t f'|_{Y \times_X T^*_M X} : Y \times_X T^*_M X \to T^* Y$ is injective.*

If f is a closed embedding, one also says Y is clean (resp. transversal) instead of saying that f is clean (resp. transversal). If f is transversal to M, then f is clean with respect to M, but the converse is false, as shown by the example where Y is a submanifold of X, and M a submanifold of Y. If f is clean with respect to M and $f^{-1}(M) = N$, then the map $p \times \tilde{f}' : \tilde{Y}_N \to Y \times \tilde{X}_M$ is a closed embedding. This follows from the fact that $f'_N : T_N Y \to N \times_M T_M X$ is a closed embedding, the injectivity of $\tilde{Y}_N \hookrightarrow Y \times \tilde{X}_M$, $t_Y = t_X \circ \tilde{f}'$, $T_N Y = t_Y^{-1}(0)$ and $T_M X = t_X^{-1}(0)$.

If f is transversal to M and $f^{-1}(M) = N$, then $f' : T_N Y \to Y \times_X T_M X$ is an isomorphism, and the square (4.1.12) is Cartesian.

In Chapter VII we shall need the following generalization of Definition 4.1.5.

Definition 4.1.6. *Let $f_1 : Y_1 \to X$ and $f_2 : Y_2 \to X$ be two morphisms of manifolds. One says f_1 and f_2 are transversal (resp. clean) if the map $(f_1, f_2) : Y_1 \times Y_2 \to X \times X$ is transversal (resp. clean) with respect to the diagonal of $X \times X$.*

Note that if f_2 is the embedding of a submanifold Y_2 into X, then this is equivalent to saying that f_1 is transversal (resp. clean) with respect to Y_2.

4.2. Specialization

Let M be a closed submanifold of X, as in §1. In all this book we shall keep the notations introduced in §1, especially those in (4.1.5).

Let $F \in \mathrm{Ob}(\mathbf{D}^b(X))$.

Lemma 4.2.1. *There is a natural isomorphism:*

$$
s^{-1} Rj_* \tilde{p}^{-1} F \simeq s^! j_! \tilde{p}^! F \ .
$$

Proof. Consider the distinguished triangle:

$$(4.2.1) \qquad (p^{-1}F)_\Omega \longrightarrow Rj_*\tilde{p}^{-1}F \longrightarrow R\Gamma_{\{t=0\}}((p^{-1}F)_\Omega)[1] \xrightarrow[+1]{} .$$

Applying s^{-1} we get:

$$\begin{aligned} s^{-1}Rj_*\tilde{p}^{-1}F &\simeq s^!(p^{-1}F)_\Omega[1] \\ &\simeq s^!j_!\tilde{p}^!F , \end{aligned}$$

since $\tilde{p}^! \simeq \tilde{p}^{-1}[1]$. □

Definition 4.2.2. *Let $F \in \mathrm{Ob}(\mathbf{D}^b(X))$. One sets:*

$$\nu_M(F) = s^{-1}Rj_*\tilde{p}^{-1}F \simeq s^!j_!\tilde{p}^!F ,$$

and says that $\nu_M(F)$ is the specialization of F along M.

Theorem 4.2.3. *Let $F \in \mathrm{Ob}(\mathbf{D}^b(X))$. Then:*

(i) $\nu_M(F) \in \mathrm{Ob}(\mathbf{D}^b{}_{\mathbb{R}^+}(T_MX))$, *and* $\mathrm{supp}(\nu_M(F)) \subset C_M(\mathrm{supp}(F))$.

(ii) *Let V be a conic open subset of T_MX. Then:*

$$H^j(V;\nu_M(F)) = \varinjlim_U H^j(U;F) ,$$

where U ranges through the family of open subsets of X such that $C_M(X\setminus U) \cap V = \varnothing$; (cf. Figure 4.2.a).
In particular if $v \in T_MX$, then:

$$H^j(\nu_M(F))_v = \varinjlim_U H^j(U;F) ,$$

where U ranges through the family of open subsets of X such that $v \notin C_M(X\setminus U)$.

(iii) *Let A be a closed conic subset of T_MX. Then:*

$$H^j_A(T_MX;\nu_M(F)) = \varinjlim_{Z,U} H^j_{Z\cap U}(U;F) ,$$

where U ranges through the family of open neighborhoods of M in X, and Z through the family of closed subsets of X such that $C_M(Z) \subset A$; (cf. Figure 4.2.b).

(iv) *One has the isomorphisms:*

$$\nu_M(F)|_M \simeq R\tau_*(\nu_M(F)) \simeq F|_M ,$$

$$(R\Gamma_M(\nu_M(F)))|_M \simeq R\tau_!(\nu_M(F)) \simeq R\Gamma_M(F)|_M .$$

(v) $R\dot{\tau}_*(\nu_M(F)|_{\dot{T}_MX}) \simeq R\Gamma_{X\setminus M}(F)|_M$.

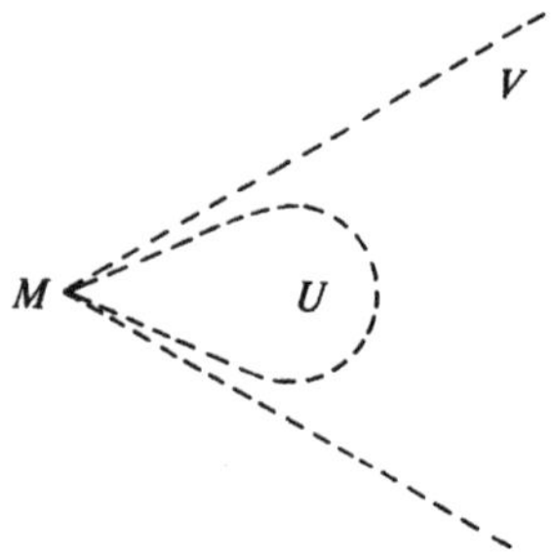

Fig. 4.2a

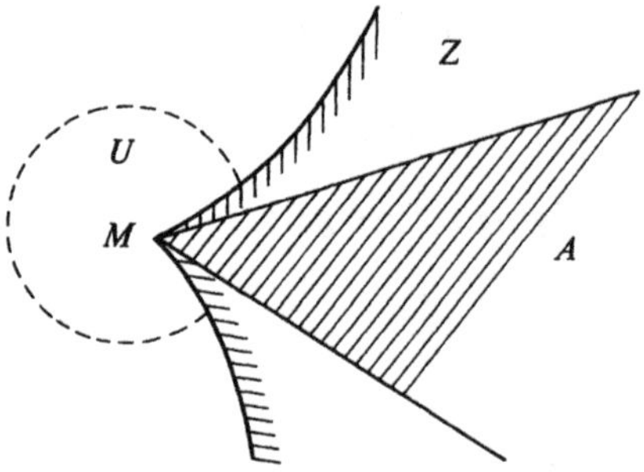

Fig. 4.2b

Proof. (i) Since $\tilde{p}^{-1}F$ is locally constant with respect to the action of $\mathbb{R}^+$, the same property will hold for $Rj_*\tilde{p}^{-1}F$, then for $s^{-1}Rj_*\tilde{p}^{-1}F$. The last statement is obvious.

(ii) Let U be an open subset of X such that $V \cap C_M(X \setminus U) = \emptyset$. We have the chain of morphisms:

$$(4.2.2)\qquad \begin{cases} R\Gamma(U;F) \to R\Gamma(p^{-1}(U);p^{-1}F) \\ \qquad \to R\Gamma(p^{-1}(U) \cap \Omega;p^{-1}F) \\ \qquad \to R\Gamma(\tilde{p}^{-1}(U) \cup V;Rj_*j^{-1}p^{-1}F) \\ \qquad \to R\Gamma(V;\nu_M(F)) \ , \end{cases}$$

where the third arrow exists since $p^{-1}(U) \cup V$ is a neighborhood of V in $\bar{\Omega}$ (Proposition 4.1.3).

Thus we obtain the morphism:

$$\varinjlim_U H^k(U;F) \to H^k(V;\nu_M(F)) \ .$$

Let us show that this is an isomorphism. We have:

$$\begin{aligned} H^k(V;\nu_M(F)) &\simeq \varinjlim_W H^k(W;Rj_*j^{-1}p^{-1}F) \\ &\simeq \varinjlim_W H^k(W \cap \Omega;p^{-1}F) \ , \end{aligned}$$

where W ranges through a neighborhood system of V. Hence we may assume by Proposition 4.1.4 that $p: W \cap \Omega \to p(W \cap \Omega)$ has connected fibers, that is, fibers homeomorphic to $\mathbb{R}$. Then we apply Proposition 3.3.9 to get:

$$H^k(W \cap \Omega;p^{-1}F) \simeq H^k(p(W \cap \Omega);F) \ .$$

Since $p(W \cap \Omega)$ ranges through the family of open subsets U of X such that $C_M(X \setminus U) \cap V = \emptyset$ (Proposition 4.1.3), we obtain (ii).

(iii) Let U and Z be as in the statement. We have the chain of morphisms:

$$\begin{aligned} R\Gamma_{Z\cap U}(U;F) &\to R\Gamma_{p^{-1}(Z\cap U)}(p^{-1}(U);p^{-1}F) \\ &\to R\Gamma_{p^{-1}(Z\cap U)\cap\Omega}(p^{-1}(U)\cap\Omega;p^{-1}F) \\ &\to R\Gamma_{(p^{-1}(Z\cap U)\cap\Omega)\cup A}(p^{-1}(U);Rj_*j^{-1}p^{-1}F) \\ &\to R\Gamma_A(T_MX;\nu_M(F)) \ . \end{aligned}$$

Here we used the fact that $(p^{-1}(Z \cap U) \cap \Omega) \cup A$ is closed in $p^{-1}(U)$. Thus we obtain the commutative diagram:

$$\begin{array}{ccccccc} \cdots \longrightarrow & \varinjlim_U H^{k-1}(U \setminus Z;F) & \longrightarrow & \varinjlim_U H^k_{Z\cap U}(U;F) & \longrightarrow & \varinjlim_U H^k(U;F) & \longrightarrow \cdots \\ & \downarrow \gamma_{k-1} & & \downarrow \alpha_k & & \downarrow \beta_k & \\ \cdots \longrightarrow & H^{k-1}(T_MX \setminus A;\nu_M(F)) & \longrightarrow & H^k_A(T_MX;\nu_M(F)) & \longrightarrow & H^k(T_MX;\nu_M(F)) & \longrightarrow \cdots . \end{array}$$

Since all the rows are exact, and all γ_k and β_k's are isomorphisms by (ii), all the α_k's are isomorphisms.

(iv) Consider the diagram (4.1.5) and denote by k the embedding of M into T_MX by the zero-section. We have the morphisms:

$$\begin{aligned} F|_M &\simeq k^{-1}s^{-1}p^{-1}F \\ &\to k^{-1}s^{-1}Rj_*j^{-1}p^{-1}F \\ &\simeq \nu_M(F)|_M \ , \end{aligned}$$

and

$$\begin{aligned} k^!\nu_M(F) &\simeq k^!s^!j_!j^!p^!F \\ &\to k^!s^!p^!F \\ &\simeq i^!F \ , \end{aligned}$$

and these morphisms are isomorphisms by (ii) and (iii). The other isomorphisms in (iv) follow from Proposition 3.7.5.

(v) We have a morphism of distinguished triangles:

$$\begin{array}{ccccccc} R\Gamma_M(F)|_M & \longrightarrow & F|_M & \longrightarrow & R\Gamma_{X\setminus M}(F)|_M & \xrightarrow{+1} & \\ \downarrow & & \downarrow & & \downarrow & & \\ R\Gamma_M(\nu_M(F))|_M & \longrightarrow & R\tau_*\nu_M(F) & \longrightarrow & R\dot{\tau}_*(\nu_M(F)|_{\dot{T}_MX}) & \xrightarrow{+1} & . \end{array}$$

Since the left and the middle vertical arrows are isomorphisms by (iv), the right arrow is also an isomorphism. □

Let $f: Y \to X$ be a morphism of manifolds, N (resp. M) a closed submanifold of Y (resp. X), and assume $f(N) \subset M$. We follow the notations of IV §1 (cf. (4.1.8), (4.1.9), (4.1.11)).

Proposition 4.2.4. *Let* $G \in \mathrm{Ob}(\mathbf{D}^b(Y))$.

(i) *There exists a commutative diagram of canonical morphisms:*

$$\begin{array}{ccc} R(T_N f)_! \nu_N(G) & \longrightarrow & \nu_M(Rf_! G) \\ \downarrow & & \downarrow \\ R(T_N f)_* \nu_N(G) & \longleftarrow & \nu_M(Rf_* G) \end{array} .$$

(ii) *Moreover if* $\mathrm{supp}(G) \to X$ *and* $C_N(\mathrm{supp}(G)) \to T_M X$ *are proper, and if* $\mathrm{supp}(G) \cap f^{-1}(M) \subset N$*, then all these morphisms are isomorphisms.*

In particular if $f^{-1}(M) = N$ *and* f *is clean with respect to* M *and proper on* $\mathrm{supp}(G)$*, then all these morphisms are isomorphisms.*

Proof. (i) We have the chains of morphisms (cf. Exercise III.9):

$$\begin{aligned} R(T_N f)_! \nu_N(G) &= R(T_N f)_! s_Y^{-1} Rj_{Y*} \tilde{p}_Y^{-1} G \\ &\simeq s_X^{-1} R\tilde{f}_!' Rj_{Y*} \tilde{p}_Y^{-1} G \\ &\to s_X^{-1} Rj_{X*} R\tilde{f}_! \tilde{p}_Y^{-1} G \\ &\simeq s_X^{-1} Rj_{X*} \tilde{p}_X^{-1} Rf_! G \\ &= \nu_M(Rf_! G) , \end{aligned}$$

and

$$\begin{aligned} \nu_M(Rf_* G) &= s_X^{-1} Rj_{X*} \tilde{p}_X^{-1} Rf_* G \\ &\simeq s_X^{-1} Rj_{X*} R\tilde{f}_* \tilde{p}_Y^{-1} G \\ &\simeq s_X^{-1} R\tilde{f}_*' Rj_{Y*} \tilde{p}_Y^{-1} G \\ &\to R(T_N f)_* s_Y^{-1} Rj_{Y*} \tilde{p}_Y^{-1} G \\ &= R(T_N f)_* \nu_N(G) . \end{aligned}$$

The commutativity follows easily by the construction.

(ii) If $\overline{\tilde{p}_Y^{-1}(\mathrm{supp}(G))}$ is proper over $\tilde{X}_M$, then all the morphisms are isomorphisms because one can replace $R\tilde{f}_*'$ and $R(T_N f)_*$ by $R\tilde{f}_!'$ and $R(T_N f)_!$, respectively. Hence it is enough to prove that for a closed subset Z of Y, $\overline{\tilde{p}_Y^{-1}(Z)}$

is proper over $\tilde{X}_M$ if Z is proper over X, if $C_N(Z)$ is proper over $T_M X$ and if $Z \cap f^{-1}(M) \subset N$. Since the fibers of $\overline{\tilde{p}_Y^{-1}(Z)} \to \tilde{X}_M$ are compact, it is enough to show that this is a closed map. Let $\{u_n\}_n$ be a sequence in $\tilde{p}_Y^{-1}(Z)$ such that $\{\tilde{f}'(u_n)\}_n$ converges. We must show that a subsequence of $\{u_n\}_n$ converges. We may assume $\{\tilde{p}_Y(u_n)\}_n$ converges. Since $\tilde{p}_Y^{-1}(Z) \setminus T_N Y \to \tilde{X}_M \setminus T_M X$ is proper, we may assume $\{\tilde{f}'(u_n)\}_n$ converges to a point of $T_M X$. Then the limit point of $\{\tilde{p}_Y(u_n)\}$ is contained in $Z \cap f^{-1}(M)$ and hence in N.

Taking local coordinate systems of X and Y as in (4.1.10), set $u_n = (y'_n, y''_n, t_n)$. Then, $t_n \underset{n}{\to} 0$, $t_n y'_n \underset{n}{\to} 0$. We may assume further $t_n > 0$ (i.e. $u_n \in \tilde{p}_Y^{-1}(z)$). Since $\{y''_n\}_n$ converges, it is enough to show that $\{|y'_n|\}_n$ is bounded. Now assuming $|y'_n| \underset{n}{\to} \infty$, we shall deduce a contradiction. By extracting a subsequence, $\{y'_n/|y'_n|\}_n$ converges to a non-zero vector v. Then $\{(y'_n/|y'_n|, y''_n, t_n|y'_n|)\}_n$ belongs to $\tilde{p}_Y^{-1}(Z)$ and converges to a point $p \in T_N Y$ which does not belong to the zero section. On the other hand, $\tilde{f}'(u_n) = \left(\frac{1}{t_n} f_1(t_n y'_n, y''_n), f_2(t_n y'_n, y''_n)\right)$ converges and hence $\left\{\frac{1}{t_n|y'_n|} f_1(t_n y'_n, y''_n)\right\}_n$ converges to zero. This implies that $T_N f(p)$ belongs to the zero section of $T_M X$. Hence $C_N(Z) \cap (T_N f)^{-1}(T_N f(p)) \supset \mathbb{R}_{\geqslant 0} p$, which contradicts the fact that $C_N(Z) \to T_M X$ is proper. $\square$

In order to study inverse images, note first that if τ_X denotes the map $T_M X \to X$, then:

$$\tau_X^! A_X \simeq \tau_X^{-1} A_X \ , \tag{4.2.3}$$

which implies $\tau_Y^{-1} f^! A_X \simeq (T_N f)^! A_{T_M X}$, or equivalently $\tau^{-1}\omega_{Y/X} \simeq \omega_{T_N Y/T_M X}$. Then:

$$\begin{cases} \omega_{T_N Y/N \times_M T_M X} \simeq \tau^{-1}\omega_{Y/X} \otimes \tau^{-1}\omega_{N/M}^{\otimes -1} \ , \\ \omega_{N \times_M T_M^* X/T_N^* Y} \simeq \pi^{-1}\omega_{N/M} \otimes \pi^{-1}\omega_{Y/X}^{\otimes -1} \ . \end{cases} \tag{4.2.4}$$

For simplicity, we shall often write $\omega_{Y/X}$ instead of $\tau^{-1}\omega_{Y/X}$ or $\pi^{-1}\omega_{Y/X}$, and similarly for $\omega_{N/M}$, $or_{Y/X}$, etc. We also write $\omega_{Y/X} \otimes \cdot$ instead of $\omega_{Y/X} \otimes^L \cdot$, etc.

Proposition 4.2.5. *Let $F \in \mathrm{Ob}(\mathbf{D}^b(X))$. Then there are canonical morphisms:*

$$\alpha : (T_N f)^{-1}\nu_M(F) \to \nu_N(f^{-1}F) \ ,$$

$$\beta : \nu_N(f^! F) \to (T_N f)^! \nu_M(F) \ ,$$

such that the diagram below commutes:

$$\begin{array}{ccc} \omega_{T_N Y/T_M X} \otimes (T_N f)^{-1}\nu_M(F) & \xrightarrow[\omega_{Y/X} \otimes \alpha]{} & \nu_N(\omega_{Y/X} \otimes f^{-1}F) \\ \Big\downarrow \gamma & & \Big\downarrow \\ (T_N f)^! \nu_M(F) & \xleftarrow[\beta]{} & \nu_N(f^! F) \end{array} \ .$$

Here the vertical arrows are deduced from (3.1.6).

All these morphisms are isomorphisms on the open set where $T_N f : T_N Y \to T_M X$ is smooth.

In particular, if $f : Y \to X$ and $f|_N : N \to M$ are smooth all these morphisms are isomorphisms.

Proof. The maps $\tilde{p}_Y$ and $\tilde{p}_X$ being smooth, we have an isomorphism $\tilde{p}_Y^{-1} f^! \simeq \tilde{f}^! \tilde{p}_X^{-1}$. We obtain the chains of morphisms:

$$\begin{aligned}(T_N f)^{-1} \nu_M(F) &= (T_N f)^{-1} s_X^{-1} Rj_{X*} \tilde{p}_X^{-1} F \\ &\simeq s_Y^{-1} \tilde{f}'^{-1} Rj_{X*} \tilde{p}_X^{-1} F \\ &\to s_Y^{-1} Rj_{Y*} \tilde{f}^{-1} \tilde{p}_X^{-1} F \\ &\simeq s_Y^{-1} Rj_{Y*} \tilde{p}_Y^{-1} f^{-1} F \\ &= \nu_N(f^{-1} F) ,\end{aligned}$$

and

$$\begin{aligned}\nu_N(f^! F) &= s_Y^{-1} Rj_{Y*} \tilde{p}_Y^{-1} f^! F \\ &\simeq s_Y^{-1} Rj_{Y*} \tilde{f}^! \tilde{p}_X^{-1} F \\ &\simeq s_Y^{-1} \tilde{f}'^! Rj_{X*} \tilde{p}_X^{-1} F \\ &\to (T_N f)^! s_X^{-1} Rj_{X*} \tilde{p}_X^{-1} F \\ &= (T_N f)^! \nu_M(F) .\end{aligned}$$

The commutativity follows from the proof.

At the points where $T_N f$ is smooth, $\tilde{f}'$ is smooth, and α, β and γ are isomorphisms. □

Finally let us study the tensor product.

Let X and Y be two manifolds, and let M and N two closed submanifolds of X and Y, respectively.

Proposition 4.2.6. *Let $F \in \mathrm{Ob}(\mathbf{D}^b(X))$ and $G \in \mathrm{Ob}(\mathbf{D}^b(Y))$. Then there is a natural morphism in $\mathbf{D}^b(T_{M \times N}(X \times Y))$:*

$$\nu_M(F) \overset{L}{\boxtimes} \nu_N(G) \to \nu_{M \times N}\left(F \overset{L}{\boxtimes} G\right) .$$

Proof. Let p_X, j_X, $\tilde{p}_X$, s_X be the natural maps associated to the normal deformation of M in X, and similarly for $p_Y, \ldots, s_Y$. We write for short p, j, $\tilde{p}$, s instead of $p_{X \times Y}, \ldots, s_{X \times Y}$, and we denote by p', j', $\tilde{p}'$, s' the maps $p_X \times p_Y$, $j_X \times j_Y$, $\tilde{p}_X \times \tilde{p}_Y$, $s_X \times s_Y$.

There is a natural closed embedding k from $\widetilde{(X \times Y)}_{M \times N}$ to $\tilde{X}_M \times \tilde{Y}_N$ which defines the commutative diagram:

$$\begin{array}{ccccc}
T_M X \times T_N Y & \xhookrightarrow{s'} & \tilde{X}_M \times \tilde{Y}_N & \xleftarrow{\tilde{j}'} & \Omega_X \times \Omega_Y \\
\| & \square & \uparrow{k} & \searrow^{p'} \nearrow^{\tilde{p}'} & \uparrow{\tilde{k}} \\
& & & X \times Y & \\
& & \nearrow^{p} \searrow^{\tilde{p}} & & \\
T_{M \times N}(X \times Y) & \xhookrightarrow{s} & \widetilde{(X \times Y)}_{M \times N} & \xleftarrow{\tilde{j}} & \Omega_{X \times Y}
\end{array}$$

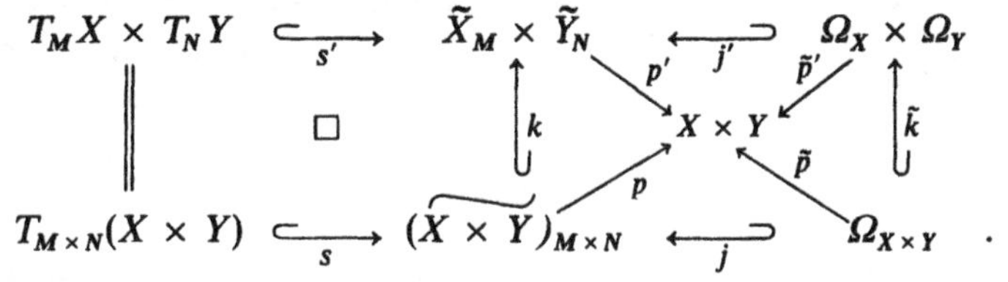

We have the morphisms:

$$v_M(F) \overset{L}{\boxtimes} v_N(G) = s_X^{-1} R j_{X*} \tilde{p}_X^{-1} F \overset{L}{\boxtimes} s_Y^{-1} R j_{Y*} \tilde{p}_Y^{-1} G$$

$$\to s'^{-1} R j'_* \tilde{p}'^{-1}\left(F \overset{L}{\boxtimes} G\right)$$

$$\simeq s^{-1} k^{-1} R j'_* \tilde{p}'^{-1}\left(F \overset{L}{\boxtimes} G\right)$$

$$\to s^{-1} R j_* \tilde{k}^{-1} \tilde{p}'^{-1}\left(F \overset{L}{\boxtimes} G\right)$$

$$\simeq s^{-1} R j_* \tilde{p}^{-1}\left(F \overset{L}{\boxtimes} G\right). \quad \square$$

Corollary 4.2.7. *Let F and G belong to $\mathrm{Ob}(\mathbf{D}^b(X))$. Then there exists a natural morphism in $\mathbf{D}^b(T_M X)$:*

$$v_M(F) \overset{L}{\otimes} v_M(G) \to v_M\left(F \overset{L}{\otimes} G\right).$$

Proof. Let Δ_X be the diagonal of $X \times X$, δ_X the embedding $\Delta_X \hookrightarrow X \times X$, and define similarly Δ_{TX}, $\Delta_{T_M X}$, δ_{TX}, $\delta_{T_M X}$. Note that, with the notations (4.1.9), $\delta_{T_M X} = T_M \delta_X$. Then applying Propositions 4.2.5 and 4.2.6, we get the morphisms:

$$v_M(F) \overset{L}{\otimes} v_M(G) \simeq \delta_{T_M X}^{-1}\left(v_M(F) \overset{L}{\boxtimes} v_M(G)\right)$$

$$\to \delta_{T_M X}^{-1}\left(v_{M \times M}\left(F \overset{L}{\boxtimes} G\right)\right)$$

$$\to v_M\left(\delta_X^{-1}\left(F \overset{L}{\boxtimes} G\right)\right)$$

$$\simeq v_M\left(F \overset{L}{\otimes} G\right). \quad \square$$

4.3. Microlocalization

Let M be a closed submanifold of X of codimension l, as in §1. We denote by $T^*_M X$ the conormal bundle to M in X, i.e. the kernel of the map $M \times_X T^*X \to T^*M$. We denote by π the projection $T^*X \to X$ or its restriction $T^*_M X \to M$, and we sometimes write π instead of $i \circ \pi$, where $i : M \to X$.

We denote by $\dot{\pi}$ the restriction of π to $\dot{T}^*X = T^*X \backslash X$.

Definition 4.3.1. *Let $F \in \mathrm{Ob}(\mathbf{D}^b(X))$. The microlocalization of F along M, denoted $\mu_M(F)$, is the Fourier-Sato transform of $\nu_M(F)$:*

$$\mu_M(F) = \nu_M(F)^\wedge \ .$$

Applying Theorem 4.2.3 and the results of III §7. We get:

Theorem 4.3.2. *Let $F \in \mathrm{Ob}(\mathbf{D}^b(X))$. Then:*

(i) $\mu_M(F) \in \mathrm{Ob}(\mathbf{D}^b_{\mathbb{R}^+}(T^*_M X))$.

(ii) *Let V be a convex open cone of $T^*_M X$. Then:*

$$H^j(V; \mu_M(F)) = \varinjlim_{U, Z} H^j_{Z \cap U}(U; F) \ ,$$

where U ranges through the family of open subsets of X such that $U \cap M = \pi(V)$, and Z through the family of closed subsets such that $C_M(Z) \subset V^\circ$. In particular let $p \in T^*_M X$. Then:

$$H^j(\mu_M(F))_p = \varinjlim_{Z} H^j_Z(F)_{\pi(p)} \ ,$$

where Z ranges through the family of closed subsets of X such that $C_M(Z)_{\pi(p)} \subset \{v \in (T_M X)_{\pi(p)}; \langle v, p \rangle > 0\} \cup \{0\}$.

(iii) *Let Z be a proper closed convex cone of $T^*_M X$ containing the zero-section M. Then:*

$$H^j_Z(T^*_M X; \mu_M(F) \otimes or_{M/X}) = \varinjlim_{U} H^{j-l}(U; F) \ ,$$

where U ranges through the family of open subsets of X such that $C_M(X \backslash U) \cap \mathrm{Int}\, Z^{\circ a} = \emptyset$.

(iv)

$$\mu_M(F)|_M \simeq R\pi_* \mu_M(F) \simeq R\Gamma_M(F)|_M \simeq i^! F \ ,$$

$$R\pi_! \mu_M(F) \simeq R\Gamma_M(\mu_M(F)) \simeq i^{-1} F \otimes \omega_{M/X} \ .$$

Note that:

$$i^{-1} F \otimes \omega_{M/X} = F|_M \otimes or_{M/X}[-l] \ .$$

Applying the last results to the distinguished triangle:

$$R\Gamma_M(\mu_M(F)) \longrightarrow R\pi_* \mu_M(F) \longrightarrow R\dot{\pi}_* \mu_M(F) \xrightarrow[+1]{} \ ,$$

we get the distinguished triangle:

$$F|_M \otimes \omega_{M/X} \longrightarrow R\Gamma_M(F)|_M \longrightarrow R\dot{\pi}_* \mu_M(F) \xrightarrow[+1]{} . \tag{4.3.1}$$

Now let $f: Y \to X$ be a morphism of manifolds, N a closed submanifold of codimension k of Y, with $f(N) \subset M$. The map Tf (cf. 4.1.8) defines the maps:

$$T^*Y \xleftarrow[{}^t f']{} Y \underset{X}{\times} T^*X \xrightarrow[f_\pi]{} T^*X , \tag{4.3.2}$$

which induce

$$T_N^*Y \xleftarrow[{}^t f'_N]{} N \underset{M}{\times} T_M^*X \xrightarrow[f_{N\pi}]{} T_M^*X . \tag{4.3.3}$$

If there is no fear of confusion, we write f_π instead of $f_{N\pi}$, and also ${}^t f'$ instead of ${}^t f'_N$. We set:

$$T_Y^*X = \operatorname{Ker}({}^t f' : Y \underset{X}{\times} T^*X \to T^*Y) = {}^t f'^{-1}(T_Y^*Y) . \tag{4.3.4}$$

Remark 4.3.3. In the literature one often writes ρ_f and ϖ_f, or simply ρ and ϖ instead of ${}^t f'$ and f_π, respectively. We shall not use these notations here.

We shall apply the Fourier-Sato functor to the morphisms of Propositions 4.2.4 and 4.2.5.

Proposition 4.3.4. *Let $G \in \mathrm{Ob}(\mathbf{D}^b(Y))$. Then, there exists a commutative diagram of canonical morphisms:*

$$\begin{array}{ccc} Rf_{N\pi!}\,{}^t f_N'^{-1} \mu_N(G) & \longrightarrow & \mu_M(Rf_! G) \\ \downarrow & & \downarrow \\ Rf_{N\pi *}({}^t f_N'^{!} \mu_N(G) \otimes \omega_{Y/X} \otimes \omega_{N/M}^{\otimes -1}) & \longleftarrow & \mu_M(Rf_* G) . \end{array}$$

If $\mathrm{supp}(G) \to X$ *and* $C_N(\mathrm{supp}(G)) \to T_M X$ *are proper and if* $f^{-1}(M) \cap \mathrm{supp}(G) \subset N$, *then these morphisms are isomorphisms.*

In particular if $f^{-1}(M) = N$, and f is clean with respect to M and proper on $\mathrm{supp}(G)$, *then all these morphisms are isomorphisms.*

Proof. Let $H = \nu_N(G)$. Applying Propositions 3.7.13 and 3.7.14 to $f' = f_\tau \circ f'_N$, we get:

$$\begin{aligned} (R(T_N f)_! H)^\wedge &\simeq Rf_{\pi!}(Rf'_{N!} H)^\wedge \\ &\simeq Rf_{\pi!}\,{}^t f_N'^{-1}(H^\wedge) , \\ (R(T_N f)_* H)^\wedge &\simeq Rf_{\pi *}(Rf'_{N*} H)^\wedge \\ &\simeq Rf_{\pi *}({}^t f_N'^{!} H^\wedge \otimes \omega^{\otimes -1}_{N \times_M T_M^*X / T_N^*Y}) . \end{aligned}$$

Then the result follows from Proposition 4.2.4 and formula (4.2.4). □

Proposition 4.3.5. *Let $F \in \mathrm{Ob}(\mathbf{D}^b(X))$.*

(i) *There exists a commutative diagram of canonical morphisms:*

$$\begin{array}{ccc} R^t f'_{N!}(\omega_{N/M} \otimes f_{N\pi}^{-1}\mu_M(F)) & \longrightarrow & \mu_N(\omega_{Y/X} \otimes f^{-1}F) \\ \downarrow & & \downarrow \\ R^t f'_{N*} f^!_{N\pi}\mu_M(F) & \longleftarrow & \mu_N(f^!F) \end{array} .$$

Here the vertical arrows are deduced from (3.1.6).

(ii) *If $f: Y \to X$ and $f|_N : N \to M$ are smooth, all these morphisms are isomorphisms.*

(iii) *If f is transverse to M and $f^{-1}(M) = N$, there is a natural morphism:*

$$R^t f'_{N*} f_{N\pi}^{-1}\mu_M(F) \to \mu_N(f^{-1}F) .$$

Proof. Set $H = \nu_M(F)$, and let us apply Propositions 3.7.13 and 3.7.14. We get:

$$\begin{aligned}(T_N f)^! A_{T_M X} \otimes ((T_N f)^{-1} H)^\wedge &\simeq (f_N'^! A_{N \times_M T_M X} \otimes f_N'^{-1}(f^!_{N\tau} A_{T_M X} \otimes f_{N\tau}^{-1} H))^\wedge \\ &\simeq R^t f'_{N!}(f^!_{N\tau} A_{T_M X} \otimes f_{N\tau}^{-1} H)^\wedge \\ &\simeq R^t f'_{N!}(f^!_{N\pi} A_{T_M^* X} \otimes f_{N\pi}^{-1}(H^\wedge)) , \\ ((T_N f)^! H)^\wedge &\simeq (f_N'^! f^!_{N\tau} H)^\wedge \\ &\simeq R^t f'_{N*} f^!_{N\pi}(H^\wedge) .\end{aligned}$$

Hence (i) follows from Proposition 4.2.5, as well as (ii).

If f is transverse to M and $f^{-1}(M) = N$, then ${}^t f'_N : T_N^* Y \to N \times_M T_M^* X$ is an isomorphism, and we have: $\omega_{N/M} \simeq (\omega_{Y/X})|_N$. □

Note that if f and $f|_N$ are smooth, then $Y \times_X T^*X$ is a sub-bundle of T^*Y and f_π induces an isomorphism $(Y \times_X T^*X) \cap T_N^*Y \xrightarrow{\sim} Y \times_X T_M^*X$.

We shall come back to the morphisms defined in Propositions 4.3.4 and 4.3.5 in Chapters V and VI where we will give new conditions in order that they are isomorphisms, with the help of the micro-support of sheaves.

Finally we study the tensor product.

Proposition 4.3.6. *In the situation of Proposition 4.2.6 there is a natural morphism:*

$$\mu_M(F) \overset{L}{\boxtimes} \mu_N(G) \to \mu_{M \times N}\left(F \overset{L}{\boxtimes} G\right) .$$

Proof. Apply Proposition 4.2.6 and Proposition 3.7.15. □

Proposition 4.3.7. *Let M be a submanifold of X and let $\gamma : T_M^*X \times_M T_M^*X \to T_M^*X$ be the morphism given by the addition. Then, for any $F, G \in \mathrm{Ob}(\mathbf{D}^b(X))$, there exists a natural morphism:*

$$(4.3.5) \qquad R\gamma_!\left(\mu_M(F)\overset{L}{\underset{M}{\boxtimes}}\mu_M(G)\right) \to \mu_M\left(F\overset{L}{\otimes}G\right)\otimes\omega_{M/X}\,.$$

Proof. Let $\delta : T_MX \hookrightarrow T_MX \times_M T_MX$ be the diagonal embedding. Then, we have:

$$\nu_M(F)\overset{L}{\otimes}\nu_M(G) \simeq \delta^{-1}\left(\nu_M(F)\underset{M}{\boxtimes}\nu_M(G)\right)\,.$$

Noting that ${}^t\delta = \gamma$ and applying Propositions 3.7.13 and 3.7.15, we obtain:

$$\begin{aligned}\left(\omega_{T_MX/T_MX\times_M T_MX}\otimes\nu_M(F)\overset{L}{\otimes}\nu_M(G)\right)^\wedge &\simeq R{}^t\delta_!\left(\nu_M(F)\overset{L}{\underset{M}{\boxtimes}}\nu_M(G)\right)^\wedge \\ &\simeq R\gamma_!\left(\mu_M(F)\overset{L}{\underset{M}{\boxtimes}}\mu_M(G)\right)\,.\end{aligned}$$

Then, we obtain the desired homomorphism by Corollary 4.2.7. □

Remark 4.3.8. The morphisms in Propositions 4.3.4 and 4.3.5 are related as follows. (Proofs are left to the reader.)

(a) For any morphism $\varphi : G \to f^!F$ in $\mathbf{D}^b(Y)$, the diagram below commutes:

$$\begin{array}{ccc} Rf_{N\pi!}{}^tf_N'^{-1}\mu_N(G) & \xrightarrow[\alpha]{} & \mu_M(Rf_!G) \\ \big\downarrow{\scriptstyle\varphi} & & \big\downarrow{\scriptstyle\varphi} \\ Rf_{N\pi!}{}^tf_N'^{-1}\mu_N(f^!F) & \xrightarrow[\beta]{} & \mu_M(F)\quad, \end{array}$$

where α is the morphism given by the first horizontal arrow in Proposition 4.3.4 and β is that given by the second horizontal arrow in Proposition 4.3.5.

(b) Similarly, for any morphism $\psi : F \to Rf_*G$ in $\mathbf{D}^b(X)$, the diagram below commutes:

$$\begin{array}{ccc} \mu_M(F) & \xrightarrow[\alpha']{} & Rf_{N\pi*}({}^tf_N'^{!}\mu_N(f^{-1}F)\otimes\omega_{Y/X}\otimes\omega_{N/M}^{\otimes -1}) \\ \big\downarrow{\scriptstyle\psi} & & \big\downarrow{\scriptstyle\psi} \\ \mu_M(Rf_*G) & \xrightarrow[\beta']{} & Rf_{N\pi*}({}^tf_N'^{!}\mu_N(G)\otimes\omega_{Y/X}\otimes\omega_{N/M}^{\otimes -1})\quad, \end{array}$$

where α' is given by the first horizontal arrow in Proposition 4.3.5 and β' is given by the second horizontal arrow in Proposition 4.3.4.

4.4. The functor μhom

Let f be a morphism from Y to X. Let $\Delta_f \subset X \times Y$ be the graph of f. We denote by Δ_X (resp. Δ_Y) the diagonal of $X \times X$ (resp. $Y \times Y$). We shall identify

$Y \times_X T^*X$ with $T^*_{\Delta_f}(X \times Y)$, by the projection $T^*(X \times Y) \to (T^*X) \times Y$, and similarly we identify T^*X (resp. T^*Y) with $T^*_{\Delta_X}(X \times X)$ (resp. $T^*_{\Delta_Y}(Y \times Y)$) by the first projection. If there is no fear of confusion we shall write Δ instead of Δ_f. Thus we have the maps (cf. (4.3.3)):

(4.4.1)
$$\begin{array}{ccccc} T^*_{\Delta_Y}(Y \times Y) & \longleftarrow & T^*_{\Delta_f}(X \times Y) & \longrightarrow & T^*_{\Delta_X}(X \times X) \\ \downarrow \wr & & \downarrow \wr & & \downarrow \wr \\ T^*Y & \underset{{}^t f'}{\longleftarrow} & Y \underset{X}{\times} T^*X & \underset{f_\pi}{\longrightarrow} & T^*X \end{array} .$$

Notice the useful formula:

(4.4.2)
$$\omega_{Y \times_X T^*X/T^*Y} \otimes \omega_{Y \times_X T^*X/T^*X} \simeq A_{Y \times_X T^*X} \, .$$

We denote by $f_1 : Y \times Y \to X \times Y$ the map (f, id_Y) and by $f_2 : X \times Y \to X \times X$ the map (id_X, f). Hence we have the commutative diagram where the square on the right-hand side is Cartesian and f_2 is transversal to Δ_X:

(4.4.3)
$$\begin{array}{ccccc} Y \times Y & \xrightarrow{f_1} & X \times Y & \xrightarrow{f_2} & X \times X \\ \uparrow & & \uparrow & \square & \uparrow \\ \Delta_Y & \xrightarrow{\sim} & \Delta & \xrightarrow{f} & \Delta_X \end{array} .$$

We denote by q_j (resp. $\tilde{q}_j$, resp. q'_j) the j-th projection ($j = 1, 2$) defined on $X \times X$ (resp. $X \times Y$, resp. $Y \times Y$).

Definition 4.4.1. *Let* $G \in \mathrm{Ob}(\mathbf{D}^b(Y))$, $F \in \mathrm{Ob}(\mathbf{D}^b(X))$. *We set:*

(i)
$$\mu hom(G \to F) = \mu_\Delta R\mathcal{H}om(\tilde{q}_2^{-1}G, \tilde{q}_1^! F) \, ,$$

(ii)
$$\mu hom(F \leftarrow G) = (\mu_\Delta R\mathcal{H}om(\tilde{q}_1^{-1}F, \tilde{q}_2^! G))^a \, .$$

(iii) *When* $Y = X$ *and* f *is the identity, we set:*

$$\mu hom(G, F) = \mu hom(G \to F) = \mu_{\Delta_X}(R\mathcal{H}om(q_2^{-1}G, q_1^! F)) \, .$$

In formula (ii), $(\cdot)^a$ denotes the inverse image by the antipodal map on $Y \times_X T^*X$.

Let π denote the projection from $T^*_\Delta(X \times Y)$ to $\Delta \simeq Y$.

Proposition 4.4.2. *We have canonical isomorphisms:*

(i)
$$R\pi_* \mu hom(G \to F) \simeq R\mathcal{H}om(G, f^! F) \, ,$$

$$R\pi_* \mu hom(F \leftarrow G) \simeq R\mathcal{H}om(f^{-1}F, G) \, .$$

In particular when $Y = X$ *and* f *is the identity:*

$$R\pi_* \mu hom(G, F) \simeq R\mathcal{H}om(G, F) \, .$$

(ii) *Assume G (resp. F) is cohomologically constructible. Then:*

$$R\pi_! \mu hom(G \to F) \simeq R\mathscr{H}om(G, A_Y) \overset{L}{\otimes} f^{-1}F \otimes \omega_{Y/X}$$

$$\left(\text{resp. } R\pi_! \mu hom(F \leftarrow G) \simeq f^{-1}R\mathscr{H}om(F, A_X) \overset{L}{\otimes} G\right) .$$

In particular when $Y = X$ and f is the identity

$$R\pi_! \mu hom(G, F) \simeq R\mathscr{H}om(G, A_X) \overset{L}{\otimes} F .$$

Proof. (i) We have the isomorphisms by Theorem 4.3.2:

$$\begin{aligned} R\pi_* \mu_\Delta R\mathscr{H}om(\tilde{q}_2^{-1}G, \tilde{q}_1^{-1}F) &\simeq R\tilde{q}_{2*} R\Gamma_\Delta R\mathscr{H}om(\tilde{q}_2^{-1}G, \tilde{q}_1^! F) \\ &\simeq R\tilde{q}_{2*} R\mathscr{H}om(\tilde{q}_2^{-1}G, R\Gamma_\Delta \tilde{q}_1^! F) \\ &\simeq R\mathscr{H}om(G, R\tilde{q}_{2*} R\Gamma_\Delta \tilde{q}_1^! F) \\ &\simeq R\mathscr{H}om(G, f^! F) . \end{aligned}$$

Similarly:

$$\begin{aligned} R\pi_* \mu_\Delta R\mathscr{H}om(\tilde{q}_1^{-1}F, \tilde{q}_2^! G) &\simeq R\tilde{q}_{2*} R\Gamma_\Delta R\mathscr{H}om(\tilde{q}_1^{-1}F, \tilde{q}_2^! G) \\ &\simeq R\tilde{q}_{2*} R\mathscr{H}om((\tilde{q}_1^{-1}F)_\Delta, \tilde{q}_2^! G) \\ &\simeq R\mathscr{H}om(R\tilde{q}_{2!}(\tilde{q}_1^{-1}F)_\Delta, G) \\ &\simeq R\mathscr{H}om(f^{-1}F, G) . \end{aligned}$$

(ii) The proof is similar, in view of Proposition 3.4.4. □

The functor μhom generalizes the functor of Sato microlocalization.

Proposition 4.4.3. *Let Y be a closed submanifold of X, j the embedding $T_Y^*X \to T^*X$ and let $F \in \mathrm{Ob}(\mathbf{D}^b(X))$. Then one has the isomorphism:*

$$\mu hom(A_Y, F) \simeq j_* \mu_Y(F) .$$

Proof. Let f be the immersion $Y \hookrightarrow X$. We have:

$$\begin{aligned} \mu_{\Delta_X} R\mathscr{H}om(q_2^{-1}A_Y, q_1^! F) &\simeq \mu_{\Delta_X}(Rf_{2*} f_2^! q_1^! F) \\ &\simeq \mu_{\Delta_X}(Rf_{2*} \tilde{q}_1^! F) . \end{aligned}$$

Let us identify $T_\Delta^*(X \times Y)$ with $Y \times_X T_{\Delta_X}^*(X \times X)$.

Applying Proposition 4.3.4 we get:

$$\mu_{\Delta_X}(Rf_{2*} \tilde{q}_1^! F) \simeq \mu_\Delta(\tilde{q}_1^! F) .$$

Then by Proposition 4.3.5 we have:

$$\mu_\Delta(\tilde{q}_1^! F) \simeq j_* \mu_Y(F) . \quad \square$$

One can describe the stalk of $\mu hom(G, F)$ using the γ-topology (cf. III§5).

Proposition 4.4.4. *Assume X is a vector space. Let F and G belong to* Ob($\mathbf{D}^b(X)$), *and let* $(x_0; \xi_0) \in T^*X$. *Then:*

$$H^j(\mu hom(G, F))_{(x_0;\xi_0)} = \varinjlim_{U,\gamma} H^j(R\Gamma(U; R\mathcal{H}om(\phi_\gamma^{-1} R\phi_{\gamma *} G_U, F)) ,$$

where U ranges through the family of open neighborhoods of x_0 *and* γ *ranges through the family of closed convex proper cones of X such that* $\gamma \subset \{v \in X; \langle v, \xi_0 \rangle < 0\} \cup \{0\}$.

Recall that ϕ_γ denotes the map $X \to X_\gamma$, where X_γ is the space X endowed with the γ-topology.

Proof. Let γ be a closed proper cone of X. Set:

$$Z_\gamma = \{(x, x') \in X \times X; x' - x \in \gamma\} . \tag{4.4.4}$$

Then:

$$H^j(\mu hom(G, F))_{(x_0;\xi_0)} = \varinjlim_{U,V,\gamma} H^j(R\Gamma_{Z_\gamma}(U \times V; R\mathcal{H}om(q_2^{-1} G, q_1^! F)) ,$$

where U and V range through the family of open neighborhoods of x_0, and γ ranges through the family of closed convex proper cones satisfying the condition of the proposition. Then we have:

$$\begin{aligned} R\Gamma_{Z_\gamma}(U \times V; R\mathcal{H}om(q_2^{-1} G, q_1^! F)) &\simeq R\Gamma(U \times X; R\mathcal{H}om((q_2^{-1} G_V)_{Z_\gamma}, q_1^! F) \\ &\simeq R\Gamma(U; R\mathcal{H}om(Rq_{1!}(q_2^{-1} G_V)_{Z_\gamma}, F) . \end{aligned}$$

Then the result follows from Proposition 3.5.4. $\square$

Let us proceed to describe some functorial properties of μhom.

Proposition 4.4.5. *In the situation of Definition* 4.4.1 *one has commutative diagrams of canonical morphisms*:

(i)

$$\begin{array}{ccc} R^t f_!' \mu hom(G \to F) & \longrightarrow & \mu hom(G, f^{-1} F \otimes \omega_{Y/X}) \\ \downarrow & & \downarrow \\ R^t f_*' \mu hom(G \to F) & \longleftarrow & \mu hom(G, f^! F) \end{array} ,$$

(ii)
$$\begin{array}{ccc} R^t f_!' \mu hom(F \leftarrow G) & \longrightarrow & \mu hom(f^! F, G \otimes \omega_{Y/X}) \\ \downarrow & & \downarrow \\ R^t f_*' \mu hom(F \leftarrow G) & \longleftarrow & \mu hom(f^{-1}F, G) \end{array} ,$$

(iii)
$$\begin{array}{ccc} Rf_{\pi!} \mu hom(G \to F) & \longrightarrow & \mu hom(Rf_* G, F) \\ \downarrow & & \downarrow \\ Rf_{\pi*} \mu hom(G \to F) & \longleftarrow & \mu hom(Rf_! G, F) \end{array} ,$$

(iv)
$$\begin{array}{ccc} Rf_{\pi!} \mu hom(F \leftarrow G) & \longrightarrow & \mu hom(F, Rf_! G) \\ \downarrow & & \downarrow \\ Rf_{\pi*} \mu hom(F \leftarrow G) & \longleftarrow & \mu hom(F, Rf_* G) \end{array} .$$

If f is smooth, all the morphisms in (i) and (ii) are isomorphisms. If f is proper on supp(G), *all the morphisms in (iii) and (iv) are isomorphisms.*

Before entering into the proof, let us recall (Theorem 3.1.5 and formula (2.6.14)) that a morphism $Rf_! G \to F$ defines a morphism $G \to f^! F$ and conversely, and a morphism $F \to Rf_* G$ defines a morphism $f^{-1}F \to G$, and conversely. We shall systematically use these remarks (with Y replaced by $Y \times_X T^*X$ and X by T^*Y or T^*X). We shall also make a systematic use of the morphisms constructed in II§6.

Proof. We keep the notations introduced at the beginning of this section. Then we have the following natural morphisms, obtained by applying Propositions 4.3.4 or 4.3.5.

(i, a)
$$\begin{aligned} \mu hom(G \to F) &= \mu_\Delta R\mathcal{H}om(\tilde{q}_2^{-1}G, \tilde{q}_1^! F) \\ &\to \mu_\Delta R\mathcal{H}om(\tilde{q}_2^{-1}G, Rf_{1*} f_1^{-1} \tilde{q}_1^! F) \\ &\simeq \mu_\Delta Rf_{1*} R\mathcal{H}om(q_2'^{-1}G, q_1'^! f^{-1}F) \\ &\to {}^t f'^! \mu_{\Delta_Y} R\mathcal{H}om(q_2'^{-1}G, q_1'^! f^{-1}F \otimes \omega_{Y/X}) . \end{aligned}$$

(i, b)
$$\begin{aligned} \mu hom(G, f^! F) &= \mu_{\Delta_Y} R\mathcal{H}om(q_2'^{-1}G, q_1'^! f^! F) \\ &\simeq \mu_{\Delta_Y} f_1^! R\mathcal{H}om(\tilde{q}_2^{-1}G, \tilde{q}_1^! F) \\ &\to R^t f_*' \mu_\Delta R\mathcal{H}om(\tilde{q}_2^{-1}G, \tilde{q}_1^! F) . \end{aligned}$$

If f is smooth, this last morphism is an isomorphism.

(ii, a)
$$\begin{aligned}\mu hom(F \leftarrow G) &= \mu_\Delta R\mathcal{H}om(\tilde{q}_1^{-1}F, \tilde{q}_2^! G)^a \\ &\to \mu_\Delta R\mathcal{H}om(Rf_{1!}f_1^!\tilde{q}_1^{-1}F, \tilde{q}_2^! G)^a \\ &\simeq \mu_\Delta Rf_{1*}R\mathcal{H}om(q_1'^{-1}f'F, q_2'^! G)^a \\ &\to {}^t f'^! \mu_{\Delta_Y} R\mathcal{H}om(q_1'^{-1}f'F, q_2'^! G \otimes \omega_{Y/X})^a .\end{aligned}$$

(ii, b)
$$\begin{aligned}\mu hom(f^{-1}F, G) &\simeq \mu_{\Delta_Y} R\mathcal{H}om(q_1'^{-1}f^{-1}F, q_2'^! G)^a \\ &\simeq \mu_{\Delta_Y} R\mathcal{H}om(f_1^{-1}\tilde{q}_1^{-1}F, f_1^!\tilde{q}_2^! G)^a \\ &\simeq \mu_{\Delta_Y} f_1^! R\mathcal{H}om(\tilde{q}_1^{-1}F, \tilde{q}_2^! G)^a \\ &\to R^t f'_* \mu_\Delta R\mathcal{H}om(\tilde{q}_1^{-1}F, \tilde{q}_2^! G)^a .\end{aligned}$$

If f is smooth, this last morphism is an isomorphism.

(iii, a)
$$\begin{aligned}\mu hom(G \to F) &\to \mu_\Delta R\mathcal{H}om(f_2^{-1}Rf_{2*}\tilde{q}_2^{-1}G, \tilde{q}_1^! F) \\ &\simeq \mu_\Delta f_2^! R\mathcal{H}om(q_2^{-1}Rf_* G, q_1^! F) \\ &\to f_\pi^! \mu_{\Delta_X} R\mathcal{H}om(q_2^{-1}Rf_* G, q_1^! F) .\end{aligned}$$

(iii, b)
$$\begin{aligned}\mu hom(Rf_! G, F) &= \mu_{\Delta_X} R\mathcal{H}om(q_2^{-1}Rf_! G, q_1^! F) \\ &\simeq \mu_{\Delta_X} Rf_{2*} R\mathcal{H}om(\tilde{q}_2^{-1}G, \tilde{q}_1^! F) \\ &\to Rf_{\pi*}\mu_\Delta R\mathcal{H}om(\tilde{q}_2^{-1}G, \tilde{q}_1^! F) .\end{aligned}$$

If f is proper on supp(G), this last morphism is an isomorphism.

(iv, a)
$$\begin{aligned}\mu hom(F \leftarrow G) &\to \mu_\Delta R\mathcal{H}om(\tilde{q}_1^{-1}F, f_2^! Rf_{2!}\tilde{q}_2^! G)^a \\ &\simeq \mu_\Delta f_2^! R\mathcal{H}om(q_1^{-1}F, q_2^! Rf_! G)^a \\ &\to f_\pi^! \mu_{\Delta_X} R\mathcal{H}om(q_1^{-1}F, q_2^! Rf_! G)^a .\end{aligned}$$

(iv, b)
$$\begin{aligned}\mu hom(F, Rf_* G) &\simeq \mu_{\Delta_X} R\mathcal{H}om(q_1^{-1}F, q_2^! Rf_* G)^a \\ &\simeq \mu_{\Delta_X} Rf_{2*} R\mathcal{H}om(\tilde{q}_1^{-1}F, \tilde{q}_2^! G)^a \\ &\to Rf_{\pi*}\mu_\Delta R\mathcal{H}om(\tilde{q}_1^{-1}F, \tilde{q}_2^! G)^a .\end{aligned}$$

If f is proper on supp(G), this last morphism is an isomorphism.

By these morphisms, we have obtained all the horizontal arrows stated in Proposition 4.4.5. The vertical arrows are the natural ones, defined by $f_! \to f_*$ or $f^{-1} \otimes \omega_{Y/X} \to f^!$. The commutativity is left to the reader. □

Proposition 4.4.6. *In the situation of Definition* 4.4.1 *one has canonical morphisms*:

(i) $$R{}^tf'_!f_\pi^{-1}\mu hom(Rf_!G,F)\to\mu hom(G,f^{-1}F\otimes\omega_{Y/X})\ ,$$

(ii) $$R{}^tf'_!f_\pi^{-1}\mu hom(F,Rf_*G)\to\mu hom(f^!F,G\otimes\omega_{Y/X})\ ,$$

(iii) $$Rf_{\pi!}{}^tf'^{-1}\mu hom(G,f^!F)\to\mu hom(Rf_*G,F)\ ,$$

(iv) $$Rf_{\pi!}{}^tf'^{-1}\mu hom(f^{-1}F,G)\to\mu hom(F,Rf_!G)\ .$$

If f is smooth and proper on supp(*G*), *then the morphisms in (iii) and (iv) are isomorphisms.*

Proof. We have the following morphisms:

(i) $$\begin{aligned} f_\pi^{-1}\mu hom(Rf_!G,F) &\to f_\pi^{-1}Rf_{\pi*}\mu hom(G\to F)\\ &\to\mu hom(G\to F)\\ &\to{}^tf'^!\mu hom(G,f^{-1}F\otimes\omega_{Y/X})\ ,\end{aligned}$$

(ii) $$\begin{aligned} f_\pi^{-1}\mu hom(F,Rf_*G) &\to f_\pi^{-1}Rf_{\pi*}\mu hom(F\leftarrow G)\\ &\to\mu hom(F\leftarrow G)\\ &\to{}^tf'^!\mu hom(f^!F,G\otimes\omega_{Y/X})\ ,\end{aligned}$$

(iii) $$\begin{aligned} {}^tf'^{-1}\mu hom(G,f^!F) &\to{}^tf'^{-1}R{}^tf'_*\mu hom(G\to F)\\ &\to\mu hom(G\to F)\\ &\to f_\pi^!\mu hom(Rf_*G,F)\ ,\end{aligned}$$

(iv) $$\begin{aligned} {}^tf'^{-1}\mu hom(f^{-1}F,G) &\to{}^tf'^{-1}R{}^tf'_*\mu hom(F\leftarrow G)\\ &\to\mu hom(F\leftarrow G)\\ &\to f_\pi^!\mu hom(F,Rf_!G)\ .\end{aligned}$$

If f is smooth, and $H\in\mathrm{Ob}(\mathbf{D}^b(Y\times_X T^*X))$ then $H\to{}^tf'^{-1}R{}^tf'_*H$ is an isomorphism. Applying Proposition 4.4.5, we obtain that if f is proper on supp(*G*), then (iii) and (iv) are isomorphisms. □

Let F_1 and F_2 belong to $\mathbf{D}^b(X)$. There are canonical morphisms:

(4.4.5)
$$\begin{array}{ccccc} & & \mu hom(f^!F_2,f^!F_1) & & \\ & \nearrow & & \searrow & \\ \mu hom(f^!F_2,f^{-1}F_1\otimes\omega_{Y/X}) & & & & \mu hom(f^{-1}F_2\otimes\omega_{Y/X},f^!F_1)\ . \\ & \searrow & & \nearrow & \\ & & \mu hom(f^{-1}F_2,f^{-1}F_1) & & \end{array}$$

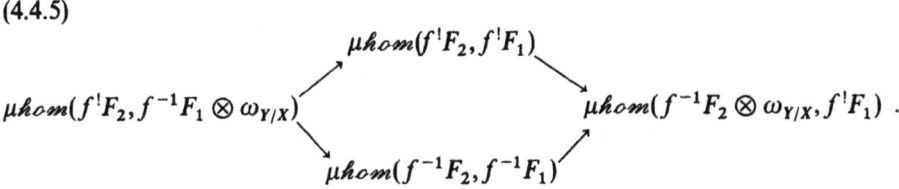

Similarly let G_1 and G_2 belong to $\mathbf{D}^b(Y)$. There are canonical morphisms:

$$
(4.4.6)\qquad
\begin{array}{ccccc}
 & & \mu hom(Rf_! G_2, Rf_! G_1) & & \\
 & \nearrow & & \searrow & \\
\mu hom(Rf_* G_2, Rf_! G_1) & & & & \mu hom(Rf_! G_2, Rf_* G_1)\ . \\
 & \searrow & & \nearrow & \\
 & & \mu hom(Rf_* G_2, Rf_* G_1) & &
\end{array}
$$

Proposition 4.4.7. *Let F_1, F_2, G_1, G_2 be as above. Then there are commutative diagrams of canonical morphisms:*

(i)
$$
\begin{array}{ccc}
R{}^tf'_! f_\pi^{-1} \mu hom(F_2, F_1) & \longrightarrow & \mu hom(f^! F_2, f^{-1}F_1 \otimes \omega_{Y/X}) \\
\downarrow & & \downarrow \\
R{}^tf'_* f_\pi^!(\mu hom(F_2, F_1) \otimes \omega_{Y/X}^{\otimes -1}) & \longleftarrow & \mu hom(f^{-1}F_2 \otimes \omega_{Y/X}, f^! F_1)\ ,
\end{array}
$$

(ii)
$$
\begin{array}{ccc}
Rf_{\pi !}{}^tf'^{-1} \mu hom(G_2, G_1) & \longrightarrow & \mu hom(Rf_* G_2, Rf_! G_1) \\
\downarrow & & \downarrow \\
Rf_{\pi *}({}^tf'^! \mu hom(G_2, G_1) \otimes \omega_{Y/X}) & \longleftarrow & \mu hom(Rf_! G_2, Rf_* G_1)\ .
\end{array}
$$

If f is smooth, the morphisms in (i) are all isomorphisms.
If f is a closed embedding, the morphisms in (ii) are all isomorphisms.

Proof. (i) By Proposition 4.3.5, we have a commutative diagram:

$$
\begin{array}{ccc}
f_\pi^{-1} \mu hom(F_2, F_1) & \longrightarrow & \mu_\Delta f_2^{-1} R\mathcal{H}om(q_2^{-1}F_2, q_1^! F_1) \\
\downarrow & & \downarrow \\
f_\pi^! \mu hom(F_2, F_1) \otimes \omega_{Y/X}^{\otimes -1} & \longleftarrow & \mu_\Delta f_2^! R\mathcal{H}om(q_2^{-1}F_2 \otimes \omega_{Y/X}, q_1^! F_1)\ .
\end{array}
$$

Using the morphism $f^{-1}R\mathcal{H}om(\cdot, \cdot) \to R\mathcal{H}om(f^!(\cdot), f^!(\cdot))$ (cf. Exercise III.9), we get:

$$
(4.4.7)\qquad
\begin{array}{ccc}
f_\pi^{-1} \mu hom(F_2, F_1) & \longrightarrow & \mu hom(f^! F_2 \to F_1) \\
\downarrow & & \downarrow \\
f_\pi^! \mu hom(F_2, F_1) \otimes \omega_{Y/X}^{\otimes -1} & \longleftarrow & \mu hom((f^{-1}F_2 \otimes \omega_{Y/X}) \to F_1)\ .
\end{array}
$$

We apply $R{}^tf'_!$ to the first horizontal arrow of (4.4.7) and $R{}^tf'_*$ to the second one. Then the result follows from Proposition 4.4.5 (i).

(ii) By Proposition 4.3.4 we have a commutative diagram:

$$\begin{array}{ccc} {}^t f'^{-1}\mu hom(G_2, G_1) & \longrightarrow & \mu_\Delta Rf_{1!} R\mathcal{H}om(q_1'^{-1}G_2, q_2'^! G_1)^a \\ \downarrow & & \downarrow \\ {}^t f'^! \mu hom(G_2, G_1) \otimes \omega_{Y/X} & \longleftarrow & \mu_\Delta Rf_{1*} R\mathcal{H}om(q_1'^{-1}G_2, q_2'^! G_1)^a \end{array},$$

from which we deduce:

$$\begin{array}{ccc} {}^t f'^{-1}\mu hom(G_2, G_1) & \longrightarrow & \mu hom(Rf_* G_2 \leftarrow G_1) \\ \downarrow & & \downarrow \\ {}^t f'^! \mu hom(G_2, G_1) \otimes \omega_{Y/X} & \longleftarrow & \mu hom(Rf_! G_2 \leftarrow G_1) \end{array}. \tag{4.4.8}$$

Now we apply $Rf_{\pi !}$ to the first horizontal arrow of (4.4.8) and $Rf_{\pi *}$ to the second one. Then the result follows from Proposition 4.4.5 (iv).

Assume f is smooth. By Proposition 4.4.5 (i) we have the isomorphism:

$$R{}^t f'_! \mu hom(f^! F_2 \to F_1) \simeq \mu hom(f^! F_2, f^{-1}F_1 \otimes \omega_{Y/X}) .$$

On the other hand we have:

$$\begin{aligned} \mu hom(f^! F_2 \to F_1) \otimes \omega_{Y/X} &\simeq \mu_\Delta R\mathcal{H}om(f_2^{-1}q_2^{-1}F_2, f_2^! q_1^! F_1) \\ &\simeq \mu_\Delta f_2^! R\mathcal{H}om(q_2^{-1}F_2, q_1^! F_1) \\ &\simeq f_\pi^! \mu_{\Delta_X} R\mathcal{H}om(q_2^{-1}F_2, q_1^! F_1) \end{aligned}$$

by Proposition 4.3.5.

This proves the isomorphisms (i).

Finally assume f is a closed embedding. By Proposition 4.4.5 (iii) we get the isomorphism:

$$Rf_{\pi !}\mu hom(G_2 \to Rf_! G_1) \xrightarrow{\sim} \mu hom(Rf_* G_2, Rf_! G_1) .$$

On the other hand we have:

$$\begin{aligned} \mu hom(G_2 \to Rf_! G_1) &\simeq \mu_\Delta R\mathcal{H}om(\tilde{q}_2^{-1}G_2, \tilde{q}_1^! Rf_! G_1) \\ &\simeq \mu_\Delta Rf_{1!} R\mathcal{H}om(q_2'^{-1}G_2, q_1'^! G_1) \\ &\simeq {}^t f'^{-1} \mu_{\Delta_Y} R\mathcal{H}om(q_2'^{-1}G_2, q_1'^! G_1) . \end{aligned}$$

This proves the isomorphisms (ii). □

Now we shall study the external tensor product for μhom (cf. Notation 2.3.12).

Let $p_X : X \to S$ and $p_Y : Y \to S$ be two morphisms of manifolds. We denote by q_1 and q_2 the first and second projection defined on $X \times Y$ or on $X \times_S Y$. We assume:

(4.4.9) $$X \underset{S}{\times} Y \text{ is a submanifold of } X \times Y \, .$$

Let j denote the embedding $X \times_S Y \hookrightarrow X \times Y$. Since $(X \times_S Y) \times_{X \times Y} T^*(X \times Y) \simeq T^*X \times_S T^*Y$, we obtain the two maps:

(4.4.10) $$T^*\left(X \underset{S}{\times} Y\right) \xleftarrow[{}^t j']{} T^*X \underset{S}{\times} T^*Y \xrightarrow[j_\pi]{} T^*X \times T^*Y \, .$$

Proposition 4.4.8. *Let F_1 and F_2 belong to $\mathbf{D}^b(X)$ and let G_1 and G_2 belong to $\mathbf{D}^b(Y)$. One has canonical morphisms:*

(i) $$R^t j'_!\left(\mu hom(F_2, F_1) \underset{S}{\overset{L}{\boxtimes}} \mu hom(G_2, G_1)\right) \to \mu hom\left(F_2 \underset{S}{\overset{L}{\boxtimes}} G_2, F_1 \underset{S}{\overset{L}{\boxtimes}} G_1\right) ,$$

(ii) $$R^t j'_!\left(\mu hom(F_2, F_1)^a \underset{S}{\overset{L}{\boxtimes}} \mu hom(G_2, G_1)\right) \to \mu hom(R\mathcal{H}om(q_1^{-1}F_1, q_2^{-1}G_2), R\mathcal{H}om(q_1^{-1}F_2, q_2^{-1}G_1)) \, .$$

Proof. (i) Let us first define the morphism when $S = \{\mathrm{pt}\}$. By Proposition 4.3.6 we get the morphisms:

$$\begin{aligned}
&\mu_{\Delta_X}(R\mathcal{H}om(q_2^{-1}F_2, q_1^! F_1)) \overset{L}{\boxtimes} \mu_{\Delta_Y}(R\mathcal{H}om(q_2^{-1}G_2, q_1^! G_1)) \\
&\quad \to \mu_{\Delta_X \times \Delta_Y}\left(R\mathcal{H}om(q_2^{-1}F_2, q_1^! F_1) \overset{L}{\boxtimes} R\mathcal{H}om(q_2^{-1}G_2, q_1^! G_1)\right) \\
&\quad \to \mu_{\Delta_X \times \Delta_Y}\left(R\mathcal{H}om\left(q_2^{-1}F_2 \overset{L}{\boxtimes} q_2^{-1}G_2, q_1^! F_1 \overset{L}{\boxtimes} q_1^! G_1\right)\right) \\
&\quad \xrightarrow{\sim} \mu hom\left(F_2 \overset{L}{\boxtimes} G_2, F_1 \overset{L}{\boxtimes} G_1\right) .
\end{aligned}$$

Now we treat the general case. Applying the preceding result and Proposition 4.4.7(i) we obtain:

$$\begin{aligned}
&R^t j'_!\left(\mu hom(F_2, F_1) \underset{S}{\overset{L}{\boxtimes}} \mu hom(G_2, G_1)\right) \\
&\quad \simeq R^t j'_! j_\pi^{-1}\left(\mu hom(F_2, F_1) \overset{L}{\boxtimes} \mu hom(G_2, G_1)\right) \\
&\quad \to R^t j'_! j_\pi^{-1} \mu hom\left(F_2 \overset{L}{\boxtimes} G_2, F_1 \overset{L}{\boxtimes} G_1\right)
\end{aligned}$$

$$\to \mu hom\left(j^{-1}\left(F_2 \overset{L}{\boxtimes} G_2\right), j^{-1}\left(F_1 \overset{L}{\boxtimes} G_1\right)\right)$$

$$\simeq \mu hom\left(F_2 \underset{S}{\overset{L}{\boxtimes}} G_2, F_1 \underset{S}{\overset{L}{\boxtimes}} G_1\right) .$$

(ii) Let us first define the morphism when $S = \{\mathrm{pt}\}$.

We denote by q_j' (resp. q_j'', resp. $\tilde{q}_j$) the j-th projection defined on $X \times X$ (resp. $Y \times Y$, resp. $(X \times Y) \times (X \times Y)$).

$$\begin{aligned}
&\mu hom(F_2, F_1)^a \overset{L}{\boxtimes} \mu hom(G_2, G_1) \\
&\quad \simeq \mu_{\Delta_X}(R\mathcal{H}om(q_1'^{-1}F_2, q_2'^{!}F_1)) \overset{L}{\boxtimes} \mu_{\Delta_Y}(R\mathcal{H}om(q_2''^{-1}G_2, q_1''^{!}G_1)) \\
&\quad \to \mu_{\Delta_X \times \Delta_Y}\left(R\mathcal{H}om(q_1'^{-1}F_2, q_2'^{!}F_1) \overset{L}{\boxtimes} R\mathcal{H}om(q_2''^{-1}G_2, q_1''^{!}G_1)\right) \\
&\quad \to \mu_{\Delta_X \times \Delta_Y}\left(R\mathcal{H}om(\tilde{q}_1^{-1}q_1^{-1}F_2, \tilde{q}_2^{-1}q_1^{-1}F_1) \overset{L}{\otimes} R\mathcal{H}om(\tilde{q}_2^{-1}q_2^{-1}G_2, \tilde{q}_1^{-1}q_2^{-1}G_1)\right) \\
&\quad\quad \otimes \omega_{X \times Y} \\
&\quad \to \mu_{\Delta_{X \times Y}}(R\mathcal{H}om(R\mathcal{H}om(\tilde{q}_2^{-1}q_1^{-1}F_1, \tilde{q}_2^{-1}q_2^{-1}G_2), R\mathcal{H}om(\tilde{q}_1^{-1}q_1^{-1}F_2, \tilde{q}_1^{-1}q_2^{-1}G_1))) \\
&\quad\quad \otimes \omega_{X \times Y} \\
&\quad \simeq \mu hom(R\mathcal{H}om(q_1^{-1}F_1, q_2^{-1}G_2), R\mathcal{H}om(q_1^{-1}F_2, q_2^{-1}G_1)) .
\end{aligned}$$

Now we treat the general case, using the preceding result and Proposition 4.4.7 (i).

$$\begin{aligned}
&{}^{t}j_!'\left(\mu hom(F_2, F_1)^a \underset{S}{\overset{L}{\boxtimes}} \mu hom(G_2, G_1)\right) \\
&\qquad \simeq {}^{t}j_!' j_\pi^{-1}\left(\mu hom(F_2, F_1)^a \overset{L}{\boxtimes} \mu hom(G_2, G_1)\right) \\
&\qquad \to {}^{t}j_!' j_\pi^{-1} \mu hom(R\mathcal{H}om(q_1^{-1}F_1, q_2^{-1}G_2), R\mathcal{H}om(q_1^{-1}F_2, q_2^{-1}G_1)) \\
&\qquad \to \mu hom(j^! R\mathcal{H}om(q_1^{-1}F_1, q_2^{-1}G_2), j^! R\mathcal{H}om(q_1^{-1}F_2, q_2^{-1}G_1)) .
\end{aligned}$$

In the last term, we can replace both $q_2^{-1}G_2$ and $q_2^{-1}G_1$ by $q_2^{!}G_2$ and $q_2^{!}G_1$. Then the result follows from Proposition 3.1.13. $\square$

Now, we shall compose the functors μhom.

Let $g: Z \to Y$ and $f: Y \to X$ be two morphisms of manifolds, $h = f \circ g$. We define q_i, q_i', q_i'' $(i = 1, 2)$ the i-th projection of $X \times Y$, $X \times Z$, $Y \times Z$ respectively, and q_{ij} the (i, j)-th projection of $X \times Y \times Z$, (cf. Diagram (3.6.1)). We use the notation $\tilde{q}_{ij}$ to denote the (i, j)-th projection on $X \times Y \times Y \times Z$, and we denote by p_{ij} (resp. $\tilde{p}_{ij}$) the (i, j)-th projection of $T^*X \times T^*Y \times T^*Z$ (resp. of $T^*X \times T^*Y \times T^*Y \times T^*Z$). We denote by r_{ij} the restriction of $\tilde{p}_{ij}$ to $\Delta_Y \times_{Y \times Y} T^*(X \times Y \times Y \times Z)$, and finally we denote by j the diagonal embedding $X \times Y \times Z \hookrightarrow X \times Y \times Y \times Z$. Note that ${}^tj'$ induces an isomorphism:

$$\text{(4.4.11)} \qquad \Delta_Y \underset{Y \times Y}{\times} T^*_{\Delta_f \times \Delta_g}(X \times Y \times Y \times Z) \underset{{}^tj'}{\xrightarrow{\sim}} T^*_{\Delta_f \times_Y \Delta_g}(X \times Y \times Z) \;,$$

and $q_{13\pi}$ induces an isomorphism:

$$\text{(4.4.12)} \qquad \left(\Delta_f \underset{Y}{\times} \Delta_g\right) \underset{\Delta_h}{\times} T^*_{\Delta_h}(X \times Z) \underset{q_{13\pi}}{\xrightarrow{\sim}} T^*_{\Delta_h}(X \times Z) \;.$$

Hence we have the commutative diagram:

(4.4.13)

$$\begin{array}{ccccc}
 & & & \nearrow^{r_{12}} & T^*_{\Delta_f}(X \times Y) \simeq Y \underset{X}{\times} T^*X \\
 & & & & \uparrow g_\pi \\
\Delta_Y \underset{Y \times Y}{\times} T^*_{\Delta_f \times \Delta_g}(X \times Y \times Y \times Z) & \underset{{}^tj'}{\xrightarrow{\sim}} & T^*_{\Delta_f \times_Y \Delta_g}(X \times Y \times Z) & \underset{{}^tq'_{13}}{\hookleftarrow} & T^*_{\Delta_h}(X \times Z) \simeq Z \underset{X}{\times} T^*X \\
 & & & & \downarrow {}^tf' \\
 & & & \searrow_{r_{34}} & T^*_{\Delta_g}(Y \times Z) \simeq Z \underset{Y}{\times} T^*Y \;.
\end{array}$$

Proposition 4.4.9. *Let F (resp. G, resp. H) belong to* $\mathrm{Ob}(\mathbf{D}^b(X))$, *(resp.* $\mathrm{Ob}(\mathbf{D}^b(Y))$, *resp.* $\mathrm{Ob}(\mathbf{D}^b(Z))$*). Then there is a canonical morphism:*

$${}^tf'^{-1}\mu hom(H \to G) \overset{L}{\otimes} g_\pi^{-1}\mu hom(G \to F) \to \mu hom(H \to F) \;.$$

Proof. Set $K_1 = R\mathcal{H}om(q_2^{-1}G, q_1^!F)$, $K_2 = R\mathcal{H}om(q_2''^{-1}H, q_1''^!G)$ and $L = g_\pi^{-1}\mu_{\Delta_f}(K_1) \otimes^L {}^tf'^{-1}\mu_{\Delta_g}(K_2)$.

We have the chain of morphisms:

$$\begin{aligned}
L &\simeq {}^tq_{13}'^{-1}\,{}^tj'_*\left((r_{12}^{-1}\mu_{\Delta_f}(K_1) \overset{L}{\otimes} r_{34}^{-1}\mu_{\Delta_g}(K_2)\right) \\
&\simeq {}^tq_{13}'^{-1}\,{}^tj'_*\, j_\pi^{-1}\left(\tilde{p}_{12}^{-1}\mu_{\Delta_f}(K_1) \overset{L}{\otimes} \tilde{p}_{34}^{-1}\mu_{\Delta_g}(K_2)\right) \\
&\to {}^tq_{13}'^{-1}\,{}^tj'_*\, j_\pi^{-1}\mu_{\Delta_f \times \Delta_g}\left(\tilde{q}_{12}^{-1}K_1 \overset{L}{\otimes} \tilde{q}_{34}^{-1}K_2\right) \\
&\to {}^tq_{13}'^{-1}\mu_{\Delta_f \times_Y \Delta_g}\left(j^{-1}\left(\tilde{q}_{12}^{-1}K_1 \overset{L}{\otimes} \tilde{q}_{34}^{-1}K_2\right)\right) \\
&\to {}^tq_{13}'^{-1}\mu_{\Delta_f \times_Y \Delta_g}(R\mathcal{H}om(q_{13}^{-1}q_2'^{-1}H, q_{13}^!q_1'^!F))
\end{aligned}$$

$$\to \mu_{\Delta_h}(Rq_{13!}R\mathcal{H}om(q_{13}^{-1}q_2'^{-1}H, q_{13}^! q_1'^! F))$$

$$\to \mu_{\Delta_h}(R\mathcal{H}om(q_2'^{-1}H, q_1'^! F)) \ .$$

Here we applied Proposition 4.3.5 to the map j, then Proposition 4.3.6, and finally Proposition 4.3.4 to the map q_{13}. □

There is a similar result for the composition of $\mu hom(\cdot \leftarrow \cdot)$.

Corollary 4.4.10. *Let F_1, F_2 and F_3 belong to $\mathbf{D}^b(X)$. Then we have a canonical morphism:*

$$\mu hom(F_1, F_2) \overset{L}{\otimes} \mu hom(F_2, F_3) \to \mu hom(F_1, F_3) \ .$$

Now let X, Y, Z be three manifolds. We denote as usual, by q_{ij} the (i, j)-th projection defined on $X \times Y \times Z$ and by p_{ij} the (i, j)-th projection defined on $T^*X \times T^*Y \times T^*Z$, and we denote by p_{ij}^a the composition of p_{ij} and the antipodal map on the j-th factor.

Proposition 4.4.11. *For K_1 and F_1 belonging to $\mathbf{D}^b(X \times Y)$ and K_2 and F_2 belonging to $\mathbf{D}^b(Y \times Z)$, there is a canonical morphism:*

$$Rp_{13!}^a\left(p_{12}^{a-1}\mu hom(K_1, F_1) \overset{L}{\otimes} p_{23}^{a-1}\mu hom(K_2, F_2)\right) \to \mu hom(K_1 \circ K_2, F_1 \circ F_2) \ , \tag{4.4.14}$$

where $K_1 \circ K_2$ and $F_1 \circ F_2$ are defined by (3.6.2).

Proof. Let $j: X \times Y \times Z \to X \times Y \times Y \times Z$ be the diagonal embedding, and consider the following commutative diagram:

(4.4.15)

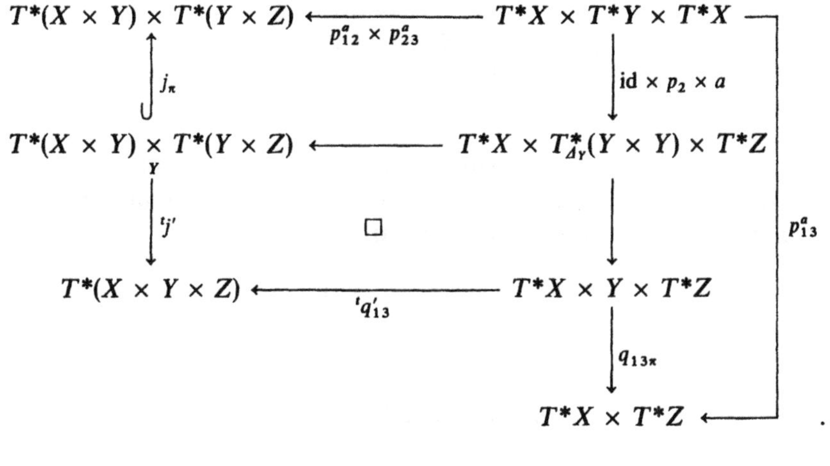

$T^*(X \times Y) \times T^*(Y \times Z)$
$p_{12}^a \times p_{23}^a$
$T^*X \times T^*Y \times T^*X$
j_π
$\mathrm{id} \times p_2 \times a$
$T^*(X \times Y) \underset{Y}{\times} T^*(Y \times Z)$
$T^*X \times T^*_{\Delta_Y}(Y \times Y) \times T^*Z$
${}^t j'$
□
p_{13}^a
$T^*(X \times Y \times Z)$
${}^t q'_{13}$
$T^*X \times Y \times T^*Z$
$q_{13\pi}$
$T^*X \times T^*Z$

Applying Proposition 4.4.8 (i) to the maps $X \times Y \to Y$ and $Y \times Z \to Y$, we get the morphism:

$$(4.4.16)\quad R^t j'_! j_\pi^{-1}\left(\mu hom(K_1, F_1) \overset{L}{\boxtimes} \mu hom(K_2, F_2)\right) \to \mu hom\left(q_{12}^{-1}K_1 \overset{L}{\otimes} q_{23}^{-1}K_2, q_{12}^{-1}F_1 \overset{L}{\otimes} q_{23}^{-1}F_2\right).$$

Applying Proposition 4.4.6 (ii) together with formula (4.4.6), we get:

$$(4.4.17)\quad Rq_{13\pi!}{}^t q_{13}'^{-1}\mu hom\left(q_{12}^{-1}K_1 \overset{L}{\otimes} q_{23}^{-1}K_2, q_{12}^{-1}F_1 \overset{L}{\otimes} q_{23}^{-1}F_2\right) \to \mu hom\left(Rq_{13!}\left(q_{12}^{-1}K_1 \overset{L}{\otimes} q_{23}^{-1}K_2\right), Rq_{13!}\left(q_{12}^{-1}F_1 \overset{L}{\otimes} q_{23}^{-1}F_2\right)\right).$$

Combining (4.4.16) and (4.4.17), we get the morphism:

$$(4.4.18)\quad Rq_{13\pi!}{}^t q_{13}'^{-1} R^t j'_! j_\pi^{-1}\left(\mu hom(K_1, F_1) \overset{L}{\boxtimes} \mu hom(K_2, F_2)\right) \to \mu hom(K_1 \circ K_2, F_1 \circ F_2).$$

Since the left-hand side of (4.4.18) is isomorphic to the left-hand side of (4.4.14), due to the commutative diagram (4.4.15), the result follows. □

Exercises to Chapter IV

Exercise IV.1. Let X be an n-dimensional manifold and M a closed submanifold of codimension l.

(i) By choosing local coordinates (x', x'', t) on $\tilde{X}_M$ and (y', y'') on X such that $p(x', x'', t) = (y', y'')$, show that dt/t^l is a well-defined section of the sheaf of relative forms $\Omega_{\tilde{X}}^{(n+1)} \otimes_{p^{-1}\mathscr{C}_X^\infty} \Omega_X^{(n)\otimes -1}$, all over $\tilde{X}_M$.

(ii) Show that dt/t^l is intrinsically defined all over $\tilde{X}_M$.

(iii) Deduce from (i) and (ii) that the relative orientation sheaf $or_{\tilde{X}/X}$ is canonically isomorphic to $A_{\tilde{X}}$.

Exercise IV.2. Let M be a closed submanifold of X, S a closed subset of X. Prove that there is a functor $\nu: \mathsf{D}^b(X\setminus S) \to \mathsf{D}^b(T_M X \setminus C_M(S))$ such that the following diagram commutes (up to an isomorphism):

$$\begin{array}{ccc} \mathsf{D}^b(X) & \xrightarrow{\nu_M} & \mathsf{D}^b(T_M X) \\ \downarrow & & \downarrow \\ \mathsf{D}^b(X\setminus S) & \xrightarrow{\nu} & \mathsf{D}^b(T_M X\setminus C_M(S)). \end{array}$$

Exercise IV.3. Let M be a closed submanifold of X and let S be a locally closed subset of X. Prove that $C_M(S) = \operatorname{supp} \nu_M(A_S)$.

Exercise IV.4. Let X be a manifold, let F and G belong to $\mathbf{D}^b(X)$, and assume F and G are cohomologically constructible. Prove:

$$\mu hom(F, G) \simeq \mu hom(\mathrm{D}G, \mathrm{D}F)^a$$

(cf. Proposition 3.4.6).

Exercise IV.5. Let $E \to Z$ be a vector bundle, $F \in \mathrm{Ob}(\mathbf{D}^b_{\mathbb{R}^+}(E))$. Prove that $\nu_Z(F) \simeq F$ and $\mu_Z \simeq F^\wedge$, where Z is identified with the zero section of E.

Exercise IV.6. Let F_2 and F_1 belong to $\mathbf{D}^b(X)$.
Let $(t; \tau)$ be the coordinates on $T^*\mathbb{R}$. Prove the isomorphisms:

$$\mu hom(F_2, F_1) \simeq \mu hom\left(F_2 \overset{L}{\boxtimes} A_{\{0\}}, F_1 \overset{L}{\boxtimes} A_{\{0\}}\right)\bigg|_{t=0,\tau=1},$$

$$\mu hom(F_2, F_1) \simeq \mu hom\left(F_2 \overset{L}{\boxtimes} A_{\mathbb{R}}, F_1 \overset{L}{\boxtimes} A_{\mathbb{R}}\right)\bigg|_{t=0,\tau=0},$$

(hint: use Proposition 4.4.7).

Exercise IV.7. Let M be a closed submanifold of X, Z a closed subset of X and let F and G belong to $\mathbf{D}^b(X)$. Construct the following natural morphisms and give examples such that they are not isomorphisms:

(i) $$A_{C_M(Z)} \to \nu_M(A_Z),$$

(ii) $$\nu_M R\mathcal{H}om(G, F) \to R\mathcal{H}om(\nu_M(G), \nu_M(F)),$$

(iii) $$\nu_M R\Gamma_Z(F) \to R\Gamma_{C_M(Z)}(\nu_M(F)).$$

Notes

To blow up singularities is a familiar operation in geometry, and the use of "polar coordinates" is classical. However, the normal deformation of a submanifold M that we perform in §1 is not so common, since the manifold that we obtain contains the whole normal vector bundle to M, and not only the sphere or projective bundle. It is the analog in the real case of an operation recently introduced in algebraic geometry (cf. Fulton [1]).

The notion of "normal cones" that we discuss in Definition 4.1.1 coincides with the classical one and is familiar to many authors. The names of Thom [1] and Whitney [1, 2], are naturally attached to it.

The specialization and microlocalization functors ν and μ have been introduced by Sato [2] in 1969 (cf. Sato-Kawai-Kashiwara [1]). Our construction of ν is slightly different from Sato's, since we use the normal deformation of M. In the algebraic case a similar construction has been performed by Verdier [5], who relates it to the famous "vanishing-cycle" functor (cf. Chapter VIII §6).

When the microlocalization functor appeared, it was first considered as a tool for defining microfunctions and analytic wave front sets. In fact, soon after Sato's work, Hörmander [1] introduced the C^∞-wave front set of distributions and Fourier integral operators, and this was the beginning of an intense activity in the field of linear partial differential equations. Then the idea of looking at singularities in the cotangent bundle (of course this idea takes its origin in the nineteenth century) spread over to other fields of mathematics, such as Feynmann integrals, analytic geometry, group representation, or sheaf theory and this constitutes what is now called "microlocal analysis".

The results of §2 and §3 are essentially contained in Sato-Kawai-Kashiwara [1]. The functor μhom of §4 has been introduced by Kashiwara-Schapira [3] as the sheaf theoretical version of the functor $\mathcal{H}om$ of $\mathcal{E}$-modules (cf. Chapter XI§4).

Chapter V. Micro-support of sheaves

Summary

On a manifold X, one can naturally associate to any object F of $\mathbf{D}^b(X)$ a closed conic subset of T^*X, the micro-support of F, denoted by $SS(F)$. Roughly speaking, $SS(F)$ describes the set of codirections of X in which F "does not propagate", and we shall prove in the subsequent chapter that $SS(F)$ is an involutive subset of T^*X.

This notion also gives a condition of commutativity of various functors in sheaf theory; e.g. when is $f^!F$ isomorphic to $\omega_{Y/X} \otimes f^{-1}F$ or when is $f^{-1}R\mathscr{H}om(F_1, F_2)$ isomorphic to $R\mathscr{H}om(f^{-1}F_1, f^{-1}F_2)$?

We first prove the equivalence of three definitions of $SS(F)$, and use the micro-support to give a criterion in order that, given two open subsets Ω_0 and Ω_1 of X with $\Omega_0 \subset \Omega_1$, the restriction morphism from $R\Gamma(\Omega_1; F)$ to $R\Gamma(\Omega_0; F)$ is an isomorphism. The γ-topology associated to a closed convex proper cone γ, introduced in Chapter III, is a natural tool in the study of the micro-support. In fact if X is affine and $\phi_\gamma : X \to X_\gamma$ is the map which weakens the topology of X, then $\phi_\gamma^{-1} R\phi_{\gamma*}$ plays the role of a cut-off functor in the following sense: $SS(\phi_\gamma^{-1} R\phi_{\gamma*}F)$ is contained in $X \times \gamma^{\circ a}$, and the morphism $\phi_\gamma^{-1} R\phi_{\gamma*}F \to F$ is "an isomorphism on $X \times \operatorname{Int} \gamma^{\circ a}$". This notion of microlocal isomorphism is defined here, and will be developed in the next chapter.

After having given some examples of micro-supports, we study the behavior of micro-supports with respect to various operations on sheaves: tensor product and $\mathscr{H}om$, direct or inverse images, Fourier-Sato transformation.

In this chapter, we always make the assumption that the morphisms are proper or "non-characteristic" with respect to the micro-support. This restriction will be removed in Chapter VI. The results we shall obtain here and in the next chapter, are very similar to many classical results in the theory of partial differential equations with analytic coefficients, and we shall show in Chapter XI, how to deduce some of these classical results from the theory of the micro-support.

We often follows quite tightly the exposition of Kashiwara-Schapira [3], but some proofs are actually simplified.

We keep convention 4.0.

5.1. Equivalent definitions of the micro-support

Let X be a manifold. If α is an integer or $\alpha = +\infty$ we shall consider functions on X of class C^α ($1 \leqslant \alpha \leqslant \infty$). If X is real analytic, a function of class C^ω means a real analytic function.

As usual, $\pi : T^*X \to X$ denotes the cotangent bundle.

Proposition 5.1.1. *Assume X is open in a vector space E, let $p = (x_\circ; \xi_\circ) \in T^*X$ and let $F \in \mathrm{Ob}(\mathbf{D}^b(X))$. Then the following conditions are equivalent, where $1 \leqslant \alpha \leqslant \infty$ or $\alpha = \omega$.*

$(1)_\alpha$ *There exists an open neighborhood U of p such that for any $x_1 \in X$ and any real function ψ of class C^α defined in a neighborhood of x_1, with $\psi(x_1) = 0$, $d\psi(x_1) \in U$, we have:*

$$(R\Gamma_{\{x;\,\psi(x)\geqslant 0\}}(F))_{x_1} = 0 \ .$$

(2) *There exist a proper closed convex cone γ in E with $0 \in \gamma$, and $F' \in \mathrm{Ob}(\mathbf{D}^b(E))$ such that:*
(a) $\gamma\setminus\{0\} \subset \{v; \langle v, \xi_\circ\rangle < 0\}$, (*i.e.* $\xi_\circ \in \mathrm{Int}(\gamma^{\circ a})$),
(b) $F'|_U \simeq F|_U$ *for a neighborhood U of $x_\circ$ in X.*
(c) $R\phi_{\gamma *}F' = 0$ (*cf.* III§5).

(3) *There exist a neighborhood U of $x_\circ$, an $\varepsilon > 0$ and a proper closed convex cone γ with $0 \in \gamma$ satisfying the condition* (a) *of* (2) *such that if we set:*

$$H = \{x; \langle x - x_\circ, \xi_\circ\rangle \geqslant -\varepsilon\} \ ,$$

$$L = \{x; \langle x - x_\circ, \xi_\circ\rangle = -\varepsilon\} \ ,$$

then $H \cap (U + \gamma) \subset X$ and we have the natural isomorphism:

$$R\Gamma(H \cap (x + \gamma); F) \xrightarrow{\sim} R\Gamma(L \cap (x + \gamma); F)$$

for every $x \in U$.

Proof. If $\xi_\circ = 0$, the three conditions are equivalent to $F = 0$ in a neighborhood of $x_\circ$. Assume $\xi_\circ \neq 0$.

$(2) \Rightarrow (1)_1$. We may assume $X = E$ and $F = F'$. We set $U = X \times \mathrm{Int}\,\gamma^{\circ a}$. For any function ψ with $\psi(x_1) = 0$, $d\psi(x_1) \in U$, there exists a γ-open set Ω such that Ω and $\{x; \psi(x) < 0\}$ coincide on a neighborhood of x_1. Moreover we may assume that, when Ω' runs over a system of γ-neighborhoods of x_1, the $\Omega'\setminus\Omega$ form a system of neighborhoods of x_1 in $X\setminus\Omega$ in the usual topology. Thus we obtain:

$$(R\Gamma_{\{x;\,\psi(x)\geqslant 0\}}(F))_{x_1} = (R\Gamma_{(X_\gamma\setminus\Omega_\gamma)}(R\phi_{\gamma *}F))_{x_1} = 0 \ .$$

$(3) \Rightarrow (2)$. We may assume X is γ-open and $U \subset H\setminus L$. Let Ω_0 and Ω_1 be two γ-open sets such that $\Omega_0 \subset \Omega_1$, $x_\circ \in \mathrm{Int}(\Omega_1\setminus\Omega_0)$, $\Omega_1\setminus\Omega_0 \subset\subset U$. We have:

$$R\Gamma((x + \gamma) \cap H; F) \simeq R\Gamma(x + \gamma; F_H) \tag{5.1.1}$$

and similarly with H replaced by L. From (5.1.1) we get that $R\Gamma(x+\gamma; F_{H\setminus L}) = 0$ for every $x \in U$, hence:

$$(\phi_\gamma^{-1} R\phi_{\gamma *} F_{H\setminus L})|_U = 0 \ . \tag{5.1.2}$$

Applying Proposition 3.5.3, we obtain:

$$\begin{aligned} R\phi_{\gamma *} R\Gamma_{\Omega_1 \setminus \Omega_0}(F_{H\setminus L}) &\simeq R\Gamma_{\Omega_{1\gamma}\setminus\Omega_{0\gamma}} R\phi_{\gamma *} F_{H\setminus L} \\ &\simeq R\Gamma_{\Omega_{1\gamma}\setminus\Omega_{0\gamma}} R\phi_{\gamma *}\phi_\gamma^{-1} R\phi_{\gamma *} F_{H\setminus L} \\ &\simeq R\phi_{\gamma *} R\Gamma_{\Omega_1\setminus\Omega_0}\phi_\gamma^{-1} R\phi_{\gamma *} F_{H\setminus L} \\ &= 0 \ . \end{aligned}$$

Therefore $R\Gamma_{\Omega_1\setminus\Omega_0}(F_{H\setminus L})$ has the required properties.

$(1)_\omega \Rightarrow (3)$. We may assume $\xi_\circ = (1, 0, \ldots, 0)$, $x_\circ = 0$ and $F \in \mathrm{Ob}(\mathbf{D}^b(E))$.

Set $x = (x_1, x')$, $x' = (x_2, \ldots, x_n)$. We take $\varepsilon > 0$ small enough and define H and L as in (3). We take $\delta > 0$ small enough and set:

$$\gamma = \{x; x_1 \leqslant -\delta|x'|\} \ . \tag{5.1.3}$$

We choose ε, δ and a γ-open subset Ω_1 such that $0 \in \Omega_1$ and $(\Omega_1 \cap H) \times \operatorname{Int}\gamma^{\circ a} \subset \mathbb{R}^+ \cdot U$. Then we shall show the isomorphism:

$$R\Gamma(H \cap (x+\gamma); F) \xrightarrow{\sim} R\Gamma(L \cap (x+\gamma); F) \ , \tag{5.1.4}$$

for every $x \in \Omega_1 \cap H$.

For each $a \in \Omega_1 \cap (H\setminus L)$, we can construct a family $\{\Omega_t(a)\}_{t\in\mathbb{R}^+}$ of open subsets of X such that:

$$\left\{\begin{array}{l} \text{(i) } \Omega_t(a) \subset a + \operatorname{Int}\gamma \ , \\ \text{(ii) } \Omega_t(a) \cap L = (a + \operatorname{Int}\gamma) \cap L \ , \\ \text{(iii) } \Omega_t(a) = \bigcup_{r<t} \Omega_r(a) \ , \\ \text{(iv) } \Omega_t(a) \text{ has a real analytic smooth boundary } \partial\Omega_t(a) \ , \\ \text{(v) } Z_t(a) = \left(\bigcap_{s>t} \overline{\Omega_s(a)\setminus\Omega_t(a)}\right) \cap H \text{ is contained} \\ \quad \text{in } \partial\Omega_t(a) \text{ and the exterior conormal of} \\ \quad \Omega_t(a) \text{ at } Z_t(a) \text{ is contained in } U \ , \\ \text{(vi) } \left(\bigcup_{t>0} \Omega_t(a)\right) \cap H = (a + \operatorname{Int}\gamma) \cap H \ , \\ \text{(vii) } \left(\bigcap_{t>0} \Omega_t(a)\right) \cap H = (a + \operatorname{Int}\gamma) \cap L \ . \end{array}\right. \tag{5.1.5}$$

For example, let:

$$\psi_t(x') = \delta^2 |x' - a'|^2 + \frac{(\delta^2 |x' - a'|^2 - (\varepsilon + a_1)^2)^2}{\sqrt{t + (\delta^2 |x' - a'|^2 - (\varepsilon + a_1)^2)^2}} .$$

Then one can take:

$$\Omega_t(a) = \{x; x_1 < a_1 \text{ and } (x_1 - a_1)^2 > \psi_t(x')\} .$$

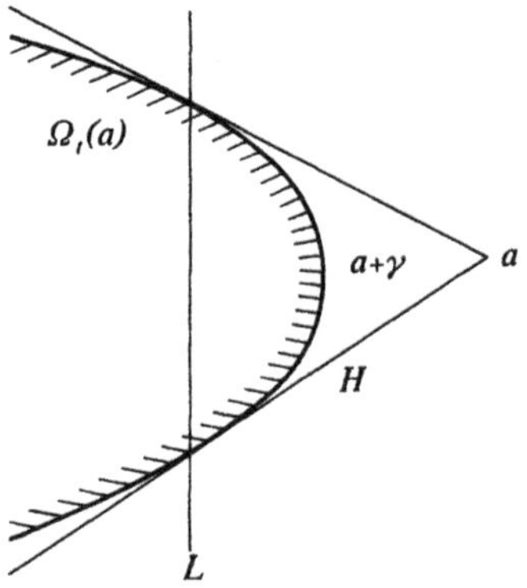

Fig. 5.1

Now we set:

(5.1.6) $$\tilde{\Omega}_t(a) = \Omega_t(a) \cup ((a + \operatorname{Int} \gamma) \setminus H) ,$$

and we shall show the isomorphism:

(5.1.7) $$R\Gamma(a + \operatorname{Int} \gamma; F) \xrightarrow{\sim} R\Gamma(\tilde{\Omega}_t(a); F) ,$$

for every $t > 0$.

Let j be the embedding $a + \operatorname{Int} \gamma \hookrightarrow X$, and let us apply the non-characteristic deformation lemma (Proposition 2.7.2). It is enough to show:

(5.1.8) $$(R\Gamma_{X \setminus \tilde{\Omega}_t(a)}(Rj_* j^{-1} F))_y = 0$$

for any $y \in Z_s(a) \setminus \tilde{\Omega}_t(a)$, any $s \leq t$. Since $Z_s(a) \setminus \tilde{\Omega}_t(a) \subset Z_t(a)$, we may assume $y \in Z_t(a) \setminus \tilde{\Omega}_t(a)$. If y belongs to $H \setminus L$ then (5.1.8) follows from the hypothesis $(1)_\omega$. Now assume $y \in Z_t(a) \cap L = L \cap \partial(a + \operatorname{Int} \gamma)$. Since $a + \operatorname{Int} \gamma$ and $\Omega_t(a)$ have smooth real analytic boundaries in a neighborhood of y, we have by $(1)_\omega$:

(5.1.9) $$(R\Gamma_{X \setminus \Omega_t(a)}(F))_y = (R\Gamma_{X \setminus (a + \operatorname{Int} \gamma)}(F))_y = 0 ,$$

thus:

(5.1.10) $$(R\Gamma_{X \setminus \Omega_t(a)}(Rj_* j^{-1} F))_y = 0 .$$

On the other hand, $(a + \operatorname{Int}\gamma)\backslash\Omega_t(a)$ is the disjoint union of $(a + \operatorname{Int}\gamma)\backslash\tilde{\Omega}_t(a)$ and $(a + \operatorname{Int}\gamma)\backslash(\Omega_t(a) \cup H)$. Hence $(R\Gamma_{X\backslash\tilde{\Omega}_t(a)}(Rj_* j^{-1}F))$ is a direct summand of $R\Gamma_{X\backslash\Omega_t(a)}(Rj_* j^{-1}F)$ and hence (5.1.10) implies (5.1.8). This shows (5.1.7) for all t.

From (5.1.7) we get the isomorphisms:

$$R\Gamma((a + \operatorname{Int}\gamma) \cap H; F|_H) \simeq R\Gamma(\tilde{\Omega}_t(a) \cap H; F|_H) \simeq R\Gamma(\Omega_t(a) \cap H; F|_H) \; . \tag{5.1.11}$$

Finally we shall show the isomorphism for any $x \in \Omega_1 \cap H$:

$$R\Gamma((x + \gamma) \cap H; F|_H) \xrightarrow{\sim} R\Gamma((x + \gamma) \cap L; F|_H) \; , \tag{5.1.12}$$

from which (5.1.4) is immediately deduced.

Let $v = (1, 0, \ldots, 0)$. Then the family $\{(x + \rho v + \operatorname{Int}\gamma) \cap H\}_{\rho>0}$ forms a neighborhood system of $(x + \gamma) \cap H$ in H, and the family $\{\Omega_t(x + \rho v) \cap L\}_{\rho>0, t>0}$ forms a neighborhood system of $(x + \gamma) \cap L$ in H. Thus (5.1.12) follows from (5.1.11). □

Definition 5.1.2. *Let X be a manifold.*

(i) *Let $F \in \mathrm{Ob}(\mathbf{D}^b(X))$. The micro-support of F, denoted by $\mathrm{SS}(F)$, is the subset of T^*X defined by:*

$$p \notin \mathrm{SS}(F) \Leftrightarrow \textit{condition } (1)_1 \textit{ of Proposition 5.1.1 is satisfied.}$$

(ii) *Let $u: F \to F'$ be a morphism in $\mathbf{D}^b(X)$, and let A be a subset of T^*X. We say that u is an isomorphism on A if u is embedded in a distinguished triangle $F \xrightarrow[u]{} F' \longrightarrow F'' \xrightarrow[+1]{}$ with $\mathrm{SS}(F'') \cap A = \emptyset$.*

Note that u is an isomorphism at p if and only if there exists a neighborhood U of p such that for any x_1 and any function ψ as in $(1)_\alpha$ of Proposition 5.1.1, the morphism:

$$(R\Gamma_{\{x;\, \psi(x) \geq 0\}}(F))_{x_1} \to (R\Gamma_{\{x;\, \psi(x) \geq 0\}}(F'))_{x_1}$$

is an isomorphism.

The following properties are immediately deduced from the definition.

Proposition 5.1.3. (i) *Let $F \in \mathrm{Ob}(\mathbf{D}^b(X))$. Then $\mathrm{SS}(F)$ is a closed conic subset of T^*X and $\mathrm{SS}(F) \cap T^*_X X = \mathrm{supp}(F)$.*

(ii) $\mathrm{SS}(F) = \mathrm{SS}(F[1])$.

(iii) *Let $F_1 \to F_2 \to F_3 \xrightarrow[+1]{}$ be a distinguished triangle in $\mathbf{D}^b(X)$. Then for $i, j, k \in \{1, 2, 3\}$:*

$$\mathrm{SS}(F_i) \subset \mathrm{SS}(F_j) \cup \mathrm{SS}(F_k) \quad \textit{for} \quad j \neq k \; , \tag{5.1.13}$$

$$(\mathrm{SS}(F_i)\backslash\mathrm{SS}(F_j)) \cup (\mathrm{SS}(F_j)\backslash\mathrm{SS}(F_i)) \subset \mathrm{SS}(F_k) \quad \textit{for} \quad k \neq i, j \tag{5.1.14}$$

We sometimes call the properties in (iii) the **triangular inequalities** for micro-supports. We shall prove in Chapter VI below that micro-supports are always involutive subsets of T^*X.

Remark 5.1.4. Let $F \in \mathrm{Ob}(\mathbf{D}^b(X))$. The inclusion $\mathrm{SS}(H^j(F)) \subset \mathrm{SS}(F)$ is not true in general, but we have $\mathrm{SS}(F) \subset \bigcup_j \mathrm{SS}(H^j(F))$ (cf. Exercises V.4 and V.6).

Remark 5.1.5. Let ϕ_X be the forgetful functor $\mathbf{D}^b(A_X) \to \mathbf{D}^b(\mathbb{Z}_X)$ and let $F \in \mathrm{Ob}(\mathbf{D}^b(A_X))$. Then:

$$\mathrm{SS}(F) = \mathrm{SS}(\phi_X(F)) \ . \tag{5.1.15}$$

In other words, $\mathrm{SS}(F)$ does not depend on the base ring A.

Remark 5.1.6. By Proposition 5.1.1, the micro-support of F depends only on the C^1-structure of X.

5.2. Propagation

Let E be a real finite-dimensional vector space, X an open subset of E, and let $F \in \mathrm{Ob}(\mathbf{D}^b(X))$.

Proposition 5.2.1. *Let U be an open subset of X and let γ be a closed proper convex cone in E with $\gamma \ni 0$. Let Ω_0 and Ω_1 be γ-open subsets of E. We assume*

$$\mathrm{SS}(F) \cap (U \times \mathrm{Int}\,\gamma^{\circ a}) = \emptyset \ , \tag{5.2.1}$$

$$\Omega_0 \subset \Omega_1 \qquad \textit{and} \qquad \Omega_1 \backslash \Omega_0 \subset U \ , \tag{5.2.2}$$

$$\textit{For any } x \in \Omega_1, (x + \gamma) \backslash \Omega_0 \textit{ is compact.} \tag{5.2.3}$$

Then we have

$$R\Gamma(\Omega_1 \cap X; F) \to R\Gamma(\Omega_0 \cap X; F) \quad \textit{is an isomorphism,} \tag{5.2.4}$$

and

$$R\phi_{\gamma *}(R\Gamma_{X \backslash \Omega_0}(F))|_{\Omega_1} = 0 \ , \tag{5.2.5}$$

where ϕ_γ is the map $X \to X_\gamma$.

We shall start the proof of the proposition by the following special case and deduce the general case afterwards.

Lemma 5.2.2. *Proposition 5.2.1 is true when Ω_0 has the form $\{x \in E; \langle x, \xi_\circ \rangle < c\}$ for some $\xi_\circ \in \mathrm{Int}\,\gamma^{\circ a}$, and*

$$\mathrm{SS}(F) \cap (U \times (\gamma^{\circ a} \backslash \{0\})) = \emptyset \ . \tag{5.2.6}$$

Proof. We endow E with a Euclidian structure and let $d(\cdot,\cdot)$ denote the distance function. We set:

$$W_t = \{(s_1,s_2) \in \mathbb{R}^2; s_1 < t\} \cup \{(s_1,s_2) \in \mathbb{R}^2; s_2 < 0\} \cup \{(s_1,s_2) \in \mathbb{R}^2; s_1 < 2t, s_2 < t, (s_1-2t)^2 + (s_2-t)^2 > t^2\} \tag{5.2.7}$$

for $0 < t$.

We set $l(x) = \langle x, \xi_\circ\rangle - c$, $H = \{x; l(x) \geqslant 0\}$, and define, for $0 < t$:

$$U(t,x_1) = \{x \in X; (d(x, x_1+\gamma), l(x)) \in W_t\} \ . \tag{5.2.8}$$

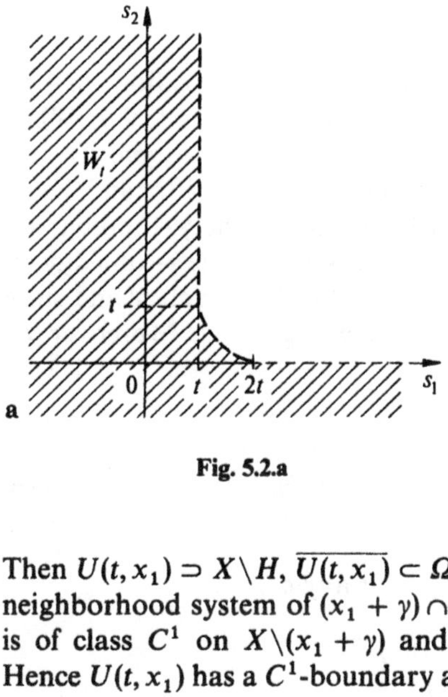

Fig. 5.2.a

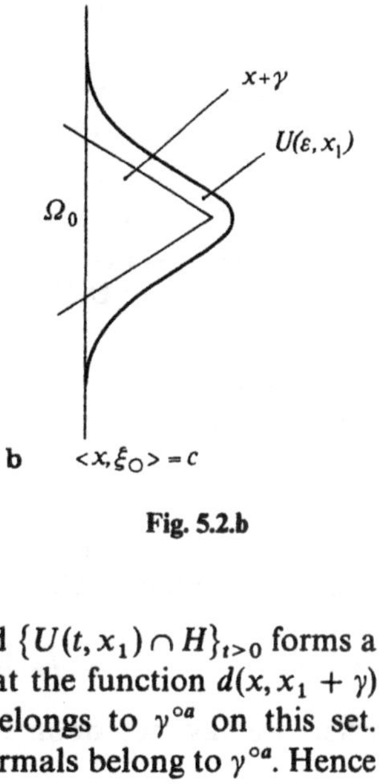

Fig. 5.2.b

Then $U(t,x_1) \supset X\backslash H$, $\overline{U(t,x_1)} \subset \Omega_1$ if $0 < t \ll 1$, and $\{U(t,x_1) \cap H\}_{t>0}$ forms a neighborhood system of $(x_1+\gamma) \cap H$. Now, note that the function $d(x, x_1+\gamma)$ is of class C^1 on $X\backslash(x_1+\gamma)$ and its differential belongs to $\gamma^{\circ a}$ on this set. Hence $U(t,x_1)$ has a C^1-boundary and its outer conormals belong to $\gamma^{\circ a}$. Hence $(R\Gamma_{(X\backslash U(t,x_1))}(F))_x = 0$, for $x \in U \cap \partial U(t,x_1)$.

We choose ε small enough such that $\overline{U(\varepsilon,x_1)}$ is contained in Ω_1. Then we may apply the non-characteristic deformation lemma (Proposition 2.7.2) to the family $\{U(t,x_1)\}_{0<t<\varepsilon}$. Setting $\tilde{F} = R\Gamma_{X\backslash\Omega_0}(F)$. we obtain the isomorphism:

$$R\Gamma(U(\varepsilon,x_1);\tilde{F}) \xrightarrow{\sim} R\Gamma(U(t,x_1);\tilde{F}) \ . \tag{5.2.9}$$

By taking the cohomology in both terms of (5.2.9) and the inductive limit for $t > 0$, we get:

$$R\Gamma(U(\varepsilon,x_1);\tilde{F}) \xrightarrow{\sim} R\Gamma(x_1+\gamma;\tilde{F}) \ . \tag{5.2.10}$$

We shall use this isomorphism to prove that for every $x \in \Omega_1$, $R\Gamma(x+\gamma;\tilde{F}) = 0$, which will complete the proof. Choose $v \in \operatorname{Int}\gamma$ and set $x_t = x + tv$. Let:

$$I = \{t \geqslant 0; R\Gamma(x_t+\gamma;\tilde{F}) = 0\} \ .$$

Then I is not empty since $x_t \in \Omega_0$ for $t \gg 0$. Let $t_\circ = \inf I$. Let $\varepsilon > 0$ be such that $\overline{U(\varepsilon, x_{t_\circ})}$ is contained in Ω_1 and let $t > t_\circ$ be such that $t \in I$ and:

$$x_{t_\circ} + \gamma \subset U(\varepsilon, x_t) \subset U(\varepsilon, x_{t_\circ}) \ .$$

The isomorphism $R\Gamma(U(\varepsilon, x_{t_\circ}); \tilde{F}) \xrightarrow{\sim} R\Gamma(x_{t_\circ} + \gamma; \tilde{F})$ factors through $R\Gamma(U(\varepsilon, x_t); \tilde{F})$ which is zero. Thus $t_\circ$ belongs to I. If $t_\circ$ would be strictly positive, we find t with $0 < t < t_\circ$ and $\varepsilon > 0$ such that:

$$x_t + \gamma \subset U(\varepsilon, x_{t_\circ}) \subset U(\varepsilon, x_t) \subset \overline{U(\varepsilon, x_t)} \subset \Omega_1 \ .$$

Since the isomorphism $R\Gamma(U(\varepsilon, x_t); \tilde{F}) \to R\Gamma(x_t + \gamma; \tilde{F})$ factors through $R\Gamma(U(\varepsilon, x_{t_\circ}); \tilde{F})$ which is zero, t belongs to I. Thus $t_\circ = 0$ which completes the proof of Lemma 5.2.2. □

End of the proof of Proposition 5.2.1. We shall reduce Proposition 5.2.1 to Lemma 5.2.2. Since (5.2.4) follows from (5.2.5), it is enough to show (5.2.5). When $\gamma = \{0\}$, $F|_U = 0$ and the proposition is obvious. Hence we may assume $\gamma \neq \{0\}$. Let γ' be a closed proper convex cone such that $\operatorname{Int}\gamma' \supset \gamma \backslash \{0\}$, and let Ω be a subset of $\Omega_1 \backslash \Omega_0$ such that Ω is open with respect to the γ'-topology on $\Omega_1 \backslash \Omega_0$ and $\Omega \subset\subset U$. Then such Ω form a base of open sets of the γ-topology on $\Omega_1 \backslash \Omega_0$. Since it is enough to show $R\Gamma(\Omega; R\Gamma_{X\backslash\Omega_0}(F)) = 0$ for such Ω and since $\operatorname{Int}\gamma^{\circ a} \supset \gamma'^{\circ a}\backslash\{0\}$, we may assume (5.2.6) from the beginning. Moreover we may assume

$$\Omega_1 \backslash \Omega_0 \subset\subset U \ . \tag{5.2.11}$$

By the non-characteristic deformation lemma (Proposition 2.7.2), in order to see (5.2.5), it is enough to show

$$R\Gamma_{H_c} R\Gamma_{\Omega_1\backslash\Omega_0}(F)_{\partial H_c} = 0 \qquad \text{for any } c \ , \tag{5.2.12}$$

where we have set for an arbitrary point $\xi_\circ \in \operatorname{Int}\gamma^{\circ a}$:

$$H_c = \{x; \langle x, \xi_\circ\rangle \geqslant c\} \qquad \text{and} \qquad \partial H_c = \{x; \langle x, \xi_\circ\rangle = c\} \ .$$

For any $x_\circ \in \partial H_c \cap U$, there exists a γ-open subset $\Omega \ni x_\circ$ such that $\Omega \cap H_c \subset U$. Then by Lemma 5.2.2, $R\phi_{\gamma *}(R\Gamma_{\Omega\cap H_c}(F)) = 0$. Then (5.2.12) follows from the fact that the $W \cap H_c$ form a neighborhood system of $x_\circ$ in H_c when W ranges through a neighborhood system of $x_\circ$ in the γ-topology. This completes the proof of Proposition 5.2.1. □

Now we shall consider the case $X = Y \times E$, where E is a finite-dimensional real vector space and Y is a manifold. Let γ be a closed convex cone in E, with $0 \in \gamma$, (we do not ask γ to be proper). We set $X_\gamma = Y \times E_\gamma$ and denote by ϕ_γ the continuous map $X \to X_\gamma$.

We shall use the γ-topology to cut off the micro-support of sheaves.

Let $F \in \operatorname{Ob}(\mathbf{D}^b(X))$.

Proposition 5.2.3 (the microlocal cut-off lemma). (i) *The micro-support* SS(F) *is contained in* $T^*Y \times (E \times \gamma^{\circ a})$ *if and only if the morphism* $\phi_\gamma^{-1}R\phi_{\gamma *}F \to F$ *is an isomorphism.*

(ii) *The morphism* $\phi_\gamma^{-1}R\phi_{\gamma *}F \to F$ *is an isomorphism on* $T^*Y \times (E \times \operatorname{Int}\gamma^{\circ a})$.

Proof. We may assume Y affine. Replacing γ by $\{0\} \times \gamma$ we may assume $X = E$ from the beginning. Assume first $\phi_\gamma^{-1}R\phi_{\gamma *}F \simeq F$. We shall prove the inclusion $SS(F) \subset X \times \gamma^{\circ a}$. Let $\xi_\circ \notin \gamma^{\circ a}$. Choose a proper convex closed cone γ' such that $\gamma' \setminus \{0\} \subset \{v; \langle v, \xi_\circ\rangle < 0\}$ and $\gamma' + \gamma = E$. Then for any γ'-open non-empty convex subset Ω of X we have:

$$R\Gamma(\Omega; F) \simeq R\Gamma(\Omega + \gamma; F) \simeq R\Gamma(X; F) \ .$$

This implies that for any pair of convex γ'-open subsets (Ω_0, Ω_1) with $\Omega_0 \subset \Omega_1$, we have:

$$R\phi_{\gamma' *}R\Gamma_{\Omega_1 \setminus \Omega_0}(F) = 0 \ ,$$

thus $SS(F) \cap (X \times \{\xi_\circ\}) = \varnothing$.

Conversely assume $SS(F) \subset X \times \gamma^{\circ a}$. In order to prove the isomorphism $\phi_\gamma^{-1}R\phi_{\gamma *}F \simeq F$, it is sufficient to show that for any relatively compact open convex subset Ω of X, the restriction morphism from $R\Gamma(\Omega + \gamma; F)$ to $R\Gamma(\Omega; F)$ is an isomorphism. Let $\mathfrak{S}$ be the set of open convex subsets V of $(\Omega + \gamma)$ such that $V \supset \Omega$ and that $R\Gamma(V; F) \to R\Gamma(\Omega; F)$ is an isomorphism. It follows from Proposition 2.7.1 that $\mathfrak{S}$ is inductively ordered. Let V be a maximal element of $\mathfrak{S}$. We shall show that $V = \Omega + \gamma$ by contradiction. If $V \neq \Omega + \gamma$, there exists $x_\circ \in (\Omega + \gamma) \setminus \bar{V}$.

Lemma 5.2.4. *Let V be a convex open subset of X, and let $x_\circ \in X$. Let λ be the cone with vertex at $x_\circ$ generated by V, let λ' be the closed cone $\bar{\lambda} - x_\circ$ with vertex at the origin and let V_1 be the interior of the convex hull of $V \cup \{x_\circ\}$. Then $V_1 \setminus V$ is locally closed for the λ'-topology.*

Proof of Lemma 5.2.4. We have:

$$V_1 = \{(1-t)x_\circ + tu; u \in V, 0 < t \leqslant 1\} = \{(1-t)x_\circ + tu; u \in V, 0 < t < 1\} \ .$$

Set:

$$V_2 = \{(1-t)x_\circ + tu; u \in V, t > 0\} \ ,$$

$$V_3 = \{(1-t)x_\circ + tu; u \in V, t > 1\} \ .$$

Then V_2 and V_3 are λ'-open, and it is enough to show that $V_1 \setminus V = V_2 \setminus V_3$. Since we have $V \subset V_1 \subset V_2$, $V \subset V_3$, $V_2 \subset V_1 \cup V_3$, it is enough to show that $V_1 \cap V_3$ is contained in V. For $x \in V_1 \cap V_3$ we may write $x = (1-t)x_\circ + tu = (1-s)x_\circ + sv$, with $0 < t \leqslant 1$, $1 < s$ and $u, v \in V$. Then we have $(s-t)x = (s-1)tu + (1-t)sv$, thus $x \in V$. □

End of the proof of Proposition 5.2.3. Let V_1, λ and λ' be as in Lemma 5.2.4. Then λ' is a proper convex cone, and $V_1 \setminus V$ is locally closed for the λ'-topology, by the preceding lemma. On the other hand we have $\operatorname{Int}\lambda' \ni v$ for some $v \in \gamma^a$. Hence $\lambda'^{\circ a} \cap \gamma^{\circ a} \subset \{0\}$. This implies, by Proposition 5.2.1:

$$(R\phi_{\lambda'*} R\Gamma_{V_1 \setminus V}(F))|_{V_1} = 0 \ .$$

Thus we obtain $V_1 \in \mathfrak{S}$, which is a contradiction.

In order to prove (ii), let $\phi_\gamma^{-1} R\phi_{\gamma*} F \longrightarrow F \longrightarrow F' \xrightarrow[+1]{}$ be a distinguished traingle. Applying the functor $R\phi_{\gamma*}$ we obtain $R\phi_{\gamma*}F' = 0$, which implies $SS(F') \cap (X \times \operatorname{Int}\gamma^{\circ a}) = \emptyset$, by Proposition 5.1.1. Note that when γ is not proper, $\operatorname{Int}\gamma^{\circ a} = \emptyset$. □

Remark 5.2.5. The cut-off of micro-support will be again discussed in VI §1.

5.3. Examples: micro-supports associated with locally closed subsets

Recall that if Z is a locally closed subset of X, the sheaf A_Z on X is the zero sheaf on $X \setminus Z$ and the constant sheaf with stalk A on Z.

Let E be a finite-dimensional real vector space, γ a closed convex cone with vertex at 0. Recall that we set $\gamma^a = -\gamma$ and $\gamma^\circ = \{\xi \in E^*; \langle v, \xi \rangle \geqslant 0 \text{ for any } v \in \gamma\}$.

Proposition 5.3.1. *One has:*

$$SS(A_\gamma) \cap \pi^{-1}(0) = \gamma^\circ \ .$$

Proof. The section $1 \in \Gamma(E; A_E)$ defines a section $1_\gamma \in \Gamma(E; A_\gamma)$ whose support is γ. Therefore γ° is contained in $\pi^{-1}(0) \cap SS(A_\gamma)$. On the other hand, the morphism $\phi_{\gamma^a}^{-1} \circ R\phi_{\gamma^a *} A_\gamma \to A_\gamma$ is an isomorphism. Hence the result follows from the microlocal cut-off lemma (Proposition 5.2.3). □

Proposition 5.3.2. *Let X be a manifold and M a closed submanifold. Then:*

$$SS(A_M) = T_M^*X \ .$$

Proof. Since the problem is local on M, we may assume that X is a vector space and M is a vector subspace. Then this is a particular case of the preceding result. □

Note that, in particular, $SS(A_X) = T_X^*X$, the zero-section of T^*X.

Proposition 5.3.3. *Let φ be a real function of class C^1, and assume* $d\varphi \neq 0$ on the set $\{x; \varphi(x) = 0\}$. *Then:*

(i) $$\mathrm{SS}(A_{\{x;\varphi(x)\geqslant 0\}}) = \{(x;\lambda d\varphi(x)); \lambda\varphi(x) = 0, \lambda \geqslant 0, \varphi(x) \geqslant 0\} ,$$

(ii) $$\mathrm{SS}(A_{\{x;\varphi(x)>0\}}) = \{(x;\lambda d\varphi(x)); \lambda\varphi(x) = 0, \lambda \leqslant 0, \varphi(x) \geqslant 0\} .$$

Proof. (i) follows from Proposition 5.3.1 by choosing a local coordinates system such that $\{x; \varphi(x) \geqslant 0\}$ is a closed half-space.

(ii) follows from (i) and the triangular inequalities (Proposition 5.1.3), by considering the exact sequence $0 \to A_{\{x;\varphi(x)>0\}} \to A_X \to A_{\{x;\varphi(x)\leqslant 0\}} \to 0$. □

Now we shall discuss some concrete examples on $\mathbb{R}^2$. We denote by (x_1, x_2) the coordinates on $\mathbb{R}^2$ and by (ξ_1, ξ_2) the dual coordinates.

Example 5.3.4. Let $Z = \{x \in \mathbb{R}^2; x_1 > 0, -x_1^{3/2} \leqslant x_2 < x_1^{3/2}\}$. Then:

$$\begin{cases} \mathrm{SS}(A_Z) = (T^*_X X \cap \bar{Z}) \cup \{(x;\xi); \xi_2 > 0, x_2 = -(2\xi_1/3\xi_2)^3, \\ \qquad\qquad\qquad\qquad\qquad\qquad x_1 = (2\xi_1/3\xi_2)^2\} . \end{cases} \tag{5.3.1}$$

In fact, set $\varphi_\pm(x) = x_2 \pm x_1^{3/2}$ for $x_1 \geqslant 0$ and $\varphi_\pm(x) = x_2$ for $x_1 \leqslant 0$, $Z_i = \{x; \varphi_i(x) \geqslant 0\}$, $(i = \pm)$. One checks easily, using Proposition 5.1.1 that for $i = \pm$, $\mathrm{SS}(A_{Z_i})$ is the closure of the set: $\{(x; \lambda d\varphi_i(x)); \lambda\varphi_i(x) = 0, \lambda \geqslant 0, \varphi_i(x) \geqslant 0\}$. Then (5.3.1) follows from Proposition 5.1.3, since $Z = Z_+ \setminus Z_-$.

Example 5.3.5. Let $Z = \{x \in \mathbb{R}^2; x_1 x_2 \geqslant 0\}$. Then:

$$\pi^{-1}(0) \cap \mathrm{SS}(A_Z) = \{\xi; \xi_1\xi_2 \leqslant 0\} . \tag{5.3.2}$$

In fact, let $Z^\pm = \{x \in Z; \pm x_1 \geqslant 0, \pm x_2 \geqslant 0\}$. We have an exact sequence; $0 \to A_Z \to A_{Z^+} \oplus A_{Z^-} \to A_{\{0\}} \to 0$. By the result of Proposition 5.3.1 and the triangular inequalities (Proposition 5.1.3), we get the inclusion $\pi^{-1}(0) \cap \mathrm{SS}(A_Z) \supset \{\xi; \xi_1\xi_2 \leqslant 0\}$. To prove the converse inclusion, denote by γ the cone Z^-. Then $\phi_\gamma^{-1} R\phi_{\gamma *} A_{Z^+} \simeq \phi_\gamma^{-1} R\phi_{\gamma *} A_{\{0\}} \simeq A_{Z^+}$, thus $\phi_\gamma^{-1} R\phi_{\gamma *} A_Z \simeq \phi_\gamma^{-1} R\phi_{\gamma *} A_{Z^-}$. Since $\mathrm{Int}\,\gamma^{\circ a} \cap \mathrm{SS}(A_{Z^-}) = \emptyset$, we get $\mathrm{Int}\,\gamma^{\circ a} \cap \mathrm{SS}(A_Z) = \emptyset$ by Proposition 5.2.3. Replacing γ by γ^a we get (5.3.2).

In general it is difficult to calculate exactly $\mathrm{SS}(A_Z)$. In order to give a bound to this set, we introduce:

Definition 5.3.6. *Let S be a subset of X, where X is a manifold. We set:*

$$N_x(S) = T_x X \setminus C_x(X \setminus S, S) ,$$

$$N_x^*(S) = (N_x(S))^\circ ,$$

$$N(S) = \bigcup_{x \in X} N_x(S) , \qquad N^*(S) = \bigcup_{x \in X} N_x^*(S) .$$

We call $N(S)$ the strict normal cone to S, and $N^*(S)$ the conormal cone to S.

(Recall that the normal cone $C(X\setminus S, S)$ has been defined in IV §1.)

It follows from Proposition 4.1.2 that a non-zero vector $\theta \in T_xX$ belongs to $N_x(S)$ if and only if, in a local chart in a neighborhood of x, there exists an open cone γ containing θ and a neighborhood U of x such that:

$$U \cap ((S \cap U) + \gamma) \subset S \ . \tag{5.3.3}$$

In particular, $N(S)$ is an open convex cone of TX, which satisfies:

$$\begin{aligned} N_x(S) = T_xX &\Leftrightarrow C_x(X\setminus S, S) = \emptyset \\ &\Leftrightarrow x \notin \overline{S} \quad \text{or} \quad x \in \operatorname{Int} S \ . \end{aligned}$$

Moreover:

$$N_x(S) = \emptyset \Leftrightarrow N_x^*(S) = T_x^*X \ ,$$

$$N_x(S) \neq \emptyset \Leftrightarrow N_x^*(S) \quad \text{is a proper closed convex cone} \ .$$

Also notice that:

$$N_x(X\setminus S) = N_x(S)^a \ ,$$

$$N_x^*(X\setminus S) = N_x^*(S)^a \ .$$

Proposition 5.3.7. *Assume X is affine, and let γ be a proper closed convex cone with* $\operatorname{Int}\gamma \neq \emptyset$. *Let Ω (resp. Z) be an open (resp. closed) subset of X. Let $x \in X$.*

(i) *If $N_x^*(\Omega) \subset \operatorname{Int}\gamma^\circ \cup \{0\}$ (resp. $N_x^*(Z)^a \subset \operatorname{Int}\gamma^\circ \cup \{0\}$), then Ω (resp. Z) is γ-open (resp. γ-closed) in a neighborhood of x.*

(ii) *Conversely if Ω (resp. Z) is γ-open (resp. γ-closed) in a neighborhood of x, then $N_x^*(\Omega) \subset \gamma^\circ$ (resp. $N_x^*(Z)^a \subset \gamma^\circ$).*

Proof. Since Ω is γ-open in a neighborhood of x if and only if $X\setminus\Omega$ is γ-closed in a neighborhood of x and $N^*(\Omega) = N^*(X\setminus\Omega)^a$, it is enough to prove the results for Ω.

(i) For each $v \in N_x(\Omega)$, $v \neq 0$, there exists a convex open cone γ_v such that Ω is γ_v-open in a neighborhood of x (cf. (5.3.3)). One may cover $\gamma\setminus\{0\}$ by a finite number of such cones γ_v, and the result follows.

(ii) follows from (5.3.3). □

Proposition 5.3.8. *Let X be a manifold, Ω an open subset and Z a closed subset. Then:*

$$\operatorname{SS}(A_\Omega) \subset N^*(\Omega)^a \ ,$$

$$\operatorname{SS}(A_Z) \subset N^*(Z) \ .$$

Proof. We may assume X affine, and it is enough to prove the result for Ω, by applying the triangular inequality to $0 \to A_{X\setminus Z} \to A_X \to A_Z \to 0$.

Let $x \in X$ and assume $N_x^*(\Omega) \neq T_x^*X$. Let γ be a closed convex proper cone such that:

$$N_x^*(\Omega) \subset \operatorname{Int}\gamma^\circ \cup \{0\} . \tag{5.3.4}$$

By the preceding proposition, Ω is γ-open in a neighborhood of x, and we may assume Ω is γ-open. Then $A_\Omega = \phi_\gamma^{-1} R\phi_{\gamma *} A_\Omega$, hence $SS(A_\Omega) \subset X \times \gamma^{\circ a}$ (Proposition 5.2.3). Since this inclusion holds for any γ satisfying (5.3.4), we get the result. □

5.4. Functorial properties of the micro-support

In this section, we study the behaviour of the micro-support under several operations such as proper direct images, inverse images in the non-characteristic case, etc. Some results obtained here will be generalized in the subsequent chapter.

Let X and Y be two manifolds. We denote as usual by q_1 and q_2 (resp. p_1 and p_2) the first and second projection defined on $X \times Y$ (resp. $T^*X \times T^*Y$). If f is a map from Y to X, we denote as in Chapter IV by ${}^t f'$ and f_π the maps:

$$T^*Y \xleftarrow{{}^t f'} Y \times_X T^*X \xrightarrow{f_\pi} T^*X .$$

Proposition 5.4.1. *Let $F \in \operatorname{Ob}(\mathbf{D}^b(X))$ and $G \in \operatorname{Ob}(\mathbf{D}^b(Y))$. Then:*

$$SS\left(F \overset{L}{\boxtimes} G\right) \subset SS(F) \times SS(G) .$$

Proof. We may assume X and Y are vector spaces. Let $(x_\circ, y_\circ; \xi_\circ, \eta_\circ) \in T^*(X \times Y)$, and assume for example that $(x_\circ; \xi_\circ) \notin SS(F)$. Take H, L, U, γ satisfying the conditions in Proposition 5.1.1 (3) for the sheaf F on X and set $\tilde{\gamma} = \gamma \times \{0\}$. For $z = (x, y) \in U \times Y$ we have:

$$\begin{aligned} R\Gamma\left(H \times Y \cap (z + \tilde{\gamma}); q_1^{-1}F \overset{L}{\otimes} q_2^{-1}G\right) &\simeq R\Gamma\left(H \cap (x + \gamma); F \overset{L}{\otimes} G_y\right) \\ &\simeq R\Gamma(H \cap (x + \gamma); F) \overset{L}{\otimes} G_y . \end{aligned} \tag{5.4.1}$$

In fact, $H \cap (x + \gamma)$ being compact, we may apply Proposition 2.6.6.

Since we have a similar formula to (5.4.1) with H replaced by L, the result follows from Proposition 5.1.1. □

Proposition 5.4.2. *Let* $F \in \mathrm{Ob}(\mathbf{D}^b(X))$ *and* $G \in \mathrm{Ob}(\mathbf{D}^b(Y))$. *Then:*

$$\mathrm{SS}(R\mathcal{H}om(q_2^{-1}G, q_1^{-1}F)) \subset \mathrm{SS}(F) \times \mathrm{SS}(G)^a .$$

Proof. It is enough to prove the similar statement with $q_1^{-1}F$ replaced by $q_1^!F$.

We may assume X and Y are vector spaces. Let $(x_\circ, y_\circ; \xi_\circ, -\eta_\circ) \notin \mathrm{SS}(F) \times \mathrm{SS}(G)^a$.

First assume $(x_\circ; \xi_\circ) \notin \mathrm{SS}(F)$. In this case the proof is similar to that of Proposition 5.4.1. Take H, L, U, γ satisfying the conditions in Proposition 5.1.1 (3) for the sheaf F on X and set $\tilde{\gamma} = \gamma \times \{0\}$. For $z = (x, y) \in U \times Y$ we have by Proposition 3.1.15:

$$(5.4.2) \qquad H^j(R\Gamma(H \times Y \cap (z + \tilde{\gamma}); R\mathcal{H}om(q_2^{-1}G, q_1^!F)))$$
$$\simeq \varinjlim_W H^j(R\Gamma((H \cap (x + \gamma)) \times W; R\mathcal{H}om(q_2^{-1}G, q_1^!F)))$$
$$\simeq \varinjlim_W H^j(R\mathrm{Hom}(R\Gamma_c(W, G), R\Gamma(H \cap (x + \gamma); F))) ,$$

where W ranges through the family of open neighborhoods of y. Since we have a similar formula with H replaced by L, the proof follows from Proposition 5.1.1 in that case.

Now assume $(y_\circ; \eta_\circ) \notin \mathrm{SS}(G)$. Take a cone γ satisfying the condition (2) of Proposition 5.1.1 for the sheaf G on Y. We may assume $R\phi_{\gamma *}G = 0$.

Lemma 5.4.3. *Let Y be a finite-dimensional real vector space, γ a closed convex proper cone with $\gamma \ni 0$, Ω a γ^a-open subset of Y such that, for any compact K of Y, $\Omega \cap (K + \gamma)$ is relatively compact, and let Ω' be a γ^a-open subset of Ω such that $\Omega \setminus \Omega'$ is relatively compact in Y. Let $G \in \mathrm{Ob}(\mathbf{D}^b(Y))$, and assume $R\phi_{\gamma *}G = 0$. Then:*

(i) $R\phi_{\gamma *}G_\Omega = 0$,
(ii) $R\Gamma_c(\Omega'; G) \simeq R\Gamma_c(\Omega; G)$.

Proof of Lemma 5.4.3. Let Ω_1 be a γ-open subset of Y and assume $\Omega_1 \cap \Omega \subset\subset Y$. We have:

$$(5.4.3) \qquad H^j(\Omega_1; G_\Omega) \simeq \varinjlim_K H^j_K(\Omega_1; G) ,$$

where K ranges through the family of closed subsets of Ω_1 contained in Ω.

For such a K, $K' = \bar{K} + \gamma^a$ is closed in Y, $\bar{K}$ denoting the closure of K in Y. Then we have:

$$(5.4.4) \qquad K' \cap \Omega_1 \subset \Omega .$$

In fact if $x = y + v \in \Omega_1$, with $y \in \bar{K}, v \in \gamma^a$, then $y = x - v$ belongs to $(\Omega_1 + \gamma) \cap \bar{K} = \Omega_1 \cap \bar{K} = K$. Therefore $x \in K + \gamma^a \subset \Omega + \gamma^a = \Omega$.

Since $K \cap \Omega_1 \subset K' \cap \Omega_1$, we get by (5.4.3):

$$H^j(\Omega_1; G_\Omega) \simeq \varinjlim_{K'} H^j_{K' \cap \Omega_1}(\Omega_1; G) ,$$

where K' ranges through the family of γ-closed subsets such that $K \cap \Omega_1 \subset \Omega$. Then $R\Gamma_{K' \cap \Omega_1}(\Omega_1; G) \simeq R\Gamma_{K' \cap \Omega_1}(\Omega_1; R\phi_{\gamma *}G) = 0$, which proves (i).

To prove the second assertion, note that $R\Gamma_c(\Omega, G) \simeq R\Gamma_c(Y, G_\Omega)$, and a similar formula holds with Ω replaced by Ω'. Then it is enough to show that $R\Gamma_c(Y; G_{\Omega \setminus \Omega'}) = 0$, and $\Omega \setminus \Omega'$ being relatively compact, it is enough to show that $R\Gamma(Y; G_{\Omega \setminus \Omega'}) = 0$. But $R\Gamma(Y; G_\Omega) \simeq R\Gamma(Y; G_{\Omega'}) = 0$ by (i). □

End of the proof of Proposition 5.4.2. By the preceding lemma and Proposition 3.1.15, we find that for any open set W in X and any pair of γ^a-open subsets Ω and Ω' of Y such that $\Omega' \subset \Omega$, $\Omega \setminus \Omega' \subset\subset Y$, we have:

$$R\Gamma(W \times \Omega; R\mathscr{H}om(q_2^{-1}G, q_1^! F)) \simeq R\Gamma(W \times \Omega'; R\mathscr{H}om(q_2^{-1}G, q_1^! F)) .$$

Then the condition (2) of Proposition 5.1.1 is satisfied for the sheaf $R\mathscr{H}om(q_2^{-1}G, q_1^! F)$, with the cone $\{0\} \times \gamma^a$. □

Proposition 5.4.4. *Let $f: Y \to X$ be a morphism of manifolds, $G \in \mathrm{Ob}(\mathbf{D}^b(Y))$, and assume f is proper on* supp(G). *Then*:

$$\mathrm{SS}(Rf_* G) \subset f_\pi({}^t f'^{-1}(\mathrm{SS}(G))) .$$

Moreover if f is a closed embedding, the inclusion above is an equality.

Proof. Let $x \in X$ and φ be a real C^1-function on X such that $\varphi(x) = 0$ and $d(\varphi \circ f)(y) \notin \mathrm{SS}(G)$ for all $y \in f^{-1}(x)$. Hence we have $R\Gamma_{\{\varphi \circ f \geq 0\}}(G)|_{f^{-1}(x)} = 0$ and

$$\begin{aligned}(R\Gamma_{\{\varphi \geq 0\}}(Rf_* G))_x &\simeq (Rf_* R\Gamma_{\{\varphi \circ f \geq 0\}}(G))_x \\ &\simeq R\Gamma(f^{-1}(x); R\Gamma_{\{\varphi \circ f \geq 0\}}(G)) \\ &= 0 .\end{aligned}$$

(Here, $\{\varphi \geq 0\}$ denotes the set $\{x \in X; \varphi(x) \geq 0\}$, and similarly for $\{f \circ \varphi \geq 0\}$.) This proves the first assertion. Now assume f is a closed immersion. We may assume X is a vector space, Y a vector subspace. Suppose $p \notin \mathrm{SS}(Rf_* G)$, and let H, L, γ, U which satisfy the condition (3) of Proposition 5.1.1 for the sheaf $Rf_* G$ on X. For $x \in U \cap Y$ we have:

$$R\Gamma(H \cap (x + \gamma); Rf_* G) \xrightarrow{\sim} R\Gamma(L \cap (x + \gamma); Rf_* G) .$$

On the other hand we have:

$$R\Gamma(H \cap (x + \gamma); Rf_* G) \simeq R\Gamma((Y \cap H) \cap (x + (Y \cap \gamma)); G)$$

and a similar formula holds with H replaced by L. Then the result follows from Proposition 5.1.1. □

Proposition 5.4.5. *Assume that $f: Y \to X$ is smooth.*

(i) *Let $F \in \mathrm{Ob}(\mathbf{D}^b(X))$. Then:*

$$\mathrm{SS}(f^{-1}F) = {}^t f'(f_\pi^{-1}(\mathrm{SS}(F))) \ .$$

(ii) *Let $G \in \mathrm{Ob}(\mathbf{D}^b(Y))$. Then the following conditions are equivalent:*

(a) *all the $H^j(G)$'s are locally constant sheaves on the fibers of f,*

(b) *locally on Y, there exists $F \in \mathrm{Ob}(\mathbf{D}^b(X))$, such that $G \simeq f^{-1}F$,*

(c) $\mathrm{SS}(G) \subset {}^t f'(Y \times_X T^*X)$.

Proof. (i) First we prove the inclusion of $\mathrm{SS}(f^{-1}F)$ in ${}^t f'(f_\pi^{-1}(\mathrm{SS}(F)))$. We may assume $Y = \mathbb{R}^n \times \mathbb{R}^l$, $X = \mathbb{R}^n$, f being the projection $(x, y) \mapsto x$. Let $p = (x_\circ, y_\circ; \xi_\circ, \eta_\circ) \notin {}^t f'(f_\pi^{-1}(\mathrm{SS}(F)))$. If $\eta_\circ \neq 0$ we choose $v \in \mathbb{R}^l$ such that $\langle v, \eta_\circ \rangle < 0$ and set $\gamma = \{(0, tv); t \geqslant 0\}$. For an arbitrary $\varepsilon > 0$, set $H = \{(x, y); \langle y, \eta_\circ \rangle \geqslant -\varepsilon\}$, $L = H \setminus \mathrm{Int}\, H$. For $z \in \mathrm{Int}\, H$ we have:

$$R\Gamma(H \cap (z + \gamma); f^{-1}F) = R\Gamma(L \cap (z + \gamma); f^{-1}F) \ ,$$

since $f^{-1}F|_{(z + \mathbb{R}(0, v))}$ is a constant sheaf. This proves $p \notin \mathrm{SS}(f^{-1}F)$ in that case.

Now assume $\eta_\circ = 0$, $(x_\circ; \xi_\circ) \notin \mathrm{SS}(F)$. We take H, L, γ, U satisfying the condition (3) of Proposition 5.1.1 for the sheaf F on X and we set $\tilde{\gamma} = \gamma \times \{0\}$. Then for any $z \in f^{-1}(U)$ we have:

$$R\Gamma(f^{-1}(H) \cap (z + \tilde{\gamma}); f^{-1}F) \simeq R\Gamma(H \cap (f(z) + \gamma); F)$$

and a similar isomorphism with H replaced by L. Therefore $p \notin \mathrm{SS}(f^{-1}F)$.

To prove the converse inclusion, we use the condition (1) of Proposition 5.1.1 and note that if φ is a real C^1-function on X and $x \in X$ we have:

$$(R\Gamma_{\{\varphi \geqslant 0\}}(F))_x = (R\Gamma_{\{\varphi \circ f \geqslant 0\}}(f^{-1}F))_y$$

for any $y \in f^{-1}(x)$.

(ii) We may assume $Y = X \times \mathbb{R}^l$. Then (a)$\Leftrightarrow$(b) by Proposition 2.7.8, (b)$\Rightarrow$(c) by (i) and, to prove that (c)$\Rightarrow$(a), we apply Proposition 5.2.3 with $\gamma = \mathbb{R}^l$. $\square$

Remark 5.4.6. In the situation of Proposition 5.4.5, let $V = {}^t f'(Y \times_X T^*X)$. This is a smooth involutive submanifold of T^*Y. Although the inclusion $\mathrm{SS}(H^j(G)) \subset \mathrm{SS}(G)$ does not hold in general (cf. Remark 5.1.4), we have the equivalence: $\mathrm{SS}(G) \subset V \Leftrightarrow \mathrm{SS}(H^j(G)) \subset V$ for any j, which follows from Proposition 5.4.5.

In order to describe the behavior of micro-supports under various operations in sheaf theory, let us introduce the notation $A + B$ for two cones A and B in T^*X, by

$$(5.4.5) \qquad (A + B) \cap \pi^{-1}(x) = \{a + b; a \in A \cap \pi^{-1}(x), b \in B \cap \pi^{-1}(x)\} \ .$$

Lemma 5.4.7. *If A and B are closed cones of T^*X and if $A \cap B^a \subset T^*_X X$, then $A + B$ is also a closed cone in T^*X.*

Proof. Let (x) be a coordinate system on X and let $(x;\xi)$ be the associated coordinates on T^*X. The condition implies $\xi+\eta\neq 0$ on $\{((x;\xi),(x;\eta))\in A\times_X B; |\xi|+|\eta|=1\}$. Hence we may assume $|\xi+\eta|\geqslant\varepsilon$ for some $\varepsilon>0$ on this set. Hence $A\times_X B$ is contained in $C=\{(x;\xi,\eta); |\xi+\eta|\geqslant\varepsilon(|\xi|+|\eta|)\}$. Since $\mu: C\to T^*X$, given by $(x;\xi,\eta)\mapsto(x;\xi+\eta)$, is a proper map, $A+B=\mu(A\times_X B)$ is closed. □

Proposition 5.4.8. *Let* $F\in\mathrm{Ob}(\mathbf{D}^b(X))$.

(a) *Let* Ω *be an open subset of* X, *and* j *the open embedding* $\Omega\hookrightarrow X$.

(i) *Assume* $SS(F)\cap N^*(\Omega)^a\subset T^*_XX$. *Then*:

$$SS(Rj_*j^{-1}F)\subset N^*(\Omega)+SS(F)\ .$$

(ii) *Assume* $SS(F)\cap N^*(\Omega)\subset T^*_XX$. *Then*:

$$SS(Rj_!j^{-1}F)\subset N^*(\Omega)^a+SS(F)\ .$$

(b) *Let* Z *be a closed subset of* X.

(i) *If* $SS(F)\cap N^*(Z)\subset T^*_XX$, *then* $SS(R\Gamma_Z(F))\subset N^*(Z)^a+SS(F)$.

(ii) *If* $SS(F)\cap N^*(Z)^a\subset T^*_XX$, *then* $SS(F_Z)\subset N^*(Z)+SS(F)$.

Proof. (a) We may assume X is a vector space. Taking $x_\circ\in X$, we shall prove the assertion at $x_\circ$.

(i) Letting $\xi_\circ\notin N^*_{x_\circ}(\Omega)+(SS(F)\cap\pi^{-1}(x_\circ))$ we shall show $(x_\circ;\xi_\circ)\notin SS(Rj_*j^{-1}F)$. By the hypothesis:

$$(N^*_{x_\circ}(\Omega)+\overline{\mathbb{R}^-\xi_\circ})\cap SS(F)^a\subset\{0\}\ ,$$

and there exists a closed convex proper cone K in $T^*_{x_\circ}X$ such that:

$$N^*_{x_\circ}(\Omega)+\overline{\mathbb{R}^-\xi_\circ}\subset \mathrm{Int}(K)\cup\{0\}\ ,$$

$$K^a\cap(SS(F)\cap\pi^{-1}(x_\circ))\subset\{0\}\ .$$

Let γ be the polar cone to K. Then $\gamma\subset\{v;\langle v,\xi_\circ\rangle<0\}\cup\{0\}$, and by Proposition 5.3.7 we may assume Ω is γ-open. Let U be a relatively compact open neighborhood of $x_\circ$ such that $(U\times\gamma^{\circ a})\cap SS(F)\subset T^*_XX$. Let Ω_0 and Ω_1 be two γ-open sets such that $\Omega_0\subset\Omega_1\subset\Omega_0\cup U$. Applying Proposition 5.2.1 we get:

$$(R\phi_{\gamma*}R\Gamma_{X\setminus\Omega_0}(F))|_{\Omega_1}=0\ . \tag{5.4.6}$$

The open set Ω being γ-open, Rj_*j^{-1} commutes with $R\phi_{\gamma*}$. Hence (5.4.6) remains true with F replaced by $Rj_*j^{-1}F$, which completes the proof by Proposition 5.1.1.

(ii) The proof is similar. Let $\xi_\circ\notin N^*_{x_\circ}(\Omega)^a+(SS(F)\cap\pi^{-1}(x_\circ))$. We may find a closed convex proper cone γ and a neighborhood U of $x_\circ$ such that $\gamma\subset\{v;\langle v,\xi_\circ\rangle<0\}\cup\{0\}$ and $(U\times\gamma^{\circ a})\cap SS(F)\subset T^*_XX$, and we may assume Ω is γ^a-open. Applying Proposition 5.2.1, we may find γ-open subsets Ω_0 and Ω_1 such that $\Omega_0\subset\Omega_1\subset\Omega_0\cup U$, $\Omega_1\setminus\Omega_0$ is a neighborhood of $x_\circ$ and $R\phi_{\gamma*}R\Gamma_{\Omega_1\setminus\Omega_0}(F)=0$.

Set $F' = R\Gamma_{\Omega_1\setminus\Omega_0}(F)$. By replacing Ω by another γ^a-open set which coincides with Ω in a neighborhood of $x_\circ$, we may assume from the beginning that $\Omega \cap (K + \gamma)$ is relatively compact in X for any compact subset K of X. Then we get by Lemma 5.4.3 that $R\phi_{\gamma *}(F'_\Omega) = 0$. This implies $(x_\circ; \xi_\circ) \notin SS(F'_\Omega)$ by Proposition 5.1.1, and F'_Ω is isomorphic to $F_\Omega = Rj_! j^{-1}F$ in a neighborhood of $x_\circ$.

(b) Setting $\Omega = X\setminus Z$, it follows from (a) by applying the triangular inequalities to $R\Gamma_Z(F) \longrightarrow F \longrightarrow Rj_* j^{-1}F \xrightarrow[+1]{}$ and $Rj_! j^{-1}F \longrightarrow F \longrightarrow F_Z \xrightarrow[+1]{}$. □

Corollary 5.4.9. *Let Z be a closed subset of X, $x \in Z$ and assume $N_x^*(Z) \neq T_x^*X$. Let $F \in \mathrm{Ob}(\mathbf{D}^b(X))$ be such that* $SS(F) \cap N_x^*(Z) \subset \{0\}$. *Then $(R\Gamma_Z(F))_x = 0$.*

Proof. We may assume X is affine. Setting $F' = R\Gamma_Z(F)$, we have:

$$SS(F') \cap \pi^{-1}(x) \subset (SS(F) \cap \pi^{-1}(x)) + N_x^*(Z)^a .$$

Since $((SS(F) \cap \pi^{-1}(x)) + N_x^*(Z)^a) \cap N_x^*(Z)$ is contained in $\{0\}$, there exists a closed convex proper cone γ such that $\mathrm{Int}\,\gamma^{\circ a} \cup \{0\}$ contains $N_x^*(Z)$ and

$$(SS(F') \cap \pi^{-1}(x)) \cap \gamma^{\circ a} \subset \{0\} .$$

By Proposition 5.3.7 we may assume Z is γ-closed. By Proposition 5.2.1, there exist γ-open subsets $\Omega_1 \supset \Omega_0$ such that $R\phi_{\gamma *}R\Gamma_{X\setminus\Omega_0}(F')|_{\Omega_1} = 0$ and $\Omega_1\setminus\Omega_0$ is a relatively compact neighborhood of x. Since $N_x^*(Z) \subset \mathrm{Int}\,\gamma^{\circ a} \cup \{0\}$, $\Omega \cap Z$ forms a neighborhood system of x in Z, where Ω ranges through the family of γ-open neighborhoods of x. Therefore

$$\begin{aligned} H^j(F')_x &= \varinjlim_{\Omega} H^j(\Omega; F') \\ &= \varinjlim_{\Omega} H^j(\Omega \cap \Omega_0; F') , \end{aligned}$$

where Ω ranges through the family of γ-open neighborhoods of x. Since $\Omega \cap \Omega_0 \cap Z = \emptyset$ for a γ-open neighborhood Ω of x such that $\Omega \cap Z \subset \Omega_1\setminus\Omega_0$, we get the result. □

Corollary 5.4.10. (i) *For a closed submanifold M of X and $F \in \mathrm{Ob}(\mathbf{D}^b(X))$,*

$$\mathrm{supp}(\mu_M(F)) \subset T_M^*X \cap SS(F) .$$

(ii) *Let F and G belong to $\mathbf{D}^b(X)$. Then:*

$$\mathrm{supp}(\mu hom(G, F)) \subset SS(G) \cap SS(F) .$$

Proof. (i) By Theorem 4.3.2(ii), for $p \in T_M^*X\setminus SS(F)$, $H^j(\mu_M(F))_p \simeq \varinjlim_{Z} H_Z^j(F)_{\pi(p)}$, where Z ranges through the family described there. But we can assume further $N_{\pi(p)}^*(Z) \subset T_M^*X\setminus SS(F)$. Then Corollary 5.4.9 implies $H_Z^j(F)_{\pi(p)} = 0$.

(ii) follows from (i) and Proposition 5.4.2. □

We shall give a bound to $SS(\mu hom(G, F))$ in the next chapter.

Corollary 5.4.11. *Let M be a closed submanifold of X, $F \in \mathrm{Ob}(\mathbf{D}^b(X))$, and assume $SS(F) \cap T^*_M X \subset T^*_X X$. Then:*

(i) $SS(F_M) \subset SS(F) + T^*_M X$,
(ii) *The natural morphism $F_M \otimes \omega_{M/X} \to R\Gamma_M(F)$ is an isomorphism.*

Recall that $\omega_{M/X} = or_{M/X}[\dim M - \dim X]$ is the relative dualizing complex.

Proof. (i) Arguing by induction on codim M, we may assume M is a hypersurface. Since the result is local on M, we also may assume M separates X in two open subsets Ω^+ and Ω^- : $X = \Omega^- \cup M \cup \Omega^+$. Let $j_\pm$ be the open embedding $\Omega_\pm \hookrightarrow X$. We have distinguished triangles:

$$Rj_{-!}j_-^{-1}F \oplus Rj_{+!}j_+^{-1}F \longrightarrow F \longrightarrow F_M \underset{+1}{\longrightarrow} . \tag{5.4.7}$$

Then the result follows from Proposition 5.4.8.

(ii) Let $i: M \hookrightarrow T^*_M X$ be the zero section and $\dot{\pi}: \dot{T}^*_M X \to M$ the projection. Then we have a distinguished triangle:

$$i^! \mu_M(F) \longrightarrow R\pi_* \mu_M(F) \longrightarrow R\dot{\pi}_*(\mu_M(F)|_{\dot{T}^*_M X}) \underset{+1}{\longrightarrow} . \tag{5.4.8}$$

On the other hand, by Theorem 4.3.2, $i^! \mu_M(F) = F_M \otimes \omega_{M/X}$ and $R\pi_* \mu_M(F) = R\Gamma_M(F)$. By Corollary 5.4.10, $\mathrm{supp}(\mu_M(F)) \subset i(M)$ and hence $R\dot{\pi}_*(\mu_M(F)|_{\dot{T}^*_M X}) = 0$. Thus, we obtain (ii). □

Definition 5.4.12. *Let $f: Y \to X$ be a morphism of manifolds, A a closed conic subset of T^*X. We say that f is non characteristic for A if:*

$$f_\pi^{-1}(A) \cap T^*_Y X \subset Y \underset{X}{\times} T^*_X X . \tag{5.4.9}$$

If $F \in \mathrm{Ob}(\mathbf{D}^b(X))$, one says f is non-characteristic for F if f is non-characteristic for $SS(F)$. If f is an embedding, one also says "Y is non-characteristic" instead of "f is non-characteristic".

Recall that $T^*_Y X$ denotes the kernel of ${}^t f': Y \times_X T^*X \to T^*Y$. In particular if f is smooth, f is non-characteristic for any conic subset of T^*X. Arguments similar to those of the proof of Lemma 5.4.7 show that if f is non-characteristic for A, then ${}^t f'(f_\pi^{-1}(A))$ is also a closed conic subset.

Proposition 5.4.13. *Let $F \in \mathrm{Ob}(\mathbf{D}^b(X))$, and assume $f: Y \to X$ is non-characteristic. Then:*

(i) $SS(f^{-1}F) \subset {}^t f'(f_\pi^{-1}(SS(F)))$,
(ii) *The natural morphism $f^{-1}F \otimes \omega_{Y/X} \to f^! F$ is an isomorphism.*

Proof. We decompose f by the graph map:

$$Y \xrightarrow{g} Y \times X \xrightarrow{h} X , \qquad f = h \circ g ,$$

where $g(y) = (y, f(y))$ and h is the second projection on $Y \times X$. Then it is enough to prove the results for h and g separately, and we may assume from the beginning that f is either smooth or is a closed embedding. Then the result follows from Proposition 5.4.5 in the smooth case and from Proposition 5.4.4 and Corollary 5.4.11 in the embedding case. □

Proposition 5.4.14. *Let F and G belong to $\mathbf{D}^b(X)$.*

(i) *Assume* $SS(F) \cap SS(G)^a \subset T^*_X X$. *Then*:

$$SS\left(F \overset{L}{\otimes} G\right) \subset SS(F) + SS(G) .$$

(ii) *Assume* $SS(F) \cap SS(G) \subset T^*_X X$. *Then*:

$$SS(R\mathscr{H}om(G, F)) \subset SS(F) + SS(G)^a .$$

If moreover G is cohomologically constructible, then $R\mathscr{H}om(G, A_X) \otimes^L F \to R\mathscr{H}om(G, F)$ is an isomorphism.

Proof. Recall that:

$$F \overset{L}{\otimes} G \simeq \delta_X^{-1}\left(F \overset{L}{\boxtimes} G\right) , \tag{5.4.10}$$

$$R\mathscr{H}om(G, F) \simeq \delta_X^! R\mathscr{H}om(q_2^{-1}G, q_1^! F) , \tag{5.4.11}$$

where $\delta_X : X \to X \times X$ is the diagonal embedding. Therefore the results follow from Propositions 5.4.13, 5.4.1, 5.4.2 and Proposition 3.4.4. □

Remark 5.4.15. In Propositions 5.4.8–5.4.13 we have always made a hypothesis of "non-charactericity". This hypothesis will be removed in Chapter VI.

Examples 5.4.16. (i) Let t be the coordinate on $\mathbb{R}$, $f: \mathbb{R} \to \mathbb{R}$ be the map $t \mapsto t^3$, and let $G = A_{\mathbb{R}}$. Then $Rf_* G \simeq A_{\mathbb{R}}$ but $f_\pi {}^t f'^{-1}(T^*_{\mathbb{R}}\mathbb{R})$ contains $T^*_{\{0\}}\mathbb{R}$. Hence the inclusion in Proposition 5.4.4 is not an equality in general. However, if f is holomorphic and finite and if G is $\mathbb{C}$-constructible (see VIII §5), we have the equality (Kashiwara [5]).

(ii) Let $X = \mathbb{R}^2$ with coordinates (t, y), $Y = \{(t, y); t = 0\}$ and let $Z = \{(t, y); t > 0, -t < y \leqslant t\}$. Then:

$$SS(A_Z) \cap \pi^{-1}(0) = \{(\tau, \eta); \tau \leqslant -|\eta|\} , \tag{5.4.12}$$

where (τ, η) denotes the dual coordinates. Thus Y is non-characteristic for the sheaf A_Z. Since $(A_Z)_Y = R\Gamma_Y(A_Z) = 0$, the inclusions in Proposition 5.4.13 are not equalities in general.

Now we shall extend Proposition 5.4.4 to a non proper situation.

It is a well-known fact from Morse theory that if Y is a compact manifold, φ a real C^1-function on Y, and φ has no critical values on an interval $]t_1, t_2[$, then for each j, the cohomology groups $H^j(\{x; \varphi(x) < t\}; A_X)$ are all isomorphic for $t \in]t_1, t_2]$. A similar result holds in the relative case, for any sheaf G on Y, the hypothesis that φ has no critical values (i.e.: $d\varphi(x) \notin T^*_X X$ when $\varphi(x) \in]t_1, t_2[$) being replaced by the hypothesis that $d\varphi(x)$ does not belong to SS(G). More precisely we have the following result.

Let $f\colon Y \to X$ be a morphism of manifolds, and let φ be a real C^1-function on Y. We set:

$$Y_t = \{y \in Y; \varphi(y) < t\}\ , \qquad \tilde{Y}_t = \{y \in Y; \varphi(y) \leqslant t\} \tag{5.4.13}$$

and we denote by j_t (resp. $\tilde{j}_t$) the embedding $Y_t \hookrightarrow Y$ (resp. $\tilde{Y}_t \hookrightarrow Y$), and by f_t (resp. $\tilde{f}_t$) the map $f|_{Y_t} = f \circ j_t$ (resp. $f \circ \tilde{j}_t$). Let $t_\circ \in \mathbb{R}$.

Proposition 5.4.17. *Let $G \in \mathrm{Ob}(\mathbf{D}^b(Y))$, and assume* $\mathrm{supp}(G) \cap \tilde{Y}_t$ *is proper over X for all $t \in \mathbb{R}$.*

(i) *Assume that, for any $y \in Y \setminus \tilde{Y}_{t_\circ}$,* we have:

$$d\varphi(y) \notin \left(\mathrm{SS}(G) + {}^t f'\left(Y \underset{X}{\times} T^*X\right)\right) .$$

Then:

(a) $$Rf_* G \simeq Rf_{t*} j_t^{-1} G \quad \text{for} \quad t > t_\circ\ ,$$

(a′) $$Rf_* G \simeq R\tilde{f}_{t*} \tilde{j}_t^{-1} G \quad \text{for} \quad t \geqslant t_\circ\ ,$$

(b) $$\mathrm{SS}(Rf_* G) \subset f_\pi({}^t f'^{-1}(\mathrm{SS}(G) \cap \pi^{-1}(\tilde{Y}_{t_\circ}))) = f_\pi({}^t f'^{-1}(\mathrm{SS}(G)))\ .$$

(ii) *Assume that, for any $y \in Y \setminus \tilde{Y}_{t_\circ}$,* we have:

$$-d\varphi(y) \notin \left(\mathrm{SS}(G) + {}^t f'\left(Y \underset{X}{\times} T^*X\right)\right) .$$

Then:

(c) $$Rf_! G \simeq Rf_{t!} j_t^{-1} G \quad \text{for} \quad t > t_\circ\ ,$$

(c′) $$Rf_! G \simeq R\tilde{f}_{t*} \tilde{j}_t^! G \quad \text{for} \quad t \geqslant t_\circ\ ,$$

(d) $$\mathrm{SS}(Rf_! G) \subset f_\pi({}^t f'^{-1}(\mathrm{SS}(G) \cap \pi^{-1}(\tilde{Y}_{t_\circ}))) \subset f_\pi({}^t f'^{-1}(\mathrm{SS}(G)))\ .$$

Proof. Let $\tilde{f}$ be the map $(f, \varphi)\colon Y \to X \times \mathbb{R}$, and let q and $\tilde{\varphi}$ denote the first and the second projections defined on $X \times \mathbb{R}$, respectively. Then $\tilde{f}$ is proper on supp(G) and $f = q \circ \tilde{f}$. Let i_t (resp. $\tilde{i}_t$) denote the embedding $X \times]-\infty, t[\hookrightarrow X \times \mathbb{R}$, (resp. $X \times]-\infty, t] \hookrightarrow X \times \mathbb{R}$), and set $q_t = q \circ i_t$, $\tilde{q}_t = q \circ \tilde{i}_t$. Then, setting $\tilde{G} = R\tilde{f}_* G$, we have:

$$Rf_{t*}j_t^{-1}G \simeq Rq_{t*}i_t^{-1}\tilde{G} , \qquad R\tilde{f}_{t*}\tilde{j}_t^{-1}G \simeq Rq_{t*}\tilde{i}_t^{-1}\tilde{G} ,$$

$$Rf_{t!}j_t^{-1}G \simeq Rq_{t!}i_t^{-1}\tilde{G} , \qquad R\tilde{f}_{t*}\tilde{j}_t^!G \simeq R\tilde{q}_{t*}\tilde{i}_t^!\tilde{G} .$$

Therefore, it is enough to prove the result for $\tilde{G}$, q, i_t, $\tilde{\varphi}$ instead of G, f, j_t, φ.

(i) The hypothesis implies:

$$\text{(5.4.14)} \qquad SS(\tilde{G}) \cap (T^*X \times \{(t;\tau); t > t_\circ\}) \subset T^*X \times \{(t;\tau); \tau \leqslant 0\} .$$

Let U be an open subset of X contained in a local chart, and convex in such a chart. By Proposition 5.2.3 and Corollary 3.5.5, we have an isomorphism for all $t > t_\circ$:

$$\text{(5.4.15)} \qquad R\Gamma(U \times \mathbb{R}; \tilde{G}) \xrightarrow{\sim} R\Gamma(U \times]-\infty, t[; \tilde{G}) .$$

In order to see this, it is enough to show the isomorphism $R\Gamma(U \times]t_\circ, \infty[; \tilde{G}) \xrightarrow{\sim} R\Gamma(U \times]t_\circ, t[; \tilde{G})$. By an isomorphism $]t_\circ, \infty[\simeq \mathbb{R}$, we may apply the propositions mentioned above. This proves (a).

Since q is proper on $\mathrm{supp}(i_{t*}i_t^{-1}\tilde{G})$, we get by Proposition 5.4.4:

$$\text{(5.4.16)} \quad SS(Rq_*Ri_{t*}i_t^{-1}\tilde{G}) \subset \{(x;\xi); (x,t';\xi,0) \in SS(Ri_{t*}i_t^{-1}\tilde{G}) \text{ for some } t'\} .$$

Since $SS(Ri_{t*}i_t^{-1}\tilde{G})$ is contained in $SS(\tilde{G}) + (T_X^*X \times \{(t;\tau); \tau \leqslant 0\})$, (Proposition 5.4.8), (5.4.16) and (5.4.14) imply for $t > t_\circ$:

$$SS(Rq_*Ri_{t*}i_t^{-1}\tilde{G}) \subset \{(x;\xi); (x,t';\xi,0) \in SS(\tilde{G}) \text{ for some } t'\}$$

which gives (b).

To get (a'), it is enough to note that, for any compact subset K of X,

$$H^k(K; R\tilde{q}_{t*}\tilde{i}_t^{-1}\tilde{G}) \xrightarrow{\sim} \varinjlim_{t'>t} H^k(K; Rq_{t'*}i_{t'}^{-1}\tilde{G}) .$$

(ii) The proof is similar, replacing "$\tau \leqslant 0$" by "$\tau \geqslant 0$" in (5.4.14) and replacing (5.4.15) by:

$$R\Gamma_{U\times]-\infty,t]}(U \times \mathbb{R} : \tilde{G}) \simeq R\Gamma_{U\times]-\infty,t']}(U \times \mathbb{R} : \tilde{G})$$

for any t, t' with $t_\circ \leqslant t \leqslant t'$. □

Remark 5.4.18. The conditions in Proposition 5.4.17 mean that f is relatively non-characteristic in the following sense. Let $T^*(Y/X)$ be the relative cotangent bundle, i.e., the cokernel of ${}^tf' : Y \times_X T^*X \to T^*Y$, and let $p : T^*Y \to T^*(Y/X)$ be the projection. Then the condition in Proposition 5.4.17 is written as $\pm p(d\varphi(y)) \notin p(SS(G))$ for any $y \in Y \setminus Y_{t_\circ}$.

As a particular case of Proposition 5.4.17, we state a useful result.

Corollary 5.4.19 (the microlocal Morse lemma). *Let* $F \in \mathrm{Ob}(\mathbf{D}^b(X))$ *and let* $\phi: X \to \mathbb{R}$ *be a* C^1*-function such that* $\phi: \mathrm{supp}(F) \to \mathbb{R}$ *is proper. Let* $a, b \in \mathbb{R}$ *with* $a < b$.

(i) *Assume* $d\phi(x) \notin SS(F)$ *for any* $x \in X$ *such that* $a \leqslant \phi(x) < b$. *Then the natural morphisms:*

$$R\Gamma(\phi^{-1}(]-\infty, b[); F) \to R\Gamma(\phi^{-1}(]-\infty, a]); F)$$
$$\to R\Gamma(\phi^{-1}(]-\infty, a[); F)$$

are isomorphisms.

(ii) *Assume* $-d\phi(x) \notin SS(F)$ *for any* $x \in X$ *such that* $a < \phi(x) \leqslant b$ *(resp.* $a \leqslant \phi(x) < b$*). Then the natural morphism:*

$$R\Gamma_{\phi^{-1}(]-\infty, a])}(X; F) \to R\Gamma_{\phi^{-1}(]-\infty, b])}(X; F)$$

(resp. $R\Gamma_c(\phi^{-1}(]-\infty, a[); F) \to R\Gamma_c(\phi^{-1}(]-\infty, b[); F)$*), is an isomorphism.*

As an easy application of the results obtained above, we shall prove the **Morse inequalities** for sheaves. For that purpose we assume, until the end of §4, that the base ring A is a field, and we denote it by k. If $V \in \mathrm{Ob}(\mathbf{D}^b(\mathfrak{Mod}^f(k))$, we set as in Exercise I.34:

$$\mathrm{b}_j(V) = \dim H^j(V) , \tag{5.4.17}$$

$$\mathrm{b}_l^*(V) = (-1)^l \sum_{j \leqslant l} (-1)^j \mathrm{b}_j(V) . \tag{5.4.18}$$

Now let $F \in \mathrm{Ob}(\mathbf{D}^b(X))$ and let $\phi: X \to \mathbb{R}$ be a C^∞-function. We set:

$$\Lambda_\phi = \{(x; d\phi(x)); x \in X\} . \tag{5.4.19}$$

Note that Λ_ϕ is a (in general non conic) Lagrangian submanifold of T^*X.

Proposition 5.4.20. *We make the hypotheses* (i), (ii), (iii) *below:*

(i) *for all* $t \in \mathbb{R}$, $\{x \in \mathrm{supp}(F); \phi(x) \leqslant t\}$ *is compact,*
(ii) *the set* $\Lambda_\phi \cap SS(F)$ *is finite, say* $\{p_1, \ldots, p_N\}$,
(iii) *setting* $x_i = \pi(p_i)$, $V_i = (R\Gamma_{\{\phi \geqslant \phi(x_i)\}}(F))_{x_i}$ *belongs to* $\mathrm{Ob}(\mathbf{D}^b(\mathfrak{Mod}^f(k)))$ *for all* $i = 1, \ldots, N$.

Then:

(a) $R\Gamma(X; F)$ *belongs to* $\mathbf{D}^b(\mathfrak{Mod}^f(k))$,
(b) *setting* $\mathrm{b}_j = \mathrm{b}_j(R\Gamma(X; F))$, $\mathrm{b}_j^* = \mathrm{b}_j^*(R\Gamma(X; F))$, $\mathrm{b}_j' = \sum_i \mathrm{b}_j(V_i)$ *and* $\mathrm{b}_j'^* = \sum_i \mathrm{b}_j^*(V_i)$ *we have:*

$$\mathrm{b}_l^* \leqslant \mathrm{b}_l'^* \quad \textit{for all } l , \tag{5.4.20}$$

(5.4.21) $$\sum_j (-1)^j b_j = \sum_j (-1)^j b'_j \ .$$

Proof. Note first that the hypothesis (i) implies that ϕ is proper on supp(F). Set $G = R\phi_* F$ and let t be the coordinate on $\mathbb{R}$. Let $\phi(\{x_1, \ldots, x_N\}) = \{t_1, \ldots, t_L\}$, with $t_i < t_{i+1}$ for all i. Then:

(i)′ For all $t \in \mathbb{R}$, $]-\infty, t[\, \cap \operatorname{supp}(G)$ is compact.
(ii)′ $SS(G) \cap \{(t; dt); t \in \mathbb{R}\}$ is contained in $\bigcup_{i=1}^L \{(t_i; dt)\}$ (this follows from Proposition 5.4.4).
(iii)′ $(R\Gamma_{\{t \geqslant t_i\}}(G))_{t_i} \simeq \bigoplus_{\phi(x_j)=t_i} V_j$.

In fact the left-hand side is isomorphic to $(R\Gamma_{\{x;\, \phi(x) \geqslant t_i\}}(F))|_{\phi^{-1}(t_i)}$ hence to $\bigoplus_{\phi(x_j)=t_i} (R\Gamma_{\{\phi(x) \geqslant t_i\}}(F))_{x_j}$, by the definition of the micro-support. Since $R\Gamma(X; F) = R\Gamma(\mathbb{R}; G)$, and G satisfies the same hypotheses as F, we may assume from the beginning that $X = \mathbb{R}$ and ϕ is the identity.

Now set $t_\circ = -\infty$, $t_{L+1} = +\infty$, $I_t =]-\infty, t[$, $Z_t =]-\infty, t]$ and write $I_j = I_{t_j}$, $Z_j = Z_{t_j}$, for short.

By the non-characteristic deformation lemma (Proposition 2.7.2), we have the isomorphisms:

$$R\Gamma(I_{j+1}; F) \xrightarrow{\sim} R\Gamma(I_t; F) \ , \qquad t_j < t \leqslant t_{j+1} \ .$$

Taking the cohomology and the inductive limit for $t > t_j$, we get:

(5.4.22) $$H^k(I_{j+1}; F) \simeq H^k(Z_j; F) \ .$$

Consider the distinguished triangles:

(5.4.23) $$(R\Gamma_{\{t \geqslant t_j\}}(F))_{t_j} \longrightarrow R\Gamma(Z_j; F) \longrightarrow R\Gamma(I_j; F) \xrightarrow[+1]{} \ .$$

Since $R\Gamma(I_1; F) = 0$, we get by induction, in view of (5.4.22), that both $R\Gamma(Z_j; F)$ and $R\Gamma(I_j; F)$ belong to $\mathbf{D}^b(\mathfrak{Mod}^f(k))$. This proves (a).

Set $b_j(Z_i) = b_j(R\Gamma(Z_i; F))$ and $b_j(I_i) = b_j(R\Gamma(I_i; F))$, for short. Applying the result of Exercise I.34, we get by (5.4.23):

$$b_l'^* \geqslant \sum_{i=1}^L (b_l^*(Z_i) - b_l^*(I_i)) \ .$$

Since

$$b_j = \sum_{i=1}^L (b_j(Z_i) - b_j(I_i))$$

by (5.4.22), we get (5.4.20).

Finally (5.4.21) follows from (5.4.20), since $b_l^* = -b_{l-1}^*$ and $b_l'^* = -b_{l-1}'^*$ for $l \gg 0$. □

Concerning the classical Morse inequalities, cf. Milnor [1]. Note that $\sum_j (-1)^j b_j$ is the **Euler-Poincaré index** of F on X. We shall come back to this index in Chapter IX.

5.5. Micro-support of conic sheaves

In this section we study the micro-support of sheaves on vector bundles.

Let $\tau : E \to Z$ be a real vector bundle over a manifold Z (cf. III §7). Let $\mu : \mathbb{R}^+ \times E \to E$ denote the action map of $\mathbb{R}^+$ on E, and let $\mathbf{e}$ denote the vector field on E describing the infinitesimal action of μ. Hence for any function φ on E we have $(\mathbf{e}(\varphi))(x) = \frac{\partial}{\partial t}\varphi(\mu(t,x))|_{t=1}$. The vector field $\mathbf{e}$ is often referred to as **the Euler vector field**. The map μ defines a map ${}^t\mu' : \mathbb{R}^+ \times T^*E \to T^*\mathbb{R}^+ \times T^*E$. When considering the restriction of ${}^t\mu'$ to $1 \in \mathbb{R}^+$ and composing with the projection $T^*\mathbb{R}^+ \times T^*E \to T^*\mathbb{R}^+ \simeq \mathbb{R}^+ \times \mathbb{R} \to \mathbb{R}$, we obtain the map:

(5.5.1) $$\theta_E : T^*E \to \mathbb{R} .$$

This map is nothing but the principal symbol of $\mathbf{e}$.

Let (z, x) be a system of coordinates (local on Z), such that (z) are coordinates of Z and (x) are linear coordinates. Let $(z, x; \zeta, \xi)$ denote the associated coordinates of T^*E. Thus:

(5.5.2) $$\begin{cases} \mathbf{e} = \left\langle x, \dfrac{\partial}{\partial x}\right\rangle = \displaystyle\sum_j x_j \frac{\partial}{\partial x_j} , \\ \theta_E = \langle x, \xi \rangle , \\ \alpha_E = \langle \zeta, dz \rangle + \langle \xi, dx \rangle . \end{cases}$$

Here, α_E is the fundamental 1-form on T^*E (cf. A§2).

Let us recall that $T^*(E/Z)$ denotes the relative cotangent bundle of E over Z. It is defined by the exact sequence of vector bundles over E:

(5.5.3) $$0 \to E \underset{Z}{\times} T^*Z \to T^*E \to T^*(E/Z) \to 0 .$$

For any $z \in Z$, the fiber $T^*(E/Z)_z$ of $T^*(E/Z)$ over z can be identified with the cotangent bundle $T^*(E_z)$ of the fiber E_z of E over z. On the other hand, $T^*(E_z)$ is isomorphic to $E_z \times E_z^*$. Therefore, for each z, we obtain a map from $T^*(E/Z)_z$ to E_z^*. This gives a map

(5.5.4) $$T^*(E/Z) \to E^* .$$

We then define a morphism over Z:

(5.5.5) $$\psi_E : T^*E \to E^*$$

as the composite of $T^*E \to T^*(E/Z) \to E^*$.

Proposition 5.5.1. (i) *There exists a unique map $\Phi_E : T^*E \to T^*E^*$ such that $\alpha_E - d\theta_E = \Phi_E^*(\alpha_{E^*})$ and that the composition $T^*E \xrightarrow[\Phi_E]{} T^*E^* \xrightarrow[\alpha]{} E^*$ is ψ_E.*

(ii) $\theta_{E^*} \circ \Phi_E = -\theta_E$.

(iii) $\Phi_{E^*} \circ \Phi_E = a^*$, *where* a^* *is the automorphism of* T^*E *induced by the antipodal map on* E.

Proof. Let (z, x) be a system of local coordinates on E as above, and let (z, y) denote the dual coordinates on E^*. Hence the canonical pairing of E and E^* over Z is given by $E \times_Z E^* \ni (z, x, y) \mapsto \langle x, y\rangle = \sum_j x_j y_j$. Let $(z, x; \zeta, \xi)$ (resp. $(z, y; \zeta, \eta)$) be the associated coordinates on T^*E (resp. T^*E^*). The map ψ_E is given by: $(z, x; \zeta, \xi) \mapsto (z; \xi)$, and we have:

$$\alpha_E - d\theta_E = \langle \zeta, dz\rangle - \langle x, d\xi\rangle \ .$$

Hence the map

$$\Phi_E : (z, x; \zeta, \xi) \mapsto (z, \xi; \zeta, -x) \tag{5.5.6}$$

satisfies the desired conditions. The map Φ_E is uniquely determined by these conditions. In fact $\pi \circ \Phi_E = \psi_E$ implies:

$$\Phi_E(z, x, \zeta, \xi) = (z, \xi; \varphi_0(z, x, \zeta, \xi), \varphi_1(z, x, \zeta, \xi)) \ ,$$

and the other condition implies:

$$\langle \varphi_0, dz\rangle + \langle \varphi_1, d\xi\rangle = \langle \zeta, dz\rangle - \langle x, d\xi\rangle \ .$$

(ii) follows from the formula (5.5.6).

(iii) Since $\Phi_{E^*}(z, y, \zeta, \eta) = (z, \eta; \zeta, -y)$, we have $\Phi_{E^*} \circ \Phi_E(z, x, \zeta, \xi) = (z, -x, \zeta, -\xi)$. □

Remark 5.5.2. The map Φ_E is a symplectic transformation (i.e.: Φ_E preserves $d\alpha_E$) but not a homogeneous symplectic transformation (i.e. Φ_E does not preserve α_E).

We denote by S_E the characteristic variety of the Euler vector field:

$$S_E = \theta_E^{-1}(0) \ . \tag{5.5.7}$$

If $(z, x; \zeta, \xi)$ are the coordinates on T^*E, then S_E is defined by the equation $\langle x, \xi\rangle = 0$.

Finally note that T^*E is endowed with two actions of $\mathbb{R}^+$, one coming from the vector bundle structure of T^*E over E, the other coming from the vector bundle structure of E over Z. In a system of local coordinates as above, these two actions are described by:

$$\begin{cases} (z, x; \zeta, \xi) \mapsto (z, x; \lambda\zeta, \lambda\xi) \ , & \lambda \in \mathbb{R}^+ \ , \\ (z, x; \zeta, \xi) \mapsto (z, \mu x; \zeta, \mu^{-1}\xi) \ , & \mu \in \mathbb{R}^+ \ . \end{cases} \tag{5.5.8}$$

We shall say that a subset of T^*E is **biconic** if it is invariant under both actions of $\mathbb{R}^+$.

Let us now study conic sheaves on E. Recall (III §7) that the category $\mathbf{D}^b_{\mathbb{R}^+}(E)$ denotes the full subcategory of $\mathbf{D}^b(E)$ consisting of objects F such that for all j, $H^j(F)$ is locally constant on the orbits of the action of $\mathbb{R}^+$.

Proposition 5.5.3. (i) *The category* $\mathbf{D}^b_{\mathbb{R}^+}(E)$ *is the full subcategory of* $\mathbf{D}^b(E)$ *consisting of objects F such that* $SS(F) \subset S_E$.

(ii) *If* $F \in \mathrm{Ob}(\mathbf{D}^b_{\mathbb{R}^+}(E))$, *then* $SS(F)$ *is biconic.*

Proof. (ii) is evident. Let us show (i).

Set $\dot{E} = E \setminus Z$, where Z is identified with the zero-section. It is enough to prove the result on $\dot{E}$. Then it follows from Proposition 5.4.5 since $\dot{E}/\mathbb{R}^+$ is a manifold and $\dot{E} \times_{\dot{E}/\mathbb{R}^+} T^*(\dot{E}/\mathbb{R}^+) = \dot{E} \times_E S_E$. $\square$

The projection $\tau : E \to Z$, and its restriction $\dot{\tau}$ to $\dot{E}$ defines the maps:

(5.5.9)
$$
\begin{array}{ccccc}
T^*E & \underset{{}^t\tau'}{\hookleftarrow} & E \underset{Z}{\times} T^*Z & \underset{\tau_\pi}{\longrightarrow} & T^*Z \\
\uparrow & & \uparrow & & \| \\
T^*\dot{E} & \underset{{}^t\dot{\tau}'}{\hookleftarrow} & \dot{E} \underset{Z}{\times} T^*Z & \underset{\dot{\tau}_\pi}{\longrightarrow} & T^*Z \, .
\end{array}
$$

Moreover the zero-section of E allows us to define the immersions:

(5.5.10)
$$T^*Z \hookrightarrow E \underset{Z}{\times} T^*Z \underset{{}^t\tau'}{\hookrightarrow} T^*E \, .$$

By this, we sometimes regard T^*Z as a subset of T^*E. Note that if A is a biconic closed subset of T^*E, then:

(5.5.11)
$$\tau_\pi {}^t\tau'^{-1}(A) = T^*Z \cap A \, .$$

Proposition 5.5.4. *Let* $F \in \mathrm{Ob}(\mathbf{D}^b_{\mathbb{R}^+}(E))$. *Then*:

(i) $SS(R\tau_*(F)) \subset T^*Z \cap SS(F)$,
(ii) $SS(R\tau_!(F)) \subset T^*Z \cap SS(F)$,
(iii) $SS(R\dot{\tau}_*(F)) \subset \dot{\tau}_\pi {}^t\dot{\tau}'^{-1}(SS(F))$,
(iv) $SS(R\dot{\tau}_!(F)) \subset \dot{\tau}_\pi {}^t\dot{\tau}'^{-1}(SS(F))$.

Proof. Locally on Z, we may choose a system of coordinates $(z;x)$, (x) denoting the fiber coordinates. Then the results follow from Proposition 5.4.17 by setting $Z = X$, $Y = E$, and $Y_t = \{(z,x); |x|^2 < t\}$ in case (i) or (ii), $Y_t = \{(z,x); t^{-1} < |x|^2 < t\}$ in case (iii) or (iv). $\square$

In Chapter III§7, we defined the Fourier-Sato transform $F^\wedge$ of a conic sheaf $F \in \mathrm{Ob}(\mathbf{D}^b_{\mathbb{R}^+}(E))$. Recall that:

$$F^\wedge = Rp_{2*} R\Gamma_P(p_1^{-1}F) \, ,$$

where p_1 and p_2 denote the first and second projection defined on $E \times_Z E^*$, and $P = \{(z, x, y) \in E \times_Z E^*; \langle x, y\rangle \geqslant 0\}$.

Theorem 5.5.5. *Let $F \in \mathrm{Ob}(\mathbf{D}^b_{\mathbb{R}^+}(E))$. Then, identifying T^*E and T^*E^* by the map Φ_E (cf. Proposition 5.5.1), we have:*

$$\mathrm{SS}(F) = \mathrm{SS}(F^\wedge) .$$

Proof. By Theorem 3.7.9, it is enough to prove the inclusion $\mathrm{SS}(F^\wedge) \subset \mathrm{SS}(F)$. We choose local coordinates (z, x) on E, and denote by (z, y) the dual coordinates on E^* and by $(z, x; \zeta, \xi)$ and $(z, y; \zeta, \eta)$ the coordinates on T^*E and T^*E^*, respectively. Then Φ_E is defined by (5.5.6). Let $(z_\circ, x_\circ; \zeta_\circ, \xi_\circ) \notin \mathrm{SS}(F)$, and let us show that $(z_\circ, \xi_\circ; \zeta_\circ, -x_\circ) \notin \mathrm{SS}(F^\wedge)$. In a first step, we assume:

$$(5.5.12) \qquad x_\circ \neq 0 , \qquad \xi_\circ \neq 0 ,$$

$$(5.5.13) \qquad R\Gamma_Z(F) = 0 .$$

Denote by j the embedding $\dot{E} \hookrightarrow E$, by $\tilde{j}$ the embedding $\dot{E} \times_Z E^* \hookrightarrow E \times_Z E^*$, and by $\dot{p}_2$ the projection $\dot{E} \times_Z E^* \to E^*$. Assuming (5.5.13), we have:

$$\begin{aligned} R\Gamma_P(p_1^{-1}F) &\simeq R\Gamma_P(p_1^{-1}Rj_*j^{-1}F) \\ &\simeq R\Gamma_P(R\tilde{j}_*j^{-1}p_1^{-1}F) \simeq R\tilde{j}_*\tilde{j}^{-1}R\Gamma_P(p_1^{-1}F) , \end{aligned}$$

hence:

$$F^\wedge \simeq R\dot{p}_{2*}\tilde{j}^{-1}R\Gamma_P(p_1^{-1}F) .$$

Assume $(z_\circ, \xi_\circ; \zeta_\circ, -x_\circ) \in \mathrm{SS}(F^\wedge)$. Applying Proposition 5.5.4 to p_2, we find some $x_1 \in \dot{E}$ such that $(z_\circ, x_1, \xi_\circ; \zeta_\circ, 0, -x_\circ) \in \mathrm{SS}(R\Gamma_P(p_1^{-1}F))$.

If $(z_\circ, x_1, \xi_\circ) \in \partial P$, the set P has a smooth boundary by (5.5.12) and its conormal at this point is the covector $\xi_\circ\, dx + x_1\, dy$. Since this point does not belong to $\mathrm{SS}(p_1^{-1}F)^a$, we may apply Proposition 5.4.14 and we find ξ_1 such that $(z_\circ, x_1, \xi_\circ; \zeta_\circ, \xi_1, 0) \in \mathrm{SS}(p_1^{-1}F)$, $(z_\circ, x_1, \xi_\circ; 0, \xi_1, x_\circ) \in \mathrm{SS}(A_P)$. Thus $x_\circ = kx_1$, $\xi_1 = k\xi_\circ$ for some $k \geqslant 0$ and $(z_\circ, x_1; \zeta_0, \xi_1) \in \mathrm{SS}(F)$. This is a contradiction because $\mathrm{SS}(F)$ is biconic.

To remove the hypothesis (5.5.13), consider the distinguished triangle:

$$R\Gamma_Z(F) \longrightarrow F \longrightarrow Rj_*j^{-1}F \xrightarrow[+1]{} .$$

Since we may apply the preceding result to $Rj_*j^{-1}F$, it is enough to show that $(z_\circ, \xi_\circ, \zeta_\circ, -x_\circ) \notin \mathrm{SS}(R\Gamma_Z(F)^\wedge)$. But this follows from Proposition 3.7.13. In fact, denote by i the embedding $Z \hookrightarrow E$. We have $R\Gamma_Z(F)^\wedge \simeq (i_! i^! F)^\wedge \simeq \pi^{-1}i^!F$, and the micro-support of $\pi^{-1}i^!F$ is contained in the set $\{(z, y; \zeta, \eta); \eta = 0\}$.

Finally we treat the general case.

Consider the sheaf $A_{\mathbb{R}\times\{0\}}$ on $\mathbb{R}^2$. Applying Propositions 3.7.15 and 3.7.12 we get:

(5.5.14) $$(F \boxtimes A_{\mathbb{R}\times\{0\}})^\wedge \simeq F^\wedge \boxtimes A_{\{0\}\times\mathbb{R}}[-1] .$$

If $(z_o, x_o; \zeta_o, \xi_o) \notin SS(F)$, then $(z_o, x_o, 1, 0; \zeta_o, \xi_o, 0, 1) \notin SS(F \boxtimes A_{\mathbb{R}\times\{0\}})$, thus $(z_o, \xi_o, 0, 1; \zeta_o, -x_o, -1, 0) \notin SS(F^\wedge \boxtimes A_{\{0\}\times\mathbb{R}})$, and this implies $(z_o, \xi_o; \zeta_o, -x_o) \notin SS(F^\wedge)$, in view of Propositions 5.4.4 and 5.4.5. □

Exercises to Chapter V

Exercise V.1. Let X be a manifold, V a closed conic subset of T^*X. One denotes by $\mathfrak{Sh}_V(X)$ (resp. $\mathbf{D}^b_V(X)$) the full subcategory of $\mathfrak{Sh}(X)$ (resp. $\mathbf{D}^b(X)$) consisting of objects F such that $SS(F) \subset V$.

(i) Assume now that X is a complex vector space endowed with linear coordinates $(z) = (z_1, \ldots, z_n)$ and let $(z; \zeta)$ be the coordinates on the complex cotangent bundle $X \times X^*$. We identify $X \times X^*$ with the real cotangent bundle by $\langle \zeta, dz \rangle + \langle \bar{\zeta}, d\bar{z} \rangle$. Let $V = \{(z; \zeta); \sum_j z_j \zeta_j = 0\}$. Prove that $\mathfrak{Sh}_V(X)$ consists of the sheaves which are locally constant on the orbits of the action of $\mathbb{C}^\times$ on X.

(ii) In the situation (i), discuss whether $\delta : \mathbf{D}^b(\mathfrak{Sh}_V(X)) \to \mathbf{D}^b_V(X)$ is an equivalence of categories.

Exercise V.2. Let $F \in Ob(\mathbf{D}^b(X))$, and let φ be a real C^1-function on X. Let $x_o \in X$ with $\varphi(x_o) = 0$, $d\varphi(x_o) \notin SS(F)$. Prove that "$\varinjlim_U$" $R\Gamma_{\{x;\varphi(x)\geq 0\}}(U; F) = 0$, where U ranges through the family of open neighborhoods of x_o in X. (For the ind-object "$\varinjlim$", see I §11.)

Exercise V.3. Let E be a real vector space, $F \in Ob(\mathbf{D}^b_{\mathbb{R}^+}(E))$. Let $\xi_o \in E^*$. Prove that $(0; \xi_o) \notin SS(F)$ if and only if there exists a fundamental system of convex conic open neighborhoods γ of ξ_o such that $R\Gamma_{\gamma^\circ}(E; F) = 0$. (Hint: use Theorem 5.5.5.)

Exercise V.4. Let $C = \{(x, y, t) \in \mathbb{R}^3; x^2 + y^2 = t^2, t > 0\}$ and let j be the embedding $C \hookrightarrow \mathbb{R}^3$. Set $F = Rj_* A_C$.

(i) By using the result of Exercise V.3 prove that $\pi^{-1}(0) \cap SS(F) \neq T^*_{\{0\}}\mathbb{R}^3$.

(ii) Prove that $H^1(F) = A_{\{0\}}$.
Note that we get an example of an $F \in Ob(\mathbf{D}^b(X))$ such that $SS(H^j(F))$ is not contained in $SS(F)$.

Exercise V.5. Let X be a vector space.

(i) Let Z be a closed convex subset of X, and $F = A_Z$. Show that for any $x \in Z$, $SS(F) \cap \pi^{-1}(x) = \gamma^\circ$, where $\gamma = C_x(Z)$.

(ii) Let Ω be an open convex subset of X and $F = A_\Omega$. Show that $SS(A_\Omega) = SS(A_{\bar{\Omega}})^a$.

Exercise V.6. Let $F \in Ob(\mathbf{D}^b(X))$. Prove the inclusion $SS(F) \subset \bigcup_j SS(H^j(F))$.

Exercise V.7. (i) Let $\{F_\lambda\}_\lambda$ be an inductive system of sheaves on X. Prove the inclusion $SS(\varinjlim_\lambda F_\lambda) \subset \overline{\bigcup_\lambda SS(F_\lambda)}$.

(ii) Let $\{F_n\}_{n \in \mathbb{N}}$ be a projective system of sheaves on X. Prove the inclusion $SS(\varprojlim_n F_n) \subset \overline{\bigcup_n SS(F_n)}$.

Exercise V.8. Let V be a finite-dimensional real vector space, Ω an open cone (not necessarily convex) in V^*. Set $G = (A_\Omega)^\wedge \otimes \omega_V$. Let $\mu : V \times V \to V$ be the map $(x, y) \mapsto y - x$ and let $K = \mu^{-1}G$. Recall the functor Φ_K of III §6.

(i) Prove that if $F \in \mathrm{Ob}(\mathbf{D}^b(V))$, then $SS(\Phi_K(F)) \subset V \times \bar{\Omega}$.

(ii) Prove that there exists a natural morphism $\Phi_K(F) \to F$ which is an isomorphism on $V \times \Omega$. (Compare with Proposition 5.2.3.)

Exercise V.9. Let $F \in \mathrm{Ob}(\mathbf{D}^b(X \times \mathbb{R}))$ satisfying $SS(F) \cap (T^*_X X \times T^*\mathbb{R}) \subset T^*_{X \times \mathbb{R}}(X \times \mathbb{R})$. Let $\varphi : \mathbb{R} \to \mathbb{R}$ be the continuous map defined by $\varphi(s) = 0$ for $s \leqslant 0$ and $\varphi(s) = s$ for $s \geqslant 0$. Prove that $SS(\varphi_X^{-1}F) \cap (T^*_X X \times T^*\mathbb{R}) \subset T^*_{X \times \mathbb{R}}(X \times \mathbb{R})$ where $\varphi_X = \mathrm{id}_X \times \varphi$.

(Hint: let $S = \{(t, s) \in \mathbb{R}^2;\ t = s \geqslant 0 \text{ or } t = 0,\ s \leqslant 0\}$. Then $\varphi_X^{-1}F \simeq Rq_{2*}(q_1^{-1}F)_S$, where q_1 and q_2 are the projections from $X \times \mathbb{R}_t \times \mathbb{R}_s$ to $X \times \mathbb{R}_t$ and $X \times \mathbb{R}_s$, respectively. Then it is enough to check that $\pm ds \notin (SS(F) \times T^*_{\mathbb{R}_s}\mathbb{R}_s + SS(A_{X \times S}))$.)

Exercise V.10. Let F_0 and F_1 belong to $\mathbf{D}^b(X)$, F_0 and F_1 both with compact supports. Let us say that F_0 and F_1 are **homotopic** if there exists $H \in \mathrm{Ob}(\mathbf{D}^b(X \times \mathbb{R}))$ such that the projection $p : X \times \mathbb{R} \to \mathbb{R}$ is proper on $\mathrm{supp}(H)$, $SS(H) \cap (T^*_X X \times T^*\mathbb{R}) \subset T^*_{X \times \mathbb{R}}(X \times \mathbb{R})$ and $H|_{X \times \{i\}} \simeq F_i$, $i = 0, 1$.

(i) Prove that if H realizes a homotopy as above, then there exists $\tilde{H}$ satisfying the same hypothesis as H and moreover $\tilde{H}|_{X \times [1,2]} \simeq F_1 \boxtimes A_{[1,2]}$. (Hint: use Exercise V.9.)

(ii) Prove that "homotopy" is an equivalence relation. (Hint: use (i) and Exercise II.22.)

(iii) Prove that if F_0 and F_1 are homotopic, then $D_X F_0$ and $D_X F_1$ are homotopic.

(iv) Prove that if F_0 and F_1 are homotopic, then for any $k \in \mathbb{Z}$, $H^k(X; F_0)$ and $H^k(X; F_1)$ are isomorphic.

(v) For $X = \mathbb{R}$, prove that $A_{[0,1]}$ is homotopic to $A_{\{0\}}$, $A_{[0,1[}$ is homotopic to zero and $A_{]0,1[}$ is homotopic to $A_{\{0\}}[-1]$.

Exercise V.11. (i) Let X be a manifold and $i : Y \hookrightarrow X$ a closed embedding of a submanifold Y. Prove that for $F \in \mathrm{Ob}(\mathbf{D}^b(X))$, $i^!F \to i^{-1}F$ is the zero morphism when $T^*_Y X \setminus SS(F) \to Y$ has a continuous section. (Hint: consider $\mu_Y(i_* i^! F) \to \mu_Y(F) \to \mu_Y(i_* i^{-1} F)$.)

(ii) Using (i) prove again Exercise III.7.

(iii) Prove that, in the situation of Exercise III.8, the composition $H^k(Y; A_Y) \to H^{k+l}(X; A_X) \to H^{k+l}(Y; A_Y)$ is zero if $\dot{T}^*_Y X \to Y$ has a continuous section.

Exercise V.12. Assume the base ring is a commutative field k. Let X be a manifold, ϕ a real C^∞-function on X, and assume:

(i) for all $t \in \mathbb{R}$; $\{x \in X; \phi(x) \leqslant t\}$ is compact,

(ii) the set $\{x; d\phi(x) = 0\}$ is finite, say $\{x_1, \dots, x_N\}$, and at each x_i, the Hessian $H_\phi(x_i)$ is non-degenerate, with ε_i negative eigenvalues.

Prove that $R\Gamma(X; k_X)$ belongs to $\mathbf{D}^b(\mathfrak{Mod}^f(k))$, and:

$$\chi(R\Gamma(X; k_X)) = \sum_i (-1)^{\varepsilon_i} ,$$

where $\chi(\cdot)$ is defined in Exercise I.32.

(Hint: use Proposition 5.4.20, a local version of Exercise III.5 and the "Morse lemma".)

Exercise V.13. Let $F \in \mathrm{Ob}(\mathbf{D}^b(X))$. Prove the formula

$$\mathrm{SS}(\mathrm{D}_X F) = \mathrm{SS}(F)^a$$

assuming condition (i) or (ii) below.

(i) F is cohomologically constructible.

(ii) The base ring A satisfies: if $M \in \mathrm{Ob}(\mathbf{D}^b(\mathfrak{Mod}(A))$ and $R\mathrm{Hom}_A(M, A) = 0$ then $M = 0$.

Observe that (ii) is true if A is either a field or $\mathbb{Z}$ (cf. Exercise I.31). The authors have no examples of a ring A which does not satisfy this property.

Exercise V.14. Set $X = \mathbb{R}$, $B = \{0\} \cup \{1/n; n \in \mathbb{N}\setminus\{0\}\}$. Let $F = G = \mathbb{Q}_B$. Prove that $\nu_{\{(0,0)\}}(F \boxtimes G) \neq \nu_{\{0\}}(F) \boxtimes \nu_{\{0\}}(G)$.

Notes

The notion of micro-support of a complex of sheaves F on a real manifold X was introduced by the authors in 1982 (Kashiwara-Schapira [2, 3]). It can be considered as a generalization of Morse theory (cf. Milnor [1]), as it appears clearly with Corollary 5.4.19, or even of the more recent "stratified Morse theory" of Goresky-MacPherson [2], (this last theory applying only to constructible sheaves, cf. Chapter VIII).

However, our motivation has come from linear partial differential equations, and especially from the study of hyperbolic systems (cf. Kashiwara-Schapira [1]). When studying such systems, one is led to make "non characteristic deformations" of the domain of definition of the solutions. It then appears that one can forget that one is working with differential equations, one can forget the complex structure of X, and the only thing to keep in mind is the characteristic variety of the system in the real cotangent bundle, that is, the micro-support of the complex of solutions of the system (cf. Theorem 11.3.3 below). As we shall show in

Chapter XI, this method is extremely fruitful when applied to the study of microdifferential equations.

The γ-topology was introduced in Kashiwara-Schapira [1] in order to make microdifferential operators operate on holomorphic functions, and Proposition 5.2.1 was proved in that paper (of course in a more restrictive context). The results of this chapter were obtained in Kashiwara-Schapira [3], with the exception of Proposition 5.4.20, which is a variant due to Schapira-Tose [1] of a result of Kashiwara [7]. (This Proposition is a generalization of the classical Morse inequalities, for which we refer to Bott [1].)

Chapter VI. Micro-support and microlocalization

Summary

Let X be a manifold, Ω a subset of T^*X. We define the triangulated category $\mathsf{D}^b(X;\Omega)$ as the localization of $\mathsf{D}^b(X)$ by the full subcategory of objects whose micro-support is disjoint from Ω. Then to work "microlocally" on Ω with a sheaf F on X gets a precise meaning: it simply means to consider F as an object of $\mathsf{D}^b(X;\Omega)$. With this new notion, we introduce the "microlocal inverse image" and the "microlocal direct image". These are pro-objects or ind-objects of the category $\mathsf{D}^b(X;p)$, the localization of $\mathsf{D}^b(X)$ at p, but we give conditions which ensure that one remains in the category $\mathsf{D}^b(X;p)$.

The localization of $\mathsf{D}^b(X)$ is related to the functor μhom by the formula:

(6.0.1) $$\mathrm{Hom}_{\mathsf{D}^b(X;p)}(G,F) = H^0(\mu hom(G,F))_p \ .$$

This formula is an essential step in the proof of Theorem 6.5.4 which asserts that $SS(F)$ is an involutive subset of T^*X.

Before getting the involutivity theorem, we study the micro-support of sheaves after various operations (direct images for an open embedding, microlocalization, etc.), extending the results of the preceding chapter to the characteristic case, or to the non-proper case. In particular we obtain the formula:

(6.0.2) $$SS(\mu hom(G,F)) \subset C(SS(F), SS(G)) \ .$$

This formulation makes use of normal cones in cotangent bundles that we study in §2.

Next we characterize "microlocally" sheaves whose micro-support is contained in an involutive submanifold. In particular, we show that if $SS(F)$ is contained in the conormal bundle to a submanifold Y of X, then F is microlocally isomorphic to the sheaf L_Y, for some A-module L.

Finally we investigate the case when the functors of inverse image and that of microlocalization commute, and obtain a sheaf-theoretical version of a result on the Cauchy problem for micro-hyperbolic systems.

6.1. The category $\mathbf{D}^b(X;\Omega)$

Let V be a subset of T^*X. We define the full subcategory $\mathbf{D}^b_V(X)$ of $\mathbf{D}^b(X)$ by setting:

$$\mathrm{Ob}(\mathbf{D}^b_V(X)) = \{F \in \mathrm{Ob}(\mathbf{D}^b(X)); \mathrm{SS}(F) \subset V\} \ . \tag{6.1.1}$$

This is a triangulated category, the distinguished triangles in $\mathbf{D}^b_V(X)$ being those of $\mathbf{D}^b(X)$ whose objects belong to $\mathbf{D}^b_V(X)$, and $\mathrm{Ob}(\mathbf{D}^b_V(X))$ is a null system in $\mathbf{D}^b(X)$. Hence we can apply the results of I §6 and localize $\mathbf{D}^b(X)$ with respect to $\mathrm{Ob}(\mathbf{D}^b_V(X))$.

Definition 6.1.1. *Let $\Omega = T^*X\setminus V$.*

(i) *We set:*

$$\mathbf{D}^b(X;\Omega) = \mathbf{D}^b(X)/\mathrm{Ob}(\mathbf{D}^b_V(X)) \ .$$

(ii) *Let F and G belong to $\mathbf{D}^b(X)$. If they are isomorphic in $\mathbf{D}^b(X;\Omega)$, we shall say that F and G are isomorphic on Ω.*

For $p \in T^*X$, we write $\mathbf{D}^b(X;p)$ instead of $\mathbf{D}^b(X;\{p\})$. Recall that $\mathrm{Ob}(\mathbf{D}^b(X;\Omega)) = \mathrm{Ob}(\mathbf{D}^b(X))$ and a morphism $u: G \to F$ in $\mathbf{D}^b(X;\Omega)$ is given by the data of two morphisms $v: G \to F'$ and $w: F \to F'$ in $\mathbf{D}^b(X)$ such that w is an isomorphism on Ω, or else by the data of two morphisms $w': G' \to G$ and $v': G' \to F$ in $\mathbf{D}^b(X)$, such that w' is an isomorphism on Ω. More precisely:

$$\begin{cases} \mathrm{Hom}_{\mathbf{D}^b(X;\Omega)}(G,F) = \varinjlim \mathrm{Hom}_{\mathbf{D}^b(X)}(G,F') \\ \qquad\qquad\qquad\qquad\quad = \varinjlim \mathrm{Hom}_{\mathbf{D}^b(X)}(G',F) \ , \end{cases} \tag{6.1.2}$$

where the inductive limit in the first (resp. second) line of (6.1.2) is taken over the category of morphisms $w: F \to F'$ (resp. $w : G' \to G$), such that w is an isomorphism on Ω, (cf. Exercise I.14). Note that if $F \in \mathrm{Ob}(\mathbf{D}^b(X;\Omega))$, then $\mathrm{SS}(F)$ is well-defined in Ω.

In IV §4, we have introduced the bifunctor μhom from $\mathbf{D}^b(X)^\circ \times \mathbf{D}^b(X)$ to $\mathbf{D}^b(T^*X)$. If the micro-support of F or G does not intersect Ω, then $\mu hom(G,F)|_\Omega = 0$ in view of Corollary 5.4.10. Thus $\mu hom(\cdot,\cdot)|_\Omega$ is a well-defined functor from $\mathbf{D}^b(X;\Omega)^\circ \times \mathbf{D}^b(X;\Omega)$ to $\mathbf{D}^b(\Omega)$. We shall still denote this functor by μhom.

$$\mu hom: \mathbf{D}^b(X;\Omega)^\circ \times \mathbf{D}^b(X;\Omega) \to \mathbf{D}^b(\Omega) \ . \tag{6.1.3}$$

The isomorphism $\mathrm{Hom}_{\mathbf{D}^b(X)}(G, F) \simeq H^0(T^*X; \mu hom(G, F))$, (Proposition 4.4.2), defines the morphism:

$$\mathrm{Hom}_{\mathbf{D}^b(X;\Omega)}(G,F) \to H^0(\Omega; \mu hom(G,F)) \ . \tag{6.1.4}$$

In fact if $w : F \to F'$ is a morphism in $\mathbf{D}^b(X)$ and w is an isomorphism on Ω, then w induces an isomorphism $\mu hom(G,F)|_\Omega \xrightarrow{\sim} \mu hom(G,F')|_\Omega$.

In general the morphism (6.1.4) is not an isomorphism (cf. Exercise VI 6). However:

Theorem 6.1.2. *Let $p \in T^*X$ and let F and G belong to $\mathbf{D}^b(X)$. Then the natural morphism* $\mathrm{Hom}_{\mathbf{D}^b(X;p)}(G,F) \to H^0(\mu hom(G,F))_p$ *is an isomorphism.*

Proof. If $p \in T^*_X X$, there is nothing to prove.

Assume X is a vector space and $p = (x_\circ;\xi_\circ) \in \dot{T}^*X$. With the same notations as for Proposition 4.4.4 we have:

$$(6.1.5) \qquad H^0(\mu hom(G,F))_p \simeq \varinjlim_{U,\gamma} H^0(R\Gamma(U;R\mathcal{H}om(\phi_\gamma^{-1}R\phi_{\gamma *}G_U,F)))$$

$$\simeq \varinjlim_{U,\gamma} \mathrm{Hom}((\phi_\gamma^{-1}R\phi_{\gamma *}G_U)_U,F) \ .$$

Let α be the morphism of Theorem 6.1.2. First we show that α is injective. Let $u \in \mathrm{Hom}_{\mathbf{D}^b(X)}(G,F)$, with $\alpha(u) = 0$. There exist U, γ such that the composite morphism $(\phi_\gamma^{-1}R\phi_{\gamma *}G_U)_U \to G \underset{u}{\to} F$ is zero. Since $(\phi_\gamma^{-1}R\phi_{\gamma *}G_U)_U \to G$ is an isomorphism in $\mathbf{D}^b(X;p)$, (Proposition 5.2.3), we get $u = 0$ in $\mathbf{D}^b(X;p)$. Let us show that α is surjective. If $v \in H^0(\mu hom(G,F))_p$, there exist U, γ and $w \in \mathrm{Hom}_{\mathbf{D}^b(X)}((\phi_\gamma^{-1}R\phi_{\gamma *}G_U)_U,F)$ which represents v. Since $(\phi_\gamma^{-1}R\phi_{\gamma *}G_U)_U \to G$ is an isomorphism in $\mathbf{D}^b(X;p)$, w defines an element of $\mathrm{Hom}_{\mathbf{D}^b(X;p)}(G,F)$ whose image by α is v. □

The following corollary is an important step in the proof of the involutivity of micro-supports.

Let $F \in \mathrm{Ob}(\mathbf{D}^b(X))$. Consider the natural morphisms:

$$\mathrm{Hom}(F,F) \xrightarrow{\sim} H^0(R\Gamma(X;R\mathcal{H}om(F,F)))$$

$$\xrightarrow{\sim} H^0(R\Gamma(T^*X;\mu hom(F,F)))$$

$$\to \Gamma(T^*X;H^0(\mu hom(F,F))) \ .$$

Denote by s the image of $\mathrm{id}_F \in \mathrm{Hom}(F,F)$ in $\Gamma(T^*X;H^0(\mu hom(F,F)))$.

Corollary 6.1.3. *One has:*

$$\mathrm{supp}(s) = \mathrm{supp}(\mu hom(F,F)) = SS(F) \ .$$

In particular, if $p \in T^*X$, one has:

$$(6.1.6) \qquad p \notin SS(F) \Leftrightarrow \mu hom(F,F)_p = 0 \ .$$

Proof. The inclusion $\mathrm{supp}(s) \subset \mathrm{supp}(\mu hom(F,F))$ is clear and the inclusion $\mathrm{supp}(\mu hom(F,F)) \subset SS(F)$ follows from Corollary 5.4.10.

Let $p \in T^*X$, $p \notin \mathrm{supp}(s)$. Then $s_p = 0$ implies that $\mathrm{id}_F \in \mathrm{Hom}_{\mathbf{D}^b(X;p)}(F, F)$ vanishes, by Theorem 6.1.2. Hence $F = 0$ in $\mathbf{D}^b(X; p)$, which means that $p \notin SS(F)$. □

In the sequel we shall discuss the microlocal operations on sheaves such as inverse images and direct images. For that purpose we need to refine the microlocal cut-off lemma of V §2.

Proposition 6.1.4 (the refined microlocal cut-off lemma). *Let $x_\circ$ be a point of a manifold X, K a proper closed convex cone of $T^*_{x_\circ}X$ and $U \subset K$ an open cone. Let $F \in \mathrm{Ob}(\mathbf{D}^b(X))$ and let W be a conic neighborhood of $K \cap SS(F) \setminus \{0\}$. Then there exist $F' \in \mathrm{Ob}(\mathbf{D}^b(X))$ and a morphism $u: F' \to F$ satisfying the following conditions:*

(6.1.7) $$u \text{ is an isomorphism on } U ,$$

(6.1.8) $$\pi^{-1}(x_\circ) \cap SS(F') \subset W \cup \{0\} .$$

Proof. By thickening K, we may assume $\{0\} \cup \mathrm{Int}\, K \supset \bar{U}$. We may also assume X is a vector space and $x_\circ = 0$. Let us take a closed proper convex cone γ such that $K^{\circ a} \subset \gamma \subset U^{\circ a}$ and $\gamma \setminus \{0\}$ has a C^1-boundary $\delta\gamma$. We take coordinates $(x) = (x_1, \ldots, x_n)$ on X such that:

(6.1.9) $$\gamma \subset \{0\} \cup \{x; x_1 < 0\} .$$

Let $(x; \xi)$ be the associated coordinates on T^*X. Since $\gamma^{\circ a} \subset K$, there exists an open neighborhood Ω of 0 such that:

(6.1.10) $$W \supset G = \{\xi \in \gamma^{\circ a} \setminus \{0\}; \text{there exists } (x; \xi) \in \pi^{-1}(\Omega) \cap SS(F)\} .$$

Take $\varepsilon > 0$ such that:

(6.1.11) $$\Omega \supset \{x \in \gamma; x_1 \geqslant -\varepsilon\} .$$

Now we shall take a closed subset Z of X with C^1-boundary δZ satisfying the following conditions (cf. Figure 6.1.1):

(6.1.12) $$0 \in \mathrm{Int}\, Z ,$$

(6.1.13) $$Z \subset \{x; x_1 \geqslant -\varepsilon\} ,$$

(6.1.14) $$\begin{cases} \delta Z \text{ and } \delta\gamma \text{ are tangent at their intersection,} \\ \text{more precisely for any } x \in \delta Z \cap \delta\gamma,\ N^*_x(Z)^a = N^*_x(\gamma) . \end{cases}$$

The construction of Z is left to the reader.

Now we set $F' = \phi_\gamma^{-1} R\phi_{\gamma *} R\Gamma_Z(F)$ and we denote by u the canonical morphism $F' \to F$. We shall prove that F' and u satisfy the desired properties. By the microlocal cut-off lemma (Proposition 5.2.3), u is an isomorphism on $\mathrm{Int}\, Z \times$

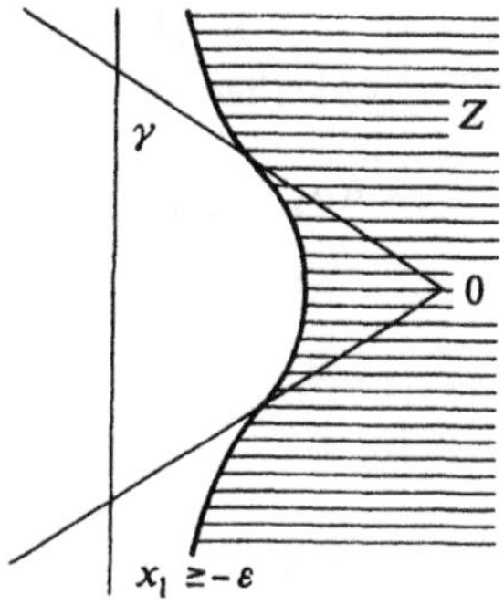

Fig. 6.1.1

$\operatorname{Int}\gamma^{\circ a}$. Since $\operatorname{Int}\gamma^{\circ a} \supset U$, (6.1.7) is satisfied. The same proposition implies $SS(F') \subset X \times \gamma^{\circ a}$ and hence it remains to prove:

(6.1.15) $\quad$ if $\xi \in \delta(\gamma^{\circ a})\backslash\{0\}$ and $(0;\xi) \in SS(F')$ then $\xi \in G$.

Let $q_i : X \times X \to X$ denote the i-th projection ($i = 1, 2$) and let $s : X \times X \to X$ be the map $s(x, x') = x - x'$. By Proposition 3.5.4 we have:

(6.1.16)
$$F' \simeq Rq_{2*}(s^{-1}A_\gamma \otimes q_1^{-1}R\Gamma_Z(F)) .$$

By the conditions (6.1.9) and (6.1.13), $q_2 : s^{-1}(\gamma) \cap q_1^{-1}Z \to X$ is proper, and we can apply Propositions 5.4.4, 5.4.5 and 5.4.14 to calculate $SS(F')$. We obtain:

(6.1.17)
$$\begin{cases} \text{if } (0;\xi) \in SS(F'), \text{then there exists} \\ (x;\xi) \in SS(A_\gamma)^a \cap SS(R\Gamma_Z(F)) . \end{cases}$$

Then by (6.1.11) and (6.1.13), x is contained in Ω. Hence the proof of (6.1.15) is reduced to that of:

(6.1.18)
$$\begin{cases} \text{if } \xi \neq 0 \text{ and } (x;\xi) \in SS(A_\gamma)^a \cap SS(R\Gamma_Z(F)) , \\ \text{then } (x;\xi) \in SS(F) . \end{cases}$$

Since $\xi \neq 0$ and $(x;\xi) \in SS(A_\gamma)^a$, we have $x \in \delta\gamma$. If $x \in \operatorname{Int} Z$, then $F \simeq R\Gamma_Z(F)$ at x, and hence $(x;\xi) \in SS(F)$. If $x \in \delta Z$, then by (6.1.14), $N_x^*(Z) = N_x^*(\gamma)^a = \mathbb{R}_{\geqslant 0}\xi$. Assume $(x;\xi) \notin SS(F)$. Then Proposition 5.4.8 shows that $\xi \in SS(R\Gamma_Z(F)) \cap \pi^{-1}(x) \subset -\mathbb{R}_{\geqslant 0}\xi + (SS(F) \cap \pi^{-1}(x))$, which implies $(x;\xi) \in SS(F)$. This is a contradiction, and this completes the proof. □

We shall need a dual statement to the preceding one. For that purpose we begin with a dual statement to Proposition 5.2.3.

Lemma 6.1.5 (The dual microlocal cut-off lemma). *Let γ be a closed convex proper cone in a vector space X, with* $\operatorname{Int}\gamma \neq \emptyset$. *Let $s : X \times X \to X$ be the map*

$(x, x') \mapsto x - x'$ *and let* $q_i : X \times X \to X$ *be the i-th projection* $(i = 1, 2)$. *Then for* $F \in \mathrm{Ob}(\mathbf{D}^b(X))$ *with compact support, there is a natural morphism*:

$$u : F \to Rq_{2!}R\Gamma_{s^{-1}(\gamma^a)}(q_1^! F) \ , \tag{6.1.19}$$

such that:

(i) $SS(Rq_{2!}R\Gamma_{s^{-1}(\gamma^a)}(q_1^! F)) \subset X \times \gamma^{\circ a}$,
(ii) *u is an isomorphism on* $X \times \mathrm{Int}\,\gamma^{\circ a}$.

Proof. We have a chain of morphisms:

$$F \simeq Rq_{2*}R\mathcal{H}om(A_{s^{-1}(0)}, q_1^! F) \to Rq_{2*}R\mathcal{H}om(s^{-1}A_{\gamma^a}, F) \ .$$

Since $A_{\gamma^a} \simeq \phi_\gamma^{-1} R\phi_{\gamma *} A_{\{0\}}$, we have by Proposition 5.2.3:

$$SS(A_{\gamma^a}) \subset X \times \gamma^{\circ a} \ , \tag{6.1.20}$$

$$A_{\gamma^a} \to A_{\{0\}} \text{ is an isomorphism on } X \times \mathrm{Int}\,\gamma^{\circ a} \ . \tag{6.1.21}$$

Then Propositions 5.4.5, 5.4.14 and 5.4.4 imply $SS(Rq_{2*}R\mathcal{H}om(s^{-1}A_{\gamma^a}, F)) \subset X \times \gamma^{\circ a}$. This shows (i).

On the other hand, (6.1.21) implies $SS(A_{\gamma \setminus \{0\}}) \subset X \times (X^* \setminus \mathrm{Int}\,\gamma^{\circ a})$. Hence, again by Propositions 5.4.5, 5.4.14 and 5.4.4 we get $SS(Rq_{2*}R\mathcal{H}om(s^{-1}A_{\gamma^a \setminus \{0\}}, F) \subset X \times (X^* \setminus \mathrm{Int}\,\gamma^{\circ a})$. This shows (ii). □

Proposition 6.1.6 (The dual refined microlocal cut-off lemma). *Let* $x_\circ$, *K*, *U*, *F* *and W be as in Proposition* 6.1.4. *Then there exists a morphism* $u : F \to F'$ *satisfying the conditions* (6.1.7) *and* (6.1.8).

Proof. We may assume F has compact support. The proof is similar to that of Proposition 6.1.4, except that we make use of the dual cut-off functor $Rq_{2!}R\Gamma_{s^{-1}(\gamma^a)}(q_1^! F_Z)$ instead of $\phi_\gamma^{-1} R\phi_{\gamma *}\,(R\Gamma_Z(F)$. More precisely, we take γ, Ω, G as in the former proof and we take $\varepsilon > 0$ such that

$$\Omega \supset \{x \in \gamma^{\circ a}; x_1 \leqslant \varepsilon\} \ . \tag{6.1.11$'$}$$

Similarly, we take a closed subset Z with C^1-boundary δZ, satisfying (6.1.12) and the following conditions:

$$Z \subset \{x; x_1 \leqslant \varepsilon\} \ , \tag{6.1.13$'$}$$

(6.1.14)′ δZ *and* $\delta \gamma^a$ *are tangent at their intersection, more precisely for any* $x \in \delta Z \cap \delta \gamma^a$, $N_x^*(Z)^a = N_x^*(\gamma^a)$.

We set:

$$F' = Rq_{2!}R\Gamma_{s^{-1}(\gamma^a)}(q_1^! F_Z) \ . \tag{6.1.16$'$}$$

Then by the preceding lemma there exists a canonical morphism $F_Z \to F'$. Let u be the composition $F \to F_Z \to F'$. The same arguments as in the former proof show that u satisfies the desired properties. We leave the details to the reader. □

Now we are ready to introduce the "microlocal operations" on sheaves. Let $f: Y \to X$ be a morphism of manifolds, $p \in Y \times_X T^*X$. We set $p_Y = {}^tf'(p)$, $p_X = f_\pi(p)$. We shall keep this notation until the end of Proposition 6.1.10. We shall make use of ind-objects and pro-objects (cf. I §11).

Definition 6.1.7. (i) *Let* $F \in \mathrm{Ob}(\mathsf{D}^b(X;p_X))$. *We denote by* $f_\mu^{-1}F$ *(resp.* $f_\mu^!F$*) the pro-object (resp. the ind-object)* $\text{“}\varprojlim_{F' \to F}\text{”}\, f^{-1}F'$ *(resp.* $\text{“}\varinjlim_{F \to F'}\text{”}\, f^!F'$*) of* $\mathsf{D}^b(Y;p_Y)$. *Here* $F' \to F$ *(resp.* $F \to F'$*) ranges over the category of morphisms in* $\mathsf{D}^b(X)$ *which are isomorphisms at* p_X. *We call* $f_\mu^{-1}F$ *the microlocal inverse image of* F *at* p.

(ii) *Let* $G \in \mathrm{Ob}(\mathsf{D}^b(Y;p_Y))$. *We denote by* $f_!^\mu G$ *(resp.* $f_*^\mu G$*) the pro-object (resp. the ind-object)* $\text{“}\varprojlim_{G' \to G}\text{”}\, Rf_!G'$ *(resp.* $\text{“}\varinjlim_{G \to G'}\text{”}\, Rf_*G'$*) of* $\mathsf{D}^b(X;p_X)$. *Here* $G' \to G$ *(resp.* $G \to G'$*) ranges over the category of morphisms in* $\mathsf{D}^b(Y)$ *which are isomorphisms at* p_Y. *We call* $f_!^\mu G$ *(resp.* $f_*^\mu G$*) the microlocal proper direct image (resp. microlocal direct image) of* G *at* p.

These four operations are related as follows.

Proposition 6.1.8. *Let* $F \in \mathrm{Ob}(\mathsf{D}^b(X;p_X))$ *and* $G \in \mathrm{Ob}(\mathsf{D}^b(Y;p_Y))$.

(i) *There are natural isomorphisms:*

$$\mathrm{Hom}_{\mathsf{D}^b(X;p_X)^\wedge}(f_!^\mu G, F) \simeq \mathrm{Hom}_{\mathsf{D}^b(Y;p_Y)^\vee}(G, f_\mu^! F)\ , \tag{6.1.22}$$

$$\mathrm{Hom}_{\mathsf{D}^b(X;p_X)^\vee}(F, f_*^\mu G) \simeq \mathrm{Hom}_{\mathsf{D}^b(Y;p_Y)^\wedge}(f_\mu^{-1}F, G)\ . \tag{6.1.23}$$

(ii) *There are canonical morphisms:*

$$f_!^\mu G \to f_*^\mu G\ , \tag{6.1.24}$$

$$\omega_{Y/X} \otimes f_\mu^{-1}F \to f_\mu^! F\ . \tag{6.1.25}$$

Proof. The proof being straightforward, we shall only prove (6.1.22). We have:

$$\begin{aligned}
\mathrm{Hom}_{\mathsf{D}^b(X;p_X)^\wedge}(f_!^\mu G, F) &\simeq \varinjlim_{G' \to G} \mathrm{Hom}_{\mathsf{D}^b(X;p_X)}(Rf_!G', F) \\
&\simeq \varinjlim_{G' \to G}\, \varinjlim_{F \to F'} \mathrm{Hom}_{\mathsf{D}^b(X)}(Rf_!G', F') \\
&\simeq \varinjlim_{G' \to G}\, \varinjlim_{F \to F'} \mathrm{Hom}_{\mathsf{D}^b(Y)}(G', f^!F') \\
&\simeq \varinjlim_{F \to F'} \mathrm{Hom}_{\mathsf{D}^b(Y;p_Y)}(G, f^!F') \\
&\simeq \mathrm{Hom}_{\mathsf{D}^b(Y;p_Y)^\vee}(G, f_\mu^! F)\ . \quad \square
\end{aligned}$$

Now we shall investigate some conditions which ensure that the microlocal inverse images or the microlocal direct images belong to $\mathbf{D}^b(Y;p_Y)$ or to $\mathbf{D}^b(X;p_X)$.

Proposition 6.1.9. *Let* $F \in \mathrm{Ob}(\mathbf{D}^b(X;p_X))$.

(i) *If* ${}^t f'^{-1}(p_Y) \cap f_\pi^{-1}(\mathrm{SS}(F)) \subset \{p\}$ *on a neighborhood of* p, *then* $f_\mu^{-1}F$ *and* $f_\mu^! F$ *belong to* $\mathbf{D}^b(Y, p_Y)$ *and the morphism* $\omega_{Y/X} \otimes f_\mu^{-1}F \to f_\mu^!$ *is an isomorphism. Moreover, for any neighborhood* W *of* p, *one has*:

$$\text{(6.1.26)}\quad \mathrm{SS}(f_\mu^{-1}F) \subset {}^t f'(W \cap f_\pi^{-1}(\mathrm{SS}(F))) \textit{ on a neighborhood of } p_Y \ .$$

(ii) *Let* $F \in \mathrm{Ob}(\mathbf{D}^b(X))$ *and assume*:
(a) f *is non-characteristic for* F,
(b) ${}^t f'^{-1}(p_Y) \cap f_\pi^{-1}(\mathrm{SS}(F)) \subset \{p\}$.
Then $f_\mu^{-1}F \simeq f^{-1}F$ *and* $f_\mu^! F \simeq f^! F$.

Proof. If $p_X \in T_X^*X$ then f is non-characteristic for F and the result follows from Proposition 5.4.13.

Assume $p_X \in \dot{T}^*X$, and let us first prove the results concerning f_μ^{-1}. Let F be as in (i) and set $x_\circ = \pi(p_X)$, $y_\circ = \pi(p_Y)$, $V = \mathrm{Ker}(T_{x_\circ}^*X \to T_{y_\circ}^*Y)$. Take a proper closed convex cone K and an open convex cone U such that $p_X \in U \subset K$, $K \cap V \subset \{0\}$ and $f_\pi {}^t f'^{-1}(p_Y) \cap \mathrm{SS}(F) \cap K \subset \{p_X\}$.

By the refined microlocal cut-off lemma, there exists $u : F' \to F$ such that u is an isomorphism at p_X and F' satisfies the conditions (ii) (a) and (ii) (b).

Moreover, if $F'' \to F'$ is an isomorphism at p_X and F' and F'' satisfy the conditions in (ii), then, by embedding this morphism into a distinguished triangle $F'' \longrightarrow F' \longrightarrow F_\circ \xrightarrow{+1}$, we find that f is non-characteristic for $F_\circ$ and ${}^t f'^{-1}(p_Y) \cap f_\pi^{-1}(\mathrm{SS}(F_\circ)) = \varnothing$. Hence $p_Y \notin \mathrm{SS}(f^{-1}F_\circ)$ by Proposition 5.4.13. This means that $f^{-1}F'' \to f^{-1}F'$ is an isomorphism in $\mathbf{D}^b(Y;p_Y)$. By the definition of $f_\mu^{-1}F$, the "inductive limit" can be taken over the category of morphisms $F' \to F$ where F' satisfies the conditions in (ii). Then all $f^{-1}F'$ are isomorphic in $\mathbf{D}^b(Y;p_Y)$, which proves the statements in (i) and (ii) concerning f_μ^{-1}. The statements concerning $f_\mu^!$ are similarly proven, using the dual refined microlocal cut-off lemma, and the isomorphism $\omega_{Y/X} \otimes f_\mu^{-1}F \xrightarrow{\sim} f_\mu^! F$ follows from (ii) and Proposition 5.4.13. □

Proposition 6.1.10. *Let* $G \in \mathrm{Ob}(\mathbf{D}^b(Y;p_Y))$.

(0) *We have*:

$$f_!^\mu G \simeq \text{"}\varprojlim_V\text{"}\, Rf_! G_V \simeq \text{"}\varprojlim_K\text{"}\, Rf_! R\Gamma_K(G) \ ,$$

$$f_*^\mu G \simeq \text{"}\varinjlim_V\text{"}\, Rf_* R\Gamma_V(G) \simeq \text{"}\varinjlim_K\text{"}\, Rf_*(G_K) \ ,$$

where V *(resp.* K*) ranges over an open (resp. a closed) neighborhood system of* $y_\circ = \pi(p_Y)$. *(Note that the isomorphisms above are defined in the categories* $\mathbf{D}^b(X;p_X)^\wedge$ *and* $\mathbf{D}^b(X;p_X)^\vee$.*)*

(i) *If* $f_\pi^{-1}(p_X) \cap {}^t f'^{-1}(\mathrm{SS}(G)) \subset \{p\}$ *on a neighborhood of* p*, then* $f_!^\mu G$ *and* $f_*^\mu G$ *belong to* $\mathbf{D}^b(X;p_X)$ *and the morphism* $f_!^\mu G \to f_*^\mu G$ *is an isomorphism. Moreover, for any neighborhood* W *of* p*, one has:*

(6.1.27) $\mathrm{SS}(f_!^\mu G) \subset f_\pi(W \cap {}^t f'^{-1}(\mathrm{SS}(G)))$ *on a neighborhood of* p_X .

(ii) *If* $\mathrm{supp}(G)$ *is proper over* X *and if* $f_\pi^{-1}(p_X) \cap {}^t f'^{-1}(\mathrm{SS}(G)) \subset \{p\}$ *then* $f_!^\mu G \simeq f_*^\mu G \simeq Rf_* G$.

Proof. The isomorphisms "$\varprojlim_V$" $Rf_! G_V \simeq$ "$\varprojlim_K$" $Rf_! R\Gamma_K(G)$ and "$\varinjlim_V$" $Rf_* R\Gamma_V(G) \simeq$ "$\varinjlim_K$" $Rf_* G_K$ are obvious, since there are natural morphisms $G_V \to R\Gamma_K(G)$, $G_K \to R\Gamma_V(G)$ if $V \subset K$, and $R\Gamma_K(G) \to G_V$, $R\Gamma_V(G) \to G_K$ if $K \subset V$.

(a) We shall first show that if G satisfies the conditions in (i), then the set of open neighborhoods V of $y_\circ$ satisfying the conditions:

(6.1.28) $$\bar{V} \to X \text{ is proper} ,$$

(6.1.29) $${}^t f'^{-1}(\mathrm{SS}(G_V)) \cap f_\pi^{-1}(p_X) \subset \{p\} ,$$

is a neighborhood system of $y_\circ$.

The question being local, we may assume Y and X are vector spaces, $p_Y = (0;\eta_\circ)$ and $p_X = (0;\xi_\circ)$.

Then there exists $\varepsilon > 0$ such that:

(6.1.30) $$\{y \in Y; f(y) = 0 \text{ and } (y; {}^t f'(y)\cdot\xi_\circ) \in \mathrm{SS}(G)\} \subset \{0\} \cup \{y; |y| > 2\varepsilon\} .$$

Hence there exists $\varepsilon_1 > 0$ such that:

(6.1.31) $$(y; {}^t f'(y)\cdot\xi_\circ + \eta) \notin \mathrm{SS}(G) \quad \text{if} \quad |y| = \varepsilon \quad \text{and} \quad |\eta| \leqslant \varepsilon_1 .$$

Now, set:

(6.1.32) $$V = \{y; \varepsilon_1 |y| < \langle f(y), \xi_\circ \rangle + \varepsilon_1 \varepsilon\} .$$

Then it is enough to show that V satisfies (6.1.29).

By (6.1.31), $\mathrm{SS}(G) \cap N^*(V) \cap \pi_Y^{-1}(f^{-1}(0)) \subset T_Y^* Y$. Hence, by Proposition 5.4.8, we have:

(6.1.33) $$\mathrm{SS}(G_V) \subset \mathrm{SS}(G) + N^*(V)^a \quad \text{over} \quad f^{-1}(0) .$$

Since ${}^t f'^{-1}\mathrm{SS}(G_V) \cap f_\pi^{-1}(p_X) \cap \pi^{-1}(V) \subset \{p\}$, it is enough to show that if $y \in \bar{V} \setminus V$, then $(y; {}^t f'(y)\cdot\xi_\circ) \notin \mathrm{SS}(G_V)$.

By (6.1.33), there exist $k \geqslant 0$ and $(y;\eta) \in \mathrm{SS}(G)$ such that ${}^t f'(y)\cdot\xi_\circ = \eta + k(-{}^t f'(y)\cdot\xi_\circ + \varepsilon_1 y/|y|)$.

Hence $\eta = (1+k){}^t f'(y)\cdot\xi_\circ - k\varepsilon_1 y/|y|$.

Since $|k\varepsilon_1 y/(1+k)|y|| \leqslant \varepsilon_1$, this contradicts (6.1.31).

(b) Let us prove the isomorphism $f_!^\mu G \simeq$ "$\varprojlim_V$" $Rf_! G_V$. Let $G' \to G$ be an isomorphism at p_Y. We embed it into a distinguished triangle $G' \longrightarrow G \longrightarrow G_\circ \xrightarrow{+1}$.

Then $p_Y \notin SS(G_\circ)$, and by (a) there exists an open neighborhood system of $y_\circ$ consisting of V such that $\bar{V} \to X$ is proper and ${}^t f'^{-1}(SS(G_{\circ V})) \cap f_\pi^{-1}(p_X) = \emptyset$. For such a V, $p_X \notin SS(Rf_! G_{\circ V})$ by Proposition 5.4.4, and hence $Rf_! G'_V \to Rf_! G_V$ is an isomorphism in $\mathbf{D}^b(X; p_X)$. Taking the "projective limit" with respect to V and G', we obtain:

$$\text{“}\varprojlim_{G'}\text{”}\, Rf_! G' \simeq \text{“}\varprojlim_{G',V}\text{”}\, Rf_! G'_V \simeq \text{“}\varprojlim_{V}\text{”}\, Rf_! G_V \ .$$

(c) If G satisfies the conditions in (ii), then taking V satisfying (6.1.28) and (6.1.29), $Rf_! G_V \to Rf_! G$ is an isomorphism in $\mathbf{D}^b(X; p_X)$. Hence we obtain $f_!^\mu G \simeq Rf_! G$.

(d) The statements concerning $f_*^\mu G$ are similarly proven by replacing condition (6.1.29) in step (a) by the condition:

$$(6.1.29)' \qquad {}^t f'^{-1}(SS(R\Gamma_V(G))) \cap f_\pi^{-1}(p_X) \subset \{p\} \ .$$

The details are left to the reader.

(e) To prove (i), we may assume by the step (a), that G satisfies the conditions in (ii). Hence the result follows from step (c) (and the corresponding result for $f_*^\mu(G)$). □

Finally let us state the following result, whose proof follows immediately from Proposition 5.4.1.

Proposition 6.1.11. *Let Y and X be two manifolds, and let $p_Y \in T^*Y$, $p_X \in T^*X$, $p = (p_X, p_Y) \in T^*(X \times Y)$.*

Let F_1 and F_2 (resp. G_1 and G_2) belong to $\mathbf{D}^b(X)$ (resp. $\mathbf{D}^b(Y)$). Then there is a canonical homomorphism:

$$(6.1.34) \quad \mathrm{Hom}_{\mathbf{D}^b(X;p_X)}(F_1, F_2) \times \mathrm{Hom}_{\mathbf{D}^b(Y;p_Y)}(G_1, G_2)$$
$$\to \mathrm{Hom}_{\mathbf{D}^b(X \times Y; p)}\left(F_1 \overset{L}{\boxtimes} G_1, F_2 \overset{L}{\boxtimes} G_2\right) .$$

6.2. Normal cones in cotangent bundles

In the next section, we shall study the behaviour of the micro-support under various operations. To formulate the results we need to introduce first some new operations on conic subsets in cotangent bundles. These constructions make use of the notion of "normal cone" introduced in IV §1 and the symplectic structure of T^*X (cf. Appendix). Let X be a manifold. We identify T^*T^*X and TT^*X by $-H$, where H is the Hamiltonian isomorphism. If $(x) = (x_1, \ldots, x_n)$ is a system of local coordinates on X, $(x; \xi)$ the associated coordinates on T^*X (hence the canonical 1-form α_X is given by $\langle \xi, dx \rangle = \sum_j \xi_j dx_j$), we have:

(6.2.1) $$-H(\langle\lambda, dx\rangle + \langle\mu, d\xi\rangle) = \left\langle \lambda, \frac{\partial}{\partial\xi}\right\rangle - \left\langle \mu, \frac{\partial}{\partial x}\right\rangle ,$$

where $\langle\lambda, dx\rangle + \langle\mu, d\xi\rangle \in T_p^*T^*X$, $\left\langle \lambda, \frac{\partial}{\partial\xi}\right\rangle - \left\langle \mu, \frac{\partial}{\partial x}\right\rangle \in T_pT^*X$, $p \in T^*X$.

If Λ is a smooth conic Lagrangian submanifold of T^*X, $-H$ induces an isomorphism $T^*\Lambda \simeq T_\Lambda T^*X$. In particular if M is a submanifold of X we have isomorphisms:

(6.2.2) $$T^*T_MX \simeq T^*T_M^*X \simeq T_{T_M^*X}T^*X .$$

Here the first isomorphism is obtained by Proposition 5.5.1 with $E = T_MX$. Let (x', x'') be a system of local coordinates on X such that $M = \{(x', x''); x' = 0\}$ and let $(x', x''; \xi', \xi'')$ denotes the associated coordinates on T^*X. With these coordinates we identify T_MX with X, and $T^*(T_MX)$ with T^*X. We also identify $T_{T_M^*X}(T^*X)$ with T^*X. Then the isomorphisms in (6.2.2) are described by:

(6.2.3)
$$\begin{array}{ccccc} T^*T_MX & \xrightarrow{\sim} & T^*T_M^*X & \xrightarrow{\sim} & T_{T_M^*X}T^*X \\ \Psi & & \Psi & & \Psi \\ (x', x''; \xi', \xi'') & \longleftrightarrow & (\xi', x''; -x', \xi'') & \longleftrightarrow & (x', x''; \xi', \xi'') . \end{array}$$

Let p denote the projection $T_M^*X \to M$, and $\dot{p}$ its restrition to $\dot{T}_M^*X$. We have the maps (cf. (5.5.9)):

$$\begin{array}{ccccc} T^*T_M^*X & \xleftarrow[{}^t p']{} & T_M^*X \underset{M}{\times} T^*M & \xrightarrow[p_\pi]{} & T^*M \\ \cup & & \cup & & \| \\ T^*\dot{T}_M^*X & \xleftarrow[{}^t \dot{p}']{} & \dot{T}_M^*X \underset{M}{\times} T^*M & \xrightarrow[\dot{p}_\pi]{} & T^*M . \end{array}$$

With the coordinates above, $T_M^*X \times_M T^*M$ is isomorphic to the subset $\{x' = 0\}$ of $T^*T_M^*X$, and p_π is given by $(0, x''; \xi', \xi'') \mapsto (x''; \xi'')$.

Recall (cf. (5.5.10)) that T^*M is embedded into $T_{T_M^*X}T^*X \simeq T^*T_M^*X$ via the zero-section of T_M^*X, and the map ${}^tp'$. Using the same coordinates as above, this embedding is described by $(x''; \xi'') \mapsto (0, x''; 0, \xi'')$.

Lemma 6.2.1. *Let (x', x'') be a system of local coordinates on X with $M = \{(x', x''); x' = 0\}$ and let $(x', x''; \xi', \xi'')$ be the associated coordinates on T^*X. Let A be a conic subset of T^*X. Then*

(i) $p_\pi {}^tp'^{-1}(C_{T_M^*X}(A)) = T^*M \cap C_{T_M^*X}(A)$.

(ii) $(x''_0; \xi''_0) \in T^*M \cap C_{T_M^*X}(A) \Leftrightarrow$ *there exists a sequence $\{(x'_n, x''_n; \xi'_n, \xi''_n)\}$ in A such that:*

$$(6.2.4)\qquad \begin{cases} (x''_n;\xi''_n) \underset{n}{\longrightarrow} (x''_0;\xi''_0)\ , \\ |x'_n| \underset{n}{\longrightarrow} 0\ , \\ |x'_n||\xi'_n| \underset{n}{\longrightarrow} 0\ . \end{cases}$$

(iii) $(x''_0;\xi''_0) \in \dot{p}_\pi {}^t\dot{p}'^{-1}(C_{\dot{T}^*_MX}(A)) \Leftrightarrow$ *there exists a sequence* $\{(x'_n, x''_n;\xi'_n,\xi''_n)\}$ *in* A *satisfying* (6.2.4) *and also*:

$$(6.2.5)\qquad |\xi'_n| \underset{n}{\longrightarrow} +\infty\ .$$

Recall that $C_{T^*_MX}(A)$ denote the normal cone to A along T^*_MX, cf. IV §1.

Proof. (i) The set $C_{T^*_MX}(A)$ is biconic. Thus (i) is a particular case of (5.5.11).

(ii), (iii) Let $(x',x'';\xi',\xi'')$ denote the coordinates on $T_{T^*_MX}(T^*X)$. The submanifold $T^*_MX \times_M T^*X$ is defined by the equations $\{x' = 0\}$, and the map p_π is given by $(0,x'';\xi',\xi'') \mapsto (x'';\xi'')$. The embedding $T^*M \hookrightarrow T_{T^*_MX}(T^*X)$ is given by $(x'';\xi'') \mapsto (0,x'';0,\xi'')$.

(a) Let $\{(x'_n,x''_n;\xi'_n,\xi''_n)\}$ be a sequence in A satisfying (6.2.4).

First assume the sequence $\{\xi'_n\}$ is bounded. By extracting a subsequence, we may assume $\xi'_n \underset{n}{\to} \xi'_0$ for some ξ'_0. Let $\{t_n\}$ be a sequence in $\mathbb{R}^+$ such that $t_n \underset{n}{\to} 0$, $t_n^{-1}x'_n \underset{n}{\to} 0$. We have:

$$\begin{cases} (x'_n,x''_n;t_n\xi_n,t_n\xi''_n) \underset{n}{\longrightarrow} (0,x''_0;0,0)\ , \\ t_n^{-1}(x'_n;t_n\xi''_n) \underset{n}{\longrightarrow} (0;\xi''_0)\ . \end{cases}$$

Hence $(x''_0;\xi''_0)$ belongs to $T^*M \cap C_{T^*_MX}(A)$.

Now assume $\{\xi'_n\}$ is not bounded. By extracting a subsequence, we may assume $|\xi'_n| \underset{n}{\to} \infty$ and $\xi'_n/|\xi'_n| \underset{n}{\to} \xi'_0 \neq 0$. Set $t_n = |\xi'_n|^{-1}$. We have:

$$\begin{cases} (x'_n,x''_n;t_n\xi'_n,t_n\xi''_n) \underset{n}{\longrightarrow} (0,x''_0;\xi'_0,0)\ , \\ t_n^{-1}(x'_n;t_n\xi''_n) \underset{n}{\longrightarrow} (0;\xi''_0)\ . \end{cases}$$

Hence $(0,x''_0;\xi'_0,\xi''_0) \in C_{\dot{T}^*_MX}(A)$ and therefore $(x''_0;\xi''_0) \in \dot{p}_\pi {}^t\dot{p}'^{-1}C_{\dot{T}^*_MX}(A)$.

(b) Let $(x''_0;\xi''_0) \in p_\pi {}^tp'^{-1}(C_{T^*_MX}(A))$.

There exists ξ'_0 such that $(0,x''_0;\xi'_0,\xi''_0) \in C_{T^*_MX}(A)$. Hence there exist sequences $\{(x'_n,x''_n;\xi'_n,\xi''_n)\}$ in A, $\{t_n\}$ in $\mathbb{R}^+$, such that:

$$\begin{cases} (x'_n,x''_n;\xi'_n,\xi''_n) \underset{n}{\longrightarrow} (0,x''_0;\xi'_0,0)\ , \\ t_n(x'_n;\xi''_n) \underset{n}{\longrightarrow} (0;\xi''_0)\ . \end{cases}$$

The sequence $\{(x'_n,x''_n;t_n\xi'_n,t_n\xi''_n)\}$ satisfies (6.2.4). If $(x''_0;\xi''_0) \in \dot{p}_\pi {}^t\dot{p}'^{-1}C_{\dot{T}^*_MX}(A)$, then we may assume $\xi'_0 \neq 0$. If $t_n \underset{n}{\to} \infty$, then $|t_n\xi'_n| \underset{n}{\to} \infty$ and (6.2.5) is satisfied. If $\{t_n\}$ is bounded, then $\xi''_0 = 0$. In this case, taking a sequence $\{s_n\}$ with $s_n|\xi''_n| \underset{n}{\to} 0$

$s_n \underset{n}{\to} \infty$, $s_n|x'_n| \underset{n}{\to} 0$, the sequence $\{(x'_n, x''_n; s_n\xi'_n, s_n\xi''_n)\}$ in A satisfies (6.2.4) and (6.2.5). □

Remark 6.2.2. We have:

$$ {}^t\dot{p}'^{-1}(C_{\dot{T}_M^*X}(A)) = ({}^t\dot{p}'^{-1}(C_{T_M^*X}(A))) \cap \dot{T}_M^*X \underset{M}{\times} T^*M \ .$$

Therefore there is no risk of confusion to denote this set by ${}^t\dot{p}'^{-1}(C_{T_M^*X}(A))$.

Now let Y and X be two manifolds and let f be a morphism from Y to X. We identify Y with the graph of f in $X \times Y$. We denote by q the projection $T_Y^*(X \times Y) \to Y$, and by $\dot{q}$ its restriction to $\dot{T}_Y^*(X \times Y)$. Note that $T_Y^*(X \times Y)$ is identified with $Y \times_X T^*X$ by the first projection, and q is identified with f_π.

Definition 6.2.3. *Let A (resp. B) be a conic subset of T^*X (resp. T^*Y). We set:*

(i)
$$C_\mu(A, B) = C_{T_Y^*(X \times Y)}(A \times B^a) \ ,$$

(this is a closed biconic subset of $T_{T_Y^*(X \times Y)}T^*(X \times Y) \simeq T^*(Y \times_X T^*X)$) ,

(ii)
$$f^\#(A, B) = q_\pi {}^t q'^{-1}(C_\mu(A, B)) \ ,$$
$$= T^*Y \cap C_\mu(A, B) \ ,$$

(iii)
$$f_\infty^\#(A, B) = \dot{q}_\pi {}^t \dot{q}'^{-1}(C_\mu(A, B)) \ .$$

(iv) *We set $f^\#(A) = f^\#(A, T_Y^*Y)$ and $f_\infty^\#(A) = f_\infty^\#(A, T_Y^*Y)$.*
(v) *If $Y = X$ and f is the identity, we set $A \hat{+} B = f^\#(A, B^a) \subset T^*X$ and $A \hat{+}_\infty B = f_\infty^\#(A, B^a) \subset T^*X$.*

Note that, when $Y = X$ and f is the identity, the definition of $C_\mu(A, B)$ in (i) coincides with $C(A, B)$.

Note also that $C_\mu(A, B)$, $f^\#(A, B)$, $f_\infty^\#(A, B)$, etc. are closed sets.

Proposition 6.2.4. (i) *Assume f is a closed embedding. Then:*

$$f^\#(A) = T^*Y \cap C_{T_Y^*X}(A) \ ,$$
$$f_\infty^\#(A) = \dot{p}_\pi {}^t\dot{p}'^{-1}(C_{\dot{T}_Y^*X}(A)) \ ,$$

where p denotes the projection $T_Y^*X \to Y$.

(ii) *Let π denote the projection $T^*X \to X$. Then if A and B are conic subsets of T^*X:*

$$A \hat{+} B = T^*X \cap C(A, B^a) \ ,$$
$$A \underset{\infty}{\hat{+}} B = \dot{\pi}_\pi {}^t\dot{\pi}'^{-1}(C(A, B^a)) \ .$$

(iii) *Let* (x) *(resp.* (y)*) be a system of local coordinates on* X *(resp.* Y*), and let* $(x;\xi)$ *(resp.* $(y;\eta)$*) be the associated coordinates on* T^*X *(resp.* T^*Y*). Then:*

(a) $(y_\circ;\eta_\circ)\in f^\#(A,B)\Leftrightarrow$ *there exists a sequence* $\{(x_n;\xi_n),(y_n;\eta_n)\}$ *in* $A\times B$ *such that:*

$$\text{(6.2.6)}\qquad \begin{cases} y_n \xrightarrow[n]{} y_\circ, x_n \xrightarrow[n]{} f(y_\circ)\ , \\ ({}^tf'(y_n)\cdot\xi_n-\eta_n)\xrightarrow[n]{}\eta_\circ, |x_n-f(y_n)||\xi_n|\xrightarrow[n]{}0\ , \end{cases}$$

(b) $(y_\circ;\eta_\circ)\in f_\infty^\#(A,B)\Leftrightarrow$ *there exists a sequence* $\{(x_n;\xi_n),(y_n;\eta_n)\}$ *in* $A\times B$ *satisfying* (6.2.6) *and also:*

$$\text{(6.2.7)}\qquad |\xi_n|\xrightarrow[n]{}+\infty\ .$$

Proof. (i) and (ii) follow immediately from the definition.

(iii) follows from Lemma 6.2.1 after the change of coordinates on $T^*(X\times Y)$:

$$(x,y;\xi,\eta)\mapsto(x-f(y),y;\xi,\eta+{}^tf'(y)\cdot\xi)\ .\quad\square$$

Remark 6.2.5. Let A be a closed conic subset of T^*X. Then:

$$\text{(6.2.8)}\qquad f^\#(A)={}^tf'f_\pi^{-1}(A)\cup f_\infty^\#(A)\ .$$

Similarly, let A and B be two closed conic subsets of T^*X. Then:

$$\text{(6.2.9)}\qquad A\mathbin{\hat{+}}B=(A+B)\cup(A\mathbin{\hat{+}}_\infty B)\ .$$

Remark 6.2.6. Assume A is a closed conic subset of T^*X. Then $f_\infty^\#(A)=\emptyset$ if and only if f is non-characteristic for A (cf. Definition 5.4.12).

Similarly $A\mathbin{\hat{+}}_\infty B=\emptyset$, for two closed conic subsets of T^*X, if and only if $A\cap B^a\subset T_X^*X$.

In view of the preceding remark, it is natural to set:

Definition 6.2.7. (i) *Let* A *be a closed conic subset of* T^*X, V *a subset of* T^*Y. *We say that* f *is non-characteristic for* A *on* V *if* $f_\infty^\#(A)\cap V=\emptyset$.

(ii) *If* $F\in\mathrm{Ob}(\mathbf{D}^b(X))$, *we say that* f *is non-characteristic for* F *on* V *if* f *is non-characteristic for* $SS(F)$ *on* V.

If f *is a closed embedding, one also says that "*Y *is non-characteristic" instead of "*f *is non-characteristic".*

Remark 6.2.8. The following particular cases of Proposition 6.2.4 will be particularly useful.

(i) Let (x',x'') be a system of local coordinates on X, $M=\{x'=0\}$, and let j denote the embedding $M\to X$. Let $(x',x'';\xi',\xi'')$ denotes the associated coordinates on T^*X. Then if A is a closed conic subset of T^*X, we have: $(x''_\circ;\xi''_\circ)\in j^\#(A)\Leftrightarrow$ there exists a sequence $\{(x'_n,x''_n;\xi'_n,\xi''_n)\}$ in A such that $x'_n\xrightarrow[n]{}0$, $x''_n\xrightarrow[n]{}x''_\circ$, $\xi''_n\xrightarrow[n]{}\xi''_\circ$ and $|x'_n||\xi'_n|\xrightarrow[n]{}0$.

(ii) Let (x) be a system of local coordinates on X, $(x;\xi)$ the associated coordinates on T^*X. Let A and B be two closed conic subsets of T^*X. Then: $(x_o;\xi_o) \in A \hat{+} B \Leftrightarrow$ there exist sequences $\{(x_n;\xi_n)\}$ in A and $\{(y_n;\eta_n)\}$ in B such that $x_n \xrightarrow[n]{} x_o$, $y_n \xrightarrow[n]{} x_o$, $\xi_n + \eta_n \xrightarrow[n]{} \xi_o$ and $|x_n - y_n||\xi_n| \xrightarrow[n]{} 0$.

Also note that $(x_o;\xi_o) \in A \hat{+}_\infty B$ if and only if there exist sequences $\{(x_n;\xi_n)\}$ and $\{(y_n;\eta_n)\}$ satisfying the conditions above and $|\xi_n| \xrightarrow[n]{} \infty$.

6.3. Direct images

In V §4 we have studied the behaviour of the micro-support under various operations such as proper direct images or non-characteristic inverse images. Here we extend these results.

We shall make use of the notions of normal cones introduced in IV §1, and the operations $f^\#$, $\hat{+}$, etc. introduced in the preceding section.

Theorem 6.3.1. *Let Ω be an open subset of X and j the embedding $\Omega \hookrightarrow X$. Let $F \in \mathrm{Ob}(\mathbf{D}^b(\Omega))$. Then:*

(i)
$$\mathrm{SS}(Rj_*F) \subset \mathrm{SS}(F) \hat{+} N^*(\Omega) \ ,$$

(ii)
$$\mathrm{SS}(Rj_!F) \subset \mathrm{SS}(F) \hat{+} N^*(\Omega)^a \ .$$

Proof. We shall prove this by moving Ω to a generic position so that F is non-characteristic for $N^*(\Omega)$.

(i) We may assume X is a vector space. Let $(x_o;\xi_o) \notin \mathrm{SS}(F) \hat{+} N^*(\Omega)$. In order to show $(x_o;\xi_o) \notin \mathrm{SS}(Rj_*F)$ we may assume $x_o \in \mathrm{supp}(F)$, $\xi_o \neq 0$, $x_o \in \partial\Omega$, thus $(x_o;\xi_o) \notin \overline{\mathrm{SS}(F)}$. Since ξ_o does not belong to $N^*_{x_o}(\Omega)$, we may assume Ω is γ-open for a closed convex proper cone γ such that $N^*_{x_o}(\Omega) \subset \mathrm{Int}\,\gamma^\circ \cup \{0\}$, $\mathrm{Int}\,\gamma \neq \emptyset$ and $\xi_o \notin \gamma^\circ$ (cf. Proposition 5.3.7).

Let us take $v \in \mathrm{Int}\,\gamma$ with $\langle \xi_o, v\rangle < 0$. For $s > 0$, $t > 0$, set:

$$H_s = \{x; \langle x - x_o, \xi_o\rangle > -s\} \ ,$$

$$\Omega_{t,s} = \{x; x - t(\langle x - x_o, \xi_o\rangle + s)v \in \Omega\} \ .$$

Then:

$$\bar{\Omega}_{t,s} \cap H_s \subset \Omega \ ,$$

$$\Omega_{t,s} \cap H_s \subset \Omega_{t',s} \qquad 0 < t' \leqslant t \ ,$$

$$\left(\bigcup_t \Omega_{t,s}\right) \cap H_s = \Omega \cap H_s \ .$$

Because $t\langle v, \xi_o\rangle < 1$, we define the transformation $(x;\xi) \leftrightarrow (y;\eta)$ by:

$$\begin{cases} y = x - t(\langle x - x_\circ, \xi_\circ\rangle + s)v \ , \\ \xi = \eta - t\langle \eta, v\rangle \xi_\circ \ . \end{cases}$$

Then we have:

$$(x;\xi) \in N^*(\Omega_{t,s}) \Leftrightarrow (y;\eta) \in N^*(\Omega) \ .$$

Now we shall prove that there exist an open neighborhood $U \times W$ of $(x_\circ;\xi_\circ)$ and $\varepsilon > 0$ such that setting $U_s = U \cap H_s$, we have for $0 < t < \varepsilon$, $0 < s < \varepsilon$:

$$\text{(6.3.1)} \qquad \begin{cases} \text{(a)} \quad SS(F) \cap N^*(\Omega_{t,s})^a \cap \pi^{-1}(U_s) \subset T_X^*X \ , \\ \text{(b)} \quad (SS(F) + N^*(\Omega_{t,s})) \cap (U_s \times W) = \emptyset \ . \end{cases}$$

In fact assume (a) or (b) is false. We find sequences $\{t_n\}$, $\{s_n\}$, $\{x_n\}$, $\{\xi_n\}$, $\{\zeta_n\}$ such that:

$$\begin{cases} t_n \xrightarrow[n]{} 0 \ , \quad s_n \xrightarrow[n]{} 0 \ , \qquad t_n > 0 \ , \quad s_n > 0 \ , \\ x_n \xrightarrow[n]{} x_\circ \ , \qquad \xi_n \in N^*_{x_n}(\Omega_{t_n,s_n}) \setminus \{0\} \ , \\ (x_n;\xi_n) \in SS(F) \ , \\ \xi_n + \zeta_n = c\tilde{\xi}_n \ , \quad \tilde{\xi}_n \to \xi_\circ \ , \qquad \text{where} \\ c = 0 \quad \text{or} \quad c = 1 \qquad (c = 0 \text{ for (a)}, c = 1 \text{ for (b)}) \ . \end{cases}$$

We define $(y_n;\eta_n) \in N^*(\Omega)$ by:

$$\begin{cases} y_n = x_n - t_n(\langle x_n - x_\circ, \xi_\circ\rangle + s_n)v \ , \\ \xi_n = \eta_n - t_n\langle \eta_n, v\rangle \xi_\circ \ , \end{cases}$$

and we set:

$$\rho_n = \zeta_n + \eta_n = c\tilde{\xi}_n + t_n\langle \eta_n, v\rangle \xi_\circ \ .$$

Then $\eta_n \in \gamma^\circ$ and we have:

$$\begin{cases} \langle \xi_\circ, v\rangle < 0 \ , \quad \langle \tilde{\xi}_n, v\rangle < 0 \ , \quad \langle \eta_n, v\rangle \geqslant 0 \ , \quad \langle \rho_n, v\rangle \leqslant 0 \ , \\ |\eta_n| \leqslant c'\langle \eta_n, v\rangle \qquad \text{for some } c' > 0 \ , \\ c''|\rho_n| \geqslant -\langle \rho_n, v\rangle \qquad \text{for some } c'' > 0 \ . \end{cases}$$

The last assertion follows from the fact that the direction of ρ_n converges to that of $\xi_\circ$. Hence we have:

$$c''|\rho_n| \geqslant -c\langle\tilde{\xi}_n, v\rangle - t_n\langle\eta_n, v\rangle\cdot\langle\xi_\circ, v\rangle$$

$$\geqslant t_n\langle\eta_n, v\rangle|\langle\xi_\circ, v\rangle|$$

$$\geqslant c'''t_n|\eta_n| \qquad \text{for some } c''' > 0 \ .$$

Note that $\rho_n \neq 0$ because $\eta_n \neq 0$.

Thus $|\eta_n|(|\rho_n|^{-1})t_n$ is bounded and $|\eta_n|(|\rho_n|^{-1})|x_n - y_n| \underset{n}{\rightarrow} 0$. Since $\rho_n/|\rho_n|$ converges to $\xi_\circ/|\xi_\circ|$ and $(x_n;\zeta_n/|\rho_n|) \in SS(F)$, $(y_n;\eta_n/|\rho_n|) \in N^*(\Omega)$, we have $(x_\circ;\xi_\circ/|\xi_\circ|) \in SS(F) \hat{+} N^*(\Omega)$. This contradicts the hypothesis and proves (6.3.1).

Now let $j_{t,s}$ denote the immersion $\Omega_{t,s} \hookrightarrow X$, and set for s being fixed, $0 < s < \varepsilon$:

$$F_t = Rj_{t,s*}(F|_{\Omega_{t,s}}) \ . \tag{6.3.2}$$

By Proposition 5.4.8 together with the non-characteristic condition (6.3.1), we have:

$$SS(F_t) \cap (U_s \times W) = \emptyset \ . \tag{6.3.3}$$

Hence by Proposition 5.2.1, there exist a closed proper cone γ' and γ'-open subsets $\Omega_0 \subset \Omega_1$, such that $\gamma' \subset \{y;\langle y;\xi_\circ\rangle < 0\} \cup \{0\}$ and $x_\circ \in \mathrm{Int}(\Omega_1\backslash\Omega_0) \subset H_s$, with $R\phi_{\gamma'*}R\Gamma_{(\Omega_1\backslash\Omega_0)}(F_t) = 0$, for all $t \in]0,\varepsilon[$.

Applying Proposition 2.7.1 we get $R\phi_{\gamma'*}R\Gamma_{(\Omega_1\backslash\Omega_0)}(F) = 0$, which implies $(x_\circ;\xi_\circ) \notin SS(F)$.

(ii) The proof is similar. Setting

$$G_t = Rj_{t,s!}(F|_{\Omega_{t,s}}) \tag{6.3.4}$$

we find:

$$SS(G_t) \cap (U_s \times W) = \emptyset \ . \tag{6.3.5}$$

Hence, there exist H, L, γ as in Proposition 5.1.1 such that

$$R\Gamma(H \cap (x + \gamma); G_t) \simeq R\Gamma(L \cap (x + \gamma); G_t)$$

holds for x near $x_\circ$ and $0 < t \ll 1$. Since $H^k(H \cap (x + \gamma); Rj_!F) = \varinjlim_t H^k(H \cap (x + \gamma); G_t)$ and $H^k(L \cap (x + \gamma); Rj_!F) = \varinjlim_t H^k(L \cap (x + \gamma); G_t)$ we have $H^k(H \cap (x + \gamma); Rj_!F) \simeq H^k(L \cap (x + \gamma); Rj_!F)$. This implies $(x_\circ;\xi_\circ) \notin SS(Rj_!F)$. □

Proposition 6.3.2. *Let M be a closed submanifold of X, $U = X\backslash M$, j the embedding $U \hookrightarrow X$, and let $F \in \mathrm{Ob}(\mathrm{D}^b(U))$. Then:*

$$SS(Rj_*F) \cap \pi^{-1}(M) \subset SS(F) \hat{+} T^*_M X \ ,$$

$$SS(Rj_!F) \cap \pi^{-1}(M) \subset SS(F) \hat{+} T^*_M X \ ,$$

$$SS((Rj_*F)|_M) \subset T^*M \cap C_{T^*_M X}(SS(F)) \ .$$

Proof. Let $p : \tilde{X}_M \to X$ be the normal deformation of X along M (cf. IV §1). Recall that there exist local coordinate systems (x', x'') on X, (x', x'', t) on $\tilde{X}_M$ such that $M = \{x' = 0\}$, $p(x', x'', t) = (tx', x'')$ and $\mathbb{R}^+$ acts on $\tilde{X}_M$ by $\lambda(x', x'', t) = (\lambda x', x'', \lambda^{-1}t)$, $(\lambda \in \mathbb{R}^+)$. Define:

$$\tilde{U} = \{(x', x'', t); x' \neq 0\} = \tilde{X}_M \backslash (M \times \mathbb{R}) ,$$

$$\tilde{U}_\pm = \tilde{U} \cap \{\pm t > 0\} , \qquad \tilde{M} = \tilde{U} \cap \{t = 0\} = \dot{T}_M X ,$$

$$X' = \tilde{U}/\mathbb{R}^+ , \qquad X'_\pm = \tilde{U}_\pm/\mathbb{R}^+ , \qquad M' = \tilde{M}/\mathbb{R}^+ .$$

Denote by γ the projection $\tilde{U} \to X'$ and by ρ the map $X' \to X$, so that $p = \rho \circ \gamma$. Then X' is a manifold, ρ is proper and γ is smooth. Moreover $M' = \rho^{-1}(M)$ is a hypersurface of X' isomorphic to the normal sphere bundle $S_M X = \dot{T}_M X/\mathbb{R}^+, X' = X'_- \sqcup M' \sqcup X'_+$ and ρ induces an isomorphism $X'_+ \simeq U$. Let us denote by i this isomorphism, and by k the embedding $X'_+ \hookrightarrow X'$:

$$\begin{array}{ccccc}
\tilde{M} & \longrightarrow & \tilde{M}/\mathbb{R}^+ = M' & \longrightarrow & M \\
\cap & & \cap & & \cap \\
\tilde{U} & \xrightarrow{\gamma} & \tilde{U}/\mathbb{R}^+ = X' & \xrightarrow{\rho} & X \\
\cup & & \uparrow k & & \uparrow j \\
\tilde{U}_+ & \longrightarrow & \tilde{U}_+/\mathbb{R}^+ = X'_+ & \xrightarrow[i]{\sim} & U .
\end{array}$$

Setting $F' = i^{-1}F$, we get:

$$R\rho_* Rk_* F' \simeq Rj_* F ,$$

$$R\rho_* Rk_! F' \simeq Rj_! F .$$

Set:

$$S = \rho_\pi {}^t\rho'^{-1}(\mathrm{SS}(F') \hat{+} T^*_{M'}X') .$$

Applying Proposition 5.4.4 and Theorem 6.3.1 we get:

$$\pi^{-1}(M) \cap \mathrm{SS}(Rj_* F) \subset S , \tag{6.3.6}$$

and a similar formula with j_* replaced by $j_!$. Since γ is smooth and surjective, we find:

$$\mathrm{SS}(F') \hat{+} T^*_{M'}X' = \gamma_\pi {}^t\gamma'^{-1}(\mathrm{SS}(\gamma^{-1}F'|_{\tilde{U}_+}) \hat{+} T^*_{\tilde{M}}\tilde{U}) .$$

Hence:

$$S = p_\pi {}^t p'^{-1}\left((\mathrm{SS}(p^{-1}F|_{\tilde{U}_+}) \hat{+} T^*_{\tilde{M}}\tilde{U}) \underset{\tilde{X}_M}{\times} \tilde{U}\right). \tag{6.3.7}$$

Let $(x', x'', t; \xi', \xi'', \tau)$ denotes the coordinates on $T^*\tilde{X}_M$. Then p_π and ${}^t p'$ are defined by $(x', x'', t; \xi', \xi'') \mapsto (tx', x''; \xi', \xi'')$ and $(x', x'', t; \xi', \xi'') \mapsto (x', x'', t; t\xi', \xi'', \langle x', \xi' \rangle)$, respectively. Hence $(0, x''_\circ; \xi'_\circ, \xi''_\circ) \in S$ implies the existence of $x'_\circ \neq 0$ such that:

$$(x'_\circ, x''_\circ, 0; 0, \xi''_\circ, \langle x'_\circ, \xi'_\circ \rangle) \in SS(p^{-1}F|_{\tilde{U}_+}) \hat{+} T^*_{\tilde{M}}\tilde{U} .$$

We get a sequence $\{(x'_n, x''_n, t_n; \xi'_n, \xi''_n, \tau_n)\}$ in $SS(p^{-1}F|_{\tilde{U}_+})$ such that:

$$\text{(6.3.8)} \qquad \begin{cases} x'_n \underset{n}{\longrightarrow} x'_\circ, \quad x''_n \underset{n}{\longrightarrow} x''_\circ, \quad t_n \underset{n}{\longrightarrow} 0, \quad \xi'_n \underset{n}{\longrightarrow} 0 , \\ \xi''_n \underset{n}{\longrightarrow} \xi''_\circ, \quad t_n \tau_n \underset{n}{\longrightarrow} 0 . \end{cases}$$

Since:

$$(x', x'', t; \xi', \xi'', \tau) \in SS(p^{-1}F|_{\tilde{U}_+}) \Leftrightarrow (tx', x''; t^{-1}\xi', \xi'') \in SS(F) ,$$

$$t > 0 , \quad x' \neq 0 , \quad t\tau = \langle x', \xi' \rangle ,$$

we find that the sequence $\{(t_n x'_n, x''_n; t_n^{-1}\xi'_n, \xi''_n)\}$ is contained in $SS(F)$ and satisfies:

$$t_n x'_n \underset{n}{\longrightarrow} 0 , \qquad x''_n \underset{n}{\longrightarrow} x''_\circ , \qquad \xi''_n \underset{n}{\longrightarrow} \xi''_\circ , \qquad |t_n x'_n| \cdot |t_n^{-1}\xi'_n| \underset{n}{\longrightarrow} 0 .$$

This shows the first two inclusions by Remark 6.2.8. In order to prove the last inclusion, let us consider the distinguished triangle:

$$Rj_! F \to Rj_* F \to (Rj_* F)_M \underset{+1}{\longrightarrow} .$$

Then by the triangular inequalities $SS((Rj_* F)_M) \subset SS(F) \hat{+} T^*_M X$. Then the third inclusion follows from Proposition 5.4.4 and Lemma 6.2.1 (i). □

In Chapter VII, we shall have to consider direct images for maps which are proper only in a "microlocal" sense. More precisely, consider the following situation.

Let $q_1 : X \times Y \to X$ and $p : T^*(X \times Y) \to (T^*X) \times Y$ denote the natural projections.

Proposition 6.3.3. *Let Ω be an open subset of T^*X and let $F \in \mathrm{Ob}(\mathbf{D}^b(X \times Y))$. Assume:*

$$\text{(6.3.9)} \qquad \textit{the projection } \overline{p(SS(F))} \cap (\Omega \times Y) \to \Omega \textit{ is proper} .$$

Then:

$$\text{(i)} \qquad SS(Rq_{1*}F) \cap \Omega \subset q_{1\pi}{}^t q_1'^{-1}(SS(F)) ,$$

$$\text{(ii)} \qquad SS(Rq_{1!}F) \cap \Omega \subset q_{1\pi}{}^t q_1'^{-1}(SS(F)) ,$$

$$\text{(iii)} \qquad Rq_{1!}F \to Rq_{1*}F \textit{ is an isomorphism in } \mathbf{D}^b(X; \Omega) .$$

Note that (6.3.9) means that for any compact subset $K \subset \Omega$ there exists a compact subset $L \subset Y$, such that:

$$\mathrm{SS}(F) \cap (K \times T^*Y) \subset K \times \left(L \underset{Y}{\times} T^*Y\right) . \tag{6.3.10}$$

Proof. By choosing a closed embedding of Y into a vector space, we may assume $Y = \mathbb{R}^n$, for some n. Since $\mathbb{R}^n$ is isomorphic to the open ball of $\mathbb{R}^n$, we may now assume Y is the open ball of $\mathbb{R}^n$. Set $\tilde{Y} = \mathbb{R}^n$, and denote by j the open embedding $X \times Y \hookrightarrow X \times \tilde{Y}$, and by $\tilde{q}_1$ the projection $X \times \tilde{Y} \to X$. Then $Rq_{1*}F \simeq R\tilde{q}_{1*}Rj_*F$ and $Rq_{1!}F \simeq R\tilde{q}_{1*}Rj_!F$. Set $Z = X \times \partial Y$. By Theorem 6.3.1, we have:

$$\mathrm{SS}(Rj_*F) \subset \mathrm{SS}(F) \cup (T_Z^*(X \times \tilde{Y}) \hat{+} \mathrm{SS}(F)) ,$$

$$\mathrm{SS}(Rj_!F) \subset \mathrm{SS}(F) \cup (T_Z^*(X \times \tilde{Y}) \hat{+} \mathrm{SS}(F)) ,$$

$$\mathrm{SS}((Rj_*F)_Z) \subset T_Z^*(X \times \tilde{Y}) \hat{+} \mathrm{SS}(F) .$$

By considering the distinguished triangle $Rj_!F \to Rj_*F \to (Rj_*F)_Z \xrightarrow{+1}$ we see that all the conclusions of the Proposition will follow from:

$$\tilde{q}_{1\pi}{}^t\tilde{q}_1'^{-1}(T_Z^*(X \times \tilde{Y}) \hat{+} \mathrm{SS}(F)) \cap \Omega = \emptyset . \tag{6.3.11}$$

Let us prove (6.3.11). Let K be a compact subset of Ω. By (6.3.10) we get:

$$\mathrm{SS}(F) \cap (K \times T^*Y) \cap \pi^{-1}(Z) = \emptyset ,$$

hence

$$K \cap \tilde{q}_{1\pi}{}^t\tilde{q}_1'^{-1}(T_Z^*(X \times \tilde{Y}) \hat{+} \mathrm{SS}(F)) = \emptyset . \quad \square$$

6.4. Microlocalization

We shall give a bound to the micro-support of the microlocalization.

Theorem 6.4.1. *Let M be a closed submanifold of X and let $F \in \mathrm{Ob}(\mathbf{D}^b(X))$. Then:*

(i) $$\mathrm{SS}(\nu_M(F)) \subset C_{T_M^*X}(\mathrm{SS}(F)) ,$$

(ii) $$\mathrm{SS}(\nu_M(F)) = \mathrm{SS}(\mu_M(F)) .$$

Here we identify $T_{T_M^*X}T^*X$, $T^*T_M^*X$ and T^*T_MX by using Proposition 5.5.1 (cf. (6.2.2) and (6.2.3)).

Proof. (ii) follows from Theorem 5.5.5.

(i) We choose a system of local coordinates (x', x'') on X such that $M = \{(x', x''); x' = 0\}$. As in IV §1, $\tilde{X}_M$ has coordinates (x', x'', t), and we denote by

$p: \tilde{X}_M \to X$ the map $(x', x'', t) \mapsto (tx', x'')$, and by Ω the open set $\{t > 0\}$ of $\tilde{X}_M$. Hence we may identify $T_M X$ to $\{t = 0\}$ in $\tilde{X}_M$. We denote by j the embedding $\Omega \hookrightarrow \tilde{X}_M$, by s the immersion $T_M X \hookrightarrow \tilde{X}$ and set $\tilde{p} = p \circ j$ (cf. diagram (4.1.5)). We denote by $(x', x'', t; \xi', \xi'', \tau)$ the coordinates of $T^*\tilde{X}_M$. By the definition, $\nu_M(F) = s^{-1}Rj_*\tilde{p}^{-1}F$. Applying Proposition 5.4.5 and Proposition 6.3.2, we get:

$$SS(\tilde{p}^{-1}F) = \{(x', x'', t; \xi', \xi'', \tau); t > 0, (tx', x''; t^{-1}\xi', \xi'') \in SS(F), t\tau - \langle x', \xi' \rangle = 0\}$$

and hence if $(x'_0, x''_0; \xi'_0, \xi''_0) \in SS(\nu_M(F))$ then there exists a sequence $\{(x'_n, x''_n, t_n; \xi'_n, \xi''_n, \tau_n)\}$ in $SS(\tilde{p}^{-1}F)$ with $t_n \underset{n}{\to} 0$, $|t_n||\tau_n| \underset{n}{\to} 0$, $(x'_n, x''_n; \xi'_n, \xi''_n) \underset{n}{\to} (x'_0, x''_0; \xi'_0, \xi''_0)$. Therefore $(t_n x'_n, x''_n; t_n^{-1}\xi'_n, \xi''_n)$ belongs to $SS(F)$. Since $SS(F)$ is conic, $(t_n x'_n, x''_n; \xi'_n, t_n\xi''_n)$ belongs to $SS(F)$. Then the definition of $C_{T_M^*X}(SS(F))$ immediately implies $(x'_0, x''_0; \xi'_0, \xi''_0) \in C_{T_M^*X}(SS(F))$. □

Note that with the same notations as in the proof of Theorem 6.4.1, we find that if $(x'_0, x''_0; \xi'_0, \xi''_0)$ belongs to $SS(\nu_M(F))$ then $\langle x'_0, \xi'_0 \rangle = 0$. (This follows from $|t_n||\tau_n| \underset{n}{\to} 0$ and $t_n\tau_n - \langle x'_n, \xi'_n \rangle = 0$, in the course of the proof.) With the notation (5.5.7), this means $SS(\nu_M(F)) \subset S_{T_M^*X}$, which is equivalent to saying that $\nu_M(F)$ is conic.

Corollary 6.4.2. *Let $f: Y \to X$ be a morphism of manifolds. Let $F \in \mathrm{Ob}(\mathbf{D}^b(X))$ and $G \in \mathrm{Ob}(\mathbf{D}^b(Y))$. Then:*

(i) $$SS(\mu hom(G \to F)) \subset C_\mu(SS(F), SS(G)) \; .$$

(ii) $$SS(\mu hom(F \leftarrow G)) \subset C_\mu(SS(F), SS(G))^a \; ,$$

where a is the antipodal map with respect to the vector bundle structure of $T^*(Y \times_X T^*X)$ over $Y \times_X T^*X$.

(iii) *Assume G is cohomologically constructible. Then the natural morphism:*

$$R\mathcal{H}om(G, A_Y) \overset{L}{\otimes} f^{-1}F \otimes \omega_{Y/X} \to R\mathcal{H}om(G, f^!F)$$

is an isomorphism on $T^*Y \setminus f^\#_\infty(SS(F), SS(G))$.

Proof. (i) and (ii) follow immediately from Theorem 6.4.1.

(iii) Consider the distinguished triangle

$$R\pi_! H \longrightarrow R\pi_* H \longrightarrow R\dot{\pi}_* H \underset{+1}{\longrightarrow} , \tag{6.4.1}$$

in which $H = \mu hom(G \to F)$. Then $SS(R\dot{\pi}_* H) \cap V = \varnothing$ by Proposition 5.5.4 and Definition 6.2.3 (iii), and the result follows from Proposition 4.4.2. □

Corollary 6.4.3. *Let F and G belong to $\mathbf{D}^b(X)$. Then:*

(i) $$SS(\mu hom(G, F)) \subset C(SS(F), SS(G)) \; .$$

(ii) *Assume G is cohomologically constructible.*
Then the natural morphism:

$$R\mathcal{H}om(G, A_X) \overset{L}{\otimes} F \to R\mathcal{H}om(G, F)$$

is an isomorphism on $T^*X \setminus (SS(F) \hat{+}_\infty SS(G)^a)$.

Proof. This is a particular case of Corollary 6.4.2. □

Corollary 6.4.4. *Let* $f: Y \to X$ *be a morphism of manifolds and let* $F \in \mathrm{Ob}(\mathbf{D}^b(X))$. *Then:*

(i) $$SS(f^{-1}F) \subset f^{\#}(SS(F)) ,$$

(ii) $$SS(f^!F) \subset f^{\#}(SS(F)) .$$

(iii) *Let V be a subset of* T^*Y *and assume f is non-characteristic for F on V. Then the natural morphism:*

$$f^{-1}F \otimes \omega_{Y/X} \to f^!F$$

is an isomorphism on V, and moreover:

$$SS(f^{-1}F) \cap V \subset {}^t f'(f_\pi^{-1}(SS(F))) .$$

Proof. Identify Y with the graph of f in $X \times Y$, and denote by q the projection $T_Y^*(X \times Y) \to Y$. Then:

$$(6.4.2) \qquad \begin{cases} f^{-1}F \otimes \omega_{Y/X} \simeq Rq_! \mu_Y(F \boxtimes \omega_Y) , \\ f^!F \simeq Rq_* \mu_Y(F \boxtimes \omega_Y) , \end{cases}$$

and there is a distinguished triangle (6.4.1) with $H = \mu_Y(F \boxtimes \omega_Y)$. Then the results follow from Proposition 5.5.4. □

Corollary 6.4.5. *Let F and G belong to* $\mathbf{D}^+(X)$. *Then:*

(i) $$SS\left(F \overset{L}{\otimes} G\right) \subset SS(F) \hat{+} SS(G) ,$$

(ii) $$SS(R\mathcal{H}om(G, F)) \subset SS(F) \hat{+} SS(G)^a .$$

Proof. Apply Propositions 5.4.1, 5.4.2 and Corollary 6.4.4, as in the proof of Proposition 5.4.14. □

Remark 6.4.6. In the situation of Corollary 6.4.4 (iii), assume V is open in T^*Y. Then the map ${}^t f' : {}^t f'^{-1}(V) \cap f_\pi^{-1}(SS(F)) \to V$ is proper. However this last condition is strictly weaker than the condition "f is non-characteristic for F on V", as shown by the following example.

Take $X = \mathbb{R}^2$ with coordinates (y, t), $Y = \{(y,t); t = 0\}$ and let $F = A_Z$ where $Z = \{(y,t); t = y^2\}$. Let $V = \{(y;\eta) \in T^*Y; \eta > 0\}$. Then $F|_Y \simeq A_{\{0\}}$ and $f_\pi^{-1}(SS(F)) \cap {}^t f'^{-1}(V) = \emptyset$.

Remark 6.4.7. The micro-support is invariant by C^1-transformations on X, but $\cdot \hat{+} \cdot$ and $f^{\#}(\cdot)$ are not, (cf. Exercise VI 7). This means that Theorem 6.3.1 and its corollaries are not the best possible results.

6.5. Involutivity and propagation

In the theory of partial differential equations, results on *propagation of singularities along bicharacteristic curves* and *involutivity of the characteristic variety* are of fundamental importance. In this section we shall prove results of this type in the framework of sheaves, as an application of Corollary 6.4.3.

Let U be an open subset of T^*X, ϕ a real C^2-function on U. We set:

$$V_0 = \{p \in U; \phi(p) = 0\} \ , \qquad V_\pm = \{p \in U; \pm\phi(p) \geqslant 0\} \ . \tag{6.5.1}$$

Let $p \in V_0$. The integral curve of the Hamiltonian vector field H_ϕ passing through p is called the *bicharacteristic curve* of V_0 (or of ϕ) issued at p, and is denoted by b_p. One also defines b_p^+, the positive half-bicharacteristic curve issued at p, as the positive half-integral curve of H_ϕ issued at p. If $(x;\xi)$ is a system of homogeneous symplectic coordinates, then b_p^+ is the curve $(x(t), \xi(t))_{t \geqslant 0}$ such that:

$$\frac{\partial x_j}{\partial t} = \frac{\partial \phi}{\partial \xi_j} \ , \qquad \frac{\partial \xi_j}{\partial t} = -\frac{\partial \phi}{\partial x_j} \ , \qquad (x(0); \xi(0)) = p \ . \tag{6.5.2}$$

Note that b_p^+ depends on V_+, not on ϕ. More precisely, replacing ϕ by $h\phi$ with $h(p) > 0$ does not affect b_p^+ in a neighborhood of p.

One defines similarly b_p^-, the negative half-bicharacteristic curve. If $d\phi(p) = 0$, we understand it as $b_p^\pm = \{p\}$.

Let S be a locally closed subset of T^*X.

Definition 6.5.1. *Let $p \in S$. We shall say that S is involutive at p if for any $\theta \in T_p^*T^*X$ such that the normal cone $C_p(S,S)$ is contained in the hyperplane $\{v \in T_pT^*X; \langle v, \theta\rangle = 0\}$ one has: $-H(\theta) \in C_p(S)$.*

If S is involutive at each $p \in S$, we shall say that S is involutive.

Of course, if S is a smooth submanifold, then S is involutive in the sense of Definition 6.5.1 if and only if T_pS is involutive in T_pT^*X for all $p \in S$, because $C_p(S,S) = C_p(S) = T_pS$.

Proposition 6.5.2. *Assume S is involutive and closed in an open subset U of T^*X. Let $\phi : U \to \mathbb{R}$ be a function of class C^2 such that $\phi|_S = 0$. Then S is a union of bicharacteristic curves of ϕ.*

Proof. Let $p \in S$. We may assume $d\phi(p) \neq 0$, otherwise the integral curve of H_ϕ issued at p is confined to $\{p\}$. Hence S is contained in the smooth hypersurface $\{\phi = 0\}$, which implies:

$$C_p(S, S) \subset \{v \in T_p T^*X; \langle v, d\phi(p)\rangle = 0\} .$$

Thus $H_\phi \in C_p(S)$, and the result will follow from the next lemma. □

Lemma 6.5.3. *Let U be a manifold, S a closed subset of U, v a C^1-vector field on U, $p_\circ \in S$, and let b^+ be the positive half-integral curve of v issued at $p_\circ$; (i.e.: $b^+ : [0, +\infty[\to U$ satisfies $b(0) = p_\circ$ and $\frac{\partial}{\partial t} b^+(t) = v(b^+(t))$). Assume that $v(p) \in C_p(S)$ for all $p \in S$. Then b^+ is contained in S.*

Proof. We may assume that U is an open subset of $\mathbb{R}^N$, $p_\circ = 0$ and $v = \frac{\partial}{\partial t}$ with respect to the coordinates $(t, x) \in \mathbb{R} \times \mathbb{R}^{N-1}$. Then $b^+(t) = (t, 0)$.

Suppose $(a, 0) \notin S$ for some $a > 0$. There exists an $\varepsilon > 0$ such that setting $B = \{x \in \mathbb{R}^{N-1}; |x| \leqslant \varepsilon\}$, we have $S \cap \{a\} \times B = \emptyset$. Let γ_t denote the convex hull of $\{(t, 0)\} \cup (\{a\} \times B)$ and let $t_\circ = \inf\{t; \gamma_t \cap S = \emptyset\}$. Then $\gamma_{t_\circ} \cap S \neq \emptyset$, $0 \leqslant t_\circ < a$ and $\gamma_t \cap S = \emptyset$ for $t_\circ < t \leqslant a$. Choose $p \in \gamma_{t_\circ} \cap S$. Then $\frac{\partial}{\partial t} \notin C_p(S)$. This is a contradiction. □

Theorem 6.5.4 (the involutivity theorem). *Let $F \in \mathrm{Ob}(\mathbf{D}^b(X))$. Then $SS(F)$ is involutive.*

Proof. Let $S = SS(F)$, $p \in S$, $\theta \in T_p^* T^*X$. Assume:

$$C_p(S, S) \subset \{\theta = 0\} , \tag{6.5.3}$$

$$H(\theta) \notin C_p(S) . \tag{6.5.4}$$

We shall derive a contradiction. By (6.5.4), we have $\theta \neq 0$ and there exists a closed subset Z of T^*X such that $p \in Z$, $S \subset Z$ and $\langle H(\theta), \lambda\rangle < 0$ for all $\lambda \in N_p^*(Z) \setminus \{0\}$. In fact if one chooses a local coordinate system at p, we find a convex open cone γ with vertex at p, γ containing $H(\theta)$, and such that $\gamma \cap S = \emptyset$ in a neighborhood of p; it is then enough to put $Z = T^*X \setminus \gamma$.

On the other hand, (6.5.3) implies (in view of Corollary 6.4.3) that $SS(\mu hom(F, F)) \cap \pi^{-1}(p)$ is contained in the set $\{H(\theta) = 0\}$ (here π denotes the projection $T^*T^*X \to T^*X$). Hence we get:

$$SS(\mu hom(F,F)) \cap N_p^*(Z) \subset \{0\} \ .$$

Since $\mu hom(F,F)_p = (R\Gamma_Z \mu hom(F,F))_p$ we obtain $\mu hom(F,F)_p = 0$ by Corollary 5.4.9. Thus $p \notin SS(F)$ by Corollary 6.1.3, which is a contradiction. □

Remark 6.5.5. In Definition 6.5.1, if we would have replaced the condition $C_p(V,V) \subset \{\theta = 0\}$ by the weaker condition $C_p(V) \subset \{\theta = 0\}$, then Theorem 6.5.4 would not remain true. See Exercise VI.2.

Now let U be an open subset of T^*X, ϕ a real C^2-function on U and define V_0, V_+, V_- as in (6.5.1).

Proposition 6.5.6. *Let F and G belong to $\mathbf{D}^b(X)$. Assume $SS(F) \cap U \subset V_+$ and $SS(G) \cap U \subset V_-$. Let $j \in \mathbb{Z}$ and let u be a section of the sheaf $H^j(\mu hom(G,F))$. Then $\mathrm{supp}(u)$ is contained in V_0 and is a union of positive half-bicharacteristic curves.*

Proof. Set $K = \mu hom(G,F)$. We know by Corollary 6.4.3 that, above U, $\mathrm{supp}(K)$ is contained in $V_+ \cap V_- = V_0$, and $SS(K)$ is contained in the set $C(V_+, V_-) = \{v \in TT^*X; \langle v, d\phi\rangle \geqslant 0\}$, after identifying TT^*X and T^*T^*X by the isomorphism $-H$. Hence:

$$SS(K) \subset \{\theta \in T^*T^*X; \langle \theta, H_\phi\rangle \geqslant 0\}$$

(since $\langle -H(\theta), d\phi\rangle = \langle \theta, H_\phi\rangle$).

Then the result follows from the next lemma.

Lemma 6.5.7. *Let Z be a manifold, v a C^1-vector field on Z, and let $K \in \mathrm{Ob}(\mathbf{D}^b(Z))$. Assume $SS(K) \subset \{\theta \in T^*Z; \langle \theta, v\rangle \geqslant 0\}$. Let $j \in \mathbb{Z}$ and let u be a section of the sheaf $H^j(K)$ on Z. Then $\mathrm{supp}(u)$ is a union of positive half-integral curves of v.*

Proof. Let $p \in Z$ and let b be the integral curve through p. We may assume $b \neq \{p\}$, and after a change of coordinates, we may assume $Z = \mathbb{R} \times Y$, $b = \mathbb{R} \times \{y_0\}$, $v = \dfrac{\partial}{\partial t}$, where t is the coordinate on $\mathbb{R}$.

Let γ be the cone $\{t \leqslant 0\}$ of $\mathbb{R}$, and let $\tilde{\gamma} = \gamma \times Y$. Applying Proposition 5.2.3 we get $K \simeq \phi_{\tilde{\gamma}}^{-1} R\phi_{\tilde{\gamma}*} K$. This implies $H^j(K)|_b \simeq \phi_\gamma^{-1} L$ for some $L \in \mathrm{Ob}(\mathfrak{Sh}(\mathbb{R}_\gamma))$, and if $u|_b$ is a section of $H^j(K)|_b$, we find that $\mathrm{supp}(u|_b)$ is a union of intervals $[a, +\infty[$, in view of Proposition 3.5.3. □

Example 6.5.8. Let t be the coordinate on $\mathbb{R}$, $(t;\tau)$ the coordinates on $T^*\mathbb{R}$ and let $F = A_{\{t>0\}}$, $G = A_{\{t=0\}}$. Then $\mu hom(G,F) = A_{\{t=0,\tau\leqslant 0\}}[-1]$ and if u is a non-zero section of $H^1(\mu hom(G,F))$, its support is the interval $\{t = 0, \tau \in]-\infty, 0]\}$. This is the positive half-bicharacteristic curve of $H_t = -\dfrac{\partial}{\partial \tau}$ issued at $(0,0)$, but it is not a union of negative half-bicharacteristic curves.

6.6. Sheaves in a neighborhood of an involutive manifold

Let $f: Y \to X$ be a morphism of manifolds.

Proposition 6.6.1. *Assume f is a closed embedding and identify Y with a submanifold of X. Let $p \in T_Y^*X$ and let $F \in \mathrm{Ob}(\mathbf{D}^b(X))$.*

(i) *Assume* $SS(F) \subset \pi^{-1}(Y)$ *in a neighborhood of* p. *Then there exists* $G \in \mathrm{Ob}(\mathbf{D}^b(Y))$ *such that* $F \simeq f_* G$ *in* $\mathbf{D}^b(X; p)$.

(ii) *Assume* $SS(F) \subset T_Y^*X$ *in a neighborhood of* p. *Then there exists* $M \in \mathrm{Ob}(\mathbf{D}^b(\mathfrak{Mod}(A)))$ *such that* $F \simeq M_Y$ *in* $\mathbf{D}^b(X; p)$.

Proof. (i) If $p \in T_X^*X$, there is nothing to prove. Assume $p \in \dot{T}_Y^*X$. By induction on the codimension of Y we may assume Y is a hypersurface. Let $\{\phi = 0\}$ be an equation of Y, with $p = (x_\circ; d\phi(x_\circ))$. Set $\Omega^\pm = \{x \in X; \pm\phi(x) > 0\}$ and denote by $j_\pm$ the open embeddings $\Omega^\pm \hookrightarrow X$. Applying Theorem 6.3.1 we find that $p \notin SS(Rj_{-*}j_-^{-1}(F))$. Hence $R\Gamma_{\{\phi \geqslant 0\}}(F) \to F$ is an isomorphism in $\mathbf{D}^b(X; p)$ and we may assume from the beginning that $\mathrm{supp}(F)$ is contained in $\{\phi \geqslant 0\}$. Again by Theorem 6.3.1 we find that $p \notin SS(Rj_{+!}j_+^{-1}(F))$. Hence $F \to F_Y$ is an isomorphism in $\mathbf{D}^b(X; p)$. Since $F_Y \simeq f_* f^{-1}F$, the result follows.

(ii) We have $F = f_* G$ in $\mathbf{D}^b(X; p)$, for some $G \in \mathrm{Ob}(\mathbf{D}^b(Y))$. Applying Proposition 5.4.4 we get $SS(G) \subset T_Y^*Y$ in a neighborhood of $\pi(p)$. Let g be the map $Y \to \{pt\}$. By Proposition 5.4.5, $G = g^{-1}M$, for some $M \in \mathrm{Ob}(\mathbf{D}^b(\mathfrak{Mod}(A)))$. Hence $F \simeq M_Y$ in $\mathbf{D}^b(X; p)$. □

Proposition 6.6.2. *Assume f is smooth and identify $Y \times_X T^*X$ with a submanifold of T^*Y. Let $p \in Y \times_X T^*X$ and let $G \in \mathrm{Ob}(\mathbf{D}^b(Y))$. Assume $SS(G) \subset Y \times_X T^*X$ in a neighborhood of p. Then there exists $F \in \mathrm{Ob}(\mathbf{D}^b(X))$ such that $G \simeq f^{-1}F$ in $\mathbf{D}^b(Y; p)$.*

Proof. We may argue by induction on $\dim Y - \dim X$, and assume $Y = \mathbb{R}^n$, $X = \mathbb{R}^{n-1}$, f is the projection $(x_1, x') \mapsto x'$, $p = (0; \xi_\circ)$ with $\xi_\circ = (0, \xi'_\circ)$.

If $\xi_\circ = 0$, the result has already been proved in Proposition 5.4.5. Hence we assume $\xi_\circ \neq 0$. Let H_ε be the open half-space $\{x \in \mathbb{R}^n; \langle x, \xi_\circ\rangle > -\varepsilon\}$, let γ be a proper closed convex cone in $\mathbb{R}^n$ such that $\gamma^{\circ a}$ is a neighborhood of $\xi_\circ$, and finally let U be a neighborhood of 0 which is the intersection of H_ε and a γ-open subset of $\mathbb{R}^n$. We may assume:

$$SS(G) \cap (U \times \gamma^{\circ a}) \subset Y \underset{X}{\times} T^*X . \tag{6.6.1}$$

We deduce

$$((SS(G) \backslash (U \times \gamma^{\circ a})) \hat{+} N^*(H_\varepsilon)^a) \cap (U \times \mathrm{Int}\,\gamma^{\circ a}) = \emptyset ,$$

$$((SS(G) \cap (U \times \gamma^{\circ a})) \hat{+} N^*(H_\varepsilon)^a) \cap (U \times \gamma^{\circ a}) \subset Y \underset{X}{\times} T^*X .$$

Then:

(6.6.2) $$\mathrm{SS}(G_{H_\varepsilon}) \cap (U \times \gamma^{\circ a}) \subset Y \times_X T^*X \; .$$

We set:

(6.6.3) $$G' = \phi_\gamma^{-1} R\phi_{\gamma *}(G_{H_\varepsilon}) \; .$$

It is enough to show:

(6.6.4) $$\mathrm{SS}(G') \cap (U \times \gamma^{\circ a}) \subset Y \times_X T^*X \; .$$

In fact, assumming (6.6.4), we get $\mathrm{SS}(G') \cap \pi^{-1}(U) \subset Y \times_X T^*X$, by Proposition 5.2.3, and it will remain to apply Proposition 5.4.5 to complete the proof, since $G' \simeq G$ in $\mathbf{D}^b(Y; p)$ by Proposition 5.2.3.

Finally (6.6.4) follows from (6.6.2) and the next lemma. ☐

Lemma 6.6.3. *Let E be a real finite-dimensional vector space, γ a closed convex proper cone of E with $\gamma \ni 0$, and let $G \in \mathrm{Ob}(\mathbf{D}^b(E))$, $G' = \phi_\gamma^{-1} R\phi_{\gamma *}(G)$. Let $x \in E$ and assume that for a compact neighborhood K of x, $(K + \gamma) \cap \mathrm{supp}(G)$ is compact. Let $\xi \in E^*$, with $(x + \gamma; \xi) \cap \mathrm{SS}(G) = \varnothing$. Then $(x; \xi) \notin \mathrm{SS}(G')$.*

Proof. Denote by q_1 and q_2 the first and second projections of $E \times E$ and denote by s the map:

(6.6.5) $$s: E \times E \to E \; , \qquad s(x, y) = y - x \; .$$

Then by Proposition 3.5.4,

(6.6.6) $$G' \simeq Rs_*\left(q_1^{-1} A_\gamma \overset{L}{\otimes} q_2^{-1} G\right).$$

Now we have:

$$\mathrm{SS}(G') \subset \{(x; \xi); \text{ there exists } y, (y, x + y; -\xi, \xi) \in \mathrm{SS}(A_\gamma) \times \mathrm{SS}(G)\} \; ,$$

which completes the proof. ☐

6.7. Microlocalization and inverse images

In this section we shall complete the results we obtained in IV §3 with the help of the micro-support.

Let $f: Y \to X$ be a morphism of manifolds, and let M (resp. N) be a closed submanifold of X (resp. Y), with $f(N) \subset M$. We denote as in IV §3 by ${}^t f'$, f_π, ${}^t f'_N$, $f_{N\pi}$ the maps:

$$\begin{array}{ccccc} T^*Y & \xleftarrow[{}^t f']{} & Y \underset{X}{\times} T^*X & \xrightarrow[f_\pi]{} & T^*X \ , \\ \uparrow & & \uparrow & & \uparrow \\ T^*_N Y & \xleftarrow[{}^t f'_N]{} & N \underset{M}{\times} T^*_M X & \xrightarrow[f_{N\pi}]{} & T^*_M X \ . \end{array}$$

Theorem 6.7.1. *Let V be an open subset of $T^*_N Y$ and let $F \in \mathbf{D}^b(X)$. Assume:*

(i) *f is non-characteristic for F on V (cf. Definition 6.2.7),*

(ii) *$f_{N\pi}|_{{}^t f_N'^{-1}(V)} : {}^t f_N'^{-1}(V) \to T^*_M X$ is non-characteristic for $C_{T^*_M X}(\mathrm{SS}(F))$,*

(iii) *${}^t f'^{-1}(V) \cap f_\pi^{-1}(\mathrm{SS}(F)) \subset Y \times_X T^*_M X$.*

Then the natural morphisms (cf. Proposition 4.3.5):

$$R^t f'_{N!}(\omega_{N/M} \otimes f_{N\pi}^{-1} \mu_M(F))|_V \to \mu_N(\omega_{Y/X} \otimes f^{-1}F)|_V$$

and

$$(\mu_N(f^! F))|_V \to (R^t f'_{N*} f^!_{N\pi} \mu_M(F))|_V$$

are isomorphisms.

Proof. The proof goes along the same lines as that of Propositions 4.2.5 or 4.3.5. First, note that $f_\pi^{-1}(\mathrm{SS}(F)) \to T^*Y$ is proper on a neighborhood of V by (i) and Remark 6.4.6. Hence, the vertical arrows in Proposition 4.3.5 are isomorphisms on V by (i) and (ii) in view of Corollary 6.4.4. Therefore, it is enough to show that the first morphism in the theorem is an isomorphism. Consider the diagram (4.1.11) whose notations we shall keep. Then:

$$(T_N f)^{-1} \nu_M(F) \simeq s_Y^{-1} \tilde{f}'^{-1} Rj_{X*} \tilde{p}_X^{-1} F \ ,$$

$$\begin{aligned} \nu_N(f^{-1}F) &\simeq s_Y^{-1} Rj_{Y*} \tilde{f}^{-1} \tilde{p}_X^{-1} F \simeq s_Y^{-1}(Rj_{Y*} \tilde{f}^! \tilde{p}_X^{-1} F \otimes \omega^{\otimes -1}_{\tilde{Y}_N/\tilde{X}_M}) \\ &\simeq s_Y^{-1}(\tilde{f}'^! Rj_{X*} \tilde{p}_X^{-1} F \otimes \omega^{\otimes -1}_{\tilde{Y}_N/\tilde{X}_M}) \ . \end{aligned}$$

Consider a distinguished triangle:

$$\omega_{\tilde{Y}_N/\tilde{X}_M} \otimes \tilde{f}'^{-1} Rj_{X*} \tilde{p}_X^{-1} F \xrightarrow[\alpha]{} \tilde{f}'^! Rj_{X*} \tilde{p}_X^{-1} F \longrightarrow H \xrightarrow[+1]{} \ .$$

Then H is supported by $T_N Y$ and in order to prove the theorem it is enough to show that $V \cap \mathrm{SS}((s_Y^{-1}H)^\wedge) = \emptyset$. Here we identify $T^*T_N Y$ and $T^*T^*_N Y$, as in Proposition 5.5.1. By Corollary 6.4.4, it is enough to prove:

$$(6.7.1) \quad \begin{cases} \textit{for any } q_\circ \in V, \textit{there exists } q \in T_N Y \underset{\tilde{Y}_N}{\times} T^*\tilde{Y}_N \textit{ such that} \\ q_\circ = {}^t s_Y'(q) \in V \textit{ and } \tilde{f}' \textit{ is non-characteristic for } Rj_{X*}\tilde{p}_X^{-1}F \textit{ at } q \ . \end{cases}$$

We choose local coordinate systems $(x) = (x', x'')$ on X, $(y) = (y', y'')$ on Y such that $M = \{(x', x''); x' = 0\}$, $N = \{(y', y''); y' = 0\}$. We denote by $(x', x'', t; \xi', \xi'', \tau)$

the coordinates on $\tilde{X}_M$, by $(y', y'', t; \eta', \eta'', \tau)$ the coordinates on $\tilde{Y}_N$. We write:

$$f(y', y'') = (g(y', y''), h(y', y'')) ,$$
$$\tilde{f}(y', y'', t) = (\tilde{g}(y', y'', t), \tilde{h}(y', y'', t), t) .$$

Hence:

$$t\tilde{g}(y', y'', t) = g(ty', y'') ,$$
$$\tilde{h}(y', y'', t) = h(ty', y'') .$$

Note that:

$$(6.7.2) \qquad {}^t f'(ty', y'') \cdot (\xi', t\xi'')$$
$$= (\tilde{g}_{y'}(y, t) \cdot \xi' + \tilde{h}_{y'}(y, t) \cdot \xi'', \tilde{g}_{y''}(y, t) \cdot t\xi' + \tilde{h}_{y''}(y, t) \cdot t\xi'') .$$

We set $q = (0, y''_\circ, 0; \eta'_\circ, 0, \tau_\circ)$, $x''_\circ = h(0, y''_\circ)$. Then $q_\circ = (y''_\circ; \eta'_\circ) \in V$. Let us assume $\tilde{f}'$ is characteristic for $Rj_{X*}\tilde{p}_X^{-1}F$ at q. Applying Proposition 6.2.4 we find sequences:

$$\{(y'_n, y''_n, t'_n)\} \text{ in } \tilde{Y}_N , \qquad \{(x'_n, x''_n, t_n; \xi'_n, \xi''_n, \tau_n)\} \text{ in } SS(Rj_{X*}\tilde{p}_X^{-1}F) ,$$

such that:

$$(6.7.3) \qquad \begin{cases} (y'_n, y''_n, t'_n) \xrightarrow[n]{} (0, y''_\circ, 0) , \\ (x'_n, x''_n, t_n) \xrightarrow[n]{} (0, x''_\circ, 0) , \end{cases}$$

$$(6.7.4) \qquad \tau_n + \tilde{g}_t(y'_n, y''_n, t'_n) \cdot \xi'_n + \tilde{h}_t(y'_n, y''_n, t'_n) \cdot \xi''_n \xrightarrow[n]{} \tau_\circ ,$$

$$(6.7.5) \qquad \begin{cases} \tilde{g}_{y'}(y'_n, y''_n, t'_n) \cdot \xi'_n + \tilde{h}_{y'}(y'_n, y''_n, t'_n) \cdot \xi''_n \xrightarrow[n]{} \eta'_\circ , \\ \tilde{g}_{y''}(y'_n, y''_n, t'_n) \cdot \xi'_n + \tilde{h}_{y''}(y'_n, y''_n, t'_n) \cdot \xi''_n \xrightarrow[n]{} 0 , \end{cases}$$

$$(6.7.6) \qquad |\tau_n| + |\xi'_n| + |\xi''_n| \xrightarrow[n]{} \infty ,$$

$$(6.7.7) \qquad |(x'_n, x''_n, t_n) - \tilde{f}(y'_n, y''_n, t'_n)| \cdot (|\tau_n| + |\xi'_n| + |\xi''_n|) \xrightarrow[n]{} 0 .$$

By (6.7.4) and (6.7.6) we obtain:

$$(6.7.8) \qquad |\xi'_n| + |\xi''_n| \xrightarrow[n]{} \infty .$$

Moreover (6.7.7) implies in particular:

$$(6.7.9) \qquad |t_n - t'_n| \cdot (|\tau_n| + |\xi'_n| + |\xi''_n|) \xrightarrow[n]{} 0 .$$

First assume $t_n > 0$. Then $(t_n x'_n, x''_n; \xi'_n, t_n \xi''_n) \in SS(F)$, and we deduce from (6.7.2), (6.7.5) and (6.7.9):

$$(6.7.10) \qquad {}^t f'(t_n y'_n, y''_n) \cdot (\xi'_n, t_n \xi''_n) \xrightarrow[n]{} (\eta'_\circ, 0) .$$

Since f is non-characteristic for F at $p = (0, y''_\circ; \eta'_\circ, 0)$, (hypothesis (i)), this implies:

$$\{|\xi'_n| + |t_n \xi''_n|\}_n \quad \text{is bounded} . \tag{6.7.11}$$

Thus $|\xi''_n| \underset{n}{\to} +\infty$.

Since the sequence $\{(\xi'_n, t_n\xi''_n)\}$ is bounded, we may assume after extracting a subsequence, that $\xi'_n \underset{n}{\to} \xi'_1$, $t_n\xi''_n \to \xi''_1$. Then by (6.7.10), $(0, y''_\circ; \xi'_1, \xi''_1) \in {}^t f'^{-1}(V) \times_X f_\pi^{-1}(\mathrm{SS}(F))$. By the hypothesis (iii), we get $\xi''_1 = 0$. Thus $(\xi'_n, t_n\xi''_n) \underset{n}{\to} (\xi'_1, 0)$. We may assume that $\xi''_n/|\xi''_n|$ has a limit ξ''_2. We obtain a sequence $\{(t_n x'_n, x''_n; \xi'_n, t_n\xi''_n)\}$ in SS(F) such that:

$$(t_n x'_n, x''_n) \underset{n}{\longrightarrow} (0, x''_\circ) ,$$

$$(\xi'_n, t_n\xi''_n) \underset{n}{\longrightarrow} (\xi'_1, 0) ,$$

$$\frac{1}{t_n|\xi''_n|}(t_n x'_n, t_n\xi''_n) \underset{n}{\longrightarrow} (0, \xi''_2) .$$

Then $(0, x''_\circ; \xi'_1, \xi''_2) \in C_{T^*_M X}(\mathrm{SS}(F))$, and hypothesis (ii) implies $h_{y''}(0, y''_\circ) \cdot \xi''_2 = \tilde{h}_{y''}(0, y''_\circ, 0) \cdot \xi''_2 \neq 0$. By (6.7.5) we have (recall that $|\xi''_n| \underset{n}{\to} \infty$):

$$\tilde{g}_{y''}(y'_n, y''_n, t_n) \cdot \frac{\xi'_n}{|\xi''_n|} + \tilde{h}_{y''}(y'_n, y''_n, t_n) \cdot \frac{\xi''_n}{|\xi''_n|} \underset{n}{\longrightarrow} 0 .$$

We get a contradiction, since $|\xi'_n|$ is bounded.

Finally assume $t_n = 0$. By Theorem 6.3.1 there exists a double sequence $\{(x'_{n,m}, x''_{n,m}, t_{n,m}; \xi'_{n,m}, \xi''_{n,m}, \tau_{n,m})\}$ in $\mathrm{SS}(\tilde{p}_X^{-1}F)$ such that:

$$(x'_{n,m}, x''_{n,m}, t_{n,m}; \xi'_{n,m}, \xi''_{n,m}, \tau_{n,m}) \underset{m}{\longrightarrow} (x'_n, x''_n, 0; \xi'_n, \xi''_n, \tau_n)$$

and $t_{n,m}$ is positive.

Hence we can choose a subsequence satisfying (6.7.3)–(6.7.7), and the proof is complete. □

Remark 6.7.2. We can decompose f by its graph, as follows:

$$\begin{array}{ccccccc} Y & \hookrightarrow & Y \times X & = & Y \times X & \longrightarrow & X \\ \cup & & \cup & & \cup & & \cup \\ N & = & N & \hookrightarrow & N \times M & \longrightarrow & M . \end{array}$$

Then it would have been possible to deduce Theorem 6.7.1 from the two following corollaries.

Corollary 6.7.3. *In the situation of Theorem* 6.7.1, *assume* (i) *and also*:

$$f|_N : N \to M \qquad \textit{is smooth} .$$

Then the conclusion holds.

In fact, in this case, $f_{N\pi}$ is smooth and ${}^tf'^{-1}(T^*_N Y) \subset Y \times_X T^*_M X$.

Corollary 6.7.4. *In the situation of Theorem* 6.7.1 *assume* $Y = X$, f *is the identity (hence:* $N \subset M \subset X$ *and* $N \times_M T^*_M X = T^*_N X \cap T^*_M X$). *Assume* (ii). *Then the natural morphism*:

$$\mu_N(F)|_{V \cap T^*_M X} \to (f^!_{N\pi}\mu_M(F))|_{V \cap T^*_M X}$$

is an isomorphism.

Proof. The hypothesis (ii) implies that, over V, $C_{T^*_M X}(\mathrm{SS}(F)) \cap T^*_{N \times_M T^*_M X}(T^*_M X)$ is contained in the zero-section. In particular, so is $C_{N \times_M T^*_M X}(\mathrm{SS}(F) \cap T^*_N X)$, because $T^*_{N \times_M T^*_M X}(T^*_M X) \simeq T_{N \times_M T^*_M X}(T^*_N X)$ by the identification of the tangent bundle and the cotangent bundle. Hence, there exists an open neighborhood W of $V \cap T^*_M X$ in $V \subset T^*_N X$ such that $\mathrm{SS}(F) \cap W \subset T^*_M X$. Then the hypotheses of Theorem 6.7.1 hold with V replaced by W. □

Corollary 6.7.5. *Let* $f : Y \to X$ *be a morphism of manifolds,* V *a subset of* T^*Y, $F \in \mathrm{Ob}(\mathbf{D}^b(X))$ *and* $G \in \mathrm{Ob}(\mathbf{D}^b(Y))$. *Assume* f *is non-characteristic for* F *on* $V \cap \mathrm{SS}(G)$. *Then the natural morphisms (cf. Proposition* 4.4.5): $\mu hom(G, f^!F) \to R^tf'_*\mu hom(G \to F)$ *and* $\mu hom(f^{-1}F, G) \to R^tf'_*\mu hom(F \leftarrow G)$ *are isomorphisms on* V.

Proof. Use Corollary 6.7.3 in the proof of Proposition 4.4.5. □

Corollary 6.7.6. *In the situation of Corollary* 6.7.5 *assume moreover that* f *is a closed embedding. Then the natural morphisms (cf. Proposition* 4.4.6): $R^tf'_! f_\pi^{-1}\mu hom(Rf_! G, F) \to \mu hom(G, f^!F)$ *and* $R^tf'_! f_\pi^{-1}\mu hom(F, Rf_* G) \to \mu hom(f^{-1}F, G)$ *are isomorphisms on* V.

Proof. Use Corollary 6.7.5 and Proposition 4.4.5 as in the proof of Proposition 4.4.6 (i) and (ii). □

Exercises to Chapter VI

Exercise VI.1. Let X be an open subset of a vector space E, γ a proper closed convex cone of E and let $F \in \mathrm{Ob}(\mathbf{D}^b(X))$. Assume $\mathrm{SS}(F) \subset X \times \gamma^{\circ a}$.

(i) Prove that for any $x \in X$ there exist a neighborhood U of x and $G \in \mathrm{Ob}(\mathbf{D}^b(E_\gamma))$ such that $F|_U \simeq (\phi_\gamma^{-1} G)|_U$.
(ii) Prove that $SS(H^j(F)) \subset X \times \gamma^{\circ a}$ for all j.

Exercise VI.2. Let $Z = \{(x, y) \in \mathbb{R}^2; x^2 \geqslant y > -x^2\}$ and let $F = A_Z$.
(i) Prove that $SS(F) = \{(x, y; \xi, \eta);\ x^2 \geqslant y \geqslant -x^2,\ \xi = \eta = 0$ or $y = -x^2$, $\xi = 2x\eta$, $\eta \leqslant 0$ or $y = x^2$, $\xi = -2x\eta$, $\eta \leqslant 0\}$.
(ii) Let $p = (0, 0; 0, 0)$. Prove that $C_p(SS(F)) = \{y = 0, \xi = 0, \eta \leqslant 0\}$.
(iii) Let $\theta = dy$. Show that $C_p(SS(F)) \subset \{\theta^{-1}(0)\}$ but $-H_\theta \notin C_p(SS(F))$. (Cf. Remark 6.5.5.)

Exercise VI.3. Let $F \in \mathrm{Ob}(\mathbf{D}^b(X))$ and assume $H^j(F) = 0$ for $j < 0$. Let $u \in H^0(X; F) = \Gamma(X; H^0(F))$, let $x \in X$ and let γ be a closed convex proper cone in $T_x X$, with $\gamma \ni 0$. Assume $(SS(F) \cap \pi^{-1}(x)) \cap \gamma^{\circ a} \subset \{0\}$ and $C_x(\mathrm{supp}(u)) \cap \gamma \subset \{0\}$. Prove that $x \notin \mathrm{supp}(u)$. (Hint: use Proposition 5.2.1.)

Exercise VI.4. Let $f: Y \to X$ be a morphism of manifolds and let F_1 and F_2 belong to $\mathbf{D}^b(X)$.
(i) Assume f is non-characteristic for $SS(F_1) \hat{+} SS(F_2)^a$. Prove the isomorphism:

$$f^{-1} R\mathcal{H}om(F_2, F_1) \simeq R\mathcal{H}om(f^{-1}F_2, f^{-1}F_1) \ .$$

(ii) Assume f is non-characteristic for $SS(F_1) \hat{+} SS(F_2)$. Prove the isomorphism:

$$f^{-1}F_1 \overset{L}{\otimes} f^! F_2 \simeq f^!\left(F_1 \overset{L}{\otimes} F_2\right) \ .$$

Exercise VI.5. Let M be a closed submanifold of X, $\tilde{X}$ a covering of $X \setminus M$ and ρ the map $\tilde{X} \to X$ (i.e.: $\rho = i \circ p$ where $i: X \setminus M \hookrightarrow X$, and p is the projection $\tilde{X} \to X \setminus M$).

To $F \in \mathrm{Ob}(\mathbf{D}^b(X))$ one associates:

$$\begin{aligned} \rho_M(F) &= R\rho_* \rho^{-1} F \\ &\simeq R\mathcal{H}om(R\rho_! A_{\tilde{X}}, F) \ . \end{aligned}$$

(i) Prove that:

$$SS(\rho_M(F)) \subset SS(F) \cup (SS(F) \hat{+} T_M^* X) \ .$$

(ii) Let $f: Y \to X$ be a morphism of manifolds transversal to M and set $N = f^{-1}(M)$. Assume f is non-characteristic for F and for $SS(F) \hat{+} T_M^* X$. Prove the isomorphism:

$$f^{-1} \rho_M(F) \simeq \rho_N(f^{-1} F) \ .$$

Here ρ_N is similarly defined by using the covering $\tilde{X} \times_X Y$ of $Y \setminus N$.

Exercise VI.6. Let $X = \mathbb{R}^2$ and let (x, y) be the coordinates on X, $(x, y; \xi, \eta)$ the associated coordinates on T^*X. Set $\Omega = \{(x, y; \xi, \eta); \eta > 0\}$, $\Omega' = \Omega \setminus \{(x, y; \xi, \eta); \eta > 0, x \leqslant 0, \xi = 0\}$.

(i) Prove that for $F \in \mathrm{Ob}(\mathbf{D}^b(X))$, $SS(F) \cap \Omega' = \varnothing$ implies $SS(F) \cap \Omega = \varnothing$.

(ii) Prove that for any $F, G \in \mathrm{Ob}(\mathbf{D}^b(X))$,

$$\mathrm{Hom}_{\mathbf{D}^b(X;\,\Omega)}(F, G) \to \mathrm{Hom}_{\mathbf{D}^b(X;\,\Omega')}(F, G)$$

is an isomorphism.

(iii) Prove that $H^0(\Omega'; \mu hom(A_{\{0\}}, A_{\{0\}})) \simeq A^2$ but $\mathrm{Hom}_{\mathbf{D}^b(X;\,\Omega')}(A_{\{0\}}, A_{\{0\}}) \simeq A$. (Hint: use the involutivity theorem.)

Exercise VI.7. Let γ be the curve $\{(x, y) \in \mathbb{R}^2; y = 0, x \leqslant 0\} \cup \{(x, y) \in \mathbb{R}^2; y = x^\lambda, x \geqslant 0\}$, where $1 < \lambda < 2$. Let $\Lambda = T_\gamma^*\mathbb{R}^2$. Prove that

$$\Lambda \mathbin{\hat{+}} \Lambda = \Lambda \cup T^*_{\{0\}}\mathbb{R}^2 \ .$$

Exercise VI.8. Let $\gamma = \{(x, y) \in \mathbb{R}^2;\ xy = 0,\ x \geqslant 0,\ y \geqslant 0\}$. Show that if $F \in \mathrm{Ob}(\mathbf{D}^b(X))$ and $SS(F) \subset SS(A_\gamma)$, then there exists $M \in \mathrm{Ob}(\mathbf{D}^b(\mathfrak{Mod}(A)))$, with $F \simeq M_\gamma$ in $\mathbf{D}^b(\mathbb{R}^2)$.

Exercise VI.9. We keep the notations of Proposition 6.1.8. Prove the isomorphisms:

$$\mathrm{Hom}_{\mathbf{D}^b(X;p_X)^\wedge}(f_!^\mu G, F) \simeq \mathrm{Hom}_{\mathbf{D}^b(Y;p_Y)^\vee}(G, f_\mu^! F) \simeq \mu hom(G \to F)_p \ ,$$

$$\mathrm{Hom}_{\mathbf{D}^b(X;p_X)^\vee}(F, f_*^\mu G) \simeq \mathrm{Hom}_{\mathbf{D}^b(Y;p_Y)^\wedge}(f_\mu^{-1} F, G) \simeq \mu hom(F \leftarrow G)_p \ .$$

Notes

The study of normal cones in cotangent bundles (i.e. the results of §2) was initiated in Kashiwara-Schapira [1]. It was motivated by the remark that the classical notion of hyperbolicity for partial differential equations can naturally be formulated in terms of such cones, and in fact, Theorems 6.3.1, 6.4.1 and 6.7.1 are sheaf-theoretical versions of results on micro-hyperbolic systems, cf. Propositions 11.5.4 and 11.5.8 below and see D'Agnolo-Schapira [1] for further developments.

The operation $\hat{+}$, introduced first in Kashiwara-Schapira [2], appears now as a natural tool when studying products (e.g. product of distributions, cf. Lebeau [1]), or when studying stratifications (cf. Chapter VIII).

The involutivity theorem (Theorem 6.5.4) is a wide generalization of the corresponding result for systems of linear differential equations, first proved using analytical methods, by Sato-Kawai-Kashiwara [1] (after the fundamental paper of Guillemin-Quillen-Sternberg [1]). It is interesting to note that the involutivity theorem for differential equations has now three radically different proofs: the first one is analytic, as mentioned above, the second one is purely algebraic and due to Gabber [1], and the last one purely "geometrical" (and "real"), using Theorem 6.5.4 (cf. Theorem 11.3.3 below). This theorem was first obtained in Kashiwara-Schapira [2, 3], in a less precise form, and with a different proof.

Proposition 6.5.6 is very useful in the study of propagation of analytic singularities of microdifferential equations, and particularly of diffractive problems, and is due to Schapira [3]. All the other results of this chapter are due to the authors, and many of them were already proved in Kashiwara-Schapira [3].

Chapter VII. Contact transformations and pure sheaves

Summary

In this chapter, we perform contact transformations for sheaves. We begin by extending the notion of kernel introduced in III §6 to a microlocal situation and we develop the microlocal calculus of kernels.

Let X and Y be two manifolds, Ω_X and Ω_Y two open subsets of T^*X and T^*Y respectively. We show that under suitable hypotheses, the functors Φ_K and Ψ_K are well-defined from $\mathbf{D}^b(Y;\Omega_Y)$ to $\mathbf{D}^b(X;\Omega_X)$ and from $\mathbf{D}^b(X;\Omega_X)$ to $\mathbf{D}^b(Y;\Omega_Y)$ respectively, and give equivalences of categories. Moreover, these equivalences are compatible with the functor μhom. Next, if $\chi : \Omega_X \xrightarrow{\sim} \Omega_Y$ is a contact transformation, we show that it is always possible after shrinking Ω_X and Ω_Y to construct an equivalence $\mathbf{D}^b(X;\Omega_X) \xrightarrow{\sim} \mathbf{D}^b(Y;\Omega_Y)$, using these kernels. Now let M and N be two hypersurfaces of X and Y respectively, and assume the contact transformation χ interchanges $T^*_M X \cap \Omega_X$ and $T^*_N Y \cap \Omega_Y$. If the graph of χ is associated to the conormal bundle to a hypersurface S of $X \times Y$, and if one chooses the sheaf A_S as kernel K, then one proves that $\Phi_K(A_N) \simeq A_M[d]$ in $\mathbf{D}^b(X;p)$, $(p \in \Omega_X)$, where d is a shift that we calculate using the inertia index.

This calculation leads to the notion of *pure sheaves* along a smooth Lagrangian manifold Λ, a sheaf-theoretical analogue of the notion of *Fourier distributions* of Hörmander [2], [4], or of that of *simple holonomic systems* of Sato-Kawai-Kashiwara [1]. In case $\Lambda = T^*_M X$, for a closed submanifold M of X, a pure sheaf F along Λ at p is nothing but the image in $\mathbf{D}^b(X;p)$ of $L_M[d]$, where L is an A-module (hence L_M is the sheaf on X supported by M and constant on M with stalk L) and d is a shift. (In such a case one says F is pure with shift $d + \frac{1}{2}\operatorname{codim} M$.) When the rank of the projection $\pi : \Lambda \to X$ is not constant any more, the shift of F may "jump" and its calculation requires the full machinery of the inertia index. We end this chapter by calculating the shift of the composite of two kernels, and the shift of a pure sheaf, after taking its direct or inverse image.

The contents of this chapter are not necessary for the understanding of the rest of the book, with the exception of Chapter X §3 and Chapter XI §4.

We keep convention 4.0. Moreover, unless otherwise specified, all submanifolds of cotangent bundles are supposed to be locally conic.

7.1. Microlocal kernels

In this section, we shall "microlocalize" the constructions of III §6.

Let X and Y be two manifolds. We denote by q_1 and q_2 the projections $X \times Y \to X$ and $X \times Y \to Y$ respectively, and by p_1 and p_2 the projections $T^*(X \times Y) \to T^*X$ and $T^*(X \times Y) \to T^*Y$ respectively. We also set $p_j^a = p_j \circ a$, where "a" is the antipodal map. If Z is a third manifold, we denote by q_{ij} the projection from $X \times Y \times Z$ to the (i, j)-th factor. For example q_{13} is the projection to $X \times Z$. One defines similarly the projection p_{ij} from $T^*(X \times Y \times Z) \simeq T^*X \times T^*Y \times T^*Z$. In this chapter, $T^*(X \times Y) \times_{T^*Y} T^*(Y \times Z)$ means the fiber product by the projections $p_1 : T^*(Y \times Z) \to T^*Y$ and $p_2^a : T^*(X \times Y) \to T^*Y$. We identify this set with $T^*(X \times Y \times Z)$ by:

$$(((x;\xi),(y;-\eta)),((y;\eta),(z;\zeta))) \leftrightarrow ((x;\xi),(y;\eta),(z;\zeta)) \ .$$

Let Ω_X, Ω_Y, Ω_Z be open subsets of T^*X, T^*Y, T^*Z respectively. One sets $\Omega_X^a = a(\Omega_X)$, where a is the antipodal map, and similarly for Ω_Y^a, Ω_Z^a, etc.

Definition 7.1.1. *One denotes by* $\mathbf{N}(X, Y; \Omega_X, \Omega_Y)$ *the full subcategory of* $\mathbf{D}^b(X \times Y; \Omega_X \times T^*Y)$ *consisting of objects K satisfying:*

(i) $SS(K) \cap (\Omega_X \times T^*Y) \subset \Omega_X \times \Omega_Y^a$,

(ii) $p_1 : SS(K) \cap (\Omega_X \times T^*Y) \to \Omega_X$ *is proper.*

If there is no risk of confusion, we write $\mathbf{N}(\Omega_X, \Omega_Y)$ *instead of* $\mathbf{N}(X, Y; \Omega_X, \Omega_Y)$.

Of course if Ω'_X and Ω'_Y are open subsets of T^*X and T^*Y respectively, with $\Omega'_X \subset \Omega_X$ and $\Omega_Y \subset \Omega'_Y$, then $\mathrm{Ob}(\mathbf{N}(\Omega_X, \Omega_Y))$ is contained in $\mathrm{Ob}(\mathbf{N}(\Omega'_X, \Omega'_Y))$. If $Y = \{\mathrm{pt}\}$, then:

$$\mathbf{N}(\Omega_X, \{\mathrm{pt}\}) \simeq \mathbf{D}^b(X; \Omega_X) \ . \tag{7.1.1}$$

Let $K \in \mathrm{Ob}(\mathbf{D}^b(X \times Y))$, $L \in \mathrm{Ob}(\mathbf{D}^b(Y \times Z))$. Recall, III§6, that $K \circ L$ is the object of $\mathrm{Ob}(\mathbf{D}^b(X \times Z))$ defined by:

$$K \circ L = Rq_{13!}\left(q_{12}^{-1}K \overset{L}{\otimes} q_{23}^{-1}L\right) \ . \tag{7.1.2}$$

Proposition 7.1.2. *Assume* $K \in \mathrm{Ob}(\mathbf{N}(\Omega_X, \Omega_Y))$. *Then:*

(i) *the natural morphism* $K \circ L \to Rq_{13*}(q_{12}^{-1}K \otimes^L q_{23}^{-1}L)$ *is an isomorphism in* $\mathbf{D}^b(X \times Z; \Omega_X \times T^*Z)$,

(ii) $SS(K \circ L) \cap (\Omega_X \times T^*Z) \subset p_{13}((SS(K) \cap (\Omega_X \times \Omega_Y^a)) \times_{T^*Y} (SS(L) \cap (\Omega_Y \times T^*Z)))$,

(iii) *if* $L \in \mathrm{Ob}(\mathbf{N}(\Omega_Y, \Omega_Z))$, *then* $K \circ L \in \mathrm{Ob}(\mathbf{N}(\Omega_X, \Omega_Z))$.

In particular, $(K, L) \mapsto K \circ L$ *is a bifunctor from* $\mathbf{N}(\Omega_X, \Omega_Y) \times \mathbf{N}(\Omega_Y, \Omega_Z)$ *to* $\mathbf{N}(\Omega_X, \Omega_Z)$.

Proof. First notice that:

$$(7.1.3)\qquad (p_{12}^{-1}(\mathrm{SS}(K)) \underset{\infty}{\hat{+}} p_{23}^{-1}(\mathrm{SS}(L)) \cap (\Omega_X \times T^*Y \times T^*Z) = \emptyset \ .$$

In fact, take a sequence $\{(x_n, y_n; \xi_n, \eta_n)\}$ in $\mathrm{SS}(K)$ and a sequence $\{(y'_n, z_n; \eta'_n, \zeta_n)\}$ in $\mathrm{SS}(L)$ such that $(x_n; \xi_n) \underset{n}{\to} (x_\circ, \xi_\circ) \in \Omega_X$, $(z_n; \zeta_n) \underset{n}{\to} (z_\circ; \zeta_\circ)$, $y_n \underset{n}{\to} y_\circ$, $y'_n \underset{n}{\to} y_\circ$, $\eta_n + \eta'_n \underset{n}{\to} \eta_\circ$. Then the sequence $\{\eta_n\}$ is bounded, since $K \in \mathrm{Ob}(\mathbf{N}(\Omega_X, \Omega_Y))$. This implies (7.1.3). By Corollary 6.4.5 (cf. Remark 6.2.5) we get:

$$(7.1.4)\quad \mathrm{SS}\left(q_{12}^{-1}K \overset{L}{\otimes} q_{23}^{-1}L\right) \cap \Omega_X \times T^*Y \times T^*Z \subset p_{12}^{-1}(\mathrm{SS}(K)) + p_{23}^{-1}(\mathrm{SS}(L)) \ .$$

The hypothesis on K also implies that for any compact subset A of $\Omega_X \times T^*Z$, there exists a compact subset B of Y such that:

$$(7.1.5)\quad (x, y, z; \xi, \eta, \zeta) \in (p_{12}^{-1}(\mathrm{SS}(K)) + p_{23}^{-1}(\mathrm{SS}(L))) \cap p_{13}^{-1}(A) \Rightarrow y \in B \ .$$

By (7.1.4) and (7.1.5) we may apply Proposition 6.3.3 to $q_{12}^{-1}K \otimes^L q_{23}^{-1}L$ on $\Omega_X \times T^*Z$ for the map q_{13}. We obtain (i) and also:

$$\mathrm{SS}(K \circ L) \cap (\Omega_X \times T^*Z) \subset \{(x, z; \xi, \zeta);\ \text{there exists } (y; \eta) \in T^*Y \text{ with}$$
$$(x, y; \xi, -\eta) \in \mathrm{SS}(K),\ (x; \xi) \in \Omega_X \text{ and } (y, z; \eta, \zeta) \in \mathrm{SS}(L)\} \ .$$

This proves (ii), and (iii) follows. □

Recall that in III §6 we associated to $K \in \mathrm{Ob}(\mathbf{D}^b(X \times Y))$ the functors $\Phi_K\colon \mathbf{D}^b(Y) \to \mathbf{D}^b(X)$ and $\Psi_K\colon \mathbf{D}^b(X) \to \mathbf{D}^b(Y)$ by setting:

$$(7.1.6)\qquad \begin{cases} \Phi_K(G) = Rq_{1!}\left(K \overset{L}{\otimes} q_2^{-1}G\right) \simeq K \circ G \ , \\ \Psi_K(F) = Rq_{2*}R\mathcal{H}om(K, q_1^! F) \ . \end{cases}$$

Definition 7.1.3. *Let $K \in \mathrm{Ob}(\mathbf{N}(\Omega_X, \Omega_Y))$. One defines the functor $\Phi_K\colon \mathbf{D}^b(Y;\Omega_Y) \to \mathbf{D}^b(X;\Omega_X)$ by setting $\Phi_K(G) = K \circ G$.*

This definition makes sense in view of Proposition 7.1.2 applied with $Z = \{\mathrm{pt}\}$, and is compatible with (7.1.6).

We shall denote by $r\colon X \times Y \to Y \times X$ the canonical map:

$$(7.1.7)\qquad r(x, y) = (y, x) \ .$$

Proposition 7.1.4. *Assume $K \in \mathrm{Ob}(\mathbf{D}^b(X \times Y))$ satisfies: $r_* K \in \mathbf{N}(\Omega_Y^a, \Omega_X^a)$. Then:*

(i) *Ψ_K induces a well-defined functor from $\mathbf{D}^b(X;\Omega_X)$ to $\mathbf{D}^b(Y;\Omega_Y)$.*

(ii) *The natural morphism* $Rq_{2!}R\mathcal{H}om(K, q_1^! F) \to \Psi_K(F)$ *is an isomorphism in* $\mathbf{D}^b(Y; \Omega_Y)$.

(iii) $SS(\Psi_K(F)) \cap \Omega_Y \subset p_2^a(SS(K) \cap p_1^{-1}(SS(F) \cap \Omega_X))$.

Since the proof is almost the same as that of Proposition 7.1.2, we shall not repeat it.

Let W be a fourth manifold, Ω_W an open subset of T^*W.

Proposition 7.1.5. *The two functors from* $\mathbf{N}(\Omega_X, \Omega_Y) \times \mathbf{N}(\Omega_Y, \Omega_Z) \times \mathbf{N}(\Omega_Z, \Omega_W)$ *to* $\mathbf{N}(\Omega_X, \Omega_W)$ *given by* $(K, L, M) \mapsto (K \circ L) \circ M$ *and* $(K, L, M) \mapsto K \circ (L \circ M)$ *are isomorphic.*

The proof is obvious. □

Proposition 7.1.6. (i) *The two functors from* $\mathbf{N}(\Omega_X, \Omega_Y) \times \mathbf{N}(\Omega_Y, \Omega_Z) \times \mathbf{D}^b(Z; \Omega_Z)$ *to* $\mathbf{D}^b(X; \Omega_X)$ *given by* $(K, L, H) \mapsto \Phi_{K \circ L}(H)$ *and* $(K, L, H) \mapsto \Phi_K(\Phi_L(H))$ *are isomorphic.*

(ii) *The two functors from* $\mathbf{N}(\Omega_Z^a, \Omega_Y^a) \times \mathbf{N}(\Omega_Y^a, \Omega_X^a) \times \mathbf{D}^b(X; \Omega_X)$ *to* $\mathbf{D}^b(Z; \Omega_Z)$ *given by* $(L, K, F) \mapsto \Psi_{r^{-1}(K \circ L)}(F)$ *and* $(L, K, F) \mapsto \Psi_{r^{-1}L}(\Psi_{r^{-1}K}(F))$ *are isomorphic.*

Here r denotes one of the canonical maps $X \times Y \to Y \times X$, $Y \times Z \to Z \times Y$, $X \times Z \to Z \times X$.

Proof. (i) is a particular case of Proposition 7.1.5.

(ii) follows from Proposition 3.6.4. □

Let q_{ij} denote the (i, j)-th projection defined on $X \times Z \times Y \times Z$. We denote by i_Z the functor from $\mathbf{N}(\Omega_X, \Omega_Y)$ to $\mathbf{N}(\Omega_X \times \Omega_Z, \Omega_Y \times \Omega_Z)$ given by:

$$i_Z : K \mapsto q_{13}^{-1} K \overset{L}{\otimes} q_{24}^{-1} A_{\Delta_Z} \,. \tag{7.1.8}$$

Consider the diagram:

(7.1.9)

$$\begin{array}{ccc} \mathbf{N}(\Omega_X, \Omega_Y) \times \mathbf{N}(\Omega_Y, \Omega_Z) & \xrightarrow{\quad\beta_1\quad} & \mathbf{N}(\Omega_X, \Omega_Z) \\ \downarrow{\scriptstyle \alpha_1} & & \downarrow{\scriptstyle \alpha_2} \\ \mathbf{N}(\Omega_X \times \Omega_Z, \Omega_Y \times \Omega_Z) \times \mathbf{D}^b(Y \times Z; \Omega_Y \times T^*Z) & \xrightarrow{\quad\beta_2\quad} & \mathbf{D}^b(X \times Z; \Omega_X \times T^*Z) \end{array}$$

where $\alpha_1(K, L) = (i_Z(K), L)$, $\alpha_2(H) = H$, $\beta_1(K, L) = K \circ L$ and $\beta_2(H, G) = \Phi_H(G)$.

Proposition 7.1.7. *The diagram* (7.1.9) *commutes, i.e.* $\beta_2 \circ \alpha_1 \simeq \alpha_2 \circ \beta_1$.

The proof is obvious. □

Proposition 7.1.8. *Let* $K \in \mathrm{Ob}(\mathbf{D}^b(X \times Y))$. *Assume* $K \in \mathrm{Ob}(\mathbf{N}(\Omega_X, \Omega_Y))$ *and* $r^{-1}K \in \mathrm{Ob}(\mathbf{N}(\Omega_Y^a, \Omega_X^a))$. *Then the functors* $\Phi_K : \mathbf{D}^b(Y;\Omega_Y) \to \mathbf{D}^b(X;\Omega_X)$ *and* $\Psi_K : \mathbf{D}^b(X;\Omega_X) \to \mathbf{D}^b(Y;\Omega_Y)$ *are adjoint functors.*

Proof. We know that (cf. Exercise I.14):

$$\mathrm{Hom}_{\mathbf{D}^b(X;\,\Omega_X)}(\Phi_K(G), F) = \varinjlim \mathrm{Hom}_{\mathbf{D}^b(X)}(\Phi_K(G'), F')\ ,$$

where the inductive limit is taken over the category of morphisms $G' \to G$ and $F \to F'$ such that $G' \simeq G$ in $\mathbf{D}^b(Y;\Omega_Y)$ and $F \simeq F'$ in $\mathbf{D}^b(X;\Omega_X)$. Similarly:

$$\mathrm{Hom}_{\mathbf{D}^b(Y;\,\Omega_Y)}(G, \Psi_K(F)) = \varinjlim \mathrm{Hom}_{\mathbf{D}^b(Y)}(G', \Psi_K(F'))\ ,$$

where the inductive limit is taken over the same category. Hence the result follows from Proposition 3.6.2. □

Proposition 7.1.9. *Let* $K \in \mathrm{Ob}(\mathbf{D}^b(X \times Y))$. *Assume* K *is cohomologically constructible and assume* $K \in \mathrm{Ob}(\mathbf{N}(\Omega_X, \Omega_Y))$. *Set* $K^* = r_* R\mathcal{H}om(K, \omega_{X \times Y/Y})$. *Then* $r^{-1}K^* \in \mathrm{Ob}(\mathbf{N}(\Omega_X^a, \Omega_Y^a))$ *and* $\Phi_K \simeq \Psi_{K^*}$ *as functors from* $\mathbf{D}^b(Y;\Omega_Y)$ *to* $\mathbf{D}^b(X;\Omega_X)$.

Proof. Since $SS(r^{-1}K^*) = SS(K)^a$, $r^{-1}K^*$ belongs to $\mathbf{N}(\Omega_X^a, \Omega_Y^a)$. Let $G \in \mathrm{Ob}(\mathbf{D}^b(Y))$. By (7.1.3) we know that:

$$\left(SS(K^*) \underset{\infty}{\hat{+}} SS(q_2^! G)\right) \cap (\Omega_X \times T^*Y) = \emptyset\ .$$

Applying Corollary 6.4.3 we get the isomorphism in $\mathbf{D}^b(X \times Y;\Omega_X \times T^*Y)$:

$$R\mathcal{H}om(K^*, q_2^! G) \simeq R\mathcal{H}om(K^*, q_2^! A_Y) \overset{L}{\otimes} q_2^{-1}G\ .$$

Thus by Proposition 7.1.2:

$$\begin{aligned}\Psi_{K^*}(G) &\simeq Rq_{1*}\left(K \overset{L}{\otimes} q_2^{-1}G\right)\\ &\simeq Rq_{1!}\left(K \overset{L}{\otimes} q_2^{-1}G\right)\end{aligned}$$

in $\mathbf{D}^b(X;\Omega_X)$. □

Proposition 7.1.10. *Let* $K \in \mathrm{Ob}(\mathbf{N}(\Omega_X, \Omega_Y))$ *and* $L \in \mathrm{Ob}(\mathbf{N}(\Omega_Y, \Omega_X))$. *Assume* $K \circ L \simeq A_{\Delta_X}$ *in* $\mathbf{D}^b(X \times X;\Omega_X \times T^*X)$ *and* $L \circ K \simeq A_{\Delta_Y}$ *in* $\mathbf{D}^b(Y \times Y;\Omega_Y \times T^*Y)$. *Then* $\Phi_K : \mathbf{D}^b(Y;\Omega_Y) \to \mathbf{D}^b(X;\Omega_X)$ *and* $\Phi_L : \mathbf{D}^b(X;\Omega_X) \to \mathbf{D}^b(Y;\Omega_Y)$ *are equivalences of categories, inverse to each other.*

Proof. This immediately follows from Proposition 7.1.6. □

Proposition 7.1.11. *Let* $K \in \mathrm{Ob}(\mathbf{D}^b(X \times Y))$, $F \in \mathrm{Ob}(\mathbf{D}^b(X;\Omega_X))$, $G \in \mathrm{Ob}(\mathbf{D}^b(Y;\Omega_Y))$.

(i) *Assume* $K \in \mathrm{Ob}(\mathbf{N}(\Omega_X, \Omega_Y))$. *Then there is a natural isomorphism in* $\mathbf{D}^b(\Omega_X)$:

$$Rp_{1*}\mu hom(K, R\mathcal{H}om(q_2^{-1}G, q_1^! F)) \simeq \mu hom(\Phi_K(G), F) \ .$$

(ii) *Assume* $r_* K \in \mathrm{Ob}(\mathbf{N}(\Omega_Y^a, \Omega_X^a))$. *Then there is a natural isomorphism in* $\mathbf{D}^b(\Omega_Y)$:

$$Rp_{2*}^a\mu hom(K, R\mathcal{H}om(q_2^{-1}G, q_1^! F)) \simeq \mu hom(G, \Psi_K(F)) \ .$$

Proof. (i) Consider the maps:

$$(7.1.10)\qquad \begin{array}{ccccc} X \times X & \xleftarrow{q} & X \times X \times Y & \overset{j}{\hookrightarrow} & X \times Y \times X \times Y \\ \cup & & \cup & & \cup \\ \Delta_X & \longleftarrow & \Delta_X \times Y & \xrightarrow{\sim} & \Delta_{X \times Y} \ , \end{array}$$

and the associated maps:

$$(7.1.11)\qquad \begin{array}{ccc} T^*_{\Delta_X}(X \times X) & \xleftarrow{q_\pi} & Y \times T^*_{\Delta_X}(X \times X) \\ \wr\vert & & \wr\vert \\ T^*X & \longleftarrow & Y \times T^*X \\ T^*_{\Delta_X \times Y}(X \times X \times Y) & \xleftarrow{{}^t j'} & T^*_{\Delta_{X\times Y}}(X \times Y \times X \times Y) \\ \wr\vert & & \wr\vert \\ (T^*X) \times Y & \longleftarrow & T^*(X \times Y) \end{array} \ .$$

One denotes, as usual, by q_j the j-th projection defined on $X \times Y \times X \times Y$, and by q_{ij} the (i, j)-th projection. Set:

$$\begin{aligned} H &= R\mathcal{H}om(q_{34}^{-1}K, R\mathcal{H}om(q_2^{-1}G, q_1^! F)) \\ &\simeq R\mathcal{H}om\left(q_{34}^{-1}K \overset{L}{\otimes} q_2^{-1}G, q_1^! F\right) \ . \end{aligned}$$

Then, $\mu hom(K, R\mathcal{H}om(q_2^{-1}G, q_1^! F)) \simeq \mu_{\Delta_X \times Y}(H)$.

The map j is non-characteristic for H on $\Omega_X \times T^*Y \subset T^*_{\Delta_X \times Y}(X \times X \times Y)$. In fact, consider a sequence $\{(x_n, y_n, x'_n, y'_n; \xi_n, \eta_n, \xi'_n, \eta'_n)\}$ in $SS(H)$, with $(x_n, y_n, x'_n, y'_n) \underset{n}{\to} (x_\circ, y_\circ, x_\circ, y_\circ)$, $(\xi_n, \xi'_n, \eta_n + \eta'_n) \underset{n}{\to} (\xi_\circ, -\xi_\circ, \eta_\circ)$, $(x_\circ; \xi_\circ) \in \Omega_X$. Then the sequence $\{\eta'_n\}$ is bounded, by the hypothesis on K.

Applying Corollary 6.7.3, we get the isomorphism on $\Omega_X \times T^*Y$:

$$R^t j'_* \mu hom(K, R\mathcal{H}om(q_2^{-1}G, q_1^! F)) \simeq \mu_{\Delta_X \times Y} j^! R\mathcal{H}om\left(q_{34}^{-1}K \overset{L}{\otimes} q_2^{-1}G, q_1^! F\right)$$

$$\simeq \mu_{\Delta_X \times Y} R\mathcal{H}om\left(q_{23}^{-1}K \overset{L}{\otimes} q_3^{-1}G, q_1^! F\right) .$$

Now we take the direct image of both sides by q_π. By Proposition 4.3.4 there is a natural morphism:

$$Rq_{\pi *} R^t j'_* \mu hom(K, R\mathcal{H}om(q_2^{-1}G, q_1^! F)) \leftarrow \mu_{\Delta_X} Rq_* R\mathcal{H}om\left(q_{23}^{-1}K \overset{L}{\otimes} q_3^{-1}G, q_1^! F\right)$$

$$\simeq \mu_{\Delta_X} R\mathcal{H}om(q_2^{-1}\Phi_K(G), q_1^! F)$$

$$= \mu hom(\Phi_K(G), F) ,$$

and this morphism is an isomorphism if G has compact support. Since $q_\pi \circ {}^t j' = p_1$, we may summarize as follows.

There is a natural morphism:

$$\mu hom(\Phi_K(G), F) \to Rp_{1*} \mu hom(K, R\mathcal{H}om(q_2^{-1}G, q_1^! F)) \tag{7.1.12}$$

and this morphism is an isomorphism as soon as G has compact support.

To end the proof of (i), we note that for each compact subset A of Ω_X there exists a compact subset B of Y such that $\operatorname{supp}(G) \cap B = \emptyset$ implies the following conditions:

(a) $\operatorname{supp}(Rp_{1*} \mu hom(K, R\mathcal{H}om(q_2^{-1}G, q_1^! F))) \cap A = \emptyset$,
(b) $\operatorname{supp}(\mu hom(\Phi_K(G), F)) \cap A = \emptyset$.

In fact, since K belongs to $\mathbf{N}(\Omega_X, \Omega_Y)$, (a) is clear, and (b) follows from Proposition 7.1.2 (iii). To prove that (7.1.12) is an isomorphism, it remains to replace G with G' such that $G'|_B = G|_B$ and $\operatorname{supp}(G')$ is compact.

(ii) The proof is similar. □

In the next section, we shall give sufficient conditions in order that Φ_K and Ψ_K are equivalences of categories.

7.2. Contact transformations for sheaves

Assume to be given a closed conic subset $\Lambda \subset \Omega_X \times \Omega_Y^a$, where Ω_X and Ω_Y are open subsets of T^*X and T^*Y respectively, as in §1. We shall assume:

$$p_1|_\Lambda : \Lambda \to \Omega_X \quad \text{and} \quad p_2^a|_\Lambda : \Lambda \to \Omega_Y \qquad \text{are homeomorphisms .} \tag{7.2.1}$$

Let us denote by χ the map $p_1|_\Lambda \circ (p_2^a|_\Lambda)^{-1}$ from Ω_Y to Ω_X. If Λ is smooth and Lagrangian, and if the p_j's are diffeomorphisms, then χ is a contact transformation.

Theorem 7.2.1. *Let* $K \in \mathrm{Ob}(\mathbf{D}^b(X \times Y))$. *Assume* (7.2.1) *and also*:

(7.2.2) *K is cohomologically constructible,*

(7.2.3) $(p_1^{-1}(\Omega_X) \cup p_2^{a-1}(\Omega_Y)) \cap \mathrm{SS}(K) \subset \Lambda$,

(7.2.4) *the natural morphism $A_\Lambda \to \mu hom(K,K)|_\Lambda$ is an isomorphism in $\mathbf{D}^b(\Lambda)$.*

Then $\Phi_K : \mathbf{D}^b(Y;\Omega_Y) \to \mathbf{D}^b(X;\Omega_X)$ and $\Psi_K : \mathbf{D}^b(X;\Omega_X) \to \mathbf{D}^b(Y;\Omega_Y)$ are equivalences of categories, inverse to each other.

Moreover if G_1 and G_2 belong to $\mathbf{D}^b(Y;\Omega_Y)$, there is a natural isomorphism in $\mathbf{D}^b(\Omega_X)$:

(7.2.5) $\chi_* \mu hom(G_2, G_1) \simeq \mu hom(\Phi_K(G_2), \Phi_K(G_1))$.

Proof. By the hypotheses (7.2.1) and (7.2.3), K belongs to $\mathbf{N}(\Omega_X, \Omega_Y)$ and r_*K belongs to $\mathbf{N}(\Omega_Y^a, \Omega_X^a)$, r denoting the map (7.1.7) from $X \times Y$ to $Y \times X$. Consider the Cartesian square:

$$\begin{array}{ccc} X \times Y & \xrightarrow[\tilde{j}]{} & X \times Y \times Y \\ \Big\downarrow{\scriptstyle q_2} & \square & \Big\downarrow{\scriptstyle q_{23}} \\ Y & \xrightarrow[j]{} & Y \times Y \end{array},$$

where j and $\tilde{j}$ are the diagonal embeddings. Set $E = R\mathcal{H}om(q_{12}^{-1}K, q_{13}^!K)$. This is an object of $\mathbf{D}^b(X \times Y \times Y)$. By Proposition 7.1.9, setting $K^* = r^{-1}R\mathcal{H}om(K, \omega_{X\times Y/X})$, we have $K^* \circ K \simeq Rq_{23*}E$ in $\mathbf{N}(\Omega_Y, \Omega_Y)$.

On the other hand we have $\tilde{j}^!E \simeq R\mathcal{H}om(\tilde{j}^{-1}q_{12}^{-1}K, \tilde{j}^!q_{13}^!K) \simeq R\mathcal{H}om(K,K)$. Hence we obtain the canonical morphisms:

$$A_{X\times Y} \to R\mathcal{H}om(K,K) \to \tilde{j}^!E \ ,$$

which induce:

$$A_Y \to Rq_{2*}A_{X\times Y} \to Rq_{2*}\tilde{j}^!E \simeq j^!Rq_{23*}E \ .$$

We thus have gotten the morphism:

(7.2.6) $$\alpha : A_{\Delta_Y} \to K^* \circ K \quad \text{in} \quad \mathbf{N}(\Omega_Y, \Omega_Y) \ .$$

We shall prove that α is an isomorphism. Let Z be another manifold, let $F \in \mathrm{Ob}(\mathbf{D}^b(X \times Z; \Omega_X \times T^*Z))$, $G \in \mathrm{Ob}(\mathbf{D}^b(Y \times Z; \Omega_Y \times T^*Z))$. Let $i_Z(K)$ denote the object of $\mathbf{N}(\Omega_X \times T^*Z, \Omega_Y \times T^*Z)$ constructed in (7.1.8). By Proposition 7.1.11 we have a natural isomorphism in $\mathbf{D}^b(\Omega_X \times T^*Z)$:

(7.2.7) $\chi_*(\mu hom(G, \Psi_{i_Z(K)}(F))|_{\Omega_Y \times T^*Z}) \simeq \mu hom(\Phi_{i_Z(K)}(G), F)|_{\Omega_X \times T^*Z}$.

Choose $Z = Y$, $G = A_{\Delta_Y}$ and $F = K$. We get the isomorphism:

$$\chi_* \mu hom(A_{\Delta_Y}, \Psi_{i_Y(K)}(K)) \simeq \mu hom(K, K) \ .$$

Since $\Psi_{i_Y(K)}(K) \simeq K^* \circ K$ by Proposition 7.1.9, we have obtained the isomorphism on $\Omega_Y \subset T^*_{\Delta_Y}(Y \times Y)$:

$$\mu_\Delta(\alpha) : \mu_{\Delta_Y}(A_{\Delta_Y}) \simeq \mu_{\Delta_Y}(K^* \circ K) \ . \tag{7.2.8}$$

Since $SS(K^* \circ K) \cap (\Omega_Y \times T^*Y) \subset T^*_{\Delta_Y}(Y \times Y)$, Proposition 6.6.1 and the isomorphism (7.2.8) imply that α is an isomorphism in $\mathbf{N}(\Omega_Y, \Omega_Y)$. One proves similarly the isomorphism $A_{\Delta_X} \simeq K \circ K^{*\prime}$ in $\mathbf{N}(\Omega_X, \Omega_X)$ where $K^{*\prime} = r^{-1} R\mathcal{H}om(K, \omega_{X \times Y/Y})$. By Proposition 7.1.10, they imply that $\Phi_K : \mathbf{D}^b(Y; \Omega_Y) \to \mathbf{D}^b(X; \Omega_X)$ is an equivalence of categories. Since Ψ_K is the adjoint functor of Φ_K, Ψ_K is a quasi-inverse of Φ_K. Then (7.2.5) follows from (7.2.7) with $Z = \{\mathrm{pt}\}$. □

We shall show that if χ is a contact transformation between Ω_Y and Ω_X, after shrinking Ω_Y and Ω_X, one can construct equivalences of categories between $\mathbf{D}^b(Y; \Omega_Y)$ and $\mathbf{D}^b(X; \Omega_X)$, using Theorem 7.2.1.

Let Ω_X and Ω_Y be two open subsets of T^*X and T^*Y respectively, $\chi : \Omega_Y \to \Omega_X$ a contact transformation. We set

$$\Lambda = \{(x, y; \xi, \eta) \in \Omega_X \times \Omega_Y^a; (x; \xi) = \chi(y; -\eta)\} \ . \tag{7.2.9}$$

This is a conic Lagrangian manifold, closed in $\Omega_X \times \Omega_Y^a$. Let $p_Y \in \Omega_Y$, $p_X = \chi(p_Y) \in \Omega_X$.

Corollary 7.2.2. *There exist open neighborhoods X' of $\pi(p_X)$, Y' of $\pi(p_Y)$, Ω'_X of p_X, Ω'_Y of p_Y with $\Omega'_X \subset T^*X' \cap \Omega_X$, $\Omega'_Y \subset T^*Y' \cap \Omega_Y$, and there exists $K \in$* $\mathrm{Ob}(\mathbf{D}^b(X' \times Y'))$ *such that:*

(a) *χ induces a contact transformation $\Omega'_Y \simeq \Omega'_X$,*
(b) $((\Omega'_X \times T^*Y') \cup (T^*X' \times \Omega'^a_Y)) \cap SS(K) \subset \Lambda \cap (\Omega'_X \times \Omega'^a_Y)$,
(c) *$\Phi_K : \mathbf{D}^b(Y'; \Omega'_Y) \to \mathbf{D}^b(X'; \Omega'_X)$ is an equivalence of categories,*
(d) *for G_1 and G_2 which belong to $\mathbf{D}^b(Y'; \Omega'_Y)$, we have the isomorphism* (7.2.5) *in* $\mathbf{D}^b(\Omega'_X)$.

Proof. By Corollary A.2.7, after shrinking Ω_Y and Ω_X we may decompose χ as $\chi_2 \circ \chi_1$ where each χ_i $(i = 1, 2)$ is a contact transformation and the Lagrangian manifold Λ_i associated with χ_i by (7.2.9) is the conormal bundle to a hypersurface. By Propositions 7.1.2 and 7.1.6, if $K_1 \in \mathrm{Ob}(\mathbf{D}^b(X' \times Z))$ and $K_2 \in \mathrm{Ob}(\mathbf{D}^b(Z \times Y'))$ satisfy conditions (b), (c), (d) of the corollary, then $K_2 \circ K_1$ will also satisfy these conditions. Therefore we may assume from the beginning that there exists a hypersurface $S \subset X \times Y$ such that the Lagrangian manifold Λ defined in (7.2.9) is contained in $T^*_S(X \times Y)$. Since Λ is $\mathbb{R}^+$-conic, and S is a hypersurface, there exist open neighborhoods Ω'_X, Ω'_Y, X', Y' of p_X, p_Y, $\pi(p_X)$, $\pi(p_Y)$ respectively, such that $((\Omega'_X \times T^*Y') \cup (T^*X' \times \Omega'^a_Y)) \cap T^*_S(X \times Y) \subset \Lambda$.

Then all the hypotheses of Theorem 7.2.1 are satisfied for $K = A_{S\cap(X'\times Y')}$ in $\mathsf{D}^b(X' \times Y')$. □

Definition 7.2.3. *In the situation of Theorem 7.2.1, we say that Φ_K and Ψ_K are extended contact transformations above χ.*

We shall show that all extended contact transformations above the identity come from equivalences of categories in $\mathsf{D}^b(\mathfrak{Mod}(A))$. In that sense Φ_K is essentially unique. Assume χ is the identity in a neighborhood of $p \in T^*X$. By Proposition 6.6.1 we find that $K \simeq M_{\Delta_X}$ in $\mathsf{D}^b(X \times X; (p, p^a))$, with $M \in \mathrm{Ob}(\mathsf{D}^b(\mathfrak{Mod}(A)))$. If $G \in \mathrm{Ob}(\mathsf{D}^b(X))$, then:

$$\Phi_K(G) = M_X \overset{L}{\otimes} G \qquad \text{in} \qquad \mathsf{D}^b(X;p) \ .$$

Proposition 7.2.4. *Assume that the functor $M_X \otimes^L \cdot$ defines an equivalence of categories in $\mathsf{D}^b(X;p)$. Then the functor $M \otimes^L \cdot$ defines an equivalence of categories in $\mathsf{D}^b(\mathfrak{Mod}(A))$.*

Proof. Let Y be a submanifold of X such that $p \in T^*_YX$. Set $\Lambda = T^*_YX$. Then $M_X \otimes^L \cdot$ induces an equivalence of categories on $\mathsf{D}^b_\Lambda(X;p)$, and this last category is equivalent to $\mathsf{D}^b(\mathfrak{Mod}(A))$ by Proposition 6.6.1. □

Example 7.2.5. Let X and Y be two copies of $\mathbb{R}^n$, endowed with systems of linear coordinates (x) and (y), respectively. Let $(x;\xi)$ and $(y;\eta)$ denote the associated coordinates on the cotangent bundles, and consider the contact transformation $\chi: \dot{T}^*Y \simeq \dot{T}^*X$:

$$\chi: (y;\eta) \mapsto (x;\xi) = (y + \eta/|\eta|; \eta) \ , \tag{7.2.10}$$

where we have set $|\eta| = (\sum_j \eta_j^2)^{1/2}$. Let $S = \{(x,y) \in X \times Y; \sum_j (x_j - y_j)^2 \geqslant 1\}$, and let $\Lambda = SS(A_S) \cap \dot{T}^*(X \times Y)$. Then:

$$\Lambda = \left\{(x,y;\xi,\eta); \sum_j (x_j - y_j)^2 = 1, \xi = -\eta = \lambda(x-y), \lambda > 0\right\} .$$

Thus:

$$(x;\xi) = \chi(y;\eta) \Leftrightarrow (x,y;\xi,-\eta) \in \Lambda \ .$$

Let $K = A_S$. Then all conditions of Theorem 7.2.1 are satisfied, and Φ_K: $\mathsf{D}^b(Y;\dot{T}^*Y) \to \mathsf{D}^b(X;\dot{T}^*X)$ is an equivalence of categories.

If $G \in \mathrm{Ob}(\mathsf{D}^b(Y))$, we have $\Phi_K(G) \simeq Rq_{1!}(q_2^{-1}G)_S$. In particular we find: $\Phi_K(A_{\{0\}}) \simeq A_{\{\sum_j x_j^2 \geqslant 1\}}$, $\Phi_K(A_Y) \simeq 0$, $\Phi_K(A_{\{y\neq 0\}}) \simeq A_{\{\sum_j x_j^2 \geqslant 1\}}[-1]$ in $\mathsf{D}^b(X;\dot{T}^*X)$.

Example 7.2.6. Let us consider a similar example to 7.2.5, on complex manifolds.

Let X and Y be two copies of $\mathbb{C}^n$ endowed with $\mathbb{C}$-linear coordinates (z) and (w) respectively, with $z = x + \sqrt{-1}y$, $w = u + \sqrt{-1}v$. Let $(z;\zeta)$ and $(w;\theta)$

denote the associated coordinates on the complex cotangent bundles. If $X^{\mathbb{R}}$ and $Y^{\mathbb{R}}$ denote the real underlying manifolds to X and Y, then the canonical 1-forms on $T^*X^{\mathbb{R}}$ and $T^*Y^{\mathbb{R}}$ are given by $2\,\mathrm{Re}(\sum_j \zeta_j dz_j)$, and $2\,\mathrm{Re}(\sum_j \theta_j dw_j)$. Let $\Omega_X = \{(z;\zeta); \zeta^2 \notin \mathbb{R}^+ \cup \{0\}\}$, where we set $\zeta^2 = \sum_j \zeta_j^2$, and similarly let $\Omega_Y = \{(w;\theta); \theta^2 \notin \mathbb{R}^+ \cup \{0\}\}$. Then $(-\theta^2)^{1/2}$ is a holomorphic function on Ω_Y, well-defined by $\mathrm{Re}((-\theta^2)^{1/2}) > 0$, and we may consider the holomorphic contact transformation χ from Ω_Y to Ω_X given by:

$$\chi : (w;\theta) \mapsto (z;\zeta) = (w + \theta/(-\theta^2)^{1/2}; \theta) \ . \tag{7.2.11}$$

Let $Z = \{(z, w); (z - w)^2 = -1\}$. Consider the Lagrangian manifolds:

$$\Lambda^{\pm} = \{(z, w; \zeta, \theta); \zeta^2 \notin \mathbb{R}^+ \cup \{0\}, \zeta = -\theta, z = w \pm \zeta/(-\zeta^2)^{1/2}\} \ .$$

Then $(z;\zeta) = \chi(w;\theta) \Leftrightarrow (z, w; \zeta, -\theta) \in \Lambda^+$, and $T^*_Z(X \times Y) \cap (\Omega_X \times \Omega_Y^a) = \Lambda^+ \sqcup \Lambda^-$. Set $Z_+ = \{(z, w); \mathrm{Im}((z - w)^2) = 0, \mathrm{Re}((z - w)^2) < -1\}$ and let $K = A_{Z_+}$. Then:

$$\begin{aligned} SS(K) = \{&(z, w; \zeta, \theta); (z, w) \in \bar{Z}_+, \zeta = -\theta = k(z - w), k \in \mathbb{C}, \mathrm{Re}\, k \geqslant 0, \\ &(1 + \mathrm{Re}((z - w)^2)) \cdot \mathrm{Re}\, k = 0\} \ . \end{aligned}$$

This gives:

$$SS(K) \cap (\Omega_X \times T^*Y) = SS(K) \cap (T^*X \times \Omega_Y^a) = \Lambda^+ \ .$$

(Note that $SS(A_Z) \cap (\Omega_X \times T^*Y) = SS(A_Z) \cap (T^*X \times \Omega_Y^a) = \Lambda^+ \sqcup \Lambda^-$, and K and $A_Z[-1]$ are isomorphic in $\mathbf{D}^b(X \times Y; \Lambda_+)$.)

Hence Φ_K defines an equivalence of categories $\mathbf{D}^b(Y;\Omega_Y) \xrightarrow{\sim} \mathbf{D}^b(X;\Omega_X)$.

Define the real submanifolds $N = \{w \in Y; \mathrm{Im}\, w = 0\}$ and $M = \{z \in X; (\mathrm{Im}\, z)^2 = 1\}$. Let us calculate $\Phi_K(A_N)$.

For $z = x + \sqrt{-1}y, (x \in \mathbb{R}^n, y \in \mathbb{R}^n)$, we have $\Phi_K(A_N)_z = R\Gamma_c(S_z; A_Y)$, where:

$$S_z = \{w \in N; (z, w) \in Z_+\} = \{w \in \mathbb{R}^n; \langle x - w, y\rangle = 0, (x - w)^2 < y^2 - 1\} \ .$$

Hence $S_z = \emptyset$ for $y^2 \leqslant 1$ and S_z is homeomorphic to the $(n-1)$-dimensional open ball for $y^2 > 1$. Therefore we obtain:

$$\Phi_K(A_N) \simeq A_{\{y^2 > 1\}}[1 - n] \ .$$

7.3. Microlocal composition of kernels

Let X, Y and Z be three manifolds and let p_X, p_Y and p_Z be a point of T^*X, T^*Y and T^*Z respectively. We set $x_\circ = \pi_X(p_X)$, $y_\circ = \pi_Y(p_Y)$ and $z_\circ = \pi_Z(p_Z)$. We keep the same notations p_{ij}, q_{ij}, p_{ij}^a, etc. as in §1.

For a pair (K_1, K_2) of an object K_1 of $\mathbf{D}^b(X \times Y; (p_X, p_Y^a))$ and an object K_2 of $\mathbf{D}^b(Y \times Z; (p_Y, p_Z^a))$, we say that (K_1, K_2) is **microlocally composable** (at (p_X, p_Y, p_Z)) if it satisfies

$$(7.3.1) \qquad \begin{cases} \mathrm{SS}(K_1) \underset{T^*Y}{\times} \mathrm{SS}(K_2) \cap p_{13}^{a-1}((p_X, p_Z^a)) \subset \{((p_X, p_Y^a), (p_Y, p_Z^a))\} \\ \text{on a neighborhood of } ((p_X, p_Y^a), (p_Y, p_Z^a)) . \end{cases}$$

Note that, if F belongs to $\mathbf{D}^b(X; p)$ then the germ of $\mathrm{SS}(F)$ at p is well-defined. By this remark, one sees that (7.3.1) makes sense.

Proposition 7.3.1. *Let $(K_1, K_2) \in \mathrm{Ob}(\mathbf{D}^b(X \times Y)) \times \mathrm{Ob}(\mathbf{D}^b(Y \times Z))$ be a microlocally composable pair at (p_X, p_Y, p_Z).*

(i) *"$\varprojlim$" $K_1' \circ K_2'$ belongs to $\mathbf{D}^b(X \times Z; (p_X, p_Z^a))$, where $K_1' \to K_1$ ranges over the category of isomorphisms at (p_X, p_Y^a) and $K_2' \to K_2$ ranges over the category of isomorphisms at (p_Y, p_Z^a).*

Moreover, for any neighborhood W of (p_X, p_Y, p_Z^a) in $T^*X \times T^*Y \times T^*Z$, we have:

$$(7.3.2) \qquad \begin{cases} \mathrm{SS}(\text{"}\varprojlim\text{"}\, K_1' \circ K_2') \subset q_{13}^a\left(W \cap \left(\mathrm{SS}(K_1) \underset{T^*Y}{\times} \mathrm{SS}(K_2)\right)\right) \\ \textit{on a neighborhood of } (p_X, p_Z^a) . \end{cases}$$

(ii) *If moreover (K_1, K_2) satisfies the following conditions:*

$$(7.3.3) \qquad \left(\mathrm{SS}(K_1) \underset{T^*Y}{\times} \mathrm{SS}(K_2)\right) \cap \{p_X\} \times T^*_{y_\circ}Y \times \{p_Z^a\} \subset \{(p_X, p_Y, p_Z^a)\} ,$$

$$(7.3.4) \qquad \left(\mathrm{SS}(K_1) \underset{T^*Y}{\times} \mathrm{SS}(K_2)\right) \cap \{(x_\circ; 0)\} \times \dot{T}^*_{y_\circ}Y \times \{(z_\circ; 0)\} = \varnothing ,$$

then "$\varprojlim_V$" $(K_1)_{X \times V} \circ K_2$ belongs to $\mathbf{D}^b(X \times Z; (p_X, p_Z^a))$ and is isomorphic to "$\varprojlim$" $K_1' \circ K_2'$. Here, V ranges over an open neighborhood system of $y_\circ$.

(iii) *There exist morphisms $K_1' \to K_1$ and $K_2' \to K_2$ such that they are isomorphisms at (p_X, p_Y^a) and (p_Y, p_Z^a) respectively, and that (K_1', K_2') satisfies (7.3.3) and (7.3.4).*

(iv) *If $p_X \in \dot{T}^*X$, then "$\varprojlim$" $K_1' \circ K_2$ belongs to $\mathbf{D}^b(X \times Z; (p_X, p_Z^a))$ and it is isomorphic to "$\varprojlim$" $K_1' \circ K_2'$.*

Definition 7.3.2. *For a pair $(K_1, K_2) \in \mathrm{Ob}(\mathbf{D}^b(X \times Y; (p_X, p_Y^a))) \times \mathrm{Ob}(\mathbf{D}^b(Y \times Z; (p_Y, p_Z^a)))$, the pro-object "$\varprojlim$" $K_1' \circ K_2'$ of $\mathbf{D}^b(X \times Z; (p_X, p_Z^a))$ given in the preceding proposition is denoted by $K_1 \circ_\mu K_2$ and called the microlocal composition of K_1 and K_2.*

In order to prove Proposition 7.3.1, let us begin with the following lemma.

Lemma 7.3.3. *The notations being as in Proposition* 7.3.1, *assume* $p_X \in \dot{T}^*X$. *Let* S *be a closed conic subset of* $T^*(Y \times Z)$ *satisfying*:

$$\text{(7.3.5)}\quad SS(K_1) \underset{T^*Y}{\times} S \subset \{(p_X, p_Y, p_Z^a)\} \text{ on a neighborhood of } (p_X, p_Y, p_Z^a)\ .$$

Then there exists a morphism $\varphi : K_1' \to K_1$ *in* $\mathbf{D}^b(X \times Y)$ *such that* φ *is an isomorphism at* (p_X, p_Z^a) *and* K_1' *satisfies the following conditions*:

$$\text{(7.3.6)}\qquad SS(K_1') \underset{T^*Y}{\times} S \cap (\{p_X\} \times T_{y_\circ}^*Y \times \{p_Z^a\}) \subset \{(p_X, p_Y, p_Z^a)\}\ ,$$

$$\text{(7.3.7)}\qquad SS(K_1') \cap ((T_X^*X)_{x_\circ} \times \dot{T}_{y_\circ}^*Y) = \emptyset\ .$$

Proof. Let us take a compact neighborhood L of p_Y in $T_{y_\circ}^*Y$ such that $(SS(K_1) \times_{T^*Y} S) \cap (\{p_X\} \times L \times \{p_Z^a\}) \subset \{(p_X, p_Y, p_Z^a)\}$. Let G be the image of $S \cap (T_{y_\circ}^*Y \times \{p_Z^a\})$ by $p_1 : T^*(Y \times Z) \to T^*Y$. Then we have:

$$\text{(7.3.8)}\qquad (\{p_X\} \times (G \cap L)^a) \cap SS(K_1) \subset \{(p_X, p_Y^a)\}\ .$$

Take a proper closed convex cone γ in $T_{(x_\circ, y_\circ)}^*(X \times Y)$ such that we have:

$$\text{(7.3.9)}\qquad (\{p_X\} \times T_{Y_\circ}^*Y) \cap \operatorname{Int}\gamma \subset \{p_X\} \times L^a \quad\text{and}\quad (p_X, p_Y^a) \in \operatorname{Int}\gamma\ ,$$

$$\text{(7.3.10)}\qquad \gamma \cap (\{(x_\circ; 0)\} \times \dot{T}_{y_\circ}^*Y) = \emptyset\ .$$

Here $(x_\circ; 0)$ denotes the origin of $T_{x_\circ}^*X$.

Then, take an open conic set U such that

$$\operatorname{Int}\gamma \supset U \ni (p_X, p_Y^a)\ .$$

Now, applying the refined microlocal cut-off lemma (Proposition 6.1.4), we obtain a morphism $\varphi : K_1' \to K_1$ such that:

$$\text{(7.3.11)}\qquad \varphi \text{ is an isomorphism on } U\ ,$$

$$\text{(7.3.12)}\qquad SS(K_1') \subset U \cup (\gamma \setminus (\{p_X\} \times G^a))\ .$$

Then φ satisfies the desired properties. □

Proof of Proposition 7.3.1. First, we shall show that (7.3.1), (7.3.3) and (7.3.4) imply:

$$\text{(7.3.13)}\qquad \begin{cases} \left(SS(K_1) \underset{T^*Y}{\times} SS(K_2)\right) \cap (\{p_X\} \times T^*Y \times \{p_Z^a\}) \subset \{(p_X, p_Y, p_Z^a)\} \\ \text{on a neighborhood of } (x_\circ, y_\circ, z_\circ)\ . \end{cases}$$

If (7.3.13) is false then there exists a sequence $\{(y_n; \eta_n)\} \subset T^*Y \setminus \{p_Y\}$ such that $y_n \underset{n}{\to} y$, $(p_X, (y_n; -\eta_n)) \in SS(K_1)$, $((y_n; \eta_n), p_Z^a) \in SS(K_2)$. If $\{\eta_n\}$ is bounded, then

$\{(y_n;\eta_n)\}$ converges to p_Y by (7.3.3), which contradicts (7.3.1). If $\{\eta_n\}$ is unbounded, we may assume that $\eta_n/|\eta_n|$ converges to $\eta \neq 0$. Then $((x_\circ;0),(y_\circ;-\eta)) \in SS(K_1)$ and $((y_\circ;\eta),(z_\circ;0)) \in SS(K_2)$, which contradicts (7.3.4).

Now we shall prove the proposition in several steps.

(a) First we assume $p_X \in \dot{T}^*X$. Then (iii) follows from Lemma 7.3.3. Moreover, in order to prove (i), (ii), (iv), we may assume that (K_1, K_2) satisfies (7.3.3) and (7.3.4) by using the same lemma.

Then, by (7.3.4), $SS(q_{12}^{-1}K_1) \cap SS(q_{23}^{-1}K_2)^a \subset T^*_{X\times Y\times Z}(X \times Y \times Z)$ on a neighborhood of $(x_\circ, y_\circ, z_\circ)$. Hence, by Proposition 5.4.14, $G = q_{12}^{-1}K_1 \otimes^L q_{23}^{-1}K_2$ satisfies:

$$(7.3.14)\qquad \begin{cases} SS(G) \subset S = SS(K_1) \times T^*_Z Z + T^*_X X \times SS(K_2) \\ \text{on a neighborhood of}\,(x_\circ, y_\circ, z_\circ)\ . \end{cases}$$

Since $({}^t q'_{13})^{-1}(S)$ is isomorphic to the image of $SS(K_1) \times_{T^*Y} SS(K_2)$ to $T^*X \times Y \times T^*Z$, (7.3.13) and (7.3.14) imply:

$$(7.3.15)\qquad \begin{cases} ({}^t q'_{13})^{-1}(SS(G)) \cap q_{13\pi}^{-1}((p_X, p_Z^a)) \subset \{(p_X, y_\circ, p_Z^a)\} \\ \text{on a neighborhood of}\,(p_X, y_\circ, p_Z^a)\ . \end{cases}$$

Hence, we can apply Proposition 6.1.10 to conclude that $\text{“}\varprojlim_V\text{''}\, Rf_*(G_{X\times V\times Y}) = \text{“}\varprojlim_V\text{''}\,(K_1)_{X\times V} \circ K_2$ belongs to $\mathbf{D}^b(X \times Z;(p_X, p_Z^a))$. Now let $K'_1 \to K_1$ and $K'_2 \to K_2$ be isomorphisms at (p_X, p_Y^a) and (p_Y, p_Z^a), respectively. There exists an isomorphism $K''_1 \to K'_1$ at (p_X, p_Y^a) such that (K''_1, K'_2) and (K''_1, K_2) satisfy the conditions (7.3.3), (7.3.4). Then $q_{12}^{-1}K''_1 \otimes^L q_{23}^{-1}K'_2 \to q_{12}^{-1}K''_1 \otimes^L q_{23}^{-1}K_2 \to q_{12}^{-1}K_1 \otimes^L q_{23}^{-1}K_2$ are isomorphisms at $(p_X, y_\circ, p_Z^a)$ by Proposition 5.4.14. Hence we have by Proposition 6.1.10:

$$(7.3.16)\qquad \text{“}\varprojlim_V\text{''}\,(K''_1)_{X\times V} \circ K'_2 \xrightarrow{\sim} \text{“}\varprojlim_V\text{''}\,(K_1)_{X\times V} \circ K_2\ .$$

Taking the “projective limit” with respect to K''_1 we get:

$$(7.3.17)\qquad \text{“}\varprojlim_{K'_1}\text{''}\, K'_1 \circ K'_2 \xrightarrow{\sim} \text{“}\varprojlim_V\text{''}\,(K_1)_{X\times V} \circ K_2\ .$$

Setting $K'_2 = K_2$ we obtain:

$$\text{“}\varprojlim_{K'_1}\text{''}\, K'_1 \circ K_2 \xrightarrow{\sim} \text{“}\varprojlim_V\text{''}\,(K_1)_{X\times V} \circ K_2\ ,$$

and taking the “projective limit” with respect to K'_2 in (7.3.17), we obtain $\text{“}\varprojlim_{K'_1,K'_2}\text{''}\, K'_1 \circ K'_2 \xrightarrow{\sim} \text{“}\varprojlim_V\text{''}(K_1)_{X\times V} \circ K_2$.

(b) $p_Z \in \dot{T}^*Z$. The proof is similar to that of (a).

(c) Assume $p_X \in T^*_X X$, $p_Z \in T^*_Z Z$ and $p_Y \in \dot{T}^*Y$. Since $(p_X, tp_Y, p_Z^a) \notin SS(K_1) \times_{T^*Y} SS(K_2)$ for $t > 0$, we have either $(p_X, p_Y^a) \notin SS(K_1)$ or $(p_Y, p_Z^a) \notin SS(K_2)$, and the results follow.

(d) $(p_X, p_Y, p_Z) \in T_X^*X \times T_Y^*Y \times T_Z^*Z$. Since $\mathrm{supp}(K_1) \times_Y \mathrm{supp}(K_2) \cap \{x_\circ\} \times Y \times \{z_\circ\} \subset \{(x_\circ, y_\circ, z_\circ)\}$ on a neighborhood of $(x_\circ, y_\circ, z_\circ)$, (K_1, K_2) satisfies the conditions (7.3.3) and (7.3.4), and $\text{“}\varinjlim_{K_1', K_2'}\text{”}\, K_1' \circ K_2' \xrightarrow{\sim} \text{“}\varinjlim_{V}\text{”}\, (K_1)_{X \times V} \circ K_2$ belongs to $\mathbf{D}^b(X \times Z; (p_X, p_Z^a))$. □

Proposition 7.3.4. *Let q_1' and q_2' denote the projection from $(X \times Z) \times (Y \times Z)$ to $X \times Z$ and $Y \times Z$ respectively and let i denote the diagonal embedding $X \times Y \times Z \hookrightarrow (X \times Z) \times (Y \times Z)$. Let $(K_1, K_2) \in \mathrm{Ob}(\mathbf{D}^b(X \times Y; (p_X, p_Y^a))) \times \mathrm{Ob}(\mathbf{D}^b(Y \times Z; (p_Y, p_Z^a)))$ be a microlocally composable pair at (p_X, p_Y, p_Z). Then $(i_*(K_1 \boxtimes A_Z), K_2) \in \mathrm{Ob}(\mathbf{D}^b(X \times Z \times Y \times Z; (p_X, p_Z^a, p_Y^a, p_Z))) \times \mathrm{Ob}(\mathbf{D}^b(Y \times Z \times \{\mathrm{pt}\}; (p_Y, p_Z^a, \mathrm{pt})))$ is microlocally composable at $((p_X, p_Z^a), (p_Y, p_Z^a), \mathrm{pt})$ and:*

$$K_1 \underset{\mu}{\circ} K_2 \simeq i_*(K_1 \boxtimes A_Z) \underset{\mu}{\circ} K_2 \ .$$

The proof is straightforward.

Proposition 7.3.5. *Let X, Y, Z and W be four manifolds, $p_X \in T^*X$, $p_Y \in T^*Y$, $p_Z \in T^*Z$, $p_W \in T^*W$. Let $K_1 \in \mathrm{Ob}(\mathbf{D}^b(X \times Y; (p_X, p_Y^a)))$, $K_2 \in \mathrm{Ob}(\mathbf{D}^b(Y \times Z; (p_Y, p_Z^a)))$, $K_3 \in \mathrm{Ob}(\mathbf{D}^b(Z \times W; (p_Z, p_W^a)))$. Assume:*

$$\text{(7.3.18)} \quad \begin{cases} SS(K_1) \underset{T^*Y}{\times} SS(K_2) \underset{T^*Z}{\times} SS(K_3) = \{(p_X, p_Y^a), (p_Y, p_Z^a), (p_Z, p_W^a)\} \\ \textit{in a neighborhood of this point} \ . \end{cases}$$

Then (K_1, K_2), (K_2, K_3), $(K_1 \circ_\mu K_2, K_3)$ and $(K_1, K_2 \circ_\mu K_3)$ are microlocally composable and:

$$\text{(7.3.19)} \quad \left(K_1 \underset{\mu}{\circ} K_2\right) \underset{\mu}{\circ} K_3 \simeq K_1 \underset{\mu}{\circ} \left(K_2 \underset{\mu}{\circ} K_3\right) \text{ in } \mathbf{D}^b(X \times W; (p_X, p_W^a)) \ .$$

The proof is straightforward.

Proposition 7.3.6. *Let X, Y, Z, X', Y', Z' be six manifolds and $p_W \in T^*W$ ($W = X, Y, Z, X', Y', Z'$).*

Let $K_1 \in \mathrm{Ob}(\mathbf{D}^b(X \times Y; (p_X, p_Y^a)))$, $K_2 \in \mathrm{Ob}(\mathbf{D}^b(Y \times Z; (p_Y, p_Z^a)))$, $K_1' \in \mathrm{Ob}(\mathbf{D}^b(X' \times Y'; (p_{X'}, p_{Y'}^a)))$ and $K_2' \in \mathrm{Ob}(\mathbf{D}^b(Y' \times Z'; (p_{Y'}, p_{Z'}^a)))$. Assume that (K_1, K_2) and (K_1', K_2') are microlocally composable. Then $(K_1 \boxtimes^L K_1', K_2 \boxtimes^L K_2')$ is microlocally composable and we have:

$$\text{(7.3.20)} \quad \left(K_1 \overset{L}{\boxtimes} K_1'\right) \underset{\mu}{\circ} \left(K_2 \overset{L}{\boxtimes} K_2'\right) \simeq \left(K_1 \underset{\mu}{\circ} K_2\right) \overset{L}{\boxtimes} \left(K_1 \underset{\mu}{\circ} K_2'\right) .$$

The proof is straightforward.

To end this section, we define a class of kernels which are microlocally composable with any kernels.

Definition 7.3.7. *One denotes by* $\mathbf{N}(X, Y; p_X, p_Y)$ *the full triangulated subcategory of* $\mathbf{D}^b(X \times Y; (p_X, p_Y^a))$ *consisting of objects K such that*

(7.3.21) $SS(K) \cap (\{p_X\} \times T^*Y) \subset \{(p_X, p_Y^a)\}$ *on a neighborhood of* (p_X, p_Y^a) .

Then one immediately sees:

Proposition 7.3.8. *For any* $K \in \mathrm{Ob}(\mathbf{N}(X, Y; p_X, p_Y))$ *and* $L \in \mathrm{Ob}(\mathbf{D}^b(Y \times Z; (p_Y, p_Z^a)))$, (K, L) *is microlocally composable at* (p_X, p_Y, p_Z), *and the microlocal composition of kernels induces functors*:

(7.3.22) $\cdot \underset{\mu}{\circ} \cdot : \mathbf{N}(X, Y; p_X, p_Y) \times \mathbf{D}^b(Y \times Z; (p_Y, p_Z^a)) \to \mathbf{D}^b(X \times Z; (p_X, p_Z^a))$,

(7.2.23) $\cdot \underset{\mu}{\circ} \cdot : \mathbf{N}(X, Y; p_X, p_Y) \times \mathbf{N}(Y, Z; p_Y, p_Z) \to \mathbf{N}(X, Z; p_X, p_Z)$.

7.4. Integral transformations for sheaves associated with submanifolds

Let us begin with an elementary result of differential geometry. Let $f: Y \to X$ be a morphism of manifolds and N (resp. M) a closed submanifold of Y (resp. X). We assume f is smooth and we identify $Y \times_X T^*X$ with a closed involutive submanifold of T^*Y. We set:

(7.4.1) $$V = Y \underset{X}{\times} \dot{T}^*X \ , \qquad \Lambda = \dot{T}^*_N Y \ .$$

Let $p \in \Lambda \cap V$. We shall assume:

(7.4.2) *the intersection* $\Lambda \cap V$ *is clean at p.*

Recall the maps ${}^t f'$ and f_π, from $Y \times_x T^*X$ to T^*Y and T^*X, respectively. By (7.4.2) the map $f_\pi|_{\Lambda \cap V} : \Lambda \cap V \to T^*X$ has constant rank (cf. Exercise A.5), and for W a sufficiently small open neighborhood of p, $f_\pi(W \cap \Lambda \cap V)$ is a Lagrangian manifold. We shall assume that there exists a submanifold M of X such that:

(7.4.3) $f_\pi(W \cap \Lambda \cap V) = T^*_M X$ in a neighborhood of $f_\pi(p)$.

We set $y_\circ = \pi_Y(p)$, $x_\circ = \pi_X(f_\pi(p))$ and $\lambda_\circ(p) = T_p \pi_Y^{-1}(y_\circ)$.

Proposition 7.4.1. *In the preceding situation* (f *smooth*, (7.4.2), (7.4.3)), *we assume moreover that N is a hypersurface of Y. Then there exist local coordinate systems* $(x) = (x_1, x', x'')$ *on X and* (x, t, u) *on Y, at* $x_\circ$ *and* $y_\circ$ *respectively, such that*: $p = (0; dx_1)$, $f(x, t, u) = x$, $M = \{x \in X; x_1 = x' = 0\}$, $N = \{(x, t, u) \in Y;\ x_1 = q(t) + \langle x', u \rangle\}$, *where* $q(t)$ *is a quadratic form, and* $x_1 \in \mathbb{R}$, $x', u \in \mathbb{R}^r$, $t \in \mathbb{R}^n$, $x'' \in \mathbb{R}^m$ *for some r, n, m* $\in \mathbb{N}$. *Moreover if one replaces the hypothesis* (7.4.2) *by*:

(7.4.4) *the intersection* $\Lambda \cap V$ *is transversal at* p,

then the quadratic form $q(t)$ *is non-degenerate.*

Proof. The corank of the projection $\Lambda \cap V \to Y$ at p is equal to:

$$\dim(\lambda_\circ(p) \cap T_p(V \cap \Lambda)) = \dim(\lambda_\circ(p) \cap T_pV \cap T_p\Lambda) = 1 \ .$$

Since this formula is also true on a neighborhood of p in $V \cap \Lambda$, $\pi_Y(\Lambda \cap V)$ is a smooth submanifold L of Y. Moreover $L \simeq (\Lambda \cap V)/\mathbb{R}^+$. (Note that L is the discriminant locus of $f|_N : N \to X$.) Consider the diagram:

$$\begin{array}{ccccc} \dot{T}^*_N Y & \hookleftarrow & \Lambda \cap V & \xrightarrow[\tilde{f}_L]{} & \dot{T}^*_M X \\ \downarrow & \square & \downarrow & & \downarrow \\ N & \hookleftarrow & L & \xrightarrow[f_L]{} & M \end{array} ,$$

where we set $f_L = f|_L$, $\tilde{f}_L = f_\pi|_{\Lambda \cap V}$.

We choose coordinates (x_1, x', x'') on X such that $M = \{x_1 = x' = 0\}$, and $x_\circ = 0$, $f_\pi(p) = dx_1$. Let $(x_1, x', x''; \xi_1, \xi', \xi'')$ be the associated coordinates on $\dot{T}^*X$, with $\xi_1 \neq 0$. Since $L \simeq (\Lambda \cap V)/\mathbb{R}^+$, and $\tilde{f}_L$ is smooth, $u = \xi'/\xi_1 \circ \tilde{f}_L$ and $x'' \circ f_L$ define coordinates on L. We write x'' instead of $x'' \circ f_L$. Then we complete the coordinates (x'', u) on L by a coordinate system (x'', u, t'') on L, that we extend to Y; (note that if the intersection $\Lambda \cap V$ is transversal, $\tilde{f}_L$ is an isomorphism and (x'', u) is a coordinate system on L). Then we have the coordinates (x_1, x', x'', t'', u) on Y that we complete with t' vanishing on L. We set $t = (t', t'')$. Then:

$$f(x_1, x', x'', t', t'', u) = (x_1, x', x'') \ ,$$

$$L = \{x_1 = x' = t' = 0\} \ ,$$

$$N = \{x_1 = g(x', x'', t', t'', u)\} \quad \text{for some function } g \ .$$

Since $\partial_{x'} g = \xi'/\xi_1$ on L, we may write:

$$g = \langle x', u \rangle + h(x', x'', t', t'', u) \ ,$$

with:

(7.4.5) $$\partial_{x'} h = 0 \quad \text{on} \quad \{x' = t' = 0\} \ .$$

The inclusion $L \subset N$ implies:

(7.4.6) $$h = 0 \quad \text{on} \quad \{x' = t' = 0\} \ ,$$

and the inclusion $T^*_N Y \times_N L \subset V$ implies:

(7.4.7) $$\partial_{t'} h = 0 \quad \text{on} \quad \{x' = t' = 0\} \ .$$

Let us assume for a while:

(7.4.8) $$\partial^2_{t't'}h(0) \quad \text{is non-degenerate} \ .$$

Then by the Morse Lemma (cf. Hörmander [4, III Appendix C]) we may write, by replacing $t' = (t_1, \ldots, t_r)$ with other coordinates:

$$h(0, x'', t', t'', u) = \sum_{j=1}^{r} \pm t_j^2 \ ,$$

and we get:

$$h(x', x'', t', t'', u) = q(t') + \langle x', \varphi(x', x'', t', t'', u) \rangle \ ,$$

where $q(t')$ is a non-degenerate quadratic form (with respect to the t'-variable). By (7.4.5), $\varphi(0, x'', 0, t'', u) = 0$. Then replacing u by $u - \varphi$, we get the result.

Finally we prove (7.4.8). Let $(x, t, u; \xi, \tau, v)$ denote the coordinates on T^*Y, with $\xi = (\xi_1, \xi', \xi'')$, $\tau = (\tau', \tau'')$. We have:

$$\begin{aligned} \Lambda = \{ & x_1 = \langle x', u \rangle + h, -\xi'/\xi_1 = u + \partial_{x'} h, -\xi''/\xi_1 = \partial_{x''} h, \\ & -\tau/\xi_1 = \partial_t h, -v/\xi_1 = x' + \partial_u h \} \ , \\ V = \{ & \tau = v = 0 \} \ . \end{aligned}$$

By (7.4.5)–(7.4.7) we get:

$$\begin{aligned} T_p\Lambda = \{ & x_1 = 0, -\xi' = u + \partial^2_{x'x'} h(0) \cdot x' + \partial^2_{x't'} h(0) \cdot t', \\ & -\tau' = \partial^2_{t't'} h(0) \cdot t' + \partial^2_{t'x'} h(0) \cdot x', \xi'' = \tau'' = 0, -v = x' \} \ , \\ T_pV = \{ & \tau = v = 0 \} \ . \end{aligned}$$

Hence:

$$\begin{aligned} T_p\Lambda \cap T_pV = \{ & x_1 = x' = \tau = v = \xi'' = 0, \\ & -\xi' = u + \partial^2_{x't'} h(0) \cdot t', 0 = \partial^2_{t't'} h(0) \cdot t' \} \ . \end{aligned}$$

Since the image of $T_p\Lambda \cap T_pV$ by the map $T_p(T^*Y) \to T_{y_0}Y$ is contained in $T_{y_0}L$, we have: $(x, t, u; \xi, \tau, v) \in T_p\Lambda \cap T_pV \Rightarrow t' = 0$. Thus $\partial^2_{t't'}h(0)$ is non-degenerate. This completes the proof. □

Now we keep the coordinate systems given by Proposition 7.4.1 and we set:

(7.4.9) $$N^+ = \{ (x, t, u); x_1 \geqslant q(t) + \langle x', u \rangle \} \ .$$

Let n_+ (resp. n_-) be the number of positive (resp. negative) eigenvalues of $q(t)$, and define n_0 by the relation:

(7.4.10) $$n_+ + n_0 + n_- = \dim Y/X - \operatorname{codim} M + 1 \ .$$

In other words, n_0 is the dimension of the space of the totally isotropic space of q.
In particular, $n_0 = 0$ if q is non-degenerate.

Proposition 7.4.2. *We keep the hypotheses of Proposition* 7.4.1, *and let* $L \in \mathrm{Ob}(\mathbf{D}^b(\mathfrak{Mod}(A)))$.

(i) *There exists a morphism in the category of pro-objects of* $\mathbf{D}^b(X)$:

$$\text{"}\varprojlim_{U}\text{"}\, Rf_!(L_{N^+\cap U}) \to L_M[\delta]$$

with $\delta = 1 - \operatorname{codim} M - (n_0 + n_-) = n_+ - \dim Y/X$. *Here U ranges through the family of open neighborhoods of* $y_\circ$.

(ii) $f_!^\mu L_{N^+} \simeq \text{"}\varprojlim_{U}\text{"}\, Rf_!(L_{N^+\cap U})$ *exists in* $\mathbf{D}^b(X; f_\pi(p))$ *and the morphism in* (i) *is an isomorphism in* $\mathbf{D}^b(X; f_\pi(p))$ (*cf. Definition* 6.1.7 *and Proposition* 6.1.10).

Note that (i) and (ii) follow from the next two statements that we are going to prove.

(a) For an open neighborhood U of $y_\circ$, there is a morphism $Rf_!(L_{N^+\cap U}) \to L_M[\delta]$ in $\mathbf{D}^b(X)$.

(b) There exists an open neighborhood system of $y_\circ$ in U such that, for any U' in such a system, the morphism $Rf_!(L_{N^+\cap U'}) \to L_M[\delta]$ is an isomorphism in $\mathbf{D}^b(X; f_\pi(p))$.

Also note that "$\varprojlim_{U}$" $Rf_!(L_{N^+\cap U})$ does not exist in $\mathbf{D}^b(X)$.

Proof. We decompose the coordinates (t) as $(t) = (s', s'', t')$ such that

$$q(t) = s'^2 - s''^2$$

and we decompose f as follows:

$$f: Y \xrightarrow[f_1]{} Y_1 \xrightarrow[f_2]{} Y_2 \xrightarrow[f_3]{} Y_3 \xrightarrow[f_4]{} X$$

with $f_1(x, s', s'', t', u) = (x, s', t', u)$, $f_2(x, s', t', u) = (x, t', u)$, $f_3(x, t', u) = (x, u)$, $f_4(x, u) = (x)$.

Let $\eta > 0$, $\varepsilon > 0$ with $\eta \ll \varepsilon$. Set:

$$U_0 = \{(x, s', s'', t', u); |x| < \eta, |t'| < \varepsilon, |u| < \varepsilon, s'^2 < \varepsilon, s''^2 < \varepsilon\}\ .$$

(a) Let us calculate $Rf_{1!}L_{N^+\cap U_0}$. We have for $|x| < \eta$, $|t''| < \varepsilon$, $|u| < \varepsilon$, $s'^2 < \varepsilon$:

$$f_1^{-1}(x, s', t', u) \cap N_+ \cap U_0 = \{s''; \langle x', u\rangle + s'^2 - x_1 \leqslant s''^2 < \varepsilon\}\ .$$

This set is empty for $\langle x', u\rangle + s'^2 - x_1 \geqslant \varepsilon$, it is an open ball for $\langle x', u\rangle + s'^2 - x_1 \leqslant 0$, and it is homeomorphic to the set (with $0 < a < b$):

$$S_a^b = \{a \leqslant s''^2 < b\} \qquad \text{for} \qquad 0 < \langle x', u\rangle + s'^2 - x_1 < \varepsilon\ .$$

Since $R\Gamma_c(S_a^b; L_{\mathbb{R}^{n_-}}) = 0$, we get:

$$Rf_{1!}L_{N^+\cap U_0} \simeq L_{N_1^+\cap U_1}[-n_-] ,$$

where $U_1 = \{|x| < \eta, |t'| < \varepsilon, |u| < \varepsilon, s'^2 < \varepsilon\}$ and $N_1^+ = \{x_1 \geqslant s'^2 + \langle x', u\rangle\}$.

(b) Let us calculate $Rf_{2!}L_{N_1^+\cap U_1}$. We have for $|x| < \eta$, $|t'| < \varepsilon$, $|u| < \varepsilon$:

$$f_2^{-1}(x, t', u) \cap N_1^+ \cap U_1 = \{s'; s'^2 < \varepsilon, s'^2 \leqslant x_1 - \langle x', u\rangle\} .$$

For $0 \leqslant x_1 - \langle x', u\rangle < \varepsilon$, this set is a closed ball or $\{0\}$. For $x_1 - \langle x', u\rangle < 0$, this set is empty. Since $|x| < \eta$ and $\eta \ll \varepsilon$, we need not to consider the case where $x_1 - \langle x', u\rangle \geqslant \varepsilon$. Hence:

$$Rf_{2!}L_{N_1^+\cap U_1} \simeq L_{N_2^+\cap U_2} ,$$

where $U_2 = \{|x| < \eta, |t'| < \varepsilon, |u| < \varepsilon\}$ and $N_2^+ = \{x_1 \geqslant \langle x', u\rangle\}$.

(c) Let us calculate $Rf_{3!}L_{N_2^+\cap U_2}$. We find immediately:

$$Rf_{3!}L_{N_2^+\cap U_2} = L_{N_3^+\cap U_3}[-n_0] ,$$

where $U_3 = \{|x| < \eta, |u| < \varepsilon\}$ and $N_3^+ = \{x_1 \geqslant \langle x', u\rangle\}$.

(d) Finally, we calculate $Rf_{4!}L_{N_3^+\cap U_3}$. Consider the distinguished triangle:

$$L_{\{|u| < \varepsilon, \langle x', u\rangle > x_1\}} \longrightarrow L_{\{|u| < \varepsilon\}} \longrightarrow L_{\{|u| < \varepsilon, \langle x', u\rangle \leqslant x_1\}} \xrightarrow[+1]{} .$$

The set $\{u; |u| < \varepsilon, \langle x', u\rangle > x_1\}$ is empty for $x_1 \geqslant \varepsilon|x'|$ and is a non-empty open convex set otherwise.

Hence $(Rf_{4!}(L_{N_3^+\cap U_3}))_x$ is isomorphic to $(Rf_{4!}L_{U_3})_x$ if $x_1 \geqslant \varepsilon|x'|$ and is zero otherwise:

$$\begin{aligned} Rf_{4!}L_{N_3^+\cap U_3} &\simeq (Rf_{4!}L_{\{|u| < \varepsilon\}})_{\{x_1 \geqslant \varepsilon|x'|\}\cap U_4} \\ &\simeq L_{N_4^+\cap U_4}[-r] , \end{aligned}$$

where $U_4 = \{|x| < \eta\}$, $N_4^+ = \{x_1 \geqslant \varepsilon|x'|\}$ and r is the dimension of the space of the u-variables, i.e. $r = \operatorname{codim} M - 1$.

(e) We have obtained:

$$Rf_!L_{N^+\cap U_0} \simeq L_{N_4^+\cap U_4}[\delta] ,$$

where $\delta = 1 - \operatorname{codim} M - n_0 - n_-$, and U_0, U_4, N_4^+ are defined above. To complete the proof it remains to note that the natural morphism $L_{\{x_1 \geqslant \varepsilon|x'|\}} \to L_M$ is an isomorphism in $\mathbf{D}^b(X; f_\pi(p))$. (Recall that $f_\pi(p) = (0; dx_1)$.) □

One can interpret the shift δ which appears in Proposition 7.4.2 using the inertia index. We refer to the appendix for the properties concerning this index.

We introduce some notations. In the situation of Proposition 7.4.1, we set:

$$(7.4.11) \quad \begin{cases} E_p = T_pT^*Y , & \lambda_N(p) = T_pT_N^*Y , \\ \lambda_0(p) = T_p\pi_Y^{-1}\pi_Y(p) , & \lambda_1(p) = T_pf_\pi^{-1}\pi_X^{-1}\pi_X(f_\pi(p)) . \end{cases}$$

Proposition 7.4.3. *In the situation of Proposition* 7.4.2, *we have*:

$$1 - \operatorname{codim} M - (n_0 + n_-)$$
$$= \tfrac{1}{2}[1 + \dim M - \dim Y - \dim((T_pV)^{\perp} \cap \lambda_N(p)) - \tau] \ ,$$

where $\tau = \tau_{E_p}(\lambda_\circ(p), \lambda_N(p), \lambda_1(p))$.

(Recall that $\tau_{E_p}(\cdot,\cdot,\cdot)$ is the inertia index of three Lagrangian linear subspaces in the symplectic vector space E_p.)

Proof. We make use of the coordinate systems (x,t,u) on Y, (x) on X given by Proposition 7.4.1, so that $f(x,t,u) = x$, $M = \{x_1 = x' = 0\}$, $N = \{x_1 = \langle x', u\rangle + q(t)\}$, with $q(t) = \frac{1}{2}\langle At, t\rangle$ for a symmetric matrix A. We denote by $(x,t,u;\xi,\tau,v)$ the associated coordinates on T^*Y.

Then:

$$\lambda_\circ(p) = \{x = t = u = 0\} \ ,$$
$$\lambda_N(p) = \{x_1 = \xi'' = 0, \xi' = -u, \tau = -At, v = -x'\} \ ,$$
$$\lambda_1(p) = \{x = \tau = v = 0\} \ .$$

Consider the isotropic space ρ of E_p defined by $\rho = \{x = 0\}^{\perp}$. We may identify E_p^{ρ} to the space of the $(t,u;\tau,v)$-variables. Since $\rho^{\perp} \supset \lambda_\circ(p) + \lambda_1(p)$, we have by Theorem A.3.2:

$$\tau = \tau_{E_p^{\rho}}(\{t = u = 0\}, \{v = 0, \tau = -At\}, \{\tau = v = 0\}) \ .$$

Consider the isotropic space $\{v = 0\}^{\perp}$. We get:

$$\tau = \tau(\{t = 0\}, \{\tau = -At\}, \{\tau = 0\}) \ ,$$

where the inertia index is now calculated in the space of the (t,τ)-variables. By Proposition A.3.6, we get:

$$\tau = -\operatorname{sgn}(A) \ , \tag{7.4.12}$$

where $\operatorname{sgn}(A) = n_+ - n_-$ is the number of positive eigenvalues minus the number of negative eigenvalues of q.

On the other hand, we have:

$$n_0 = \dim(T_pV^{\perp} \cap \lambda_N(p)) \ .$$

By writing $n_0 + n_- = \frac{1}{2}(n_0 + n_- + n_+ + n_0 - (n_+ - n_-))$, we find by (7.4.10):

$$1 - \operatorname{codim} M - \tfrac{1}{2}[\dim Y/X - \operatorname{codim} M + 1 + \dim(T_pV^{\perp} \cap \lambda_N(p)) + \tau]$$
$$= \tfrac{1}{2}[1 - \operatorname{codim} M - \dim Y/X - \dim(T_pV^{\perp} \cap \lambda_N(p)) - \tau] \ ,$$

which is the desired result. □

We shall apply the preceding results to compose kernels associated with hypersurfaces. First we need an elementary lemma.

Let $\Lambda_1 \subset T^*(X \times Y)$ and $\Lambda_2 \subset T^*(Y \times Z)$ be two (conic) Lagrangian submanifolds, let $(p_X, p_Y^a) \in \Lambda_1$, $(p_Y, p_Z^a) \in \Lambda_2$. We make the hypothesis:

$$\text{(7.4.13)} \qquad \begin{cases} p_2^a|_{\Lambda_1} : \Lambda_1 \to T^*Y \quad \text{and} \quad p_1|_{\Lambda_2} : \Lambda_2 \to T^*Y \\ \text{are transversal} \ . \end{cases}$$

As in §1, $T^*(X \times Y) \times_{T^*Y} T^*(Y \times Z)$ means the fiber product of p_2^a: $T^*(X \times Y) \to T^*Y$ and $p_1 : T^*(Y \times Z) \to T^*Y$.

Lemma 7.4.4. *Assume* (7.4.13). *Then* $\Lambda_1 \times_{T^*Y} \Lambda_2$ *is a smooth submanifold of* $T^*(X \times Y) \times_{T^*Y} T^*(Y \times Z)$. *Moreover, replacing* Λ_1 *and* Λ_2 *by* $\Lambda_1 \cap U$ *and* $\Lambda_2 \cap V$, *where* U *and* V *are sufficiently small open neighborhoods of* (p_X, p_Y^a) *and* (p_Y, p_Z^a) *respectively, one has*:

(i) p_{13} *induces an isomorphism of* $\Lambda_1 \times_{T^*Y} \Lambda_2$ *with a smooth Lagrangian submanifold* Λ *of* $T^*(X \times Z)$.
(ii) *Let* δ *denote the diagonal embedding* $X \times Y \times Z \hookrightarrow X \times Y \times Y \times Z$. *Then* δ_π *is transversal to* $\Lambda_1 \times \Lambda_2$ *and* $\Lambda_{12} = {}^t\delta'\delta_\pi^{-1}(\Lambda_1 \times \Lambda_2)$ *is a Lagrangian submanifold of* $T^*(X \times Y \times Z)$. *Moreover* ${}^tq'_{13}$ *is transversal to* Λ_{12} *and* $q_{13\pi}{}^tq'^{-1}_{13}(\Lambda_{12}) = \Lambda$.

Proof. The hypothesis implies that $\Lambda_1 \times_{T^*Y} \Lambda_2$ is a smooth submanifold of $T^*(X \times Y \times Z)$ with dimension $\dim(X \times Y) + \dim(Y \times Z) - \dim T^*Y = \dim(X \times Z)$.

(i) We set:

$$\text{(7.4.14)} \qquad \begin{cases} E_X = T_{p_X}T^*X \ , \quad E_Y = T_{p_Y}T^*Y \ , \quad E_Z = T_{p_Z}T^*Z \ , \\ \lambda_1 = T_{(p_X, p_Y^a)}\Lambda_1 \ , \quad \lambda_2 = T_{(p_Y, p_Z^a)}\Lambda_2 \ . \end{cases}$$

If E is a symplectic vector space, E^a denotes the space E endowed with the opposite skew-symmetric form.

We shall denote by "a" the map from E to E^a given by $x \mapsto x$. Then $E_Y^a \simeq T_{p_Y^a}(T^*Y)$, $E_Z^a \simeq T_{p_Z^a}(T^*Z)$ by a^*, and we still denote by λ_1 and λ_2 the Lagrangian planes of $E_X \oplus E_Y^a$ and $E_Y \oplus E_Z^a$ corresponding to λ_1 and λ_2 by the isomorphisms $T_{(p_X, p_Y^a)}T^*(X \times Y) \simeq E_X \oplus E_Y^a$ and $T_{(p_Y, p_Z^a)}T^*(Y \times Z) \simeq E_Y \oplus E_Z^a$, respectively.

Then by Proposition A.1.4 we have:

$$\lambda_1 \circ \lambda_2 = p_{13}\left(\lambda_1 \underset{E_Y}{\times} \lambda_2\right) ,$$

where p_{13} is the projection $(E_X \oplus E_Y^a) \times_{E_Y} (E_Y \oplus E_Z^a)$ to $E_X \oplus E_Z^a$.

Since $\lambda_1 \circ \lambda_2$ is a Lagrangian subspace of $E_X \oplus \hat{E}_Z^a$ and $\dim(\lambda_1 \times_{E_Y} \lambda_2) = \dim \lambda_1 + \dim \lambda_2 - \dim E_Y = \dim(X \times Z) = \dim \lambda_1 \circ \lambda_2$, we see that p_{13}: $\Lambda_1 \times_{T^*Y} \Lambda_2 \to T^*(X \times Z)$ is an embedding and $p_{13}(\Lambda_1 \times_{T^*Y} \Lambda_2)$ is Lagrangian.

(ii) It is enough to show that

$$\delta_\pi : T_{(p_X, p_Y^a, p_Y, p_Z^a)}\left(X \times Y \times Z \underset{X \times Y \times Y \times Z}{\times} T^*(X \times Y \times Y \times Z)\right)$$

$$\hookrightarrow T_{(p_X, p_Y^a, p_Y, p_Z^a)}(X \times Y \times Y \times Z)$$

is transversal to $\lambda_1 \times \lambda_2 \simeq T_{(p_X, p_Y^a, p_Y, p_Z^a)}(\Lambda_1 \times \Lambda_2)$ and

$${}^t q'_{13} : T_{(y_\circ, p_X, p_Z^a)}(Y \times T^*(X \times Z)) \hookrightarrow T_{(p_X, y_\circ, p_Z^a)} T^*(X \times Y \times Z)$$

is transversal to $\operatorname{Im}(\delta_\pi^{-1}(\lambda_1 \times \lambda_2) \xrightarrow{{}^t\delta} T_{(p_X, y_\circ, p_Z^a)} T^*(X \times Y \times Z))$, where $y_\circ = \pi(p_Y) \in T_Y^* Y$. Now, let $\rho' \subset \rho \subset T_{(p_Y^a, p_Y)}(T^*(Y \times Y))$ be the isotropic subspaces defined by

$$\rho'^{\perp} = T_{(p_Y^a, p_Y)}(Y \times_{Y \times Y} T^*(Y \times Y)) \text{ and } \rho = T_{(p_Y^a, p_Y)}(T_Y^*(Y \times Y)) \ .$$

Then we have:

$$T_{(p_X, p_Y^a, p_Y, p_Z^a)}\left(X \times Y \times Z \underset{X \times Y \times Y \times Z}{\times} T^*(X \times Y \times Y \times Z)\right) \simeq E_Y \oplus \rho'^{\perp} \oplus E_Z^a \ ,$$

$$T_{(p_X, y_\circ, p_Z^a)} T^*(X \times Y \times Z) \simeq (E_X \oplus E_Y^a \oplus E_Y \oplus E_Z^a)^{(0 \oplus \rho' \oplus 0)} \ ,$$

and

$$T_{(y_\circ, p_X, p_Z^a)}(Y \times T^*(X \times Z)) \simeq (E_X \oplus \rho \oplus E_Z^a)^{(0 \oplus \rho' \oplus 0)} \ .$$

Therefore the problem is reduced to

$$(\lambda_1 \oplus \lambda_2) + (E_X \oplus \rho'^{\perp} \oplus E_Z^a) = E_X \oplus E_Y^a \oplus E_Y \oplus E_Z^a \quad \text{and}$$

$$(\lambda_1 \oplus \lambda_2)^{(0 \oplus \rho' \oplus 0)} + (E_X \oplus \rho \oplus E_Z^a)^{(0 \oplus \rho' \oplus 0)} = (E_X \oplus E_Y^a \oplus E_Y \oplus E_Z^a)^{(0 \oplus \rho' \oplus 0)} \ .$$

This follows immediately from:

$$(\lambda_1 \oplus \lambda_2) \cap (0 \oplus \rho' \oplus 0) \subset (\lambda_1 \oplus \lambda_2) \cap (0 \oplus \rho \oplus 0)$$

$$\simeq \operatorname{Ker}\left(\lambda_1 \underset{E_Y}{\times} \lambda_2 \to E_X \oplus E_Z^a\right) = 0 \ . \quad \square$$

Definition 7.4.5. *In the situation of Lemma* 7.4.4, *one sets* $\Lambda = \Lambda_1 \circ \Lambda_2$.

Of course, $\Lambda_1 \circ \Lambda_2$ is only defined in a neighborhood of (p_X, p_Z^a). More precisely $\Lambda_1 \circ \Lambda_2$ is a germ of set at (p_X, p_Z^a). Now let $S_1 \subset X \times Y$ and $S_2 \subset Y \times Z$ be two hypersurfaces. We set:

$$\Lambda_1 = T_{S_1}^*(X \times Y) \cap (\dot{T}^*X \times \dot{T}^*Y) \ , \tag{7.4.15}$$

$$\Lambda_2 = T_{S_2}^*(Y \times Z) \cap (\dot{T}^*Y \times \dot{T}^*Z) \ .$$

Let $(p_X, p_Y^a) \in \Lambda_1$, $(p_Y, p_Z^a) \in \Lambda_2$. We set $x_\circ = \pi(p_X)$, $y_\circ = \pi(p_Y)$, $z_\circ = \pi(p_Z)$. Then $p_X \notin T_X^*X$ and $p_Z \notin T_Z^*Z$ imply:

(7.4.16) $$q_2|_{S_1}: S_1 \to Y \quad \text{and} \quad q_1|_{S_2}: S_2 \to Y \text{ are smooth} .$$

We assume (7.4.13) and the following:

(7.4.17) $$\begin{cases} \Lambda_1 \circ \Lambda_2 = T_S^*(X \times Z) \text{ on a neighborhood of } (p_X, p_Z^a), \\ \text{for a submanifold } S \text{ of } X \times Z . \end{cases}$$

We keep the notations (7.4.14) and we also set:

(7.4.18) $$\begin{cases} \lambda_\circ(p_X) = T_{p_X}\pi^{-1}(x_\circ) , \qquad \lambda_\circ(p_Y) = T_{p_Y}\pi^{-1}(y_0) , \\ \lambda_\circ(p_Z) = T_{p_Z}\pi^{-1}(z_\circ) . \end{cases}$$

We shall also write for short $\lambda_{\circ X}$ (resp. $\lambda_{\circ Y}$, $\lambda_{\circ Z}$) instead of $\lambda_\circ(p_X)$ (resp. $\lambda_\circ(p_Y)$, $\lambda_\circ(p_Z)$). These are Lagrangian subspaces of E_X, E_Y, E_Z, respectively. Finally we set:

(7.4.19) $$\begin{cases} \lambda_1(p_Y) = p_2^a(\lambda_1 \cap p_1^{-1}(\lambda_\circ(p_X))) \\ \qquad = T_{p_Y}(p_2^a(\Lambda_1 \cap \pi^{-1}(x_\circ) \times T^*Y)) = \lambda_\circ(p_X) \circ \lambda_1^a , \\ \lambda_2(p_Y) = p_1(\lambda_2 \cap p_2^{a-1}(\lambda_\circ(p_Z))) \\ \qquad = T_{p_Y}(p_1(\Lambda_2 \cap T^*Y \times \pi^{-1}(z_\circ))) = \lambda_2 \circ \lambda_\circ(p_Z) . \end{cases}$$

Here p_2^a is the projection $E_X \oplus E_Y^a \to E_Y^a$ or $E_Y \oplus E_Z^a \to E_Z^a$ and we used the notation in Proposition A.1.4 and the one given after Proposition A.3.8. Note that $\lambda_1(p_Y)$ and $\lambda_2(p_Y)$ are Lagrangian subspaces of E_Y.

Proposition 7.4.6. *Let $(p_X, p_Y, p_Z) \in \dot{T}^*X \times \dot{T}^*Y \times \dot{T}^*Z$ and let S_1 and S_2 be two hypersurfaces of $X \times Y$ and $Y \times Z$, as above.*

Let L' and L'' belong to $\mathbf{D}^b(\mathfrak{Mod}(A))$ and let $L = L' \otimes^L L''$. Assume the transversality condition (7.4.13) and assume (7.4.17). Then (L'_{S_1}, L''_{S_2}), is microlocally composable at (p_X, p_Y, p_Z) and there is an isomorphism in $\mathbf{D}^b(X \times Z; (p_X, p_Z^a))$:

(7.4.20) $$L'_{S_1} \underset{\mu}{\circ} L''_{S_2} \simeq L_S[\delta] ,$$

with

(7.4.21) $$\begin{cases} \delta = 1 - \frac{1}{2}[\dim Y + \operatorname{codim} S + \tau] , \\ \tau = \tau_{E_Y}(\lambda_\circ(p_Y), \lambda_2(p_Y), \lambda_1(p_Y)) . \end{cases}$$

Proof. The first statement (7.4.21) follows immediately from (7.4.20). By (7.4.16), $q_{12}^{-1}S_1$ and $q_{23}^{-1}S_2$ intersect transversally in $X \times Y \times Z$. Set:

(7.4.22) $$S_{12} = q_{12}^{-1}S_1 \cap q_{23}^{-1}S_2 , \qquad W = q_{12}^{-1}S_1 , \qquad f = q_{13}|_W .$$

By (7.4.16) f is smooth. Moreover, S_{12} is a hypersurface of W, ${}^t f'$ is transversal to $T_{S_{12}}^*W$, and f_π induces an isomorphism ${}^t f'^{-1} T_{S_{12}}^*W \simeq T_S^*(X \times Z)$. Applying

Propositions 7.4.2 and 7.4.3 we obtain the isomorphism in $\mathbf{D}^b(X \times Z;(p_X, p_Z^a))$:

$$\text{(7.4.23)} \qquad \text{“}\varprojlim_U\text{”}\, Rf_!(L_{S_{12}\cap U}) \simeq L_S[\delta]$$

where U ranges through the family of open neighborhoods of $(x_\circ, y_\circ, z_\circ)$ in W and the shift δ will be calculated latter. Let U_1, U_2, U_3 be open neighborhoods of $x_\circ$, $y_\circ$, $z_\circ$ in X, Y, Z respectively, $U = (U_1 \times U_2 \times U_3) \cap W$, $U_{13} = U_1 \times U_3$. We have $Rf_!(L_{S_{12}\cap U}) \simeq (Rq_{13!}(L'_{S_1} \overset{L}{\otimes} L''_{S_2})_{U_2})_{U_{13}}$.

By Lemma 7.3.3 we get that “$\varprojlim_{U_{13}}$”$(L'_{S_1} \circ L''_{S_2})_{U_2}$ exists and is isomorphic to $L_S[\delta]$ in $\mathbf{D}^b(X \times Z;(p_X, p_Z^a))$.

By Proposition 7.3.1 (ii), we have:

$$L'_{S_1} \underset{\mu}{\circ} L''_{S_2} \simeq L_S[\delta] \qquad \text{in} \qquad \mathbf{D}^b(X \times Z;(p_X, p_Z^a)) \ .$$

Finally we calculate the shift δ. Let $\tilde{p} = (p_X, p_Y^a, p_Y, p_Z^a) \in T^*_{S_1\times S_2}(X \times Y \times Y \times Z)$, $p' = {}^t i'(\tilde{p}) \in T^*_{S_{12}}(X \times Y \times Z)$, where i denotes the diagonal embedding $X \times Y \times Z \hookrightarrow X \times Y \times Y \times Z$, and let $p'' = {}^t j'(p')$, where j denotes the embedding $W \hookrightarrow X \times Y \times Z$. Set

$$E'' = T_{p''}T^*W \ , \qquad \lambda''_\circ = T_{p''}\pi^{-1}\pi(p'') \ ,$$

$$\lambda''_1 = T_{p''}f_\pi^{-1}(\pi^{-1}(x_\circ, z_\circ)) \ , \qquad \lambda''_2 = T_{p''}T^*_{S_{12}}W \ .$$

By Proposition 7.4.3, we have:

$$\delta = \tfrac{1}{2}[1 + \dim S - (\dim(X \times Y \times Z) - 1) - \tau'']$$

with $\tau'' = \tau_{E''}(\lambda''_\circ, \lambda''_2, \lambda''_1)$. Since $\frac{1}{2}[1 + \dim S - (\dim(X \times Y \times Z) - 1)] = 1 - \frac{1}{2}(\dim Y + \operatorname{codim} S)$, it remains to show that $\tau = \tau''$.

Set $E' = T_{p'}T^*(X \times Y \times Z)$, $\lambda'_\circ = T_{p'}\pi^{-1}\pi(p')$, $\lambda'_1 = T_{p'}T^*_{\{x_\circ\}\times Y\times\{z_\circ\}}(X \times Y \times Z)$, $\lambda'_2 = T_{p'}T^*_{S_{12}}(X \times Y \times Z)$. Note that $p' \in T^*X \times T^*_YY \times T^*Z$. Consider the isotropic space ρ' of E' given by $\rho' = (T_{p'}(W \times_{X\times Y\times Z} T^*(X \times Y \times Z)))^\perp$. Then $(E')^{\rho'} \simeq E''$, $\rho' \subset \lambda'_\circ \cap \lambda'_2$ and $(\lambda'_i)^{\rho'} \simeq \lambda''_i$ $(i = 0, 1, 2)$. Hence:

$$\tau'' = \tau' \underset{\text{def}}{=} \tau_{E'}(\lambda'_\circ, \lambda'_2, \lambda'_1) \ .$$

Now set $\tilde{E} = T_{\tilde{p}}T^*(X \times Y \times Y \times Z)$, $\tilde{\lambda}_\circ = T_{\tilde{p}}\pi^{-1}\pi(\tilde{p})$, $\tilde{\lambda}_1 = T_{\tilde{p}}T^*_{\{x_\circ\}\times Y\times Y\times\{z_\circ\}}(X \times Y \times Y \times Z)$, $\tilde{\lambda}_2 = T_pT^*_{S_1\times S_2}(X \times Y \times Y \times Z)$. Consider the isotropic space $\tilde{\rho}$ of $\tilde{E}$ given by $\tilde{\rho} = \{0\} \times (T_{(p_Y^a, p_Y)}(T_Y^*(Y \times Y) \cap \pi^{-1}\pi(p_Y^a, p_Y))) \times \{0\}$. Then $\tilde{E}^{\tilde{\rho}} \simeq E'$, $\tilde{\rho} \subset \tilde{\lambda}_\circ \cap \tilde{\lambda}_1$, $(\tilde{\lambda}_i)^{\tilde{\rho}} \simeq \lambda'_i$, $(i = 0, 1, 2)$. Hence:

$$\tau' = \tilde{\tau} \underset{\text{def}}{=} \tau_{\tilde{E}}(\tilde{\lambda}_\circ, \tilde{\lambda}_2, \tilde{\lambda}_1) \ .$$

Now we make use of notations (7.4.18), (7.4.19) and we identify $\tilde{E}$ with $E_X \oplus E_Y^a \oplus E_Y \oplus E_Z^a$. Then:

$$\tilde{\tau} = \tau(\lambda_{\circ,X} \oplus \lambda_{\circ,Y} \oplus \lambda_{\circ,Y} \oplus \lambda_{\circ,Z}, \lambda_1 \oplus \lambda_2, \lambda_{\circ,X} \oplus \Delta \oplus \lambda_{\circ,Z})$$

where Δ is the diagonal of $E_Y^a \oplus E_Y$.

Thus:

$$\tilde{\tau} = \tau_{E_Y^a \oplus E_Y}(\lambda_{o,Y} \oplus \lambda_{o,Y}, p_2^a(\lambda_1 \cap p_1^{-1}(\lambda_{o,X})) \oplus p_1(\lambda_2 \cap p_2^{a-1}(\lambda_{o,Z})), \Delta) .$$

Applying Proposition A.3.9, we get:

$$\begin{aligned}\tilde{\tau} &= \tau_{E_Y}(\lambda_{o,Y}, \lambda_{o,Y}, \lambda_2(p_Y), \lambda_1(p_Y)) \\ &= \tau_{E_Y}(\lambda_{o,Y}, \lambda_2(p_Y), \lambda_1(p_Y)) \\ &= \tau . \quad \square\end{aligned}$$

Corollary 7.4.7. *Let X and Y be two manifolds of the same dimension n, S a hypersurface of $X \times Y$. Set $\Lambda = T_S^*(X \times Y)$, and let $p = (p_X, p_Y^a) \in \Lambda \cap (\dot{T}X \times \dot{T}Y)$. Assume*

$$(7.4.23) \qquad \begin{cases} p_1|_\Lambda : \Lambda \to T^*X \quad \text{and} \quad p_2^a|_\Lambda : \Lambda \to T^*Y \quad \text{are local} \\ \text{isomorphisms in a neighborhood of } p . \end{cases}$$

Set $K = A_S$, $L = r_* A_S[n-1]$, where r is the canonical map $X \times Y \to Y \times X$. Then

(i) $K \in \mathrm{Ob}(\mathbf{N}(X, Y; p_X, p_Y))$ *and* $L \in \mathrm{Ob}(\mathbf{N}(Y, X; p_Y, p_X))$,
(ii) $K \circ_\mu L \simeq A_{\Delta_X}$ *in* $\mathbf{N}(X, X; p_X, p_X)$ *and* $L \circ_\mu K = A_{\Delta_Y}$ *in* $\mathbf{N}(Y, Y; p_Y, p_Y)$.

Proof. Set $x_o = \pi(p_X)$, $y_o = \pi(p_Y)$. Set $\Lambda' = T^*_{r(S)}(Y \times X)$, $\Lambda'' = \Lambda \circ \Lambda'$. Then, $(x, x'; \xi, \xi') \in \Lambda''$ if and only if there exists $(y; \eta) \in T^*Y$ such that $(x, y; \xi, -\eta) \in \Lambda$ and $(y, x'; \eta, \xi') \in \Lambda'$. By (7.4.23) this implies $x = x'$, $\xi = -\xi'$. Thus $\Lambda'' = T^*_{\Delta_X}(X \times X)$. Similarly, $\Lambda' \circ \Lambda = T^*_{\Delta_Y}(Y \times Y)$. Hence all the hypotheses of Proposition 7.4.6 are satisfied and we obtain:

$$K \underset{\mu}{\circ} L \simeq A_{\Delta_X}[n - 1 + \delta] \quad \text{in} \quad \mathbf{D}^b(X \times X; (p_X, p_X^a)) ,$$

with $\delta = 1 - \frac{1}{2}(2n + \tau) = 1 - n + \frac{1}{2}\tau$,

$$\tau = \tau_{T_{p_Y} T^*Y}(\lambda_o(p_Y), \lambda_2(p_Y), \lambda_1(p_Y)) ,$$

$$\lambda_o(p_Y) = T_{p_Y}\pi^{-1}\pi(p_Y) ,$$

$$\lambda_2(p_Y) = p_1(T_{(p_Y, p_X^a)} T^*_{r(S)}(Y \times X) \cap p_2^{a-1} T_{p_X}\pi^{-1}(x_o)) ,$$

$$\lambda_1(p_Y) = p_2^a(T_{(p_X, p_Y^a)} T_S^*(X \times Y) \cap p_1^{-1} T_{p_X}\pi^{-1}(x_o)) .$$

Since $\lambda_1(p_Y) = \lambda_2(p_Y)$, $\tau = 0$ and $K \circ_\mu L \simeq A_{\Delta_X}$ in $\mathbf{N}(X, X; p_X, p_X)$. Similarly one proves $L \circ_\mu K \simeq A_{\Delta_Y}$. $\square$

Remark 7.4.8. There exist open neighborhoods Ω_X of p_X and Ω_Y of p_Y, such that hypotheses (7.2.1) and (7.2.3) of Theorem 7.2.1 are satisfied with $K = A_S$.

Hence Φ_K and Ψ_K give equivalences of categories inverse to each other between $\mathbf{D}^b(X;\Omega_X)$ and $\mathbf{D}^b(Y;\Omega_Y)$, and (7.2.5) is satisfied.

By Proposition 7.4.6, we have obtained a more direct proof of the equivalence $\mathbf{D}^b(X;p_X) \simeq \mathbf{D}^b(Y;p_Y)$, and we are now able to control the shift δ which appears in contact transformations (cf. Examples 7.2.5, 7.2.6). The study of these shifts is the subject of the next section.

7.5. Pure sheaves

Let Λ be a (conic) Lagrangian submanifold of T^*X, $p \in \Lambda$ and let φ be a real function (of class C^∞) on X. We set:

$$\Lambda_\varphi = \{(x;d\varphi(x)); x \in X\} \ . \tag{7.5.1}$$

Note that Λ_φ is a Lagrangian manifold of T^*X, but Λ_φ is not conic in general.

Definition 7.5.1. *We say that φ is transversal to Λ at p if $\varphi(\pi(p)) = 0$ and the manifolds Λ and Λ_φ intersect transversally at p.*

Note that, when Λ is the conormal bundle of the submanifold M of X, this condition is equivalent to saying that $\pi(p)$ is a non-degenerate critical point of $\varphi|_M$, (i.e. its Hessian is non-degenerate).

We set for $p \in \Lambda_\varphi \cap \Lambda$,

$$\lambda_\circ(p) = T_p\pi^{-1}\pi(p) \ , \qquad \lambda_\Lambda(p) = T_p\Lambda \ , \qquad \lambda_\varphi(p) = T_p\Lambda_\varphi \ , \tag{7.5.2}$$

$$\tau_\varphi(p) = \tau_{T_pT^*X}(\lambda_\circ(p), \lambda_\Lambda(p), \lambda_\varphi(p)) \ . \tag{7.5.3}$$

If there is no risk of confusion, we write $\lambda_\circ, \tau_\varphi, \ldots$, instead of $\lambda_\circ(p)$, $\tau_\varphi(p)$, etc.

If φ is transversal to Λ at $p \in \dot{T}^*X$, then $d\varphi(p) \neq 0$, and the set $\{\varphi = 0\}$ is a smooth manifold. In such a case we set:

$$\lambda_{\{\varphi=0\}}(p) = T_p T^*_{\{\varphi=0\}}X \ . \tag{7.5.4}$$

Then:

$$\tau_\varphi = \tau(\lambda_\circ, \lambda_\Lambda, \lambda_{\{\varphi=0\}}) \ . \tag{7.5.5}$$

In fact denote by $\rho(p)$ the line of T_pT^*X generated by the Euler vector field at p; (recall that this vector field is defined by $\sum_j \xi_j \frac{\partial}{\partial \xi_j}$ in a system of homogeneous symplectic coordinates $(x;\xi)$). Then $\rho(p)$ is an isotropic space contained in $\lambda_\circ(p) \cap \lambda_\Lambda(p)$, $(\lambda_\varphi(p))^{\rho(p)} = (\lambda_{\{\varphi=0\}}(p))^{\rho(p)}$, and (7.5.5) follows from Theorem A.3.2.

Note that:

$$(7.5.6) \qquad \begin{cases} \lambda_\varphi(p) \ \ \textit{and} \ \ \lambda_\Lambda(p) \ \ \textit{intersect transversally if an only if} \\ \lambda_{\{\varphi=0\}} \cap \lambda_\Lambda(p) = \rho(p) \ . \end{cases}$$

Lemma 7.5.2. *Let X and X' be two manifolds of the same dimension, $S \subset X' \times X$ a hypersurface, $\Lambda_S = \dot{T}^*_S(X' \times X)$. Let $\tilde{p} = (p', p^a) \in \Lambda_S$, and assume that $p_1|_{\Lambda_S}$ and $p_2^a|_{\Lambda_S}$ are local isomorphisms in a neighborhood of $\tilde{p}$. Denote by $\chi: T^*X \to T^*X'$ the contact transformation defined in a neighborhood of p, given by $\chi = p_1|_{\Lambda_S} \circ (p_2^a|_{\Lambda_S})^{-1}$.*

Let φ be a function on X such that $\Lambda_\varphi \ni p$. Assume $\chi(T^*_{\{\varphi=0\}}X) = T^*_{\{\psi=0\}}X'$, for a real function ψ on X'. We choose ψ such that $d\psi(\pi(p')) = p'$ and $\psi(\pi(p')) = 0$. Let $F \in \mathrm{Ob}(\mathbf{D}^b(X))$, and set $K = A_S$.

Then, for any Lagrangian plane λ of $T_p(T^*X)$ and any $j \in \mathbb{Z} + \tau_\varphi/2$, we have:

$$R\Gamma_{\{\varphi\geqslant 0\}}(F)_{\pi(p)}[j + \tau_\varphi/2] \simeq R\Gamma_{\{\psi\geqslant 0\}}(\Phi_K F)_{\pi(p')}[j + (\tau_\psi/2) + d]$$

with $d = \frac{1}{2}(n-1) + \frac{1}{2}\tau(\lambda_0(p), \lambda, \lambda_1(p))$ and $\lambda_1(p) = \chi(\lambda_0(p')) \subset T_p(T^*X)$, $\tau_\varphi = \tau(\lambda_0(p), \lambda, \lambda_\varphi(p))$ and $\tau_\psi = \tau(\lambda_0(p'), \chi(\lambda), \lambda_\psi(p')) = \tau(\lambda_1(p), \lambda, \lambda_\varphi(p))$.

Proof. First note that:

$$R\Gamma_{\{\varphi\geqslant 0\}}(F)_{\pi(p)} = \mu hom(A_{\{\varphi=0\}}, F)_p \ .$$

Applying Theorem 7.2.1 and Proposition 7.4.6, we find:

$$R\Gamma_{\{\varphi\geqslant 0\}}(F)_{\pi(p)} = R\Gamma_{\{\psi\geqslant 0\}}(\Phi_K(F))_{\pi(p')}[-\delta] \ ,$$

where δ is defined by the relation $\Phi_K(A_{\{\varphi=0\}}) \simeq A_{\{\psi=0\}}[\delta]$ in $\mathbf{D}^b(X'; p')$, that is, (Proposition 7.4.6):

$$\delta = 1 - \tfrac{1}{2}[\dim X + 1 + \tau(\lambda_0(p), \lambda_{\{\varphi=0\}}(p), \lambda_1(p))] \ .$$

Let us calculate $d = \frac{1}{2}\tau_\varphi - \delta - \frac{1}{2}\tau_\psi$. We get by the cocycle condition (Theorem A.3.2):

$$\begin{aligned} 2d &= \tau(\lambda_0, \lambda, \lambda_\varphi) - (1 - n - \tau(\lambda_0, \lambda_\varphi, \lambda_1)) - \tau(\lambda_1, \lambda, \lambda_\varphi) \\ &= (n-1) + \tau(\lambda_0, \lambda, \lambda_1) \ . \quad \square \end{aligned}$$

Proposition 7.5.3. *Assume φ is transversal to Λ at $p \in T^*X$, and let $F \in \mathrm{Ob}(\mathbf{D}^b(X))$. We assume $\mathrm{SS}(F) \subset \Lambda$ in a neighborhood of p. Let j be a number such that $j - \frac{1}{2}(n + \dim(\lambda_0 \cap \lambda_\Lambda)) \in \mathbb{Z}$. Then $R\Gamma_{\{\varphi\geqslant 0\}}(F)_{\pi(p)}[j + \tau_\varphi(p)/2]$ does not depend on φ. (Cf. (7.5.3) for the definition of $\tau_\varphi(p)$.)*

Proof. (a) First assume that $\Lambda = T^*_M X$, for a submanifold M of X. Let us choose a coordinate system $x = (x', x'')$ with $x' = (x_1, \dots, x_l)$, such that $M = \{x'' = 0\}$ and $\pi(p) = 0$. Let $(x; \xi)$ denote the associated coordinates on T^*X, with $\xi = (\xi'; \xi'')$. Then:

$$\Lambda_\varphi = \{(x;\xi); \xi_j = \partial\varphi/\partial x_j\}$$

$$T_p\Lambda_\varphi = \left\{(x;\xi); \xi_j = \sum_k \partial^2_{x_jx_k}\varphi(0)\cdot x_k\right\}$$

$$T_pT^*_MX = \{(x;\xi); x'' = \xi' = 0\} .$$

The intersection $T_p\Lambda_\varphi \cap T_pT^*_MX$ is $\{0\}$ if and only if the matrix $(\partial^2_{x_jx_k}\varphi(0))_{1\leqslant j,k\leqslant l}$ is non-degenerate. Hence by the Morse lemma (cf. Hörmander [4, III, Appendix C]) we may assume after a change of coordinates that $\varphi|_M = \sum_{1\leqslant j\leqslant l} a_jx_j^2$, with $a_j \in \mathbb{R}$, $a_j \neq 0$. We obtain:

$$\begin{aligned}\tau_\varphi &= \tau(\{x=0\},\{x''=\xi'=0\},\{\xi=\partial^2_{x,x}\varphi(0)\cdot x\})\\ &= \tau(\{x'=0\},\{\xi'=0\},\{\xi'=\partial^2_{x',x'}\varphi(0)\cdot x'\}) ,\end{aligned}$$

where the index in the last line is calculated in the space of the $(x';\xi')$-variables. By Proposition A.3.6 we get:

$$\begin{aligned}\tau_\varphi &= -\operatorname{sgn}(\partial^2_{x',x'}\varphi(0))\\ &= \#\{j; 1\leqslant j\leqslant l, a_j<0\} - \#\{j, 1\leqslant j\leqslant l, a_j>0\} .\end{aligned} \tag{7.5.7}$$

On the other hand, by Proposition 6.6.1, there exists $L \in \mathrm{Ob}(\mathbf{D}^b(\mathfrak{Mod}(A)))$ such that $F \simeq L_M$ in $\mathbf{D}^b(X;p)$. We get:

$$\begin{aligned}R\Gamma_{\{\varphi\geqslant 0\}}(F)_{\pi(p)}[j+\tau_\varphi/2] &\simeq R\Gamma_{\{\sum_{1\leqslant i\leqslant l}a_ix_i^2\geqslant 0\}}(L_{\mathbb{R}^l})_0[j+\tau_\varphi/2]\\ &\simeq L[j+\tau_\varphi/2-q] ,\end{aligned}$$

where $q = \#\{j; 1\leqslant j\leqslant l, a_j<0\}$. (This is a particular case of the calculation performed in the course of the proof of Proposition 7.4.2.) Since $\tau_\varphi - 2q = -l$, we get the result in this case.

(b) We treat the general case. We may assume $p \in \dot{T}^*X$ (otherwise $\Lambda = T^*_MX$, cf. Exercise A.2). Let φ_1 and φ_2 be functions transversal to Λ at p. By Corollary A.2.7, we may find a contact transformations $\chi\colon T^*X \simeq T^*X'$, defined in a neighborhood of p, such that:

(i) $\chi = p_1|_{\Lambda_S}\circ(p_2^a|_{\Lambda_S})^{-1}$, $\Lambda_S = T^*_S(X'\times X)$, S being a hypersurface of $X'\times X$, and $\chi(\Lambda) = T^*_{M'}X'$ for a submanifold M' of X'.
(ii) $\chi(\dot{T}^*_{\{\varphi_i=0\}}X) = \dot{T}^*_{\{\psi_i=0\}}X'$ for some function ψ_i $(i=1,2)$.

Then the result follows from the first part of the proof and Lemma 7.5.2, since the shift d obtained in this lemma does not depend on i. □

Definition 7.5.4. *Let Λ be a Lagrangian submanifold of T^*X, $p\in\Lambda$ and let $F\in\mathrm{Ob}(\mathbf{D}^b(X))$. Assume $\mathrm{SS}(F)\subset\Lambda$ in a neighborhood of p and let φ be a function transversal to Λ at p. Let d be a number satisfying:*

$$d \equiv \tfrac{1}{2}\dim(\lambda_0(p)\cap\lambda_\Lambda(p)) \bmod \mathbb{Z} , \tag{7.5.8}$$

let $\tau_\varphi(p) = \tau(\lambda_0(p),\lambda_\Lambda(p),\lambda_\varphi(p))$ (cf. (7.5.3)), and let $L\in\mathrm{Ob}(\mathbf{D}^b(\mathfrak{Mod}(A)))$. If there *is an isomorphism*:

$$R\Gamma_{\{\varphi\geqslant 0\}}(F)_{\pi(p)}[-d+\tfrac{1}{2}\dim X+\tfrac{1}{2}\tau_\varphi(p)]\simeq L\ ,$$

then one says F is of type L with shift d at p.

If $H^j(L)=0$ for $j\neq 0$, one says F is pure at p. If moreover L is a free A-module of rank one, one says F is simple at p.

In the preceding situation, the isomorphy class of L is called the type of F with shift d at p. Note that the type of F does not depend on φ by Proposition 7.5.3.

Examples 7.5.5. (i) If F has type L with shift d, then F has type $L[-k]$ with shift $d+k$ and $F[k]$ has type L with shift $d+k$.

(ii) Let M be a closed submanifold of X. Then A_M is a simple sheaf with shift $\frac{1}{2}\operatorname{codim} M$ at each $p\in T^*_MX$. In fact choose coordinates $x=(x',x'')$ with $M=\{x''=0\}$, $x'=(x_1,\dots,x_l)$, $p=(0;\lambda\,dx_n)$. Take $\varphi(x)=\lambda x_n+\sum_{j=1}^{l}x_j^2$. Then φ is transversal to T^*_MX at p and we have:

$$\begin{aligned}(H^k_{\{\varphi\geqslant 0\}}(A_M))_0&=A\quad \text{for } k=0\ ,\\ &=0\quad \text{otherwise}\ .\end{aligned}$$

We get $d=\frac{1}{2}(\dim X-l)$ by (7.5.7).

(iii) Let ψ be a real function on X, $Z=\{x;\psi(x)\geqslant 0\}$, $U=\{x;\psi(x)<0\}$, $p=(x_\circ;d\psi(x_\circ))$, with $\psi(x_\circ)=0$, $d\psi(x_\circ)\neq 0$. Then A_Z (resp. A_U) is simple with shift $\frac{1}{2}$ (resp. $-\frac{1}{2}$) at p. In fact $A_Z\simeq A_{\{\psi=0\}}\simeq A_U[1]$ in $\mathsf{D}^b(X;p)$.

(iv) Recall Example 5.3.4: let $Z=\{x\in\mathbb{R}^2;x_1>0,-x_1^{3/2}\leqslant x_2<x_1^{3/2}\}$, and $\Lambda=\{(x;\xi);\xi_2>0,x_2=-(2\xi_1/3\xi_2)^3,x_1=(2\xi_1/3\xi_2)^2\}$. By (iii) the sheaf A_Z has shift $\frac{1}{2}$ on $\Lambda\cap\{\xi_1>0\}$ and shift $-\frac{1}{2}$ on $\Lambda\cap\{\xi_1<0\}$. Let us calculate the shift of A_Z at $(0;dx_2)$. Choose $\varphi(x)=x_2$. Then φ is transversal to Λ at $(0;dx_2)$ and $\tau_\varphi=0$. Since $(H^j_{\{x_2\geqslant 0\}}(A_Z))_0=A$ for $j=1$, and 0 otherwise, the shift is 0.

We shall study the behavior of the shift under particular contact transformations.

Proposition 7.5.6. *Assume F is of type L with shift d along Λ at p. Consider a contact transformation $\chi: T^*X \simeq T^*X'$ defined in a neighborhood of p, and assume $\chi=p_1|_{\Lambda_S}\circ(p_2^a|_{\Lambda_S})^{-1}$, $\Lambda_S=T_S^*(X'\times X)$, S is a hypersurface. Set $K=A_S$. Then $\Phi_K(F)$ is of type L with shift d′ along $\chi(\Lambda)$ at $\chi(p)$, with*:

$$d'=d-\tfrac{1}{2}(n-1)-\tfrac{1}{2}\tau(\lambda_\circ(p),\lambda_\Lambda(p),\lambda_1(p))\quad \textit{and}$$

$$\begin{aligned}\lambda_1(p)&=p_2^a(\lambda_{\Lambda_S}(\chi(p),p^a)\cap p_1^{-1}(\lambda_\circ(\chi(p))))\\ &=\chi^{-1}(\lambda_\circ(\chi(p)))=\lambda_\circ(\chi(p))\circ\lambda_{\Lambda_S}(\chi(p),p^a)^a\ .\end{aligned}$$

Proof. We may find a function φ on X such that φ is transversal to Λ at p, and $\chi(T^*_{\{\varphi=0\}}X)=T^*_{\{\psi=0\}}X'$, for a smooth function ψ. Hence the result follows from Lemma 7.5.2. ☐

Corollary 7.5.7. *Let U be an open subset of T^*X, $L \in \mathrm{Ob}(\mathbf{D}^b(\mathfrak{Mod}(A)))$, $F \in \mathrm{Ob}(\mathbf{D}^b(X))$ and assume $\mathrm{SS}(F) \cap U \subset \Lambda$. Then, the set of points p in $\Lambda \cap U$ such that F is of type L at p is open and closed in $\Lambda \cap U$.*

Proof. Let $p \in \Lambda \cap U$. We may find a contact transformation χ satisfying the hypothesis of Proposition 7.5.6 and such that $\chi(\Lambda) = T_N^*X'$, for a hypersurface N of X'. Then $\phi_K(F) \simeq L'_N$ in $\mathbf{D}^b(X'; \chi(p))$, for some $L' \in \mathrm{Ob}(\mathbf{D}^b(\mathfrak{Mod}(A)))$. This implies $\phi_K(F) \simeq L'_N$ in $\mathbf{D}^b(X'; \Omega)$ for an open neighborhood Ω of $\chi(p)$. By Proposition 7.5.6, we can reduce to the case $\Lambda = T_N^*X$. Then the corollary is obvious. □

Remark 7.5.8. By the proof of the last Proposition, we see that if F is of type L at p, then there exists an object G simple at p such that $F \simeq L_X \otimes G$ in $\mathbf{D}^b(X; p)$.

The calculation of the shift is not so easy, as shown by Example 7.5.5 (iv), but we use the following result to bring the situation into the generic one.

Consider a connected topological space S and a continuous map $p: S \to \Lambda$. We assume to be given a continuous family of Lagrangian planes $\mu(s)$ of $T_{p(s)}T^*X$, $s \in S$, satisfying:

$$\mu(s) \cap \lambda_0(p(s)) = \mu(s) \cap \lambda_\Lambda(p(s)) = \{0\} \ . \tag{7.5.9}$$

Proposition 7.5.9. *Assume that the function $d(s) - \frac{1}{2}\tau(\lambda_0(p(s)), \lambda_\Lambda(p(s)), \mu(s))$ is constant on S. Then the type of F with shift $d(s)$ at $p(s)$ is constant.*

Proof. (a) First assume $\Lambda = T_M^*X$, for a submanifold M of X. Then $d(s)$ is locally constant, as well as $\tau(\lambda_0(p(s)), \lambda_\Lambda(p(s)), \mu(s))$, since $\dim(\lambda_0(p(s)) \cap \lambda_\Lambda(p(s)))$ is constant.

(b) To treat the general case, we choose a contact transformation χ defined in a neighborhood of $p(s_0)$ satisfying the hypothesis of Proposition 7.5.6 and such that $\chi(\Lambda) = T_N^*X'$ for a submanifold N of X', and $\chi(\mu(s_0))$ is transversal to $\lambda_0(\chi(p(s)))$. Set $p'(s) = \chi(p(s))$, and write p or p' instead of $p(s)$ or $p'(s)$, for short.

Let us define $d'(s)$ by $d'(s) = d(s) - \frac{1}{2}(n-1) - \frac{1}{2}\tau(\lambda_0(p), \lambda_\Lambda(p), \lambda_1(p))$ where $\lambda_1(p) = \chi^{-1}(\lambda_0(p'))$.

Then by Proposition 7.5.6 the type of F with shift $d(s)$ at $p(s)$ is the type of $\Phi_K(F)$ with shift $d'(s)$ at $p'(s)$. We have:

$$\tau(\lambda_0(p'), \lambda_{\chi(\Lambda)}(p'), \chi(\mu(s))) = \tau(\lambda_1(p), \lambda_\Lambda(p), \mu(s)) \ .$$

Then by (a), it is enough to show that

$$d(s) - \tfrac{1}{2}\tau(\lambda_0(p), \lambda_\Lambda(p), \lambda_1(p)) - \tfrac{1}{2}\tau(\lambda_1(p), \lambda_\Lambda(p), \mu(s)) = d(s) - \tfrac{1}{2}\tau(\lambda_0(p), \mu(s), \lambda_1(p))$$
$$- \tfrac{1}{2}\tau(\lambda_0(p), \lambda_\Lambda(p), \mu(s))$$

is locally constant. Since $\mu(s)$ is transversal both to $\lambda_0(p)$ and $\lambda_1(p)$, it remains to check that $\dim(\lambda_0(p) \cap \lambda_1(p))$ is locally constant, (Theorem A.3.2). We have:

$$\lambda_0(p) \cap \lambda_1(p) = \lambda_0(p) \cap p_2^a(p_1^{-1}(\lambda_0(p')) \cap \lambda_{T_S^*(X' \times X)}(p', p^a))$$
$$\simeq \lambda_{T_S^*(X' \times X)}(p', p^a) \cap \lambda_0(p', p^a)$$

and the dimension of this space is 1 since S is a hypersurface. This completes the proof. □

Let us give an application of Proposition 7.5.9.

Let X_1 and X_2 be two manifolds, Λ_i Lagrangian submanifold of T^*X_i, $p_i \in \Lambda_i$ $(i = 1, 2)$. We denote by q_i the projection $X_1 \times X_2 \to X_i$.

Proposition 7.5.10. *Let $F_i \in \mathrm{Ob}(\mathbf{D}^b(X_i))$, and assume F_i is of type L_i with shift d_i at p_i along Λ_i, $(i = 1, 2)$.*

(i) *$F_1 \boxtimes^L F_2$ is of type $L_1 \otimes^L L_2$ with shift $d_1 + d_2$ at (p_1, p_2) along $\Lambda_1 \times \Lambda_2$.*
(ii) *$R\mathcal{H}om(q_1^{-1}F_1, q_2^{-1}F_2)$ is of type $R\mathcal{H}om(L_1, L_2)$ with shift $d_2 - d_1$ at (p_1^a, p_2) along $\Lambda_1^a \times \Lambda_2$.*

Proof. (i) At generic points p_i' of Λ_i, $F_i \simeq (L_i)_{M_i}[d(p_i') - \frac{1}{2}\operatorname{codim} M_i]$, $i = 1, 2$, for submanifolds M_i of X_i. Hence $F_1 \boxtimes^L F_2 \simeq (L_1 \otimes L_2)_{M_1 \times M_2}[d(p_1') + d(p_2') - \frac{1}{2}\operatorname{codim} M_1 \times M_2]$. Thus $F_1 \boxtimes^L F_2$ is of type $L_1 \otimes^L L_2$ at generic points of $\Lambda_1 \times \Lambda_2$, hence at each point of $\Lambda_1 \times \Lambda_2$ by Corollary 7.5.7. To calculate the shift at (p_1, p_2) let us choose two families of Lagrangian spaces $\mu_i(p_i') \subset T_{p_i'}T^*X_i$, $(i = 1, 2)$, $p_i' \in \Lambda_i$, which satisfy (7.5.9). Let $d(p_1', p_2')$ denote the shift of $F_1 \boxtimes^L F_2$ at (p_1', p_2'). By Proposition 7.5.9, we have:

$$
\begin{aligned}
2(d(p_1', p_2') - d(p_1, p_2)) &= \tau(\lambda_0(p_1', p_2'), \lambda_{\Lambda_1 \times \Lambda_2}(p_1', p_2'), \mu(p_1') \oplus \mu(p_2')) \\
&\quad - \tau(\lambda_0(p_1, p_2), \lambda_{\Lambda_1 \times \Lambda_2}(p_1, p_2), \mu(p_1) \oplus \mu(p_2)) \ , \\
2(d(p_i') - d(p_i)) &= \tau(\lambda_0(p_i'), \lambda_{\Lambda_i}(p_i'), \mu(p_i')) \\
&\quad - \tau(\lambda_0(p_i), \lambda_{\Lambda_i}(p_i), \mu(p_i)) \ , \qquad i = 1, 2 \ .
\end{aligned}
$$

Since $d(p_1', p_2') = d(p_1') + d(p_2')$ for generic p_i', and

$$
\begin{aligned}
\tau(\lambda_0(p_1', p_2'), \lambda_{\Lambda_1 \times \Lambda_2}(p_1', p_2'), \mu(p_1') \oplus \mu(p_2')) &= \tau(\lambda_0(p_1'), \lambda_{\Lambda_1}(p_1'), \mu(p_1')) \\
&\quad + \tau(\lambda_0(p_2'), \lambda_{\Lambda_2}(p_2'), \mu(p_2'))
\end{aligned}
$$

for any (p_1', p_2') we get $d(p_1, p_2) = d(p_1) + d(p_2)$.

(ii) The proof is similar. □

To end this section, we shall extend Propositions 7.4.6 and 7.5.6 to a more general situation. Let X, Y, Z be three manifolds, $\Lambda_1 \subset T^*(X \times Y)$, $\Lambda_2 \subset T^*(Y \times Z)$ two Lagrangian manifolds. Let $\tilde{p}_1 = (p_X, p_Y^a) \in \Lambda_1$ and $\tilde{p}_2 = (p_Y, p_Z^a) \in \Lambda_2$. We keep notations (7.4.14) and (7.4.19), that is, we set $\lambda_i = T_{\tilde{p}_i}\Lambda_i$, $i = 1, 2$, $\lambda_0(p_W) = T_{p_W}\pi^{-1}\pi(p_W)$ for $W = X, Y, Z$, $\lambda_1(p_Y) = \lambda_0(p_X) \circ \lambda_1^a$, $\lambda_2(p_Y) = \lambda_2 \circ \lambda_0(p_Z)$. We introduce the notation (cf. Exercise A.9):

$$
\tau(\lambda_1 : \lambda_2) = \tau(\lambda_0(p_Y), \lambda_2(p_Y), \lambda_1(p_Y)) \ . \tag{7.5.10}
$$

Theorem 7.5.11. *Let* $K_1 \in \mathrm{Ob}(\mathbf{D}^b(X \times Y; (p_X, p_Y^a)))$ *and* $K_2 \in \mathrm{Ob}(\mathbf{D}^b(Y \times Z; (p_Y, p_Z^a)))$, *with* $\mathrm{SS}(K_i) \subset \Lambda_i$ *in a neighborhood of* $\tilde{p}_i$ $(i = 1, 2)$. *Assume* $p_2^a|_{\Lambda_1} : \Lambda_1 \to T^*Y$ *and* $p_1|_{\Lambda_2} : \Lambda_2 \to T^*Y$ *are transversal and* K_i *is of type* L_i *with shift* d_i *along* Λ_i *at* $\tilde{p}_i$, $i = 1, 2$.

Then (K_1, K_2) *is microlocally composable,* $\mathrm{SS}(K_1 \circ_\mu K_2) \subset \Lambda \underset{\mathrm{def}}{=} \Lambda_1 \circ \Lambda_2$ *in a neighborhood of* (p_X, p_Z^a) *and* $K_1 \circ_\mu K_2$ *is of type* $L_1 \otimes^L L_2$ *with shift* $d_1 + d_2 - \frac{1}{2}(\dim Y + \tau(\lambda_1 : \lambda_2))$ *along* Λ *at* (p_X, p_Z^a).

Proof. (a) Note that if Λ_1 and Λ_2 are the conormal bundles to hypersurfaces and Λ is the conormal bundle to a submanifold then we recover Proposition 7.4.6.

If $Z = \{\mathrm{pt}\}$, and Λ_1 is the conormal bundle to a hypersurface associated with a contact transformation, then we recover Proposition 7.5.6.

(b) Let Δ denote the diagonal of $\mathbb{R} \times \mathbb{R}$ and let $q \in \dot{T}^*\mathbb{R}$. Then by Proposition 7.3.6, $(K_1 \boxtimes A_\Delta) \circ_\mu (K_2 \boxtimes A_\Delta) \simeq (K_1 \circ_\mu K_2) \boxtimes A_\Delta$.

Replacing W, p_W with $W \times \mathbb{R}, (p_W, q), (W = X, Y, Z)$ and K_1, K_2 with $K_1 \boxtimes A_\Delta$, $K_2 \boxtimes A_\Delta$, we may assume from the beginning that $p_W \in \dot{T}^*W$ $(W = X, Y, Z)$.

(c) It is possible to reduce the theorem to the case where $Z = \{\mathrm{pt}\}$ by the following procedure. With the same notations as in Proposition 7.3.4, $K_1 \circ_\mu K_2 \simeq i_*(K_1 \boxtimes A_Z) \circ_\mu K_2$. Since $i_*(K_1 \boxtimes A_Z)$ has type L_1 with shift $d_1 + \frac{1}{2}\dim Z$, and $\tau(\lambda_1 : \lambda_2) = \tau(\lambda_1' : \lambda_2)$, with $\lambda_1' = \lambda_1 \times T_Z^*(Z \times Z)$, the theorem for (K_1, K_2) is equivalent to the theorem for $(i_*(K_1 \boxtimes A_Z), K_2)$.

(d) Let W be another manifold and let Λ_3 be a Lagrangian submanifold of $T^*(Z \times W)$, and $K_3 \in \mathrm{Ob}(\mathbf{D}^b(Z \times W; (p_Z, p_W^a)))$. Assume that K_3 has type L_3 with shift d_3 along Λ_3 and assume moreover that the maps $\Lambda_1 \circ \Lambda_2 \to T^*Z$ and $\Lambda_3 \to T^*Z$ are transversal and also that $\Lambda_2 \to T^*Z$ and $\Lambda_3 \to T^*Z$ are transversal. Hence, $\Lambda_1 \to T^*Y$ and $\Lambda_2 \circ \Lambda_3 \to T^*Y$ are transversal. If the theorem is valid for (K_1, K_2), (K_2, K_3) and $(K_1 \circ_\mu K_2, K_3)$ (resp. $(K_1, K_2 \circ_\mu K_3)$) then it is valid for $(K_1, K_2 \circ_\mu K_3)$ (resp. $(K_1 \circ_\mu K_2, K_3)$). This follows immediately from the isomorphism $(K_1 \circ_\mu K_2) \circ_\mu K_3 \simeq K_1 \circ_\mu (K_2 \circ_\mu K_3)$ and the formula (cf. Exercise A.9):

$$\tau(\lambda_1 : \lambda_2) + \tau(\lambda_1 \circ \lambda_2 : \lambda_3) = \tau(\lambda_1 : \lambda_2 \circ \lambda_3) + \tau(\lambda_2 : \lambda_3) \ .$$

(e) If there exists a subset Ω of $\Lambda_1 \times_{T^*Y} \Lambda_2$ such that (p_X, p_Y, p_Z) belongs to $\bar{\Omega}$ and if the theorem is valid at each point of Ω, then the theorem is valid at (p_X, p_Y, p_Z). In fact this will follow from Proposition 7.5.9.

Let $p : S \to \Lambda_1 \times_{T^*Y} \Lambda_2$ be a continuous map, with $p(s_\circ) = (p_X, p_Y, p_Y)$ for an $s_\circ \in S$. We write $p(s) = (p_X(s), p_Y(s), p_Z(s))$, and we set:

$$\lambda_1(s) = T_{(p_X(s), p_Y(s)^a)}\Lambda_1 \ , \qquad \lambda_2(s) = T_{(p_Y(s), p_Z(s)^a)}\Lambda_2 \qquad \text{and}$$

$$E_W(s) = T_{p_W(s)}T^*W \ , \qquad \lambda_{\circ W}(s) = \lambda_\circ(p_W(s)) \qquad \text{for } W = X, Y, Z \ .$$

We choose Lagrangian planes $\mu_W(s) \subset E_W(s)$ depending continuously on s, and such that $\mu_W(s) \cap \lambda_{\circ W}(s) = 0$ $(W = X, Y, Z)$, and that $(\mu_X(s) \oplus \mu_Y(s)^a) \cap \lambda_1(s) = 0$, $(\mu_Y(s) \oplus \mu_Z(s)^a) \cap \lambda_2(s) = 0$, $(\mu_X(s) \oplus \mu_Z(s)^a) \cap (\lambda_1(s) \circ \lambda_2(s)) = 0$. Then we choose functions $d_1(s)$, $d_2(s)$, $d(s)$ with $d_i(s_\circ) = d_i$, $(i = 1, 2)$ such that the following functions are locally constant:

$$d_1(s) - \tfrac{1}{2}\tau_{E_X(s)\oplus E_Y(s)^a}(\lambda_{\circ X}(s) \oplus \lambda_{\circ Y}(s)^a, \lambda_1(s), \mu_X(s) \oplus \mu_Y(s)^a) \ ,$$

$$d_2(s) - \tfrac{1}{2}\tau_{E_Y(s)\oplus E_Z(s)^a}(\lambda_{\circ Y}(s) \oplus \lambda_{\circ Z}(s)^a, \lambda_2(s), \mu_Y(s) \oplus \mu_Z(s)^a) \ ,$$

$$d(s) - \tfrac{1}{2}\tau_{E_X(s)\oplus E_Z(s)^a}(\lambda_{\circ X}(s) \oplus \lambda_{\circ Z}(s)^a, \lambda_1(s) \circ \lambda_2(s), \mu_X(s) \oplus \mu_Y(s)^a) \ .$$

Then by Proposition 7.5.9, K_1 has type L_1 with shift $d_1(s)$ at $(p_X(s), p_Y(s)^a)$ and K_2 has type L_2 with shift $d_2(s)$ at $(p_Y(s), p_Z(s)^a)$.

Assume $K_1 \circ_\mu K_2$ has type $L_1 \otimes^L L_2$ with shift $d = d_1(s_1) + d_2(s_1) - \frac{1}{2}[\dim Y + \tau(\lambda_{\circ Y}(s_1), \lambda_2(s_1) \circ \lambda_{\circ Z}(s_1), \lambda_{\circ X}(s_1) \circ \lambda_1(s_1)^a)]$ at some point $(p_X(s_1), p_Z(s_1)^a)$. If $d(s_1) = d$, then $K_1 \circ_\mu K_2$ has type $L_1 \otimes^L L_2$ with shift $d(s)$ at $(p_X(s), p_Z(s)^a)$ by Proposition 7.5.9. Hence, in order to prove that the theorem is valid at $s = s_\circ$, it is enough to show that $d_1(s) + d_2(s) - d(s) - \frac{1}{2}\tau(\lambda_{\circ Y}(s), \lambda_2(s) \circ \lambda_{\circ Z}(s), \lambda_{\circ X}(s) \circ \lambda_1(s)^a)$ is locally constant in s. This is equivalent to proving that the function $\tau(s)$ defined below is locally constant:

$$\begin{aligned}\tau(s) = {} & \tau_{E_X(s)\oplus E_Y(s)^a}(\lambda_{\circ X}(s) \oplus \lambda_{\circ Y}(s)^a, \lambda_1(s), \mu_X(s) \oplus \mu_Y(s)^a) \\ & + \tau_{E_Y(s)\oplus E_Z(s)^a}(\lambda_{\circ Y}(s) \oplus \lambda_{\circ Z}(s)^a, \lambda_2(s), \mu_Y(s) \oplus \mu_Z(s)^a) \\ & - \tau_{E_X(s)\oplus E_Z(s)^a}(\lambda_{\circ X}(s) \oplus \lambda_{\circ Z}(s)^a, \lambda_1(s) \circ \lambda_2(s), \mu_X(s) \oplus \mu_Z(s)^a) \\ & - \tau_{E_Y(s)}(\lambda_{\circ Y}(s), \lambda_2(s) \circ \lambda_{\circ Z}(s), \lambda_{\circ X}(s) \circ \lambda_1(s)^a) \ .\end{aligned}$$

By Exercise A.10, we have:

$$\tau(s) = \tau_{E_Y(s)}(\mu_Y(s), \mu_X(s) \circ \lambda_1(s)^a, \lambda_2(s) \circ \mu_Z(s)) \ .$$

Since $\mu_Y(s)$, $\mu_X(s) \circ \lambda_1(s)^a$ and $\lambda_2(s) \circ \mu_Z(s)$ are transversal to each other, $\tau(s)$ is locally constant.

Choosing $S = \bar{\Omega}$ and p the identity map, the result follows.

(f) We shall prove the theorem when Λ_1 is the graph of a contact transformation from T^*X to T^*Y. By (c) we may assume from the beginning that $Z = \{\mathrm{pt}\}$. Then we decompose the contact transformation $T^*X \to T^*Y$ into $T^*X \to T^*X' \to T^*Y$, such that the graph of $T^*X \to T^*X'$ (resp. $T^*X' \to T^*Y$) is associated to the conormal bundle to a hypersurface $S \subset X \times X'$ (resp. $S' \subset X' \times Y$). By Remark 7.5.8 we may assume $K_1 = A_S \circ A_{S'}$. The theorem is valid for $(A_S, A_{S'})$ at any point where $\Lambda_1 \to X \times Y$ has a constant rank, by (a). Hence it is valid for $(A_S, A_{S'})$ everywhere by (e). Applying (d) and Proposition 7.5.6, we get (f).

(g) Now we shall prove the theorem in the general case.

By (c) we may assume $Z = \{\mathrm{pt}\}$. Then there exist contact transformations χ_1 on T^*X and χ_2 on T^*Y such that $(\chi_1 \times \chi_2^{-1})(\Lambda_1)$, $\chi_2(\Lambda_2)$ and $\chi_1(\Lambda_1 \circ \Lambda_2)$ are conormal bundles to hypersurfaces (Proposition A.2.6). Let F_i and G_i be simple objects whose micro-supports are contained in the Lagrangian manifolds associated with χ_i and χ_i^{-1}, respectively $(i = 1, 2)$, and satisfying $G_1 \circ_\mu F_1 \simeq A_{\Delta_X}$,

$G_2 \circ_\mu F_2 \simeq A_{\Delta_Y}$. Then the micro-supports of $F_1 \circ_\mu K_1 \circ_\mu G_2$ and $F_2 \circ_\mu K_2$ are contained in conormal bundles to hypersurfaces. Hence by (a) the theorem is valid for $(F_1 \circ_\mu K_1 \circ_\mu G_2, F_2 \circ_\mu K_2)$. By (f), the theorem is valid for $(G_2, F_2 \circ_\mu K_2)$ and $(F_1 \circ_\mu K_1, G_2)$. Hence by (d) it is valid for $(F_1 \circ_\mu K_1, K_2)$. Again by (f), the theorem is valid for $(G_1, F_1 \circ_\mu K_1)$ and $(G_1, F_1 \circ_\mu K_1 \circ_\mu K_2)$. Hence by (f) it is valid for $(G_1 \circ_\mu F_1 \circ_\mu K_1, K_2)$, that is, (K_1, K_2). □

Note that in the proof of (g) it was not possible to apply Proposition 7.5.6, since this result only applies when $Z = \{\mathrm{pt}\}$.

As an application of Theorem 7.5.11, we obtain:

Corollary 7.5.12. *Let $f: Y \to X$ be a morphism of manifolds, $p \in Y \times_X T^*X$, $p_Y = {}^t f'(p)$, $p_X = f_\pi(p)$. Let Λ_Y be a Lagrangian submanifold of T^*Y such that ${}^t f'$ is transversal to Λ_Y at p_Y. Let $G \in \mathrm{Ob}(\mathbf{D}^b(Y))$ and assume that $\mathrm{SS}(G) \subset \Lambda_Y$ on a neighborhood of p_Y, and G has type L with shift d along Λ_Y at p_Y.*

(i) *For a sufficiently small open neighborhood W of p, $\Lambda_X = f_\pi({}^t f'^{-1}(\Lambda_Y) \cap W)$ is a Lagrangian manifold isomorphic to ${}^t f'^{-1}(\Lambda_Y) \cap W$ by f_π.*

(ii) *$f_!^\mu G = \text{“}\varinjlim_U\text{”}\, Rf_* G_U$ exists in $\mathbf{D}^b(X; p_X)$ (cf. Proposition 6.1.10) and $f_!^\mu G$ has type L with shift*

$$d - \tfrac{1}{2}(\dim Y/X + \tau(\lambda_\circ(p_X), \lambda_{\Lambda_X}(p_X), f_\pi {}^t f'^{-1}(\lambda_\circ(p_Y))))\ .$$

Here U ranges over an open neighborhood system of $\pi_Y(p_Y)$ and we wrote $f_\pi {}^t f'^{-1}(\lambda_\circ(p_Y))$ instead of $(df_\pi)(d^t f')^{-1}(\lambda_\circ(p_Y))$.

Proof. (i) follows from Exercise A.5.

(ii) Apply Theorem 7.5.11 with $K_1 = A_\Delta$, $K_2 = G$, where $\Delta \subset X \times Y$ is the graph of f. □

Corollary 7.5.13. *Let $f: Y \to X$ be a morphism of manifolds, $p \in Y \times_X T^*X$, $p_Y = {}^t f'(p)$, $p_X = f_\pi(p)$. Let Λ_X be a Lagrangian submanifold of T^*X such that f_π is transversal to Λ_X at p_X. Let $F \in \mathrm{Ob}(\mathbf{D}^b(X))$ and assume that $\mathrm{SS}(F) \subset \Lambda_X$ on a neighborhood of p_X and F has type L with shift d along Λ_X at p_X.*

(i) *For a sufficiently small open neighborhood W of p, $\Lambda_Y = {}^t f'(f_\pi^{-1}(\Lambda_X) \cap W)$ is a Lagrangian manifold, isomorphic to $f_\pi^{-1}(\Lambda_X) \cap W$ by ${}^t f'$.*

(ii) *$f_\mu^{-1} F = \text{“}\varinjlim_{F' \to F}\text{”}\, f^{-1} F'$ and $f_\mu^! F = \text{“}\varinjlim_{F \to F'}\text{”}\, f^! F'$ exist in $\mathbf{D}^b(Y; p_Y)$, where $F' \to F$ (resp. $F \to F'$) ranges over the category of isomorphisms at p_X,* and *$\omega_{Y/X} \otimes f_\mu^{-1} F \simeq f_\mu^! F$ (cf. Proposition 6.1.9).*

(iii) *$f_\mu^{-1} F$ has type L with shift d.*

Proof. (i) follows from Exercise A.5.

(ii) is a consequence of Proposition 6.1.9.

(iii) Apply Theorem 7.5.11 with $K_1 = A_\Delta$ and $K_2 = F$, where $\Delta \subset Y \times X$ is the graph of f. □

Exercises to Chapter VII

Exercise VII.1. Let V be a real (resp. complex) finite-dimensional vector space, V^* its dual. Set $S_V = (V\setminus\{0\})/\mathbb{R}^+$ (resp. $(V\setminus\{0\})/\mathbb{C}^\times$) and let $Z = \{(x, y) \in S_V \times S_{V^*}; \langle x, y\rangle = 0\}$. Let $K = A_Z$, $\Omega = \dot{T}^*S_V$, $\Omega' = \dot{T}^*S_{V^*}$. Prove that the hypotheses of Theorem 7.2.1 are satisfied for K on (Ω, Ω') (cf. Brylinski [2] for a detailed study of such a contact transformation in the framework of complex manifolds and constructible sheaves).

Exercise VII.2. Let $\tau: E \to Z$ be a vector bundle over a manifold Z. Identify T^*E and T^*E^* as in Proposition 5.5.1. Let F_1 and F_2 belong to $\mathbf{D}^b_{\mathbb{R}^+}(E)$. Prove the isomorphism:

$$\mu hom(F_2, F_1) \simeq \mu hom(F_2^\wedge, F_1^\wedge) \ .$$

(Hint: using Exercises III 17 and IV 6, show that it is equivalent to proving a similar formula for F_1 and F_2 in $\mathbf{D}^b_{\mathbb{R}_+}(\dot{E})$ or else in $\mathbf{D}^b(\dot{E}/\mathbb{R}^+)$. Then use Theorem 7.2.1.)

Exercise VII.3. Let (x) and (y) denote two systems of linear coordinates on two copies X and Y of $\mathbb{R}^n$, respectively, and let $(x; \xi)$ and $(y; \eta)$ be the associated coordinates. Consider the contact transformation χ (**partial Legendre transformation**) defined for $\xi_n \neq 0$, $\eta_n \neq 0$ by $y_j = \xi_j\xi_n^{-1}$ $(p < j < n)$, $y_n = (\sum_{j=p+1}^n x_j\xi_j)\xi_n^{-1}$, $y_k = x_k$ $(k \leqslant p)$, $\eta_j = -x_j\xi_n$ $(p < j < n)$, $\eta_n = \xi_n$, $\eta_k = \xi_k$ $(k \leqslant p)$.

(i) Prove that χ is associated to the conormal bundle to the submanifold $S = \{(x, y); x_k - y_k = 0\ 1 \leqslant k \leqslant p, x_n - y_n + \sum_{j=p+1}^{n-1} x_jy_j = 0\}$.

(ii) By applying Theorem 7.2.1, prove that microlocally, $\psi_K(A_M) \simeq A_N$, where $N = \{y_n = 0\}$, $M = \{x_{p+1} = \cdots = x_n = 0\}$ and $K = A_S[d]$ for a shift d to be calculated.

Exercise VII.4. Let X be a complex manifold and let Λ be a complex connected Lagrangian submanifold of T^*X. Let $F \in \mathrm{Ob}(\mathbf{D}^b(X))$, and assume $SS(F) \subset \Lambda$ in a neighborhood of Λ. Prove that if F is pure with shift d at $p \in \Lambda$, then F is pure with shift d at any $p \in \Lambda$. (Hint: use Proposition 7.5.9 and Exercise A.7.)

Exercise VII.5. In the situation of Theorem 7.5.11, assume X, Y, Z and Λ_i are complex manifolds, and K_i is of type L_i with shift 0 $(i = 1, 2)$. Prove that $K_1 \circ_\mu K_2$ has type $L_1 \otimes^L L_2$ with shift $-\dim_{\mathbb{C}} Y$.

Notes

Symplectic geometry goes back to Hamilton and Jacobi, and physicists are used to work in the "phase space", that is, in the cotangent bundle (cf. the excellent introduction of Guillemin-Sternberg [1]). However, it is only quite recently that mathematical tools appeared, which make it possible to pass beyond the purely geometrical point of view, and to work in the cotangent bundle (one says "microlocally") with classical objects attached to a manifold X, such as for example distributions or differential operators. The first step was the introduction of "pseudo-differential operators" (we shall not review this theory here, and refer to Hörmander [4]), then came the "canonical operator" of Maslov [1], and finally the theory of microfunctions of Sato [2] and the theory of Fourier integral operators of Hörmander [2], which in particular, allow us to make contact transformations operate on those classical objects.

The theory we present in this chapter is a sheaf-theoretical version of the theory of Fourier integral operators, or its complex equivalent, the theory of simple holonomic modules of Sato-Kawai-Kashiwara [1]. It was first introduced by a different method, in Kashiwara-Schapira [3].

Chapter VIII. Constructible sheaves

Summary

We begin this chapter by explaining the notion of constructible sheaves on a simplicial complex, then by recalling some basic facts about Hironaka's subanalytic sets.

Next we introduce the following definition, a modification of that of a Whitney stratification. A stratification $X = \bigsqcup_{\alpha \in A} X_\alpha$ of a real analytic manifold X is a μ-stratification if the X_α's are subanalytic submanifolds, and for each pair (α, β) with $X_\beta \cap \overline{X_\alpha} \neq \emptyset$, we have $X_\beta \subset \overline{X_\alpha}$ and also:

$$(T^*_{X_\beta}X \mathbin{\hat{+}} T^*_{X_\alpha}X) \cap \pi^{-1}(X_\beta) \subset T^*_{X_\beta}X \ .$$

(Recall the operation $\hat{+}$ of VI §2.) One proves that given a closed conic subanalytic isotropic subset Λ of T^*X, there exists a μ-stratification $X = \bigsqcup_\alpha X_\alpha$ such that $\Lambda \subset \bigsqcup_\alpha T^*_{X_\alpha}X$.

Next we introduce the following definitions. An object F of $\mathbf{D}^b(X)$ is weakly $\mathbb{R}$-constructible (w-$\mathbb{R}$-constructible, for short) if there exists a locally finite covering $X = \bigcup_j X_j$ by subanalytic subsets such that for all k and all j, the sheaves $H^k(F)|_{X_j}$ are locally constant. If moreover the complexes F_x are perfect for all $x \in X$, one says F is $\mathbb{R}$-constructible.

Using the existence of μ-stratifications, we prove that F is w-$\mathbb{R}$-constructible if and only if $SS(F)$ is contained in a closed conic subanalytic isotropic set, or equivalently if $SS(F)$ is subanalytic and Lagrangian. In other words, to be w-$\mathbb{R}$-constructible is a microlocal property. Then we can make full use of the results of the preceding chapters to obtain quite immediately various functorial properties of w-$\mathbb{R}$ and $\mathbb{R}$-constructible objects. Using the existence of triangulations for subanalytic sets, we also prove that the derived category of $\mathbb{R}$-constructible sheaves is equivalent to the full subcategory of $\mathbf{D}^b(X)$ consisting of objects with $\mathbb{R}$-constructible cohomology.

When X is a complex manifold, one can also introduce the notions of w-$\mathbb{C}$ and $\mathbb{C}$-constructible objects. The definitions are similar, replacing "subanalytic submanifold" by "complex analytic submanifold". A useful result asserts that F is w-$\mathbb{C}$-constructible if and only if F is w-$\mathbb{R}$-constructible (on the real underlying manifold) and moreover $SS(F)$ is invariant by the action of $\mathbb{C}^\times$ on T^*X. By this theorem, many properties of $\mathbb{C}$-constructible sheaves are deduced from those of $\mathbb{R}$-constructible sheaves.

Finally we introduce the famous "nearby-cycle functor" and "vanishing-cycle functor" and compare them to the specialization functor and the microlocalization functor.

Convention 8.0. We keep conventions 4.0. In particular the base ring A satisfies $\mathrm{gld}(A) < \infty$. Moreover, unless otherwise specified, all manifolds and morphisms of manifolds, are real analytic and countable at infinity.

One denotes by $\dim X$ (resp. $\dim_{\mathbb{C}} X$) the dimension of a real (resp. complex) analytic subset X.

8.1. Constructible sheaves on a simplicial complex

Definition 8.1.1. *A simplicial complex* $\mathbf{S} = (S, \Delta)$ *is the data consisting of a set* S *and a set* Δ *of subsets of* S, *satisfying the following axioms.*

(S.1) *Any* $\sigma \in \Delta$ *is a finite and non-empty subset of* S.
(S.2) *If* τ *is a non-empty subset of an element* σ *of* Δ, *then* τ *belongs to* Δ.
(S.3) *For any* $p \in S$, $\{p\}$ *belongs to* Δ.
(S.4) *For any* $p \in S$, *the set* $\{\sigma \in \Delta; p \in \sigma\}$ *is finite.*

An element of Δ is called a **simplex** and an element of S is called a **vertex**.

Let $\mathbb{R}^S$ denote the set of maps from S to $\mathbb{R}$. Recall that an element $x \in \mathbb{R}^S$ is nothing but a family $x(p) \in \mathbb{R}$, indexed by $p \in S$. We equip $\mathbb{R}^S$ with the product topology. To $\sigma \in \Delta$ we associate the subset $|\sigma|$ of $\mathbb{R}^S$ by:

$$(8.1.1)\quad |\sigma| = \left\{x \in \mathbb{R}^S; x(p) = 0 \text{ for } p \notin \sigma,\ x(p) > 0 \text{ for } p \in \sigma \text{ and } \sum_p x(p) = 1\right\} .$$

Note that the $|\sigma|$'s are disjoint to each other. One also sets:

$$(8.1.2)\qquad |\mathbf{S}| = \bigcup_{\sigma \in \Delta} |\sigma| ,$$

$$(8.1.3)\qquad U(\sigma) = \bigcup_{\tau \in \Delta, \tau \supset \sigma} |\tau| ,$$

and for $x \in |\mathbf{S}|$:

$$(8.1.4)\qquad U(x) = U(\sigma(x)) ,$$

where $\sigma(x)$ is the unique simplex such that $x \in |\sigma(x)|$, i.e. $\sigma(x) = \{p; x(p) \neq 0\}$.

$$(8.1.5)\qquad U(\sigma) = \{x \in |\mathbf{S}|; x(p) > 0 \text{ for } p \in \sigma\} ,$$

$$(8.1.6)\qquad U(x) = \{y \in |\mathbf{S}|; y(p) > 0 \text{ for any } p \text{ such that } x(p) > 0\} .$$

Example 8.1.2. Take $S = \{1,2,3\}$, and let Δ be the family of non-empty subsets of S. One can visualize the sets $|\sigma|$, $U(\sigma)$ and $U(x)$ by the following pictures, where $\sigma = \{1,2\}$, $x = (1,0,0)$.

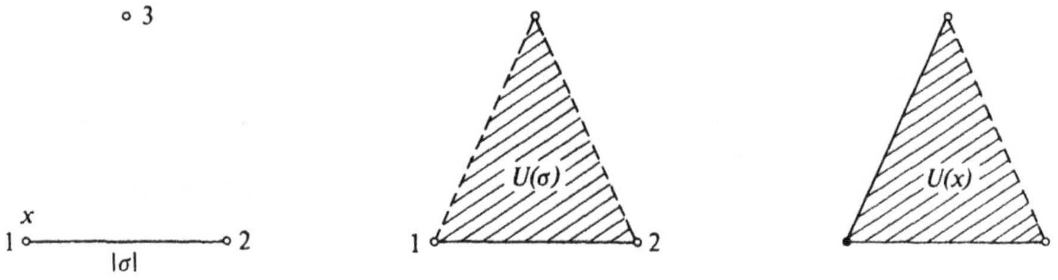

Fig. 8.1.1

We endow $|\mathbf{S}|$ with the induced topology from $\mathbb{R}^S$. Then $U(\sigma)$ and $U(x)$ are open subsets of $|\mathbf{S}|$ as easily seen by (8.1.5), (8.1.6). Define:

$$S(\sigma) = \{p \in S; \{p\} \cup \sigma \in \Delta\} \ . \tag{8.1.7}$$

Then $S(\sigma)$ is a finite subset of S by the axiom (S.4) and $U(\sigma)$ is contained in $\mathbb{R}^{S(\sigma)}$. Hence $U(\sigma)$ is homeomorphic to a locally closed subset of $\mathbb{R}^l$, with $l = \# S(\sigma)$. Since $|\sigma| = \{x \in U(\sigma);\ x(p) = 0 \text{ for } p \notin \sigma\}$, $|\sigma|$ is a closed subset of $U(\sigma)$, thus a locally closed subset of $|\mathbf{S}|$. One also easily verify that:

$$\begin{aligned} \overline{|\sigma|} &= \left\{x \in \mathbb{R}^S;\ x(p) = 0 \text{ if } p \notin \sigma,\ x(p) \geqslant 0 \text{ and } \sum_p x(p) = 1\right\} \\ &= \bigcup_{\tau \subset \sigma} |\tau| \ , \end{aligned} \tag{8.1.8}$$

where $\overline{|\sigma|}$ denotes the closure of $|\sigma|$ in $|\mathbf{S}|$. Also note that for σ and τ in Δ:

$$\begin{cases} U(\sigma) \cap U(\tau) = U(\sigma \cup \tau) & \text{if } \sigma \cup \tau \in \Delta \ , \\ \qquad\qquad\quad\ = \varnothing & \text{if } \sigma \cup \tau \notin \Delta \ . \end{cases} \tag{8.1.9}$$

$$U(\sigma) \subset U(\tau) \qquad \text{if and only if } \tau \subset \sigma \ . \tag{8.1.10}$$

The axiom (S.4) together with (8.1.9) imply that the family $\{U(\sigma)\}_{\sigma \in \Delta}$ is a locally finite open covering of $|\mathbf{S}|$. Therefore $|\mathbf{S}|$ is a paracompact topological space.

Definition 8.1.3. *Let* $F \in \mathrm{Ob}(\mathbf{D}^b(|\mathbf{S}|))$.

(i) *One says F is weakly* **S**-*constructible* (w-**S**-*constructible, for short*) *if for all* $j \in \mathbb{Z}$ *and all* $\sigma \in \Delta$, *the sheaves* $H^j(F)|_{|\sigma|}$ *are constant.*
(ii) *If F is* w-**S**-*constructible and moreover for each* $x \in X$, F_x *is a perfect complex, one says F is* **S**-*constructible.*

(For the notion of perfect complexes, cf. Exercise I 30 and recall that if A is Noetherian, F_x is perfect iff for all $j \in \mathbb{Z}$, $H^j(F)_x$ is finitely generated. Recall that we assumed $\mathrm{gld}(A) < \infty$.)

A sheaf F is called w-S-constructible (resp. S-constructible) if it is so, considered as an object of $\mathbf{D}^b(|\mathbf{S}|)$. One denotes by $\mathbf{D}^b_{w\text{-}\mathbf{S}\text{-}c}(|\mathbf{S}|)$ (resp. $\mathbf{D}^b_{\mathbf{S}\text{-}c}(|\mathbf{S}|)$ the full triangulated subcategory of $\mathbf{D}^b(|\mathbf{S}|)$ consisting of w-S-constructible (resp. S-constructible) objects.

One denotes by w-$\mathfrak{Cons}(\mathbf{S})$ (resp. $\mathfrak{Cons}(\mathbf{S})$) the full subcategory of $\mathfrak{Mod}(A_{|\mathbf{S}|})$ consisting of w-S-constructible (resp. S-constructible) sheaves.

Let $u: F \to G$ be a morphism of sheaves on $|\mathbf{S}|$, F and G being w-S-constructible. Then one immediately proves that $\mathrm{Ker}\, u$, $\mathrm{Im}\, u$, $\mathrm{Coker}\, u$ are w-S-constructible. Moreover if $0 \to F' \to F \to F'' \to 0$ is an exact sequence of sheaves on $|\mathbf{S}|$ and F' and F'' are w-S-constructible then so is F. Hence w-$\mathfrak{Cons}(\mathbf{S})$ is an abelian category.

If the base ring A is Noetherian, the same results hold for $\mathfrak{Cons}(\mathbf{S})$.

Proposition 8.1.4. *Assume F is a w-S-constructible sheaf. Then for any $\sigma \in \Delta$ and any $x \in |\sigma|$ we have the isomorphisms:*

(i) $$H^0(U(\sigma); F) \xrightarrow{\sim} H^0(|\sigma|; F) \xrightarrow{\sim} F_x \ ,$$

(ii) $$H^j(U(\sigma); F) = H^j(|\sigma|; F) = 0 \qquad \textit{for } j \neq 0 \ .$$

Proof. For $0 < \varepsilon < 1$, set $I_\varepsilon = \{t \in \mathbb{R}; \varepsilon \leqslant t \leqslant 1\}$ and define the map $\pi_\varepsilon : I_\varepsilon \times U(\sigma) \to U(\sigma)$ by:

(8.1.11) $$\pi_\varepsilon(t, y)(p) = ty(p) + (1 - t)x(p) \ , \qquad \text{for } p \in S \ .$$

The map π_ε is continuous and surjective, $\pi_\varepsilon(1, \cdot)$ is the identity $\{1\} \times U(\sigma) \simeq U(\sigma)$, and $\pi_\varepsilon(\varepsilon, \cdot)$ is a homeomorphism $\{\varepsilon\} \times U(\sigma) \xrightarrow{\sim} \pi_\varepsilon(\{\varepsilon\} \times U(\sigma))$.

Denote by q_2 the projection from $I_\varepsilon \times U(\sigma)$ to $U(\sigma)$. Since $y \in |\tau|$ implies $\pi_\varepsilon(t, y) \in |\tau|$, we get that $\pi_\varepsilon^{-1}F$ is a constant sheaf on the fibers of q_2, thus $\pi_\varepsilon^{-1}F = q_2^{-1}G$, for some sheaf G on $U(\sigma)$. (Take $G = q_{2*}\pi_\varepsilon^{-1}F$.) Applying Corollary 2.7.7, we obtain that for each $t \in I_\varepsilon$, the morphism $R\Gamma(I_\varepsilon \times U(\sigma); \pi_\varepsilon^{-1}F) \to R\Gamma(\{t\} \times U(\sigma); \pi_\varepsilon^{-1}F)$ is an isomorphism. Hence:

(8.1.12) $$\begin{aligned} R\Gamma(U(\sigma); F) &\simeq R\Gamma(\{\varepsilon\} \times U(\sigma); \pi_\varepsilon^{-1}F) \\ &\simeq R\Gamma(\pi_\varepsilon(\{\varepsilon\} \times U(\sigma)); F) \ . \end{aligned}$$

Now we note that the family $\{\pi_\varepsilon(\{\varepsilon\} \times U(\sigma))\}_{\varepsilon > 0}$ forms a neighborhood system of x in X. Taking the cohomology of both sides of (8.1.12) and the inductive limit with respect to $\varepsilon > 0$, we get that $H^j(U(\sigma); F)$ is zero for $j > 0$ and is isomorphic to F_x for $j = 0$. If we apply the same argument to the simplicial complex σ and the sheaf $F|_{|\sigma|}$, we obtain the same result with $U(\sigma)$ replaced by $|\sigma|$, which completes the proof. □

Corollary 8.1.5. (i) *The functor $\Gamma(U(\sigma);\cdot)$ is exact on the category* w-$\mathfrak{Cons}$(S).

(ii) *If F is a w-S-constructible sheaf and $\Gamma(U(\sigma);F)=0$ for all $\sigma\in\Delta$, then* $F=0$.

Let us say that a sheaf F on $|\mathbf{S}|$ is S-acyclic if it satisfies $H^k(U(\sigma);F)=0$ for all $k>0$ and all $\sigma\in\Delta$. Then w-S-constructible sheaves are S-acyclic, in view of Proposition 8.1.4. Of course, flabby sheaves on $|\mathbf{S}|$ are also S-acyclic.

Let F be a sheaf on $|\mathbf{S}|$. We shall associate functorially to F a w-S-constructible sheaf $\beta(F)$ and a morphism $s(F):\beta(F)\to F$. For that purpose define first

$$\alpha(F)=\bigoplus_{\sigma\in\Delta}(F(U(\sigma)))_{U(\sigma)}\ . \tag{8.1.13}$$

If M is an A-module, then $\mathrm{Hom}(M_{U(\sigma)},F)=\mathrm{Hom}(M,\Gamma(U(\sigma);F))$. Therefore there exists a natural morphism $i(F):\alpha(F)\to F$, and this morphism is functorial with respect to F. Let F' denote the kernel of $i(F)$ and let $\beta(F)$ denote the cokernel of the morphism $\alpha(F')\to\alpha(F)$ obtained by the composition $\alpha(F')\to F'\to\alpha(F)$. We get a morphism $s(F):\beta(F)\to F$ which is functorial with respect to F.

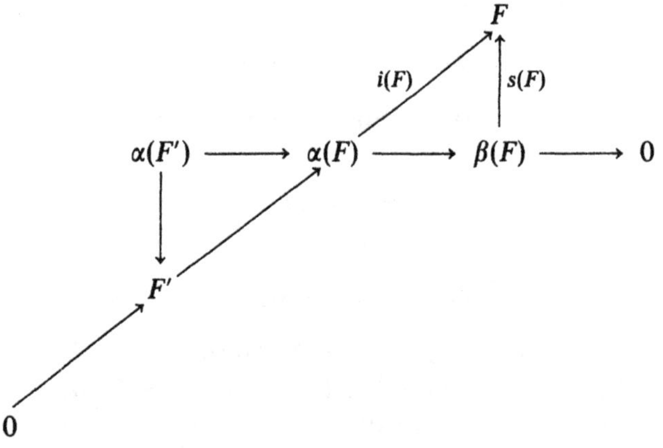

Note that $\alpha(F)$ and $\beta(F)$ are w-S-constructible.

Lemma 8.1.6. *For $\sigma\in\Delta$, the natural morphism $\Gamma(U(\sigma);\beta(F))\to\Gamma(U(\sigma);F)$ is an isomorphism.*

Proof. Consider the diagram:

$$\begin{array}{ccccccccc}
 & & \alpha(F')(U(\sigma)) & \longrightarrow & \alpha(F)(U(\sigma)) & \longrightarrow & \beta(F)(U(\sigma)) & \longrightarrow & 0\\
 & & \downarrow & & \| & & \downarrow & & \\
0 & \longrightarrow & F'(U(\sigma)) & \longrightarrow & \alpha(F)(U(\sigma)) & \longrightarrow & F(U(\sigma)) & \longrightarrow & 0\ .
\end{array}$$

Since $\Gamma(U(\sigma); \cdot)$ is an exact functor on w-$\mathfrak{Cons}(\mathrm{S})$, the top row is exact. Since the morphism $\alpha(F)(U(\sigma)) \to F(U(\sigma))$ is surjective, the bottom row is exact. Then the surjectivity of $\alpha(F')(U(\sigma)) \to F'(U(\sigma))$ implies the desired result. □

Lemma 8.1.7. *Assume F is weakly* S*-constructible. Then $s(F): \beta(F) \to F$ is an isomorphism.*

Proof. The sheaves $\mathrm{Ker}\, s(F)$ and $\mathrm{Coker}\, s(F)$ are w-S-constructible and their sections over $U(\sigma)$ are zero by Lemma 8.1.6 and Corollary 8.1.5 (i). Then these sheaves are zero in view of Corollary 8.1.5 (ii). □

By this lemma, $\beta: \mathfrak{Mod}(A_{|\mathrm{S}|}) \to w$-$\mathfrak{Cons}(\mathrm{S})$ is a right adjoint functor to the inclusion functor w-$\mathfrak{Cons}(\mathrm{S}) \to \mathfrak{Mod}(A_{|\mathrm{S}|})$.

Proposition 8.1.8. (i) *The functor β from $\mathfrak{Sh}(|\mathrm{S}|)$ to w-$\mathfrak{Cons}(\mathrm{S})$ is left exact.*

(ii) *$\Gamma(U(\sigma); R^k\beta(F)) = H^k(U(\sigma); F)$ for all $\sigma \in \Delta$, $k \geqslant 0$, F a sheaf on $|\mathrm{S}|$.*

(iii) *$R^k\beta(F) = 0$ for any $k > 0$ if and only if F is* S*-acyclic.*

Proof. (i) Let $0 \to F' \to F \to F'' \to 0$ be an exact sequence of sheaves on $|\mathrm{S}|$. The sheaves $\beta(F')$, $\beta(F)$, $\beta(F'')$ being w-S-constructible, in order to see that the sequence $0 \to \beta(F') \to \beta(F) \to \beta(F'')$ is exact, it is enough to show, in view of Corollary 8.1.5 (ii) that for all $\sigma \in \Delta$, the sequence $0 \to \Gamma(U(\sigma); \beta(F')) \to \Gamma(U(\sigma); \beta(F)) \to \Gamma(U(\sigma); \beta(F''))$ is exact. But this follows from Lemma 8.1.6 and the left exactness of the functor $\Gamma(U(\sigma); \cdot)$.

(ii) Let $F^{\cdot}$ be an injective resolution of F. Then $R^k\beta(F) = H^k(\beta(F^{\cdot}))$. Moreover:

$$\begin{aligned} \Gamma(U(\sigma); R^k\beta(F)) &= \Gamma(U(\sigma); H^k(\beta(F^{\cdot}))) \\ &= H^k(\Gamma(U(\sigma); \beta(F^{\cdot}))) \\ &= H^k(\Gamma(U(\sigma); F^{\cdot})) \\ &= H^k(U(\sigma); F) \ . \end{aligned}$$

(We have used the fact that w-S-constructible sheaves are S-acyclic, and Lemma 8.1.6.)

(iii) follows from (ii) and Corollary 8.1.5 (ii). □

Proposition 8.1.9. *Let $F^{\cdot}$ be a complex bounded from below of* S*-acyclic sheaves. Assume that for all $n \in \mathbb{Z}$, $H^n(F^{\cdot})$ is w-*S*-constructible. Then the morphism $\beta(F^{\cdot}) \to F^{\cdot}$ is a quasi-isomorphism.*

Proof. Let $d^{\cdot}$ denote the differential of the complex $F^{\cdot}$, (i.e.: $d^n: F^n \to F^{n+1}$), and set $Z^n(F^{\cdot}) = \mathrm{Ker}\, d^n$, $B^n(F^{\cdot}) = \mathrm{Im}\, d^{n-1}$. We get the exact sequences:

$$0 \to Z^{n-1}(F^{\cdot}) \to F^{n-1} \to B^n(F^{\cdot}) \to 0 \ , \tag{8.1.14}$$

(8.1.15) $$0 \to B^n(F^\cdot) \to Z^n(F^\cdot) \to H^n(F^\cdot) \to 0 \ .$$

Assume $Z^{n-1}(F^\cdot)$ is S-acyclic. Applying the functor β to (8.1.14) we find $R^k\beta(B^n(F^\cdot)) = 0$ for $k > 0$ and this implies $B^n(F^\cdot)$ is S-acyclic by Proposition 8.1.8. The same argument applied to the sequence (8.1.15) gives that $Z^n(F^\cdot)$ is S-acyclic. By induction this proves that all the $Z^n(F^\cdot)$ and $B^n(F^\cdot)$ are S-acyclic.

The functor β being left exact, we have $\beta(Z^n(F^\cdot)) = Z^n(\beta(F^\cdot))$. Since $R^1\beta(Z^{n-1}(F^\cdot)) = 0$, the sequence (8.1.14) yields $\beta(B^n(F^\cdot)) = B^n(\beta(F^\cdot))$. Similarly, since $R^1\beta(B^n(F^\cdot)) = 0$, the sequence (8.1.15) yields $\beta(H^n(F^\cdot)) = H^n(\beta(F^\cdot))$. By the hypothesis and Lemma 8.1.7, $\beta(H^n(F^\cdot)) \simeq H^n(F^\cdot)$. Hence we have proved the isomorphism $H^n(\beta(F^\cdot)) \simeq H^n(F^\cdot)$. □

In order to work in the category $\mathbf{D}^b(|\mathbf{S}|)$ we shall make the following hypothesis.

(8.1.16) $$\begin{cases} \textit{There exists an integer } n \textit{ such that} \\ \#\sigma \leqslant n+1 \textit{ for all } \sigma \in \Delta \ . \end{cases}$$

We call the smallest n satisfying (8.1.16) the **dimension** of S. Under the assumption (8.1.16), we have $R^k\beta = 0$ for $k > n$, in view of Propositions 8.1.8 and 3.2.2. Then consider the functors (where δ is the natural functor):

$$w\text{-}\mathfrak{Cons}(\mathbf{S}) \underset{\beta}{\overset{\delta}{\rightleftarrows}} \mathfrak{Mod}(A_{|\mathbf{S}|}) \ .$$

They induce the functors

(8.1.17) $$\mathbf{D}^b(w\text{-}\mathfrak{Cons}(\mathbf{S})) \underset{R\beta}{\overset{\delta}{\rightleftarrows}} \mathbf{D}^b_{w\text{-}\mathbf{S}\text{-}c}(|\mathbf{S}|) \ .$$

Theorem 8.1.10. *Assume* (8.1.16). *Then the functors δ and $R\beta$ in* (8.1.17) *are equivalences of categories, inverse to each other.*

Proof. The isomorphism $R\beta \circ \delta \simeq id$ in $\mathbf{D}^b(w\text{-}\mathfrak{Cons}(\mathbf{S}))$ follows from the isomorphism $\beta \circ \delta \simeq id$ in $w\text{-}\mathfrak{Cons}(\mathbf{S})$.

The isomorphism $\delta \circ R\beta \simeq id$ in $\mathbf{D}^b_{w\text{-}\mathbf{S}\text{-}c}(|\mathbf{S}|)$ follows from Proposition 8.1.9. □

Theorem 8.1.11. *Assume* (8.1.16) *and assume the base ring A is Noetherian. Then the natural functor $\delta : \mathbf{D}^b(\mathfrak{Cons}(\mathbf{S})) \to \mathbf{D}^b_{\mathbf{S}\text{-}c}(|\mathbf{S}|)$ is an equivalence of categories.*

Proof. By Proposition 1.7.11 (and Remark 1.7.12) and Theorem 8.1.10, it is enough to prove the following lemma.

Lemma 8.1.12. *Let $u : F \to G$ be an epimorphism of w-S-constructible sheaves, and assume G is S-constructible. Then there exists an S-constructible sheaf H and a morphism $v : H \to F$ such that $u \circ v$ is an epimorphism.*

Proof. For any $\sigma \in \Delta$, $F(U(\sigma)) \to G(U(\sigma))$ is surjective by Corollary 8.1.5. Since $G(U(\sigma)) \xrightarrow{\sim} G_x$ for $x \in |\sigma|$, by Proposition 8.1.4, $G(U(\sigma))$ is finitely generated. Hence there exists a finitely generated free module $H(\sigma)$ and an epimorphism $H(\sigma) \to G(U(\sigma))$. This morphism splits and we get the morphisms $H(\sigma) \to F(U(\sigma)) \to G(U(\sigma))$. Set $H = \bigoplus_\sigma H(\sigma)_{U(\sigma)}$. Then H is an S-constructible sheaf, the morphisms $H(\sigma) \to F(U(\sigma))$ define the morphism $H \to F$ and the composite $H \to F \to G$ is an epimorphism. $\square$

Note that by this argument, one sees that, w-$\mathfrak{Cons}(\mathrm{S})$ has enough projectives and if A is Noetherian, the category $\mathfrak{Cons}(\mathrm{S})$ has also enough projectives.

8.2. Subanalytic sets

In this section we recall the definition of Hironaka's subanalytic sets, and their main properties, without proofs. References are made to Hironaka [1, 2], and Hardt [1, 2] or Bierstorne-Milman [1].

Let X be a manifold, (recall that all manifolds are supposed real analytic), and let Z be a subset of X.

Definition 8.2.1. *One says Z is subanalytic at $x \in X$ if there exist an open neighborhood U of x, compact manifolds Y_j^i $(i = 1, 2, 1 \leqslant j \leqslant N)$ and morphisms $f_j^i : Y_j^i \to X$ such that:*

$$Z \cap U = U \cap \bigcup_{j=1}^{N} (f_j^1(Y_j^1) \setminus f_j^2(Y_j^2)) \ .$$

If Z is subanalytic at each $x \in X$, one says Z is subanalytic in X.

Subanalytic sets inherit the following properties.

Proposition 8.2.2. (i) *Assume Z is subanalytic in X. Then $\bar{Z}$ and* $\mathrm{Int}(Z)$ *are subanalytic in X. Moreover the connected components of Z are locally finite and subanalytic.*

(ii) *Assume Z_1 and Z_2 are subanalytic in X. Then $Z_1 \cup Z_2$, $Z_1 \setminus Z_2$, $Z_1 \cap Z_2$ are subanalytic.*

(iii) *Let $f : Y \to X$ be a morphism of manifolds. If $Z \subset X$ is subanalytic in X then $f^{-1}(Z)$ is subanalytic in Y. If $W \subset Y$ is subanalytic in Y and f is proper on $\overline{W}$, then $f(W)$ is subanalytic in X.*

(iv) *Let Z be a closed subanalytic subset of X. Then there exist a manifold Y and a proper morphism $f : Y \to X$ such that $f(Y) = Z$.*

Now we state the **curve selection lemma**.

Proposition 8.2.3. *Let Z be a subanalytic subset of X and let $x_\circ \in \bar{Z}$. Then there exists an analytic curve $t \mapsto x(t)$, $]-1,1[\to X$, such that $x(0) = x_\circ$ and $x(t) \in Z$ for $t \neq 0$.*

There is a **desingularization theorem** in the real analytic case.

Proposition 8.2.4. *Let $\varphi : X \to \mathbb{R}$ be a real analytic function, not identically zero on each connected component of X. Set $Z = \{x \in X; \varphi(x) = 0, d\varphi(x) = 0\}$. Then there exists a proper morphism of manifolds $f : Y \to X$ which induces an isomorphism $Y \backslash f^{-1}(Z) \simeq X \backslash Z$ and such that in a neighborhood of each $y_\circ \in f^{-1}(Z)$ there exists a local coordinate system $(y_1, \ldots, y_n)$ with $\varphi \circ f = \pm y_1^{r_1} \ldots y_n^{r_n}$, the r_j's being non-negative integers.*

Let Z be a subanalytic subset of X. One defines Z_{reg} as the subset of points $x \in Z$ such that there exists an open neighborhood U of x in X such that $U \cap Z$ is a closed submanifold of U. One sets $Z_{\text{sing}} = Z \backslash Z_{\text{reg}}$. Then the sets Z_{reg} and Z_{sing} are subanalytic in X and $\overline{Z_{\text{reg}}}$ contains Z.

If $x \in Z_{\text{reg}}$, the dimension of Z at x, denoted $\dim_x(Z)$, is well defined. One sets:

$$\dim(Z) = \sup_{x \in Z_{\text{reg}}} \dim_x(Z) . \tag{8.2.1}$$

The **triangulation theorem** below will reduce the study of constructible sheaves along subanalytic stratifications to the study of constructible sheaves on simplicial complexes.

Proposition 8.2.5. *Let $X = \bigsqcup_{\alpha \in A} X_\alpha$ be a locally finite partition of X by subanalytic subsets. Then there exist a simplicial complex $\mathbf{S} = (S, \Delta)$ and a homeomorphism $i : |\mathbf{S}| \xrightarrow{\sim} X$ such that:*

(i) *for any $\sigma \in \Delta$, $i(|\sigma|)$ is a subanalytic submanifold of X,*
(ii) *for any $\sigma \in \Delta$ there exists $\alpha \in A$ such that $i(|\sigma|) \subset X_\alpha$.*

8.3. Subanalytic isotropic sets and μ-stratifications

In this section we shall collect all results concerning subanalytic isotropic sets that we shall need later, and we shall introduce the new notion of a μ-stratification.

Let S be a locally closed subanalytic subset of the manifold X. We set:

$$T_S^*X = \overline{T_{S_{\text{reg}}}^*X} \cap \pi^{-1}(S) . \tag{8.3.1}$$

Proposition 8.3.1. *The set T_S^*X is subanalytic in T^*X.*

Proof. We may assume S closed.

Let $f : Y \to X$ be a proper morphism of manifolds such that $f(Y) = S$. Set:

(8.3.2) $$Y_\circ = \{y \in f^{-1}(S_{reg});\, f'_y : T_yY \to T_{f(y)}S \text{ is surjective}\} .$$

Then $Y_\circ$ is open, subanalytic and $S_\circ = f(Y_\circ)$ is open, dense and subanalytic in S. The set $P = T^*X \setminus f_\pi {}^t f'^{-1}(\dot{T}^*Y_\circ)$ is subanalytic and we have:

$$P = \{(x;\xi) \in T^*X; \text{for any } y \in Y_\circ \cap f^{-1}(x), {}^t f'(y)\cdot\xi = 0\} .$$

Hence $P \cap \pi^{-1}(S_\circ) = T^*_{S_\circ}X$ and this set is subanalytic. Therefore $T^*_SX = \overline{T^*_{S_\circ}X}$ is subanalytic. □

Proposition 8.3.2. (i) *Let M be a closed submanifold and S a subanalytic subset of X. Then the normal cone $C_M(S)$ is subanalytic in T_MX.*

(ii) *Let S_1 and S_2 be two subanalytic subsets of X. Then the normal cone $C(S_1, S_2)$ is subanalytic in TX.*

Proof. (ii) follows from (i), and (i) follows immediately from Definition 4.1.1. □

Proposition 8.3.3. *Let S_1 and S_2 be two subanalytic subsets of $\mathbb{R}^n$ and let $v \in C_x(S_1, S_2)$. Then there exist two real analytic curves $x_i(t)$, $t \in]-1,1[$, $i = 1, 2$ such that $x_i(0) = x$, $x_i(t) \in S_i$ $(t \neq 0, i = 1,2)$ and $x_1(t) - x_2(t) = t^k v + O(t^{k+1})$ for some integer $k > 0$.*

Proof. Let p be the map $\mathbb{R}^n \times \mathbb{R}^n \times \mathbb{R} \to \mathbb{R}^n \times \mathbb{R}^n$, $(x, y, s) \mapsto (x, x - sy)$. Then identifying TX to the set $\{s = 0\}$ of $\mathbb{R}^n \times \mathbb{R}^n \times \mathbb{R}$, the cone $C(S_1, S_2)$ is identified to $\overline{p^{-1}(S_1 \times S_2) \cap \Omega} \cap TX$ where $\Omega = \{s > 0\}$.

Let $v \in C_x(S_1, S_2)$. By the curve selection lemma, there exists a curve $u(t) = (x(t), y(t), s(t))$, $-1 < t < 1$, such that $x(0) = x$, $y(0) = v$, $s(0) = 0$ and $x(t) \in S_1$, $x(t) - s(t)y(t) \in S_2$, $s(t) > 0$, for $t \neq 0$. Setting $x_1(t) = x(t)$, $x_2(t) = x(t) - s(t)y(t)$, we get $x_1(t) - x_2(t) = s(t)y(t) = t^k v + O(t^{k+1})$, for some $k > 0$. □

Proposition 8.3.4. *Let S be a subanalytic subset of X and let θ be a (real analytic) 1-form on X. Then the following conditions are equivalent:*

(i) $\theta|_{S_{reg}} = 0$,
(ii) *for any* $x \in X$, $\theta|_{C_x(S)} = 0$.

The condition (ii) means that the linear function on T_xX given by θ vanishes on $C_x(S)$.

Proof. (ii) ⇒ (i) is obvious.

Assume (i) and let $x_\circ \in \bar{S}$, $v \in C_{x_\circ}(S)$, $v \neq 0$. By the preceding result, there exists a curve $x(t)$, $-1 < t < 1$, such that $x(t) \in S_{reg}$ for $t \neq 0$ and $x(t) = x_\circ + t^k v + O(t^{k+1})$. Choose a local coordinate system and write $\theta = \sum_j a_j(x)\,dx_j$, $v = (v_1, \dots, v_n)$. Then $\sum_j a_j(x(t))(\partial x_j/\partial t) = 0$ for $t \neq 0$, thus $k \sum_j a_j(x_\circ) v_j t^{k-1} + O(t^k) = 0$, and we get $\langle \theta(x_\circ), v\rangle = 0$. □

Definition 8.3.5. *If the equivalent conditions of Proposition* 8.3.4 *are satisfied we say that θ vanishes on S and we write $\theta|_S = 0$.*

Corollary 8.3.6. (i) *Let $h: X' \to X$ be a morphism of manifolds, and let $S \subset X$, $S' \subset X'$ be subanalytic subsets with $h(S') \subset S$. Let θ be a* 1*-form on X. Then $\theta|_S = 0$ implies $h^*\theta|_{S'} = 0$.*

(ii) *Let $(S_j)_{j \in J}$ be a locally finite family of subanalytic subsets of X and let θ be a* 1*-form on X. If $\theta|_{S_j} = 0$ for all j, then $\theta|_{\bigcup_j S_j} = 0$.*

Proof. (i) For all $x' \in X'$, one has:

$$h'(C_{x'}(S')) \subset C_{h(x')}(S) \ .$$

(ii) This follows from

$$C_x\Big(\bigcup_j S_j\Big) = \bigcup_j C_x(S_j) \ . \quad \square$$

Proposition 8.3.7. *Let $h: X' \to X$ be a morphism of manifolds and let $S \subset X$ and $S' \subset X'$ be subanalytic subsets. Assume $h(S') = S$. Let θ be a* 1*-form on X. Then the following conditions are equivalent.*

(i) $\theta|_S = 0$,
(ii) $h^*\theta|_{S'} = 0$.

Proof. (i) $\Rightarrow$ (ii) by Corollary 8.3.6 (i).

(ii) $\Rightarrow$ (i) We may assume S is non-singular. Let $g: Y \to X'$ be a proper morphism such that $g(Y) = S'$. Then $g^*h^*\theta = 0$.

Set $f = g \circ h: Y \to X$ and define $Y_\circ = \{y \in Y; f'_y: T_yY \to T_{f(y)}S$ is surjective$\}$. Since $f(Y_\circ)$ is dense in S, the proof is complete. (Note that we assume Y countable at infinity.) $\square$

We need a result concerning conic subanalytic subsets in vector bundles.

Proposition 8.3.8. *Let $\tau: E \to X$ be a vector bundle.*

(i) *Let A be a conic subset of $\dot{E}$, subanalytic in $\dot{E}$. Then A is subanalytic in E.*
(ii) *If A is a subanalytic subset of E such that $\bar{A} \to X$ is proper, then $\mathbb{R}^+ \cdot A$ is subanalytic.*
(iii) *If A is a conic subanalytic subset of E, then $\tau(A)$ is subanalytic in X.*
(iv) *Let $\tau': E' \to X$ be another vector bundle, $f: E \to E'$ a morphism of bundles. If A is a conic subanalytic subset of E, then $f(A)$ is subanalytic in E'.*

Proof. Let $\gamma: \dot{E} \to \dot{E}/\mathbb{R}^+$ be the projection, $i: X \to E$ the zero-section. Let us take a submanifold S of $\dot{E}$ such that $S \to \dot{E}/\mathbb{R}^+$ is an isomorphism.

(i) Let $\mu: \mathbb{R}^+ \times E \to E$ be the multiplication map. Then $A' = \mu([0, 1] \times (S \cap A))$

is subanalytic in E, and (i) follows from the fact that $\Lambda'\backslash i(X) = \Lambda$ on a neighborhood of $i(X)$.

(ii) $\mu: [0,1] \times \bar{\Lambda} \to E$ is proper because $\bar{\Lambda}$ is proper over X. Then (ii) follows from $\mathbb{R}^+ \cdot \Lambda = \mu([0,1] \times \Lambda) \cup p(\mu^{-1}(\Lambda) \cap (]0,1] \times E))$, where p is the second projection $\mathbb{R} \times E \to E$.

(iii) is a particular case of (iv) with $E' = X$.

(iv) Let $A \subset E$ be a subanalytic open neighborhood of the zero-section of E such that $\bar{A} \to X$ is proper. Then $f(A \cap \Lambda)$ is subanalytic. Then (iv) follows from (ii) because $f(\Lambda) = \mathbb{R}^+ \cdot f(A \cap \Lambda)$. □

Now we shall study subanalytic isotropic subsets of T^*X. Recall that α_X (or simply α) denotes the canonical 1-form on T^*X.

Definition 8.3.9. *Let Λ be a conic subanalytic subset of T^*X.*

(i) *We say that Λ is isotropic if $\alpha|_\Lambda = 0$.*

(ii) *We say that Λ is Lagrangian if Λ is both isotropic and involutive* (*cf. Definition* 6.5.1).

Proposition 8.3.10. (i) *Let Λ be a closed conic subanalytic subset of T^*X. The following conditions are equivalent.*

(a) *Λ is isotropic.*

(b) *There exists a locally finite family $\{X_j\}$ of subanalytic subsets of X such that $\Lambda \subset \bigcup_j T^*_{X_j}X$.*

(c) *The same as* (b) *with a finite family of subanalytic submanifolds $X_j \subset \pi(\Lambda)$.*

(ii) *Moreover if Λ is isotropic and Y is a subanalytic subset of X, then there exists a subanalytic submanifold $Y_\circ \subset Y$ open dense in Y, such that $\Lambda \cap \pi^{-1}(Y_\circ) \subset T^*_{Y_\circ}X$.*

Proof. (i) (b) $\Rightarrow$ (a) By Corollary 8.3.6 it is enough to prove that $\alpha|_{T^*_{X_j}X} = 0$, which follows from the corresponding result for $X_{j,\mathrm{reg}}$ since $T^*_{X_j}X = \overline{T^*_{X_{j,\mathrm{reg}}}X}$.

(a) $\Rightarrow$ (c) Let d be the maximum of the rank of the projection $\pi: \Lambda_{\mathrm{reg}} \to X$. We shall argue by induction on d. Let $\Lambda_\circ = \{p \in \Lambda_{\mathrm{reg}};$ rank of π at p is $d\}$. Then $\Lambda_\circ$ is open and subanalytic in Λ. Set $X_\circ = (\pi(\Lambda))_{\mathrm{reg}}$ and $\Lambda'_\circ = \Lambda_\circ \cap \pi^{-1}(X_\circ)$. Then $\Lambda'_\circ$ is open and subanalytic in Λ, dense in $\Lambda_\circ$ and the differential of the projection $\pi: \Lambda'_\circ \to X_\circ$ is surjective at each $p \in \Lambda'_\circ$. Let us prove:

$$\Lambda'_\circ \subset T^*_{X_\circ}X \ . \tag{8.3.3}$$

Choose a local coordinate system $(x_1, \ldots, x_n) = (x', x'')$ with $x' = (x_1, \ldots, x_p)$ such that $X_\circ = \{x'' = 0\}$ and let $\xi = (\xi', \xi'')$ denote the dual coordinates. Then $\alpha|_{\pi^{-1}(X_\circ)} = \xi'\,dx'$ and $dx_1, \ldots, dx_p$ being linearly independent on $X_\circ$ and π being smooth, we get $\xi' = 0$ on $\Lambda'_\circ$, which proves (8.3.3). Now set $\Lambda' = \overline{\Lambda \backslash T^*_{X_\circ}X}$. Since $\Lambda'_\circ \cap \Lambda' = \varnothing$ by (8.3.3), the induction proceeds.

(c) $\Rightarrow$ (b) Obvious.

(ii) Replacing Λ and Y with $\Lambda \cap \pi^{-1}(\bar{Y})$ and $\bar{Y}$ respectively, we may assume from the beginning that $\Lambda \subset \pi^{-1}(Y)$. Then choosing $\{X_j\}$ as in (i) (c), it is enough to take Y_0 as the union of the connected components of $Y_{reg} \cap X_j$ open in Y_{reg}. □

Proposition 8.3.11. *Let $f: Y \to X$ be a morphism of manifolds.*

(i) *Let $\Lambda \subset T^*Y$ be a conic subanalytic isotropic set such that $f_\pi : \overline{{}^t f'^{-1}(\Lambda)} \to T^*X$ is proper. Then $f_\pi {}^t f'^{-1}(\Lambda)$ is a conic subanalytic isotropic subset of T^*X.*
(ii) *Let $\Lambda \subset T^*X$ be a conic subanalytic isotropic set. Then ${}^t f' f_\pi^{-1}(\Lambda)$ is a conic subanalytic isotropic subset of T^*Y.*

Proof. (i) First note that ${}^t f'^* \alpha_Y = f_\pi^* \alpha_X$. Next apply Proposition 8.3.7. We get the implications:

$$\alpha_Y|_\Lambda = 0 \Rightarrow ({}^t f')^* \alpha_Y|_{{}^t f'^{-1}(\Lambda)} = 0 \Leftrightarrow f_\pi^* \alpha_X|_{{}^t f'^{-1}(\Lambda)} = 0 \Rightarrow \alpha_X|_{f_\pi {}^t f'^{-1}(\Lambda)} = 0 \ .$$

(ii) The proof is similar except for the subanalyticity of ${}^t f' f_\pi^{-1}(\Lambda)$, which follows from Proposition 8.3.8. □

The following result will play an important role when proving finiteness theorems.

Proposition 8.3.12 (microlocal Bertini-Sard theorem). *Let $\varphi : X \to \mathbb{R}$ be a real analytic function and let $\Lambda \subset T^*X$ be a closed conic subanalytic isotropic subset. Let $S = \{t \in \mathbb{R}; t = \varphi(x), d\varphi(x) \in \Lambda$ for some $x \in X\}$. Assume φ is proper on $\pi(\Lambda)$. Then S is discrete.*

Proof. We have $S = \{t \in \mathbb{R}; (t; dt) \in \varphi_\pi {}^t\varphi'^{-1}(\Lambda)\}$. Since $\varphi_\pi {}^t\varphi'^{-1}(\Lambda)$ is a closed subanalytic isotropic subset, S is discrete. □

Proposition 8.3.13. *Let Λ and Λ_0 be two locally closed conic subsets of T^*X with $\Lambda \subset \Lambda_0$. Assume Λ_0 is subanalytic and isotropic and Λ is involutive and closed in Λ_0 (but we don't assume Λ is subanalytic). Then Λ is subanalytic and Lagrangian.*

Proof. We need two lemmas.

Lemma 8.3.14. *Let Λ_0 be an isotropic submanifold of X and Λ an involutive subset, closed in Λ_0. Then Λ is open in Λ_0, and* $\dim \Lambda = \dim X$.

Proof of Lemma 8.3.14. By Proposition A.2.9, Λ_0 is locally contained in a Lagrangian manifold. Hence we may assume Λ_0 Lagrangian. For any real function φ vanishing on Λ_0, Λ is a union of integral curves of H_φ by Proposition 6.5.2, and the union of the integral curves of H_φ issued at a point x is a neighborhood of x. □

Lemma 8.3.15. *Let S be a conic locally closed subanalytic isotropic subset of T^*X and let V be a locally closed involutive subset. Assume $V \subset S$ and* $\dim S < \dim X$. *Then $V = \emptyset$.*

Proof of Lemma 8.3.15. Arguing by induction on $\dim S$, it is enough to prove that $V \cap S_{\mathrm{reg}} = \emptyset$. Hence we may assume S is smooth. Let $x \in V$. By Lemma 8.3.14, $\dim_x V = \dim X$. This contradicts $\dim S < \dim X$. □

End of the proof of Proposition 8.3.13. Set $\Lambda'_\circ = (\Lambda_\circ)_{\mathrm{reg}}$, $\Lambda' = \Lambda \cap \Lambda'_\circ$. By Lemma 8.3.14, we find that Λ' is both open and closed in $\Lambda'_\circ$. Since $\Lambda'_\circ$ is subanalytic, Λ' is also subanalytic. It is then enough to show that $\Lambda = \overline{\Lambda'} \cap \Lambda_\circ$. Set $V = \Lambda \setminus \overline{\Lambda'}$. This is an involutive subset of T^*X contained in $\Lambda_\circ \setminus (\Lambda_\circ)_{\mathrm{reg}}$. Then V is empty by Lemma 8.3.15. □

We shall study the normal cone to an isotropic subset along a Lagrangian submanifold. First, we need a lemma.

Lemma 8.3.16. *Let $X = \mathbb{R}^n \times \mathbb{R}$, with coordinates (x, t), let Y be the hypersurface $\{t = 0\}$, and let Z be a subanalytic subset of X such that $Z \subset \overline{Z \setminus Y}$. Let α be a 1-form on X and let $\theta = t\alpha + b\,\mathrm{d}t$, where b is a real analytic function on X. Then $\theta|_Z = 0$ implies $\alpha|_{Z \cap Y} = 0$ and $b|_{Z \cap Y} = 0$.*

Proof. One may assume Z is closed.

Let $f: X' \to X$ be a proper morphism of manifolds such that $f(X') = Z$, and set $Y' = f^{-1}(Y)$. Let X'' be the union of connected components of X' on which $t \circ f = 0$. Since $f(X' \setminus X'') \supset Z \setminus Y$, $f(X' \setminus X'') = Z$. Hence we may assume Y' is nowhere dense by replacing X' with $X' \setminus X''$. By applying Proposition 8.2.4 to the function $t' = t \circ f$, we may assume from the beginning that in a neighborhood of each point of Y', there is a local coordinate system $(t_1, \ldots, t_N)$ such that $t' = \pm\prod_{j=1}^{l} t_j^{a_j}$, where $1 \leqslant l \leqslant N$ and the a_j's are positive integers. Set $\alpha' = f^*\alpha$, $b' = f^*b$. Then $t'\alpha' + b'\,\mathrm{d}t' = 0$. Since $\mathrm{d}t'/t' = \sum_j (a_j/t_j)\,\mathrm{d}t_j$ we get:

$$\alpha' = -b'\left(\sum_j (a_j/t_j)\,\mathrm{d}t_j\right). \tag{8.3.4}$$

By this formula we see that $t_1 \ldots t_l$ divides b', hence $b'|_{Y'} = 0$ and $b|_{Y \cap Z} = 0$.

We can also write (8.3.4) as:

$$\alpha' = (-b'/t_1 \ldots t_l)\left(\sum_j t_1 \ldots t_l (a_j/t_j)\,\mathrm{d}t_j\right).$$

Hence $\alpha'|_{Y'_{\mathrm{reg}}} = 0$, which implies $\alpha'|_{Y'} = 0$ and $\alpha|_{Y \cap Z} = 0$ by Proposition 8.3.7.
□

Theorem 8.3.17. *Let $\Lambda \subset T^*X$ be a (smooth, conic) Lagrangian submanifold and let S be a conic subanalytic isotropic subset of T^*X. We identify $T_\Lambda T^*X$ and $T^*\Lambda$*

by $-H$, *where* H *is the Hamiltonian isomorphism*: $TT^*X \simeq T^*T^*X$. *Then*:

(i) *the normal cone* $C_\Lambda(S)$ *is a conic subanalytic isotropic subset of* $T^*\Lambda$,
(ii) $C_\Lambda(S)$ *is contained in the hyperplane* $\{\alpha_X = 0\}$.

(Recall that α_X being 0 on Λ, α_X defines a linear function on $T_\Lambda T^*X$. Hence the hyperplane $\{\alpha_X = 0\}$ is well-defined in $T_\Lambda T^*X \simeq T^*\Lambda$.)

Proof. We may assume Λ is the conormal bundle to a submanifold (Proposition A.2.9 and Exercise A.2). For simplicity we shall assume $\Lambda = \{(x;\xi) \in T^*\mathbb{R}^n; x = 0\}$ (in the general case, the proof is similar). Consider the map $p: T^*X \times \mathbb{R} \to T^*X$, $((x;\xi), t) \mapsto (tx;\xi)$. Then $T_\Lambda T^*X \simeq \{t = 0\}$, $C_\Lambda(S) \simeq \overline{p^{-1}(S) \cap \{t > 0\}} \cap \{t = 0\}$ and $p^*(\alpha_X) = \sum_j \xi_j d(tx_j) = t(\sum_j \xi_j dx_j) + \langle x, \xi\rangle dt$. Since $p^*(\alpha_X)|_{p^{-1}(S)} = 0$ by Proposition 8.3.7, we get by Lemma 8.3.16:

$$\left(\sum_j \xi_j dx_j\right)\Big|_{C_\Lambda(S)} = 0 \qquad \text{and} \qquad \langle x, \xi\rangle|_{C_\Lambda(S)} = 0 \ .$$

Hence $(\sum_j x_j d\xi_j)|_{C_\Lambda(S)} = 0$, and the proof is complete, since $-\sum_j x_j d\xi_j$ is the canonical 1-form on $T^*\Lambda$. □

Corollary 8.3.18. (i) *Let* Λ_1 *and* Λ_2 *be two conic subanalytic isotropic subsets of* T^*X. *Then* $\Lambda_1 \hat{+} \Lambda_2$ *is a conic subanalytic isotropic subset of* T^*X.

(ii) *Let* $f: Y \to X$ *be a morphism of manifolds, and let* $\Lambda \subset T^*X$ *be a conic subanalytic isotropic subset. Then* $f^\#(\Lambda)$ *is a conic subanalytic isotropic subset of* T^*Y (*cf. Definition* 6.2.3).

Now we shall study stratifications of X by subanalytic manifolds. Let us first recall some basic definitions.

Let $(X_j)_{j\in J}$ be a family of subsets of X. One says this family is a covering if $X = \bigcup_{j\in J} X_j$. If moreover the X_j's are disjoint to each other, one says this covering is a partition and one writes $X = \bigsqcup_{j\in J} X_j$. One says that a covering $X = \bigcup_{i\in I} X_i$ is finer than a covering $X = \bigcup_{j\in J} X'_j$, if for each $i \in I$ there exists $j \in J$ with $X_i \subset X'_j$.

Definition 8.3.19. (i) *For a closed subanalytic subset* Y *of* X, *a partition* $Y = \bigsqcup_{\alpha\in A} X_\alpha$ *is called a subanalytic stratification of* Y, *if it is locally finite, the* X_α*'s are subanalytic submanifolds and for all pairs* $(\alpha, \beta) \in A \times A$ *such that* $\overline{X}_\alpha \cap X_\beta \neq \emptyset$ *one has* $X_\beta \subset \overline{X}_\alpha$. *Each* X_α *is called a stratum.*

(ii) *For two submanifolds* N *and* M *of* X, *we say that* (M, N) *satisfies the* μ*-condition if*:

$$(8.3.5) \qquad (T^*_M X \hat{+} T^*_N X) \cap \pi^{-1}(N) \subset T^*_N X \ .$$

(iii) *A partition* $X = \bigsqcup_{\alpha\in A} X_\alpha$ *is a* μ*-stratification if it is a subanalytic stratification and moreover for all pairs* $(\alpha, \beta) \in A \times A$ *such that* $X_\beta \subset \overline{X}_\alpha \backslash X_\alpha$, (X_α, X_β) *satisfies the* μ*-condition.*

Note that if $X = \bigsqcup_\alpha X_\alpha$ is a μ-stratification, then the set $\Lambda = \bigsqcup_\alpha T^*_{X_\alpha}X$ is a closed conic subanalytic isotropic subset of T^*X. In fact if $X_\beta \subset \overline{X}_\alpha$, then $\overline{T^*_{X_\alpha}X} \cap \pi^{-1}(X_\beta)$ is contained in $T^*_{X_\beta}X$.

Theorem 8.3.20. *Let $X = \bigcup_{j\in J} X_j$ be a locally finite covering of X by subanalytic subsets. Then there exists a μ-stratification $X = \bigsqcup_{\alpha\in A} X_\alpha$ finer than the covering $X = \bigcup_{j\in J} X_j$.*

In order to prove this result, we need a lemma.

Lemma 8.3.21. (i) *Let Y be a closed subanalytic subset of X, $Y = \bigcup_{j\in J} X_j$ a locally finite covering by subanalytic subsets. Then there exists a finer subanalytic stratification of Y.*

(ii) *Let N and M be two subanalytic submanifolds of X. Then the set Ω of points x in N such that (M, N) satisfies the μ-condition on a neighborhood of x is dense in N and subanalytic in X.*

(iii) *Let X' be a subanalytic open subset of X, $X' = \bigsqcup_{\alpha\in A} X_\alpha$ a subanalytic stratification of X', with the strata X_α subanalytic in X. Then there exists a largest open subset Ω of X' such that $\Omega = \bigsqcup_{\alpha\in A}(X_\alpha \cap \Omega)$ is a μ-stratification of Ω. Moreover Ω is subanalytic in X.*

Proof of Lemma 8.3.21. (i) Let A be the set of finite subsets of J and set, for $\alpha \in A$, $Z_\alpha = (\bigcap_{j\in\alpha} X_j)\setminus(\bigcup_{j\notin\alpha} X_j)$. Then $Y = \bigsqcup_\alpha Z_\alpha$ is a locally finite subanalytic partition. Hence we may assume from the beginning that $Y = \bigsqcup_j X_j$ is a locally finite subanalytic partition. We shall argue by induction on dim Y. Let X_j' be the union of connected components of $(X_j)_{\mathrm{reg}} \cap Y_{\mathrm{reg}}$, open in Y_{reg}. Set $Y' = Y\setminus(\bigcup_j X_j')$. Since $Y\setminus Y' = \bigsqcup_j X_j'$ is a disjoint union, it is a subanalytic stratification. Let A be the set of finite subsets of J. Then by the hypothesis of induction, there exists a subanalytic stratification $Y' = \bigsqcup_{j\in J'} Y_j$ finner than

$$Y' = \bigcup_{j\in J, \alpha\in A}\left[Y' \cap X_j \cap \left(\bigcap_{k\in\alpha}(\overline{X_k\setminus Y'})\right)\setminus\left(\bigcup_{k\in J\setminus\alpha}\overline{X_k\setminus Y'}\right)\right].$$

Then $Y = (\bigsqcup_{j\in J}(X_j\setminus Y')) \sqcup (\bigsqcup_{j\in J'} Y_j)$ is a stratification.

In order to see this, it is enough to show, for any j and for any $k \in J'$, that $\overline{X_j\setminus Y'} \cap Y_k \neq \emptyset$ implies $\overline{X_j\setminus Y'} \supset Y_k$. Then there is $\alpha \in A$ such that $Y_k \subset \bigcap_{i\in\alpha}(\overline{X_i\setminus Y'})\setminus(\bigcup_{i\in J\setminus\alpha}\overline{X_i\setminus Y'})$. Therefore $\overline{X_j\setminus Y'} \cap Y_k \neq \emptyset$ implies $j \in \alpha$. This proves the desired result.

(ii) Ω is $N\setminus\pi(Z)$, where $Z = \overline{(T^*_N X \hat{+} T^*_M X)\setminus T^*_N X}$. Hence it is subanalytic. Since $T^*_N X \hat{+} T^*_M X$ is isotropic, there is an open dense subset U of N such that $(T^*_N X \hat{+} T^*_M X) \cap \pi^{-1}(U) \subset T^*_N X$ (Proposition 8.3.10 (ii)). Then $\pi(Z) \cap U = \emptyset$.

(iii) The set Ω is the complement of $\bigcup_{(\alpha,\beta)} \overline{X_\beta \cap \pi((T^*_{X_\alpha}X \hat{+} T^*_{X_\beta}X)\setminus T^*_{X_\beta}X)}$, where $(\alpha, \beta) \in A \times A$, $X_\beta \subset \overline{X}_\alpha\setminus X_\alpha$. □

Proof of Theorem 8.3.20. We may assume $X = \bigsqcup_j X_j$ is a subanalytic stratification. By induction on dim Y, it is enough to show that if Y is a closed subanalytic subset of X such that $X \setminus Y = \bigsqcup_{j \in J} (X_j \setminus Y)$ is a μ-stratification, then there exist a nowhere dense subanalytic subset Y' of Y and a finer subanalytic stratification $X = \bigsqcup_i X'_i$, such that $X \setminus Y' = \bigsqcup_i (X'_i \setminus Y')$ is a μ-stratification. In order to prove this, let us take a stratification $Y = \bigsqcup_k Y_k$ of Y by subanalytic manifolds finer than the partition $Y = \bigsqcup_j (X_j \cap Y)$. Then $X = (\bigsqcup_j (X_j \setminus Y)) \sqcup (\bigsqcup_k Y_k)$ is a subanalytic stratification, and we may assume from the beginning that $J = J' \sqcup J''$, $X \setminus Y = \bigsqcup_{j \in J'} X_j$ and $Y = \bigsqcup_{j \in J''} X_j$. Let Ω be the largest open subset of X such that $\Omega = \bigsqcup_j (X_j \cap \Omega)$ is a μ-stratification (Lemma 8.3.21 (ii)). Then Ω is subanalytic and contains $X \setminus Y$. Hence it is enough to show that $\Omega \cap Y$ is dense in Y. But this follows immediately from Lemma 8.3.21 (ii). □

Corollary 8.3.22. *Let Λ be a closed conic subanalytic isotropic subset of T^*X. Then there exists a μ-stratification $X = \bigsqcup_\alpha X_\alpha$ such that $\Lambda \subset \bigsqcup_\alpha T^*_{X_\alpha}X$.*

Proof. By Proposition 8.3.10 there exists a locally finite covering $X = \bigcup_j X_j$ by subanalytic subsets with $\Lambda \subset \bigcup_j T^*_{X_j}X$. Applying Theorem 8.3.20, we find a μ-stratification $X = \bigsqcup_\alpha X_\alpha$ finer than the covering. Then $\bigcup_j T^*_{X_j}X$ is contained in $\bigsqcup_\alpha T^*_{X_\alpha}X$, and the proof follows. □

Proposition 8.3.23. *Let M be a submanifold of X, Λ a conic subanalytic isotropic subset of T^*X. If $T^*_MX \cap \Lambda = \emptyset$ and $(\Lambda \hat{+} T^*_MX) \cap \pi^{-1}(M) \subset T^*_MX$, then $\bar{\Lambda} \cap T^*_MX$ is nowhere dense on each fiber of $T^*_MX \to M$.*

Proof. We shall show that, for $x_0 \in M$, $\pi^{-1}(x_0) \cap \bar{\Lambda}$ is nowhere dense in T^*_MX. Let us take a local coordinate system $x = (x', x'')$ on X such that $M = \{x' = 0\}$, and x_0 is the origin. We shall use the associated coordinate systems on $T^*T^*_MX$ and $T_{T^*_MX}T^*X$ as in (6.2.3). Then the hypothesis $(\Lambda \hat{+} T^*_MX) \cap \pi^{-1}(M) \subset T^*_MX$ implies:

$$\{x' = 0\} \cap C_{T^*_MX}(\Lambda) \subset \{\xi'' = 0\} \; . \tag{8.3.6}$$

One the other hand, $C_{T^*_MX}(\Lambda)$ is isotropic by Theorem 8.3.17, and hence there exists an open dense subset U of $\pi^{-1}(x_0) \cap T^*_MX$ such that $C_{T^*_MX}(\Lambda) \times_{T^*_MX} U \subset T^*_{\pi^{-1}(x_0) \cap T^*_MX}(T^*_MX)$, by Proposition 8.3.10 (ii). Since $T^*_{\pi^{-1}(x_0) \cap T^*_MX}(T^*_MX)$ is given by $\{x'' = 0, x' = 0\}$, $C_{T^*_MX}(\Lambda) \times_{T^*_MX} U$ is contained in $\{x' = \xi'' = 0\}$, that is, in the zero-section of $T_{T^*_MX}T^*X$. This means that $\bar{\Lambda} \cap U$ is contained in T^*_MX on a neighborhood of U, and $T^*_MX \cap \Lambda$ being empty, we get $\bar{\Lambda} \cap U = \emptyset$. □

Corollary 8.3.24. *Let $X = \bigsqcup_\alpha X_\alpha$ be a μ-stratification. Then:*

$$\pi\left(T^*_{X_\alpha}X \setminus \bigsqcup_{\beta \neq \alpha} \overline{T^*_{X_\beta}X}\right) = X_\alpha \; .$$

Let us introduce another notion, similar to that of a stratification but sometime more convenient.

Definition 8.3.25. (i) *A (decreasing) filtration on a topological space X is a decreasing sequence $\{X_j\}$ of closed subsets of X such that $X_j = X$ for $j \ll 0$ and $X_j = \emptyset$ for $j \gg 0$.*

(ii) *A subanalytic filtration on a real analytic manifold X is a filtration $\{X_j\}$ such that the X_j's are subanalytic in X and $X_j \backslash X_{j+1}$ is a real analytic submanifold.*

(iii) *A subanalytic filtration is called a μ-filtration if $(X_j \backslash X_{j+1}, X_k \backslash X_{k+1})$ satisfies the μ-condition for any j and k with $j > k$.*

Similarly, one can consider increasing filtrations.

Proposition 8.3.26. *Let $X = \bigcup_\alpha X_\alpha$ be a locally finite covering of X by closed subanalytic subsets. Then there exists a subanalytic μ-filtration $\{X_j\}$ of X such that for any j, any connected component of $X_j \backslash X_{j+1}$ is contained in some X_α.*

Proof. We may assume that $X = \bigsqcup_\alpha X_\alpha$ is a μ-stratification. Set:

$$X_j = \bigcup_{\dim X_\alpha \leqslant -j} X_\alpha \ .$$

Then $X_j \backslash X_{j+1}$ is a disjoint union of $(-j)$-dimensional manifolds, and one checks immediately that the filtration $\{X_j\}$ satisfies the required properties. □

To end this section, let us show the existence of **Morse functions** with respect to an isotropic subset, a result that we shall use in Chapter X.

Proposition 8.3.27. *Let X be an n-dimensional closed submanifold of $\mathbb{R}^N$ and let Λ be a closed conic subanalytic isotropic subset of T^*X. Let $\Lambda_\circ$ be a subanalytic n-dimensional submanifold contained in Λ and assume $\dim(\Lambda \backslash \Lambda_\circ) < n$. Then there exists $x_\circ \in \mathbb{R}^N$, such that, setting $\phi(x) = |x - x_\circ|^2$ for $x \in X$ and $\Lambda_\phi = \{(x; d\phi(x)); x \in X\} \subset T^*X$, we have:*

$$\Lambda_\phi \cap \Lambda \subset \Lambda_\circ \qquad \textit{and the intersection is transversal} \ . \tag{8.3.7}$$

Proof. Let q denote the projection $X \times_{\mathbb{R}^N} T^*\mathbb{R}^N \to T^*X$. Replacing X, Λ, $\Lambda_\circ$ with $\mathbb{R}^N$, $q^{-1}\Lambda$, $q^{-1}\Lambda_\circ$, we may assume from the beginning that $X = \mathbb{R}^n$. We denote by $(x;\xi)$ the homogeneous symplectic coordinates on $T^*\mathbb{R}^n$. Setting $g(x,\xi) = x - \xi/2$, we have $\Lambda_\phi = g^{-1}(x_\circ)$. Now let f be the composite map:

$$f : \Lambda \hookrightarrow T^*\mathbb{R}^n \xrightarrow[g]{} \mathbb{R}^n \ .$$

Since $\dim(\Lambda \backslash \Lambda_\circ) \leqslant n - 1$, $f(\Lambda \backslash \Lambda_\circ)$ has measure zero. Set $G = \{p \in \Lambda_\circ; T_p f$ is not surjective$\}$. By Sard's theorem (cf. eg. Guillemin-Pollack [1]), $f(G)$ has measure zero. Hence $f(\Lambda \backslash \Lambda_\circ) \cup f(G) \neq \mathbb{R}^n$ and any $x_\circ \in \mathbb{R}^n \backslash (f(\Lambda \backslash \Lambda_\circ) \cup f(G))$ has the required properties. □

Concerning Morse functions on stratified spaces, cf. Lazzeri [1] Pignogni [1] and Goresky-MacPherson [2].

8.4. $\mathbb{R}$-constructible sheaves

In this section we shall study sheaves which are locally constant along subanalytic stratifications.

Proposition 8.4.1. *Let* $X = \bigsqcup_{\alpha \in A} X_\alpha$ be a μ-stratification and let $F \in \mathrm{Ob}(\mathbf{D}^b(X))$. *The following conditions are equivalent:*

(i) *for all* $j \in \mathbb{Z}$, *all* $\alpha \in A$, *the sheaves* $H^j(F)|_{X_\alpha}$ *are locally constant,*
(ii) $SS(F) \subset \bigsqcup_{\alpha \in A} T^*_{X_\alpha}X$.

Proof. (ii) $\Rightarrow$ (i). We know by Corollary 6.4.5 that $SS(F_{X_\alpha})$ is contained in $SS(F) \mathbin{\hat{+}} T^*_{X_\alpha}X$ on a neighborhood of X_α. Thus $SS(F_{X_\alpha}) \cap \pi^{-1}(X_\alpha)$ is contained in $T^*_{X_\alpha}X$ and we get $SS(F|_{X_\alpha}) \subset T^*_{X_\alpha}X_\alpha$ by Proposition 5.4.4. This implies that $H^j(F)|_{X_\alpha} = H^j(F|_{X_\alpha})$ is locally constant, by Proposition 5.4.5.

(i) $\Rightarrow$ (ii). The problem being local on X, we may assume the stratification is finite. Let B be a subset of A such that $Y = \bigsqcup_{\alpha \in B} X_\alpha$ is closed in X, and assume we already know that $SS(F) \cap \pi^{-1}(X \setminus Y)$ is contained in $\bigsqcup_{\alpha \in A} T^*_{X_\alpha}X$. Let $\alpha_0 \in B$ be such that X_{α_0} is open in Y and set $Y' = Y \setminus X_{\alpha_0}$. Arguing by induction, it is enough to prove that $SS(F) \cap \pi^{-1}(X \setminus Y')$ is contained in $\bigsqcup_{\alpha \in A} T^*_{X_\alpha}X$. Hence we may assume $Y = X_{\alpha_0}$. Let j denote the open embedding $X \setminus X_{\alpha_0} \hookrightarrow X$. We get a distinguished triangle:

$$Rj_!j^{-1}F \longrightarrow F \longrightarrow F_{X_{\alpha_0}} \xrightarrow[+1]{} .$$

By the hypothesis and Proposition 5.4.4, $SS(F_{X_\alpha}) \subset T^*_{X_\alpha}X$. By Proposition 6.3.2 we have:

$$\begin{aligned} SS(Rj_!j^{-1}F) \cap \pi^{-1}(X_{\alpha_0}) &\subset \left(\bigsqcup_{\alpha \in A} T^*_{X_\alpha}X\right) \mathbin{\hat{+}} T^*_{X_{\alpha_0}}X \\ &= \bigcup_{\alpha \in A} (T^*_{X_\alpha}X \mathbin{\hat{+}} T^*_{X_{\alpha_0}}X) \\ &\subset \bigsqcup_{\alpha \in A} T^*_{X_\alpha}X . \end{aligned}$$

Then the result follows. $\square$

Theorem 8.4.2. *Let* $F \in \mathrm{Ob}(\mathbf{D}^b(X))$. *The following conditions are equivalent.*

(i) *There exists a locally finite covering* $X = \bigcup_{i \in I} X_i$ *by subanalytic subsets such that for all* $j \in \mathbb{Z}$, *all* $i \in I$, *the sheaves* $H^j(F)|_{X_i}$ *are locally constant.*
(ii) $SS(F)$ *is contained in a closed conic subanalytic isotropic subset.*
(iii) $SS(F)$ *is a closed conic subanalytic Lagrangian subset.*

Proof. (i) $\Rightarrow$ (ii) by Propositions 8.3.20, 8.4.1 and 8.3.10.

(ii) $\Rightarrow$ (i) by Corollary 8.3.22 and Proposition 8.4.1.

(iii) ⇒ (ii) is obvious.

(ii) ⇒ (iii) by the involutivity theorem (Theorem 6.5.4) and Proposition 8.3.13.
□

Definition 8.4.3. *Let F be an object of* $\mathbf{D}^b(X)$.

(i) *One says F is weakly ℝ-constructible (w-ℝ-constructible, for short) if F satisfies the equivalent conditions of Theorem* 8.4.2.

(ii) *If F is w-ℝ-constructible and moreover for each* $x \in X$, F_x *is a perfect complex, one says that F is ℝ-constructible.*

Note that when the base ring is Noetherian, the condition (ii) is equivalent to saying that all cohomology groups $H^j(F_x)$ are finitely generated.

One denotes by $\mathbf{D}^b_{w-\mathbb{R}-c}(X)$ (resp. $\mathbf{D}^b_{\mathbb{R}-c}(X)$) the full triangulated subcategory of $\mathbf{D}^b(X)$ consisting of w-ℝ-constructible (resp. ℝ-constructible) objects.

A sheaf F on X is called w-ℝ-constructible (resp. ℝ-constructible) if it is so, considered as an object of $\mathbf{D}^b(X)$. We denote by w-ℝ-$\mathfrak{Cons}(X)$ (resp. ℝ-$\mathfrak{Cons}(X)$) the full subcategory of $\mathfrak{Mod}(A_X)$ consisting of w-ℝ-constructible (resp. ℝ-constructible) sheaves.

Let $u: F \to G$ be a morphism of sheaves on X, F and G being w-ℝ-constructible. One proves immediately that $\operatorname{Ker} u$, $\operatorname{Im} u$, and $\operatorname{Coker} u$ are w-ℝ-constructible. Moreover if $0 \to F' \to F \to F'' \to 0$ is an exact sequence of sheaves on X and F' and F'' are w-ℝ-constructible, then so is F. Hence w-ℝ-$\mathfrak{Cons}(X)$ is an abelian category.

If the base ring A is Noetherian, the same results hold with w-ℝ-$\mathfrak{Cons}(X)$ replaced by ℝ-$\mathfrak{Cons}(X)$.

Example 8.4.4. Let Z be a locally closed subanalytic subset of X. Then the sheaf A_Z is ℝ-constructible.

Theorem 8.4.5. (i) *The natural functor* $\mathbf{D}^b(w\text{-}\mathbb{R}\text{-}\mathfrak{Cons}(X)) \to \mathbf{D}^b_{w-\mathbb{R}-c}(X)$ *is an equivalence of categories.*

(ii) *Assume the base ring A is Noetherian. Then the natural functor* $\mathbf{D}^b(\mathbb{R}\text{-}\mathfrak{Cons}(X)) \to \mathbf{D}^b_{\mathbb{R}-c}(X)$ *is an equivalence of categories.*

Proof. (i) We must prove the assertions (a) and (b) below:

(a) for any $F \in \operatorname{Ob}(\mathbf{D}^b_{w-\mathbb{R}-c}(X))$, there exists $G \in \operatorname{Ob}(\mathbf{D}^b(w\text{-}\mathbb{R}\text{-}\mathfrak{Cons}(X)))$ which is isomorphic to F in $\mathbf{D}^b(X)$,

(b) for any objects F and G in $\mathbf{D}^b(w\text{-}\mathbb{R}\text{-}\mathfrak{Cons}(X))$, we have the isomorphism:

$$\operatorname{Hom}_{\mathbf{D}^b(w-\mathbb{R}-\mathfrak{Cons}(X))}(F, G) \xrightarrow{\sim} \operatorname{Hom}_{\mathbf{D}^b(X)}(F, G) \, . \tag{8.4.1}$$

Let us prove (a).

We choose a locally finite covering by subanalytic sets $X = \bigcup_{j \in J} X_j$, such that $H^k(F)|_{X_j}$ is locally constant for all k, all j. Applying Proposition 8.2.5, we find a simplicial complex $\mathbf{S} = (S, \Delta)$ and a homeomorphism $i: |\mathbf{S}| \xrightarrow{\sim} X$ such that:

(8.4.2) $\qquad i(|\sigma|)$ is subanalytic for every $\sigma \in \Delta$,

(8.4.3) $\qquad$ for any $\sigma \in \Delta$, there exists $j \in J$ with $i(|\sigma|) \subset X_j$.

Therefore $i^{-1}(F)$ is an object of $\mathbf{D}^b_{w-\mathrm{S}-c}(|\mathrm{S}|)$, and by Theorem 8.1.10 there exists $G \in \mathrm{Ob}(\mathbf{D}^b(w\text{-}\mathfrak{Cons}(\mathrm{S})))$ isomorphic to $i^{-1}(F)$. Then i_*G is an object of $\mathbf{D}^b(w\text{-}\mathbb{R}\text{-}\mathfrak{Cons}(X))$ isomorphic to F.

Let us prove (b).

Let $F^\cdot$ and $G^\cdot$ be two bounded complexes of w-$\mathbb{R}$-constructible sheaves. There exist a simplicial complex $\mathrm{S} = (S, \Delta)$ and a homeomorphism $i : |S| \xrightarrow{\sim} X$ satisfying (8.4.2) and also:

(8.4.4) $\qquad \left\{ \begin{array}{l} i^{-1}(F^\cdot) \textit{ and } i^{-1}(G^\cdot) \textit{ are complexes of w-S-constructible} \\ \textit{sheaves} \ . \end{array} \right.$

Consider the diagram:

$$\begin{array}{ccc} \mathrm{Hom}_{\mathbf{D}^b(w-\mathfrak{Cons}(\mathrm{S}))}(i^{-1}F^\cdot, i^{-1}G^\cdot) & \xrightarrow[v]{} & \mathrm{Hom}_{\mathbf{D}^b(w-\mathbb{R}-\mathfrak{Cons}(X))}(F^\cdot, G^\cdot) \\ \Big\downarrow u & & \Big\downarrow w \\ \mathrm{Hom}_{\mathbf{D}^b(|\mathrm{S}|)}(i^{-1}F^\cdot, i^{-1}G^\cdot) & \xrightarrow{\sim} & \mathrm{Hom}_{\mathbf{D}^b(X)}(F^\cdot, G^\cdot) \end{array} \tag{8.4.5}$$

By Theorem 8.1.10, u is an isomorphism. Hence w is surjective. Let us prove that w is injective. Let $\varphi \in \mathrm{Hom}_{\mathbf{D}^b(w-\mathbb{R}-\mathfrak{Cons}(X))}(F^\cdot, G^\cdot)$ with $w(\varphi) = 0$. One represents φ by a quasi-isomorphism $G^\cdot \xrightarrow{\sim} G'^\cdot$ and a morphism $\varphi' : F^\cdot \to G'^\cdot$. Replacing φ by φ' we may assume from the beginning that φ is given by a morphism from $F^\cdot$ to $G^\cdot$.

Then $\varphi = v \circ i^{-1}(\varphi)$ and $w(\varphi) = 0$ implies $i^{-1}(\varphi) = 0$, hence $\varphi = 0$.

(ii) The proof is similar, with the help of Theorem 8.1.11. □

By Theorem 8.4.2, we see that being w-$\mathbb{R}$-constructible is a microlocal property. Hence we can make full use of the results of Chapters V and VI to study the functorial properties of these objects.

Proposition 8.4.6. (i) *Let $f : Y \to X$ be a morphism of manifolds.*

(i)$_a$ *Let $F \in \mathrm{Ob}(\mathbf{D}^b_{w-\mathbb{R}-c}(X))$. Then $f^{-1}F$ and $f^!F$ belong to $\mathbf{D}^b_{w-\mathbb{R}-c}(Y)$.*

(i)$_b$ *Let $G \in \mathrm{Ob}(\mathbf{D}^b_{w-\mathbb{R}-c}(Y))$ and assume f is proper on* supp(G). *Then $Rf_*G \in \mathrm{Ob}(\mathbf{D}^b_{w-\mathbb{R}-c}(X))$.*

(ii) *Let F and G belong to $\mathbf{D}^b_{w-\mathbb{R}-c}(X)$. Then $G \otimes^L F$ and $R\mathcal{H}om(G, F)$ belong to $\mathbf{D}^b_{w-\mathbb{R}-c}(X)$ and $\mu hom(G, F)$ belongs to $\mathbf{D}^b_{w-\mathbb{R}-c}(T^*X)$.*

(iii) *Let $E \to X$ be a vector bundle and let $F \in \mathrm{Ob}(\mathbf{D}^b_{\mathbb{R}^+}(E))$. Assume F is w-$\mathbb{R}$-constructible. Then $F^\wedge$, its Fourier-Sato transform, is w-$\mathbb{R}$-constructible.*

Proof. Using Theorem 8.3.17 and Proposition 8.3.11, (i)$_a$ follows from Corollary 6.4.4; (i)$_b$ follows from Proposition 5.4.4, (ii) follows from Corollary 6.4.5 and Corollary 6.4.3, and (iii) follows from Theorem 5.5.5. □

In order to study ℝ-constructible sheaves we need a lemma. First we introduce:

$$B_\varepsilon = \{x \in \mathbb{R}^n; |x| < \varepsilon\} ,$$

$$B_{\varepsilon',\varepsilon} = \{x \in \mathbb{R}^n; \varepsilon' < |x| < \varepsilon\} ,$$

$$S_\varepsilon = \{x \in \mathbb{R}^n; |x| = \varepsilon\} ,$$

and we denote by $\bar{B}_\varepsilon$ and $\bar{B}_{\varepsilon',\varepsilon}$ the closure of B_ε and $B_{\varepsilon',\varepsilon}$, respectively.

Lemma 8.4.7. *Let $F \in \mathrm{Ob}(\mathbf{D}^b_{w-\mathbb{R}-c}(X))$ and let $\varphi : X \to \mathbb{R}^n$ be a real analytic function. Assume $\varphi|_{\mathrm{supp}(F)}$ is proper. Then we have the natural isomorphisms:*

(i) $R\Gamma(\varphi^{-1}(\bar{B}_\varepsilon); F) \xrightarrow{\sim} R\Gamma(\varphi^{-1}(B_\varepsilon); F) \xrightarrow{\sim} R\Gamma(\varphi^{-1}(0); F)$, *for* $0 < \varepsilon \ll 1$.

(ii) $R\Gamma_{\varphi^{-1}(0)}(X; F) \xrightarrow{\sim} R\Gamma_{\varphi^{-1}(\bar{B}_{\varepsilon})}(X; F) \xrightarrow{\sim} R\Gamma_c(\varphi^{-1}(B_\varepsilon); F)$, *for* $0 < \varepsilon' < \varepsilon \ll 1$.

(iii) $R\Gamma(\varphi^{-1}(B_{0,\varepsilon}); F) \xrightarrow{\sim} R\Gamma(\varphi^{-1}(B_{\varepsilon'',\varepsilon'}); F) \xrightarrow{\sim} R\Gamma(S_{\varepsilon'''}; F)$,

for $0 \leqslant \varepsilon'' < \varepsilon''' < \varepsilon' \leqslant \varepsilon \ll 1$.

Proof. By replacing φ with $|\varphi|^2$, we may assume $n = 1$. Then apply the microlocal Bertini-Sard theorem (Proposition 8.3.12) with $\Lambda = SS(F)$, and the microlocal Morse lemma (Corollary 5.4.19). □

Note that this lemma applies in particular when X is open in $\mathbb{R}^n$, and $\varphi(x) = x - x_\circ$ for some $x_\circ \in X$.

Proposition 8.4.8. *Let $f: Y \to X$ be a morphism of manifolds, and let $G \in \mathrm{Ob}(\mathbf{D}^b_{\mathbb{R}-c}(Y))$. Assume f is proper on* supp(G). *Then $Rf_*G \in \mathrm{Ob}(\mathbf{D}^b_{\mathbb{R}-c}(X))$.*

Proof. By Proposition 8.4.6, we only have to prove that for each $x \in X$, $R\Gamma(f^{-1}(x); G|_{f^{-1}(x)})$ is a perfect complex. By decomposing f by the graph map, we may assume f is smooth, hence we may assume $X = \{\mathrm{pt}\}$. (We use the obvious remark that if $S \subset Y$ is a submanifold, $G|_S$ belongs to $\mathbf{D}^b_{\mathbb{R}-c}(S)$.) By a closed embedding $Y \hookrightarrow \mathbb{R}^n$, we may assume $Y = \mathbb{R}^n$.

Finally, arguing by induction, we may assume $n = 1$. In other words, it remains to prove that if $Y = \mathbb{R}$ and G has compact support, $R\Gamma(Y; G)$ is perfect. Since $SS(G)$ is subanalytic and isotropic, there exists a finite sequence $t_1 < \cdots < t_N$ in ℝ such that G is supported by $[t_1, t_N]$ and $G|_{]t_{i-1}, t_i[}$ is constant for all $i \leqslant N$. By considering the distinguished triangles:

$$G_{]-\infty, t_i[} \longrightarrow G \longrightarrow G_{[t_i, +\infty[} \xrightarrow[+1]{} \quad \text{and} \quad G_{]t_i, +\infty[} \longrightarrow G_{[t_i, +\infty[} \longrightarrow G_{\{t_i\}} \xrightarrow[+1]{} ,$$

and arguing by induction, it is enough to prove that $R\Gamma(Y; G_{]t_{i-1}, t_i[})$ is perfect. Since this complex is isomorphic to $R\Gamma_{\{t\}}(Y; G)$ for any $t \in]t_{i-1}, t_i[$, the result follows from the distinguished triangle

$$R\Gamma_{\{t\}}(Y; G) \longrightarrow R\Gamma(]t - \varepsilon, t + \varepsilon[; G) \longrightarrow R\Gamma(]t - \varepsilon, t[\cup]t, t + \varepsilon[; G) \xrightarrow[+1]{} . \quad \square$$

Proposition 8.4.9. *Let* $F \in \mathrm{Ob}(\mathbf{D}^b_{w-\mathbb{R}-c}(X))$. *Consider the following conditions*:

(i) *F is $\mathbb{R}$-constructible.*
(ii) *F is cohomologically constructible.*
(iii) *For any $x \in X$, $R\Gamma_{\{x\}}(X;F)$ is perfect.*
(iv) *$DF = R\mathcal{H}om(F, \omega_X)$ is $\mathbb{R}$-constructible.*

Then (i) $\Leftrightarrow$ (ii) $\Rightarrow$ (iii) $\Rightarrow$ (iv). *If the base ring A is a field or $\mathbb{Z}$, these four conditions are equivalent.*

Proof. (i) $\Rightarrow$ (ii) By Lemma 8.4.7, it remains to prove that (i) $\Rightarrow$ (iii). With the notations of Lemma 8.4.7, after having choosen a coordinate system in a neighborhood of $x_\circ \in X$, we have the distinguished triangle for $0 < \varepsilon \ll 1$:

$$R\Gamma_{\{x_\circ\}}(X;F) \longrightarrow R\Gamma(B_\varepsilon;F) \longrightarrow R\Gamma(B_\varepsilon \backslash \{x_\circ\};F) \underset{+1}{\longrightarrow} ,$$

and the isomorphisms $R\Gamma(B_\varepsilon;F) \simeq F_{x_\circ}$, $R\Gamma(B_\varepsilon\backslash\{x_\circ\};F) \simeq R\Gamma(S_{\varepsilon'};F)$ $(0 < \varepsilon' < \varepsilon)$. The complex $R\Gamma(S_{\varepsilon'};F)$ being perfect, by Proposition 8.4.8, the result follows.

(ii) $\Rightarrow$ (i) and (ii) $\Rightarrow$ (iii) are obvious.

(iii) $\Rightarrow$ (iv) By the same proof as for Proposition 3.4.3 we see that

$$(\mathrm{D}F)_x \simeq R\mathrm{Hom}(R\Gamma_{\{x\}}(X;F), A) .$$

The complex $(\mathrm{D}F)_x$ being the dual of a perfect complex, is itself perfect.

(iv) $\Rightarrow$ (i) For $x \in X$, $R\Gamma_{\{x\}}(X;\mathrm{D}F) = R\mathrm{Hom}(F_x, A)$. Since this complex is perfect, F_x is perfect (see Exercise I.31 for $A = \mathbb{Z}$). □

Proposition 8.4.10. (i) *Let $f: Y \to X$ be a morphism of manifolds and let $F \in \mathrm{Ob}(\mathbf{D}^b_{\mathbb{R}-c}(X))$. Then $f^{-1}F$ and $f^!F$ belong to $\mathbf{D}^b_{\mathbb{R}-c}(Y)$.*

(ii) *Let F and G belong to $\mathbf{D}^b_{\mathbb{R}-c}(X)$. Then $G \otimes^L F$ and $R\mathcal{H}om(G,F)$ belong to $\mathbf{D}^b_{\mathbb{R}-c}(X)$.*

Proof. By Proposition 8.4.6 we already know that the objects $f^{-1}F$, $f^!F$, etc. are w-$\mathbb{R}$-constructible.

(i) Since $(f^{-1}F)_y = F_{f(y)}$, this complex is perfect. (We have already made this remark in the course of the proof of Proposition 8.4.8.)

We have $f^!\mathrm{D}_X\mathrm{D}_XF = f^!R\mathcal{H}om(\mathrm{D}_XF, \omega_X) \simeq R\mathcal{H}om(f^{-1}\mathrm{D}_XF, \omega_Y) \simeq \mathrm{D}_Y(f^{-1}\mathrm{D}_XF)$. Hence $f^!F$ is $\mathbb{R}$-constructible by Proposition 8.4.9.

(ii) It is clear that $G \otimes^L F$ is $\mathbb{R}$-constructible.

Since $R\mathcal{H}om(G,F) = \mathrm{D}(\mathrm{D}F \otimes^L G)$, (Proposition 3.4.6), this object is $\mathbb{R}$-constructible by Proposition 8.4.9. □

Corollary 8.4.11. *Let $F \in \mathrm{Ob}(\mathbf{D}^b_{\mathbb{R}-c}(X))$.*

(i) *Let K be a compact subanalytic subset of X. Then $R\Gamma_K(X;F)$ and $R\Gamma(K;F)$ are perfect complexes.*
(ii) *Let Ω be a relatively compact open subanalytic subset of X. Then $R\Gamma(\Omega;F)$ and $R\Gamma_c(\Omega;F)$ are perfect complexes.*

Proof. Apply Propositions 8.4.8 and 8.4.10. (Recall the isomorphisms of Exercises II.19.) □

Proposition 8.4.12. (i) *Let $\tau : E \to Z$ be vector bundle and let $F \in \mathrm{Ob}(\mathbf{D}^b_{\mathbb{R}^+}(E))$. Assume F is ℝ-constructible. Then $F^\wedge$, its Fourier-Sato transform, is ℝ-constructible.*

(ii) *Let F and G belong to $\mathbf{D}^b_{\mathbb{R}-c}(X)$. Then $\mu hom(G, F)$ belongs to $\mathbf{D}^b_{\mathbb{R}-c}(T^*X)$.*

Proof. (i) We keep the notations of III §7. Then $F^\wedge = Rp_{2*}R\Gamma_P(p_1^{-1}F)$. Denote by i the zero-embedding $E^* \hookrightarrow E \times_Z E^*$. Since the object $R\Gamma_P(p_1^{-1}F)$ is conic on the vector bundle $E \times_Z E^* \to E^*$, we also have $F^\wedge \simeq i^{-1}R\Gamma_P(p_1^{-1}F)$ in view of Proposition 3.7.5. Hence $F^\wedge$ is ℝ-constructible by Proposition 8.4.10.

(ii) By Proposition 8.4.10 and the preceding result, it remains to prove that if M is a submanifold of X, then $\nu_M(F)$ is ℝ-constructible.

By the construction of the specialization $\nu_M(F)$ it is then enough to prove that $Rj_*j^{-1}F$ is ℝ-constructible when j is the open embedding $\Omega \hookrightarrow X$ and $\Omega = \{x \in X; \varphi(x) > 0\}$ for a real function φ with $d\varphi \neq 0$ on X. This follows from $Rj_*j^{-1}F \simeq R\mathcal{H}om(A_\Omega, F)$ and Proposition 8.4.10 (ii). □

To end this section we shall study some relations between the functor of duality and the functor μhom.

Proposition 8.4.13. *Let M be a submanifold of X and let $F \in \mathrm{Ob}(\mathbf{D}^b_{\mathbb{R}-c}(X))$. We then have natural isomorphisms:*

(i) $\nu_M(\mathrm{D}_X F) \simeq \mathrm{D}_{T_M X}(\nu_M(F))$,

(ii) $\mu_M(\mathrm{D}_X F) \simeq \mathrm{D}_{T_M^* X}(\mu_M(F))^a \otimes \omega_{M/X}$.

(Recall that "a" denotes the direct image by the antipodal map on T_M^*X.)

Proof. We use the notations of IV §2, and we apply Lemma 4.2.1. Then:

$$\begin{aligned}\nu_M(\mathrm{D}_X F) &= s^{-1}Rj_*\tilde{p}^{-1}R\mathcal{H}om(F, \omega_X)\\ &\simeq s^{-1}Rj_*R\mathcal{H}om(\tilde{p}^!F, \tilde{p}^!\omega_X)\\ &\simeq s^{-1}R\mathcal{H}om(j_!\tilde{p}^!F, \omega_{\tilde{X}_M})\\ &\simeq R\mathcal{H}om(s^!j_!\tilde{p}^!F, \omega_{T_M X})\\ &\simeq \mathrm{D}_{T_M X}(\nu_M(F)) .\end{aligned}$$

In the course of the proof we use the fact that $G = R\mathcal{H}om(j_!\tilde{p}^!F, \omega_{\tilde{X}_M})$ is ℝ-constructible, which follows from the proof of Proposition 8.4.12, and the formula $s^{-1}\mathrm{D}G \simeq \mathrm{D}s^!G$, (cf. Exercise VIII.3).

(ii) We shall apply (i), formula (3.7.9) and Proposition 3.7.12. Then:

$$\begin{aligned}\mu_M(\mathrm{D}F) &= (\nu_M(\mathrm{D}F))^{\vee a} \otimes \omega_{T_M^*X/M}^{\otimes -1} \\ &\simeq (\mathrm{D}\nu_M(F))^{\vee a} \otimes \omega_{M/X} \\ &\simeq \mathrm{D}\mu_M(F)^a \otimes \omega_{M/X} \; .\end{aligned}$$

Proposition 8.4.14. *Assume F and G belong to $\mathbf{D}^b_{\mathbb{R}-c}(X)$. Then we have the natural isomorphisms:*

(i) $\mu hom(F, G) \simeq \mu hom(\mathbf{D}_X G, \mathbf{D}_X F)^a$,
(ii) $\mathbf{D}_{T^*X}(\mu hom(F, G)) \simeq \mu hom(G, F) \otimes \omega_X$.

Proof. (i) In fact this formula is true under the weaker hypothesis that F and G are cohomologically constructible (cf. Exercise IV. 4) and follows immediately from Proposition 3.4.6.

(ii) By Proposition 8.4.13, we have:

$$\begin{aligned}\mathrm{D}_{T^*X}(\mu hom(F, G)) &\simeq \mathrm{D}_{T^*X}\mu_\Delta(G \boxtimes \mathrm{D}F) \\ &\simeq \mu_\Delta(\mathrm{D}_{X \times X}(G \boxtimes \mathrm{D}F))^a \otimes \omega_{\Delta/X \times X}^{\otimes -1} \\ &\simeq \mu_\Delta(\mathrm{D}G \boxtimes F)^a \otimes \omega_X \\ &\simeq \mu_\Delta(F \boxtimes \mathrm{D}G) \otimes \omega_X \\ &\simeq \mu hom(G, F) \otimes \omega_X \; . \quad \square\end{aligned}$$

8.5. ℂ-constructible sheaves

In this section we shall consider complex analytic manifolds. Since complex geometry is out of the scope of this book, we shall be rather brief, and we shall not give all proofs with details.

If X is a complex manifold, we denote by $X^{\mathbb{R}}$ the real underlying manifold, and we refer to XI§1 for a study of the relations between X and $X^{\mathbb{R}}$, the isomorphism between $(T^*X)^{\mathbb{R}}$ and $T^*(X^{\mathbb{R}})$, etc. If there is no risk of confusion we shall write X instead of $X^{\mathbb{R}}$.

Let S be a locally closed subset of X. We say that S is **ℂ-analytic** if $\bar{S}$ and $\bar{S} \setminus S$ are complex analytic subsets. (We refer to Cartan [2] for the main notions of analytic geometry.) In particular if S is ℂ-analytic in X, S is subanalytic in $X^{\mathbb{R}}$.

Let Λ be a subset of T^*X. We say that Λ is **$\mathbb{R}^+$-conic** if Λ has the corresponding property in $T^*(X^{\mathbb{R}})$. We say that Λ is **locally $\mathbb{C}^\times$-conic** if it is locally invariant by the action of $\mathbb{C}^\times$, that is, $\Lambda \cap S$ is open in S for any $\mathbb{C}^\times$-orbit S. We say that Λ is **$\mathbb{C}^\times$-conic** if Λ is a union of $\mathbb{C}^\times$-orbits.

If Λ is a complex analytic submanifold then Λ is involutive (resp. Lagrangian, resp. isotropic) if $T_p\Lambda$ is a complex involutive (resp. Lagrangian, resp. isotropic) subspace of T^*X at each $p \in \Lambda$. Assume Λ is a closed $\mathbb{C}$-analytic subset of T^*X. Then Λ is $\mathbb{R}^+$-conic iff Λ is $\mathbb{C}^\times$-conic, and we simply say Λ is conic in this case.

Note also that if Λ is complex analytic and Λ_{reg} is Lagrangian, then Λ is Lagrangian (cf. Exercise VIII 8).

Proposition 8.5.1. *Let S and Y be two $\mathbb{C}$-analytic subsets of X. Then the normal cone $C(Y, S)$ (which is defined in $TX^{\mathbb{R}}$, using the real structure of X) is $\mathbb{C}$-analytic in TX.*

For the proof we refer to Whitney [2].

Note that, in particular, $C(Y, S)$ is invariant by the action of $\mathbb{C}^\times$.

Proposition 8.5.2. *Let Ω be an open subset of T^*X and let Λ be a locally $\mathbb{C}^\times$-conic involutive closed subset of Ω. Assume that Λ is contained in a closed $\mathbb{R}^+$-conic subanalytic isotropic subset of Ω. Then Λ is a complex analytic set.*

Proof. By Proposition 8.3.13, Λ is subanalytic and Lagrangian in Ω. For any $p \in \Lambda_{reg}$, $T_p\Lambda$ is a real Lagrangian plane of T_pT^*X. Since Λ is locally $\mathbb{C}^\times$-conic, $T_p\Lambda = (T_p\Lambda)^\perp$ contains $\mathbb{C} \cdot H(\alpha)$, where α is the complex 1-form on T^*X. Hence $\alpha|_{\Lambda_{reg}} = 0$, which implies $d\alpha|_{\Lambda_{reg}} = 0$ and $\mathrm{Re}(d\alpha)|_{T_p\Lambda + \sqrt{-1}T_p\Lambda} = 0$, which reads as $\sqrt{-1}T_p\Lambda \subset (T_p\Lambda)^\perp = T_p\Lambda$. This shows that $T_p\Lambda$ is a complex vector subspace for any $p \in \Lambda_{reg}$. Thus Λ_{reg} is a complex manifold. Set $n = \dim_{\mathbb{C}} X$. Then $\dim_{\mathbb{R}}(\Lambda \setminus \Lambda_{reg}) \leqslant 2n - 1$. Let S be the union of the $(2n - 1)$-dimensional connected components of $(\Lambda \setminus \Lambda_{reg})_{reg}$, and next, let S' be the largest open subset of S where (Λ_{reg}, S) satisfies the μ-condition. Since S' is dense in S, $\dim((\Lambda \setminus \Lambda_{reg}) \setminus S') \leqslant 2n - 2$.

For any $p \in S'$, choose a sequence $\{p_n\}$ in Λ_{reg} converging to p such that $T_{p_n}\Lambda_{reg}$ converges to $\tau \subset T_pT^*X$, with $\tau \supset T_pS'$ (cf. Exercise VIII.12). Since $T_{p_n}\Lambda_{reg}$ is a complex vector space, so is τ. Hence we obtain:

$$\dim_{\mathbb{C}}(T_pS' + \sqrt{-1}T_pS') = n , \qquad \text{for any } p \in S' . \tag{8.5.1}$$

Let $S'_{\mathbb{C}}$ be a complexification of S'. The embedding $S' \hookrightarrow T^*X$ extends to a map $S'_{\mathbb{C}} \to T^*X$, and, shrinking $S'_{\mathbb{C}}$ if necessary, this map has constant rank by (8.5.1). Hence its image is a complex submanifold on a neighborhood of S'. Let us denote it by Z. The same argument shows that any complex submanifold containing S' contains Z, on a neighborhood of S'. Note that Z is isotropic and hence Lagrangian. Now we shall proceed as follows. We shall prove:

(a) $S' \cap \overline{\Lambda_{reg} \setminus Z}$ is nowhere dense in S',
(b) $S = \emptyset$,
(c) Λ is complex analytic.

(a) Let us argue by contradiction. Otherwise, $S' \cap \overline{\Lambda_{\mathrm{reg}} \backslash Z}$ contains an non-empty open subset of S'. Hence we may assume $\overline{\Lambda_{\mathrm{reg}} \backslash Z} \supset S'$. Since $\overline{\Lambda_{\mathrm{reg}} \backslash Z}$ is subanalytic, there exists a proper real analytic map $f: W \to T^*X$ such that $f(W) = \overline{\Lambda_{\mathrm{reg}} \backslash Z}$. We may assume $f^{-1}(\Lambda_{\mathrm{reg}} \backslash Z)$ is open and dense in W. Now let us take a complexification $W_{\mathbb{C}}$ of W and extend f to a holomorphic map $\tilde{f}: W_{\mathbb{C}} \to T^*X$. For w in an open dense subset of $f^{-1}(\Lambda_{\mathrm{reg}} \backslash Z)$, $\mathrm{Im}(T_w W_{\mathbb{C}} \to T_{f(w)} T^*X)$ coincides with $T_{f(w)}(\Lambda_{\mathrm{reg}})$, and we get:

(8.5.2) *the rank of $\tilde{f}$ is equal or less than n* .

Since $f(W) \supset S'$, there exists $w \in W$ such that the image of $T_w W \to T_{f(w)} T^*X$ contains $T_{f(w)} S$. Hence, at such a point w, the rank of $\tilde{f}$ is n, and $\tilde{f}$ has constant rank in a neighborhood of w. For a sufficiently small neighborhood U of w, $\tilde{f}(U)$ is an n-dimensional complex submanifold containing S on a neighborhood of $f(w)$. Hence $\tilde{f}(U) = Z$ on a neighborhood of $f(w)$, and $\emptyset \neq \tilde{f}(U \cap f^{-1}(\Lambda_{\mathrm{reg}} \backslash Z)) \subset Z$, which is a contradiction.

(b) Assume $S \neq \emptyset$. Then there exists $p \in S' \backslash (\overline{\Lambda_{\mathrm{reg}} \backslash Z})$. On a neighborhood of p, $\Lambda_{\mathrm{reg}} \subset Z$, and since $\Lambda = \overline{\Lambda_{\mathrm{reg}}}$, $\Lambda \subset Z$. By Lemma 8.3.14, Λ is open in Z. Thus $\Lambda = Z$ in a neighborhood of p, which contradicts $p \in S' \subset \Lambda \backslash \Lambda_{\mathrm{reg}}$.

(c) By (b), $\dim_{\mathbb{R}}(\Lambda \backslash \Lambda_{\mathrm{reg}}) \leqslant 2n - 2$. Applying a theorem of Remmert-Stein [1] on the extension of complex analytic set, we obtain that $\overline{\Lambda_{\mathrm{reg}}}$ is complex analytic. □

As an application of the last result, consider a closed $\mathbb{C}$-analytic subset S of X. The set T_S^*X is defined by (8.3.1). Locally on X, S is defined as $\{x \in X; f_j(x) = 0, j = 1, \ldots, p\}$ where the f_j's are holomorphic functions. Hence

$$T_S^*X = \overline{\left\{\left(x; \sum_{j=1}^{p} \lambda_j df_j(x)\right); \lambda_j \in \mathbb{C}, f_j(x) = 0\, \forall j\right\}}$$

and by Propositions 8.3.1 and 8.5.2, we see that T_S^*X is a closed conic $\mathbb{C}$-analytic and $\mathbb{C}$-Lagrangian subset of T^*X.

Now we can extend to the complex case most of the results of §3.

Proposition 8.5.3. *Let Λ be a closed conic $\mathbb{C}$-analytic subset of T^*X. Then there exists a finite family $\{X_j\}$ of closed $\mathbb{C}$-analytic subsets of X such that $\Lambda \subset \bigcup_j T_{X_j}^*X$. Moreover for any $\mathbb{C}$-analytic subset Y of X there exists a $\mathbb{C}$-analytic manifold $Y_\circ \subset Y$, open and dense in Y, such that $\Lambda \cap \pi^{-1}(Y_\circ) \subset T_{Y_\circ}^*X$.*

The proof is the same as for Proposition 8.3.10.

Proposition 8.5.4. *Let $X = \bigcup_{j \in J} X_j$ be a locally finite covering of X by $\mathbb{C}$-analytic subsets. Then there exists a μ-stratification $X = \bigsqcup_{\alpha \in A} X_\alpha$ finer than the covering and such that for all $\alpha \in A$ the X_α's are complex manifolds.*

The proof is similar to that of Theorem 8.3.20, using Propositions 8.5.3 and 8.5.1.

Now we can state the main result of this section.

Theorem 8.5.5. *Let $F \in \mathrm{Ob}(\mathbf{D}^b(X))$. The following conditions are equivalent.*

(i) *There exists a locally finite covering $X = \bigcup_{j \in J} X_j$ by $\mathbb{C}$-analytic subsets such that for all $j \in J$, all $k \in \mathbb{Z}$, the sheaves $H^k(F)|_{X_j}$ are locally constant.*
(ii) $SS(F)$ *is contained in a closed $\mathbb{C}^\times$-conic subanalytic $\mathbb{R}$-isotropic subset Λ.*
(iii) $SS(F)$ *is a closed conic $\mathbb{C}$-analytic Lagrangian subset.*
(iv) *$F \in \mathrm{Ob}(\mathbf{D}^b_{w-\mathbb{R}-c}(X))$ and $SS(F)$ is $\mathbb{C}^\times$-conic.*

Proof. (i) ⇒ (ii) Apply Propositions 8.5.4 and 8.4.1.
(ii) ⇒ (iii) Apply the involutivity theorem 6.5.4 and Proposition 8.5.2.
(iii) ⇔ (iv) by Proposition 8.5.2.
(iii) ⇒ (i) by Propositions 8.5.3, 8.5.4 and 8.4.1. □

Definition 8.5.6. *Let $F \in \mathrm{Ob}(\mathbf{D}^b(X))$.*

(i) *One says F is weakly $\mathbb{C}$-constructible (w-$\mathbb{C}$-constructible, for short) if F satisfies the equivalent conditions of Theorem 8.5.5.*
(ii) *If F is w-$\mathbb{C}$-constructible and moreover for each $x \in X$, F_x is a perfect complex, one says F is $\mathbb{C}$-constructible.*

One denotes by $\mathbf{D}^b_{w-\mathbb{C}-c}(X)$ (resp. $\mathbf{D}^b_{\mathbb{C}-c}(X)$) the full triangulated sub-category of $\mathbf{D}^b(X)$ consisting of w-ℂ-constructible (resp. ℂ-constructible) objects.

A sheaf F on X is called w-ℂ-constructible (resp. ℂ-constructible) if it is so, considered as an object of $\mathbf{D}^b(X)$. One denotes by w-ℂ-$\mathfrak{Cons}(X)$ (resp. ℂ-$\mathfrak{Cons}(X)$) the category of such sheaves. Note that the natural morphism $\mathbf{D}^b(\text{w-}\mathbb{C}\text{-}\mathfrak{Cons}(X)) \to \mathbf{D}^b_{w-\mathbb{C}-c}(X)$ is not an equivalence in general. (The proof breaks since the triangulation theorem does not hold in the complex case.)

Proposition 8.5.7. (i) *Let $f: Y \to X$ be a morphism of complex manifolds.*

(a) *Let $F \in \mathrm{Ob}(\mathbf{D}^b_{\mathbb{C}-c}(X))$. Then $f^{-1}F$ and $f^!F$ belong to $\mathbf{D}^b_{\mathbb{C}-c}(Y)$.*
(b) *Let $G \in \mathrm{Ob}(\mathbf{D}^b_{\mathbb{C}-c}(Y))$ and assume f is proper on $\mathrm{supp}(G)$. Then $Rf_*G \in \mathrm{Ob}(\mathbf{D}^b_{\mathbb{C}-c}(X))$.*

(ii) *Let F and G belong to $\mathbf{D}^b_{\mathbb{C}-c}(X)$. Then $G \otimes^L F$ and $R\mathcal{H}om(G, F)$ belong to $\mathbf{D}^b_{\mathbb{C}-c}(X)$ and $\mu hom(G, F)$ belongs to $\mathbf{D}^b_{\mathbb{C}-c}(T^*X)$.*

(iii) *Let $E \to X$ be a complex vector bundle and let $F \in \mathrm{Ob}(\mathbf{D}^b_{\mathbb{R}^+}(E))$. Assume F is $\mathbb{C}$-constructible. Then $F^\wedge$, its Fourier-Sato transform, is $\mathbb{C}$-constructible.*

Moreover (i), (ii), (iii) still hold replacing everywhere $\mathbf{D}^b_{\mathbb{C}-c}(\cdot)$ by $\mathbf{D}^b_{w-\mathbb{C}-c}(\cdot)$.

Proof. By Propositions 8.4.8, 8.4.10, 8.4.12 it is enough to prove the results for weakly $\mathbb{C}$-constructible objects.

(i) By Corollary 6.4.4, $SS(f^{-1}F)$ is contained in $f^{\#}(SS(F))$, and this last set is $\mathbb{R}^+$-conic, and isotropic. Moreover, since $SS(F)$ is $\mathbb{C}^\times$-conic, this set is $\mathbb{C}^\times$-conic and $f^{-1}F$ is w-$\mathbb{C}$-constructible by Theorem 8.5.5. The same proof holds for $f^!F$. (ii) and (iii) are similarly proven. □

One can also treat non-proper direct images.

Let $f: Y \to X$ be a morphism of complex manifolds and let $\varphi: Y \to \mathbb{R}$ be a real analytic function. Set as in (5.4.13):

$$Y_t = \{y \in Y; \varphi(y) < t\}\ , \qquad \tilde{Y}_t = \{y \in Y; \varphi(y) \leqslant t\}\ . \tag{8.5.3}$$

Denote by j_t (resp. $\tilde{j}_t$) the embedding $Y_t \hookrightarrow Y$ (resp. $\tilde{Y}_t \hookrightarrow Y$) and set $f_t = f \circ j_t$, $\tilde{f}_t = f \circ \tilde{j}_t$. Let $t_0 \in \mathbb{R}$.

Proposition 8.5.8. *Let $G \in \mathrm{Ob}(\mathbf{D}^b_{\mathbb{C}-c}(Y))$ and assume:*

(i) *$\mathrm{supp}(G) \cap \tilde{Y}_t$ is proper over X for all t,*
(ii) *for all $y \in Y \setminus Y_{t_0}$, we have:*

$$d\varphi(y) \notin SS(G) + {}^t f'\left(Y \underset{X}{\times} T^*X\right)\ .$$

Then conclusions (a), (a′), (b), (c), (c′), (d) *of Proposition* 5.4.17 *hold, and moreover Rf_*G and $Rf_!G$ belong to $\mathbf{D}^b_{\mathbb{C}-c}(X)$.*

Proof. Since $SS(G)$ is $\mathbb{C}^\times$-conic, we also have for $y \in Y \setminus Y_{t_0}$:

$$-d\varphi(y) \notin SS(G) + {}^t f'\left(Y \underset{X}{\times} T^*X\right)\ .$$

Hence we may apply Proposition 5.4.17 (i) and (ii). Moreover, since $Rf_*G \simeq Rf_*(G_{\tilde{Y}_t})$ and $Rf_!G \simeq Rf_*(R\Gamma_{\tilde{Y}_t}(G))$, these objects belong to $\mathbf{D}^b_{\mathbb{R}-c}(X)$, by Proposition 8.4.8. Since their micro-supports are contained in the set $f_\pi {}^t f'^{-1}(SS(G) \cap \pi^{-1}(\tilde{Y}_{t_0}))$, and this set is subanalytic, isotropic and $\mathbb{C}^\times$-conic, it remains to apply Theorem 8.5.5. □

When $\dim X = 1$, the hypotheses of Proposition 8.5.8 are always satisfied "locally". More precisely one has:

Proposition 8.5.9. *Let $f: Y \to X$ be a morphism of complex manifolds, with $\dim_{\mathbb{C}} X = 1$, and let $G \in \mathrm{Ob}(\mathbf{D}^b_{\mathbb{C}-c}(Y))$. Let $x_0 \in X$ and let K be a compact subset of $f^{-1}(x_0)$. There exist an open neighborhood U (resp. V) of x_0 (resp. K), with $V \subset f^{-1}(U)$, such that, denoting by f_V the morphism $f_V: V \to U$ induced by f, one has:*

(i) $Rf_{V!}G$ *and* $Rf_{V*}G$ *belong to* $\mathbf{D}^b_{\mathbb{C}-c}(U)$,
(ii) $\mathrm{SS}(Rf_{V!}G) \cup \mathrm{SS}(Rf_{V*}G) \subset f_{V\pi}{}^t f_V'^{-1}(\mathrm{SS}(G))$.

Moreover when Y *is an open subset of* $\mathbb{C}^n$ *and* $K = \{0\}$, *then we can take the open ball centered at* 0 *with radius* $\varepsilon \ll 1$ *as* V.

Proof. By decomposing f as $Y \hookrightarrow Y \times X \to X$, we may assume from the beginning that $Y = Z \times X$, for a complex manifold Z, and f is the second projection. Set $Z_\circ = Z \times \{x_\circ\}$ $(= f^{-1}(x_\circ))$, and denote by j the embedding $Z_\circ \hookrightarrow Y$. Set $\Lambda = \mathrm{SS}(G)$ and introduce:

$$\Lambda_\circ = j^{\#}\Lambda \qquad \text{(cf. VI §2)} . \tag{8.5.4}$$

Lemma 8.5.10. *One has:*

$$ {}^t j' j_\pi^{-1}\overline{\left(\Lambda + {}^t f'\left(Y \underset{X}{\times} T^*X\right)\right)} = \Lambda_\circ \ .$$

Proof of Lemma 8.5.10. We choose a local coordinate system (z,x) on $Z \times X$ and denote by $(z,x;\zeta,\xi)$ the associated coordinates on $T^*(Z \times X)$. Then $(z_\circ;\zeta_\circ)$ belongs to ${}^t j' j_\pi^{-1}(\overline{\Lambda + {}^t f'(Y \times_X T^*X)})$ if and only if there exists a sequence $\{(z_n, x_n; \zeta_n, \xi_n)\}$ in Λ with:

$$(z_n, x_n, \zeta_n) \underset{n}{\longrightarrow} (z_\circ, x_\circ, \zeta_\circ) \ .$$

Hence it is enough to show that for such a sequence one also has

$$|\xi_n||x_n - x_\circ| \underset{n}{\longrightarrow} 0 \ . \tag{8.5.5}$$

Assume (8.5.5) is not satisfied. By the curve selection lemma (in the holomorphic situation), we find a holomorphic map

$$\theta: \{t \in \mathbb{C}; 0 < |t| < 1\} \to \Lambda$$

such that, setting $\theta(t) = (z(t), x(t); \zeta(t), \xi(t))$, one has when $t \to 0$:

$$\begin{cases} (z(t); \zeta(t)) \to (z_\circ; \zeta_\circ) \\ x(t) - x_\circ \sim t^s \\ \xi(t) \sim t^{-r} \ , \quad 1 \leqslant s \leqslant r \ . \end{cases}$$

Since Λ is isotropic, we also have:

$$\left\langle \zeta(t), \frac{dz(t)}{dt} \right\rangle + \xi(t)\frac{dx(t)}{dt} = 0 \ .$$

Therefore $\xi(t)\dfrac{dx(t)}{dt}$ is bounded, which is a contradiction. $\square$

End of the proof of Proposition 8.5.9. Let $\varphi : Z_\circ \to \mathbb{R}$ be a real analytic function, and assume φ is positive, and proper. Set $s_\circ = \sup_K |\varphi|$, and using Proposition 8.3.12, choose real numbers $s_\circ < s_1 < s_2$ such that:

$$d\varphi(z) \notin \Lambda_\circ \quad \text{for} \quad s_1 \leqslant \varphi(z) \leqslant s_2 \ . \tag{8.5.6}$$

Set $Z_\circ^s = \{z \in Z_\circ; \varphi(z) < s\}$. We claim that for U a sufficiently small open neighborhood of $x_\circ$ in X, the hypotheses of Proposition 8.5.8 are satisfied on $V = Z_0^{s_2} \times U$, with φ regarded as a function on $Z \times X$ (i.e.: setting $\varphi(z, x) = \varphi(z)$), and with $t_\circ = s_1$. In fact otherwise we find a sequence $\{(z_n, x_n)\}$ in $Z \times X$ such that

$$\begin{cases} (z_n, x_n; d\varphi(z_n), 0) \in \Lambda + Z \times T^*X \\ x_n \underset{n}{\longrightarrow} x_\circ, \quad s_1 \leqslant \varphi(z_n) \leqslant s_2 \ , \end{cases}$$

and we may assume $z_n \underset{n}{\to} z_\circ$. Then by Lemma 8.5.10, $(z_\circ; d\varphi(z_\circ)) \in \Lambda_\circ$ which is a contradiction. □

Remark 8.5.11. Proposition 8.5.9 is not true if one releases the hypothesis that $\dim X = 1$. We refer to Henry-Merle-Sabbah [1] for a geometrical study of this problem when $\dim X > 1$.

8.6. Nearby-cycle functor and vanishing-cycle functor

In Chapter IV, we defined the functors ν_Y and μ_Y of specialization and microlocalization, respectively. When X and Y are complex manifolds and Y is a hypersurface, one can formulate them differently.

Let X be a complex manifold, $f : X \to \mathbb{C}$ a holomorphic map. Set $Y = f^{-1}(0)$ and denote by i the embedding $Y \hookrightarrow X$.

Let $\tilde{\mathbb{C}}^*$ be the universal covering of $\mathbb{C}^* = \mathbb{C}\backslash\{0\}$, and $p : \tilde{\mathbb{C}}^* \to \mathbb{C}$ the projection; (e.g. take $\tilde{\mathbb{C}}^* = \mathbb{C}$ and $p(z) = \exp(2\pi\sqrt{-1}z)$). Set $\tilde{X}^* = X \times_{\mathbb{C}} \tilde{\mathbb{C}}^*$ and denote by $\tilde{p}$ the projection $\tilde{X}^* \to X$ associated to p (i.e. $\tilde{p} = \mathrm{id} \times_{\mathbb{C}} p$):

$$\begin{array}{ccccc} & & \tilde{X}^* & \longrightarrow & \tilde{\mathbb{C}}^* \\ & & \downarrow{\scriptstyle \tilde{p}} & \square & \downarrow{\scriptstyle p} \\ Y & \underset{i}{\hookrightarrow} & X & \underset{f}{\longrightarrow} & \mathbb{C} \ . \end{array} \tag{8.6.1}$$

Note that $\tilde{p}_!$ is an exact functor.

Definition 8.6.1. *For* $F \in \mathrm{Ob}(\mathbf{D}^b(X))$, *one sets:*

$$\psi_f(F) = i^{-1}R\tilde{p}_*\tilde{p}^{-1}(F) \ ,$$

and calls ψ_f *the nearby-cycle functor.*

Note that $\psi_f(F)$ belongs to $\mathbf{D}^b(Y)$ and depends only on $F|_{X \setminus Y}$. Since $\tilde{p}^{-1} \simeq \tilde{p}^!$, we get by the Poincaré-Verdier duality:

$$\tilde{p}_* \tilde{p}^{-1} F \simeq R\mathcal{H}om(\tilde{p}_! A_{\tilde{X}^*}, F)$$
$$\simeq R\mathcal{H}om(f^{-1} p_! A_{\tilde{\mathbb{C}}^*}, F) ,$$

hence:

(8.6.2) $$\psi_f(F) \simeq i^{-1} R\mathcal{H}om(f^{-1} p_! A_{\tilde{\mathbb{C}}^*}, F) .$$

Now the action of $1 \in \mathbb{Z}$ on $\tilde{\mathbb{C}}^*$ induces an automorphism T of $p_! A_{\tilde{\mathbb{C}}^*}$, hence an automorphism of $\psi_f(F)$. This automorphism is called the **monodromy** of $\psi_f(F)$ and will be denoted by M. Next, consider the complex:

(8.6.3) $$K : 0 \longrightarrow p_! A_{\tilde{\mathbb{C}}^*} \underset{\mathrm{tr}}{\longrightarrow} A_{\mathbb{C}} \longrightarrow 0$$

where $A_{\mathbb{C}}$ stands in degree zero and the differential tr is the trace morphism $p_! A_{\tilde{\mathbb{C}}^*} \simeq p_! p^! A_{\mathbb{C}} \to A_{\mathbb{C}}$.

Definition 8.6.2. *For $F \in \mathrm{Ob}(\mathbf{D}^b(X))$, one sets:*

$$\phi_f(F) = i^{-1} R\mathcal{H}om(f^{-1} K, F) ,$$

and call ϕ_f the vanishing-cycle functor.

The action of T on $p_! A_{\tilde{\mathbb{C}}^*}$ and the identity on $A_{\mathbb{C}}$ induce an automorphism of K, hence an automorphism of $\phi_f(F)$. This automorphism is called the monodromy of $\phi_f(F)$ and will be still denoted by M.

Consider the exact sequences of complexes (represented by the rows):

(8.6.4) $$\begin{array}{ccccccccc} 0 & \longrightarrow & 0 & \longrightarrow & p_! A_{\tilde{\mathbb{C}}^*} & \xrightarrow{\mathrm{id}} & p_! A_{\tilde{\mathbb{C}}^*} & \longrightarrow & 0 \\ & & \downarrow & & \downarrow{\scriptstyle \mathrm{tr}} & & \downarrow & & \\ 0 & \longrightarrow & A_{\mathbb{C}} & \xrightarrow{\mathrm{id}} & A_{\mathbb{C}} & \longrightarrow & 0 & \longrightarrow & 0 \end{array}$$

and:

(8.6.5) $$\begin{array}{ccccccccc} 0 & \longrightarrow & p_! A_{\tilde{\mathbb{C}}^*} & \xrightarrow{1-T} & p_! A_{\tilde{\mathbb{C}}^*} & \xrightarrow{\mathrm{tr}} & A_{\mathbb{C}^*} & \longrightarrow & 0 \\ & & \downarrow & & \downarrow{\scriptstyle \mathrm{tr}} & & \downarrow & & \\ 0 & \longrightarrow & 0 & \longrightarrow & A_{\mathbb{C}} & \xrightarrow{\mathrm{id}} & A_{\mathbb{C}} & \longrightarrow & 0 . \end{array}$$

Note that the exactness of the top row of (8.6.5) follows from the fact that for $a \in \mathbb{C}^*$,

$$(p_! A_{\tilde{\mathbb{C}}^*})_a \simeq A^{(\mathbb{Z})} ,$$

(where $A^{(\mathbb{Z})}$ denotes the set of sequences $\{a_n\}_{n\in\mathbb{Z}}$ in A such that $a_n = 0$ except for finitely many n), and T is given by $\{a_n\}_n \to \{a_{n+1}\}_n$, while $\mathrm{tr}: (p_! A_{\tilde{\mathbb{C}}^*})_a \to (A_{\mathbb{C}^*})_a$ is given by $\{a_n\}_n \to \sum_n a_n$.

Thus we obtain two distinguished triangles in $\mathbf{D}^b(A_{\mathbb{C}})$:

$$\begin{cases} A_{\mathbb{C}} \longrightarrow K \longrightarrow p_! A_{\tilde{\mathbb{C}}^*}[1] \underset{+1}{\longrightarrow} , \\ p_! A_{\tilde{\mathbb{C}}^*}[+1] \longrightarrow K \longrightarrow A_{\{0\}} \underset{+1}{\longrightarrow} \end{cases} \tag{8.6.6}$$

(noting that $A_{\{0\}}$ is isomorphic to the complex $A_{\mathbb{C}^*} \to A_{\mathbb{C}}$ in $\mathbf{D}^b(A_{\mathbb{C}})$), which induce the distinguished triangles in $\mathbf{D}^b(Y)$:

$$\begin{cases} \psi_f(F)[-1] \overset{\mathrm{can}}{\longrightarrow} \phi_f(F) \longrightarrow i^{-1}F \underset{+1}{\longrightarrow} , \\ i^! F \longrightarrow \phi_f(F) \overset{\mathrm{var}}{\longrightarrow} \psi_f(F)[-1] \underset{+1}{\longrightarrow} . \end{cases} \tag{8.6.7}$$

Note that by this construction:

$$\begin{cases} \mathrm{can} \circ \mathrm{var} = 1 - \mathrm{M} & \text{in } \mathrm{End}(\phi_f(F)) , \\ \mathrm{var} \circ \mathrm{can} = 1 - \mathrm{M} & \text{in } \mathrm{End}(\psi_f(F)) . \end{cases} \tag{8.6.8}$$

Let us investigate the relations between ϕ_f, ψ_f, ν_Y and μ_Y when Y is non-singular.

In this case, df defines a function

$$\tilde{f}: T_Y X \to \mathbb{C} . \tag{8.6.9}$$

We denote by s the section of $T_Y X \to Y$ given by $\tilde{f}^{-1}(1)$ and by s' the section of $T_Y^* X \to Y$ given by df.

Proposition 8.6.3. *Assume Y is non-singular and let $F \in \mathrm{Ob}(\mathbf{D}^b_{w-\mathbb{R}-c}(X))$. Choose a point in $p^{-1}(1) \in \tilde{\mathbb{C}}^*$ and identify Y with the zero-section of $T_Y X$. Then we have the isomorphisms:*

$$\begin{cases} \psi_f(F) \simeq \psi_{\tilde{f}}(\nu_Y(F)) , \\ \phi_f(F) \simeq \phi_{\tilde{f}}(\nu_Y(F)) . \end{cases} \tag{8.6.10}$$

Moreover if $F \in \mathrm{Ob}(\mathbf{D}^b_{w-\mathbb{C}-c}(X))$, we have the isomorphisms:

$$\begin{cases} \psi_f(F) \simeq s^{-1}\nu_Y(F) , \\ \phi_f(F) \simeq s'^{-1}\mu_Y(F) . \end{cases} \tag{8.6.11}$$

Note that it follows that if F is (weakly) $\mathbb{C}$-constructible then so are $\psi_f(F)$ and $\phi_f(F)$, and

$$\mathrm{supp}(\phi_f(F)) \subset \{x \in Y; df(x) \in \mathrm{SS}(F)\} . \tag{8.6.12}$$

Proof. (a) Assume first that $X = \mathbb{C} \times Y$ and f is the first projection. Consider the diagram:

$$\begin{array}{ccccc} \mathbb{R}^+\times\tilde{\mathbb{C}}^*\times Y & \xhookrightarrow{\;j'\;} & \mathbb{R}\times\tilde{\mathbb{C}}^*\times Y & \xhookleftarrow{\;k'\;} & \tilde{\mathbb{C}}^*\times Y \\ \downarrow{\scriptstyle\pi'} & \square & \downarrow{\scriptstyle\pi} & \square & \downarrow{\scriptstyle\pi''} \\ \mathbb{R}^+\times\mathbb{C}^*\times Y & \xhookrightarrow{\;j\;} & \mathbb{R}\times\mathbb{C}^*\times Y & \xhookleftarrow{\;k\;} & \mathbb{C}^*\times Y \\ & \searrow{\scriptstyle h'} & \downarrow{\scriptstyle h} & & \\ & & \mathbb{C}\times Y = X & & \end{array} \tag{8.6.13}$$

Here j' is the inclusion map, k' is the inclusion map defined by $k'(z, y) = (0, z, y)$, and h is the map $(t, z, y) \mapsto (tz, y)$. The maps π', π, π'' are those induced by $p : \tilde{\mathbb{C}}^* \to \mathbb{C}$.

By the definition of ν_Y we have:

$$\nu_Y(F)|_{\dot{T}_Y X} \simeq k^{-1} Rj_* h'^{-1} F \ . \tag{8.6.14}$$

The choice of $p^{-1}(1) \in \tilde{\mathbb{C}}^*$ allows us to identify $\tilde{\mathbb{C}}^*$ with $\mathbb{C}$, and $p : \tilde{\mathbb{C}}^* \to \mathbb{C}$ with $p(z) = \exp(2\pi\sqrt{-1}z)$.

We obtain a Cartesian diagram:

$$\begin{array}{ccc} \mathbb{R}^+\times\tilde{\mathbb{C}}^*\times Y & \xrightarrow{\;h''\;} & \tilde{\mathbb{C}}^*\times_{\mathbb{C}} X \simeq \tilde{\mathbb{C}}^*\times Y \\ \downarrow{\scriptstyle\pi'} & \square & \downarrow{\scriptstyle\tilde{p}} \\ \mathbb{R}^+\times\mathbb{C}^*\times Y & \xrightarrow{\;h'\;} & X \end{array} ,$$

where $h''(t, z, y) = (z + (1/2\pi\sqrt{-1}) \log t, y)$.

Since h' and h'' are smooth, we have for $F \in \mathrm{Ob}(\mathbf{D}^b(X))$:

$$\begin{aligned} h'^{-1} R\tilde{p}_* \tilde{p}^{-1} F &\simeq \pi'_* h''^{-1} \tilde{p}^{-1} F \\ &\simeq \pi'_* \pi'^{-1} h'^{-1} F \ . \end{aligned}$$

Coming back to diagram (8.6.13), we get:

$$\begin{aligned} k^{-1} Rj_* h'^{-1} R\tilde{p}_* \tilde{p}^{-1} F &\simeq k^{-1} Rj_* \pi'_* \pi'^{-1} h'^{-1} F \\ &\simeq k^{-1} \pi_* Rj'_* \pi'^{-1} h'^{-1} F \\ &\simeq k^{-1} \pi_* \pi^{-1} Rj_* h'^{-1} F \ . \end{aligned}$$

Set $G = Rj_* h'^{-1} F$ and recall that we assumed F weakly $\mathbb{R}$-constructible.

We are going to prove

$$\text{(8.6.15)} \qquad \begin{cases} k^{-1}R\pi_*\pi^{-1}G \to R\pi''_*k'^{-1}\pi^{-1}G \simeq R\pi''_*\pi''^{-1}k^{-1}G \\ \text{is an isomorphism} . \end{cases}$$

In fact, any $x \in \mathbb{C}^* \times Y$ has a fundamental system of open simply connected neighborhoods U such that $G_x \simeq R\Gamma(U; G)$ (because G is weakly $\mathbb{R}$-constructible). Since $\pi^{-1}(U) \simeq U \times \mathbb{Z}$, we have:

$$\begin{aligned} R\Gamma(U; \pi_*\pi^{-1}G) &\simeq R\Gamma(\pi^{-1}(U); \pi^{-1}G) \\ &\simeq R\Gamma(U; G)^{\mathbb{Z}} . \end{aligned}$$

Taking the inductive limit, we obtain:

$$(\pi_*\pi^{-1}G)_x \simeq (G_x)^{\mathbb{Z}} .$$

Similarly one proves:

$$(\pi''_*\pi''^{-1}k^{-1}G)_x \simeq ((k^{-1}G)_x)^{\mathbb{Z}} .$$

This gives (8.6.15).

Combining (8.6.15) and (8.6.14), we get:

$$\begin{aligned} \nu_Y(\tilde{p}_*\tilde{p}^{-1}F)|_{\dot{T}_YX} &\simeq \pi''_*\pi''^{-1}k^{-1}Rj_*h'^{-1}F \\ &\simeq \pi''_*\pi''^{-1}(\nu_Y(F)|_{\dot{T}_YX}) . \end{aligned}$$

Applying $R\dot{\tau}_*$ we get $\psi_f(F)$ on the left-hand side and $\psi_{f'}(\nu_Y(F))$ on the right-hand side (cf. Theorem 4.2.3).

This proves the first isomorphism in (8.6.10). To get the second one, it is enough to consider the commutative diagram of distinguished triangles:

$$\begin{array}{ccccccc} \psi_f(F)[-1] & \xrightarrow{\text{can}} & \phi_f(F) & \longrightarrow & i^{-1}F & \xrightarrow[+1]{} & \\ \downarrow \wr & & \downarrow & & \downarrow \wr & & \\ \psi_{f'}(\nu_Y(F))[-1] & \xrightarrow{\text{can}} & \phi_{f'}(\nu_Y(F)) & \longrightarrow & i^{-1}(\nu_Y(F)) & \xrightarrow[+1]{} & . \end{array}$$

Now assume F is weakly $\mathbb{C}$-constructible and we shall prove (8.6.11). In this case $\tilde{p}^{-1}\nu_Y(F)$ is constant on the fibers of $\tau \circ \tilde{p} : \tilde{\mathbb{C}}^* \times_{\mathbb{C}} T_YX \to Y$. Let $\tilde{s}$ be a section of the map $\tilde{\mathbb{C}}^* \times_{\mathbb{C}} T_YX \to Y$ such that $\tilde{p} \circ \tilde{s} = s$. Applying Corollary 2.7.7, we get:

$$\begin{aligned} \psi_{f'}(\nu_Y(F)) &\simeq R\dot{\tau}_*(\tilde{p}_*\tilde{p}^{-1}\nu_Y(F)|_{\dot{T}_YX}) \\ &\simeq \tilde{s}^{-1}\tilde{p}^{-1}\nu_Y(F) \\ &\simeq s^{-1}\nu_Y(F) . \end{aligned}$$

Identifying T_YX with $\mathbb{C} \times Y$ and T_Y^*X with $\mathbb{C} \times Y$, we have $s'^{-1}\mu_Y(F) \simeq R\pi_*R\mathcal{H}om(A_{U \times Y}, \mu_Y(F))$, because $\mu_Y(F)$ is $\mathbb{C}^\times$-conic. Here $U = \{z \in \mathbb{C}; \operatorname{Re} z > 0\}$.

By Theorem 3.7.9, we get:

$$s'^{-1}\mu_Y(F) \simeq R\tau_* R\mathcal{H}om(A^\vee_{U\times Y}, \nu_Y(F))$$
$$\simeq R\tau_* R\mathcal{H}om(A_{Z\times Y}, \nu_Y(F)) \ ,$$

where $Z = \{z \in \mathbb{C}; \operatorname{Im} z = 0, \operatorname{Re} z \geqslant 0\}$.

Then we have a commutative diagram, by using the inclusion $\mathbb{C}\setminus Z \hookrightarrow \tilde{\mathbb{C}}^*$ over $\mathbb{C}$:

$$\begin{array}{ccc} A_{\mathbb{C}\setminus Z} & \longrightarrow & A_{\mathbb{C}} \\ \downarrow & & \downarrow \\ p_! A_{\tilde{\mathbb{C}}^*} & \longrightarrow & A_{\mathbb{C}} \ . \end{array}$$

Since the complex $A_{\mathbb{C}\setminus Z} \to A_{\mathbb{C}}$ is isomorphic to A_Z, it is enough to show:

(8.6.16) $R\tau_* R\mathcal{H}om(p_! A_{\tilde{\mathbb{C}}^*} \boxtimes A_Y, \nu_Y(F)) \xrightarrow{\sim} R\tau_* R\mathcal{H}om(A_{\mathbb{C}\setminus Z} \boxtimes A_Y, \nu_Y(F))$:

Since $\nu_Y(F)$ is locally constant on the fiber of $(\mathbb{C}\setminus Z) \times Y \to Y$, we have $s^{-1}\nu_Y(F) \simeq R\tau_* R\mathcal{H}om(A_{\mathbb{C}\setminus Z} \boxtimes A_Y, \nu_Y(F))$. This shows (8.6.16) and we obtain $s'^{-1}\mu_Y(F) \simeq \phi_f(F)$.

In the general case, embed X in $\mathbb{C} \times X$ by the graph map, $f: X \underset{g}{\hookrightarrow} \mathbb{C} \times X \underset{t}{\to} \mathbb{C}$ and note that:

$$\psi_f(F) \simeq \psi_t(g_* F) \ ,$$
$$\phi_f(F) \simeq \phi_t(g_* F) \ ,$$

(cf. Exercise VIII.15). □

One can recover $SS(F)$ from the vanishing-cycle functor.

Proposition 8.6.4. *Let $F \in \operatorname{Ob}(\mathbf{D}^b_{w\text{-}\mathbb{C}\text{-}c}(X))$ and let $p \in T^*X$. Then the following two conditions are equivalent.*

(i) $p \notin SS(F)$.
(ii) *There exists an open neighborhood U of p such that for any $x \in X$ and any holomorphic function f defined in a neighborhood of x with $f(x) = 0$ and $df(x) \in U$, one has $\phi_f(F)_x = 0$.*

Proof. (i) implies (ii) by Proposition 8.6.3 and Corollary 5.4.10. Conversely assume (ii) and let us prove that $U \cap SS(F) = \varnothing$. By Theorem 8.5.5, $\Lambda = SS(F)$ is a complex analytic Lagrangian subset of T^*X. At a generic point p' of $\Lambda \cap U$, F is isomorphic in $\mathbf{D}^b(X; p')$ to L_Y for some $L \in \operatorname{Ob}(\mathbf{D}^b(\mathfrak{Mod}(A))$ and a complex submanifold Y of X (Proposition 6.6.1). We shall show $L = 0$. If $p' \in T^*_X X$, then $Y = X$ and we have $L = 0$ (take $f = 0$). Otherwise choose a local coordinate system $(z_1,\ldots,z_n)$ on X such that $p' = (0; dz_1)$ and $Y = \{z_1 = \cdots = z_l = 0\}$. Let $f(z) = z_1 + \sum_{j=l+1}^{n} z_j^2$. Then $\phi_f(F)_0 \simeq \mu_{\{f=0\}}(F)_{p'} \simeq \mu_{\{f=0\}}(L_Y)_{p'} \simeq \phi_f(L_Y)_0 \simeq \phi_{f|Y}(L_Y)_0 \simeq L[l-n]$, (cf. Exercises VIII.14 and VIII.15). This completes the proof. □

Exercises to Chapter VIII

Exercise VIII.1. Let $\mathbf{S} = (S, \Delta)$ be a simplicial complex. Define the category $\mathbf{\Delta}$ by setting $\mathrm{Ob}(\mathbf{\Delta}) = \{\sigma; \sigma \in \Delta\}$ and $\mathrm{Hom}(\sigma,\tau) = \emptyset$ if $\sigma \not\subset \tau$, $\mathrm{Hom}(\sigma,\tau) = \{\mathrm{pt}\}$ if $\sigma \subset \tau$. Let $\mathbf{\Delta}(A)$ denote the category of covariant functors from $\mathbf{\Delta}$ to $\mathfrak{Mod}(A)$ and let δ be the functor from w-$\mathfrak{Cons}(\mathbf{S})$ to $\mathbf{\Delta}(A)$ given by $F \mapsto \{\sigma \mapsto F(U(\sigma))\}$. Prove that δ is an equivalence of categories.

Exercise VIII.2. Let $X = \mathbb{R}^3$ endowed with the coordinates (t, x, y) and let $Z = \{y^2 - t^2x^2 - x^3 = 0\}$, $Z_2 = \{x = y = 0\}$, $Z_1 = Z \setminus Z_2$, $Z_0 = X \setminus Z$. Prove that $X = Z_0 \sqcup Z_1 \sqcup Z_2$ is a subanalytic stratification, but is not a μ-stratification.

Exercise VIII.3. Let $f: Y \to X$ be a morphism of manifolds, $F \in \mathrm{Ob}(\mathbf{D}^b(X))$, $G \in \mathrm{Ob}(\mathbf{D}^b(Y))$.

(i) Prove the isomorphisms:

$$f^!(\mathrm{D}_X F) \simeq \mathrm{D}_Y(f^{-1}F) ,$$

$$Rf_*(\mathrm{D}_Y G) \simeq \mathrm{D}_X(Rf_! G) .$$

(ii) Assume F is $\mathbb{R}$-constructible. Prove the isomorphism:

$$f^{-1}(\mathrm{D}_X F) \simeq \mathrm{D}_Y(f^! F) .$$

(iii) Assume G and $Rf_!(\mathrm{D}_Y G)$ are $\mathbb{R}$-constructible. Prove the isomorphism:

$$Rf_!(\mathrm{D}_Y G) \simeq \mathrm{D}_X(Rf_* G) ,$$

and prove that $Rf_* G$ is $\mathbb{R}$-constructible.

Exercise VIII.4. In the situation of Proposition 5.4.17 prove that if G is $\mathbb{R}$-constructible then $Rf_* G$ (resp. $Rf_! G$) is $\mathbb{R}$-constructible.

Exercise VIII.5. Let $X = \bigsqcup_\alpha X_\alpha$ be a μ-stratification and set $\Lambda = \bigsqcup_\alpha T^*_{X_\alpha}X$. Let $F \in \mathrm{Ob}(\mathbf{D}^b(X))$. Prove that $SS(F) \subset \Lambda$ if and only if $SS(H^j(F)) \subset \Lambda$ for all j.

Exercise VIII.6. Let Ω be a subset of T^*X. One defines the full subcategory $\mathbf{D}^b_{w-\mathbb{R}-c}(X;\Omega)$ (resp. $\mathbf{D}^b_{\mathbb{R}-c}(X;\Omega)$) of $\mathbf{D}^b(X;\Omega)$ consisting of objects F satisfying the following condition: at any point $p \in \Omega$ there exists $F' \in \mathrm{Ob}(\mathbf{D}^b_{w-\mathbb{R}-c}(X))$ (resp. $F' \in \mathrm{Ob}(\mathbf{D}^b_{\mathbb{R}-c}(X))$) such that F and F' are isomorphic in $\mathbf{D}^b(X;p)$.

(i) Prove that $F \in \mathrm{Ob}(\mathbf{D}^b(X;\Omega))$ belongs to $\mathbf{D}^b_{w-\mathbb{R}-c}(X;\Omega)$ if and only if there exists an open neighborhood U of Ω such that $\mathrm{SS}(F) \cap U$ is contained in a closed subanalytic isotropic subset of U.

(ii) In the situation of Proposition 6.3.3 prove that if F is $\mathbb{R}$-constructible then $Rq_{1!}F$ and $Rq_{1*}F$ belong to $\mathbf{D}^b_{\mathbb{R}-c}(X;\Omega)$.

Exercise VIII.7. In the situation of Theorem 7.2.1 assume K is $\mathbb{R}$-constructible. Prove that Φ_K induces an equivalence $\mathbf{D}^b_{\mathbb{R}-c}(Y;\Omega_Y) \simeq \mathbf{D}^b_{\mathbb{R}-c}(X;\Omega_X)$. (Hint: use Exercise VIII.6.)

Exercise VIII.8. Let X be a complex manifold, Λ a closed conic $\mathbb{C}$-analytic subset of T^*X. Assume Λ_{reg} is involutive. Prove that Λ is involutive in $T^*X^{\mathbb{R}}$ (in the sense of Definition 6.5.1). (Hint: use the result of Kashiwara-Monteiro-Fernandes [1].)

Exercise VIII.9. Let $\mathbf{S} = (S, \Delta)$ be a simplicial complex with finite dimension (i.e. satisfying (8.1.16)) and let $X = |\mathbf{S}|$ denote the associated topological space. Set:

$$X_k = \{x \in X;\, x \in |\sigma| \text{ for some } \sigma \in \Delta \text{ with } \#\sigma \leqslant -k+1\}$$

(i) Prove that $H^j_{X_k \setminus X_{k+1}}(\omega_X) = 0$ for $j \neq k$.

(ii) Using Exercise II.21, prove that ω_X is isomorphic to the complex $\{H^k_{X_k \setminus X_{k+1}}(\omega_X)\}$ in $\mathbf{D}^b(X)$, and describe this complex.

Exercise VIII.10. Let X be a complex manifold, S a closed $\mathbb{C}$-analytic subset, $U = X \setminus S$ and j the open embedding $U \hookrightarrow X$. Let $F \in \mathrm{Ob}(\mathbf{D}^b_{\mathbb{C}-c}(U))$ with $\mathrm{SS}(F) \subset T^*_U U$. Prove that $Rj_!F$ and Rj_*F belong to $\mathbf{D}^b_{\mathbb{C}-c}(X)$.

Exercise VIII.11. Let $X = \bigsqcup_\alpha X_\alpha$ be a subanalytic stratification of a real analytic manifold X. Prove that if $\Lambda = \bigsqcup_\alpha T^*_{X_\alpha}X$ is closed, then Λ is involutive (in the sense of Definition 6.5.1).

Exercise VIII.12. Let N and M be subanalytic submanifolds of $\mathbb{R}^n$, with $N \subset \overline{M} \setminus M$. Consider the following **conditions (a) and (b)** due to Whitney.

(a) For any sequence $\{x_n\}$ in M converging to $x \in N$, if $\{T_{x_n}M\}$ converges to a plane τ, then $\tau \supset T_xN$.

(b) For any sequence $\{x_n\}$ in M and any sequence $\{y_n\}$ in N, both converging to $x \in N$, if the sequence of lines $\{\mathbb{R}(x_n, y_n)\}$ converges to l and the sequence $\{T_{x_n}M\}$ converges to τ, then $\tau \supset l$.

Prove that (b) implies (a) and that the μ-condition implies (b). (Hint: use the curve selection lemma.)

Exercise VIII.13. Let f be a holomorphic function on a complex manifold X, and let $F \in \mathrm{Ob}(\mathbf{D}^b_{w-\mathbb{C}-c}(X))$. Prove that there is a canonical isomorphism $R\Gamma_{\{\mathrm{Re}\, f \geqslant 0\}}(F)|_{f^{-1}(0)} \xrightarrow{\sim} \phi_f(F)$.

Exercise VIII.14. Let $X = \mathbb{C}^n$, $f(x) = \sum_j x_j^2$, and let $M \in \mathrm{Ob}(\mathbf{D}^b(\mathfrak{Mod}(A)))$. Prove the isomorphism $\phi_f(M_X) \simeq M_{\{0\}}[-n]$.

Exercise VIII.15. Let $f: Y \to X$ be a morphism of complex manifolds, and $t: X \to \mathbb{C}$ a holomorphic function on X. Let $G \in \mathbf{D}^b_{w-\mathbb{C}-c}(Y)$, and assume f is proper on supp(G). Denoting by $f_\circ$ the restriction of f to $\{t \circ f = 0\}$, prove that:

$$\psi_t(Rf_* G) \simeq Rf_{\circ *}\psi_{t\circ f}(G) \ ,$$

$$\phi_t(Rf_* G) \simeq Rf_{\circ *}\phi_{t\circ f}(G) \ .$$

Notes

The theory of subanalytic sets has been initiated by Gabrielov [1] and Hironaka [1, 2], and originates in the work of Lojasiewicz [1, 2] on semi-analytic sets. The main results are due to Hironaka (loc. cit.) but there are many other important contributions, in particular Hardt [1, 2], Tamm [1], and Teissier [1]. This theory which appeared for long as technically difficult, has been considerably simplified (e.g. cf. Denkowska-Lojasiewicz-Stasica [1]) and there is now a short paper by Bierstone-Milman [1] which contains the main theorems, with complete proofs.

The first time the notion of stratified space appeared seems to be in 1955, in Cartan-Chevalley [1], and it is Whitney [1, 2] who first introduced regularity conditions on the strata, (the famous "conditions (a) and (b) of Whitney"), and proved that for a given stratification, there exists a finer stratification satisfying these conditions, (a result similar to Theorem 8.3.20). There are various conditions of regularity for real stratifications (cf. Trotman [1] for a review), and in particular there is the notion of w-stratification introduced by Verdier [3] in 1976, a variant of another condition due to Kuo [1]. In his paper (loc. cit.), Verdier proved, first, that the w-condition is stronger than Whitney's conditions, and second, that there always exist w-stratifications. This last result is in fact equivalent to Theorem 8.3.20 since Trotman [2] has recently proved that the notions of w-stratifications and of μ-stratifications are equivalent. Let us also mention a paper by Delort [1] who gives a generalization of the operation $\hat{+}$.

In the complex relative case, the A_f-condition of Thom appears as an efficient tool for the study of direct images (cf. Henry-Merle-Sabbah [1]), and Hironaka [3] proved that if the target space has dimension one, this condition is "locally" satisfied: this is essentially equivalent to Proposition 8.5.9.

The microlocal study of stratifications (i.e. the idea of treating stratifications by using associated Lagrangian subsets, what is now called "conormal geometry") was initiated in Kashiwara [3, 5], then developed in Kashiwara-Schapira [1, 3].

It is Grothendieck (in [SGA 4], exposé IX by Artin) who first introduced the notion of constructible sheaves and studied their functorial properties (in the context of étale cohomology). This theory has regained new interest after

Kashiwara [3] made the link with holonomic $\mathscr{D}$-modules (cf. Chapter XI §3 below), and the discovery of intersection cohomology by Goresky-MacPherson [1] and perverse sheaves by Gabber and Beilinson-Bernstein-Deligne [1] (cf. Brylinski [1] for a review).

The theory of nearby-cycle and vanishing-cycle functors (cf. §6) is due to Grothendieck (cf. [SGA 7] exposé I) and Deligne [2]. It originates in the study of vanishing cycles and monodromy by Picard and Lefschetz.

As shown in Proposition 8.6.4, the micro-support of $\mathbb{C}$-constructible sheaves on a complex manifold may be obtained using the vanishing-cycle functor. As noted by Brylinski (loc. cit.) this fact is already implicit in Kashiwara [5].

The results of §1, as well as Theorem 8.4.5, are extracted from Kashiwara [6]. Most of the results of §3, 4, 5 were first obtained by similiar methods in Kashiwara-Schapira [3], but of course the operations on constructible sheaves have been known for long (cf. e.g. Verdier [4]). However, Definition 8.3.19 and Proposition 8.4.14 are new, as well as the proof of Proposition 8.4.1. (Note that we do not make use of the Thom-Mather's isotopy theorem in this proof.)

Chapter IX. Characteristic cycles

Summary

On a real analytic manifold X, we first introduce the notions of subanalytic chains and subanalytic cycles, with the help of the dualizing complex ω_X. Then we define the intersection of two cycles.

If F is an $\mathbb{R}$-constructible object of $\mathbf{D}^b(X)$, we construct the characteristic cycle of F, $\mathrm{CC}(F)$, as the natural image of $\mathrm{id}_F \in \mathrm{Hom}(F, F)$ in $H^0_{SS(F)}(T^*X; \pi^{-1}\omega_X)$. This is a "Lagrangian cycle". For example, if M is a closed submanifold of X and if $V \in \mathrm{Ob}(\mathbf{D}^b(\mathfrak{Mod}^f(k)))$ (the base ring A is now a field k of characteristic zero) then $\mathrm{CC}(V_M) = m[T^*_M X]$, where $[T^*_M X]$ is the Lagrangian cycle associated to the conormal bundle $T^*_M X$ to M in X and $m = \chi(V) = \sum_j (-1)^j \dim H^j(V)$. We study some functorial properties of characteristic cycles, such as non-characteristic inverse images or proper direct images, and we prove in particular that if F has compact support, the Euler-Poincaré index $\chi(X; F) = \sum_j (-1)^j \dim H^j(X; F)$ can be obtained as the intersection number of $\mathrm{CC}(F)$ and the cycle associated to the zero-section of T^*X. We also give a local Euler-Poincaré index formula. These indices can be calculated using a "Morse function with respect to $SS(F)$".

Next we study a version of the "Lefschetz fixed point formula". If f is an endomorphism of X, and if a morphism $\varphi \in \mathrm{Hom}(f^{-1}F, F)$ is given, one defines a characteristic class $\mathrm{C}(\varphi) \in H^0(X; \omega_X)$ whose degree calculates the trace $\mathrm{tr}(\varphi)$ of $\Gamma(X; \varphi) \in \mathrm{Hom}(R\Gamma(X; F), R\Gamma(X, F))$, (one assumes F has compact support). When f has only finitely many fixed points and is transversal to the identity, we show how to deduce $\mathrm{tr}(\varphi)$ from a local formula.

Finally we study the Grothendieck group of $\mathbf{D}^b_{\mathbb{R}-c}(X)$. We prove that this group is isomorphic to the group of Lagrangian cycles on T^*X (the isomorphism being defined by $F \mapsto \mathrm{CC}(F)$) as well as to the group of $\mathbb{R}$-constructible functions on X (the isomorphism is defined by $F \mapsto \chi(F)(x) = \chi(F_x)$). This gives rise to a new calculus on the algebra of constructible functions on X.

We keep Conventions 8.0. Moreover in Section 9.1, and from Section 9.4 until the end of the chapter, the base ring A will be a (commutative) field of characteristic zero, and we will denote it by k.

Remark 9.0. The commutativity of diagrams is carefully checked in §1 and serves as a prototype for this kind of proof in the rest of the chapter where we will sometimes take the liberty of leaving similar verifications to the reader. Moreover it would have made this book too heavy to follow the signs all over the calculations especially in this chapter. In other words we do not always make the difference between a commutative and an anti-commutative diagram. This does not affect the understanding of the contents. Usually for applications, the right sign in a formula may be easily obtained by looking at simple examples.

9.1. Index formula

In this section, we assume the base ring A is a commutative field of characteristic zero, and we denote it by k. If $V \in \mathrm{Ob}(\mathbf{D}^b(\mathfrak{Mod}^f(k)))$, its index $\chi(V)$ is defined by:

(9.1.1) $$\chi(V) = \sum_j (-1)^j \dim H^j(V) \ .$$

Now let X be a (real analytic) manifold and let $F \in \mathbf{D}^b_{\mathbf{R}-c}(X)$. Let $x \in X$.

Definition 9.1.1. *One sets:*

$$\chi(F)(x) = \chi(F_x) \ ,$$

$$\chi_c(F)(x) = \chi(R\Gamma_{\{x\}}(X;F)) = \chi(\mathrm{D}(F))(x) \ .$$

If $R\Gamma(X;F)$ (resp. $R\Gamma_c(X;F)$) belongs to $\mathbf{D}^b(\mathfrak{Mod}^f(k))$, one sets:

$$\chi(X;F) = \chi(R\Gamma(X;F))$$

$$(\text{resp. } \chi_c(X;F) = \chi(R\Gamma_c(X;F))) \ .$$

The integer $\chi(X;F)$ is called the Euler-Poincaré index of F and the function $\chi(F)(x)$ is called the local Euler-Poincaré index of F.

In this section we shall introduce ȧ first tool to calculate these indices.

The identity $\mathrm{Hom}(F,F) = \mathrm{Hom}(k_X, R\mathcal{H}om(F,F))$ defines the morphism:

(9.1.2) $$k_X \to R\mathcal{H}om(F,F) \ .$$

The dualizing functor D_X being defined as $R\mathcal{H}om(\cdot, \omega_X)$, by contraction we get a morphism:

(9.1.3) $$\mathrm{tr}_X : F \otimes \mathrm{D}_X F \to \omega_X \ .$$

We call tr_X the "trace morphism". Let $\delta_X : X \to X \times X$ denote the diagonal embedding. (If there is no risk of confusion, we write δ instead of δ_X, and similarly, D instead of D_X, etc.) Then there is a unique morphism:

(9.1.4) $$\delta^! \to \delta^{-1}$$

such that the following diagram commutes:

$$(9.1.5)\qquad \begin{array}{ccccc} \delta^{-1}\delta_!\delta^! & \longrightarrow & \delta^{-1} & \xrightarrow{\sim} & \delta^!\delta_!\delta^{-1} \\ \wr\downarrow & & \downarrow & & \downarrow\wr \\ \delta^{-1}\delta_*\delta^! & \xrightarrow{\sim} & \delta^! & \longrightarrow & \delta^!\delta_*\delta^{-1} \end{array} .$$

Hence we get the chain of morphisms:

$$k_X \longrightarrow R\mathscr{H}om(F,F) \simeq \delta^!(F \boxtimes DF) \longrightarrow \delta^{-1}(F \boxtimes DF) \simeq F \otimes DF \underset{\mathrm{tr}_X}{\longrightarrow} \omega_X .$$

Definition 9.1.2. *The image of* $1 \in \Gamma(X;k_X)$ *in* $H^0_{\mathrm{supp}(F)}(X;\omega_X)$ *by the chain of morphisms above is called the characteristic class of* F *and denoted* $\mathrm{C}(F)$. *For a closed subset* $S \supset \mathrm{supp}(F)$, *we also denote by the same notation* $\mathrm{C}(F)$ *the image of the characteristic class of* F *by the morphism* $H^0_{\mathrm{supp}(F)}(X;\omega_X) \to H^0_S(X;\omega_X)$.

When X is a point, then $F \in \mathrm{Ob}(\mathbf{D}^b(\mathfrak{Mod}^f(k)))$ and $\mathrm{C}(F) = \chi(F)$ (cf. Exercise I.32).

We shall study direct images of characteristic classes.

Let $f: Y \to X$ be a morphism of manifolds, let $G \in \mathrm{Ob}(\mathbf{D}^b_{\mathbb{R}\text{-}c}(Y))$ and assume f is proper on $\mathrm{supp}(G)$. Consider the diagram below:

(9.1.6)

$$\begin{array}{ccccccccc} Rf_*k_Y & \to & Rf_*R\mathscr{H}om(G,G) & \simeq & Rf_!\delta_Y^!(G \boxtimes D_YG) & \to & Rf_!(G \otimes D_YG) & \to & Rf_!\omega_Y \\ \uparrow & 1 & \downarrow & 2 & \downarrow & 3 & \uparrow & 4 & \downarrow \\ k_X & \to & R\mathscr{H}om(Rf_!G, Rf_!G) & \simeq & \delta_X^!(Rf_!G \boxtimes D_XRf_!G) & \to & (Rf_!G) \otimes (D_XRf_!G) & \to & \omega_X \end{array} .$$

Proposition 9.1.3. *The diagram* (9.1.6) *is commutative.*

Proof. We shall omit to write "R" for the derived functor, for the sake of simplicity.

(i) The square 1 in (9.1.6) is obviously commutative.

(ii) Consider the commutative diagram of maps:

$$(9.1.7)\qquad \begin{array}{ccccc} Y \times Y & \xrightarrow{f_1} & X \times Y & \xrightarrow{f_2} & X \times X \\ & \nwarrow^{\delta_Y} & \uparrow\delta & \square & \uparrow\delta_X \\ & & Y & \xrightarrow{f} & X \end{array} .$$

in which the square is Cartesian.

Then the squares 2 and 3 in (9.1.6) decompose as in diagram (9.1.8) below.

$$
\begin{array}{ccccccccc}
f_*\mathscr{H}om(G,G) & \overset{\sim}{-\!\!-} & f_*\delta_Y^!\mathscr{H}om(A_Y\boxtimes G, G\boxtimes\omega_Y) & \overset{\sim}{\longleftarrow} & f_*\delta_Y^!(G\boxtimes \mathrm{D}_Y G) & \longrightarrow & f_*\delta_Y^{-1}(G\boxtimes \mathrm{D}_Y G) & \overset{\sim}{\longleftarrow} & f_!(G\boxtimes \mathrm{D}_Y G) \\
\downarrow & & \downarrow & & \downarrow & & \uparrow & & \uparrow \\
f_*\mathscr{H}om(G,f^!f_!G) & \overset{\sim}{-\!\!-} & f_*\delta_Y^!\mathscr{H}om(A_Y\boxtimes G, f^!f_!G\boxtimes\omega_Y) & \longleftarrow & f_*\delta_Y^!(f^!f_!G\boxtimes \mathrm{D}_Y G) & & f_*\delta^{-1}f_{1*}(G\boxtimes \mathrm{D}_Y G) & & f_!(f^{-1}f_!G\boxtimes \mathrm{D}_Y G) \\
\Big| & & \wr\,\Big| & & \wr\,\Big| & \text{(a)} & \uparrow & & \wr\,\Big\downarrow \\
\Big| & & f_*\delta_Y^! f_1^!\mathscr{H}om(A_X\boxtimes G, f_!G\boxtimes\omega_Y) & \overset{\sim}{\longleftarrow} & f_*\delta_Y^! f_1^!(f_!G\boxtimes \mathrm{D}_Y G) & \longrightarrow & f_*\delta^{-1}f_{1!}(G\boxtimes \mathrm{D}_Y G) & & \Big\downarrow \\
\Big| & & \wr\,\Big| & & \wr\,\Big| & & \wr\,\Big| & & \\
\Big| & & f_*\delta^!\mathscr{H}om(A_X\boxtimes G, f_!G\boxtimes\omega_Y) & \longleftarrow & f_*\delta^!(f_!G\boxtimes \mathrm{D}_Y G) & \longrightarrow & f_*\delta^{-1}(f_!G\boxtimes \mathrm{D}_Y G) & \overset{\sim}{\longleftarrow} & f_*(f^{-1}f_!G\otimes \mathrm{D}_Y G) \\
\wr\,\Big| & & \wr\,\Big| & & \wr\,\Big| & \text{(b)} & \uparrow & & \uparrow \\
\Big| & & \delta_X^! f_{2*}\mathscr{H}om(A_X\boxtimes G, f_2^!(f_!G\boxtimes\omega_X)) & \longleftarrow & \delta_X^! f_{2*}(f_!G\boxtimes \mathrm{D}_Y G) & \longrightarrow & \delta_X^{-1}f_{2*}(f_!G\boxtimes \mathrm{D}_Y G) & & \Big| \\
\Big| & & \Big| & & \wr\,\uparrow & & \uparrow & & \Big| \\
\Big| & & \wr\,\Big| & & \delta_X^!(f_!G\boxtimes f_*\mathrm{D}_Y G) & \longrightarrow & \delta_X^{-1}(f_!G\boxtimes f_*\mathrm{D}_Y G) & \overset{\sim}{-\!\!-} & f_!G\boxtimes f_*\mathrm{D}_Y G \\
\Big\downarrow & & \Big| & & \wr\,\uparrow & & \wr\,\uparrow & & \wr\,\uparrow \\
\mathscr{H}om(f_!G,f_!G) & \overset{\sim}{-\!\!-} & \delta_X^!\mathscr{H}om(A_X\boxtimes f_!G, f_!G\boxtimes\omega_X) & \overset{\sim}{\longleftarrow} & \delta_X^!(f_!G\boxtimes \mathrm{D}_X f_!G) & \longrightarrow & \delta_X^{-1}(f_!G\boxtimes \mathrm{D}_X f_!G) & \overset{\sim}{-\!\!-} & f_!G\otimes \mathrm{D}_X f_!G
\end{array}
$$

Diagram (9.1.8)

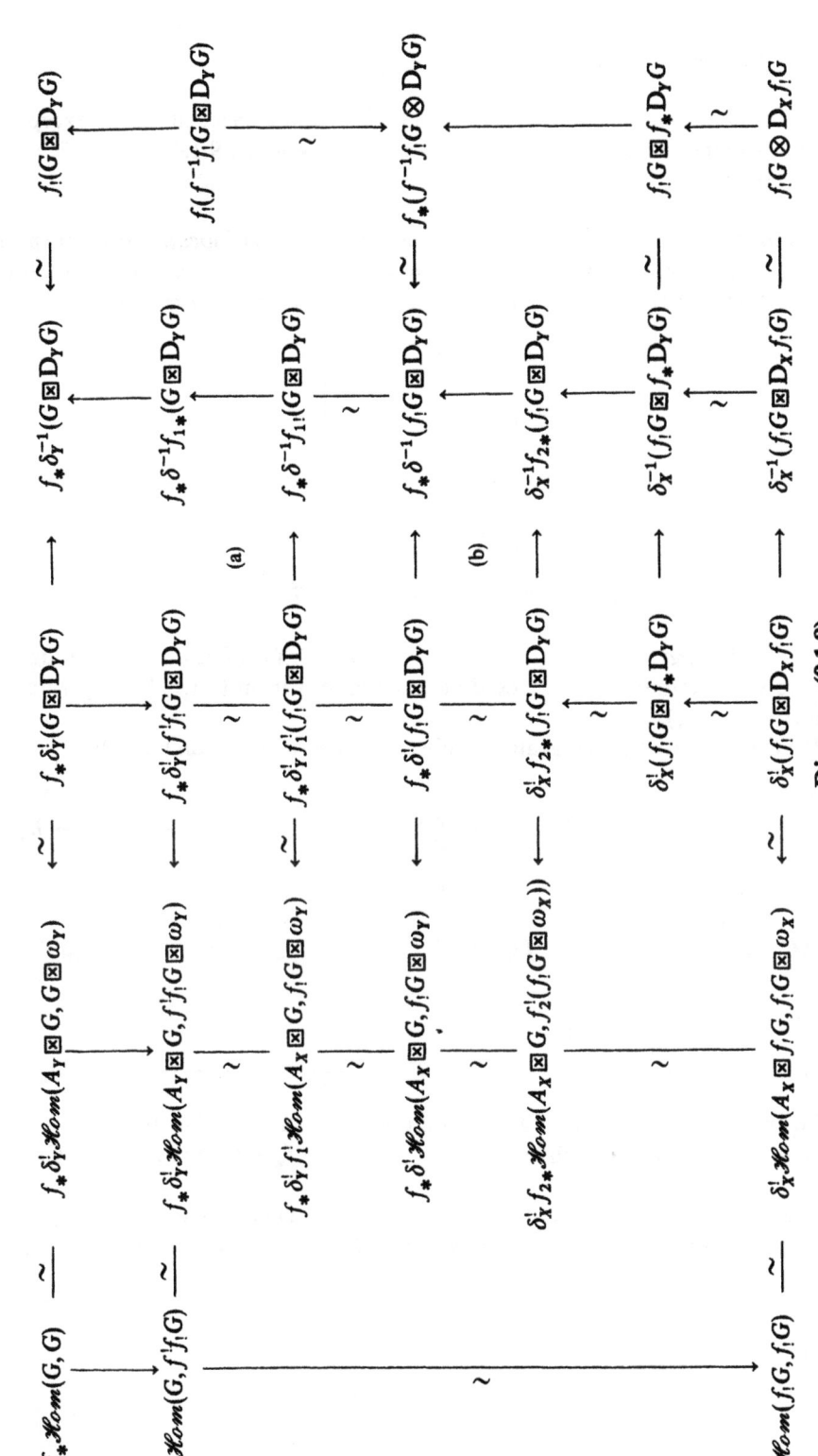

The commutativity of all diagrams in (9.1.8) is easily checked, at the exception of the squares (a) and (b), which follow from Lemmas 9.1.4 and 9.1.5 below respectively.

Lemma 9.1.4. *Let $Z \xrightarrow{g} Y \xrightarrow{f} X$ be a chain of morphisms of locally compact spaces with finite c-soft dimension and set $h = f \circ g$. Assume that g and h are closed embeddings. Then if $G \in \mathrm{Ob}(\mathbf{D}^b(Y))$, the following diagram commutes.*

$$
(9.1.9)\qquad
\begin{array}{ccc}
g^!G & \longrightarrow & g^{-1}G \\
\downarrow & & \uparrow \\
g^!f^!f_!G & & g^{-1}f^{-1}f_*G \\
\downarrow\wr & & \uparrow\wr \\
h^!f_!G & \longrightarrow & h^{-1}f_!G
\end{array}
$$

Proof of Lemma 9.1.4. Since $h_! = h_* : \mathbf{D}^b(Z) \to \mathbf{D}^b(X)$ is faithful, it is enough to prove the commutativity of the diagram deduced from (9.1.9) by applying the functor $h_!(\simeq h_*)$.

Then the resulting diagram is embedded into the diagram (9.1.10):

$$
(9.1.10)\qquad
\begin{array}{ccccccccccc}
h_!g^!G & \xrightarrow{\sim} & f_!g_!g^!G & \longrightarrow & f_!G & \longrightarrow & f_*G & \longrightarrow & f_*g_*g^{-1}G & \longrightarrow & h_*g^{-1}G \\
\downarrow & & \downarrow & & \downarrow & & \uparrow & & \uparrow & & \uparrow \\
h_!g^!f^!f_!G & \simeq & f_!g_!g^!f^!f_!G & \rightarrow & f_!f^!f_!G & & f_*f^{-1}f_*G & \rightarrow & f_*g_*g^{-1}f^{-1}f_*G & \simeq & h_*g^{-1}f^{-1}f_*G \\
\downarrow\wr & & & & \downarrow & \text{(c)} & \uparrow & & & & \uparrow\wr \\
h_!h^!f_!G & & \longrightarrow & & f_!G & \longrightarrow & f_*G & & \longrightarrow & & h_*h^{-1}f_*G
\end{array}
$$

The commutativity of (9.1.10) then follows from the fact that the composition $f_!G \to f_!f^!f_!G \to f_!G$ as well as $f_*G \to f_*f^{-1}f_*G \to f_*G$, is the identity. □

Lemma 9.1.5. *Consider the Cartesian square* (9.1.11) *of locally compact spaces with finite c-soft dimension, where the map g is a closed embedding:*

$$
(9.1.11)\qquad
\begin{array}{ccc}
Y' & \xhookrightarrow{g'} & Y \\
{\scriptstyle f'}\downarrow & \square & \downarrow{\scriptstyle f} \\
X' & \xhookrightarrow{g} & X .
\end{array}
$$

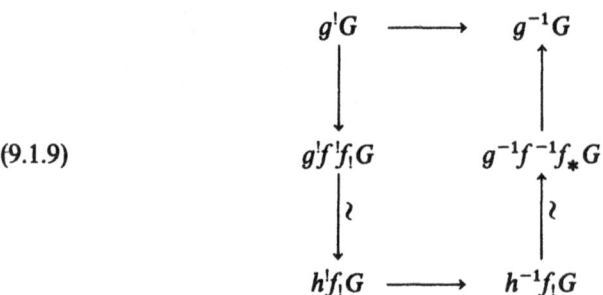

Then for $G \in \mathrm{Ob}(\mathbf{D}^b(Y))$, *the diagram* (9.1.12) *below commutes.*

$$(9.1.12)\qquad \begin{array}{ccc} f'_*g''^!G & \longrightarrow & f'_*g'^{-1}G \\ \wr\big| & & \big\uparrow \\ g^!f_*G & \longrightarrow & g^{-1}f_*G \end{array}.$$

Proof of Lemma 9.1.5. It is enough to prove the commutativity of the diagram obtained by applying the faithful functor $g_! \simeq g_*$. The resulting diagram is embedded into the following commutative diagram, which completes the proof.

(9.1.13)

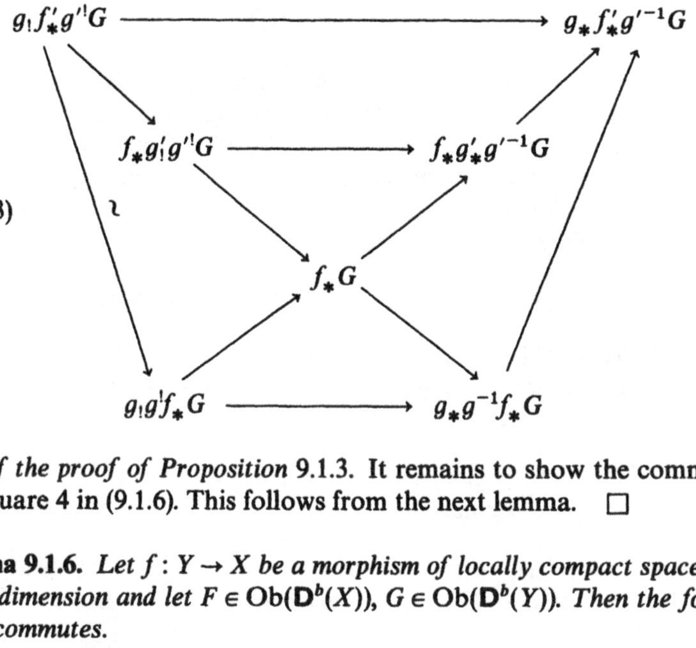

□

End of the proof of Proposition 9.1.3. It remains to show the commutativity of the square 4 in (9.1.6). This follows from the next lemma. □

Lemma 9.1.6. *Let* $f: Y \to X$ *be a morphism of locally compact spaces with finite c-soft dimension and let* $F \in \mathrm{Ob}(\mathbf{D}^b(X))$, $G \in \mathrm{Ob}(\mathbf{D}^b(Y))$. *Then the following diagram commutes.*

$$\begin{array}{ccccc} f_!(G \otimes \mathcal{H}om(G, f^!F)) & \longleftarrow & f_!G \otimes f_*\mathcal{H}om(G, f^!F) & \longrightarrow & f_!G \otimes \mathcal{H}om(f_!G, F) \\ \big\downarrow & & & & \big\downarrow \\ f_!f^!F & & \longrightarrow & & F \end{array}$$

The proof is straightforward.
By Proposition 9.1.3 we get:

Theorem 9.1.7. *Let* $f: Y \to X$ *be a morphism of manifolds, let* $G \in \mathrm{Ob}(\mathbf{D}^b_{\mathbb{R}-c}(Y))$, *and assume* f *is proper on* supp(G). *Then* $\mathrm{C}(Rf_*G)$ *is the image of* $\mathrm{C}(G)$ *by the morphism* $H^0_{\mathrm{supp}(G)}(Y; \omega_Y) \to H^0_{f(\mathrm{supp}(G))}(X; \omega_X)$.

In particular if $F \in \mathbf{D}^b_{\mathbf{R}-c}(X)$ has compact support, we find (by considering the morphism $a_X : X \to \{\mathrm{pt}\}$):

$$\chi(X;F) = \textstyle\int_X \mathrm{C}(F) \ , \tag{9.1.14}$$

where $\int_X$ is the map $H^0_c(X;\omega_X) \to k$ defined in III §3. In the subsequent sections we shall generalize these constructions in two directions: on one hand, we shall give a microlocal version of C(F). On the other hand we shall extend the index formula to a trace formula. For that purpose, we need some preparation.

9.2. Subanalytic chains and subanalytic cycles

Singular homology is defined through singular chains. In this book we restrict ourselves to the subanalytic case. Let X be a manifold. (Recall that all manifolds and morphisms of manifolds are real analytic.) For each integer p, let $CS'_p(X)$ be the A-module generated by the symbols $[S]$, where S ranges through the family of subanalytic p-dimensional oriented submanifolds of X, with the relations (9.2.1), (9.2.2), (9.2.3) below:

$$[S_1 \sqcup S_2] = [S_1] + [S_2] \ , \qquad (S_1 \text{ and } S_2 \text{ are disjoint}) \ . \tag{9.2.1}$$

$$\begin{cases} [S] = [S'] \textit{ if } S' \textit{ is an open dense} \\ \textit{subanalytic subset of } S, \textit{ endowed with the induced} \\ \textit{orientation} \ . \end{cases} \tag{9.2.2}$$

$$\begin{cases} [S^a] = -[S], \textit{ where } S^a \textit{ is } S \textit{ endowed with the} \\ \textit{opposite orientation} \ . \end{cases} \tag{9.2.3}$$

We denote by $\mathscr{CS}^X_p$ (or $\mathscr{CS}_p$, if there is no risk of confusion) the sheaf on X associated to the presheaf $U \mapsto CS'_p(U)$. If F is a sheaf on X, we also set:

$$\mathscr{CS}_p(F) = \mathscr{CS}_p \otimes F \ .$$

Definition 9.2.1. *The sheaf $\mathscr{CS}_p(F)$ is called the sheaf of subanalytic p-chains with values in F.*

In order to describe $\mathscr{CS}_p(F)$, we need some complementary results on the dualizing complex ω_X.

For a subanalytic subset S of X denote by j_S the embedding $S \hookrightarrow X$. If S has dimension $\leqslant p$, $S_{\mathrm{reg},p}$ will denote the union of p-dimensional connected components of S_{reg}. If S is locally closed, we set:

$$\partial S = \bar{S} \backslash S \ . \tag{9.2.4}$$

This is a closed set.

Proposition 9.2.2. *Let S be a closed subanalytic subset of dimension* $\leqslant p$.

(i) *For any sheaf F on S one has:*

$$H_c^j(S;F) = H^j(S;F) = 0 \qquad \text{for} \qquad j > p \ .$$

(ii) $H^j(\omega_S) = 0$ *for* $j < -p$.

(iii) $H^{-p}(S;\omega_S) \simeq H^0(S;H^{-p}(\omega_S)) \simeq \mathrm{Hom}(H_c^p(S;A_S), A)$.

Proof. (i) Choose a filtration $S = S_p \supset \cdots \supset S_0$, where S_k is a closed subanalytic subset of S and $S_k \backslash S_{k-1}$ is smooth with dimension k. Then:

$$H^j(S;F_{S_k\backslash S_{k-1}}) = \varinjlim_U H^j(S_k\backslash S_{k-1};F_U)$$

where U ranges through the family of open subsets of $S_k\backslash S_{k-1}$ such that $\bar{U} \cap S_{k-1} = \emptyset$. By Proposition 3.2.2 we find $H^j(S;F_{S_k\backslash S_{k-1}}) = 0$ for $j > k$. Then we argue by induction on p and consider the long exact sequence:

$$\cdots \to H^j(S;F_{S\backslash S_{p-1}}) \to H^j(S;F) \to H^j(S;F_{S_{p-1}}) \to \cdots \ .$$

Since $H^j(S;F_{S_{p-1}}) = H^j(S_{p-1};F_{S_{p-1}})$, all terms vanish for $j > p$, and we obtain $H^j(S;F) = 0$ for $j > p$. The vanishing of $H_c^j(S;F)$ for $j > p$ follows from the equality $H_c^j(S;F) = \varinjlim_U H^j(S;F_U)$, where U ranges through the family of relatively compact open subsets of S.

(ii) and (iii) By (3.1.8) we have:

$$R\Gamma(S;\omega_S) \simeq R\mathrm{Hom}(R\Gamma_c(S;A_S), A) \ .$$

Since $R\Gamma_c(S;A_S)$ is concentrated in degree $\leqslant p$, $R\Gamma(S;\omega_S)$ is concentrated in degree $\geqslant -p$, and $H^{-p}(R\Gamma(S;\omega_S)) \simeq H^{-p}(R\mathrm{Hom}(R\Gamma_c(S;A_S);A))$ $\simeq \mathrm{Hom}(H_c^p(S;A_S), A)$. □

Proposition 9.2.3. *Let S be a closed subanalytic subset of X of dimension* $\leqslant p$, *and let* $S_\circ$ *be a locally closed subanalytic subset of S. Then:*

(i) $j_{S_\circ *}H^{-p}(\omega_{S_\circ}) \simeq H^0_{S_\circ}(j_{S*}H^{-p}(\omega_S))$,

(ii) *if* $S_\circ$ *is closed and* $S\backslash S_\circ$ *has dimension* $<p$, *then* $j_{S_\circ *}H^{-p}(\omega_{S_\circ}) \simeq j_{S*}H^{-p}(\omega_S)$,

(iii) *if* $S_\circ$ *is open and dense in* $S_{\mathrm{reg},p}$, *then there is an exact sequence:*

$$0 \to j_{S*}H^{-p}(\omega_S) \to j_{S_\circ *}or_{S_\circ} \to j_{S\backslash S_\circ *}H^{-p+1}(\omega_{S\backslash S_\circ}) \ .$$

Proof. (i) We have $Rj_{S_\circ *}\omega_{S_\circ} \simeq R\Gamma_{S_\circ}(\omega_S)$. Since $\omega_{S_\circ}$ and ω_S are concentrated in degree $\geqslant -p$, we get (i).

(ii) follows from the fact that $\omega_S|_{S\backslash S_\circ}$ is concentrated in degree $> -p$.

(iii) Consider the distinguished triangle:

$$R\Gamma_{S\backslash S_\circ}(\omega_S) \longrightarrow \omega_S \longrightarrow Rj_{S_\circ *}j_{S_\circ}^{-1}\omega_S \xrightarrow[+1]{} \ .$$

It gives rise to the long sequence (after noting that $\omega_{S\setminus S_0} \simeq R\Gamma_{S\setminus S_0}(\omega_S)$ and $\omega_S|_{S_0} \simeq or_{S_0}[-p]$):

$$H^{-p}(\omega_{S\setminus S_0}) \to H^{-p}(\omega_S) \to j_{S_0*} or_{S_0} \to H^{-p+1}(\omega_{S\setminus S_0}) .$$

Since $S\setminus S_0$ has dimension $<p$, the first term vanishes. □

We shall now return to the study of the sheaves $\mathscr{CS}_p$. For a non-negative integer p, let $LCLS_p(X)$ denote the set of locally closed subanalytic subsets of X of dimension $\leqslant p$. We introduce a partial order $<$ on $LCLS_p(X)$ as follows. For S_1 and S_2 in $LCLS_p(X)$, $S_1 < S_2$ if and only if there exists a subanalytic subset S of $S_1 \cap S_2$ such that:

$$\text{(9.2.5)} \qquad \begin{cases} S \textit{ is open in } S_1 \textit{ and closed in } S_2 \textit{ and } S_1\setminus S \\ \textit{has dimension } <p . \end{cases}$$

Note that $LCLS_p(X)$ is a directed ordered set, i.e. for S_1 and S_2 in $LCLS_p(X)$, there always exists S with $S_1 < S$ and $S_2 < S$. In fact, choose $S = (S_1\setminus\partial S_2) \cup (S_2\setminus\partial S_1)$. Note that $S < S_{\mathrm{reg},p}$ for any $S \in LCLS_p(X)$. If $S_1 < S_2$, we have a canonical morphism:

$$\text{(9.2.6)} \qquad j_{S_1*}H^{-p}(\omega_{S_1}) \to j_{S_2*}H^{-p}(\omega_{S_2}) .$$

This morphism is obtained as the composition:

$$j_{S_1*}H^{-p}(\omega_{S_1}) \to j_{S*}H^{-p}(\omega_S) \to j_{S_2*}H^{-p}(\omega_{S_2}) ,$$

where S satisfies (9.2.5). This does not depend on the choice of S. In fact if S' is another set satisfying (9.2.5), then $S'' = S \cap S'$ still satisfies (9.2.5) and the morphism $j_{S*}H^{-p}(\omega_S) \to j_{S''*}H^{-p}(\omega_{S''})$ is well-defined, and is an isomorphism since S'' is closed and open in S and $S\setminus S''$ has dimension $<p$.

Lemma 9.2.4. *We have the isomorphims:*

$$\mathscr{CS}_p^X \simeq \varinjlim_{S \in LCLS_p(X)} j_{S*}H^{-p}(\omega_S) \simeq \varinjlim_{S'} j_{S'*} or_{S'} ,$$

where S' in the last inductive limit ranges through the family of p-dimensional subanalytic submanifolds of X.

Proof. The last isomorphism is obvious, since for $S \in LCLS_p(X)$ one has $S < S_{\mathrm{reg},p}$.

Let U be an open subset of X and let V be an open subanalytic subset of X with $\bar{V} \subset U$. We define:

$$\alpha : CS_p'(U) \to \varinjlim_S \Gamma(V; j_{S*} or_S) \to \Gamma\left(V; \varinjlim_S j_{S*} or_S\right)$$

as follows.

If S is an oriented p-dimensional subanalytic submanifold, the orientation of S defines a section s of $or_{S\cap V}$ over V. Then α is given by $[S] \mapsto s \in \Gamma(V; j_{S*}or_S)$. Taking the inductive limit with respect to U and V, we obtain the morphism $\mathscr{CS}_p^X \to \varinjlim_S j_{S*}or_S$. This is clearly an isomorphism. $\square$

Let $CLS_p(X)$ denote the set of closed subanalytic subsets of X of dimension $\leqslant p$. We order $CLS_p(X)$ by the inclusion relation. If S_1 and S_2 belong to $CLS_p(X)$ and $S_1 \subset S_2$, we have an injective morphism, by Proposition 9.2.3.

$$j_{S_1*}H^{-p}(\omega_{S_1}) \to j_{S_2*}H^{-p}(\omega_{S_2}) \ . \tag{9.2.7}$$

Definition 9.2.5. *One sets:*

$$\mathscr{ZS}_p^X = \varinjlim_{S \in CLS_p(X)} j_{S*}H^{-p}(\omega_S) \ ,$$

and one calls $\mathscr{ZS}_p^X$ the sheaf of subanalytic p-cycles on X (over the base ring A).

If there is no risk of confusion, we write $\mathscr{ZS}_p$ instead of $\mathscr{ZS}_p^X$.

For an A_X-module F on X, one sets $\mathscr{ZS}_p(F) = \mathscr{ZS}_p \otimes F$, and calls it the sheaf of subanalytic p-cycles with values in F. Since $CLS_p(X)$ is contained in $LCLS_p(X)$, we have an exact sequence:

$$0 \to \mathscr{ZS}_p \to \mathscr{CS}_p. \tag{9.2.8}$$

Let $S \in LCLS_p(X)$. We have a distinguished triangle:

$$\omega_{\partial S} \longrightarrow \omega_{\overline{S}} \longrightarrow Rj_{S*}\omega_S \xrightarrow[+1]{} \ ,$$

which gives rise to the long exact sequence:

$$0 \to H^{-p}(\omega_{\overline{S}}) \to j_{S*}H^{-p}(\omega_S) \to H^{1-p}(\omega_{\partial S}) \ .$$

Passing through the inductive limit, we obtain an exact sequence:

$$0 \to \mathscr{ZS}_p \to \mathscr{CS}_p \to \mathscr{ZS}_{p-1} \ , \tag{9.2.9}$$

and the first arrow coincides with the morphism in (9.2.8). Then we define the **boundary operator**:

$$\partial_p : \mathscr{CS}_p \to \mathscr{CS}_{p-1} \tag{9.2.10}$$

as the composite $\mathscr{CS}_p \to \mathscr{ZS}_{p-1} \to \mathscr{CS}_{p-1}$. If there is no risk of confusion, we write ∂ instead of ∂_p.

Proposition 9.2.6. (i) *$\mathscr{CS}_\cdot$ is a complex of sheaves (cf. Notations* 1.3.9).

(ii) $\mathscr{ZS}_p \simeq \operatorname{Ker} \partial_p$.

(iii) *For any sheaf F on X, $\mathscr{CS}_p(F)$ is a soft sheaf.*

(iv) *There is a canonical morphism $\omega_X \to \mathscr{CS}_\cdot$ in $\mathbf{D}^b(X)$.*

(v) *If Ω is a open subset of X then $\mathscr{CS}_p^X|_\Omega \simeq \mathscr{CS}_p^\Omega$* and $\mathscr{ZS}_p^X|_\Omega \simeq \mathscr{ZS}_p^\Omega$.

Proof. (i) and (ii) are clear.

(iii) For an open subanalytic subset W of X, let ρ_W denote the endomorphism of the sheaf $\mathscr{C}\mathscr{S}_p$ defined on $j_{S*}H^{-p}(\omega_S)$ by the morphism:

$$j_{S*}H^{-p}(\omega_S) \to j_{S\cap W*}H^{-p}(\omega_{S\cap W}) \ .$$

Then $\rho_W|_W = \mathrm{id}$, $\rho_W^2 = \rho_W$ and $\rho_W = 0$ on $X\setminus\overline{W}$. Now if Z is a closed subset of X and $s \in \Gamma(Z; \mathscr{C}\mathscr{S}_p(F))$, there exists an open neighborhood U of Z and a section $\tilde{s} \in \Gamma(U; \mathscr{C}\mathscr{S}_p(F))$ whose restriction to Z is s. Let W be a subanalytic open subset of X with $Z \subset W \subset \overline{W} \subset U$. Then $\rho_W(\tilde{s}) \in \Gamma_{\overline{W}}(X; \mathscr{C}\mathscr{S}_p(F))$ is a global extension of s.

(v) is proved by a similar argument.

(iv) Since $H^{-n}(\mathscr{C}\mathscr{S}_\cdot) = \mathrm{Ker}\,\partial_n = \mathscr{Z}\mathscr{S}_n = H^{-n}(\omega_X)$, we obtain the morphism $H^{-n}(\omega_X)[n] \to \mathscr{C}\mathscr{S}_\cdot$. □

Proposition 9.2.7. *Let L be a locally free sheaf of finite rank on X.*

(i) *For any $\alpha \in \Gamma(X; \mathscr{C}\mathscr{S}_p(L))$, $\mathrm{supp}(\alpha)$ is a closed subanalytic subset of pure dimension p.*

(ii) *Let $S \in CLS_p(X)$. Then:*

$$\Gamma_S(X; \mathscr{Z}\mathscr{S}_p(L)) \simeq H^{-p}(S; \omega_S \otimes L) \ .$$

(iii) *Let $S \in LCLS_p(X)$. Then:*

$$\Gamma(X; j_{S*}H^{-p}(\omega_S \otimes L)) \simeq H^{-p}(S; \omega_S \otimes L)$$
$$\simeq \{\alpha \in \Gamma_{\overline{S}}(X; \mathscr{C}\mathscr{S}_p(L)); \mathrm{supp}(\partial\alpha) \subset \partial S\} \ .$$

Proof. (i) Locally, the section α belongs to $j_{S*}or_S$ for some p-dimensional subanalytic submanifold S.

(ii) Let $S' \in CLS_p(X)$ with $S \subset S'$. Then $\Gamma_S(j_{S'*}H^{-p}(\omega_{S'})) \simeq j_{S*}H^{-p}(\omega_S)$. This implies (ii).

(iii) If $\alpha \in H^{-p}(S; \omega_S \otimes L)$, then clearly $\mathrm{supp}(\alpha) \subset \overline{S}$ and $\mathrm{supp}(\partial\alpha) \subset \partial S$. We have to prove the converse. The problem being local, we may assume $L = A_X$ and $\alpha \in \Gamma(S'; H^{-p}(\omega_{S'}))$ for some $S' \in LCLS_p(X)$. Then $\Gamma_{\overline{S}}(j_{S'*}H^{-p}(\omega_{S'})) \simeq j_{\overline{S}\cap S'*}H^{-p}(\omega_{\overline{S}\cap S'})$. Hence we may assume $S' \subset \overline{S}$. Replacing S' by $(S \cap S')_{\mathrm{reg},p} \cup (S\setminus\overline{S'})_{\mathrm{reg},p}$, we may even assume S' is open in S and $\dim(S\setminus S') < p$. The distinguished triangle on S: $\omega_{S\setminus S'} \to \omega_S \to \omega_{S'} \xrightarrow[+1]{}$ yields the exact sequence:

$$0 \to H^{-p}(\omega_S) \to j_{S'*}H^{-p}(\omega_{S'})|_S \to j_{S\setminus S'*}H^{-p+1}(\omega_{S\setminus S'})|_S \ .$$

Since $\partial\alpha|_S = 0$, α belongs to $H^{-p}(\omega_S)$. □

Corollary 9.2.8. *Let L be a locally free sheaf of finite rank on X. Then:*

$$\Gamma(X; \mathscr{C}\mathscr{S}_p(L)) \simeq \varinjlim_{S \in LCLS_p(X)} \Gamma(S; H^{-p}(\omega_S) \otimes L) \ ,$$

$$\Gamma(X;\mathscr{Z}\mathscr{S}_p(L)) \simeq \varinjlim_{S\in CLS_p(X)} \Gamma(S;H^{-p}(\omega_S)\otimes L) .$$

In order to prove Theorem 9.2.10 below, let us study some functorial operations on subanalytic chains.

Let $f: Y \to X$ be a morphism of manifolds and let $\mathscr{C}\mathscr{S}_\cdot^Y$ and $\mathscr{C}\mathscr{S}_\cdot^X$ denote the complexes of subanalytic chains on Y and X respectively. We have:

$$f_!\mathscr{C}\mathscr{S}_p^Y = \varinjlim_{S} j_{S*}H^{-p}(\omega_S) , \tag{9.2.11}$$

where S belongs to $LCLS_p(Y)$ and f is proper on $\bar{S}$. For such an S, the set $S' = f(\bar{S})\backslash f(\partial S)$ belongs to $LCLS_p(X)$ and the map $S \cap f^{-1}(S') \to S'$ is proper. Hence we obtain:

$$f_! j_{S*}H^{-p}(\omega_S) \to j_{S'*}H^{-p}(\omega_{S'}) \to \mathscr{C}\mathscr{S}_p^X ,$$

where the first arrow is obtained by taking the p-th cohomology of the morphism $Rf_!\omega_{\bar{S}} \to \omega_{f(\bar{S})}$. Passing through the inductive limit, we get the morphism $f_!\mathscr{C}\mathscr{S}_p^Y \to \mathscr{C}\mathscr{S}_p^X$. Since this morphism commutes with the differential, we get the morphism of complexes:

$$f_!\mathscr{C}\mathscr{S}_\cdot^Y \to \mathscr{C}\mathscr{S}_\cdot^X . \tag{9.2.12}$$

If F is a sheaf on X, this defines:

$$f_!\mathscr{C}\mathscr{S}_\cdot^Y(f^{-1}F) \to \mathscr{C}\mathscr{S}_\cdot^X(F) . \tag{9.2.13}$$

If $\alpha \in \Gamma(Y;\mathscr{C}\mathscr{S}_p(f^{-1}F))$ and f is proper on supp(α), we denote by $f_*(\alpha)$ the image of α by the morphism (9.2.13) and call it the direct image of α.

The external product $\alpha \boxtimes \beta$ of two cycles, $\alpha \in \Gamma(X;\mathscr{C}\mathscr{S}_p^X(F))$ and $\beta \in \Gamma(Y;\mathscr{C}\mathscr{S}_q^Y(G))$ is defined by the morphism:

$$j_{S_1*}H^{-p}(\omega_{S_1}) \boxtimes j_{S_2*}H^{-q}(\omega_{S_2}) \to j_{S_1\times S_2*}H^{-p-q}(\omega_{S_1\times S_2})$$

where $S_1 \in LCLS_p(X)$ and $S_2 \in LCLS_q(Y)$. Since these morphisms commute with the differentials we get a morphism:

$$\mathscr{C}\mathscr{S}_\cdot^X(F) \boxtimes \mathscr{C}\mathscr{S}_\cdot^Y(G) \to \mathscr{C}\mathscr{S}_\cdot^{X\times Y}(F\boxtimes G) . \tag{9.2.14}$$

One can also define a notion of homotopy for subanalytic cycles.

Definition 9.2.9. *Let γ_0 and γ_1 belong to $\Gamma(X;\mathscr{Z}\mathscr{S}_p^X)$. One says γ_0 and γ_1 are homotopic if there exist $\tau \in \Gamma_{X\times[0,1]}(X\times\mathbb{R};\mathscr{C}\mathscr{S}_{p+1}^{X\times\mathbb{R}})$ such that:*

$$\partial\tau = i_{0*}\gamma_0 - i_{1*}\gamma_1 , \tag{9.2.15}$$

where i_t is the inclusion $X \hookrightarrow X\times\mathbb{R}$, $x \mapsto (x,t)$.

Let q denote the projection $X \times \mathbb{R} \to X$ and let S denote the projection of $\mathrm{supp}(\tau)$ by q. Applying q_* to (9.2.15) we get $\gamma_0 - \gamma_1 = \partial q_* \tau$. Hence γ_0 and γ_1 will have the same image in $H_p(\Gamma_S(X; \mathscr{CS}_{\cdot}^X))$. Now we can prove:

Theorem 9.2.10. *The canonical morphism (cf. Proposition 9.2.6), $\omega_X \to \mathscr{CS}_{\cdot}^X$ is an isomorphism in $\mathbf{D}^b(X)$.*

Proof. Let $n = \dim X$. Since $H^{-n}(\mathscr{CS}_{\cdot}^X) = \mathscr{ZS}_n = H^{-n}(\omega_X)$, it is enough to show that $H^{-p}(\mathscr{CS}_{\cdot}^X) = 0$ for $0 \leqslant p < n$. Let $x \in X$ and let $\alpha \in (\mathscr{ZS}_p)_x$, with $0 \leqslant p < n$. Then, shrinking X if necessary, there is a closed subanalytic subset S of dimension $\leqslant p$ such that α is the image of a section (that we still denote by α) of $H^{-p}(S; \omega_S)$. We may find an $(n-1)$-dimensional manifold Y such that $X \simeq \mathbb{R} \times Y$ in a neighborhood of x, and f denoting the projection $X \to Y$, f is proper on S. (This point is left as an exercise.) Let $\beta \in H^0_{[0,\infty[}(\mathbb{R}; \mathscr{CS}_1^{\mathbb{R}})$ be the canonical element whose boundary $\partial\beta$ is $[\{0\}]$ and set $\gamma = \beta \boxtimes \alpha$. Then $\gamma \in \Gamma(\mathbb{R} \times X; \mathscr{CS}_{p+1})$ and $\partial\gamma = \partial\beta \boxtimes \alpha = i_*\alpha$, where i is the map $X \hookrightarrow \mathbb{R} \times X$, $x \mapsto (0, x)$. Let φ denote the map $\mathbb{R} \times X \to X$, $(t, (s, y)) \mapsto (t + s, y)$. Then $\varphi: \mathbb{R} \times S \to X$ is proper and hence $\varphi|_{\mathrm{supp}(\gamma)}$ is proper. Therefore $\varphi_*\gamma \in \Gamma(X; \mathscr{CS}_{p+1})$, and $\partial\varphi_*\gamma = \varphi_*\partial\gamma = \varphi_* i_*\alpha = \alpha$. This proves that $H^{-p}(\mathscr{CS}_{\cdot}^X) = 0$ for $0 \leqslant p < n$. □

Corollary 9.2.11. (i) *The stalks of $\mathscr{CS}_p$ and $\mathscr{ZS}_p$ are flat A-modules, and for any sheaf F on X, $\mathscr{ZS}_p(F)$ is the kernel of the morphism $\mathscr{CS}_p(F) \to \mathscr{CS}_{p-1}(F)$.*

(ii) *There is a natural isomorphism $\omega_X \otimes F \simeq \mathscr{CS}_{\cdot}(F)$ in $\mathbf{D}^b(X)$.*

Proof. (i) The stalk of $\mathscr{CS}_p$ is an inductive limit of free A-modules, hence it is flat.

Consider the exact sequence $0 \to \mathscr{ZS}_p \to \mathscr{CS}_p \to \mathscr{ZS}_{p-1} \to 0$, for $p \leqslant \dim X$. Then if $\mathscr{ZS}_{p-1}$ is flat, it follows that $\mathscr{ZS}_p$ is flat and we get an exact sequence $0 \to \mathscr{ZS}_p(F) \to \mathscr{CS}_p(F) \to \mathscr{ZS}_{p-1}(F) \to 0$. Therefore we get (i) by induction on p.

(ii) follows from (i) by Theorem 9.2.10. □

Finally, let us define the intersection of two cycles. Let S_j be a closed subanalytic subset of dimension p_j $(j = 1, 2)$ of the n-dimensional manifold X. Let L_1 and L_2 be two locally free sheaves of finite rank. Consider the chain of morphisms:

$$
\begin{aligned}
(9.2.16) \qquad & H_{S_1}^{-p_1}(X; \omega_X \otimes L_1) \otimes H_{S_2}^{-p_2}(X; \omega_X \otimes L_2) \\
& \simeq H_{S_1}^{n-p_1}(X; or_X \otimes L_1) \otimes H_{S_2}^{n-p_2}(X; or_X \otimes L_2) \\
& \to H_{S_1 \cap S_2}^{2n-p_1-p_2}(X; or_X \otimes L_1 \otimes or_X \otimes L_2) \\
& \simeq H_{S_1 \cap S_2}^{n-p_1-p_2}(X; \omega_X \otimes L_1 \otimes L_2 \otimes or_X) \,.
\end{aligned}
$$

Definition 9.2.12. *Let $C_j \in H_{S_j}^{-p_j}(X; \omega_X \otimes L_j)$ $(j = 1, 2)$. The image in $H_{S_1 \cap S_2}^{n-p_1-p_2}(X; \omega_X \otimes L_1 \otimes L_2 \otimes or_X)$ of $C_1 \otimes C_2$ by the morphisms (9.2.16) is called the intersection of C_1 and C_2 and denoted $C_1 \cap C_2$. If $S_1 \cap S_2$ is compact,*

$n = p_1 + p_2$ and $L_1 \otimes L_2 \simeq or_X$, the scalar $\int_X C_1 \cap C_2$ is called the intersection number of C_1 and C_2 and denoted $\#(C_1 \cap C_2)$.

Recall that $\int_X$ is the morphism $H_c^0(X;\omega_X) \to A$.

Remark 9.2.13. Let $C_j \in H_{S_j}^{-p_j}(X;\omega_X)$, $j = 1, 2$. Then:

$$C_1 \cap C_2 = (-1)^{(n-p_1)(n-p_2)} C_2 \cap C_1 \tag{9.2.17}$$

in $H_{S_1 \cap S_2}^{n-p_1-p_2}(X;\omega_X \otimes or_X)$. This follows from the fact that if K_1 and K_2 are two complexes, the diagram:

$$\begin{array}{ccc} H^{p_1}(K_1) \otimes H^{p_2}(K_2) & \longrightarrow & H^{p_1+p_2}(K_1 \otimes K_2) \\ \downarrow & & \downarrow \\ H^{p_2}(K_2) \otimes H^{p_1}(K_1) & \longrightarrow & H^{p_1+p_2}(K_2 \otimes K_1) \end{array}$$

commutes or anticommutes, according that $p_1 p_2$ is even or odd (cf. Remark 1.10.16).

If $S_1 \cap S_2$ has dimension $\leqslant p_1 + p_2 - n$, the morphism (9.2.16) defines (cf. Proposition 9.2.6):

(9.2.18)

$$\Gamma_{S_1}(X;\mathscr{LS}_{p_1}(L_1)) \otimes \Gamma_{S_2}(X;\mathscr{LS}_{p_2}(L_2)) \to \Gamma_{S_1 \cap S_2}(X;\mathscr{LS}_{p_1+p_2-n}(L_1 \otimes L_2 \otimes or_X)) .$$

We still call this morphism the intersection morphism, and we still denote it by $\cap$.

Remark 9.2.14. Applying (9.2.12) to $a_X : X \to \{pt\}$, we obtain $a_{X!}\mathscr{CS}^X_\cdot \to A$, and then this gives a morphism $\mathscr{CS}^X_\cdot \to \omega_X$ in $\mathbf{D}^b(X)$. This coincides with the isomorphism constructed in Theorem 9.2.10 (up to sign).

9.3. Lagrangian cycles

In T^*X, cycles supported by subanalytic isotropic subsets are of special interest. Let $n = \dim X$.

Definition 9.3.1. *We set:*

$$\mathscr{L}_X = \varinjlim_{\Lambda} H_\Lambda^0(\pi^{-1}\omega_X)$$

where Λ ranges through the family of all closed conic subanalytic isotropic subsets of T^*X. We call $\mathscr{L}_X$ the sheaf of Lagrangian cycles on T^*X.

Since $\pi^{-1}\omega_X \simeq \omega_{T^*X} \otimes or_{T^*X/X}[-n]$, a Lagrangian cycle is an n-cycle with values in $or_{T^*X/X}$. As a particular case of the results of §2, we get:

(a) let Λ be a conic locally closed subanalytic isotropic subset. Then:

$$(9.3.1) \qquad H^0_\Lambda(\mathscr{L}_X) \simeq H^0_\Lambda(\pi^{-1}\omega_X)$$
$$\simeq j_{\Lambda *}H^{-n}(\omega_\Lambda \otimes or_{T^*X/X}) \simeq H^0_\Lambda(\mathscr{L}\mathscr{S}_n(or_{T^*X/X})) .$$

In particular if Λ is smooth of dimension n:

$$H^0_\Lambda(\mathscr{L}_X)|_\Lambda \simeq or_\Lambda \otimes or_{T^*X/X} .$$

(b) Let s be a section of $\mathscr{L}_X$ on an open subset U of T^*X. Then supp(s) is a closed conic subanalytic isotropic subset of pure dimension n. We do not know if supp(s) is involutive in the sense of Definition 6.5.1.

Let us study some functorial properties of Lagrangian cycles. Let Y be another manifold and let $m = \dim Y$. Let Λ_X (resp. Λ_Y) be a closed conic subanalytic isotropic subset of T^*X (resp. T^*Y). One defines the external product of two Lagrangian cycles by considering the chain of morphisms:

$$H^0_{\Lambda_X}(\pi^{-1}\omega_X) \boxtimes H^0_{\Lambda_Y}(\pi^{-1}\omega_Y) \to H^0_{\Lambda_X \times \Lambda_Y}(\pi^{-1}\omega_X \boxtimes \pi^{-1}\omega_Y) \simeq H^0_{\Lambda_X \times \Lambda_Y}(\pi^{-1}\omega_{X \times Y}) .$$

Taking the inductive limit with respect to Λ_X and Λ_Y, we get:

$$(9.3.2) \qquad \mathscr{L}_X \boxtimes \mathscr{L}_Y \to \mathscr{L}_{X \times Y}$$

Now let $f: Y \to X$ be a morphism of manifolds. Recall the maps ${}^tf'$: $Y \times_X T^*X \to T^*Y$ and $f_\pi : Y \times_X T^*X \to T^*X$ (cf. (4.3.2)). To avoid confusions, we sometimes denote by π_Y (resp. π_X, resp. π) the projection $T^*Y \to Y$ (resp. $T^*X \to X$, resp. $Y \times_X T^*X \to Y$).

Let Λ_X and Λ_Y be closed conic subanalytic isotropic subsets of T^*X and T^*Y respectively, as precedingly.

Proposition 9.3.2. (i) *Assume f_π is proper on ${}^tf'^{-1}(\Lambda_Y)$. Then there is a natural morphism*:

$$(9.3.3) \qquad H^0_{\Lambda_Y}(T^*Y; \pi_Y^{-1}\omega_Y) \to H^0_{f_\pi {}^tf'^{-1}(\Lambda_Y)}(T^*X; \pi_X^{-1}\omega_X) .$$

(ii) *Assume ${}^tf'$ is proper on $f_\pi^{-1}(\Lambda_X)$. Then there is a natural morphism*:

$$(9.3.4) \qquad H^0_{\Lambda_X}(T^*X; \pi_X^{-1}\omega_X) \to H^0_{{}^tf' f_\pi^{-1}(\Lambda_X)}(T^*Y; \pi_Y^{-1}\omega_Y) .$$

Note first that with the hypothesis (i), $f_\pi {}^tf'^{-1}(\Lambda_Y)$ is a closed conic subanalytic subset of T^*X and this set is isotropic by Proposition 8.3.11. Similarly, with the hypothesis (ii), ${}^tf' f_\pi^{-1}(\Lambda_X)$ is a closed conic subanalytic isotropic subset of T^*Y.

Proof. (i) Letting $\pi : Y \times_X T^*X \to Y$ be the projection, consider the morphisms:

$$H^0_{\Lambda_Y}(T^*Y; \pi_Y^{-1}\omega_Y) \to H^0_{{}^tf'^{-1}(\Lambda_Y)}\left(Y \underset{X}{\times} T^*X; \pi^{-1}\omega_Y\right)$$
$$\to H^0_{f_\pi {}^tf'^{-1}(\Lambda_Y)}(T^*X; Rf_{\pi !}\pi^{-1}\omega_Y)$$

and use the morphism:

$$Rf_{\pi !}\pi^{-1}\omega_Y \simeq \pi_X^{-1}Rf_!\omega_Y \to \pi_X^{-1}\omega_X \ .$$

(ii) Consider the morphisms:

$$H^0_{\Delta_X}(T^*X;\pi_X^{-1}\omega_X) \to H^0_{f_\pi^{-1}(\Delta_X)}\left(Y \underset{X}{\times} T^*X;\pi^{-1}f^{-1}\omega_X\right)$$

$$\to H^0_{{}^tf' f_\pi^{-1}(\Delta_X)}(T^*Y; R{}^tf'_!\pi^{-1}f^{-1}\omega_X) \ .$$

To conclude it remains to construct the morphism:

$$R{}^tf'_!\pi^{-1}f^{-1}\omega_X \to \pi_Y^{-1}\omega_Y \ . \tag{9.3.5}$$

We proceed as follows. Consider the diagram of maps:

$$\begin{array}{ccc} Y \times Y & \xrightarrow{f_1} & Y \times X \\ \uparrow \delta_Y & & \uparrow \delta \\ \Delta_Y & \xrightarrow{\sim} & \Delta_f \end{array}$$

and note that ${}^tf'_1 : T^*_{\Delta_f}(Y \times X) \to T^*_{\Delta_Y}(Y \times Y)$ is isomorphic to ${}^tf'$. Applying Proposition 4.3.5 to the sheaf $\delta_* f^{-1}\omega_X$, we get:

$${}^tf'_!\mu_{\Delta_f}(\delta_* f^{-1}\omega_X) \to \mu_{\Delta_Y}(\omega_{Y/X} \otimes f_1^{-1}\delta_* f^{-1}\omega_X) \to \mu_{\Delta_Y}(\delta_{Y*}\omega_Y)$$

and (9.3.5) follows. □

Note that the morphism:

$$\pi^{-1}f^{-1}\omega_X \to {}^tf'^{!}\pi_Y^{-1}\omega_Y$$

deduced from (9.3.5) is an isomorphism.

Definition 9.3.3. *In the situation of Proposition* 9.3.2, *one denotes by* f_* *the morphism in* (9.3.3) *and by* f^* *the morphism in* (9.3.4).

Examples 9.3.4. (i) If X is the 0-dimensional manifold $\{\mathrm{pt}\}$, then $\mathscr{L}_X = A$. One denotes by $[\mathrm{pt}]$ the cycle corresponding to $1 \in A$.

(ii) Let a_X be the morphism $X \to \{\mathrm{pt}\}$. One sets:

$$[T_X^*X] = \mathrm{a}_X^*[\mathrm{pt}] \ . \tag{9.3.6}$$

(iii) Let $f : Y \hookrightarrow X$ be a closed embedding. One sets:

$$[T_Y^*X] = f_*[T_Y^*Y] \ . \tag{9.3.7}$$

(iv) Let $f: Y \to X$ be a morphism of manifolds and assume f is transversal to a closed submanifold Z of X. Then:

$$f^*[T_Z^*X] = [T_{f^{-1}(Z)}^*Y] \ . \tag{9.3.8}$$

Let us define a notion of intersection for Lagrangian cycles.

Let $\sigma : X \to T^*X$ be a continuous section of π (i.e.: $\pi \circ \sigma = \mathrm{id}_X$). We have an isomorphism:

$$\begin{aligned} H^0_{\sigma(X)}(T^*X;\pi^! A_X) &\simeq H^0(X;\sigma^! \pi^! A_X) \\ &= H^0(X;A_X) \ . \end{aligned} \tag{9.3.9}$$

Note that $\pi^! A_X \simeq \omega_{T^*X} \otimes \pi^{-1} or_X[-n]$, where $n = \dim X$. If one chooses an orientation on T^*X, then $\pi^! A_X \simeq \pi^{-1}\omega_X$.

Definition 9.3.5. (i) *The image in* $H^0_{\sigma(X)}(T^*X;\pi^! A_X)$ *of* $1 \in H^0(X;A_X)$ *by the isomorphism* (9.3.9) *is denoted by* $[\sigma]$.

(ii) *One denotes by* $[\sigma_0]$ *the cycle associated to the zero-section.*

Let S_1 and S_2 be two closed subsets of T^*X. Consider the chain of morphisms:

$$\begin{aligned} & H^0_{S_1}(T^*X;\pi^! A_X) \otimes H^0_{S_2}(T^*X;\pi^{-1}\omega_X) \\ & \to H^0_{S_1 \cap S_2}(T^*X;\pi^! A_X \otimes \pi^{-1}\omega_X) \\ & \to H^0_{S_1 \cap S_2}(T^*X;\pi^! \omega_X) \\ & \simeq H^0_{S_1 \cap S_2}(T^*X;\omega_{T^*X}) \ . \end{aligned} \tag{9.3.10}$$

As in Definition 9.2.12, if $\gamma \in H^0_{S_1}(T^*X;\pi^! A_X)$ and $\delta \in H^0_{S_2}(T^*X;\pi^{-1}\omega_X)$, we denote by $\gamma \cap \delta$ the image of $\gamma \otimes \delta$ by the chain of morphisms in (9.3.10) and we call it the intersection of γ and δ. If $S_1 \cap S_2$ is compact, one denotes by $\#(\gamma \cap \delta)$ the scalar $\int_{T^*X} \gamma \cap \delta$. If p is an isolated point of $S_1 \cap S_2$, one denotes by $\#(\gamma \cap \delta)_p$ the scalar $\int_U \gamma \cap \delta$, where U is a sufficiently small neighborhood of p.

Let S be a closed subset of T^*X such that π is proper on S. The morphism $R\pi_! \omega_{T^*X} \to \omega_X$ defines the morphism:

$$\alpha : H^0_S(T^*X;\omega_{T^*X}) \to H^0_{\pi(S)}(X;\omega_X) \ . \tag{9.3.11}$$

On the other hand, let Λ be a closed conic set. The morphism $R\pi_* \pi^{-1}\omega_X \xrightarrow{\sim} \omega_X$ defines the morphism:

$$\beta : H^0_\Lambda(T^*X;\pi^{-1}\omega_X) \to H^0_{\pi(\Lambda)}(X;\omega_X) \ . \tag{9.3.12}$$

Proposition 9.3.6. *Let σ be a continuous section of π and let λ be a Lagrangian cycle supported by Λ. Then:*

$$\alpha([\sigma] \cap \lambda) = \beta(\lambda) \ .$$

Proof. It is enough to consider the commutative diagram (9.3.13) below, where R is omitted. Here the commutativity of the left bottom square follows from $\sigma^! \pi^! \simeq \mathrm{id}$ and that of the right square follows from the fact that $\pi_* \pi^{-1} \to \mathrm{id}$ and $\pi_* \pi^{-1} \to \pi_* \sigma_* \sigma^{-1} \pi^{-1} \simeq \mathrm{id}$ are equal. $\square$

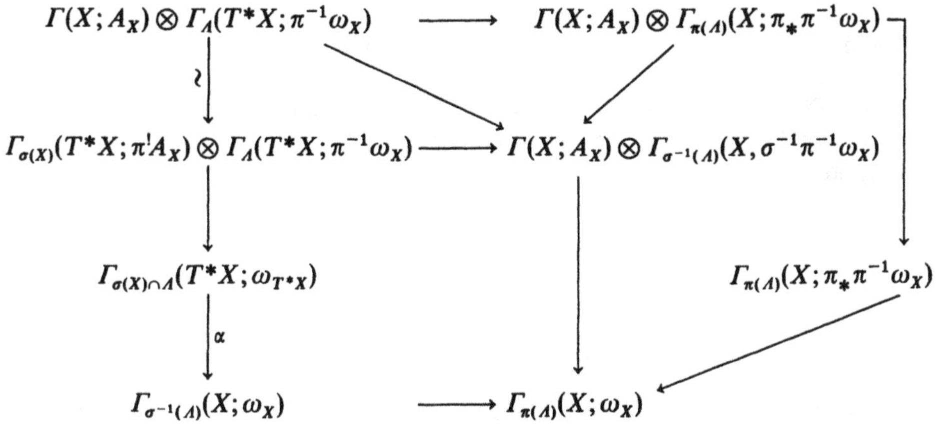

Diagram (9.3.13)

9.4. Characteristic cycles

From now on, and until the end of this chapter, we assume the base ring A is a field of characteristic zero, and we denote it by k. Let $F \in \mathrm{Ob}(\mathbf{D}^b_{\mathbb{R}-c}(X))$. Consider the chain of morphisms:

$$\begin{aligned} R\mathcal{H}om(F,F) &\simeq R\pi_* \mu hom(F,F) \\ &\simeq R\pi_* R\Gamma_{SS(F)} \mu hom(F,F) \\ &\simeq R\pi_* R\Gamma_{SS(F)} \mu_\Delta(F \boxtimes DF) \\ &\to R\pi_* R\Gamma_{SS(F)} \mu_\Delta(\delta_*(F \otimes DF)) \\ &\to R\pi_* R\Gamma_{SS(F)} \mu_\Delta(\delta_* \omega_X) \\ &\simeq R\pi_* R\Gamma_{SS(F)}(\pi^{-1}\omega_X) \ . \end{aligned}$$

Definition 9.4.1. *The image of* $\mathrm{id}_F \in \mathrm{Hom}(F,F)$ *in* $H^0_{SS(F)}(T^*X; \pi^{-1}\omega_X) \subset H^0(T^*X; \mathcal{L}_X)$ *is called the characteristic cycle of* F *and denoted* $\mathrm{CC}(F)$.

Let us compare the operations on $\mathbb{R}$-constructible sheaves and the operations on Lagrangian cycles.

Let Y be another manifold and let $G \in \mathrm{Ob}(\mathrm{D}^b_{\mathbb{R}-c}(Y))$. Using Proposition 4.4.8, one checks easily the formula:

(9.4.1) $$\mathrm{CC}(F \boxtimes G) = \mathrm{CC}(F) \boxtimes \mathrm{CC}(G)\ .$$

Now let $f: Y \to X$ be a morphism of manifolds.

Proposition 9.4.2. *Assume f is proper on* supp(G). *Then*:

(9.4.2) $$\mathrm{CC}(Rf_* G) = f_* \mathrm{CC}(G)\ .$$

Proof. Denote by δ_X(resp. δ_Y) the diagonal embedding $X \hookrightarrow X \times X$ (resp. $Y \hookrightarrow Y \times Y$) and by δ the graph embedding $Y \hookrightarrow X \times Y$, and consider the diagram (cf. (4.4.3)):

(9.4.3) $$\begin{array}{ccccc} Y \times Y & \xrightarrow[f_1]{} & X \times Y & \xrightarrow[f_2]{} & X \times X \\ \uparrow \delta_Y & & \uparrow \delta & & \uparrow \delta_X \\ \Delta_Y & \xrightarrow{\sim} & \Delta_f & \xrightarrow[f]{} & \Delta_X \end{array}\ .$$

Denote by π_Y (resp. π, resp. π_X) the projection $T^*Y \to Y$ (resp. $Y \times_X T^*X \to Y$, resp. $T^*X \to X$), and set for short $S = \mathrm{SS}(G)$, $S' = {}^t f'^{-1}(S)$, $S'' = f_\pi(S')$. There are natural morphisms (cf. IV §3):

$$ {}^t f_1' \circ \mu_{\Delta_Y} \to \mu_{\Delta_f} \circ Rf_{1!}\ , \qquad f_{2\pi!} \circ \mu_{\Delta_f} \to \mu_{\Delta_Y} \circ Rf_{2!}$$

which give rise to the commutative diagram (9.4.4) below, in which we omit to write R for the derived functor, for short.

Therefore the image of $\mathrm{CC}(G) \in \Gamma_S(T^*Y; \pi_Y^{-1}\omega_Y)$ in $\Gamma_{S''}(T^*X; \pi_X^{-1}\omega_X)$ via $\Gamma_{S'}(Y \times_X T^*X; \pi^{-1}\omega_Y)$ coincides with $\mathrm{CC}(Rf_! G)$. □

Proposition 9.4.3. *Assume ${}^t f'$ is proper on $f_\pi^{-1}(\mathrm{SS}(F))$, (that is: f is non-characteristic for F). Then*:

(9.4.5) $$\mathrm{CC}(f^{-1}F) = f^* \mathrm{CC}(F)\ .$$

Proof. We set $S = SS(F)$, $S' = f_\pi^{-1}(S)$, $S'' = {}^t f'(S')$.
Consider the diagram:

$$\begin{array}{ccccc} Y \times Y & \xrightarrow{f_1} & Y \times X & \xrightarrow{f_2} & X \times X \\ \uparrow \delta_Y & & \uparrow \delta & & \uparrow \delta_X \\ \Delta_Y & \xrightarrow{\sim} & \Delta_f & \xrightarrow[f]{} & \Delta_X \end{array}\ .$$

and denote as above, by π_Y, π, π_X the projections $T^*Y \to Y$, $Y \times_X T^*X \to Y$ and $T^*X \to X$, respectively. There are natural morphisms:

$$f_{2\pi}^{-1} \circ \mu_{\Delta_X} \to \mu_{\Delta_f} \circ f_2^{-1}\ , \qquad {}^t f_{1!}' \circ \mu_{\Delta_f} \to \mu_{\Delta_Y} \circ (\omega_{Y/X} \otimes f_1^{-1})\ .$$

$$
\begin{array}{ccccc}
\mathrm{Hom}(G,G) & & \longrightarrow & & \mathrm{Hom}(f_!G,f_!G) \\
\uparrow\wr & & & & \uparrow\wr \\
\Gamma_{\Delta_Y}(Y\times Y; G\boxtimes \mathrm{D}G) & \longrightarrow & \Gamma_{\Delta_f}(X\times Y; f_!G\boxtimes \mathrm{D}G) & \longrightarrow & \Gamma_{\Delta_X}(X\times X; f_!G\boxtimes f_*\mathrm{D}G) \\
\downarrow & & \downarrow & & \downarrow \\
\Gamma_S(T^*Y;\mu_{\Delta_Y}(G\boxtimes \mathrm{D}G)) & \longrightarrow & \Gamma_{S'}(Y\times_X T^*X;\mu_{\Delta_f}(f_!G\boxtimes \mathrm{D}G)) & \longrightarrow & \Gamma_{S''}(T^*X;\mu_{\Delta_X}(f_!G\boxtimes f_*\mathrm{D}G)) \\
\downarrow & & \downarrow & & \downarrow \\
\Gamma_S(T^*Y;\mu_{\Delta_Y}(\delta_{Y*}(G\otimes \mathrm{D}G))) & \longrightarrow & \Gamma_{S'}(Y\times_X T^*X;\mu_{\Delta_f}(\delta_*(f^{-1}f_!G\otimes \mathrm{D}G))) & \longrightarrow & \Gamma_{S''}(T^*X;\mu_{\Delta_X}(\delta_{X*}(f_!G\otimes \mathrm{D}f_!G))) \\
\downarrow & & \downarrow & & \downarrow \\
\Gamma_S(T^*Y;\pi_Y^{-1}\omega_Y) & \longrightarrow & \Gamma_{S'}(Y\times_X T^*X;\pi^{-1}\omega_Y) & \longrightarrow & \Gamma_{S''}(T^*X;\pi_X^{-1}\omega_X)
\end{array}
$$

Diagram (9.4.4)

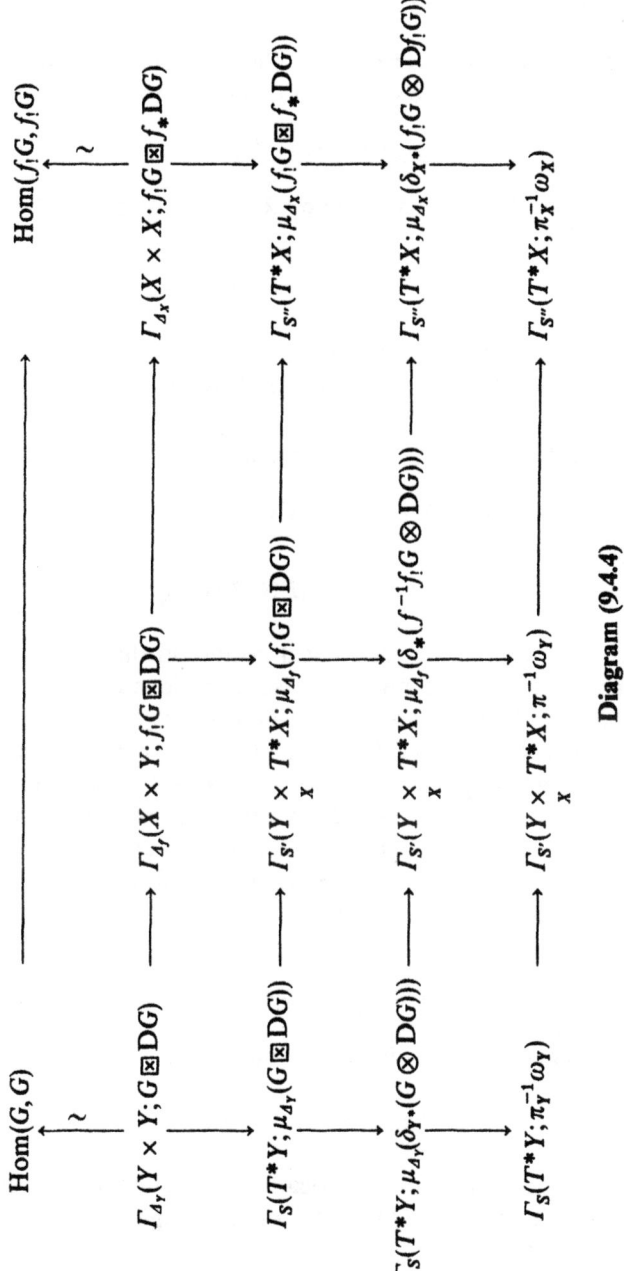

Since ${}^t f'$ is proper on S', we have a commutative diagram (9.4.6) below. Hence $CC(f^{-1}F)$ is the image of $CC(F) \in H^0_S(T^*X;\pi_X^{-1}\omega_X)$ in $H^0_{S''}(T^*Y;\pi_Y^{-1}\omega_Y)$ via $H^0_{S'}(Y \times_X T^*X; f_\pi^{-1}\pi_X^{-1}\omega_X)$. □

Let "a" denote the antipodal map on T^*X. Since "a" sends subanalytic isotropic subsets to subanalytic isotropic subsets, it defines a morphism:

$$a_* : a^{-1}\mathscr{L}_X \to \mathscr{L}_X \ .$$

If λ is a Lagrangian cycle, we denotes by λ^a its image by a_*. Note that for a submanifold Y of X,

$$[T_Y^*X]^a = (-1)^{\dim Y}[T_Y^*X] \ . \tag{9.4.7}$$

In fact $R\Gamma_{T_Y^*X}(\pi^{-1}\omega_X) \simeq \omega_{T_Y^*X/T^*X} \otimes \pi^{-1}\omega_X$, $a^{-1}\omega_{T_Y^*X/T^*X} \to \omega_{T_Y^*X/T^*X}$ and $a^!\omega_{T_Y^*X/T^*X} \to \omega_{T_Y^*X/T^*X}$ are equal up to $(-1)^{\dim Y}$ after identifying $a^!$ and a^{-1}.

Proposition 9.4.4. *Let* $F \in \mathrm{Ob}(\mathbf{D}^b_{\mathbb{R}-c}(X))$. Then:

$$CC(D_X F) = (CC(F))^a \ .$$

Proof. One has $\mu_\Delta(DF \boxtimes DDF) \simeq a^{-1}\mu_\Delta(F \boxtimes DF)$. □

Let us note that $CC(F)$ is a microlocal notion, in the following sense. If Ω is an open subset of T^*X and F and F' are isomorphic in $\mathbf{D}^b(X;\Omega)$, then we have $CC(F)|_\Omega = CC(F')|_\Omega$. This follows from $\mu hom(F,F)|_\Omega \simeq \mu hom(F',F')|_\Omega$. With the use of this remark, we shall calculate $CC(F)$ more explicitly. If X is a point, we have $CC(F) = \chi(F) = \sum_j(-1)^j \dim H^j(F)$. If Y is a closed submanifold and if $H^j(F)$ is a locally constant sheaf of rank m_j for all j, then $CC(F) = \sum_j(-1)^j m_j[T_Y^*X]$. In fact let $i: Y \hookrightarrow X$ denote the embedding of Y, and let $a_Y : Y \to \{pt\}$. Then $F = Ri_* a_Y^{-1}(V)$ for some $V \in \mathrm{Ob}(\mathbf{D}^b(\mathfrak{Mod}^f(k)))$, and the result follows by Propositions 9.4.2 and 9.4.3. In particular, we have for a closed submanifold Y of X:

$$CC(k_Y) = [T_Y^*X] \ . \tag{9.4.8}$$

To treat the general case, we introduce the sheaf:

$$\mathscr{CL}_X = \varinjlim_\Lambda H^0_\Lambda(\pi^{-1}\omega_X) \ ,$$

where Λ ranges through the family of locally closed subanalytic conic isotropic subsets of T^*X. We call $\mathscr{CL}_X$ the sheaf of **Lagrangian chains**. We have the inclusions:

$$\mathscr{L}_X \subset \mathscr{CL}_X \subset \mathscr{CS}_n^{T^*X}(or_{T^*X/X}) \ .$$

$$
\begin{array}{ccccccc}
\mathrm{Hom}(F,F) & \multicolumn{5}{c}{\xrightarrow{\hspace{20em}}} & \mathrm{Hom}(f^{-1}F,f^{-1}F) \\
\uparrow\wr & & & & & & \uparrow\wr \\
\Gamma_{\Delta_X}(X\times X;F\boxtimes \mathrm{D}F) & \longrightarrow & \Gamma_{\Delta_f}(Y\times X;f^{-1}F\boxtimes \mathrm{D}F) & \multicolumn{3}{c}{\xrightarrow{\hspace{10em}}} & \Gamma_{\Delta_Y}(Y\times Y;f^{-1}F\boxtimes \mathrm{D}(f^{-1}F)) \\
\downarrow & & \downarrow & & & & \downarrow \\
\Gamma_S(T^*X;\mu_{\Delta_X}(F\boxtimes \mathrm{D}F)) & \longrightarrow & \Gamma_{S'}(Y\underset{X}{\times}T^*X;\mu_{\Delta_f}(f^{-1}F\boxtimes \mathrm{D}F)) & \longrightarrow & \Gamma_{S''}(T^*Y;\mu_{\Delta_Y}(f^{-1}F\boxtimes\omega_{Y/X}\otimes f^{-1}\mathrm{D}F)) & \longrightarrow & \Gamma_{S''}(T^*Y;\mu_{\Delta_Y}(f^{-1}F\boxtimes f^{!}\mathrm{D}F)) \\
\downarrow & & \downarrow & & \downarrow & & \downarrow \\
\Gamma_S(T^*X;\mu_{\Delta_X}(\delta_{X*}(F\otimes \mathrm{D}F))) & \rightarrow & \Gamma_{S'}(Y\underset{X}{\times}T^*X;\mu_{\Delta_f}(\delta_*(f^{-1}F\otimes f^{-1}\mathrm{D}F))) & \rightarrow & \Gamma_{S''}(T^*Y;\mu_{\Delta_Y}(\delta_{Y*}(f^{-1}F\otimes\omega_{Y/X}\otimes f^{-1}\mathrm{D}F))) & \rightarrow & \Gamma_{S''}(T^*Y;\mu_{\Delta_Y}(\delta_{Y*}(f^{-1}F\otimes f^{!}\mathrm{D}F))) \\
\downarrow & & \downarrow & & \downarrow & & \downarrow \\
\Gamma_S(T^*X;\mu_{\Delta_X}(\delta_{X*}\omega_X)) & \longrightarrow & \Gamma_{S'}(Y\underset{X}{\times}T^*X;\mu_{\Delta_f}(\delta_*f^{-1}\omega_X)) & \longrightarrow & \Gamma_{S''}(T^*Y;\mu_{\Delta_Y}(\delta_{Y*}(\omega_{Y/X}\otimes f^{-1}\omega_X))) & \longrightarrow & \Gamma_{S''}(T^*Y;\mu_{\Delta_Y}(\delta_{Y*}\omega_Y)) \\
\downarrow & & \downarrow & & \downarrow & & \downarrow \\
\Gamma_S(T^*X;\pi_X^{-1}\omega_X) & \longrightarrow & \Gamma_{S'}(Y\underset{X}{\times}T^*X;\pi^{-1}f^{-1}\omega_X) & \longrightarrow & \Gamma_{S''}(T^*Y;\pi_Y^{-1}(\omega_{Y/X}\otimes f^{-1}\omega_X))) & \longrightarrow & \Gamma_{S''}(T^*Y;\pi_Y^{-1}\omega_Y)
\end{array}
$$

Diagram (9.4.6)

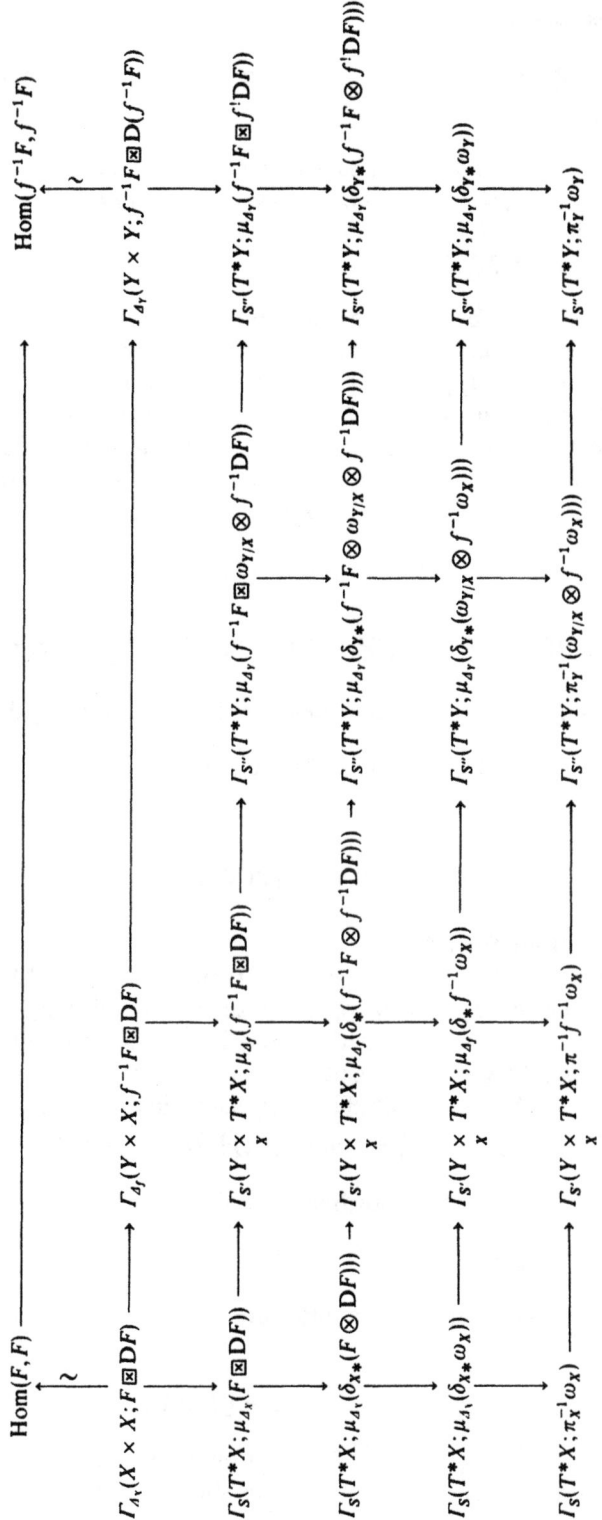

Now let $F \in \mathrm{Ob}(\mathbf{D}^b_{\mathbb{R}-c}(X))$ and let $\Lambda = SS(F)$. There exists an open dense subset $\Lambda_\circ \subset \Lambda$ such that:

$$(9.4.9)\qquad \begin{cases} \Lambda_\circ \textit{ is a subanalytic submanifold and denoting by} \\ \{\Lambda_\alpha\}_\alpha \textit{ the connected components of } \Lambda_\circ, \textit{for each} \\ \alpha \textit{ there is a submanifold } X_\alpha \textit{ of } X \textit{ such that} \\ \Lambda_\alpha \subset T^*_{X_\alpha}X \ . \end{cases}$$

For such a Λ_α, we denote by $[\Lambda_\alpha]$ the chain supported by $\overline{\Lambda_\alpha}$ which is $[T^*_{X_\alpha}X]$ on Λ_α. If U is an open subset such that $\Lambda_\alpha = T^*_{X_\alpha}X \cap U$, we also sometimes write $[T^*_{X_\alpha}X] \cap U$ instead of $[\Lambda_\alpha]$. Let $p \in \Lambda_\alpha$. By Proposition 6.6.1(ii) there exists $V \in \mathrm{Ob}(\mathbf{D}^b\mathfrak{Mod}^f(k)))$ such that $F \simeq V_{X_\alpha}$ in $\mathbf{D}^b(X;p)$. Hence $CC(F) = m_\alpha[\Lambda_\alpha]$ in a neighborhood of p, with $m_\alpha = \chi(V)$. Since Λ_α is connected, we find $CC(F) = m_\alpha[\Lambda_\alpha]$ in a neighborhood of Λ_α, and if Ω is an open dense subset of T^*X such that $\Omega \cap \Lambda = \Lambda_\circ$, we get $CC(F)|_\Omega = \sum_\alpha m_\alpha[\Lambda_\alpha]|_\Omega$. Hence:

$$(9.4.10)\qquad CC(F) = \sum_\alpha m_\alpha[\Lambda_\alpha] \quad \text{in} \quad \mathscr{CL}_X \ .$$

Proposition 9.4.5. (i) *Let $F \in \mathrm{Ob}(\mathbf{D}^b_{\mathbb{R}-c}(X))$. Then $CC(F)$ is a Lagrangian cycle over the ring $\mathbb{Z}$.*

(ii) *Let $F' \longrightarrow F \longrightarrow F'' \xrightarrow[+1]{}$ be a distinguished triangle in $\mathbf{D}^b_{\mathbb{R}-c}(X)$. Then:*

$$(9.4.11)\qquad CC(F) = CC(F') + CC(F'') \ .$$

In particular:

$$CC(F[k]) = (-1)^k CC(F).$$

Proof. (i) follows from (9.4.10).

(ii) Let Λ be a closed conic subanalytic isotropic subset of T^*X containing $SS(F') \cup SS(F'')$ and let $\Lambda_\circ \subset \Lambda$ an open dense subset satisfying (9.4.9). Then $CC(F) = \sum_\alpha m_\alpha[\Lambda_\alpha]$ and similarly with F' and F'', replacing m_α by m'_α and m''_α, respectively. Let $p \in T^*_{X_\alpha}X$. By Proposition 6.6.1 (ii) there exists a distinguished triangle $V' \longrightarrow V \longrightarrow V'' \xrightarrow[+1]{}$ in $\mathbf{D}^b(\mathfrak{Mod}^f(k))$ such that $F' \longrightarrow F \longrightarrow F'' \xrightarrow[+1]{}$ is isomorphic to $V'_{X_\alpha} \longrightarrow V_{X_\alpha} \longrightarrow V''_{X_\alpha} \xrightarrow[+1]{}$ in $\mathbf{D}^b(X;p)$. Hence $m_\alpha = m'_\alpha + m''_\alpha$. □

Example 9.4.6. Denote by t the coordinate on $X = \mathbb{R}$ and consider the sets:

$$Z_\pm = \{\pm t \geqslant 0\} \ , \qquad U_\pm = \{\pm t > 0\} \ .$$

Let $(t;\tau)$ denote the associated coordinates on T^*X. Consider the Lagrangian chains:

$$\alpha_\pm = [T^*_X X] \cap \{\pm t > 0\} \ , \qquad \beta_\pm = [T^*_{\{0\}}X] \cap \{\pm \tau > 0\} \ .$$

(Here and in the sequel, if φ is a real function on X, we often write $\{\varphi > 0\}$ to denote the subset $\pi^{-1}\{x;\varphi(x) > 0\}$ of T^*X, for short.) Then:

$$H^0_{\{t\tau=0\}}(T^*X;\mathscr{L}_X) = \mathbb{Z}[T^*_X X] \oplus \mathbb{Z}[T^*_{\{0\}}X] \oplus \mathbb{Z}(\alpha_+ + \beta_+) \ .$$

(We consider subanalytic chains with coefficients in $\mathbb{Z}$.) Now we have:

(i) $CC(k_X) = [T^*_X X] = \alpha_+ + \alpha_-$,
(ii) $CC(k_{\{0\}}) = [T^*_{\{0\}}X] = \beta_+ + \beta_-$,
(iii) $CC(k_{Z_\pm}) = \alpha_\pm + \beta_\pm$,
(iv) $CC(k_{U_\pm}) = \alpha_\pm - \beta_\pm$.

In fact the first two formulas have already been proved (cf. (9.4.8)). Let us calculate $CC(k_{U_+})$. This cycle is supported by the set $\{t \geqslant 0, \tau = 0\} \cup \{t = 0, \tau \leqslant 0\}$. On $\{t > 0, \tau = 0\}$, $k_{U_+} \simeq k_X$. On $\{t = 0, \tau < 0\}$, $k_{U_+} \simeq k_{\{0\}}[-1]$. This gives $CC(k_{U_+}) = \alpha_+ - \beta_-$. The other formulas are similarly proven.

Example 9.4.7. Let $(t, x) = (t, x_1, \ldots, x_n)$ be the coordinates on $X = \mathbb{R}^{1+n}$. Consider the sets:

$$Z_\pm = \{\pm t \geqslant |x|\} \ , \qquad Z_0 = \{|t| \leqslant |x|\} \ ,$$

$$U_\varepsilon = \operatorname{Int} Z_\varepsilon \ , \qquad (\varepsilon = +, -, 0) \ , \qquad S_\pm = \{\pm t = |x| > 0\} \ ,$$

where $|x|^2 = \sum_j x_j^2$.

Let $(t, x; \tau, \xi)$ denote the associated coordinates on T^*X. Define the Lagrangian chains:

$$\sigma_\pm = [T^*_X X] \cap U_\pm \ ,$$

$$\tau_{\varepsilon_1\varepsilon_2} = [T^*_{S_{\varepsilon_1}} X] \cap \{\varepsilon_2 \tau > 0\} \ , \qquad \varepsilon_1 = \pm 1 \ , \quad \varepsilon_2 = \pm 1 \ ,$$

$$\gamma_\pm = [T^*_{\{0\}}X] \cap \{\pm\tau > |\xi|\} \ ,$$

$$\gamma_0 = [T^*_{\{0\}}X] \cap \{|\tau| < |\xi|\} \ .$$

Then we have:

(i) $CC(k_X) = \sigma_+ + \sigma_0 + \sigma_-$
(ii) $CC(k_{\{0\}}) = \gamma_+ + \gamma_0 + \gamma_-$
(iii) $CC(k_{Z_\pm}) = \sigma_\pm + \tau_{\pm\pm} + \gamma_\pm$
(iv) $CC(k_{U_\pm}) = \sigma_\pm - \tau_{\pm\mp} + (-1)^{n+1}\gamma_\mp$
(v) $CC(k_{Z_0}) = \sigma_0 + \tau_{+-} + \tau_{-+} + (-1)^n(\gamma_+ + \gamma_-)$
(vi) $CC(k_{U_0}) = \sigma_0 - (\tau_{++} + \tau_{--}) + \gamma_0$
(vii) $CC(k_{S_\pm}) = \tau_{\pm+} + \tau_{\pm-} - \gamma_0 + ((-1)^n - 1)\gamma_\mp$.

Let us check example (iii). We have:

$$SS(k_{Z_\pm}) \subset \operatorname{supp}(\sigma_\pm) \cup \operatorname{supp}(\tau_{\pm\pm}) \cup \operatorname{supp}(\gamma_\pm) \ .$$

Moreover, $k_{Z_\pm}$ is microlocally isomorphic to k_X (resp. $k_{S_\pm}$, resp. $k_{\{0\}}$) at generic points of $\operatorname{supp}(\sigma_\pm)$ (resp. $\operatorname{supp}(\tau_{\pm\pm})$, resp. $\operatorname{supp}(\gamma_\pm)$).

The other examples can be similarly treated.

9.5. Microlocal index formulas

Let X be a manifold and let $F \in \mathrm{Ob}(\mathbf{D}^b_{\mathbb{R}-c}(X))$. We shall calculate $\chi(X;F)$ (and $\chi(F)(x)$, etc.) by an intersection formula. Recall the morphisms α and β of formulas (9.3.11) and (9.3.12):

$$\alpha : H^0_S(T^*X;\omega_{T^*X}) \to H^0_{\pi(S)}(X;\omega_X)$$

(S is closed and π is proper on S) .

$$\beta : H^0_\Lambda(T^*X;\pi^{-1}\omega_X) \to H^0_{\Lambda\cap X}(X;\omega_X)$$

(Λ is closed and conic) .

Proposition 9.5.1. *Let σ be a continuous section of π. We have in $H^0_{\mathrm{supp}(F)}(X;\omega_X)$:*

$$\begin{aligned}\mathrm{C}(F) &= \alpha([\sigma]\cap \mathrm{CC}(F))\\ &= \beta(\mathrm{CC}(F)) .\end{aligned}$$

Proof. By Proposition 9.3.6 it is enough to prove that $\mathrm{C}(F) = \beta(\mathrm{CC}(F))$. This follows from the commutative diagram (9.5.1) below, where we set $\Lambda = \mathrm{SS}(F)$.
□

$$\begin{array}{ccc}
\mathrm{Hom}(F,F) & \xrightarrow{\sim} & \Gamma(T^*X;\mu hom(F,F)) \\
\wr\big| & & \big|\wr \\
\Gamma_{X\cap\Lambda}(X;\delta^!(F\boxtimes DF)) & \xrightarrow{\sim} & \Gamma_\Lambda(T^*X;\mu_\Delta(F\boxtimes DF)) \\
\big\downarrow & & \big\downarrow \\
\Gamma_{X\cap\Lambda}(X;F\otimes DF) & \xrightarrow{\sim} & \Gamma_{\pi^{-1}(X\cap\Lambda)}(T^*X;\mu_\Delta(\delta_*(F\otimes DF))) \\
\big\downarrow & & \big\downarrow \\
\Gamma_{X\cap\Lambda}(X;\omega_X) & \underset{\beta}{\overset{\sim}{\rightleftarrows}} & \Gamma_{\pi^{-1}(X\cap\Lambda)}(T^*X;\pi^{-1}\omega_X)
\end{array}$$

Diagram (9.5.1)

Corollary 9.5.2. *Assume F has compact support. Then:*

$$\chi(X;F) = \#([\sigma]\cap \mathrm{CC}(F)) . \tag{9.5.2}$$

Proof. Apply Proposition 9.5.1 and formula (9.1.3). □

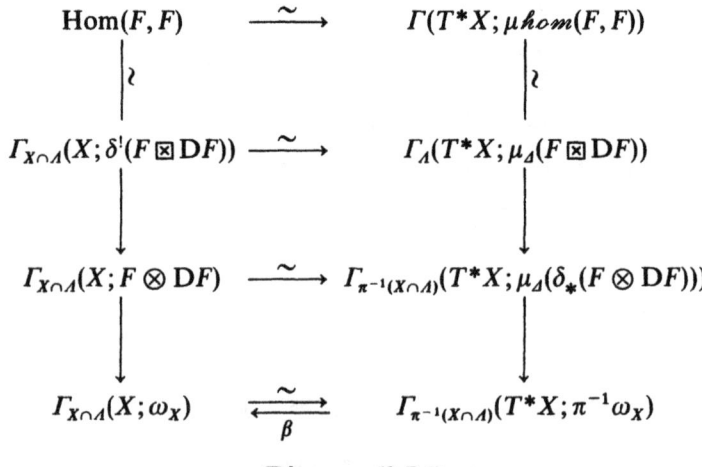

Note that this Corollary may also be deduced from (9.1.14) by using the map $a : X \to \{\mathrm{pt}\}$.

We shall generalize this last result. Let I be an open interval of $\mathbb{R}$, let $\varphi : X \to I$ be a real analytic function and let σ_φ denote the section $x \mapsto d\varphi(x)$. Let $\Lambda_\varphi = \sigma_\varphi(X) = \{(x; d\varphi(x)); x \in X\}$. Hence to φ is associated a cycle (cf. Definition 9.3.5):

$$(9.5.3) \qquad [\sigma_\varphi] \in H^0_{\Lambda_\varphi}(T^*X; \pi^! k_X) .$$

Theorem 9.5.3. *Let* $F \in \mathrm{Ob}(\mathbf{D}^b_{\mathbb{R}-c}(X))$. *Assume*:

$$(9.5.4) \qquad \{x \in \mathrm{supp}(F); \varphi(x) \leqslant t\} \quad \text{is compact for all } t \in I ,$$

$$(9.5.5) \qquad \mathrm{SS}(F) \cap \Lambda_\varphi \quad \text{is compact} .$$

Then for all $j \in \mathbb{Z}$, *the spaces* $H^j(X; F)$ *are finite-dimensional and*:

$$\chi(X; F) = \#([\sigma_\varphi] \cap \mathrm{CC}(F)) .$$

Proof. Let $\Omega_t = \{x \in X; \varphi(x) < t\}$, let j_t denote the open embedding $\Omega_t \hookrightarrow X$ and set $F_t = Rj_{t*} j_t^{-1} F$. Applying Corollary 5.4.19, we get $R\Gamma(X; F) \xrightarrow{\sim} R\Gamma(X; F_t)$ for $t \gg 0$. Hence the spaces $H^j(X; F)$ are finite-dimensional and $\chi(X; F) = \chi(X; F_t) = \#([\sigma_\varphi] \cap \mathrm{CC}(F_t))$, by Corollary 9.5.2. Since $\mathrm{SS}(F) \cap \pi^{-1}(\Omega_t) = \mathrm{SS}(F_t) \cap \pi^{-1}(\Omega_t)$ and $\mathrm{SS}(F) \cap \Lambda_\varphi \subset \pi^{-1}(\Omega_t)$ for $t \gg 0$, by the hypothesis (9.5.5), it remains to show:

$$(9.5.6) \qquad \mathrm{SS}(F_t) \cap \Lambda_\varphi \subset \pi^{-1}(\Omega_t) \quad \text{for} \quad t \gg 0 .$$

Let $K = \pi(\mathrm{SS}(F) \cap \Lambda_\varphi)$. Then $x \in \mathrm{supp}(F) \backslash K$ implies $d\varphi(x) \neq 0$ and we get by Proposition 9.5.8:

$$\mathrm{SS}(F_t) \subset \mathrm{SS}(F) + \mathbb{R}_{\leqslant 0} \Lambda_\varphi , \quad \text{for} \quad t \gg 0 .$$

This inclusion, together with (9.5.5) implies (9.5.6). □

Corollary 9.5.4. *Let* $F \in \mathrm{Ob}(\mathbf{D}^b_{\mathbb{R}-c}(X))$. *Assume* (9.5.4) *and*:

$$(9.5.7) \qquad \mathrm{SS}(F)^a \cap \Lambda_\varphi \quad \text{is compact} .$$

Then for all j, *the spaces* $H^j_c(X; F)$ *are finite-dimensional and*:

$$\begin{aligned}\chi_c(X; F) &= \#([\sigma_\varphi] \cap \mathrm{CC}(F)^a) \\ &= \#([\sigma_{-\varphi}] \cap \mathrm{CC}(F)) .\end{aligned}$$

Proof. We have $\mathrm{Hom}(R\Gamma_c(X; F), k) \simeq \mathrm{Hom}(F, \omega_X) \simeq R\Gamma(X; D_X F)$. Hence $\chi_c(X; F) = \chi(X; D_X F)$, and the result follows from Proposition 9.4.4 and Theorem 9.5.3. □

Remark 9.5.5. Let $\varphi : X \to \mathbb{R}$ be a real analytic function and assume $\varphi(x_\circ) = 0$, $d\varphi(x_\circ) = 0$ and the Hessian of φ at $x_\circ$ is positive-definite. (This implies that in a local chart in a neighborhood of $x_\circ$, one can write $\varphi(x) = \sum_{j=1}^n x_j^2$.) Applying the microlocal Bertini-Sard theorem (Proposition 8.3.12), one gets that $x_\circ$ is an isolated point of $SS(F) \cap \Lambda_\varphi$. Then by Theorem 9.5.3 we find:

$$\chi(F)(x_\circ) = \#([\sigma_\varphi] \cap CC(F))_{x_\circ} \ . \tag{9.5.8}$$

and similarly, by Corollary 9.5.4:

$$\chi_c(F)(x_\circ) = \#([\sigma_{-\varphi}] \cap CC(F))_{x_\circ} \ . \tag{9.5.9}$$

Here and in the sequel, $\#(c)_x$, for $c \in H_S^0(X; \omega_X)$ and an isolated point x of the closed subset S, means the image of c by the maps: $H_S^0(X; \omega) \to H_{\{x\}}^0(X; \omega) \to k$.

We shall in fact now prove a more general result.

Theorem 9.5.6. *Let $x_\circ \in X$, $F \in \mathrm{Ob}(\mathbf{D}^b_{\mathbb{R}-c}(X))$, and let φ be a real analytic function on X. Assume:*

$$\Lambda_\varphi \cap SS(F) \subset \{d\varphi(x_\circ)\} \ . \tag{9.5.10}$$

Then we have:

$$\chi(R\Gamma_{\{\varphi \geq \varphi(x_\circ)\}}(F))(x_\circ) = \#([\sigma_\varphi] \cap CC(F)) \ .$$

Proof. We may assume $X = \mathbb{R}^n$, $x_\circ = 0$, $\varphi(x_\circ) = 0$. Set $\Lambda = SS(F)$, $\psi(x) = \sum_{j=1}^n x_j^2$. If θ is a real function on $\mathbb{R}$, let us write, when there is no risk of confusion, $\{\theta > 0\}$ (for example) to denote the set $\pi^{-1}(\{x \in X; \theta(x) > 0\})$ (as we already did in Examples 9.4.6, 9.4.7). By the microlocal Bertini-Sard theorem (Proposition 8.3.12), there exists $\delta_\circ > 0$ such that:

$$\Lambda \cap \Lambda_\psi \cap \{0 < |x| \leq \delta_\circ\} = \varnothing \ , \tag{9.5.11}$$

$$(\Lambda \mathbin{\hat{+}} T^*_{\{\varphi=0\}}X) \cap \Lambda_\psi \cap \{0 < |x| \leq \delta_\circ\} = \varnothing \ . \tag{9.5.12}$$

(Recall that $\Lambda \mathbin{\hat{+}} T^*_{\{\varphi=0\}}X$ is isotropic by Corollary 8.3.18.) Moreover, there exists $\delta_1 > 0$ such that:

$$k \geq 0 \ , \qquad 0 < |x| \leq \delta_1 \ , \qquad \varphi(x) > 0 \Rightarrow (x; k\,d\psi(x) + d\varphi(x)) \notin \Lambda \ . \tag{9.5.13}$$

In fact, if (9.5.13) is not satisfied, there exists a real analytic curve $t \mapsto (x(t), a(t), b(t)) \in X \times \mathbb{R} \times \mathbb{R}$ such that $a(t) \geq 0$, $b(t) > 0$, $x(0) = 0$, and $\varphi(x(t)) > 0$ for $t \neq 0$, and $(x(t); a(t)\,d\psi(x(t)) + b(t)\,d\varphi(x(t))) \in \Lambda$. Since Λ is isotropic, this implies:

$$a(t)\frac{d}{dt}\psi(x(t)) + b(t)\frac{d}{dt}\varphi(x(t)) \equiv 0 \ .$$

Since $\frac{d}{dt}\psi \geqslant 0$ and $\frac{d}{dt}\varphi > 0$ for $0 < t \ll 1$, this is a contradiction. This proves (9.5.13).

Set $B(\delta) = \{x \in X; |x| < \delta\}$. Since $R\Gamma_{\{\varphi \geqslant 0\}}(F)$ is $\mathbb{R}$-constructible, there exists $\delta > 0$ with $\delta < \inf(\delta_\circ, \delta_1)$ such that:

$$R\Gamma(B(\delta); R\Gamma_{\{\varphi \geqslant 0\}}(F)) \xrightarrow{\sim} (R\Gamma_{\{\varphi \geqslant 0\}}(F))_0 \ .$$

We fix such a δ, we denote by j the open embedding $B(\delta) \hookrightarrow X$ and we set $F' = Rj_* j^{-1}F$,

$$\Lambda' = \Lambda \cup (\Lambda + T^*_{\partial B(\delta)}X) \ .$$

Note that:

$$\text{(9.5.14)} \qquad SS(F') \subset SS(F) + \mathbb{R}_{\leqslant 0}\Lambda_\psi$$

and in particular $SS(F') \subset \Lambda'$. Since Λ' is isotropic, there exists, by the microlocal Bertini-Sard theorem, an $\varepsilon > 0$ such that:

$$\text{(9.5.15)} \qquad \Lambda' \cap \Lambda_\varphi \cap \{0 < |\varphi(x)| \leqslant \varepsilon\} = \emptyset \ .$$

Applying the microlocal Morse lemma (Corollary 5.4.19), we obtain:

$$\text{(9.5.16)} \qquad R\Gamma_{\{\varphi \geqslant 0\}}(X; F') \simeq R\Gamma_c(X \cap \{\varphi > -\varepsilon\}; F') \ .$$

By (9.5.15), $SS(F') \cap \Lambda_\varphi \cap \{0 > \varphi > -\varepsilon\} = \emptyset$. Hence we get by Corollary 9.5.4:

$$\text{(9.5.17)} \qquad \chi_c(X \cap \{\varphi > -\varepsilon\}; F') = \#([\sigma_\varphi] \cap CC(F'|_{\{\varphi > -\varepsilon\}})) \ .$$

(The right-hand side of (9.5.17) denotes the intersection number of $[\sigma_\varphi]$ and $CC(F')$ in $T^*(X \cap \{\varphi > -\varepsilon\})$, which makes sense since $\Lambda_\varphi \cap SS(F') \cap \{\varphi > -\varepsilon\}$ is compact.)

We have thus obtained:

$$\chi(R\Gamma_{\{\varphi \geqslant 0\}}(F))(x_\circ) = \#([\sigma_\varphi] \cap CC(F'|_{\{\varphi > -\varepsilon\}})) \ .$$

It remains to show:

$$\Lambda_\varphi \cap SS(F') \cap \{\varphi > -\varepsilon\} \cap \{|x| = \delta\} = \emptyset \ .$$

This relation is satisfied on the set $\{-\varepsilon < \varphi < 0\}$ by (9.5.15), and on the set $\{\varphi > 0\}$ by (9.5.13). Assume it is not satisfied on $\{\varphi = 0\}$. We find $p \in SS(F)$ and $c \geqslant 0$ such that $p - c\,d\psi = d\varphi$, $|\pi(p)| = \delta$. If $c = 0$ this contradicts the hypothesis. If $c > 0$ this implies $(\Lambda \backslash \Lambda_\varphi) \cap \{\varphi = 0\} \cap \Lambda_\psi \neq \emptyset$ which contradicts (9.5.12). Hence the proof is complete. □

Examples 9.5.7. Let Y be a closed submanifold of dimension l, let $x_0 \in Y$ and let φ be a real analytic function such that $\varphi(x_\circ) = 0$ and the manifolds $T^*_Y X$ and Λ_φ

intersect transversally at $p = d\varphi(x_\circ)$. By the Morse lemma there exists a local coordinate system $(x_1, \dots, x_l)$ on Y such that $\varphi|_Y = \sum_{j=1}^{l} a_j x_j^2$, with $a_j \neq 0\ \forall j$. Let $q = \#\{j; a_j < 0\}$. Since $R\Gamma_{\{\varphi \geq 0\}}(k_Y)_{x_\circ} \simeq k[-q]$ (cf. VII §5) we get by Theorem 9.5.6:

$$\#([\sigma_\varphi] \cap [T_Y^*X]) = (-1)^q \ . \tag{9.5.18}$$

More generally let $F \in \mathrm{Ob}(\mathbf{D}^b_{\mathbb{R}-c}(X))$ and let $\phi : X \to \mathbb{R}$ be a real analytic function. Set $\Lambda = SS(F)$, assume (9.5.4) and also:

(9.5.19) *$\Lambda_\phi \cap \Lambda$ is finite, contained in Λ_{reg}, and the intersection is transversal.*

It is then natural to call ϕ "a Morse function with respect to Λ". In fact if $F = k_X$, one recovers the classical notion of a Morse function. Let $\Lambda_\phi \cap \Lambda = \{p_1, \dots, p_N\}$ and assume:

(9.5.20) *F is pure with shift d_i and multiplicity m_i at p_i, $i = 1, \dots, N$.*

From Theorem 9.5.6 and the definition of pure sheaves we deduce:

$$\#([\sigma_\phi] \cap CC(F))_{p_i} = (-1)^{d_i'} \cdot m_i \tag{9.5.21}$$

with:

$$d_i' = d_i - \tfrac{1}{2}\dim X - \tfrac{1}{2}\tau(\lambda_\circ(p_i), \lambda_\Lambda(p_i), \lambda_\phi(p_i)) \ . \tag{9.5.22}$$

Using Theorem 9.5.3, we get:

$$\chi(X; F) = \sum_{i=1}^{N} (-1)^{d_i'} \cdot m_i \ . \tag{9.5.23}$$

Note that this last result can also be obtained, and even refined (replacing this equality by "Morse inequalities") using Proposition 5.4.20.

Remark 9.5.8. Since $\mathscr{L}_X$ is a subsheaf of $\mathscr{L}\mathscr{S}_n \otimes \mathit{or}_{T^*X/X}$, a Lagrangian cycle may be considered as an $\mathit{or}_{T^*X/X}$-valued subanalytic cycle.

We shall describe Lagrangian cycles in this language.

For an n-dimensional manifold M, and a nowhere vanishing real n-form v on X, we denote by $\mathrm{sgn}(v)$ the orientation given by v. Hence $\mathrm{sgn}(v)$ is considered as a section of or_X. Now, let $(x_1, \dots, x_n)$ be a coordinate system of X and let $(x; \xi)$ be the associated coordinate system of T^*X. We identify $\mathit{or}_{T^*X/X}$ and or_X by $\mathrm{sgn}(d\xi) \leftrightarrow \mathrm{sgn}\, dx$, where $d\xi = d\xi_1 \wedge \cdots \wedge d\xi_n$, and $dx = dx_1 \wedge \cdots \wedge dx_n$. Then, by a suitable identification, we have:

$$\begin{cases} [T^*_{\{x_1 = \cdots = x_l = 0\}}(X)] \quad \text{is the cycle} \\ \{(x;\xi); x_1 = \cdots = x_l = \xi_{l+1} = \cdots = \xi_n = 0\} \quad \text{with the orientation} \\ (-1)^l \,\mathrm{sgn}(d\xi_1 \wedge \cdots \wedge d\xi_l \wedge dx_{l+1} \wedge \cdots \wedge dx_n) \otimes \mathrm{sgn}(d\xi) \ . \end{cases} \tag{9.5.24}$$

(9.5.25) $\begin{cases} \text{For a function } \varphi \text{ on } X, [\sigma_\varphi] \text{ is the cycle } \Lambda_\varphi \\ \text{with the orientation induced by that of } X \text{ via the} \\ \text{projection } \Lambda_\varphi \xrightarrow{\sim} X \ . \end{cases}$

We define the intersection of two n-cycles in T^*X as follows. Let Z_1 and Z_2 be n-dimensional submanifolds of T^*X intersecting transversally at $p \in T^*X$. Let θ_i $(i = 1, 2)$ be n-forms such that $\theta_1|_{Z_2} = 0$ and $\mathrm{sgn}(\theta_1 \wedge \theta_2) = \mathrm{sgn}(d\xi \wedge dx) = (-1)^{n(n-1)/2}\,\mathrm{sgn}((d\alpha_X)^n)$. If α_i is the n-cycle $[Z_i]$ with the orientation $\mathrm{sgn}(\theta_i|_{Z_i})$, then $\#(\alpha_1 \cap \alpha_2) = 1$. By this convention, Corollary 9.5.2, Theorem 9.5.3, Corollary 9.5.4 as well as (9.5.8) and (9.5.9) remain true.

9.6. Lefschetz fixed point formula

Consider two morphisms of manifolds:

$$Y \underset{g}{\overset{f}{\rightrightarrows}} X$$

and set $h = (f, g) : Y \to X \times X$. Let $F \in \mathrm{Ob}(\mathbf{D}^b_{\mathbb{R}-c}(X))$ and assume to be given:

(9.6.1) $$\varphi \in \mathrm{Hom}(f^{-1}F, g^!F) \ .$$

We set:

(9.6.2) $$S = \mathrm{supp}(F) \ , \qquad T = f^{-1}(S) \cap g^{-1}(S) \ ,$$

$$Z = T \cap h^{-1}(\Delta_X) = \{y \in Y; f(y) = g(y) \in S\} \ ,$$

and we assume:

(9.6.3) $$T \text{ is compact} \ .$$

Note that the support of φ is contained in T. The natural morphisms: $\mathrm{id} \to Rf_* f^{-1}$ and $Rg_! g^! \to \mathrm{id}$, define the morphisms:

$$R\Gamma(X; F) \to R\Gamma_{f^{-1}(S)}(Y; f^{-1}F) \ , \qquad R\Gamma_T(Y; g^!F) \to R\Gamma(X, F) \ .$$

Combining with φ we get a morphism, that we still denote by φ, for short:

(9.6.4) $$\varphi : R\Gamma(X; F) \to R\Gamma(X; F) \ .$$

The aim of this section is to calculate $\mathrm{tr}(\varphi)$, the trace of φ, where:

$$\mathrm{tr}(\varphi) = \sum_j (-1)^j \mathrm{tr}(H^j(\varphi)) \ ,$$

$H^j(\varphi)$ denoting the endomorphism of $H^j(X; F)$ deduced from (9.6.4). Since $H^j(\varphi)$ decomposes into $H^j(X; F) \to H^j_{g(T)}(X; F) \to H^j(X; F)$ and $H^j_{g(T)}(X; F)$ is finite-dimensional, $\mathrm{tr}(H^j(\varphi))$ has a meaning (cf. Exercise I.33). Let δ denote the diagonal embedding $X \hookrightarrow X \times X$, and consider the chain of morphisms (here we don't need to assume (9.6.3)):

$$\begin{aligned} R\mathcal{H}om(F,F) &\xleftarrow{\sim} \delta^!(F \boxtimes DF) \\ &\longrightarrow \delta^! Rh_* h^{-1}(F \boxtimes DF) \\ &\xrightarrow{\sim} R\Gamma_\Delta Rh_* R\Gamma_T(f^{-1}F \otimes Dg^! F) \\ &\xrightarrow{\sim} Rh_* R\Gamma_Z(f^{-1}F \otimes Dg^! F) \\ &\xrightarrow[\varphi]{} Rh_* R\Gamma_Z(g^! F \otimes Dg^! F) \\ &\xrightarrow[\mathrm{tr}]{} Rh_* R\Gamma_Z(\omega_Y) . \end{aligned}$$

This induces:

$$\mathrm{Hom}(F,F) \to H_Z^0(Y;\omega_Y) . \tag{9.6.5}$$

Definition 9.6.1. *The image of* id_F *in* $H_Z^0(Y;\omega_Y)$ *is called the characteristic class of* φ *and denoted* $\mathrm{C}(\varphi)$.

Proposition 9.6.2 (the Lefschetz fixed point theorem). *Let* f, g, φ *be as above, and assume* (9.6.3) *and* $\mathrm{supp}(F)$ *is compact. Then*:

$$\mathrm{tr}(\varphi) = \int_Y \mathrm{C}(\varphi) .$$

Proof. Consider the commutative diagram (9.6.6) below. Since φ decomposes as $R\Gamma(X;F) \xrightarrow{\alpha} R\Gamma_T(Y;g^!F) \xrightarrow{\beta} R\Gamma(X;F)$, the trace of φ is equal to the trace of $\beta \circ \alpha$, and it is the image of id_F in k by the chain of morphisms of the column on the right-hand side. This completes the proof. □

Note that the proposition above also holds without the condition that $\mathrm{supp}(F)$ is compact (cf. Exercise IX.9).

Now let $y \in Z$ and assume y is an isolated point of Z. Set $Z' = Z \setminus \{y\}$. Then:

$$H_Z^0(Y;\omega_Y) \simeq H_{\{y\}}^0(Y;\omega_Y) \oplus H_{Z'}^0(Y;\omega_Y) .$$

Definition 9.6.3. *In the preceding situation, one denotes by* $\mathrm{C}_y(\varphi)$ *the image of* $\mathrm{C}(\varphi)$ *by the morphism*:

$$H_Z^0(Y;\omega_Y) \to H_{\{y\}}^0(Y;\omega_Y) \simeq k .$$

Also, if $y \notin Z$, *we set* $\mathrm{C}_y(\varphi) = 0$.

Of course, if Z is finite, then:

$$\mathrm{tr}(\varphi) = \sum_{y \in Z} \mathrm{C}_y(\varphi) .$$

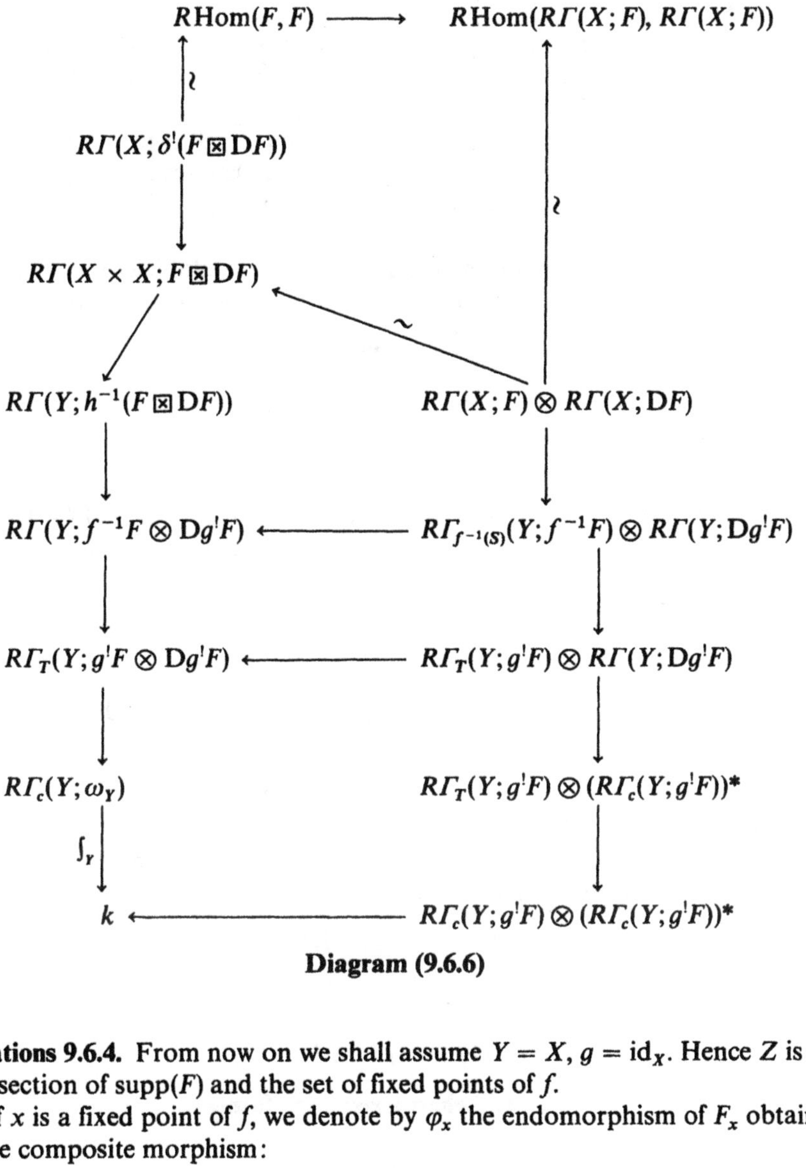

Diagram (9.6.6)

Notations 9.6.4. From now on we shall assume $Y = X$, $g = \mathrm{id}_X$. Hence Z is the intersection of supp(F) and the set of fixed points of f.

If x is a fixed point of f, we denote by φ_x the endomorphism of F_x obtained as the composite morphism:

$$F_x \simeq (f^{-1}F)_x \xrightarrow[\varphi]{} F_x \ .$$

When x is a fixed point of f and x is an isolated point of $f^{-1}(x)\cap T = f^{-1}(x)\cap \mathrm{supp}(F)\cap f^{-1}(\mathrm{supp}(F))$, we denote by $\Gamma_{\{x\}}(\varphi)$ the endomorphism of $R\Gamma_{\{x\}}(X;F)$ obtained as the composite morphism:

$$R\Gamma_{\{x\}}(X;F) \longrightarrow R\Gamma_{\{f^{-1}(x)\}}(X;f^{-1}F) \xrightarrow[\varphi]{} R\Gamma_{\{f^{-1}(x)\}\cap T}(X;F) \longrightarrow R\Gamma_{\{x\}}(X;F) \ .$$

One shall take care that, $\mathrm{C}_x(\varphi)$, $\mathrm{tr}(\varphi_x)$ and $\mathrm{tr}(\Gamma_{\{x\}}(\varphi))$ are different in general.

Example 9.6.5. Let $f:\mathbb{R}\to\mathbb{R}$ be an analytic diffeomorphism with only two fixed points, t_0 and t_1, $t_0<t_1$, so that f induces an homeomorphism of the interval $[t_0,t_1]$. Let $F=k_{[t_0,t_1]}$ and let φ denote the canonical isomorphism $f^{-1}k_{[t_0,t_1]}\xrightarrow{\sim} k_{[t_0,t_1]}$. Then $\mathrm{tr}(\varphi)=\chi(\mathbb{R};F)=1=\mathrm{C}_{t_0}(\varphi)+\mathrm{C}_{t_1}(\varphi)$. On the other hand φ_{t_i} acts as the identity on F_{t_i} $(i=0,1)$. Hence $\mathrm{tr}(\varphi_{t_i})=1$ and $\mathrm{C}_{t_i}(\varphi)\neq\mathrm{tr}(\varphi_{t_i})$ for $i=0$ or $i=1$. There is a similar argument for $\Gamma_{\{t_i\}}(\varphi)$. As we shall see later, $\mathrm{C}_{t_i}(\varphi)=0$ or 1 according that $f(t)-t$ is increasing or decreasing at t_i.

Example 9.6.6. Let X be an n-dimensional compact manifold, let $F=k_X$ and let φ be the canonical morphism $f^{-1}k_X\xrightarrow{\sim}k_X$. Consider the chain of morphisms:

$$\begin{aligned} k_X &\simeq R\mathcal{H}om(k_X,k_X)\\ &\simeq \delta^!(k_X\boxtimes\omega_X)\\ &\to \delta^!h_*\omega_X\ . \end{aligned}$$

Let $\Gamma_f=\{(f(x),x);x\in X\}$ and let $j:\Gamma_f\hookrightarrow X\times X$ the embedding of Γ_f. Then by its construction, $\mathrm{C}(\varphi)$ is the image of $1\in\Gamma(X;k_X)$ by the morphisms:

$$\begin{aligned} \Gamma(X;k_X) &\xrightarrow{\alpha} H^0_\Delta(X\times X;k_X\boxtimes\omega_X)\\ &\xrightarrow{\beta} H^0_{j^{-1}(\Delta)}(\Gamma_f;j^{-1}(k_X\boxtimes\omega_X))\\ &\xrightarrow{\sim} H^0_{j^{-1}(\Delta)}(\Gamma_f;\omega_{\Gamma_f})\\ &\xrightarrow{\sim} H^0_{j^{-1}(\Delta)}(X\times X;\omega_{X\times X})\ . \end{aligned}$$

Let $[\Gamma_f]\in H^0_{\Gamma_f}(X\times X;\mathcal{LS}_n(k_X\boxtimes or_X))$ and $[\Delta]\in H^0(X\times X;\mathcal{LS}_n(k_X\boxtimes or_X))$ be the cycles associated to Γ_f and Δ, respectively. Then, identifying $(or_X\boxtimes k_X)|_\Delta$ and $(k_X\boxtimes or_X)|_\Delta$, we have $\alpha(1)=[\Delta]$. Then we get:

$$\beta([\Delta])=[\Gamma_f]\cap[\Delta]\ ,$$

which gives:

$$\mathrm{C}(\varphi)=[\Gamma_f]\cap[\Delta]\ . \tag{9.6.7}$$

Hence we obtain the famous **Lefschetz fixed point formula**:

$$\sum_j(-1)^j\,\mathrm{tr}(H^j(\varphi))=\#([\Gamma_f]\cap[\Delta])\ ,$$

(cf. Remark 9.0).

In other words, the trace of φ is calculated as the intersection number of the graph of f and the diagonal.

Remark 9.6.7. Let x be an isolated point of Z. Then $\mathrm{C}_x(\varphi)$ is a local invariant in the following sense. Let F and F' belong to $\mathbf{D}^b_{\mathbb{R}-c}(X)$, $\varphi:f^{-1}F\to F$, $\varphi':f^{-1}F'\to F'$ and assume there is a neighborhood W of x and an isomorphism $\psi:F|_W\to F'|_W$

such that the diagram below commutes:

$$\begin{array}{ccc}
f^{-1}F|_{W\cap f^{-1}(W)} & \xrightarrow{\varphi} & F|_{W\cap f^{-1}(W)} \\
{\scriptstyle f^{-1}(\psi)}\downarrow & & \downarrow{\scriptstyle \psi} \\
f^{-1}F'|_{W\cap f^{-1}(W)} & \xrightarrow{\varphi'} & F'|_{W\cap f^{-1}(W)} .
\end{array}$$

Then $C_x(\varphi) = C_x(\varphi')$. The proof is immediate.

We shall now explain how to calculate $C(\varphi)$, in some cases. We begin by proving the homotopy invariance of $C(\varphi)$.

Let $I = [0, 1]$, $f: X \times I \to X$ the restriction of a morphism of manifolds $X \times \mathbb{R} \to X$, $p: X \times I \to X$ the projection. Let $F \in \mathrm{Ob}(\mathbf{D}^b_{\mathbb{R}-c}(X))$ and let $\varphi: f^{-1}F \to p^{-1}F$ be a morphism. For $t \in I$, let $i_t: X \hookrightarrow X \times I$ be the injection $x \mapsto (x, t)$. Set $f_t = f \circ i_t$. $\varphi_t = i_t^{-1}(\varphi): f_t^{-1}F \to F$, and finally set:

$$\tilde{Z} = \{(x, t) \in \mathrm{supp}(F) \times I; f(x, t) = x\} .$$

Proposition 9.6.8. *In the preceding situation assume $\tilde{Z}$ is compact. Then the image of $C(\varphi_t)$ in $H^0_{p(\tilde{Z})}(X; \omega_X)$ does not depend on t.*

Proof. Consider the commutative diagram:

$$\begin{array}{ccc}
R\mathrm{Hom}(F, F) & & \\
\downarrow & & \\
R\Gamma_\Delta(X \times X; F \boxtimes DF) & \longrightarrow & R\Gamma_{\Delta\times I}(X \times X \times I; F \boxtimes DF \boxtimes k_I) \\
\downarrow & & \downarrow \\
R\Gamma_{p(\tilde{Z})}(X; f_t^{-1}F \otimes DF) & \xleftarrow[i_t^\#]{} & R\Gamma_{\tilde{Z}}(X \times I; f^{-1}F \otimes p^{-1}DF) \\
{\scriptstyle \varphi_t}\downarrow & & \downarrow{\scriptstyle \varphi} \\
R\Gamma_{p(\tilde{Z})}(X; F \otimes DF) & \xleftarrow[i_t^\#]{} & R\Gamma_{\tilde{Z}}(X \times I; p^{-1}F \otimes p^{-1}DF) \\
\downarrow & & \downarrow \\
R\Gamma_{p(\tilde{Z})}(X; \omega_X) & \xleftarrow[i_t^\#]{} & R\Gamma_{\tilde{Z}}(X \times I; p^{-1}\omega_X)
\end{array} ,$$

where $i_t^\#$ is the morphism induced by $\mathrm{id} \to i_{t*} \circ i_t^{-1}$. Then $C(\varphi_t)$ is the image of $c \in H^0_{\tilde{Z}}(X \times I; p^{-1}\omega_X)$ by $i_t^\#$. Since the morphisms

$$H^0_{p(\tilde{Z})}(X; \omega_X) \xrightarrow{p^\#} H^0_{p^{-1}p(\tilde{Z})}(X \times I; p^{-1}\omega_X) \xrightarrow{i_t^\#} H^0_{p(\tilde{Z})}(X; \omega_X)$$

are isomorphisms and $i_t^\# \circ p^\# = \mathrm{id}$, $i_t^\#$ does not depend on t. □

Now let x be an isolated point of Z and let V be a locally closed subanalytic neighborhood of x, satisfying:

(9.6.8) $$V \cap f^{-1}(V) \text{ is closed in } V \text{ and open in } f^{-1}(V) \ .$$

Let $\Gamma_V(\varphi)$ denote the morphism from $f^{-1}R\Gamma_V(F)$ to $R\Gamma_V(F)$ obtained as the composite:

$$\begin{aligned} f^{-1}R\Gamma_V(F) &\longrightarrow R\Gamma_{f^{-1}(V)}(f^{-1}F) \underset{\varphi}{\longrightarrow} R\Gamma_{f^{-1}(V)}(F) \\ &\longrightarrow R\Gamma_{f^{-1}(V)\cap V}(F) \longrightarrow R\Gamma_V(F) \ . \end{aligned}$$

Proposition 9.6.9. *Assume* (9.6.8). *Then* $C_x(\varphi) = C_x(\Gamma_V(\varphi))$. *In particular if* V *is relatively compact and* $Z \cap \bar{V} = \{x\}$, $C_x(\varphi) = \mathrm{tr}(\Gamma_V(\varphi))$.

Proof. This follows from the local invariance (Remark 9.6.7) and the Lefschetz Theorem (Proposition 9.6.2). □

We also have the following similar result.

Let V be a locally closed subanalytic neighborhood of x, satisfying:

(9.6.9) $$V \cap f^{-1}(V) \text{ is open in } V \text{ and closed in } f^{-1}(V) \ .$$

Let φ_V be the morphism from $f^{-1}F_V$ to F_V obtained as the composite:

$$f^{-1}F_V \longrightarrow (f^{-1}F)_{f^{-1}(V)} \underset{\varphi}{\longrightarrow} F_{f^{-1}(V)} \longrightarrow F_{V \cap f^{-1}(V)} \longrightarrow F_V \ .$$

Proposition 9.6.10. *Assume* (9.6.9). *Then* $C_x(\varphi) = C_x(\varphi_V)$. *In particular if* V *is relatively compact and* $Z \cap \bar{V} = \{x\}$, *then* $C_x(\varphi) = \mathrm{tr}(\varphi_V)$.

Next we calculate the trace of the specialization of φ at a fixed point x of f. We shall write $\nu_x F$ instead of $\nu_{\{x\}}(F)$, for short. We denote by $\nu_x\varphi$ the composite morphism (cf. Proposition 4.2.5):

(9.6.10) $$\nu_x\varphi : (T_x f)^{-1}\nu_x F \longrightarrow \nu_x f^{-1}F \underset{\varphi}{\longrightarrow} \nu_x F \ .$$

Assume:

(9.6.11) $$\Gamma_f \text{ and } \Delta \text{ intersect transversally at } x \in X \ .$$

By this hypothesis, 0 is the only fixed point of the map $T_x f : T_x X \to T_x X$.

Proposition 9.6.11. *Assume* (9.6.11). *Then* $C_x(\varphi) = C_0(\nu_x\varphi)$.

Proof. It is enough to consider the commutative diagram (9.6.12) below, in which we use the isomorphism $D\nu_x F \simeq \nu_x DF$ of Proposition 8.4.13. □

$$
\begin{array}{ccccc}
R\mathrm{Hom}(F,F) & & \longrightarrow & & R\mathrm{Hom}(\nu_x F,\nu_x F) \\
\uparrow \wr & & & & \uparrow \wr \\
R\Gamma_\Delta(X\times X; F\boxtimes DF) & \longrightarrow & R\Gamma_\Delta(T_xX\times T_xX;\nu_xF\boxtimes\nu_xDF) & \longleftarrow & R\Gamma_\Delta(T_xX\times T_xX;\nu_xF\boxtimes D\nu_xF) \\
\downarrow & & \downarrow & & \downarrow \\
R\Gamma_Z(X;f^{-1}F\otimes DF) & \longrightarrow & R\Gamma_{\{x\}}(T_xX;(T_xf)^{-1}\nu_xF\otimes\nu_xDF) & \longleftarrow & R\Gamma_{\{x\}}(T_xX;(T_xf)^{-1}\nu_xF\otimes D\nu_xF) \\
\varphi\downarrow & & \downarrow \nu_x(\varphi) & & \downarrow \nu_x(\varphi) \\
R\Gamma_Z(X;F\otimes DF) & \longrightarrow & R\Gamma_{\{x\}}(T_xX;\nu_xF\otimes\nu_xDF) & \longleftarrow & R\Gamma_{\{x\}}(T_xX,\nu_xF\otimes D\nu_xF) \\
\downarrow & & \downarrow & & \downarrow \\
R\Gamma_Z(X;\omega_X) & \longrightarrow & R\Gamma_{\{x\}}(T_xX;\nu_x\omega_X) & \xleftarrow{\sim} & R\Gamma_{\{x\}}(T_xX;\omega_{T_xX}) \\
\downarrow & & & & \downarrow \\
R\Gamma_{\{x\}}(X;\omega_X) & & \longrightarrow & & k
\end{array}
$$

Diagram (9.6.12)

By this proposition, in order to calculate $C_x(\varphi)$ when (9.6.11) is satisfied, it is enough to consider the following situation.

Let V be a real finite-dimensional vector space, $u: V \to V$ a linear endomorphism, F a conic object of $\mathbf{D}^b_{\mathbb{R}-c}(V)$, and let $\varphi \in \mathrm{Hom}(u^{-1}F, F)$. We shall assume:

$$\text{(9.6.13)} \qquad 1 \textit{ is not an eigenvalue of } u \ .$$

Let $V^{\mathbb{C}}$ denote the complexified space $V \otimes_{\mathbb{R}} \mathbb{C}$, and for $\lambda \in \mathbb{C}$, denote by $V^{\mathbb{C}}_\lambda$ the generalized eigenspace of $V^{\mathbb{C}}$ with generalized eigenvalue λ, i.e. $x \in V^{\mathbb{C}}_\lambda$ iff there exists $m > 0$ with $(u - \lambda)^m(x) = 0$. We shall say that a linear subspace V_s of V is a "shrinking space" if:

$$\text{(9.6.14)} \qquad \begin{cases} u(V_s) \subset V_s \ , \\ u|_{V_s} \quad \textit{has no eigenvalues } \lambda \in \mathbb{R} \textit{ with } \lambda > 1 \ , \\ u|_{V/V_s} \quad \textit{has no eigenvalues } \lambda \in \mathbb{R} \textit{ with } 0 \leqslant \lambda < 1 \ . \end{cases}$$

This is equivalent to:

$$\text{(9.6.15)} \qquad \begin{cases} u(V_s) \subset V_s \ , \\ \bigoplus_{0 \leqslant \lambda < 1} V^{\mathbb{C}}_\lambda \subset V_s \otimes_{\mathbb{R}} \mathbb{C} \subset \bigoplus_{\lambda \notin]1,+\infty[} V^{\mathbb{C}}_\lambda \ . \end{cases}$$

This condition implies $u^{-1}(V_s) = V_s$.

Similarly, we shall say that a linear subspace V_e is an "expanding space" if:

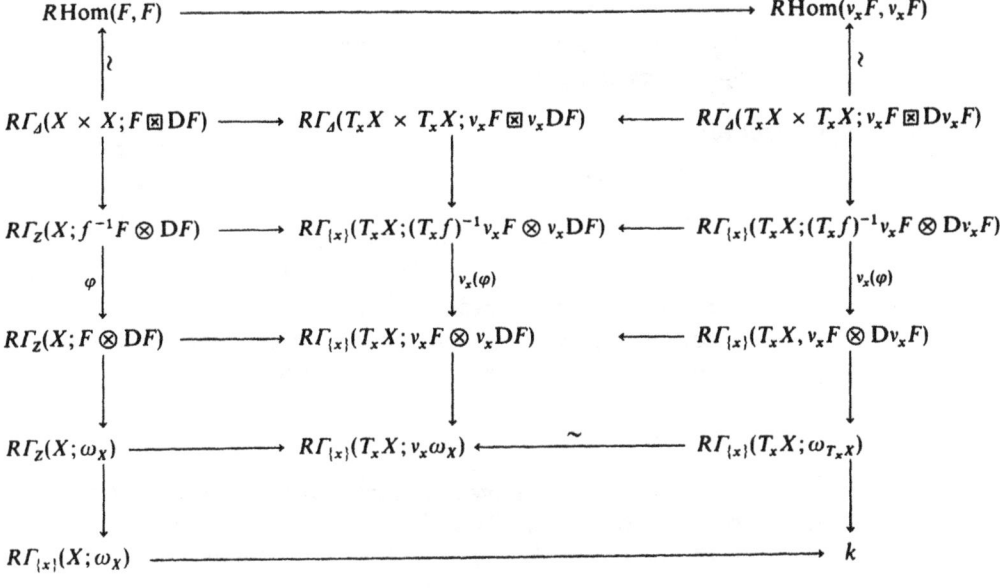

$$(9.6.16) \qquad \begin{cases} u(V_e) \subset V_e \ , \\ u|_{V_e} \quad \textit{has no eigenvalues } \lambda \in \mathbb{R} \textit{ with } 0 \leqslant \lambda < 1 \ , \\ u|_{V/V_e} \quad \textit{has no eigenvalues } \lambda \in \mathbb{R} \textit{ with } \lambda > 1 \ . \end{cases}$$

This is equivalent to:

$$(9.6.17) \qquad \begin{cases} u(V_e) \subset V_e \ , \\ \bigoplus_{1<\lambda} V_\lambda^{\mathbb{C}} \subset V_e \otimes_{\mathbb{R}} \mathbb{C} \subset \bigoplus_{\lambda \notin [0,1]} V_\lambda^{\mathbb{C}} \ . \end{cases}$$

We shall concentrate our study on expanding spaces.

Let V_e be an expanding space. Then $u|_{V_e}: V_e \to V_e$ is an isomorphism. Let $\Gamma_c(\varphi_{V_e})$ be the endomorphism obtained as the composition:

$$R\Gamma_c(V_e; F|_{V_e}) \to R\Gamma_c(V_e; u^{-1}F|_{V_e}) \xrightarrow{\varphi} R\Gamma_c(V_e; F|_{V_e}) \ .$$

Proposition 9.6.12. *Let $u: V \to V$ be a linear map without non-trivial fixed point, and V_e an expanding subspace of V(cf. (9.6.13)).*

Let F be a conic object of $\mathbf{D}^b_{\mathbb{R}\text{-}c}(V)$ and let $\varphi \in \mathrm{Hom}(u^{-1}F, F)$. Then $C_0(\varphi) = \mathrm{tr}(\Gamma_c(\varphi_{V_e}))$.

Proof. (a) Assume first that u has no eigenvalue $\lambda \in \mathbb{C}$ with $|\lambda| = 1$. Consider the decomposition $V = V_+ \oplus V_-$ with:

$$V_+ = \left(\bigoplus_{|\lambda|>1} V_\lambda^{\mathbb{C}} \right) \cap V \ , \qquad V_- = \left(\bigoplus_{|\lambda|<1} V_\lambda^{\mathbb{C}} \right) \cap V$$

and choose a metric on V such that there exist constants c_1 and c_2 with $0 < c_1 < 1 < c_2$ and $|u(x)| \geqslant c_2|x|$ on V_+, $|u(x)| \leqslant c_1|x|$ on V_-.

Set $Z_{ab} = \{x \in V_+; |x| < a\} \times \{x \in V_-; |x| \leqslant b\}$. Then $u^{-1}(Z_{ab}) \cap Z_{ab}$ is open in Z_{ab} and closed in $u^{-1}(Z_{ab})$. Applying Proposition 9.6.9, we find:

$$(9.6.18) \qquad C_0(\varphi) = \mathrm{tr}(\Gamma(\varphi_{Z_{ab}})) \ .$$

Here $\Gamma(\varphi_{Z_{a,b}})$ is the endomorphism of $R\Gamma(X; F_{Z_{ab}}) \simeq R\Gamma_c(Z_{ab}; F|_{Z_{ab}})$ induced by $\varphi_{Z_{ab}}$.

Lemma 9.6.13. *For any $b > 0$, there exists $a_0 > 0$ such that $R\Gamma_c(Z_{a,b}; F) \to R\Gamma_c(Z_{\infty,b}; F)$ is an isomorphism for $a \geqslant a_0$.*

Proof. Set $Y = \{(t, x_+) \in \mathbb{R} \times V_+; t^2 + |x_+|^2 = 1\}$ and embed V_+ into Y by $x_+ \mapsto \left(\dfrac{1}{\sqrt{1+|x_+|^2}}, \dfrac{x_+}{\sqrt{1+|x_+|^2}} \right)$.

Then V_+ is an open subset of Y defined by $t > 0$. Let $j: V_+ \times V_- \to Y \times V_-$ be the open embedding. Then $F' = Rj_! F_{\{|x_-| \leqslant b\}}$ is w-$\mathbb{R}$-constructible. Now, we can apply Lemma 8.4.7, with $\varphi = t$. In fact, the assertion is equivalent to saying that $R\Gamma(\{|t| \leqslant \varepsilon\}; F') = 0$ for $0 < \varepsilon \ll 1$. □

End of the proof of Proposition 9.6.12. Let $q: V \to V_-$ be the projection. Then, since $Rq_!F$ is conic, we have for any $b > 0$:

$$(9.6.19) \qquad H_c^k(V_+;F) \simeq H^k(Rq_!F)_0 \simeq H^k(\{x_- \in V_-; |x_-| \leqslant b\}; Rq_!F)$$
$$\simeq \varinjlim_{a \to \infty} H_c^k(Z_{a,b}; F) \; .$$

Then by the preceding lemma, this is isomorphic to $H_c^k(Z_{a,b}; F)$ for $a \gg 1$. Hence we obtain a chain of isomorphisms $R\Gamma_c(V_+;F) \leftarrow\!\!\!\sim R\Gamma_c(Z_{\infty,b};F) \leftarrow\!\!\!\sim R\Gamma_c(Z_{a,b};F)$.

This proves the result in the particular case where $V_e = V_+$ and there is no eigenvalue λ with $|\lambda| = 1$.

(b) Replacing u by tu, $t \in \mathbb{R}$, $|1 - t| \ll 1$, does not affect $C_0(\varphi)$ by Proposition 9.6.8 (cf. Remark 9.6.15 below) and does not affect $\mathrm{tr}(\Gamma_c(\varphi_{V_e}))$ by Proposition 2.7.5. Hence we may assume that all eigenvalues λ satisfy $|\lambda| \neq 1$.

(c) Let $W = \bigoplus_{\lambda > 1} (V_\lambda^{\mathbb{C}} \cap V)$ and let V_e be an expanding space. By (a) it is enough to prove:

$$(9.6.20) \qquad \mathrm{tr}(\Gamma_c(\varphi_W)) = \mathrm{tr}(\Gamma_c(\varphi_{V_e})) \; ,$$

since (9.6.20) will in particular imply $\mathrm{tr}(\Gamma_c(\varphi_W)) = \mathrm{tr}(\Gamma_c(\varphi_{V_+}))$. To prove (9.6.20), let us denote by V' the space V_e/W, by u' the map $V' \to V'$ induced by u and by $p: V_e \to V_e/W$ the projection. Replacing V by V', u by u' and F by $Rp_!F$, it is enough to prove:

$$(9.6.21) \qquad \mathrm{tr}(\Gamma_c(\varphi_V)) = \mathrm{tr}(\varphi_0)$$

under the assumption that there is no eigenvalues $\lambda \in [0, \infty[$.

Since $R\Gamma_c(V;F) \leftarrow\!\!\!\sim R\Gamma_{\{0\}}(V;F)$ and $R\Gamma(V;F) \xrightarrow{\sim} F_0$, (9.6.21) is equivalent to saying that the trace of φ acting on $R\Gamma(V\backslash\{0\};F)$ is zero. Since F is conic, this trace is also that of $\gamma_*\varphi$ acting on $R\gamma_*F$, where γ is the map $V\backslash\{0\} \to S_V \underset{\mathrm{def}}{=} (V\backslash\{0\})/\mathbb{R}^+$. By the hypothesis, γ_*u (the map on S_V induced by u) has no fixed points. Hence $\mathrm{tr}(\gamma_*\varphi) = 0$ by Proposition 9.6.2. □

Similarly:

Proposition 9.6.14. *Let $u: V \to V$ be a linear map without non-trivial fixed point and let V_s be a shrinking space (cf. (9.6.14)). Let F be a conic object of $\mathbf{D}^b_{\mathbb{R}-c}(V)$ and let $\varphi \in \mathrm{Hom}(u^{-1}F, F)$. Then $C_0(\varphi) = \mathrm{tr}(\Gamma_{V_s}(\varphi))$.*

Since the proof is similar to that of Proposition 9.6.13, we shall not repeat it. Note that $R\Gamma_{V_s}(V;F)$ and $R\Gamma_c(V_e;F)$ are not necessarily isomorphic.

Remark 9.6.15. Let V be a real finite-dimensional vector space and denote as usual by p the projection $V \times \mathbb{R} \to V$ and by i_t the embedding $V \hookrightarrow V \times \mathbb{R}$, $x \mapsto (x, t)$. Let $u: V \times \mathbb{R} \to V$ be a continuous map, linear with respect to $x \in V$, let F be a conic object of $\mathbf{D}^b_{\mathbb{R}-c}(V)$ and let φ be a morphism from $u^{-1}F$ to $p^{-1}F$. Set $u_t = i_t \circ u$ and let $\varphi_t = i_t^{-1} \circ \varphi : u_t^{-1}F \to F$. Then if u_t does not have 1 as eigenvalue, $C_0(\varphi_t)$ does not depend on t. This follows from Proposition 9.6.8.

Corollary 9.6.16. *Assume X is a complex manifold and $F \in \mathrm{Ob}(\mathbf{D}^b_{\mathbb{C}-c}(X))$. Then assuming* (9.6.11), *we have $C_x(\varphi) = \mathrm{tr}(\varphi_x)$. If moreover 0 is not an eigenvalue of $f'(x)$ (i.e: f is a local isomorphism at x), then we also have $C_x(\varphi) = \mathrm{tr}(\Gamma_{\{x\}}(\varphi))$.*

Proof. Replacing F by $\nu_x F$, we may assume from the beginning that X is a complex vector space, f is linear and F is $\mathbb{C}^\times$-conic in the sense that F is locally constant on the $\mathbb{C}^\times$-orbits in X. Replacing f by λf with $\lambda \in \mathbb{C}$, $|\lambda - 1| \ll 1$, we may assume that any non-zero eigenvalues of f is not real. Then we take $V_e = \{0\}$. If 0 is not an eigenvalue we may take $V_e = T_x X$ and then we use the fact that $R\Gamma_c(T_x X; \nu_x F) \simeq R\Gamma_{\{x\}}(X; F)$. □

Example 9.6.17. Let X be an n-dimensional manifold, $F = k_X$ and let $\varphi : f^{-1}F \to F$ be the canonical map. Assume Γ_f and Δ intersect transversally at $x_\circ \in X$. Then:

$$C_{x_\circ}(\varphi) = \mathrm{sgn}(\det(1 - f'(x_\circ))) \ . \tag{9.6.22}$$

In fact, by Proposition 9.6.11, we may assume X is a vector space and f is linear. Take $V_e = \bigoplus_{\lambda > 1}(V_\lambda^{\mathbb{C}} \cap V)$ and apply Proposition 9.6.12. Then $\det_{V_e}(f|_{V_e}) > 0$. By Proposition 3.2.3 (iv), φ_{V_e} acts as the identity on $R\Gamma_c(V_e; k_X) \simeq k[-\dim V_e]$. Hence $C_0(\varphi) = \mathrm{tr}(\Gamma_c(\varphi_{V_e})) = (-1)^{\dim V_e} = \mathrm{sgn}(\det(1 - f'(x_\circ))$. (The last equality follows from: $\det_V(1 - f'(x_\circ)) = \det_{V_e}(1 - f'(x_\circ)) \cdot \det_{V/V_e}(1 - f'(x_\circ))$ and $\det_{V/V_e}(1 - f'(x_\circ)) > 0$, $\det_{V_e}(f'(x_\circ) - 1) > 0$.)

Example 9.6.18. Let f, F, φ be as in Example 9.6.5 and assume $f'(t_0) \neq 1$. Then:

$$C_{t_0}(\varphi) = \begin{cases} 0 & \text{if } f'(t_0) > 1 \ , \\ 1 & \text{if } 0 \leqslant f'(t_0) < 1 \ . \end{cases} \tag{9.6.23}$$

In fact, $C_{t_0}(\varphi) = \mathrm{tr}(\Gamma_c(\varphi_{V_e}))$, and we take $V_e = T_x X$ if $f'(t_0) > 1$ and $V_e = \{0\}$ if $0 \leqslant f'(t_0) < 1$. Since $\nu_{t_0} F \simeq k_{\{t \geqslant 0\}}$, we have $R\Gamma_c(T_{\{t_0\}}X; \nu_{t_0}F) = 0$ and $R\Gamma_c(\{0\}; \nu_{t_0}F) \simeq k$. Then (9.6.23) follows.

9.7. Constructible functions and Lagrangian cycles

A **constructible function** φ on X is a $\mathbb{Z}$-valued function such that:

$$\begin{cases} \textit{for each } m \in \mathbb{Z}, \varphi^{-1}(m) \textit{ is subanalytic} \\ \textit{and the family } \{\varphi^{-1}(m)\}_{m \in \mathbb{Z}} \textit{ is locally finite} \ . \end{cases} \tag{9.7.1}$$

By the results of Chapter VIII, a $\mathbb{Z}$-valued function φ is constructible if and only if there exists a μ-stratification $X = \bigsqcup_\alpha X_\alpha$ such that $\varphi|_{X_\alpha}$ is constant for each α. Let $\mathbf{1}_A$ denote the characteristic function of a subset A of X. By the triangulation theorem, a function φ is constructible if and only if there exists a locally finite covering $X = \bigcup_\alpha X_\alpha$, where the X_α's are compact, subanalytic, and contractible, and there exist integers m_α, with $\varphi = \sum_\alpha m_\alpha \mathbf{1}_{X_\alpha}$.

The set of constructible functions on X is naturally endowed with a structure of an algebra and we denote by $\mathrm{CF}(X)$ this algebra. The presheaf $U \mapsto \mathrm{CF}(U)$, (U open in X) is clearly a sheaf (of groups). We denote if by $\mathscr{CF}_X$, and we call it the sheaf of ($\mathbb{Z}$-valued) constructible functions on X. Note that $\mathscr{CF}_X$ is soft.

Now let $F \in \mathrm{Ob}(\mathbf{D}^b_{\mathbb{R}-c}(X))$, (the base ring is again a field k of characteristic zero). Its local Euler-Poincaré index $\chi(F)(x) = \sum_j (-1)^j \dim H^j(F)_x$ is clearly a constructible function. Moreover:

$$\left\{\begin{aligned} &\chi(F \oplus G) = \chi(F) + \chi(G) \\ &\chi(F \otimes G) = \chi(F) \cdot \chi(G) \end{aligned}\right. \tag{9.7.2}$$

and if $F' \to F \to F'' \xrightarrow[+1]{}$ is a distinguished triangle, then:

$$\chi(F) = \chi(F') + \chi(F'') \tag{9.7.3}$$

(cf. Exercise I. 34).

We shall denote by $\mathrm{K}_{\mathbb{R}-c}(X)$ the Grothendieck group of $\mathbf{D}^b_{\mathbb{R}-c}(X)$ (cf. Exercise I.27). This group is obtained as the quotient of the free abelian group generated by $\mathrm{Ob}(\mathbf{D}^b_{\mathbb{R}-c}(X))$ by the relations $F = F' + F''$ if there exists a distinguished triangle $F' \to F \to F'' \xrightarrow[+1]{}$. By Theorem 8.1.11, $\mathrm{K}_{\mathbb{R}-c}(X)$ is also the Grothendieck group of the abelian category $\mathbb{R}$-$\mathfrak{Cons}(X)$. By (9.7.3), the local Euler-Poincaré index χ induces a group homomorphism, that we still denote by χ:

$$\chi : \mathrm{K}_{\mathbb{R}-c}(X) \to \mathrm{CF}(X) \tag{9.7.4}$$

Theorem 9.7.1. *The morphism* χ *in* (9.7.4) *is an isomorphism.*

Proof. (a) Let $\varphi \in \mathrm{CF}(X)$. Choose a subanalytic stratification $X = \bigsqcup_{\alpha \in A} X_\alpha$ such that $\varphi = \sum_{\alpha \in A} m_\alpha \mathbf{1}_{X_\alpha}$, $m_\alpha \in \mathbb{Z}$. Let $\varepsilon_\alpha = \mathrm{sgn}(m_\alpha)$, for $m_\alpha \neq 0$ and define:

$$F = \bigoplus_{\alpha \in A'} k_{X_\alpha}^{|m_\alpha|}\left[\frac{1-\varepsilon_\alpha}{2}\right] \tag{9.7.5}$$

where $A' = \{\alpha \in A; m_\alpha \neq 0\}$.

Clearly F belongs to $\mathbf{D}^b_{\mathbb{R}-c}(X)$ and $\chi(F) = \varphi$. This proves the surjectivity of χ. (Note that F satisfies: $\mathrm{supp}(F) = \mathrm{supp}(\varphi)$. We shall use this remark latter.)

(b) Let us prove that χ is injective. An element u of $\mathrm{K}_{\mathbb{R}-c}(X)$ is represented by a finite sum $u = \sum_j a_j[F_j]$, with $a_j \in \mathbb{Z}$ and $[F_j]$ denotes the image of $F_j \in \mathrm{Ob}(\mathbf{D}^b_{\mathbb{R}-c}(X))$ in $\mathrm{K}_{\mathbb{R}-c}(X)$. Setting, for $F \in \mathrm{Ob}(\mathbf{D}^b_{\mathbb{R}-c}(X))$, $F^k = F \oplus \cdots \oplus F$ (k

times) for $k \in \mathbb{N}$ and $F^k = F^{-k}[1]$ for $k < 0$, we also have: $u = [\bigoplus_j F^{a_j}]$. Hence any $u \in \mathrm{K}_{\mathbb{R}-c}(X)$ is represented by a single object F of $\mathbf{D}^b_{\mathbb{R}-c}(X)$.

Let $X = \bigsqcup_\alpha Z_\alpha$ be a subanalytic stratification such that $H^j(F)|_{Z_\alpha}$ is a constant for all j, all α. Let X_k denote the union of the k-codimensional strata. From the distinguished triangle $F_{X_0} \longrightarrow F \longrightarrow F_{X \setminus X_0} \underset{+1}{\longrightarrow}$, we deduce:

$$[F] = [F_{X_0}] + [F_{X \setminus X_0}] \ .$$

Continuing this procedure, we get $[F] = \sum_k [F_{X_k}]$, thus (cf. Exercise I.27):

$$[F] = \sum_{j,k} (-1)^j [H^j(F)_{X_k}] \ .$$

Hence $\chi(F) = 0$ implies that for any α:

$$\dim \bigoplus_{j \text{ even}} H^j(F)_{Z_\alpha} = \dim \bigoplus_{j \text{ odd}} H^j(F)_{Z_\alpha} \ .$$

Thus:

$$\bigoplus_{j \text{ even}} H^j(F)_{Z_\alpha} \simeq \bigoplus_{j \text{ odd}} H^j(F)_{Z_\alpha}$$

and

$$\bigoplus_{j \text{ even}} H^j(F)_{X_k} \simeq \bigoplus_{j \text{ odd}} H^j(F)_{X_k}$$

(because X_k is a disjoint union of Z_α's). This shows $[F] = 0$. □

Now we shall study some natural operations on constructible functions. Let X and Y be two manifolds.

(a) One defines the "external product":

$$\mathscr{CF}_X \boxtimes \mathscr{CF}_Y \to \mathscr{CF}_{X \times Y} \tag{9.7.6}$$

by setting $(\varphi \otimes \psi)(x, y) = \varphi(x) \cdot \psi(y)$. Clearly, if $\varphi = \chi(F)$, $\psi = \chi(G)$, then $\varphi \boxtimes \psi = \chi(F \boxtimes G)$.

(b) Let $f: Y \to X$ be a morphism of manifolds. One defines the "inverse image":

$$f^*: f^{-1}\mathscr{CF}_X \to \mathscr{CF}_Y \tag{9.7.7}$$

by setting $(f^*\varphi)(y) = \varphi(f(y))$. If $\varphi = \chi(F)$, then clearly $f^*\varphi = \chi(f^{-1}F)$.

(c) Let $\varphi \in \Gamma_c(X; \mathscr{CF}_X)$. One defines the "integral" of φ, denoted $\int_X \varphi$, as follows. Choose $F \in \mathrm{Ob}(\mathbf{D}^b_{\mathbb{R}-c}(X))$ with compact support such that $\chi(F) = \varphi$. One sets:

$$\int_X \varphi = \chi(X; F) \ .$$

This number depends only on φ by the additivity of the functor $\chi(X; \cdot)$ with respect to distinguished triangles (and Theorem 9.7.1).

Note that $\int_X \varphi$ may also be calculated as follows. Choose a finite family $\{X_\alpha\}$ of relatively compact locally closed subanalytic subsets such that $\varphi = \sum_\alpha m_\alpha \mathbf{1}_{X_\alpha}$, $m_\alpha \in \mathbb{Z}$. Then, we have:

$$\int_X \varphi = \sum_\alpha m_\alpha \cdot \chi(X; k_{X_\alpha}) \ . \tag{9.7.8}$$

In particular, if one chooses the X_α's compact and contractible, $\int_X \varphi = \sum_\alpha m_\alpha$.

(d) Let $f: Y \to X$ be a morphism of manifolds. One defines the "direct image":

$$f_*: f_!\mathscr{CF}_Y \to \mathscr{CF}_X \tag{9.7.9}$$

by setting:

$$(f_*\psi)(x) = \int_Y \psi \cdot \mathbf{1}_{f^{-1}(x)} \ .$$

To check that $f_*\psi$ is constructible, one applies Theorem 9.7.1 and represents ψ by $\chi(G)$ with $G \in \mathrm{Ob}(\mathbf{D}^b_{\mathbb{R}-c}(Y))$, such that f is proper on $\mathrm{supp}(G)$. Then

$$f_*\chi(G) = \chi(Rf_!G)$$

since $\chi(Rf_!G)(x) = \chi(Y; G \otimes k_{f^{-1}(x)})$.

(e) Finally if $\varphi \in \mathscr{CF}_X$, we shall introduce its "dual function", denoted $\mathrm{D}_X\varphi$ (or simply $\mathrm{D}\varphi$). We represent φ as $\varphi = \chi(F)$ and we set $\mathrm{D}_X\varphi = \chi(\mathrm{D}_X F)$. Since D_X is a triangulated functor and χ is additive, this definition makes sense. Note that $(\mathrm{D}_X\varphi)(x) = \chi_c(F)(x)$. Hence $(\mathrm{D}_X\varphi)(x)$ may also be calculated as follows. Choose a local coordinate system centered at x, and denote by $B(x, \varepsilon)$ the open ball with center x and radius ε. For $\varepsilon > 0$ small enough, we have by Lemma 8.4.7:

$$\begin{aligned} R\Gamma_{\{x\}}(X; F) &\simeq R\Gamma_c(B(x; \varepsilon); F) \\ &\simeq R\Gamma(X; F \otimes k_{B(x,\varepsilon)}) \ . \end{aligned}$$

Hence:

$$(\mathrm{D}_X\varphi)(x) = \int_X \mathbf{1}_{B(x,\varepsilon)}\varphi \tag{9.7.10}$$

for $0 < \varepsilon \ll 1$. Note that $(\mathrm{D}_X\varphi)(x)$ depends only on the germ of φ at x. Hence D_X is a sheaf endomorphism of $\mathscr{CF}_X$:

$$\mathrm{D}_X: \mathscr{CF}_X \to \mathscr{CF}_X \ . \tag{9.7.11}$$

Proposition 9.7.2. (i) *Let $\varphi \in \mathrm{CF}(X)$. Then $\mathrm{D}_X \circ \mathrm{D}_X\varphi = \varphi$.*

(ii) *Let $f: Y \to X$ be a morphism of manifolds and let $\psi \in \Gamma(X; f_!\mathscr{CF}_Y)$. Then $f_*\mathrm{D}_Y\psi = \mathrm{D}_X f_*\psi$.*

Proof. This follows from the corresponding properties for sheaves, and Theorem 9.7.1. □

Example 9.7.3. Let Z be a locally closed subanalytic subset of X and assume that for any $x \in \bar{Z}$ there is a fundamental system of open neighborhoods U of x in X such that $U \cap Z$ is homeomorphic to $\mathbb{R}^d$. Then one proves easily that $D_X k_Z \simeq k_{\bar{Z}}[d]$. Hence:

$$D_X \mathbf{1}_Z = (-1)^d \mathbf{1}_{\bar{Z}} \ . \tag{9.7.12}$$

Assume moreover that $\bar{Z}$ is compact and $\int_X \mathbf{1}_{\bar{Z}} = 1$. Set $\partial Z = \bar{Z} \backslash Z$. Since $\int_X \mathbf{1}_{\partial Z} = \int_X \mathbf{1}_{\bar{Z}} - \int_X \mathbf{1}_Z$ and $\int_X \mathbf{1}_Z = \int_X D_X \mathbf{1}_Z$, we get:

$$\chi(\partial Z; k_{\partial Z}) = \int_X \mathbf{1}_{\partial Z} = 1 - (-1)^d \ . \tag{9.7.13}$$

This is a generalization of the Euler formula.

With these operations in hands, we can now construct the analogous operations for constructible functions to those we have defined for sheaves.

Let $f: Y \to X$ be a morphism of manifolds. We define:

$$f^! : f^{-1}\mathscr{CF}_X \to \mathscr{CF}_Y \tag{9.7.14}$$

by:

$$f^! = D_Y \circ f^* \circ D_X \ .$$

If $\varphi = \chi(F)$, $F \in \mathrm{Ob}(\mathbf{D}^b_{\mathbb{R}-c}(X))$, we get that $f^!\chi(F) = \chi(f^!F)$. Similarly, we define:

$$hom : \mathscr{CF}_X \times \mathscr{CF}_X \to \mathscr{CF}_X \tag{9.7.15}$$

by setting:

$$hom(\psi, \varphi) = D_X(\psi \cdot (D_X \varphi)) \ .$$

By Proposition 3.4.6 we get for F and G in $\mathbf{D}^b_{\mathbb{R}-c}(X)$, that

$$hom(\chi(G), \chi(F)) = \chi(R\mathscr{H}om(G, F)) \ .$$

In order to define the microlocalization of constructible functions, we need to study homogeneous constructible functions on a vector bundle.

Let $\tau: E \to Z$ be a vector bundle. One denotes by $\mathscr{CF}_{E/\mathbb{R}^+}$ the subsheaf of $\mathscr{CF}_E$ consisting of functions which are constant on the orbits of $\mathbb{R}^+$. One sets:

$$\mathrm{CF}_{\mathbb{R}^+}(E) = \Gamma(E; \mathscr{CF}_{E/\mathbb{R}^+}) \tag{9.7.16}$$

Let $\mathbf{D}^b_{\mathbb{R}-c,\,\mathbb{R}^+}(E)$ denote the full triangulated subcategory of $\mathbf{D}^b_{\mathbb{R}-c}(E)$ consisting of conic objects, and let $\mathrm{K}_{\mathbb{R}-c,\,\mathbb{R}^+}(E)$ denote its Grothendieck group.

Lemma 9.7.4. *The local Euler Poincaré index χ induces an isomorphism:*

$$\chi : \mathrm{K}_{\mathbb{R}-c,\,\mathbb{R}^+}(E) \xrightarrow{\sim} \mathrm{CF}_{\mathbb{R}^+}(E)$$

Proof. Let $i: Z \hookrightarrow E$ denote the zero-embedding, and let $j: \dot{E} \hookrightarrow E$ and $\gamma: \dot{E} \to \dot{E}/\mathbb{R}^+$ denote the natural maps. Set for short $S_E = \dot{E}/\mathbb{R}^+$. Then we have a commutative diagram:

$$\begin{array}{ccccccccc} 0 & \longrightarrow & \mathrm{K}_{\mathbb{R}-c}(Z) & \xrightarrow[\alpha]{} & \mathrm{K}_{\mathbb{R}-c,\,\mathbb{R}^+}(E) & \xrightarrow[\beta]{} & \mathrm{K}_{\mathbb{R}-c}(S_E) & \longrightarrow & 0 \\ & & \downarrow & & \downarrow & & \downarrow & & \\ 0 & \longrightarrow & \mathrm{CF}(Z) & \xrightarrow[\alpha']{} & \mathrm{CF}_{\mathbb{R}^+}(E) & \xrightarrow[\beta']{} & \mathrm{CF}(S_E) & \longrightarrow & 0 \end{array} \tag{9.7.17}$$

where α is defined by the functor $Ri_* : \mathbf{D}^b_{\mathbb{R}-c}(Z) \to \mathbf{D}^b_{\mathbb{R}-c,\,\mathbb{R}^+}(E)$, β by the functor $R\gamma_* \circ j^{-1} : \mathbf{D}^b_{\mathbb{R}-c,\,\mathbb{R}^+}(E) \to \mathbf{D}^b_{\mathbb{R}-c}(S_E)$, α' is the "extension by zero" morphism and β' is the natural one. Then the first and third vertical arrows in (9.7.17) are isomorphisms by Theorem 9.7.1, and the bottom row is clearly an exact sequence. It is thus enough to show that the top row is an exact sequence. Let $\beta'' : \mathrm{K}_{\mathbb{R}-c}(S_E) \to \mathrm{K}_{\mathbb{R}-c,\,\mathbb{R}^+}(E)$ be the map derived from $F \mapsto Rj_!\gamma^{-1}F$ (cf. Proposition 8.3.8(i)) and $\alpha'' : \mathrm{K}_{\mathbb{R}-c,\,\mathbb{R}^+}(E) \to \mathrm{K}_{\mathbb{R}-c}(Z)$ the map derived from $F \mapsto i^{-1}F$. Then, for $F \in \mathbf{D}^b_{\mathbb{R}-c,\,\mathbb{R}^+}(E)$, we have a distinguished triangle $Rj_!\gamma^{-1}R\gamma_*F \longrightarrow F \longrightarrow i_*i^{-1}F \xrightarrow{+1}$. Hence we obtain $1 = \beta'' \circ \beta + \alpha \circ \alpha''$, $\alpha'' \circ \beta'' = 0$. This shows the exactness of the first horizontal line. □

We then define the morphisms:

$$\tau_! \quad \text{and} \quad \tau_* : \tau_*\mathscr{CF}_{E/\mathbb{R}^+} \to \mathscr{CF}_Z \tag{9.7.18}$$

by setting:

$$\tau_!\varphi = i^!\varphi \quad \text{and} \quad \tau_*\varphi = i^*\varphi \tag{9.7.19}$$

where i is the zero-embedding $Z \hookrightarrow E$.

We can describe $\tau_!$ and τ_* in another way. Choose a metric on the vector bundle E and let $B(\varepsilon)$ be the open set of E such that for each $z \in Z$, $B(\varepsilon) \cap \tau^{-1}(z)$ is the open ball of radius ε and center 0 for the induced norm. Let $\varphi \in \mathrm{CF}_{\mathbb{R}^+}(E)$. It follows from Proposition 3.7.5 that:

$$\tau_!\varphi = \tau_*(\varphi \cdot \mathbf{1}_{B(\varepsilon)}) \quad \text{and} \quad \tau_*\varphi = \tau_*(\varphi \cdot \mathbf{1}_{\overline{B(\varepsilon)}}) \,. \tag{9.7.20}$$

We can now define the Fourier transform of constructible functions. We follow the notations of III §7 and denote by $\pi: E^* \to Z$ the dual vector bundle and by p_1 and p_2 the projections from $E \times_Z E^*$ to E and E^*, respectively. We set $P' = \{(x, y) \in E \times_Z E^*; \langle x, y \rangle \leqslant 0\}$ and we denote by i_E the zero-embedding $E^* \simeq Z \times_Z E^* \hookrightarrow E \times_Z E^*$.

Definition 9.7.5. *Let $\varphi \in \mathrm{CF}_{\mathbb{R}^+}(E)$.*

One sets:

$$\varphi^\wedge = p_{2!}((p_1^*\varphi) \cdot \mathbf{1}_{P'}) \,,$$

$$\varphi^\vee = (-1)^n \varphi^{\wedge a} ,$$

where n is the fiber dimension of E, and $\varphi^{\wedge a} = a^*\varphi^\wedge$ *is the image of* $\varphi^\wedge$ *by the antipodal map.*

One calls $\varphi^\wedge$ *(resp.* $\varphi^\vee$*) the Fourier transform (resp. inverse Fourier transform) of* φ.

Note that one also has:

(9.7.21)
$$\varphi^\wedge = i_E^!((p_1^*\varphi)\cdot \mathbf{1}_{P'}) .$$

By the construction of $^\wedge$, if $F \in \mathrm{Ob}(\mathbf{D}^b_{\mathbf{R}-c}(E))$ and F is conic, then:

(9.7.22)
$$\chi(F)^\wedge = \chi(F^\wedge) .$$

Hence by Lemma 9.7.4 and Theorem 3.7.9, we get:

Proposition 9.7.6. *Let* $\varphi \in \mathrm{CF}_{\mathbf{R}^+}(E)$. *Then* $\varphi^\wedge$ *and* $\varphi^\vee$ *belong to* $\mathrm{CF}_{\mathbf{R}^+}(E^*)$ *and moreover*:

$$\varphi^{\wedge\vee} = \varphi^{\vee\wedge} = \varphi .$$

It is an easy exercise (left to the reader) to "translate" Propositions 3.7.13, 3.7.14 and 3.7.15 to obtain similar results for constructible functions.

Let us only note that the Fourier transformation commutes to base changes. In particular one can calculate $\varphi^\wedge$ fiberwise, that is, if $\varphi \in \mathrm{CF}_{\mathbf{R}^+}(E)$ and $z \in Z$, then:

(9.7.23)
$$(\varphi^\wedge)|_{\pi^{-1}(z)} = (\varphi|_{\tau^{-1}(z)})^\wedge .$$

Examples 9.7.7. (i) Assume for simplicity E is a vector space. Let $y \in E^*$. Then by its definition:

$$\varphi^\wedge(y) = \int_E (\varphi(x)\cdot \mathbf{1}_{|x|<\varepsilon}\cdot \mathbf{1}_{\langle x,y\rangle \leqslant 0}) .$$

(ii) Let U be an open convex cone of E and let γ be a closed convex proper cone of E containing Z. Then:

$$(\mathbf{1}_U)^\wedge = (-1)^n \mathbf{1}_{U^{\circ a}} ,$$

$$(\mathbf{1}_\gamma)^\wedge = \mathbf{1}_{\mathrm{Int}\,\gamma^\circ} .$$

This follows from Lemma 3.7.10, (or this can be calculated directly).

Finally we shall define the specialization and microlocalization of constructible functions.

Let M be a closed submanifold of X, i the embedding $M \hookrightarrow X$ and recall the maps p, j, s and diagram (4.1.5) of Chapter IV.

Definition 9.7.8. (i) *Let* $\varphi \in \mathrm{CF}(X)$.
One sets:

$$\nu_M(\varphi) = -s^!((p^*\varphi)\cdot \mathbf{1}_\Omega) \ ,$$

$$\mu_M(\varphi) = \nu_M(\varphi)^\wedge \ .$$

(ii) *Let* φ *and* ψ *belong to* $\mathrm{CF}(X)$.
One sets:

$$\mu hom(\psi, \varphi) = \mu_\Delta(\varphi \boxtimes \mathrm{D}_X\psi) \ .$$

Hence we have defined the morphisms:

$$\nu_M : i^{-1}\mathscr{CF}_X \to \tau_*\mathscr{CF}_{T_MX/\mathbb{R}^+} \ ,$$

$$\mu_M : i^{-1}\mathscr{CF}_X \to \pi_*\mathscr{CF}_{T^*_MX/\mathbb{R}^+} \ ,$$

$$\mu hom : \mathscr{CF}_X \times \mathscr{CF}_X \to \pi_*\mathscr{CF}_{T^*X/\mathbb{R}^+} \ .$$

Of course these morphisms "commute" to χ. For example

(9.7.24) $$\chi(\mu hom(G, F)) = \mu hom(\chi(G), \chi(F))$$

for F and G in $\mathbf{D}^b_{\mathbb{R}-c}(X)$.

Example 9.7.9. (i) Let (x', x'') be a local coordinate system on X such that $M = \{x' = 0\}$. Let t be the coordinate on $\mathbb{R}$ and let j denote the embedding $X \times \mathbb{R}^+ \hookrightarrow X \times \mathbb{R}$. Then:

$$(\nu_M(\varphi))(x', x'') = -\mathrm{D}_X((\mathrm{D}_{X\times\mathbb{R}}(\varphi(tx', x'')\cdot \mathbf{1}_{\{t>0\}}))|_{t=0}) \ .$$

(ii) Let (x, y) denote the coordinates on $\mathbb{R}^2$ and set $M = \{0\}$, $A = \{y = 0, x \geqslant 0\} \cup \{y = x^2, x \geqslant 0\}$. Then $C_M(A) = \{y = 0, x \geqslant 0\}$ but $\nu_M(\mathbf{1}_A) = 2\mathbf{1}_{C_M}(A) - \mathbf{1}_{\{0\}}$.

Hence, in general, $\nu_M(\mathbf{1}_A) \neq \mathbf{1}_{C_M(A)}$. (However the support of $\nu_M(\mathbf{1}_A)$ is always contained in $C_M(A)$.)

(iii) Let $\tau : E \to Z$ be a vector bundle and let $\varphi \in \mathrm{CF}_{\mathbb{R}^+}(E)$. Then $\nu_Z(\varphi) = \varphi$.

Let us now study the group of Lagrangian cycles on T^*X.

Let $\mathscr{L}_X$ denote the sheaf on T^*X of Lagrangian cycles over the ring $\mathbb{Z}$. By Proposition 9.4.5, the characteristic cycle $\mathrm{CC}(\cdot)$ defines a homomorphism (still denoted CC):

(9.7.25) $$\mathrm{CC} : \mathrm{K}_{\mathbb{R}-c}(X) \to H^0(T^*X; \mathscr{L}_X) \ .$$

Theorem 9.7.10. *The homomorphism* CC *in* (9.7.25) *is an isomorphism.*

Proof. (a) Let λ be a Lagrangian cycle. We shall construct $F \in \mathrm{Ob}(\mathbf{D}^b_{\mathbb{R}-c}(X))$ with $\mathrm{CC}(F) = \lambda$. Let $\Lambda = \mathrm{supp}(\lambda)$. This is a conic subanalytic isotropic subset of T^*X.

We shall argue by induction on $\dim \pi(\Lambda)$. There exists a subanalytic submanifold Y of X such that Y is an open subset of $\pi(\Lambda)$, $\dim(\pi(\Lambda)\setminus Y) < \dim \pi(\Lambda)$ and $\Lambda \cap \pi^{-1}(Y) \subset T_Y^*X$. Hence $\Gamma_{T_Y^*X}(\mathcal{L}_X)$ is locally constant of rank one on $\pi^{-1}(Y)$. Therefore there exists $F \in \mathrm{Ob}(\mathbf{D}^b_{\mathbb{R}\text{-}c}(X))$ such that $\mathrm{supp}(F) \subset \bar{Y}$, $H^j(F)|_Y$ is locally constant, and $\mathrm{CC}(F) = \lambda$ on $\pi^{-1}(Y)$. Setting $\mu = \lambda - \mathrm{CC}(F)$, we have $\dim \pi(\mathrm{supp}(\mu)) < \dim \pi(\Lambda)$ and, replacing λ with μ, the induction proceeds.

(b) Let $F \in \mathrm{Ob}(\mathbf{D}^b_{\mathbb{R}\text{-}c}(X))$, and assume $\mathrm{CC}(F) = 0$. We shall prove that $F = 0$ in $K_{\mathbb{R}\text{-}c}(X)$. By Theorem 9.7.1, it is enough to prove that $\chi(F)(x) = 0$ for all x. But this follows from Remark 9.5.5. □

One defines an "Euler morphism":

(9.7.26) $$Eu : \pi_* \mathcal{L}_X \to \mathcal{CF}_X$$

as follows. Let $x \in X$ and let ϕ satisfying $\phi(x) = 0$, $d\phi(x) = 0$ and the Hessian of ϕ at x is positive definite. Let λ be a section of $\pi_* \mathcal{L}_X$. One sets

$$Eu(\lambda)(x) = \#([\sigma_\phi] \cap \lambda)_x \, .$$

By Theorem 9.7.10 and Remark 9.5.5, Eu is well-defined, and moreover we obtain:

Theorem 9.7.11. *The diagram*:

$$\begin{array}{ccc} & \mathrm{K}_{\mathbb{R}\text{-}c}(X) & \\ {\scriptstyle \mathrm{CC}} \swarrow & & \searrow {\scriptstyle \chi} \\ H^0(T^*X; \mathcal{L}_X) & \xrightarrow[Eu]{} & H^0(X; \mathcal{CF}_X) \end{array}$$

is commutative, and the arrows are isomorphisms.

Exercises to Chapter IX

Exercise IX.1. Let F be an $\mathbb{R}$-constructible sheaf on a manifold X. Prove that $\mathcal{H}om(F, \mathcal{CS}^X)$ is isomorphic to $R\mathcal{H}om(F, \omega_X)$ in $\mathbf{D}^b(X)$. (Hint: use Exercise IX.2.)

Exercise IX.2. Let S be a closed subanalytic subset of a manifold X. A continuous real valued function φ on S is called subanalytic if its graph is subanalytic in $X \times \mathbb{R}$.

(i) Prove that the sheaf $\mathcal{S}_S$ of subanalytic functions on S is a soft sheaf of rings and the morphism $\mathcal{S}_X \to \mathcal{S}_S$ is surjective.

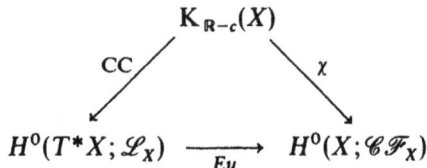

(ii) A ringed space $(X; \mathcal{S})$ is called subanalytic if it is locally isomorphic to a subanalytic subset of a real analytic manifold, endowed with the sheaf of rings of subanalytic functions. Define subanalytic subsets in X, $\mathbb{R}$-constructible sheaves, the complex of sheaves of subanalytic cycles $\mathscr{CS}^X_\cdot$ and prove that $\mathscr{CS}^X_\cdot$ is isomorphic to ω_X in $\mathbf{D}^b(X)$. Prove also the similar statement to that of Exercise IX.1. (Hint: use Hironaka's result in Exercise IX.3.)

Exercise IX.3. Let X be a manifold of dimension n, and we assume the base ring A is $\mathbb{C}$. Prove that there exists a morphism of complexes:

$$\mathscr{CS}^X_\cdot \xrightarrow{\psi} \mathscr{D}b^\cdot_X \otimes or_X[n] \ , \tag{9.E.1}$$

which is a quasi-isomorphism. Here, $\mathscr{D}b^\cdot_X$ is the de Rham complex with distribution coefficients. (Hint: one shall use the following result due to Hironaka – the authors thank him to allow them to insert it here – :

"let S be a locally closed real analytic submanifold of X of dimension p and subanalytic in X. Then there exists a p-dimensional real analytic manifold Y, a morphism of real analytic manifold $f: Y \to X$ and an open embedding $i: S \hookrightarrow Y$ such that $i(S)$ is subanalytic in Y, f is proper on $\overline{i(S)}$ and $f \circ i$ coincides with the inclusion $S \hookrightarrow X$. (One can even assume that $i(S)$ is locally defined by inequalities $\{y_1 > 0, \ldots, y_k > 0\}$ in Y, where $(y_1, \ldots, y_p)$ is a local coordinate system on Y)".

Then define ψ in (9.E.1) as follows. For a p-dimensional oriented subanalytic submanifold S and $\theta \in \Gamma_c(X; \mathscr{C}_X^{\infty,(p)})$ set: $\int \psi([S])\theta = \int_S \theta$.)

Exercise IX.4. Let $f: Y \to X$ be a morphism of manifolds and let $\varphi: Y \to \mathbb{R}$ be a real analytic function. For $\lambda \in \mathbb{R}$ define the map $\rho_\lambda: Y \times_X T^*X \to T^*Y$ by $p \mapsto {}^t f'(p) + \lambda\, d\varphi(\pi(p))$. If S is a closed subset of T^*Y, define:

$$\alpha_\lambda : H^0_S(T^*Y; \pi_Y^{-1}\omega_Y) \to H^0_{\rho_\lambda^{-1}(S)}\left(Y \underset{X}{\times} T^*X; \pi^{-1}\omega_Y\right)$$

by using the morphism $\rho_\lambda^{-1}\pi_Y^{-1}\omega_Y \simeq \pi^{-1}\omega_Y$. If T is a closed subset of $Y \times_X T^*X$ such that f_π is proper on T, define:

$$\beta : H^0_T\left(Y \underset{X}{\times} T^*X; \pi^{-1}\omega_Y\right) \to H^0_{f_\pi(T)}(T^*X; \pi_X^{-1}\omega_X)$$

by using the morphisms: $Rf_{\pi!}\pi^{-1}\omega_X \simeq \pi_X^{-1}Rf_!\omega_Y \to \pi_X^{-1}\omega_X$.

Now let $G \in \mathrm{Ob}(\mathbf{D}^b_{\mathbb{R}-c}(Y))$, and assume f is proper on $\mathrm{supp}(G) \cap \overline{Y}_t$ for all t, where $Y_t = \{\varphi < t\}$. Let $t_\circ \in \mathbb{R}$, and set for some $\lambda > 0$ fixed:

$$T_\pm = \bigcup_{0 \leqslant \pm\mu \leqslant \lambda} \rho_\mu^{-1}(\mathrm{SS}(G) \cap \pi^{-1}(\overline{Y}_{t_\circ})) \ .$$

(i) Assume that for $y \in Y \setminus Y_{t_\circ}$,

$$d\varphi(y) \notin \mathrm{SS}(G) + {}^t f'\left(Y \underset{X}{\times} T^*X\right) .$$

Prove that $Rf_*G \in \mathrm{Ob}(\mathbf{D}^b_{\mathbb{R}-c}(X))$, $\rho_\lambda^{-1}(SS(G))$ is contained in T_+ and $CC(Rf_*G) = \beta \circ \alpha_\lambda(CC(G))$ in $H^0_{f_\pi(T_+)}(T^*X; \pi_X^{-1}\omega_X)$.

(ii) Prove a similar result, replacing $d\varphi(y)$ by $-d\varphi(y)$, Rf_*G by $Rf_!G$, α_λ by $\alpha_{-\lambda}$ and T_+ by T_-.

Exercise IX.5. Let $\tau: E \to Z$ be a vector bundle, let θ be the Euler vector field on E and set $P_\pm = \{\pm\theta \geqslant 0\} \subset T^*E$. Let Λ be a closed conic subset of $P_+ \cap P_-$ and denote by i the embedding $T^*Z \hookrightarrow P_\pm \subset T^*E$.

(i) Define the morphisms:

$$\varphi_\pm : R\Gamma_\Lambda(\pi_E^{-1}\omega_E) \to i_*\pi_Z^{-1}\omega_Z .$$

(Hint: consider the morphisms: $R\Gamma_\Lambda(\pi_E^{-1}\omega_E) \simeq R\Gamma_\Lambda R\Gamma_{P_\pm}(\pi_E^{-1}\omega_E) \to R\Gamma_\Lambda(A_{P_\mp} \otimes R\Gamma_{P_\pm}(\pi_E^{-1}\omega_E))$ and use $A_{P_\mp} \otimes R\Gamma_{P_\pm}(\pi_E^{-1}\omega_E) \simeq i_*\pi_Z^{-1}\omega_Z$.)

(ii) Define: $\psi_\pm : H^0_\Lambda(T^*E; \pi_E^{-1}\omega_E) \to H^0_{i^{-1}(\Lambda)}(T^*Z; \pi_Z^{-1}\omega_Z)$.

(iii) Let F be a conic object of $\mathbf{D}^b_{\mathbb{R}-c}(E)$. Prove that:

$$CC(R\tau_*F) = \psi_-(CC(F)) ,$$

$$CC(R\tau_!F) = \psi_+(CC(F)) .$$

(Hint: use Exercise IX 4.)

Exercise IX.6. Let X be a compact manifold with odd dimension. Prove that $\chi(X; k_X) = 0$.

Exercise IX.7. Let F be a conic $\mathbb{R}$-constructible object of $\mathbf{D}^b(E)$, where E is a vector bundle. Using Exercise VII.2, prove that $CC(F) = CC(F^\wedge)$, after identifying T^*E and T^*E^* as in Proposition 5.5.1, and after having identified $\pi_E^{-1}\omega_E$ and $\pi_{E^*}^{-1}\omega_{E^*}$.

Exercise IX.8. (Hopf index theorem). Let X be a compact manifold, v a real analytic vector field on X with finite many zeroes $x_1, \ldots, x_N$. Assume each x_k ($k = 1, \ldots, N$) is non degenerate (i.e.: in a local chart in a neighborhood of x_k, $v = \sum_{ij} a^k_{ij} x_i \frac{\partial}{\partial x_j} + O(|x|^2)$, the matrix $A_k = (a^k_{ij})$ being non degenerate). Set $\varepsilon_k = \operatorname{sgn}\det(A_k)$. Then prove:

$$\chi(X; \mathbb{Q}_X) = \sum_{k=1}^{N} \varepsilon_k .$$

(Hint: consider the section s of $TX \to X$ given by v, and calculate the intersection number of s and the zero section.)

Exercise IX.9. In Proposition 9.6.2, remove the hypothesis that $\mathrm{supp}(F)$ is compact and prove the same conclusion.

Exercise IX.10. Let E be a finite-dimensional vector space and let $\mathscr{A} = \Gamma_c(E; \mathscr{CF}_E)$ be the group of compactly supported constructible functions on E.

(i) Define the convolution operation on $\mathscr{A}$ by the formula $\varphi * \psi = s_*(\varphi \boxtimes \psi)$, where s is the map $E \times E \to E$, $(x, y) \mapsto x + y$.

(ii) Prove that $(\mathscr{A}, *)$ is a commutative algebra, with unit $\mathbf{1}_{\{0\}}$.

(iii) Let Z be a convex compact subanalytic subset of E. Prove:

$$\mathbf{1}_Z * D(\mathbf{1}_{-Z}) = \mathbf{1}_{\{0\}} \ .$$

(iv) Prove that $\mathrm{D}(\varphi * \psi) = \mathrm{D}\varphi * \mathrm{D}\psi$ and $\int_E (\varphi * \psi) = (\int_E \varphi)(\int_E \psi)$.

Exercise IX.11. Let k be a field, let V be a real vector space, V^* the dual space, q the projection $T^*V \simeq V \times V^* \to V^*$. Let $F \in \mathrm{Ob}(\mathbf{D}^b_{\mathbb{R}-c}(k_V))$ with compact support and assume that: (i) $F \simeq k^m_{\{x_\circ\}}$ in $\mathbf{D}^b(V; p)$, where $p = (x_\circ; \xi_0) \in T^*V$, and (ii) $q^{-1}(\xi_\circ) \cap \mathrm{SS}(F) = \{p\}$.

Prove that $q(\mathrm{SS}(F)) = V^*$.

(Hint: prove that $\mathrm{supp}(q_* \mathrm{CC}(F)) = V^*$.)

Exercise IX.12. Let X be a complex manifold.

(i) Let $F \in \mathrm{Ob}(\mathbf{D}^b_{\mathbb{C}-c}(X))$. Prove that $\chi(F_x) = \chi(R\Gamma_{\{x\}}(X; F))$ for any $x \in X$ (cf. Sullivan [1]).

(ii) Let ϕ be a $\mathbb{C}$-constructible function on X, i.e. there exists a complex stratification $X = \bigsqcup_\alpha X_\alpha$ such that $\phi|_{X_\alpha}$ is constant. Prove that $\mathrm{D}_X(\phi) = \phi$.

(Hint: using the specialization functor, reduce to the case where X is a complex vector space and F is locally constant on the orbits of $\mathbb{C}^\times$. Then use the fact that the Euler-Poincaré index of $\mathbf{S}^1$ is zero.)

Notes

The notion of "cycle" is basic in algebraic geometry, and we refer to the book of Fulton [1] for a review. In the real analytic case, the constructions of §2 are the cohomological version of results which are well-known by the specialists of homology theory, especially after the works of Borel-Moore [1], Borel-Haefliger [1], Herrera [1], Bloom-Herrera [1], Poly [1] and Verdier [4].

A Riemann-Roch formula for constructible sheaves on complex varieties was conjectured by Grothendieck. This was obtained by MacPherson [1] who introduced the so-called "local Euler obstruction" (cf. also the pioneering work of M-H. Schwartz [1]), and proved that the isomorphism between the group of cycles and that of constructible functions commutes with direct images. Then Sabbah [1] and Ginsburg [1,2] constructed the functorial operations on Lagrangian cycles, and obtained results in the complex case similar to those of §3. However, our method in §3 is essentially different.

The index theorem was first proved for holonomic $\mathscr{D}$-modules by Kashiwara [2, 5] who introduced an invariant which turned out to be equivalent to that of

MacPherson (cf. Brylinski-Dubson-Kashiwara [1]). Then Dubson [1] in the complex case, and Kashiwara [7] in the real case showed that the index formula could be interpreted as an intersection formula. However the construction of the characteristic cycle of §4 is more recent (the former one used formula (9.4.10)), and is due to Kashiwara [8].

We shall not review here the Lefschetz formula, but only recall that it is Verdier who first saw that the duality formalism of Grothendieck could be applied to derive general trace formulas for correspondences between constructible sheaves, in the étale context (cf. [SGA 5], exposé III). Apart from that, the results of §6 are essentially new and were announced by Kashiwara [8] in 88. However, it should be mentioned that a Lefschetz formula for manifolds with boundary was first proved by Brenner-Shubin [1], and the notions of expanding or shrinking spaces already appeared in their paper.

Finally note that the construction of the caracteristic class has recently been extended by Schapira-Schneiders [1] (cf. also Schapira [4]) to a wider class of sheaves including $\mathscr{D}$-modules.

Chapter X. Perverse sheaves

Summary

Although perverse sheaves have a short history, they play an important role in various branches of mathematics, such as algebraic geometry or group representation. This theory is now well understood, but it is difficult to find in the literature a systematic treatment of it in the analytical case.

In this chapter, we define perverse sheaves on a real analytic manifold X and show that they form an abelian subcategory of $\mathbf{D}^b_{\mathbb{R}-c}(X)$. Then we study the case where X is a complex manifold. We prove that perversity is a microlocal property, that is, an object of $\mathbf{D}^b_{\mathbb{C}-c}(X)$ is perverse if and only if it is pure of shift $-\dim_{\mathbb{C}} X$ at generic points of its micro-support. Because Morse theory is well-adapted to the microlocal characterization, we can prove a vanishing theorm for perverse sheaves on Stein manifolds. Then we prove that perversity is preserved by various operations and in particular by specialization, microlocalization and Fourier-Sato transformation. To achieve this program we need some complements of homological algebra. In section 1, we explain the notion of t-structures which permits to construct abelian subcategories in triangulated categories.

References are made to Beilinson-Bernstein-Deligne [1] and Goresky–MacPherson [2, 3], (cf. also Borel et al. [1] and Gelfand-Manin [1]).

We keep conventions 8.0.

10.1. t-Structures

Let $\mathbf{D}$ be a triangulated category (cf. I §5).

Definition 10.1.1. *Let $\mathbf{D}^{\leqslant 0}$ and $\mathbf{D}^{\geqslant 0}$ be full subcategories of $\mathbf{D}$. We say that $(\mathbf{D}^{\leqslant 0}, \mathbf{D}^{\geqslant 0})$ is a t-structure on $\mathbf{D}$ if the following conditions are satisfied. Here, $\mathbf{D}^{\leqslant n} = \mathbf{D}^{\leqslant 0}[-n]$ and $\mathbf{D}^{\geqslant n} = \mathbf{D}^{\geqslant 0}[-n]$.*

(i) $\mathbf{D}^{\leqslant -1} \subset \mathbf{D}^{\leqslant 0}$ *and* $\mathbf{D}^{\geqslant 1} \subset \mathbf{D}^{\geqslant 0}$.

(ii) $\mathrm{Hom}_{\mathbf{D}}(X, Y) = 0$ *for* $X \in \mathrm{Ob}(\mathbf{D}^{\leqslant 0})$ *and* $Y \in \mathrm{Ob}(\mathbf{D}^{\geqslant 1})$.

(iii) *For any* $X \in \mathrm{Ob}(\mathbf{D})$, *there exists a distinguished triangle* $X_0 \longrightarrow X \longrightarrow X_1 \xrightarrow{+1}$ *in* $\mathbf{D}$ *with* $X_0 \in \mathrm{Ob}(\mathbf{D}^{\leqslant 0})$ *and* $X_1 \in \mathrm{Ob}(\mathbf{D}^{\geqslant 1})$.

The full subcategory $\mathscr{C} = \mathbf{D}^{\leqslant 0} \cap \mathbf{D}^{\geqslant 0}$ is called the **heart** of the t-structure.

Remark 10.1.2. (i) The notion of t-structure is self-dual, i.e. if $(\mathbf{D}^{\leqslant 0}, \mathbf{D}^{\geqslant 0})$ is a t-structure on $\mathbf{D}$ then $((\mathbf{D}^{\geqslant 0})^\circ, (\mathbf{D}^{\leqslant 0})^\circ)$ is a t-structure on the opposite triangulated category $\mathbf{D}^\circ$.

(ii) If $(\mathbf{D}^{\leqslant 0}, \mathbf{D}^{\geqslant 0})$ is a t-structure on $\mathbf{D}$, then $(\mathbf{D}^{\leqslant n}, \mathbf{D}^{\geqslant n})$ is also a t-structure on $\mathbf{D}$.

Let us give some examples.

Examples 10.1.3. (i) Let $\mathscr{C}$ be an abelian category and let $\mathbf{D} = \mathbf{D}(\mathscr{C})$ denote its derived category. Denote by $\mathbf{D}^{\leqslant 0}(\mathscr{C})$ (resp. $\mathbf{D}^{\geqslant 0}(\mathscr{C})$) the full subcategory of $\mathbf{D}$ consisting of objects X satisfying $H^j(X) = 0$ for $j > 0$ (resp. $j < 0$). Then $(\mathbf{D}^{\leqslant 0}(\mathscr{C}), \mathbf{D}^{\geqslant 0}(\mathscr{C}))$ is a t-structure on $\mathbf{D}(\mathscr{C})$. In fact the axiom (iii) in Definition 10.1.1 is verified by considering the distinguished triangle $\tau^{\leqslant 0}X \longrightarrow X \longrightarrow \tau^{\geqslant 1}X \xrightarrow[+1]{}$. In this case, the heart of $\mathbf{D}(\mathscr{C})$ is equivalent to $\mathscr{C}$.

(ii) Let A be a Noetherian commutative ring with $\mathrm{gld}(A) < \infty$, and let X be a real analytic manifold. Then $\mathbf{D}^{\leqslant 0}_{\mathbb{R}-c}(X) = \mathbf{D}^{\leqslant 0}(A_X) \cap \mathbf{D}^b_{\mathbb{R}-c}(X)$ and $\mathbf{D}^{\geqslant 0}_{\mathbb{R}-c}(X) = \mathbf{D}^{\geqslant 0}(A_X) \cap \mathbf{D}^b_{\mathbb{R}-c}(X)$ give a t-structure on $\mathbf{D}^b_{\mathbb{R}-c}(X)$. Let D_X be the duality functor (Definition 3.1.16). Then the images of $\mathbf{D}^{\geqslant 0}_{\mathbb{R}-c}(X)$ and $\mathbf{D}^{\leqslant 0}_{\mathbb{R}-c}(X)$ by D_X form a t-structure on $\mathbf{D}^b_{\mathbb{R}-c}(X)$.

In the rest of X§1, $(\mathbf{D}^{\leqslant 0}, \mathbf{D}^{\geqslant 0})$ is a t-structure on a triangulated category $\mathbf{D}$.

Proposition 10.1.4. (i) *The inclusion* $\mathbf{D}^{\leqslant n} \to \mathbf{D}$ *(resp.* $\mathbf{D}^{\geqslant n} \to \mathbf{D}$*) has a right adjoint functor* $\tau^{\leqslant n} : \mathbf{D} \to \mathbf{D}^{\leqslant n}$ *(resp. a left adjoint functor* $\tau^{\geqslant n} : \mathbf{D} \to \mathbf{D}^{\geqslant n}$*), i.e. there exists a morphism* $\tau^{\leqslant n} \to \mathrm{id}_{\mathbf{D}}$ *(resp.* $\mathrm{id}_{\mathbf{D}} \to \tau^{\geqslant n}$*) such that:*

$$\mathrm{Hom}_{\mathbf{D}^{\leqslant n}}(X, \tau^{\leqslant n} Y) \to \mathrm{Hom}_{\mathbf{D}}(X, Y)$$

is an isomorphism for any $X \in \mathrm{Ob}(\mathbf{D}^{\leqslant n})$ *and* $Y \in \mathrm{Ob}(\mathbf{D})$ *(resp.* $\mathrm{Hom}_{\mathbf{D}^{\geqslant n}}(\tau^{\geqslant n}X, Y) \to \mathrm{Hom}_{\mathbf{D}}(X, Y)$ *is an isomorphism for any* $X \in \mathrm{Ob}(\mathbf{D})$ *and* $Y \in \mathrm{Ob}(\mathbf{D}^{\geqslant n})$*).*

(ii) *There exists a unique morphism* $\mathrm{d} : \tau^{\geqslant n+1}(X) \to \tau^{\leqslant n}(X)[1]$ *such that* $\tau^{\leqslant n}(X) \to X \to \tau^{\geqslant n+1}(X) \xrightarrow{\mathrm{d}} \tau^{\leqslant n}(X)[1]$ *is a distinguished triangle. Moreover* d *is a morphism of functors* $\tau^{\geqslant n+1} \to [1] \circ \tau^{\leqslant n}$.

The functors $\tau^{\leqslant n}$ and $\tau^{\geqslant n}$ are called the truncation functors with respect to the t-structure.

Proof. We may assume $n = 0$. By Definition 10.1.1 (iii), for $X \in \mathrm{Ob}(\mathbf{D})$, there exists a distinguished triangle $X_0 \longrightarrow X \longrightarrow X_1 \xrightarrow[+1]{}$ with $X_0 \in \mathrm{Ob}(\mathbf{D}^{\leqslant 0})$ and $X_1 \in \mathrm{Ob}(\mathbf{D}^{\geqslant 1})$.

We shall show that $\mathrm{Hom}_{\mathbf{D}}(Y, X_0) \to \mathrm{Hom}_{\mathbf{D}}(Y, X)$ is an isomorphism for any $Y \in \mathrm{Ob}(\mathbf{D}^{\leqslant 0})$, (resp. $\mathrm{Hom}_{\mathbf{D}}(X_1, Z) \to \mathrm{Hom}_{\mathbf{D}}(X, Z)$ is an isomorphism for any $Z \in \mathrm{Ob}(\mathbf{D}^{\geqslant 1})$). This follows from the long exact sequence:

$$\mathrm{Hom}(Y, X_1[-1]) \to \mathrm{Hom}(Y, X_0) \to \mathrm{Hom}(Y, X) \to \mathrm{Hom}(Y, X_1)$$

(resp.

$$\mathrm{Hom}(X_0[1], Z) \to \mathrm{Hom}(X_1, Z) \to \mathrm{Hom}(X, Z) \to \mathrm{Hom}(X_0, Z)) \ .$$

This gives (i) with $X_0 \simeq \tau^{\leqslant 0}X$ and $X_1 \simeq \tau^{\geqslant 1}X$ (cf. Exercise I.2). The first part of (ii) follows from Definition 10.1.1 (iii) and the next lemma.

Lemma 10.1.5. *Let* $\mathbf{D}$ *be a triangulated category and let* $X \xrightarrow{f} Y \xrightarrow{g} Z \xrightarrow{h_i} X[1]$ $(i = 1, 2)$ *be two distinguished triangles. If* $\mathrm{Hom}_{\mathbf{D}}(X[1], Z) = 0$, *then* $h_1 = h_2$.

Proof. By (TR4) (cf. Proposition 1.4.4), we have a morphism of triangles:

$$\begin{array}{ccccccc} X & \xrightarrow{f} & Y & \xrightarrow{g} & Z & \xrightarrow{h_1} & X[1] \\ \downarrow{\scriptstyle \mathrm{id}} & & \downarrow{\scriptstyle \mathrm{id}} & & \downarrow{\scriptstyle \phi} & & \downarrow{\scriptstyle \mathrm{id}} \\ X & \xrightarrow{f} & Y & \xrightarrow{g} & Z & \xrightarrow{h_2} & X[1] . \end{array}$$

Hence $g = \phi \circ g$ and $h_1 = h_2 \circ \phi$. Since $(\mathrm{id}_Z - \phi) \circ g = 0$, by Proposition 1.5.3 there exists $\psi : X[1] \to Z$ such that $\mathrm{id}_Z - \phi = \psi \circ h_1$. By the hypothesis, ψ must be zero and hence $\phi = \mathrm{id}_Z$, $h_1 = h_2$. □

Finally let us prove that d is a morphism of functors. For $f : X \to Y$, there exists by (TR4) a morphism of distinguished triangles:

$$\begin{array}{ccccccc} \tau^{\leqslant n}X & \longrightarrow & X & \longrightarrow & \tau^{\geqslant n+1}X & \xrightarrow{\mathrm{d}(X)} & \tau^{\leqslant n}X[1] \\ \downarrow{\scriptstyle \tau^{\leqslant n}f} & & \downarrow{\scriptstyle f} & & \downarrow{\scriptstyle \phi} & & \downarrow{\scriptstyle \tau^{\leqslant n}f[1]} \\ \tau^{\leqslant n}Y & \longrightarrow & Y & \longrightarrow & \tau^{\geqslant n+1}Y & \xrightarrow{\mathrm{d}(Y)} & \tau^{\leqslant n}Y[1] . \end{array}$$

The commutativity of the middle square implies $\phi = \tau^{\geqslant n+1}f$. □

Note that we have:

$$(10.1.1) \qquad \begin{cases} \tau^{\leqslant n}(X[m]) \simeq \tau^{\leqslant n+m}(X)[m] , \\ \tau^{\geqslant n}(X[m]) \simeq \tau^{\geqslant n+m}(X)[m] . \end{cases}$$

We shall also write $\tau^{>n}$ and $\tau^{<n}$ instead of $\tau^{\geqslant n+1}$ and $\tau^{\leqslant n-1}$, respectively, and similarly for $\mathbf{D}^{<n}$ and $\mathbf{D}^{>n}$.

Proposition 10.1.6. (i) *If* $X \in \mathrm{Ob}(\mathbf{D}^{\leqslant n})$ (*resp.* $X \in \mathrm{Ob}(\mathbf{D}^{\geqslant n})$) *then the morphism* $\tau^{\leqslant n}X \to X$ (*resp.* $X \to \tau^{\geqslant n}X$) *is an isomorphism.*

(ii) *Let* $X \in \mathrm{Ob}(\mathbf{D})$. *Then* $X \in \mathrm{Ob}(\mathbf{D}^{\leqslant n})$ (*resp.* $\mathrm{Ob}(\mathbf{D}^{\geqslant n})$) *if and only if* $\tau^{>n}X = 0$ (*resp.* $\tau^{<n}X = 0$).

Proof. (i) follows from Proposition 10.1.4.

(ii) follows from the distinguished triangle $\tau^{\leqslant n}X \longrightarrow X \longrightarrow \tau^{>n}X \xrightarrow[+1]{}$. □

Proposition 10.1.7. *Let* $X' \longrightarrow X \longrightarrow X'' \xrightarrow[+1]{}$ *be a distinguished triangle in* $\mathbf{D}$. *If* X' *and* X'' *belong to* $\mathbf{D}^{\geqslant 0}$ (*resp.* $\mathbf{D}^{\leqslant 0}$) *then so does* X.

Proof. It is enough to prove the statement for $\mathbf{D}^{\geqslant 0}$. Assume X' and X'' belong to $\mathbf{D}^{\geqslant 0}$. Then $\mathrm{Hom}(\tau^{<0}X, X') = \mathrm{Hom}(\tau^{<0}X, X'') = 0$, by Definition 10.1.1 (i) and (ii). Hence $\mathrm{Hom}(\tau^{<0}X, X) = 0$. This gives $\tau^{<0}X = 0$. Then apply Proposition 10.1.6 (ii). □

Proposition 10.1.8. *Let a and b be two integers.*

(i) *If* $b \geqslant a$, *then* $\tau^{\geqslant b} \circ \tau^{\geqslant a} \simeq \tau^{\geqslant a} \circ \tau^{\geqslant b} \simeq \tau^{\geqslant b}$, *and* $\tau^{\leqslant b} \circ \tau^{\leqslant a} \simeq \tau^{\leqslant a} \circ \tau^{\leqslant b} \simeq \tau^{\leqslant a}$.
(ii) *If* $a > b$, *then* $\tau^{\leqslant b} \circ \tau^{\geqslant a} = \tau^{\geqslant a} \circ \tau^{\leqslant b} = 0$.
(iii) $\tau^{\geqslant a} \circ \tau^{\leqslant b} \simeq \tau^{\leqslant b} \circ \tau^{\geqslant a}$. *More precisely, for* $X \in \mathrm{Ob}(\mathbf{D})$, *there exists a unique morphism* $\phi : \tau^{\geqslant a} \circ \tau^{\leqslant b}X \to \tau^{\leqslant b} \circ \tau^{\geqslant a}X$ *such that the diagram*:

$$
\begin{array}{ccccc}
\tau^{\leqslant b}X & \longrightarrow & X & \longrightarrow & \tau^{\geqslant a}X \\
\big\downarrow & & & & \big\uparrow \\
\tau^{\geqslant a}\tau^{\leqslant b}X & & \xrightarrow{\quad\phi\quad} & & \tau^{\leqslant b}\tau^{\geqslant a}X
\end{array}
\tag{10.1.2}
$$

commutes, and moreover ϕ *is an isomorphism.*

Proof. (i) $\tau^{\geqslant a} \circ \tau^{\geqslant b} \simeq \tau^{\geqslant b}$ is a consequence of Proposition 10.1.6 (i). For any $X \in \mathrm{Ob}(\mathbf{D})$, $Y \in \mathrm{Ob}(\mathbf{D}^{\geqslant b})$, we have $\mathrm{Hom}_{\mathbf{D}^{\geqslant b}}(\tau^{\geqslant b}\tau^{\geqslant a}X, Y) \simeq \mathrm{Hom}_{\mathbf{D}}(\tau^{\geqslant a}X, Y) \simeq \mathrm{Hom}_{\mathbf{D}}(X, Y) \simeq \mathrm{Hom}_{\mathbf{D}^{\geqslant b}}(\tau^{\geqslant b}X, Y)$ and hence $\tau^{\geqslant b}\tau^{\geqslant a}X \simeq \tau^{\geqslant b}X$.

The dual statement give the other isomorphisms.

(ii) follows immediately from Proposition 10.1.6 (ii).

(iii) By (ii), we may assume $b \geqslant a$. By (i) there exist distinguished triangles:

$$
\left\{
\begin{array}{l}
\tau^{\leqslant b}\tau^{\geqslant a}X \longrightarrow \tau^{\geqslant a}X \longrightarrow \tau^{>b}X \xrightarrow[+1]{} , \\
\tau^{<a}X \longrightarrow \tau^{\leqslant b}X \longrightarrow \tau^{\geqslant a}\tau^{\leqslant b}X \xrightarrow[+1]{} .
\end{array}
\right.
\tag{10.1.3}
$$

Therefore $\tau^{\leqslant b}\tau^{\geqslant a}X$ and $\tau^{\geqslant a}\tau^{\leqslant b}X$ belong to $\mathbf{D}^{\geqslant a} \cap \mathbf{D}^{\leqslant b}$ by the preceding proposition. This gives the existence and the uniqueness of ϕ. Now we shall show that ϕ is an isomorphism. Apply the octahedral axiom TR5 to the morphisms $\tau^{<a}X \to \tau^{\leqslant b}X \to X$, (cf. (10.1.4)).

(10.1.4)

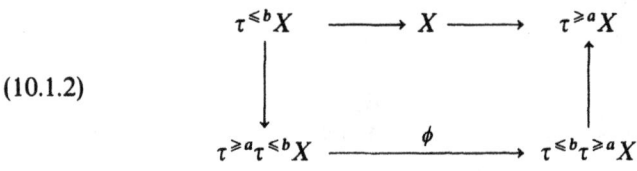

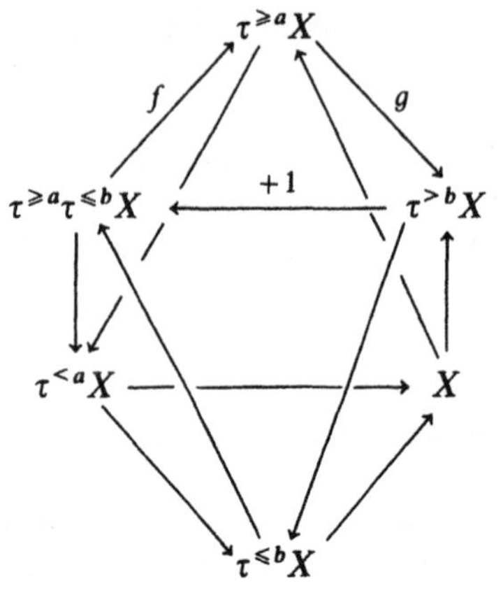

We get the distinguished triangle:

$$\tau^{\geqslant a}\tau^{\leqslant b}X \xrightarrow{f} \tau^{\geqslant a}X \xrightarrow{g} \tau^{>b}X \xrightarrow[+1]{} .$$

(Note that f and g must be unique.) Hence $\tau^{\geqslant a}\tau^{\leqslant b}X \simeq \tau^{\leqslant b}\tau^{\geqslant a}X$. □

Let $\mathscr{C}$ be the heart of $\mathbf{D}$.

Definition 10.1.9. *We define the functor* $H^0 : \mathbf{D} \to \mathscr{C}$ *by:*

$$H^0(X) = \tau^{\geqslant 0}\tau^{\leqslant 0}X \simeq \tau^{\leqslant 0}\tau^{\geqslant 0}X \ . \tag{10.1.5}$$

We also set:

$$H^n(X) = H^0(X[n]) \simeq (\tau^{\geqslant n}\tau^{\leqslant n}X)[n] \ . \tag{10.1.6}$$

Proposition 10.1.10. *Let* $X \in \mathrm{Ob}(\mathbf{D})$ *and assume that* X *belongs to* $\mathbf{D}^{\geqslant a}$ *(resp.* $\mathbf{D}^{\leqslant a}$*) for some* a*. Then* X *belongs to* $\mathbf{D}^{\geqslant 0}$ *(resp.* $\mathbf{D}^{\leqslant 0}$*) if and only if* $H^n(X) = 0$ *for* $n < 0$ *(resp.* $n > 0$*).*

Proof. It is enough to show that if $X \xrightarrow{\sim} \tau^{\geqslant a}X$ and $H^a(X) = 0$, then $X \xrightarrow{\sim} \tau^{>a}X$. This immediately follows from the distinguished triangle:

$$H^a(X)[-a] \longrightarrow \tau^{\geqslant a}X \to \tau^{>a}X \xrightarrow[+1]{} . \quad \square$$

Proposition 10.1.11. (i) *The heart* $\mathscr{C} = \mathbf{D}^{\leqslant 0} \cap \mathbf{D}^{\geqslant 0}$ *is an abelian category.*

(ii) *If* $X' \longrightarrow X \longrightarrow X'' \xrightarrow[+1]{}$ *is a distinguished triangle in* $\mathbf{D}$ *and if* X' *and* X'' *belong to* $\mathscr{C}$*, then* X *belongs to* $\mathscr{C}$*.*

(iii) *If* $0 \to X \to Y \to Z \to 0$ *is an exact sequence in* $\mathscr{C}$*, then there exists a unique* $h : Z \to X[1]$ *such that* $X \to Y \to Z \xrightarrow[h]{} X[1]$ *is a distinguished triangle in* $\mathbf{D}$*.*

Proof. (ii) follows from Proposition 10.1.7.

(i) By considering the distinguished triangle $X \longrightarrow X \oplus Y \longrightarrow Y \xrightarrow[+1]{}$, we get by (ii) that $\mathscr{C}$ is an additive category. Let $f : X \to Y$ be a morphism in $\mathscr{C}$, and let us embed f into a distinguished triangle $X \xrightarrow{f} Y \longrightarrow Z \xrightarrow[+1]{}$. Then, by Proposition 10.1.7, Z belongs to $\mathbf{D}^{\leqslant 0} \cap \mathbf{D}^{\geqslant -1}$. We shall prove:

$$\begin{cases} H^0(Z) \simeq \tau^{\geqslant 0}Z \simeq \mathrm{Coker}\, f \qquad \text{and} \\ H^0(Z[-1]) \simeq \tau^{\leqslant 0}(Z[-1]) \simeq \mathrm{Ker}\, f \ . \end{cases} \tag{10.1.7}$$

For that purpose take $W \in \mathrm{Ob}(\mathscr{C})$ and consider the long exact sequences:

$$\mathrm{Hom}(X[1], W) \to \mathrm{Hom}(Z, W) \to \mathrm{Hom}(Y, W) \to \mathrm{Hom}(X, W) \qquad \text{and}$$

$$\mathrm{Hom}(W, Y[-1]) \to \mathrm{Hom}(W, Z[-1]) \to \mathrm{Hom}(W, X) \to \mathrm{Hom}(W, Y) \ .$$

Since $\mathrm{Hom}(X[1], W) = \mathrm{Hom}(W, Y[-1]) = 0$, and $\mathrm{Hom}(Z, W) \simeq \mathrm{Hom}(\tau^{\geqslant 0}Z, W)$, $\mathrm{Hom}(W, Z[-1]) \simeq \mathrm{Hom}(W, \tau^{\leqslant 0}(Z[-1]))$, (because $W \in \mathrm{Ob}(\mathscr{C})$), we get (10.1.7).

Now let us prove that the canonical morphism $\operatorname{Coim} f \to \operatorname{Im} f$ is an isomorphism.

We embed $Y \to \tau^{\geqslant 0}Z$ into a distinguished triangle $I \longrightarrow Y \longrightarrow \tau^{\geqslant 0}Z \xrightarrow[+1]{}$. Then by Proposition 10.1.7, I belongs to $\mathbf{D}^{\geqslant 0}$.

We apply the octahedral axiom to $Y \to Z \to \tau^{\geqslant 0}Z$, (cf. (10.1.8)).

(10.1.8)

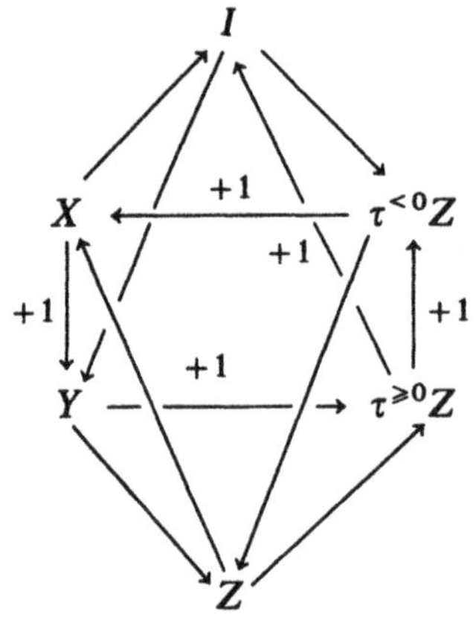

Thus we obtain the distinguished triangle $\tau^{\leqslant 0}(Z[-1]) \longrightarrow X \longrightarrow I \xrightarrow[+1]{}$. Hence I belongs to $\mathbf{D}^{\leqslant 0}$ and hence $I \in \operatorname{Ob}(\mathscr{C})$. Since $\tau^{\leqslant 0}(Z[-1]) \simeq \operatorname{Ker} f$, we find by applying (10.1.7) to the distinguished triangle $\operatorname{Ker} f \longrightarrow X \longrightarrow I \xrightarrow[+1]{}$ that $I \simeq \operatorname{Coim} f$. Similarly, applying (10.1.7) to the distinguished triangle $I \longrightarrow Y \longrightarrow \operatorname{Coker} f \xrightarrow[+1]{}$, we get $I \simeq \operatorname{Im} f$. This proves (ii).

(iii) follows from Lemma 10.1.5 and (10.1.7). □

Proposition 10.1.12. *The functor* $H^0 : \mathbf{D} \to \mathscr{C}$ (*cf. Definition* 10.1.9) *is a cohomological functor.*

Proof. Let $X \longrightarrow Y \longrightarrow Z \xrightarrow[+1]{}$ be a distinguished triangle in $\mathbf{D}$. We have to show that $H^0(X) \to H^0(Y) \to H^0(Z)$ is exact. We decompose the proof into three steps.

(a) Assume X, Y, Z belong to $\mathbf{D}^{\geqslant 0}$, and let us show that $0 \to H^0(X) \to H^0(Y) \to H^0(Z)$ is exact.

Let $W \in \operatorname{Ob}(\mathscr{C})$. Then $\operatorname{Hom}_{\mathscr{C}}(W, H^0(X)) \simeq \operatorname{Hom}_{\mathbf{D}}(W, \tau^{\geqslant 0}X) \simeq \operatorname{Hom}_{\mathbf{D}}(W, X)$. Hence, $\operatorname{Hom}_{\mathbf{D}}(W, Z[-1]) = 0$ and the long exact sequence

$$\operatorname{Hom}_{\mathbf{D}}(W, Z[-1]) \to \operatorname{Hom}_{\mathbf{D}}(W, X) \to \operatorname{Hom}_{\mathbf{D}}(W, Y) \to \operatorname{Hom}_{\mathbf{D}}(W, Z)$$

give the desired result.

(b) Here we only assume $Z \in \operatorname{Ob}(\mathbf{D}^{\geqslant 0})$, and we shall prove that $0 \to H^0(X) \to H^0(Y) \to H^0(Z)$ is exact. For any $W \in \operatorname{Ob}(\mathbf{D}^{<0})$, $\operatorname{Hom}_{\mathbf{D}}(W, Z) = \operatorname{Hom}_{\mathbf{D}}(W, Z[-1]) = 0$, which gives $\operatorname{Hom}_{\mathbf{D}}(W, X) \simeq \operatorname{Hom}_{\mathbf{D}}(W, Y)$. Hence $\tau^{<0}X \to \tau^{<0}Y$ is an isomorphism. Applying the octahedral axiom (cf. (10.1.9)), we obtain a distinguished triangle $\tau^{\geqslant 0}X \longrightarrow \tau^{\geqslant 0}Y \longrightarrow Z \xrightarrow[+1]{}$, and it is enough to apply (a).

(10.1.9)

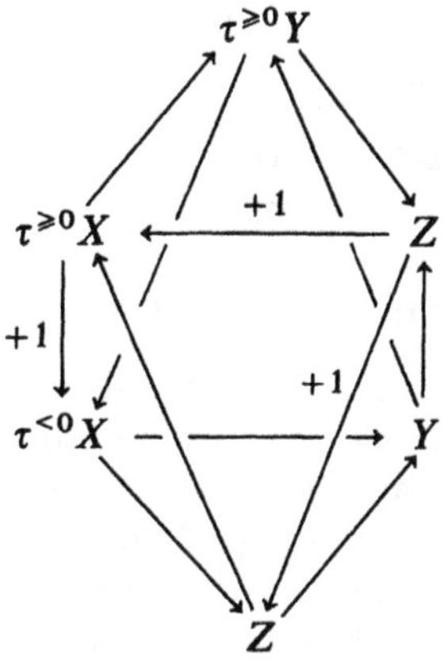

(c) The dual statement to (b) gives that $H^0(X) \to H^0(Y) \to H^0(Z) \to 0$ is exact if X belongs to $\mathbf{D}^{\leqslant 0}$.

(d) We treat the general case. The octahedral axiom applied to $\tau^{\leqslant 0}X \to X \to Y$ (cf. (10.1.10)), gives the distinguished triangles: $\tau^{\leqslant 0}X \longrightarrow Y \longrightarrow W \xrightarrow[+1]{}$ and $\tau^{>0}X \longrightarrow W \longrightarrow Z \xrightarrow[+1]{}$.

(10.1.10)

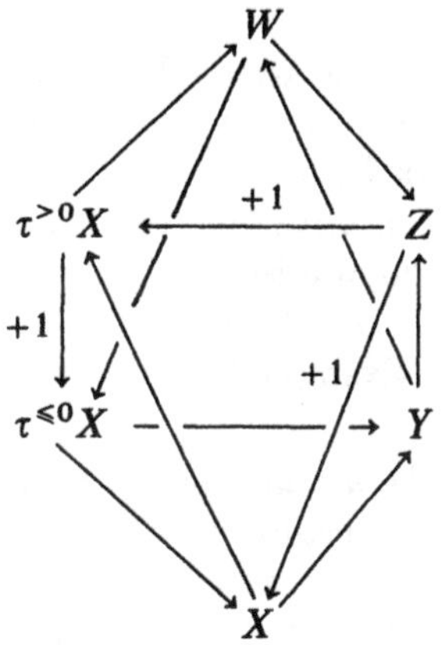

Applying (c) to the first triangle, we find that $H^0(X) \to H^0(Y) \to H^0(W)$ is exact. Applying (b) to the triangle $W \longrightarrow Z \longrightarrow \tau^{>0}X[1] \xrightarrow[+1]{}$ we find that $0 \to H^0(W) \to H^0(Z)$ is exact. Hence $H^0(X) \to H^0(Y) \to H^0(Z)$ is exact. □

Definition 10.1.13. *Let $\mathbf{D}_i$ $(i = 1, 2)$ be two triangulated categories endowed with t-structures $(\mathbf{D}_i^{\leqslant 0}, \mathbf{D}_i^{\geqslant 0})$, and let $\mathscr{C}_i$ be the heart of $\mathbf{D}_i$, $\varepsilon_i : \mathscr{C}_i \to \mathbf{D}_i$ the inclusion functor. Let $F : \mathbf{D}_1 \to \mathbf{D}_2$ be a functor of triangulated categories.*

(i) *One says F is left (resp. right) t-exact if $F(\mathbf{D}_1^{\geqslant 0}) \subset \mathbf{D}_2^{\geqslant 0}$ (resp. $F(\mathbf{D}_1^{\leqslant 0}) \subset \mathbf{D}_2^{\leqslant 0}$), and one says F is t-exact if it is both right and left t-exact.*
(ii) *One sets:*

$$ {}^pF = H^0 \circ F \circ \varepsilon_1 : \mathscr{C}_1 \to \mathscr{C}_2 \ . $$

Proposition 10.1.14. *Let $\mathbf{D}_1$, $\mathbf{D}_2$ and F be as in Definition* 10.1.13, *and assume F is left (resp. right) t-exact. Then*:

(i) *for* $X \in \mathrm{Ob}(\mathbf{D}^{\geqslant 0})$ *(resp.* $\mathbf{D}^{\leqslant 0}$), $H^0(F(X)) \simeq {}^pF(H^0(X))$,
(ii) ${}^pF : \mathscr{C}_1 \to \mathscr{C}_2$ *is left (resp. right) exact.*

Proof. It is enough to prove the results for left t-exact functors.

(i) Let $X \in \mathrm{Ob}(\mathbf{D}^{\geqslant 0})$, and apply F to the distinguished triangle $H^0(X) \longrightarrow X \longrightarrow \tau^{>0}X \xrightarrow[+1]{}$. We get the distinguished triangle $F(H^0(X)) \longrightarrow F(X) \longrightarrow F(\tau^{>0}X) \xrightarrow[+1]{}$, and $F(\tau^{>0}X)$ belongs to $\mathbf{D}_2^{>0}$. Then the result follows by applying the cohomological functor $H^0(\cdot)$ to this triangle.
(ii) Let $0 \to X \to Y \to Z \to 0$ be an exact sequence in $\mathscr{C}_1$. It gives rise to a distinguished triangle $X \longrightarrow Y \longrightarrow Z \xrightarrow[+1]{}$ in $\mathbf{D}_1$, by Proposition 10.1.11. Applying F, we get a distinguished triangle $F(X) \longrightarrow F(Y) \longrightarrow F(Z) \xrightarrow[+1]{}$. Since $F(X)$, $F(Y)$, $F(Z)$ belong to $\mathbf{D}_2^{\geqslant 0}$, we get the exact sequence: $0 \to H^0(F(X)) \to H^0(F(Y)) \to H^0(F(Z))$. Then it remains to apply (i). □

Remark 10.1.15. If $F : \mathbf{D}_1 \to \mathbf{D}_2$ is t-exact, then F sends $\mathscr{C}_1$ to $\mathscr{C}_2$ and $F|_{\mathscr{C}_1}$ is exact. Moreover $F|_{\mathscr{C}_1} \simeq {}^pF$ and $F(H^n(X)) \simeq H^n(F(X))$ for any $X \in \mathrm{Ob}(\mathbf{D}_1)$.

Remark 10.1.16. (i) Let $\mathbf{D}$ be the triangulated category with the t-structure given in Example 10.1.3 (i). Then the truncation functors $\tau^{\leqslant n}$ and $\tau^{\geqslant n}$ coincide with the ones given in Remark 1.7.6.

(ii) Let $F : \mathscr{C}_1 \to \mathscr{C}_2$ be an additive functor of abelian categories, and assume F is left exact and $\mathscr{C}_1$ has enough injective. Then $RF : \mathbf{D}^+(\mathscr{C}_1) \to \mathbf{D}^+(\mathscr{C}_2)$ is left t-exact, where the t-structure on $\mathbf{D}^+(\mathscr{C}_i)$ is the natural one, induced by the t-structure on $\mathbf{D}(\mathscr{C}_i)$ given by Example 10.1.3 (i).

Proposition 10.1.17. *Let $\tilde{\mathbf{D}}_i$ be a triangulated category, $\mathbf{D}_i$ a full triangulated subcategory, $(\mathbf{D}_i^{\leqslant 0}, \mathbf{D}_i^{\geqslant 0})$ a t-structure on $\mathbf{D}_i$ $(i = 1,2)$. Let $f : \tilde{\mathbf{D}}_1 \to \tilde{\mathbf{D}}_2$ and $g : \tilde{\mathbf{D}}_2 \to \tilde{\mathbf{D}}_1$ be functors of triangulated categories with f a left adjoint to g. Assume $f(\mathbf{D}_1) \subset \mathbf{D}_2$ (resp. $g(\mathbf{D}_2) \subset \mathbf{D}_1$) and $f|_{\mathbf{D}_1}$ is right t-exact (resp. $g|_{\mathbf{D}_2}$ is left t-exact). Then for any $Y \in \mathrm{Ob}(\mathbf{D}_2^{\geqslant 0})$ such that $g(Y) \in \mathrm{Ob}(\mathbf{D}_1)$, (resp. $X \in \mathrm{Ob}(\mathbf{D}_1^{\leqslant 0})$ such that $f(X) \in \mathrm{Ob}(\mathbf{D}_2)$), one has $g(Y) \in \mathrm{Ob}(\mathbf{D}_1^{\geqslant 0})$, (resp. $f(X) \in \mathrm{Ob}(\mathbf{D}_2^{\leqslant 0})$).*

Proof. For any $X \in \mathrm{Ob}(\mathbf{D}_1^{<0})$ (resp. $Y \in \mathrm{Ob}(\mathbf{D}_2^{>0})$), one has:

$$\mathrm{Hom}_{\mathbf{D}_1^{\geqslant 0}}(X, \tau^{<0}g(Y)) \simeq \mathrm{Hom}_{\tilde{\mathbf{D}}_1}(X, g(Y)) \simeq \mathrm{Hom}_{\tilde{\mathbf{D}}_2}(f(X), Y) = 0 \ ,$$

(resp. $\mathrm{Hom}_{\mathbf{D}_2^{\geqslant 0}}(\tau^{>0}f(X), Y) \simeq \mathrm{Hom}_{\tilde{\mathbf{D}}_2}(f(X), Y) \simeq \mathrm{Hom}_{\tilde{\mathbf{D}}_1}(X, g(Y)) = 0$). Hence $\tau^{<0}g(Y) = 0$, (resp. $\tau^{>0}f(X) = 0$). □

Corollary 10.1.18. *Let $\mathbf{D}_i$ be a triangulated category with a t-structure $(i = 1,2)$ and let $f : \mathbf{D}_1 \to \mathbf{D}_2$ and $g : \mathbf{D}_2 \to \mathbf{D}_1$ be functors of triangulated categories, with f a left adjoint to g. Then f is right t-exact if and only if g is left t-exact.*

10.2. Perverse sheaves on real manifolds

Let p be a map from $\mathbb{Z}$ to $\mathbb{Z}$. The dual map p^* is defined by:

(10.2.1) $$p^*(n) = -p(n) - n \; .$$

If both p and p^* are decreasing, we call p a **perversity**. Hence p is a perversity iff:

(10.2.2) $$p(n) - p(n+1) = 0 \text{ or } 1 \qquad \text{for any } n \; ,$$

or equivalently:

(10.2.3) $$0 \leqslant p(n) - p(m) \leqslant m - n \qquad \text{for any } n, m \text{ with } n \leqslant m \; .$$

We define $p[k]$ by:

(10.2.4) $$p[k](n) = p(k+n) \; .$$

Then:

(10.2.5) $$p^*[k]^*(n) = p[k](n) + k \; .$$

Now let X be a (real analytic) manifold. Recall that $\mathbf{D}^b_{w-\mathbb{R}-c}(X)$ denotes the full subcategory of weakly $\mathbb{R}$-constructible objects of $\mathbf{D}^b(X)$. For a subset S of X, we shall denote by i_S the inclusion map $S \hookrightarrow X$. If S is subanalytic, its dimension, $\dim S$, is well-defined (cf. VIII §2). If $S = \varnothing$, we set $\dim S = -\infty$.

Definition 10.2.1. *Let p be a perversity.*

(i) ${}^p\mathbf{D}^{\leqslant 0}_{w-\mathbb{R}-c}(X)$ *is the full subcategory of* $\mathbf{D}^b_{w-\mathbb{R}-c}(X)$ *consisting of objects F satisfying*:

(10.2.6) $$\dim(\operatorname{supp}(H^j(F))) < k \qquad \textit{for any } j \textit{ and } k \textit{ with } j > p(k) \; ,$$

(*or equivalently* $\dim(\operatorname{supp} \tau^{>p(k)}(F)) < k$).

(ii) ${}^p\mathbf{D}^{\geqslant 0}_{w-\mathbb{R}-c}(X)$ *is the full subcategory of* $\mathbf{D}^b_{w-\mathbb{R}-c}(X)$ *consisting of objects F satisfying*:

(10.2.7) $$\begin{cases} H^j(i_S^!(F)) = 0 & \textit{for any locally closed} \\ \textit{subanalytic subset } S \textit{ and any } j \textit{ with} \\ j < p(\dim S) \; . \end{cases}$$

We set:

$$ {}^p\mathbf{D}^{\geqslant n}_{w-\mathbb{R}-c}(X) = {}^p\mathbf{D}^{\geqslant 0}_{w-\mathbb{R}-c}(X)[-n] \; ,$$

$$ {}^p\mathbf{D}^{\leqslant n}_{w-\mathbb{R}-c}(X) = {}^p\mathbf{D}^{\leqslant 0}_{w-\mathbb{R}-c}(X)[-n] \; ,$$

$$ {}^p\mathbf{D}^{0}_{w-\mathbb{R}-c}(X) = {}^p\mathbf{D}^{\leqslant 0}_{w-\mathbb{R}-c}(X) \cap {}^p\mathbf{D}^{\geqslant 0}_{w-\mathbb{R}-c}(X) \; .$$

We also set:

$$ {}^p\mathbf{D}^{\geqslant n}_{\mathbb{R}-c}(X) = {}^p\mathbf{D}^{\geqslant n}_{w-\mathbb{R}-c}(X) \cap \mathbf{D}^b_{\mathbb{R}-c}(X) $$

and similarly for ${}^p\mathbf{D}^{\leqslant n}_{\mathbb{R}-c}(X)$, ${}^p\mathbf{D}^{0}_{\mathbb{R}-c}(X)$.

Remark 10.2.2. (i) The definition of ${}^p\mathbf{D}^{\geqslant 0}_{w-\mathbb{R}-c}(X)$ and ${}^p\mathbf{D}^{\leqslant 0}_{w-\mathbb{R}-c}(X)$ depends only on the values of $p(n)$ for $0 \leqslant n \leqslant \dim X$.

(ii) If $p = 0$, then ${}^p\mathbf{D}^{\leqslant 0}_{w-\mathbb{R}-c}(X) = \mathbf{D}^{\leqslant 0}_{w-\mathbb{R}-c}(X) = \{F \in \mathbf{D}^b_{w-\mathbb{R}-c}(X);\ H^j(F) = 0$ for any $j > 0\}$ and ${}^p\mathbf{D}^{\geqslant 0}_{w-\mathbb{R}-c}(X) = \mathbf{D}^{\geqslant 0}_{w-\mathbb{R}-c}(X) = \{F \in \mathbf{D}^b_{w-\mathbb{R}-c}(X); H^j(F) = 0$ for any $j < 0\}$.

(iii) If p' is another perversity satisfying, $p(n) \leqslant p'(n)$ for any n, then

$$ {}^p\mathbf{D}^{\leqslant 0}_{w-\mathbb{R}-c}(X) \subset {}^{p'}\mathbf{D}^{\leqslant 0}_{w-\mathbb{R}-c}(X) \quad \text{and} \quad {}^{p'}\mathbf{D}^{\geqslant 0}_{w-\mathbb{R}-c}(X) \subset {}^p\mathbf{D}^{\geqslant 0}_{w-\mathbb{R}-c}(X) . $$

In particular, if $a \leqslant p(n) \leqslant b$ for any n, then:

$$ (10.2.8) \qquad \begin{cases} \mathbf{D}^{\leqslant a}_{w-\mathbb{R}-c}(X) \subset {}^p\mathbf{D}^{\leqslant 0}_{w-\mathbb{R}-c}(X) \subset \mathbf{D}^{\leqslant b}_{w-\mathbb{R}-c}(X) & \text{and} \\ \mathbf{D}^{\geqslant b}_{w-\mathbb{R}-c}(X) \subset {}^p\mathbf{D}^{\geqslant 0}_{w-\mathbb{R}-c}(X) \subset \mathbf{D}^{\geqslant a}_{w-\mathbb{R}-c}(X) . \end{cases} $$

(iv) Assume $F \in \mathrm{Ob}(\mathbf{D}^b(X))$ has locally constant cohomology objects and moreover, these objects are flat A-modules. Then $F \in \mathrm{Ob}({}^p\mathbf{D}^{\geqslant 0}_{w-\mathbb{R}-c}(X))$ if $H^j(F) = 0$ for $j < p(\dim X)$. In fact, for any locally closed subanalytic set S, $i_S^!F \simeq i_S^!\omega_X \otimes i_S^{-1}F \otimes or_X[-\dim X] \simeq \omega_S \otimes i_S^{-1}F \otimes or_X[-\dim X]$ and $H^j(\omega_S) = 0$ for $j < -\dim S$ by Proposition 9.2.2. Hence $H^j(i_S^!F) = 0$ for $j < p(\dim S) \leqslant p(\dim X) + \dim X - \dim S$.

Remark 10.2.3. This definition can also be generalized to the case where X is a subanalytic space (cf. Exercise IX.2) and most of the results stated here remain true under this generalization.

We shall prove later that $({}^p\mathbf{D}^{\leqslant 0}_{w-\mathbb{R}-c}(X), {}^p\mathbf{D}^{\geqslant 0}_{w-\mathbb{R}-c}(X))$ forms a t-structure. For that purpose we need some preliminary results.

Proposition 10.2.4. *Let $F \in \mathrm{Ob}(\mathbf{D}^b_{w-\mathbb{R}-c}(X))$ and let $X = \bigsqcup_\alpha X_\alpha$ be a subanalytic stratification consisting of equidimensional strata.*

(i) *Assume $i_{X_\alpha}^{-1}F$ has locally constant cohomologies for all α. Then F belongs to ${}^p\mathbf{D}^{\leqslant 0}_{w-\mathbb{R}-c}(X)$ if and only if $H^j(i_{X_\alpha}^{-1}F) = 0$ for any α and j with $j >$ p(dim X_α).*

(ii) *Assume $i_{X_\alpha}^!F$ has locally constant cohomologies for all α. Then F belongs to ${}^p\mathbf{D}^{\geqslant 0}_{w-\mathbb{R}-c}(X)$ if and only if $H^j(i_{X_\alpha}^!F) = 0$ for any α and j with $j < p(\dim X_\alpha)$.*

Proof. (i) follows from:

$$ \dim(\mathrm{supp}(H^j(F))) = \sup\{\dim X_\alpha; H^j(F)|_{X_\alpha} \neq 0\} . $$

(ii) If F belongs to ${}^p\mathbf{D}^{\geqslant 0}_{w-\mathbb{R}-c}(X)$, then $H^j(i_{X_\alpha}^!(F)) = 0$ for $j < p(\dim X_\alpha)$, by the definition. Let us prove the converse.

Let S be a subanalytic closed subset. By the assumption, $i_{X_\alpha}^!(F)$ is concentrated to degree $\geqslant p(\dim X_\alpha)$ and has locally constant cohomology sheaves. Hence by Remark 10.2.2 (iv), $i_{X_\alpha}^!(F)$ belongs to ${}^{\tilde p}\mathbf{D}^{\geqslant 0}_{w-\mathbb{R}-c}(X_\alpha)$, where:

$$ \tilde p(n) = -n + p(\dim X_\alpha) + \dim X_\alpha . $$

By the definition this implies that $i^!_{S\cap X_\alpha}(F) = i^!_{S\cap X_\alpha} i^!_{X_\alpha} F$ is concentrated to degree $\geqslant \tilde{p}(\dim(S\cap X_\alpha)) = -\dim(S\cap X_\alpha) + p(\dim X_\alpha) + \dim X_\alpha \geqslant p(\dim(S\cap X_\alpha)) \geqslant p(\dim S)$.

Setting $X_k = \bigsqcup_{\dim X_\alpha \leqslant k} X_\alpha$, we have:

$$H^j_{X_k\cap S}(F)|_{(X_k\cap S\setminus X_{k-1}\cap S)} \simeq \bigoplus_{\dim X_\alpha = k} H^j_{X_\alpha\cap S}(F)$$

and this is 0 for $j < p(\dim S)$. Thus we can apply the result of Exercise II.25 to conclude that $H^j_S(F) = 0$ for $j < p(\dim S)$. □

The conditions (10.2.6) and (10.2.7) do not appear symmetric at the first glance, but we are going to show that in fact, they are symmetric.

For that purpose we introduce a definition. We set:

(10.2.9) $$\operatorname{cosupp}^j(F) = \{x; H^j(i^!_x F) \neq 0\} .$$

Note that if the base ring A is a field, then $i_x^{-1}H^j(\mathrm{D}_X F) \simeq H^{-j}(i^!_x F)^*$, which implies:

$$\overline{\operatorname{cosupp}^j(F)} = \operatorname{supp}(H^{-j}(\mathrm{D}_X F)) .$$

Proposition 10.2.5. *Let $F \in \mathrm{Ob}(\mathbf{D}^b_{w-\mathbf{R}-c}(X))$. Then $\operatorname{cosupp}^j(F)$ is a subanalytic subset of X and F belongs to ${}^p\mathbf{D}^{\geqslant 0}_{w-\mathbf{R}-c}(X)$ if and only if:*

(10.2.10) $$\dim(\operatorname{cosupp}^j(F)) < k \text{ for any } j \text{ and } k \text{ with } j < p(k) + k .$$

Proof. Let us take a stratification $X = \bigsqcup_\alpha X_\alpha$ such that $i^!_{X_\alpha}F$ has locally constant cohomologies.

Then we have:

(10.2.11) $$i^!_x F \simeq i^!_x i^!_{X_\alpha} F \simeq (or_{X_\alpha} \otimes i^!_{X_\alpha} F)_x [-\dim X_\alpha] .$$

Hence:

$$\operatorname{cosupp}^j(F) = \bigsqcup_\alpha \operatorname{supp}(H^{j-\dim X_\alpha}(i^!_{X_\alpha}F)) .$$

This shows immediately that $\operatorname{cosupp}^j(F)$ is subanalytic as well as the equivalence of (10.2.7) and (10.2.10) by Proposition 10.2.4 (ii). □

Note that the condition $j < p(k) + k$ is equivalent to $-j > p^*(k)$.

Corollary 10.2.6. *Let $F \in \mathrm{Ob}(\mathbf{D}^b_{w-\mathbf{R}-c}(X))$ and let $X = \bigsqcup_\alpha X_\alpha$ be a μ-stratification such that $SS(F) \subset \bigsqcup_\alpha T^*_{X_\alpha}X$. Let Y be a d-dimensional submanifold transversal to all strata X_α. Assume F belongs to ${}^p\mathbf{D}^{\geqslant 0}_{w-\mathbf{R}-c}(X)$ (resp. ${}^p\mathbf{D}^{\leqslant 0}_{w-\mathbf{R}-c}(X)$). Then $i_Y^{-1}F$ (resp. $i^!_Y F$) belongs to ${}^{p[d]}\mathbf{D}^{\geqslant 0}_{w-\mathbf{R}-c}(Y)$ (resp. ${}^{p[d]}\mathbf{D}^{\leqslant 0}_{w-\mathbf{R}-c}(Y)$).*

Proposition 10.2.7. *Let $F \in \mathrm{Ob}({}^p\mathbf{D}^{\leqslant 0}_{w-\mathbf{R}-c}(X))$ and $G \in \mathrm{Ob}({}^p\mathbf{D}^{\geqslant 0}_{w-\mathbf{R}-c}(X))$. Then:*

(10.2.12) $$H^j(R\mathcal{H}om(F, G)) = 0 \qquad \text{for } j < 0 ,$$

(10.2.13) $U \mapsto \mathrm{Hom}_{\mathbf{D}(U)}(F|_U, G|_U)$ *is a sheaf* .

Proof. Set $S = \bigcup_{j<0} \mathrm{supp}(H^j(R\mathcal{H}om(F,G)))$, and assume $S \neq \emptyset$. We shall derive a contradiction.

For $j < 0$, we have:

$$i_S^! H^j(R\mathcal{H}om(F,G)) \simeq H^j(i_S^! R\mathcal{H}om(F,G))$$
$$\simeq H^j(R\mathcal{H}om(i_S^{-1}F, i_S^! G)) .$$

Let $k = \dim S$. Then $i_S^! G$ is concentrated to degrees $\geqslant p(k)$. On the other hand the hypothesis on F implies that $S' \underset{\text{def}}{=} \bigcup_{j>p(k)} \mathrm{supp}(H^j(i_S^{-1}F))$ has dimension $< k$. Therefore $S \backslash S' \neq \emptyset$. Since $i_S^{-1}F|_{S\backslash S'}$ is concentrated to degrees $\leqslant p(k)$, $R\mathcal{H}om(i_S^{-1}F, i_S^! G)|_{S\backslash S'}$ is concentrated to degrees $\geqslant 0$. This is a contradiction.

The property (10.2.13) follows from (10.2.12) since $\mathrm{Hom}_{D(U)}(F|_U, G|_U) \simeq H^0(U; R\mathcal{H}om(F,G)) \simeq \Gamma(U; H^0(R\mathcal{H}om(F,G)))$. $\square$

We can now prove the main result of this section.

Theorem 10.2.8. *For any perversity p,* $({}^p\mathbf{D}^{\leqslant 0}_{w-\mathbb{R}-c}(X), {}^p\mathbf{D}^{\geqslant 0}_{w-\mathbb{R}-c}(X))$ *is a t-structure on* $\mathbf{D}^b_{w-\mathbb{R}-c}(X)$. *Moreover if A is Noetherian,* $({}^p\mathbf{D}^{\leqslant 0}_{\mathbb{R}-c}(X), {}^p\mathbf{D}^{\geqslant 0}_{\mathbb{R}-c}(X))$ *is a t-structure on* $\mathbf{D}^b_{\mathbb{R}-c}(X)$.

Proof. First we treat the case of $\mathbf{D}^b_{w-\mathbb{R}-c}(X)$. We have to show:

(a) for $F \in \mathrm{Ob}({}^p\mathbf{D}^{\leqslant 0}_{w-\mathbb{R}-c}(X))$ and $G \in \mathrm{Ob}({}^p\mathbf{D}^{\geqslant 1}_{w-\mathbb{R}-c}(X))$, $\mathrm{Hom}_{\mathbf{D}^b(X)}(F,G) = 0$,
(b) ${}^p\mathbf{D}^{\leqslant 0}_{w-\mathbb{R}-c}(X) \subset {}^p\mathbf{D}^{\leqslant 1}_{w-\mathbb{R}-c}(X)$ and ${}^p\mathbf{D}^{\geqslant 0}_{w-\mathbb{R}-c}(X) \supset {}^p\mathbf{D}^{\geqslant 1}_{w-\mathbb{R}-c}(X)$,
(c) for any F belonging to $\mathbf{D}^b_{w-\mathbb{R}-c}(X)$ there exists a distinguished triangle $F' \longrightarrow F \longrightarrow F'' \underset{+1}{\longrightarrow}$, with $F' \in \mathrm{Ob}({}^p\mathbf{D}^{\leqslant 0}_{w-\mathbb{R}-c}(X))$ and $F'' \in \mathrm{Ob}({}^p\mathbf{D}^{\geqslant 1}_{w-\mathbb{R}-c}(X))$.

Since (a) follows from the preceding proposition and (b) is obvious it remains to prove (c).

Let us choose an increasing μ-filtration $\{X_k\}_k$ of X (cf. Definition 8.3.25) such that:

(10.2.14) $\begin{cases} X_k \backslash X_{k-1} \text{ is a } k\text{-dimensional manifold for any } k \\ \text{and } F|_{X_k \backslash X_{k-1}} \text{ has locally constant cohomology} \end{cases}$.

Consider the condition:

(10.2.15)$_k$ $\begin{cases} \text{there exists a distinguished triangle in } \mathbf{D}^b_{w-\mathbb{R}-c}(X\backslash X_k)\text{:} \\ F' \longrightarrow i^{-1}_{X\backslash X_k}F \longrightarrow F'' \underset{+1}{\longrightarrow} \text{ with } F' \in \mathrm{Ob}({}^p\mathbf{D}^{\leqslant 0}_{w-\mathbb{R}-c}(X\backslash X_k)) \\ \text{and } F'' \in \mathrm{Ob}({}^p\mathbf{D}^{\geqslant 1}_{w-\mathbb{R}-c}(X\backslash X_k)), \text{ and } F'|_{X_j\backslash X_{j-1}} \text{ and} \\ F''|_{X_j\backslash X_{j-1}} \text{ have locally constant cohomology for any } j \geqslant k \end{cases}$.

Since (10.2.15)$_k$ is satisfied for $k \gg 0$, it is enough to prove (10.2.15)$_{k-1}$ when assuming (10.2.15)$_k$. Let $F' \longrightarrow i^{-1}_{X\backslash X_k}F \longrightarrow F'' \underset{+1}{\longrightarrow}$ be a distinguished triangle as in (10.2.15)$_k$. Let $j : X\backslash X_k \hookrightarrow X\backslash X_{k-1}$ be the open embedding and $i : X_k\backslash$

$X_{k-1} \hookrightarrow X \setminus X_{k-1}$ the closed embedding. The morphism $F' \to i^{-1}_{X \setminus X_k} F$ gives $j_! F' \to i^{-1}_{X \setminus X_{k-1}} F$. We embed this morphism into a distinguished triangle:

(10.2.16) $$j_! F' \longrightarrow i^{-1}_{X \setminus X_{k-1}} F \to G \xrightarrow[+1]{} \ .$$

Now we have a chain of morphisms $\tau^{\leqslant p(k)} i_* i^! G \to i_* i^! G \to G$. We embed the composition in a distinguished triangle:

(10.2.17) $$\tau^{\leqslant p(k)} i_* i^! G \longrightarrow G \longrightarrow \tilde{F}'' \xrightarrow[+1]{} \ .$$

Finally we embed the composition $i^{-1}_{X \setminus X_{k-1}} F \to G \to \tilde{F}''$ into a distinguished triangle:

(10.2.18) $$\tilde{F}' \longrightarrow i^{-1}_{X \setminus X_{k-1}} F \longrightarrow \tilde{F}'' \xrightarrow[+1]{} \ .$$

We shall show:

(10.2.19) $$\tilde{F}'' \in \mathrm{Ob}({}^p\mathbf{D}^{\geqslant 1}_{w-\mathbb{R}-c}(X \setminus X_{k-1})) \ ,$$

(10.2.20) $$\tilde{F}' \in \mathrm{Ob}({}^p\mathbf{D}^{\leqslant 0}_{w-\mathbb{R}-c}(X \setminus X_{k-1})) \ .$$

By the construction, $j^{-1}\tilde{F}'' \simeq F''$ and $j^{-1}\tilde{F}' \simeq F'$. Hence by Proposition 10.2.4, it is enough to show:

(10.2.21) $i^! \tilde{F}''$ is concentrated to degrees $\geqslant p(k) + 1$,

(10.2.22) $i^{-1} \tilde{F}'$ is concentrated to degrees $\leqslant p(k)$.

By applying the functor $i^!$ to the distinguished triangle (10.2.17) we get the distinguished triangle:

(10.2.23) $$\tau^{\leqslant p(k)} i^! G \longrightarrow i^! G \longrightarrow i^! \tilde{F}'' \xrightarrow[+1]{} \ .$$

Hence $i^! \tilde{F}'' \simeq \tau^{\geqslant p(k)+1} i^! G$, which shows (10.2.21).

Now let us apply the octahedral axiom (I §4) to the distinguished triangles (10.2.16), (10.2.17), (10.2.18). (cf. (10.2.24)):

(10.2.24)

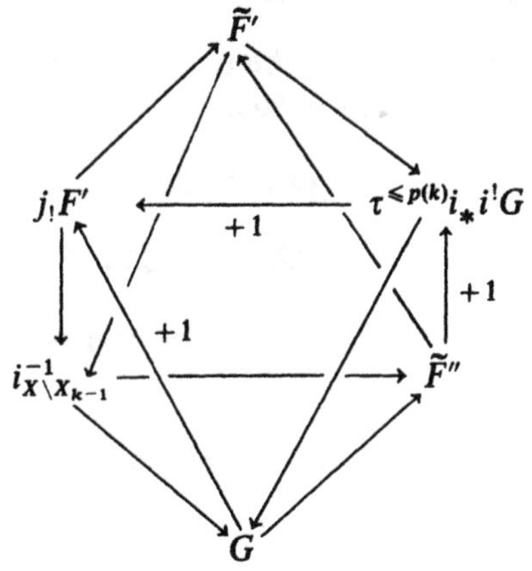

We get a distinguished triangle:

$$j_!F' \longrightarrow \tilde{F}' \longrightarrow \tau^{\leqslant p(k)} i_* i^! G \xrightarrow[+1]{} . \tag{10.2.25}$$

Hence $i^{-1}\tilde{F}' \simeq \tau^{\leqslant p(k)} i^! G$. This shows (10.2.22).

Since one checks easily that $\tilde{F}'$ and $\tilde{F}''$ have locally constant cohomology on $X_j \setminus X_{j-1}$ for any $j \geqslant k$, the induction proceeds. If A is Noetherian, the same proof holds for $\mathbf{D}^b_{\mathbb{R}-c}(X)$. □

The functor $U \mapsto {}^p\mathbf{D}^0_{w-\mathbb{R}-c}(U)$, ($U$ open in X), behaves like a sheaf. More precisely, we have:

Proposition 10.2.9. *${}^p\mathbf{D}^0_{w-\mathbb{R}-c}(X)$ is a* **stack**, *i.e. given an open covering $X = \bigcup_{i \in I} U_i$, $F_i \in \mathrm{Ob}({}^p\mathbf{D}^0_{w-\mathbb{R}-c}(U_i))$ and isomorphisms: $f_{ij} : F_j|_{U_{ij}} \simeq F_i|_{U_{ij}}$ satisfying the cocycle condition $(f_{ij}|_{U_{ijk}}) \circ (f_{jk}|_{U_{ijk}}) = f_{ik}|_{U_{ijk}}$ for any i, j, k, there exist $F \in \mathrm{Ob}({}^p\mathbf{D}^0_{w-\mathbb{R}-c}(X))$ and $f_i : F|_{U_i} \simeq F_i$, such that $f_{ij} \circ f_j|_{U_{ij}} = f_i|_{U_{ij}}$.*

Moreover the family $(F, \{f_i\}_i)$ is unique up to isomorphism. (Here we set $U_{ij} = U_i \cap U_j$ and $U_{ijk} = U_i \cap U_j \cap U_k$.)

Proof. (a) The uniqueness follows from Proposition 10.2.7.

(b) We shall first show the existence of $(F, \{f_i\}_i)$ when I is finite. Arguing by induction on $\# I$, it is enough to prove the result when $\# I = 2$, say $I = \{1, 2\}$. In this case it is enough to define F by a distinguished triangle:

$$i_{U_{12}!}(F_1|_{U_{12}}) \longrightarrow i_{U_1!}F_1 \oplus i_{U_2!}F_2 \longrightarrow F \xrightarrow[+1]{}$$

where the morphism $i_{U_{12}!}(F_1|_{U_{12}}) \to i_{U_2!}F_2$ is given by f_{21}.

(c) Now we treat the general case. By (a) and (b) we may assume $I = \mathbb{N}$ and $U_n \subset U_{n+1}$ for all n. Represent $F_n \in \mathrm{Ob}({}^p\mathbf{D}^0_{w-\mathbb{R}-c}(U_n))$ by an injective complex $I_n \in \mathrm{Ob}(\mathbf{C}^+(\mathfrak{Mod}(A_{U_n})))$. Then the isomorphisms $F_n \simeq F_{n+1}|_{U_n}$ give homomorphisms $I_n \to I_{n+1}|_{U_n}$, hence homomorphisms $\varphi_n : i_{U_n!}I_n \to i_{U_{n+1}!}I_{n+1}$. Now let $F = \varinjlim_n I_n$. It is enough to prove that the natural morphism $I_n \to i_{U_n}^{-1}F$ is a quasi-isomorphism. But this follows immediately from the fact that for $x \in U_n$, $(I_n)_x$ is quasi-isomorphic to $(I_m)_x$ for $m \geqslant n$. □

We shall denote by ${}^p\tau^{\leqslant n}$ and ${}^p\tau^{\geqslant n}$ the truncation functors associated to the t-structure $({}^p\mathbf{D}^{\leqslant 0}_{w-\mathbb{R}-c}(X), {}^p\mathbf{D}^{\geqslant 0}_{w-\mathbb{R}-c}(X))$. Hence:

$$\begin{cases} {}^p\tau^{\leqslant n} : \mathbf{D}^b_{w-\mathbb{R}-c}(X) \to {}^p\mathbf{D}^{\leqslant n}_{w-\mathbb{R}-c}(X) , \\ {}^p\tau^{\geqslant n} : \mathbf{D}^b_{w-\mathbb{R}-c}(X) \to {}^p\mathbf{D}^{\geqslant n}_{w-\mathbb{R}-c}(X) . \end{cases} \tag{10.2.26}$$

We shall denote by ${}^pH^n$ the n-cohomology with respect to the same t-structure. Hence:

$${}^pH^n : \mathbf{D}^b_{w-\mathbb{R}-c}(X) \to {}^p\mathbf{D}^0_{w-\mathbb{R}-c}(X) . \tag{10.2.27}$$

We keep the same notations for ${}^p\mathbf{D}^0_{\mathbb{R}-c}(X)$ when A is Noetherian.

To end this section, let us study functorial properties of t-structures associated to perversities.

Proposition 10.2.10. *Let Y be a locally closed subanalytic subset of X.*

(i) $i_{Y!}i_Y^{-1}$ *sends* ${}^p\mathbf{D}^{\leqslant 0}_{w-\mathbf{R}-c}(X)$ *to itself (i.e. this functor is right t-exact).*
(ii) $Ri_{Y*}i_Y^!$ *sends* ${}^p\mathbf{D}^{\geqslant 0}_{w-\mathbf{R}-c}(X)$ *to itself (i.e. this functor is left t-exact).*

Proof. (i) Since $i_{Y!}$ and i_Y^{-1} are exact functors, they commute to the functor $H^j(\cdot)$. Hence the result follows from $\mathrm{supp}(i_{Y!}i_Y^{-1}(F)) \subset \mathrm{supp}(F)$.

(ii) One has $(i_S^! i_{Y*} i_Y^! F)_{Y\cap S} \simeq i_{Y\cap S}^! F$. Hence the result follows from the definition. □

Proposition 10.2.11. *Let $f\colon Y \to X$ be a morphism of manifolds and let d be an integer. Assume that $\dim f^{-1}(x) \leqslant d$ for any $x \in X$. Then:*

(i) f^{-1} *sends* ${}^p\mathbf{D}^{\leqslant 0}_{w-\mathbf{R}-c}(X)$ *to* ${}^{p[-d]}\mathbf{D}^{\leqslant 0}_{w-\mathbf{R}-c}(Y)$.
(ii) $f^!$ *sends* ${}^p\mathbf{D}^{\geqslant 0}_{w-\mathbf{R}-c}(X)$ *to* ${}^{p[-d]}\mathbf{D}^{\geqslant -d}_{w-\mathbf{R}-c}(Y)$.

Proof. (i) follows from:

$$\begin{aligned}\dim(\mathrm{supp}(H^j(f^{-1}F))) &= \dim(f^{-1}(\mathrm{supp}(H^j(F))))\\ &\leqslant \dim(\mathrm{supp}(H^j(F))) + d\ .\end{aligned}$$

(ii) follows from Proposition 10.2.5 and the relation:

$$\begin{aligned}\mathrm{cosupp}^j(f^!F) &= \{y \in Y; H^j(i_y^! f^! F) \neq 0\}\\ &= \{y \in Y; H^j(i_{f(y)}^! F) \neq 0\}\\ &= f^{-1}(\mathrm{cosupp}^j(F))\ ,\end{aligned}$$

which implies that

$$\dim(\mathrm{cosupp}^j(f^!F)) \leqslant \dim(\mathrm{cosupp}^j(F)) + d\ . \quad □$$

Proposition 10.2.12. *Let $f\colon Y \to X$ be a morphism of manifolds and let d be an integer. Assume that $\dim f^{-1}(x) \leqslant d$ for any $x \in X$.*

(i) *Let* $G \in \mathrm{Ob}({}^p\mathbf{D}^{\leqslant 0}_{w-\mathbf{R}-c}(Y))$ *and assume that* $Rf_!G \in \mathrm{Ob}(\mathbf{D}^b_{w-\mathbf{R}-c}(X))$. *Then* $Rf_!G \in \mathrm{Ob}({}^{p[d]}\mathbf{D}^{\leqslant d}_{w-\mathbf{R}-c}(X))$.
(ii) *Let* $G \in \mathrm{Ob}({}^p\mathbf{D}^{\geqslant 0}_{w-\mathbf{R}-c}(Y))$ *and assume* $Rf_*G \in \mathrm{Ob}(\mathbf{D}^b_{w-\mathbf{R}-c}(X))$. *Then* $Rf_*G \in \mathrm{Ob}({}^{p[d]}\mathbf{D}^{\geqslant 0}_{w-\mathbf{R}-c}(X))$.

Proof. This follows immediately from the preceding proposition, Proposition 10.1.17 and the fact that f^{-1} and Rf_* as well as $Rf_!$ and $f^!$ are adjoint functors. □

Proposition 10.2.13. *Assume the base ring A is a field. Then the dual functor* D_X *(cf. Definition 3.1.16) sends* ${}^p\mathbf{D}^{\leq 0}_{\mathbb{R}-c}(X)$ *and* ${}^p\mathbf{D}^{\geq 0}_{\mathbb{R}-c}(X)$ *to* ${}^{p^*}\mathbf{D}^{\geq 0}_{\mathbb{R}-c}(X)$ *and* ${}^{p^*}\mathbf{D}^{\leq 0}_{\mathbb{R}-c}(X)$ *respectively.*

Proof. This follows from $i_x^{-1}(D_X F) \simeq \mathrm{Hom}(i_x^! F, A)$ and $i_x^!(D_X F) \simeq \mathrm{Hom}(i_x^{-1} F, A)$, and Proposition 10.2.5. □

Finally we note the following useful result.

Proposition 10.2.14. *Let Y be a closed submanifold of X and let F belong to* $\mathbf{D}^b_{w-\mathbb{R}-c}(X)$.

(i) $F \in \mathrm{Ob}({}^p\mathbf{D}^{\leq 0}_{w-\mathbb{R}-c}(X))$ *if and only if* $F|_{X\setminus Y} \in \mathrm{Ob}({}^p\mathbf{D}^{\leq 0}_{w-\mathbb{R}-c}(X\setminus Y))$ *and* $i_Y^{-1}F \in \mathrm{Ob}({}^p\mathbf{D}^{\leq 0}_{w-\mathbb{R}-c}(Y))$.
(ii) $F \in \mathrm{Ob}({}^p\mathbf{D}^{\geq 0}_{w-\mathbb{R}-c}(X))$ *if and only if* $F|_{X\setminus Y} \in \mathrm{Ob}({}^p\mathbf{D}^{\geq 0}_{w-\mathbb{R}-c}(X\setminus Y))$ *and* $i_Y^! F \in \mathrm{Ob}({}^p\mathbf{D}^{\geq 0}_{w-\mathbb{R}-c}(Y))$.

Proof. (i) follows from $\mathrm{supp}\, H^j(F) = \overline{\mathrm{supp}\, H^j(F|_{X\setminus Y})} \cup \mathrm{supp}\, H^j(i_Y^{-1}F)$.
(ii) follows from $\mathrm{cosupp}^j(F) = \mathrm{cosupp}^j(F|_{X\setminus Y}) \cup \mathrm{cosupp}^j(i_Y^! F)$. □

10.3. Perverse sheaves on complex manifolds

Let X be a complex manifold. We set:

$${}^p\mathbf{D}^{\leq 0}_{w-\mathbb{C}-c}(X) = {}^p\mathbf{D}^{\leq 0}_{w-\mathbb{R}-c}(X) \cap \mathbf{D}_{w-\mathbb{C}-c}(X) ,$$

and we define similarly ${}^p\mathbf{D}^{\geq 0}_{w-\mathbb{C}-c}(X)$, ${}^p\mathbf{D}^{0}_{w-\mathbb{C}-c}(X)$, ${}^p\mathbf{D}^{\leq 0}_{\mathbb{C}-c}(X)$, ${}^p\mathbf{D}^{\geq 0}_{\mathbb{C}-c}(X)$, ${}^p\mathbf{D}^{0}_{\mathbb{C}-c}(X)$, etc.

In Proposition 10.2.4 we can choose for the X_α's complex manifolds and hence ${}^p\mathbf{D}^{\leq 0}_{w-\mathbb{C}-c}(X)$, etc., depend only on the values of p at the even integers. In the sequel, we choose p such that $p = p^*$ on $2\mathbb{Z}$, that is:

$$p(n) = -n/2 \qquad \text{for even } n \ . \tag{10.3.1}$$

We call p the **middle perversity**.

Convention 10.3. In this section, unless otherwise specified, all manifolds and morphisms of manifolds, are complex analytic. For a complex analytic set S, $\dim S$ denotes the complex dimension of S, and we write $\dim_{\mathbb{R}}$ for the real dimension.

Hence we have for an object F of $\mathbf{D}^b_{w-\mathbb{C}-c}(X)$:

$$\begin{cases} F \in \mathrm{Ob}({}^p\mathbf{D}^{\leq 0}_{w-\mathbb{C}-c}(X)) \text{ if and only if } \dim \mathrm{supp}(H^j(F)) \leq -j \\ \text{for all } j \ . \end{cases} \tag{10.3.2}$$

$$(10.3.3)\quad \begin{cases} F \in \mathrm{Ob}({}^p\mathbf{D}^{\geq 0}_{w-\mathbb{C}-c}(X)) \text{ if and only if for any locally closed} \\ \text{complex analytic subset } S \text{ of } X, H^j_S(F)|_S = 0 \text{ for } j < -\dim S \ . \end{cases}$$

(This last condition is equivalent to $\dim \mathrm{cosupp}^j(F) \leq j$.)

Choose a complex μ-stratification $X = \bigsqcup_\alpha X_\alpha$ such that $\mathrm{SS}(F) \subset \bigsqcup_\alpha T^*_{X_\alpha}X$. Then:

$$(10.3.4)\quad \begin{cases} F \in \mathrm{Ob}({}^p\mathbf{D}^{\leq 0}_{w-\mathbb{C}-c}(X)) \text{ if and only if } F_x \in \mathrm{Ob}(\mathbf{D}^{\leq -\dim X_\alpha}(\mathfrak{Mod}(A))) \\ \text{for any } x \in X_\alpha, \text{ any } \alpha \ , \end{cases}$$

$$(10.3.5)\quad \begin{cases} F \in \mathrm{Ob}({}^p\mathbf{D}^{\geq 0}_{w-\mathbb{C}-c}(X)) \text{ if and only if} \\ (R\Gamma_{\{x\}}(F))_x \in \mathrm{Ob}(\mathbf{D}^{\geq \dim X_\alpha}(\mathfrak{Mod}(A))) \text{ for any } x \in X_\alpha, \text{ any } \alpha \ . \end{cases}$$

Definition 10.3.1. *An object of ${}^p\mathbf{D}^0_{\mathbb{C}-c}(X)$ is called a perverse sheaf.*

Note that in general, a perverse sheaf is not a sheaf, but one nevertheless still calls them "sheaves", in view of their local properties stated in Proposition 10.2.9.

Example 10.3.2. Let Y be a closed submanifold of X. Then $A_Y[\dim Y]$ is perverse.

We can translate Proposition 10.2.11 and 10.2.12 as follows.

Proposition 10.3.3. *Let d be an integer, $f: Y \to X$ a morphism of complex manifolds such that $\dim f^{-1}(x) \leq d$ for any x. Then:*

(i) f^{-1} *sends* ${}^p\mathbf{D}^{\leq 0}_{w-\mathbb{C}-c}(X)$ *to* ${}^p\mathbf{D}^{\leq d}_{w-\mathbb{C}-c}(Y)$.

(ii) $f^!$ *sends* ${}^p\mathbf{D}^{\geq 0}_{w-\mathbb{C}-c}(X)$ *to* ${}^p\mathbf{D}^{\geq -d}_{w-\mathbb{C}-c}(Y)$.

(iii) *For* $G \in \mathrm{Ob}({}^p\mathbf{D}^{\leq 0}_{w-\mathbb{C}-c}(Y))$, *if* $Rf_!G$ *belongs to* $\mathbf{D}^b_{w-\mathbb{C}-c}(X)$, *then* $Rf_!G$ *belongs to* ${}^p\mathbf{D}^{\leq d}_{w-\mathbb{C}-c}(X)$.

(iv) *For* $G \in \mathrm{Ob}({}^p\mathbf{D}^{\geq 0}_{w-\mathbb{C}-c}(Y))$, *if* Rf_*G *belongs to* $\mathbf{D}^b_{w-\mathbb{C}-c}(X)$, *then* Rf_*G *belongs to* ${}^p\mathbf{D}^{\geq -d}_{w-\mathbb{C}-c}(X)$.

Moreover the result of Theorem 10.2.8 remains valid in the complex case, by choosing the filtration $\{X_j\}$ of X in the proof of that theorem such that all X_j's are complex analytic.

Theorem 10.3.4. *$({}^p\mathbf{D}^{\leq 0}_{w-\mathbb{C}-c}(X), {}^p\mathbf{D}^{\geq 0}_{w-\mathbb{C}-c}(X))$ is a t-structure on $\mathbf{D}^b_{w-\mathbb{C}-c}(X)$. Moreover if the base ring A is Noetherian, then $({}^p\mathbf{D}^{\leq 0}_{\mathbb{C}-c}(X), {}^p\mathbf{D}^{\geq 0}_{\mathbb{C}-c}(X))$ is a t-structure on $\mathbf{D}^b_{\mathbb{C}-c}(X)$.*

The following is a consequence of Proposition 10.2.13.

Proposition 10.3.5. *If the base ring A is a field, then the duality functor D_X interchanges ${}^p\mathbf{D}^{\geq 0}_{\mathbb{C}-c}(X)$ and ${}^p\mathbf{D}^{\leq 0}_{\mathbb{C}-c}(X)$.*

The following proposition is also easy to prove.

Proposition 10.3.6. *Let X and Y be complex manifolds.*

(i) *If* $F \in \mathrm{Ob}({}^p\mathbf{D}^{\leqslant 0}_{w-\mathbb{C}-c}(X))$ *and* $G \in \mathrm{Ob}({}^p\mathbf{D}^{\leqslant 0}_{w-\mathbb{C}-c}(Y))$, *then* $F \overset{L}{\boxtimes} G \in \mathrm{Ob}({}^p\mathbf{D}^{\leqslant 0}_{w-\mathbb{C}-c}(X \times Y))$.

(ii) *If the base ring is a field and if* $F \in \mathrm{Ob}({}^p\mathbf{D}^{\geqslant 0}_{\mathbb{C}-c}(X))$ *and* $G \in \mathrm{Ob}({}^p\mathbf{D}^{\geqslant 0}_{\mathbb{C}-c}(Y))$ *then* $F \boxtimes G \in \mathrm{Ob}({}^p\mathbf{D}^{\geqslant 0}_{\mathbb{C}-c}(X \times Y))$.

(iii) *If* $F \in \mathrm{Ob}({}^p\mathbf{D}^{\leqslant 0}_{w-\mathbb{C}-c}(X))$ *and* $G \in \mathrm{Ob}({}^p\mathbf{D}^{\geqslant 0}_{w-\mathbb{C}-c}(Y))$, *then* $R\mathcal{H}om(q_1^{-1}F, q_2^!G) \in \mathrm{Ob}({}^p\mathbf{D}^{\geqslant 0}_{w-\mathbb{C}-c}(X \times Y))$.

(Here q_1 and q_2 are the projections from $X \times Y$ to X and Y, respectively.)

Proof. (i) We choose complex analytic μ-stratifications $X = \bigsqcup_\alpha X_\alpha$ and $Y = \bigsqcup_\beta Y_\beta$ such that $SS(F) \subset \bigsqcup_\alpha T^*_{X_\alpha}X$ and $SS(G) \subset \bigsqcup_\beta T^*_{Y_\beta}Y$. Then the result follows from Proposition 10.2.4, since:

$$i^{-1}_{X_\alpha \times Y_\beta}\left(F \overset{L}{\boxtimes} G\right) \simeq i^{-1}_{X_\alpha}F \overset{L}{\boxtimes} i^{-1}_{Y_\beta}G \ .$$

(ii) follows from (i) and the fact that $F \boxtimes G \simeq D_{X \times Y}(D_X F \boxtimes D_Y G)$.

(iii) One has:

$$i^!_{X_\alpha \times Y_\beta} R\mathcal{H}om(q_1^{-1}F, q_2^!G) \simeq R\mathcal{H}om(i^{-1}_{X_\alpha}F \boxtimes A_{Y_\beta}, \omega_{X_\alpha} \boxtimes i^!_{Y_\beta}G) \ .$$

By the hypothesis, $i^{-1}_{X_\alpha}F$ is concentrated to degrees $\leqslant -\dim X_\alpha$ and $\omega_{X_\alpha} \boxtimes i^!_{Y_\beta}G$ is concentrated to degrees $\geqslant -2\dim X_\alpha - \dim Y_\beta$. Thus $i^!_{X_\alpha \times Y_\beta} R\mathcal{H}om(q_1^{-1}F, q_2^!G)$ is concentrated to degrees $\geqslant -\dim(X_\alpha \times Y_\beta)$. □

Now we shall introduce a kind of microlocal perversity.

Definition 10.3.7. ${}^\mu\mathbf{D}^{\leqslant n}_{w-\mathbb{C}-c}(X)$ *(resp.* ${}^\mu\mathbf{D}^{\geqslant n}_{w-\mathbb{C}-c}(X)$*) is the full subcategory of* $\mathbf{D}^b_{w-\mathbb{R}-c}(X)$ *consisting of objects F such that the type L of F with shift 0 at any non-singular point of* $SS(F)$ *satisfies* $H^j(L) = 0$ *for* $j > n - \dim X$ *(resp.* $j < n - \dim X$*).*

By Proposition 7.5.9 and Exercise A.7 (see also Exercise VII.4), the following three conditions for $F \in \mathrm{Ob}(\mathbf{D}^b_{w-\mathbb{C}-c}(X))$ are equivalent.

(10.3.6) *F belongs to* ${}^\mu\mathbf{D}^{\leqslant 0}_{w-\mathbb{C}-c}(X)$ *(resp.* ${}^\mu\mathbf{D}^{\geqslant 0}_{w-\mathbb{C}-c}(X)$*)* .

(10.3.7) *For every non-singular point p of* $SS(F)$ *such that* $\pi : SS(F) \to X$ *has constant rank on a neighborhood of p, there exists a submanifold Y and* $L \in \mathrm{Ob}(\mathbf{D}^b(\mathfrak{Mod}(A)))$ *such that* $F \simeq L_Y[\dim Y]$ *in* $\mathbf{D}^b(X; p)$ *and* $H^j(L) = 0$ *for* $j > 0$ *(resp.* $j < 0$*)* .

(10.3.8) *The assertion of (10.3.7) is true for some point p at any irreducible component of* $SS(F)$.

One of the aims of this section is to prove that ${}^{\mu}\mathbf{D}^{\leq 0}_{w-\mathbb{C}-c}(X) = {}^{p}\mathbf{D}^{\leq 0}_{w-\mathbb{C}-c}(X)$ and ${}^{\mu}\mathbf{D}^{\geq 0}_{w-\mathbb{C}-c}(X) = {}^{p}\mathbf{D}^{\geq 0}_{w-\mathbb{C}-c}(X)$.

First we need a vanishing theorem for perverse sheaves on Stein manifolds. We shall not review here the theory of Stein manifolds (for example, cf. Hörmander [1]) but we recall that such a manifold can be holomorphically embeded into $\mathbb{C}^N$ for some N as a closed submanifold.

Theorem 10.3.8. *Let X be a Stein manifold.*

(i) *For any* $F \in \mathrm{Ob}({}^{\mu}\mathbf{D}^{\leq 0}_{w-\mathbb{C}-c}(X))$, $H^j(X;F) = 0$ *for* $j > 0$.
(ii) *For any* $F \in \mathrm{Ob}({}^{\mu}\mathbf{D}^{\geq 0}_{w-\mathbb{C}-c}(X))$, $H^j_c(X;F) = 0$ *for* $j < 0$.

The proof which relies on Morse theory, is similar to that of Proposition 5.4.20.

Proof. We embed X into $\mathbb{C}^N$ as a closed submanifold. Let $F \in \mathrm{Ob}(\mathbf{D}^b_{w-\mathbb{C}-c}(X))$, $\Lambda = SS(F)$ and let $\Lambda_\circ = \{p \in \Lambda_{\mathrm{reg}}; \pi : \Lambda_{\mathrm{reg}} \to X$ has a constant rank in a neighborhood of $p\}$. Then $\Lambda_\circ$ is a subanalytic submanifold and $\dim_{\mathbb{R}}(\Lambda \setminus \Lambda_\circ) < \dim_{\mathbb{R}} X$. Applying Proposition 8.3.27, we find $z_\circ \in \mathbb{C}^N$ such that, setting $\varphi(x) = |x - z_\circ|^2$ for $x \in X$ and $\Lambda_\varphi = \{(x; d\varphi(x)); x \in X\} \subset T^*X$, we have:

(10.3.9) $\qquad \Lambda_\varphi \cap \Lambda \subset \Lambda_\circ$ *and the intersection is transversal* .

Take a point $p \in \Lambda_\varphi \cap \Lambda$ and set $x_\circ = \pi(p)$. Since $p \in \Lambda_\circ$, $F \simeq L_Y[\dim Y]$ in $\mathbf{D}^b(X;p)$ and in $\mathbf{D}^b(X;p^a)$, for a complex submanifold Y of X and for some $L \in \mathrm{Ob}(\mathbf{D}^b(\mathfrak{Mod}(A)))$. Since $\Lambda_\circ = T^*_Y X$ on a neighborhood of p, and the intersection of Λ and Λ_φ is transversal at p, the Hessian of $\varphi|_Y$ at $x_\circ$ is non-degenerate (see VII §5). On the other hand $\partial\bar{\partial}\varphi$ is positive-definite on $T_{x_\circ}\mathbb{C}^N$, hence on $T_{x_\circ}Y$. This implies:

(10.3.10) $\quad \begin{cases} \textit{the number of positive eigenvalues of } \mathrm{Hess}(\varphi|_Y) \\ \textit{is at least } \dim Y\,. \end{cases}$

In fact (10.3.10) is a consequence of the following elementary result of linear algebra:

let V be an n-dimensional complex vector space, B a real symmetric bilinear form on the real underlying space $V^{\mathbb{R}}$, and extend B to $\tilde{B}$ a $\mathbb{C}$-bilinear form on $V^{\mathbb{R}} \otimes_{\mathbb{R}} \mathbb{C} \simeq V \oplus \bar{V}$. Then if the Hermitian form $\tilde{B}(x, \bar{x})$ is positive-definite on V, we get that B has at least n positive eigenvalues.

Hence, by choosing a real local coordinate system $(x_1, \ldots, x_{2d})$ on Y, we can write:

(10.3.11) $\qquad \varphi(x) = x_1^2 + \cdots + x_l^2 - \cdots - x_{2d}^2 + \varphi(x_\circ)$

with $d = \dim Y$ and $l \geq d$.

Then we obtain:

(10.3.12) $$(H^j_{\{x;\,\varphi(x)\geqslant\varphi(x_0)\}}(F))_{x_0} \simeq (H^{j+d}_{\{x_1^2+\cdots+x_l^2-\cdots-x_{2d}^2\geqslant 0\}}(L_{\mathbb{R}^{2d}}))_0$$
$$\simeq H^{j+d-(2d-l)}(L)$$
$$= H^{j+l-d}(L)\ ,$$

and similarly:

(10.3.13) $$(H^j_{\{x;\,\varphi(x)\leqslant\varphi(x_0)\}}(F))_{x_0} \simeq H^{j+d-l}(L)\ .$$

Since $l \geqslant d$, we get:

(10.3.14) $$\begin{cases} \textit{If } F \in \mathrm{Ob}({}^{\mu}\mathbf{D}^{\leqslant 0}_{w-\mathbb{C}-c}(X)) & \textit{then} \\ (H^j_{\{x;\,\varphi(x)\geqslant\varphi(x_0)\}}(F))_{x_0} = 0 & \textit{for } j > 0\ , \end{cases}$$

(10.3.15) $$\begin{cases} \textit{If } F \in \mathrm{Ob}({}^{\mu}\mathbf{D}^{\geqslant 0}_{w-\mathbb{C}-c}(X))\ , & \textit{then} \\ (H^j_{\{x;\,\varphi(x)\leqslant\varphi(x_0)\}}(F))_{x_0} = 0 & \textit{for } j < 0\ . \end{cases}$$

Now $\varphi(\pi(\Lambda_\varphi \cap \Lambda))$ is a discrete set by the microlocal Bertini-Sard theorem (Proposition 8.3.12), contained in $\mathbb{R}_{\geqslant 0}$. We set $t_0 = -\infty$ and $\varphi(\pi(\Lambda_\varphi \cap \Lambda)) = \{t_1, t_2, \ldots\}$.

By Proposition 5.4.4, $SS(R\varphi_* F) \subset \bigcup_i T^*_{\{t_i\}}\mathbb{R}$. Hence $R\varphi_* F$ is constant on any open interval $]t_{i-1}, t_i[$ $(i \geqslant 1)$. Setting:

$$V_i = R\Gamma(]-\infty, t_{i+1}[; R\varphi_* F)\ ,$$

$$V_i^c = R\Gamma_c(]-\infty, t_{i+1}[; R\varphi_* F)\ ,$$

we get the distinguished triangles for $i \geqslant 1$:

(10.3.16) $$(R\Gamma_{[t_i, +\infty[}(R\varphi_* F))_{t_i} \longrightarrow V_i \longrightarrow V_{i-1} \xrightarrow[+1]{}\ ,$$

(10.3.17) $$V_{i-1}^c \longrightarrow V_i^c \longrightarrow (R\Gamma_{]-\infty, t_i]}(R\varphi_* F))_{t_i} \xrightarrow[+1]{}\ .$$

Now assume $F \in \mathrm{Ob}({}^{\mu}\mathbf{D}^{\leqslant 0}_{w-\mathbb{C}-c}(X))$.

Since $(R\Gamma_{\{\varphi(x)\geqslant t_i\}}(F))|_{\varphi^{-1}(t_i)}$ is supported by $\varphi^{-1}(t_i) \cap \pi(SS(F) \cap \Lambda_\varphi)$, we have:

$$H^j(R\Gamma_{[t_i, +\infty[}(R\varphi_* F))_{t_i} \simeq \bigoplus_p H^j(R\Gamma_{\{\varphi(x)\geqslant t_i\}}(F))_p$$

where p ranges through the finite set $\varphi^{-1}(t_i) \cap \pi(SS(F) \cap \Lambda_\varphi)$. These groups vanish for $j > 0$ by (10.3.14) and the distinguished triangle (10.3.16) yields:

$$H^j(V_i) \simeq H^j(V_{i-1}) \quad \text{for} \quad j > 0 \quad \text{and}$$

$$H^0(V_i) \to H^0(V_{i-1}) \quad \text{is surjective}\ .$$

Therefore the family $\{H^{j-1}(V_i)\}_i$ satisfies the Mittag-Leffler condition for $j > 0$ and we get by Proposition 2.7.1:

$$H^j(X;F) \simeq H^j(\mathbb{R};R\varphi_* F) \simeq \varprojlim_i H^j(V_i) \simeq H^j(V_0) = 0 \quad \text{for} \quad j > 0 \ .$$

Similarly, if one assumes that F belongs to ${}^{\mu}\mathbf{D}^{\geq 0}_{w-\mathbb{C}-c}(X)$, one gets $H^j(R\Gamma_{\{t \leq t_i\}}(F))_{t_i} = 0$ for $j < 0$, hence:

$$H^j(V^c_{i-1}) \simeq H^j(V^c_i) \quad \text{for} \quad j < 0 \ .$$

This gives:

$$H^j_c(X;F) \simeq H^j_c(\mathbb{R};R\varphi_* F) \simeq \varinjlim_i H^j(V^c_i) \simeq H^j(V^c_0) = 0 \quad \text{for} \quad j < 0 \ . \quad \square$$

Lemma 10.3.9. *Let $F \in \mathrm{Ob}(\mathbf{D}^b_{w-\mathbb{C}-c}(X))$ and let $X = \bigsqcup_\alpha X_\alpha$ be a stratification such that $\mathrm{SS}(F) \subset \bigsqcup_\alpha T^*_{X_\alpha}X$. Let Y be a submanifold of X which intersects all X_α transversally.*

(i) *Assume F belongs to ${}^{\mu}\mathbf{D}^{\leq 0}_{w-\mathbb{C}-c}(X)$. Then $i_Y^{-1}F$ belongs to ${}^{\mu}\mathbf{D}^{\leq -\mathrm{codim}\, Y}_{w-\mathbb{C}-c}(Y)$ and $i_Y^!F$ belongs to ${}^{\mu}\mathbf{D}^{\leq \mathrm{codim}\, Y}_{w-\mathbb{C}-c}(Y)$.*

(ii) *Assume F belongs to ${}^{\mu}\mathbf{D}^{\geq 0}_{w-\mathbb{C}-c}(X)$. Then $i_Y^{-1}F$ belongs to ${}^{\mu}\mathbf{D}^{\geq -\mathrm{codim}\, Y}_{w-\mathbb{C}-c}(Y)$ and $i_Y^!F$ belongs to ${}^{\mu}\mathbf{D}^{\geq \mathrm{codim}\, Y}_{w-\mathbb{C}-c}(Y)$.*

Proof. By the transversality condition, $T^*_YX \cap (\bigsqcup_\alpha T^*_{X_\alpha}X) \subset T^*_XX$. Hence i_Y is non-characteristic for F, and we can apply Proposition 5.4.13 to obtain:

$$i_Y^{-1}F \otimes \omega_{Y/X} \simeq i_Y^!F \ ,$$

$$\mathrm{SS}(i_Y^{-1}F) \subset \bigsqcup_\alpha T^*_{Y \cap X_\alpha}Y \ .$$

Let $y \in Y \cap X_\alpha$. By Corollary 8.3.24, there exists $p \in Y \times_X T^*_{X_\alpha}X \setminus \bigcup_{\beta \neq \alpha} \overline{T^*_{X_\beta}X}$ with $\pi(p) = y$. Then $F \simeq L_{X_\alpha}$ in $\mathbf{D}^b(X;p)$ for some $L \in \mathbf{D}^b(\mathfrak{Mod}(A))$, and $i_Y^{-1}F \simeq L_{X_\alpha \cap Y}$ in $\mathbf{D}^b(Y;{}^t i'_Y(p))$, by Proposition 6.1.9.

Then (i) and (ii) follow by the equivalence of (10.3.6), (10.3.7) and (10.3.8). $\square$

Next we study the specialization of perverse sheaves.

Proposition 10.3.10. *Let Y be a smooth hypersurface of X and let $F \in \mathrm{Ob}(\mathbf{D}^b_{w-\mathbb{C}-c}(X))$. Assume that $F|_{X \setminus Y} \in \mathrm{Ob}({}^{\mu}\mathbf{D}^{\leq 0}_{w-\mathbb{C}-c}(X \setminus Y))$ (resp. ${}^{\mu}\mathbf{D}^{\geq 0}_{w-\mathbb{C}-c}(X \setminus Y)$). Then $\nu_Y(F)|_{\dot{T}_YX}$ belongs to ${}^{p}\mathbf{D}^{\leq 0}_{w-\mathbb{C}-c}(\dot{T}_YX)$ (resp. ${}^{p}\mathbf{D}^{\geq 0}_{w-\mathbb{C}-c}(\dot{T}_YX)$).*

Proof. We decompose the proof in three steps.

(a) Assume X is an open subset of $\mathbb{C}$ and $Y = \{0\}$. If $F \in \mathrm{Ob}(\mathbf{D}^b_{w-\mathbb{C}-c}(X))$ and $F|_{X\setminus Y}$ is concentrated to degrees $\leqslant -1$ (resp. $\geqslant -1$), then $\nu_Y(F)$ has clearly the same property.

(b) Let us prove that, under the hypotheses of the proposition, we have for any $p \in \dot{T}_Y X$:

$$\begin{cases} i_p^{-1}\nu_Y(F) \in \mathrm{Ob}(\mathbf{D}^{\leqslant -1}(\mathfrak{Mod}(A))) \ , \\ (\textit{resp. } i_p^!\nu_Y(F) \in \mathrm{Ob}(\mathbf{D}^{\geqslant 1}(\mathfrak{Mod}(A)))) \ . \end{cases} \tag{10.3.18}$$

Let us choose a local holomorphic coordinate system $(z_1, \ldots, z_n)$ on X such that $Y = \{z_1 = 0\}$ and $p = \left(0; \frac{\partial}{\partial z_1}\right)$. Let $f: X \to \mathbb{C}$ be the morphism given by the first coordinate z_1 and denote by $B(\varepsilon)$ the open ball in X centered at 0 with radius $\varepsilon > 0$, by j_ε the embedding $B(\varepsilon) \hookrightarrow X$ and by f_ε the map $f \circ j_\varepsilon$. Finally denote by $T_Y f$ the map $T_Y X \to T_{\{0\}}\mathbb{C}$. Applying Proposition 4.2.4 we get the isomorphisms:

$$\begin{cases} (T_Y f)_* \nu_Y(Rj_{\varepsilon *} j_\varepsilon^{-1} F) \simeq \nu_{\{0\}}(Rf_* Rj_{\varepsilon *} j_\varepsilon^{-1} F) \ , \\ (T_Y f)_! \nu_Y(Rj_{\varepsilon !} j_\varepsilon^{-1} F) \simeq \nu_{\{0\}}(Rf_! Rj_{\varepsilon !} j_\varepsilon^{-1} F) \ . \end{cases} \tag{10.3.19}$$

On the other-hand, we have by Lemma 8.4.7:

$$\begin{cases} i_p^{-1}\nu_Y(F) \simeq \text{“}\varinjlim_U\text{”}\, R\Gamma(U; \nu_Y(F)) \ , \\ i_p^!\nu_Y(F) \simeq \text{“}\varprojlim_U\text{”}\, R\Gamma_c(U; \nu_Y(F)) \ , \end{cases} \tag{10.3.20}$$

where U ranges through the family of open neighborhoods of p in $T_Y X$. In view of (10.3.19), this shows, setting $v = \left(0; \frac{\partial}{\partial z_1}\right) \in T_{\{0\}}\mathbb{C}$:

$$\begin{cases} i_p^{-1}\nu_Y(F) \simeq \text{“}\varinjlim_\varepsilon\text{”}\, \nu_{\{0\}}(Rf_{\varepsilon *} j_\varepsilon^{-1} F)_v \ , \\ i_p^!\nu_Y(F) \simeq \text{“}\varprojlim_\varepsilon\text{”}\, R\Gamma_{\{v\}}\nu_{\{0\}}(Rf_{\varepsilon !} j_\varepsilon^{-1} F) \ . \end{cases} \tag{10.3.21}$$

By Proposition 8.5.9 we know that for U a sufficiently small open neighborhood of $0 \in \mathbb{C}$, $Rf_{\varepsilon *} j_\varepsilon^{-1} F|_U$ and $Rf_{\varepsilon !} j_\varepsilon^{-1} F|_U$ belong to $\mathbf{D}^b_{w-\mathbb{C}-c}(U)$.

We shall prove under the hypotheses of the proposition:

$$\begin{cases} Rf_{\varepsilon *} j_\varepsilon^{-1} F \in \mathrm{Ob}({}^p\mathbf{D}^{\leqslant 0}_{w-\mathbb{C}-c}(U\setminus\{0\})) \ , \\ (\textit{resp. } Rf_{\varepsilon !} j_\varepsilon^{-1} F \in \mathrm{Ob}({}^p\mathbf{D}^{\geqslant 0}_{w-\mathbb{C}-c}(U\setminus\{0\}))) \ . \end{cases} \tag{10.3.22}$$

We choose a complex μ-stratification $X = \bigsqcup_\alpha X_\alpha$ such that $SS(F) \subset \bigsqcup_\alpha T^*_{X_\alpha}X$. Then by the microlocal Bertini-Sard theorem (Proposition 8.3.12),

$$\bigsqcup_\alpha (B(\varepsilon) \times_X T^*_{X_\alpha}X) \cap T^*_{\{f^{-1}(t)\}}X \subset T^*_X X \quad \text{for} \quad 0 < |t| \ll \varepsilon \ .$$

Applying Lemma 10.3.9 we find that if F belongs to ${}^{\mu}\mathbf{D}^{\leq 0}_{w-\mathbb{C}-c}(X)$ (resp. ${}^{\mu}\mathbf{D}^{\geq 0}_{w-\mathbb{C}-c}(X)$) then $i^!_{f_\varepsilon^{-1}(t)} j_\varepsilon^{-1} F$ (resp. $i^{-1}_{f_\varepsilon^{-1}(t)} j_\varepsilon^{-1} F$) belongs to ${}^{\mu}\mathbf{D}^{\leq 1}_{w-\mathbb{C}-c}(f_\varepsilon^{-1}(t))$ (resp. ${}^{\mu}\mathbf{D}^{\geq -1}_{w-\mathbb{C}-c}(f_\varepsilon^{-1}(t))$).

Since $f_\varepsilon^{-1}(t)$ is a Stein manifold, we can apply Theorem 10.3.8. We obtain:

$$R\Gamma(f_\varepsilon^{-1}(t); i^!_{f_\varepsilon^{-1}(t)} j_\varepsilon^{-1} F) \in \mathrm{Ob}(\mathbf{D}^{\leq 1}(\mathfrak{Mod}(A)))$$

(resp. $R\Gamma_c(f_\varepsilon^{-1}(t); i^{-1}_{f_\varepsilon^{-1}(t)} j_\varepsilon^{-1} F) \in \mathrm{Ob}(\mathbf{D}^{\geq -1}(\mathfrak{Mod}(A)))$).

Since

$$(Rf_{\varepsilon *} j_\varepsilon^{-1} F)_t \simeq (R\Gamma_{\{t\}} Rf_{\varepsilon *} j_\varepsilon^{-1} F)_t[2] \simeq R\Gamma(f_\varepsilon^{-1}(t); i^!_{f_\varepsilon^{-1}(t)} j_\varepsilon^{-1} F)[2] ,$$

and

$$(Rf_{\varepsilon !} j_\varepsilon^{-1} F)_t \simeq R\Gamma_c(f_\varepsilon^{-1}(t); i^{-1}_{f_\varepsilon^{-1}(t)} F) ,$$

we get (10.3.22). This together with (10.3.21) and the step (a) implies (10.3.18).

(c) We shall complete the proof of the proposition.

Let us choose a μ-stratification $X = \bigsqcup_{\alpha \in A} X_\alpha$ such that there exists $A_0 \subset A$ with:

$$\text{(10.3.23)} \qquad \begin{cases} SS(F) \subset \bigsqcup_{\alpha \in A} T^*_{X_\alpha} X , \\ Y = \bigsqcup_{\alpha \in A_0} X_\alpha , \\ SS(\nu_Y(F)|_{\dot{T}_Y X}) \subset \bigsqcup_{\alpha \in A_0} (T^*_{X_\alpha} Y \times \dot{T}_Y X) . \end{cases}$$

Let $x \in X_\alpha$ and take a submanifold Z of X which intersects transversally X_α at x such that $\dim Z + \dim X_\alpha = \dim X$. Applying the inverse Fourier-Sato transformation to the isomorphisms obtained in Theorem 6.7.1 (or better, by looking at the proof of this theorem), one gets the isomorphisms:

$$\text{(10.3.24)} \qquad \begin{cases} \nu_Y(F)|_{Z \times_Y T_Y X} \simeq \nu_{Z \cap Y}(i_Z^{-1} F) , \\ R\Gamma_{Z \times_Y T_Y X} \nu_Y(F) \simeq \nu_{Z \cap Y}(i_Z^! F) . \end{cases}$$

Let $p \in \{x\} \times_Y \dot{T}_Y X \simeq \{x\} \times_{Z \cap Y} \dot{T}_{Z \cap Y} Z$.

Then:

$$\text{(10.3.25)} \qquad \begin{cases} \nu_Y(F)_p \simeq \nu_{Z \cap Y}(i_Z^{-1} F)_p , \\ R\Gamma_{\{p\}} \nu_Y(F) \simeq R\Gamma_{\{p\}} \nu_{Z \cap Y}(i_Z^! F) . \end{cases}$$

Applying Lemma 10.3.9, we obtain that $i_Z^{-1}(F)|_{Z \setminus Y}$ (resp. $i_Z^!(F)|_{Z \setminus Y}$) belongs to ${}^{\mu}\mathbf{D}^{\leq -\mathrm{codim}\, Z}_{w-\mathbb{C}-c}(Z \setminus Y)$ (resp. ${}^{\mu}\mathbf{D}^{\geq \mathrm{codim}\, Z}_{w-\mathbb{C}-c}(Z \setminus Y)$). Applying step (b) we get:

$$\nu_{Z \cap Y}(i_Z^{-1} F)_p \in \mathrm{Ob}(\mathbf{D}^{-1-\dim X_\alpha}(\mathfrak{Mod}(A)))$$

(resp. $R\Gamma_{\{p\}} \nu_{Z \cap Y}(i_Z^{-1} F)_p \in \mathrm{Ob}(\mathbf{D}^{\geq 1+\dim X_\alpha}(\mathfrak{Mod}(A)))$).

Then one concludes by (10.3.25) and (10.3.4), (resp. (10.3.25) and (10.3.5)). □

Corollary 10.3.11. *Let Y be a smooth hypersurface of X defined by the equation $\{f = 0\}$, and let $F \in \mathrm{Ob}(\mathbf{D}^b_{w-\mathbb{C}-c}(X))$. Assume that $F|_{X \setminus Y}$ belongs to ${}^{\mu}\mathbf{D}^{\leq 0}_{w-\mathbb{C}-c}(X \setminus Y)$*

(resp. ${}^{\mu}\mathbf{D}^{\geq 0}_{w-\mathbb{C}-c}(X \setminus Y)$*). Then:*

(i) $\psi_f(F)[-1]$ *belongs to* ${}^{p}\mathbf{D}^{\leq 0}_{w-\mathbb{C}-c}(Y)$ *(resp.* ${}^{p}\mathbf{D}^{\geq 0}_{w-\mathbb{C}-c}(Y)$*).*

(ii) *If moreover* $i_Y^{-1}(F)$ *(resp.* $i_Y^{!}(F)$*) belongs to* ${}^{p}\mathbf{D}^{\leq 0}_{w-\mathbb{C}-c}(Y)$*, (resp.* ${}^{p}\mathbf{D}^{\geq 0}_{w-\mathbb{C}-c}(Y)$*) then* $\phi_f(F)$ *belongs to* ${}^{p}\mathbf{D}^{\leq 0}_{w-\mathbb{C}-c}(Y)$ *(resp.* ${}^{p}\mathbf{D}^{\geq 0}_{w-\mathbb{C}-c}(Y)$*).*

Proof. (i) follows from Propositions 8.6.3 and 10.3.10.

(ii) follows from (i) and the distinguished triangles (8.6.7). □

We can now prove one of the main results of this section.

Theorem 10.3.12. *One has* ${}^{\mu}\mathbf{D}^{\leq 0}_{w-\mathbb{C}-c}(X) = {}^{p}\mathbf{D}^{\leq 0}_{w-\mathbb{C}-c}(X)$ *and* ${}^{\mu}\mathbf{D}^{\geq 0}_{w-\mathbb{C}-c}(X) = {}^{p}\mathbf{D}^{\geq 0}_{w-\mathbb{C}-c}(X)$.

Proof. (a) Let $t: X \times \mathbb{C} \to \mathbb{C}$ be the projection. If F belongs to ${}^{\mu}\mathbf{D}^{\leq 0}_{w-\mathbb{C}-c}(X)$ (resp. ${}^{\mu}\mathbf{D}^{\geq 0}_{w-\mathbb{C}-c}(X)$), then $F \boxtimes A_{\mathbb{C}}[1]$ belongs to ${}^{\mu}\mathbf{D}^{\leq 0}_{w-\mathbb{C}-c}(X \times \mathbb{C})$ (resp. ${}^{\mu}\mathbf{D}^{\geq 0}_{w-\mathbb{C}-c}(X \times \mathbb{C})$). Then by Proposition 10.3.10, $\psi_t(F \boxtimes A_{\mathbb{C}}[1])[-1] \simeq F$ belongs to ${}^{p}\mathbf{D}^{\leq 0}_{w-\mathbb{C}-c}(X)$ (resp. ${}^{p}\mathbf{D}^{\geq 0}_{w-\mathbb{C}-c}(X)$).

(b) Conversely, assume F belongs to ${}^{p}\mathbf{D}^{\leq 0}_{w-\mathbb{C}-c}(X)$ (resp. ${}^{p}\mathbf{D}^{\geq 0}_{w-\mathbb{C}-c}(X)$).

Let us take a μ-stratification $X = \bigsqcup_\alpha X_\alpha$ such that $SS(F) \subset \bigsqcup_\alpha T^*_{X_\alpha}X$. Arguing by induction it is enough to show:

$$(10.3.26)\qquad \begin{cases} \textit{if } X_\alpha \textit{ is a closed stratum and if } F|_{X\setminus X_\alpha} \textit{ belongs to} \\ {}^{\mu}\mathbf{D}^{\leq 0}_{w-\mathbb{C}-c}(X\setminus X_\alpha)\textit{, then the type } L \textit{ of } F \textit{ at a generic} \\ \textit{point of } T^*_{X_\alpha}X \textit{ with shift } 0 \textit{ satisfies } H^j(L) = 0 \\ \textit{for } j > -\dim X \textit{ (resp. } j < -\dim X) \ . \end{cases}$$

We choose a submanifold Z which intersects transversally X_α at x. Since μ-perversities and p-perversities are preserved by i_Z^{-1} with the same shift (Corollary 10.2.6 and Lemma 10.3.9), we may assume, by replacing X with Z and F with $i_Z^{-1}F$, that $X_\alpha = \{x\}$. Then take a holomorphic function f such that $f(x) = 0$, $df(x) = p \in \dot{T}^*_{\{x\}}X \setminus \bigsqcup_{\beta \neq \alpha} T^*_{X_\beta}X$. Since $\operatorname{supp}(\phi_f(F))$ is contained in $\pi(SS(F) \cap \Lambda_f)$, it is concentrated to $\{x\}$. Moreover $\phi_f(F)_x \simeq \phi_f(L_{\{x\}}) = L$. Hence we obtain the desired result since by Corollary 10.3.11, $\phi_f(F)$ belongs to ${}^{p}\mathbf{D}^{\leq 0}_{w-\mathbb{C}-c}(f^{-1}(0))$ (resp. ${}^{p}\mathbf{D}^{\geq 0}_{w-\mathbb{C}-c}(f^{-1}(0))$). □

Corollary 10.3.13. *The functors* $\psi_f[-1]$ *and* ϕ_f *from* $\mathbf{D}^b_{w-\mathbb{C}-c}(X)$ *to* $\mathbf{D}^b_{w-\mathbb{C}-c}(f^{-1}(0))$ *are t-exact with respect to the t-structure given by the middle perversity.*

Proof. When $df \neq 0$, this follows from Corollary 10.3.11 and Theorem 10.3.12. Otherwise, apply Exercise VIII.15 to the graph embedding of f. □

Recall that one sets ${}^{p}H^0 = {}^{p}\tau^{\leq 0} \circ {}^{p}\tau^{\geq 0}$ and ${}^{p}H^n = {}^{p}H^0 \circ [n]$.

Corollary 10.3.14. *Let* $F \in \mathrm{Ob}(\mathbf{D}^b_{w-\mathbb{C}-c}(X))$*. Then:*

$$(10.3.27)\qquad SS({}^{p}H^j(F)) \subset SS(F) \ .$$

Proof. This follows from ${}^pH^j(\phi_f(F)) \simeq \phi_f{}^pH^j(F)$, (Corollary 10.3.13) and Proposition 8.6.4. □

We shall now study functorial operations on perverse sheaves.

Proposition 10.3.15. (The left exactness of f^{-1} and the right exactness of $f^!$). *Let $f: Y \to X$ be a morphism of manifolds.*

(i) *f^{-1} sends ${}^p\mathbf{D}^{\geq 0}_{w-\mathbb{C}-c}(X)$ to ${}^p\mathbf{D}^{\geq \dim Y/X}_{w-\mathbb{C}-c}(Y)$.*
(ii) *$f^!$ sends ${}^p\mathbf{D}^{\leq 0}_{w-\mathbb{C}-c}(X)$ to ${}^p\mathbf{D}^{\leq -\dim Y/X}_{w-\mathbb{C}-c}(Y)$.*

Proof. By decomposing f by its graph, we may consider separately the case where f is smooth and the case where f is a closed embedding.

If f is smooth the results follow from Proposition 10.3.3 since in this case $f^! \simeq f^{-1} \otimes \mathit{or}_{Y/X}[2 \dim Y/X]$.

If f is a closed embedding we argue by induction and reduce the proof to the case where Y is a hypersurface given by an equation $\{g = 0\}$. Then the results follow from the distinguished triangles (8.6.7) and Corollary 10.3.13. □

Corollary 10.3.16. *Let $F \in \mathrm{Ob}(\mathbf{D}^b_{w-\mathbb{C}-c}(X))$, and assume $f: Y \to X$ is non-characteristic for F. Then:*

(i) *f is non-characteristic for ${}^pH^j(F)$, ${}^p\tau^{\geq j}F$ and ${}^p\tau^{\leq j}F$.*
(ii) *If $F \in \mathrm{Ob}({}^p\mathbf{D}^0_{w-\mathbb{C}-c}(X))$, then $f^{-1}F \in \mathrm{Ob}({}^p\mathbf{D}^{\dim Y/X}_{w-\mathbb{C}-c}(Y))$ and $f^!F \in \mathrm{Ob}({}^p\mathbf{D}^{-\dim Y/X}_{w-\mathbb{C}-c}(Y))$.*

(iii)
$$
{}^pH^j(f^{-1}(F)) \simeq f^{-1}({}^pH^{j-\dim Y/X}(F))[\dim Y/X] ,
$$
$$
{}^pH^j(f^!(F)) \simeq f^!({}^pH^{j+\dim Y/X}(F))[-\dim Y/X] .
$$

Proof. (i) follows from Corollary 10.3.14

(ii) follows from Proposition 10.3.15 since $f^!(F) \simeq f^{-1}(F) \otimes \mathit{or}_{Y/X}[2 \dim Y/X]$.

(iii) follows easily from (i) and (ii). □

Proposition 10.3.17. *Let $f: Y \to X$ be a morphism of manifolds. Assume that any point of X has an open neighborhood U such that $f^{-1}(U)$ is a Stein manifold.*

(i) *Let $G \in \mathrm{Ob}({}^p\mathbf{D}^{\leq 0}_{w-\mathbb{C}-c}(Y))$, and assume that $Rf_*G \in \mathrm{Ob}(\mathbf{D}^b_{w-\mathbb{C}-c}(X))$. Then $Rf_*G \in \mathrm{Ob}({}^p\mathbf{D}^{\leq 0}_{w-\mathbb{C}-c}(X))$.*
(ii) *Let $G \in \mathrm{Ob}({}^p\mathbf{D}^{\geq 0}_{w-\mathbb{C}-c}(Y))$, and assume that $Rf_!G \in \mathrm{Ob}(\mathbf{D}^b_{w-\mathbb{C}-c}(X))$. Then $Rf_!G \in \mathrm{Ob}({}^p\mathbf{D}^{\geq 0}_{w-\mathbb{C}-c}(X))$.*

Proof. Since the result is clear for a closed embedding, it is enough to consider the case where f is smooth by decomposing f by its graph.

Let G be as in (i) (resp. (ii)) and let us choose a μ-stratification $X = \bigsqcup_\alpha X_\alpha$ such that $SS(Rf_*G)$ (resp. $SS(Rf_!G)$) is contained in $\bigsqcup_\alpha T^*_{X_\alpha}X$. For $x \in X_\alpha$ choose a submanifold Z of X with $\dim Z + \dim X_\alpha = \dim X$ which intersects X_α transversally at x. Then it is enough to show by (10.3.4) and (10.3.5):

$$(10.3.28)\qquad \begin{cases} i_x^{-1} i_Z^! Rf_* G \in \mathrm{Ob}(\mathbf{D}^{\leq \dim X_\alpha}(\mathfrak{Mod}(A))) \\ (\text{resp. } i_x^! i_Z^{-1} Rf_! G \in \mathrm{Ob}(\mathbf{D}^{\geq -\dim X_\alpha}(\mathfrak{Mod}(A)))) \ . \end{cases}$$

Let $\tilde{f}: f^{-1}(Z) \to Z$ be the restriction of f. Then $i_Z^! Rf_* G \simeq R\tilde{f}_* i_{f^{-1}(Z)}^! G$ and $i_Z^{-1} Rf_! G \simeq R\tilde{f}_! i_{f^{-1}(Z)}^{-1} G$. Hence we find for a sufficiently small Stein open neighborhood U of x:

$$(10.3.29)\qquad \begin{cases} i_x^{-1} i_Z^! Rf_* G \simeq R\Gamma(U; i_Z^! Rf_* G) \simeq R\Gamma(\tilde{f}^{-1}(U); i_{f^{-1}(Z)}^! G) \\ (\text{resp. } i_x^! i_Z^{-1} Rf_! G \simeq R\Gamma_c(U; i_Z^{-1} Rf_! G) \simeq R\Gamma_c(\tilde{f}^{-1}(U); i_{f^{-1}(Z)}^{-1} G)) \ . \end{cases}$$

Since $i_{f^{-1}(Z)}^! G$ belongs to ${}^p\mathbf{D}_{w-\mathbb{C}-c}^{\leq \dim X_\alpha}(f^{-1}(Z))$ (resp. $i_{f^{-1}(Z)}^{-1} G$ belongs to ${}^p\mathbf{D}_{w-\mathbb{C}-c}^{\geq -\dim X_\alpha}(f^{-1}(Z))$) by Proposition 10.3.15, (10.3.28) follows from (10.3.29) and Theorem 10.3.8. □

Proposition 10.3.18. *Let E be a (complex) vector bundle over X with fiber dimension n. Then the Fourier-Sato transformation sends ${}^p\mathbf{D}_{w-\mathbb{C}-c}^{\leq 0}(E) \cap \mathbf{D}_{\mathbb{R}^+}^b(E)$ and ${}^p\mathbf{D}_{w-\mathbb{C}-c}^{\geq 0}(E) \cap \mathbf{D}_{\mathbb{R}^+}^b(E)$ to ${}^p\mathbf{D}_{w-\mathbb{C}-c}^{\leq n}(E^*) \cap \mathbf{D}_{\mathbb{R}^+}^b(E^*)$ and ${}^p\mathbf{D}_{w-\mathbb{C}-c}^{\geq n}(E^*) \cap \mathbf{D}_{\mathbb{R}^+}^b(E^*)$, respectively.*

Proof. Let F belong to $\mathbf{D}_{w-\mathbb{C}-c}^b(E) \cap \mathbf{D}_{\mathbb{R}^+}^b(E)$. Consider the diagram:

$$\begin{array}{ccccccc} & E \times_X E^* & \xrightarrow{\ \gamma\ } & E^* \times \mathbb{C} & & & \\ \swarrow^{p_1} & & \searrow^{p_2} \quad \swarrow^{\mathrm{pr}} & & \searrow^{t} & & \\ E & & E^* & & \mathbb{C} & \xrightarrow{\mathrm{Re}} & \mathbb{R} \ , \end{array}$$

where p_1 and p_2 (resp. pr and t) denote the first and second projection defined on $E \times_X E^*$ (resp. $E^* \times \mathbb{C}$) and γ is the map $(x, y) \mapsto (y, \langle x, y \rangle)$. Then by Definition 3.7.8 one has:

$$(10.3.30)\qquad \begin{cases} F^\wedge \simeq Rp_{2*} R\Gamma_{\gamma^{-1}(E^* \times (\mathrm{Re})^{-1}(\mathbb{R}_{\geq 0}))}(p_1^{-1} F) \ , \\ \qquad \simeq Rp_{2!}((p_1^{-1} F)_{\gamma^{-1}(E^* \times (\mathrm{Re})^{-1}(\mathbb{R}_{\leq 0}))}) \ . \end{cases}$$

Using the first isomorphism in (10.3.30), we get:

$$\begin{aligned} F^\wedge &\simeq R\mathrm{pr}_* R\gamma_* R\Gamma_{\gamma^{-1}(E^* \times (\mathrm{Re})^{-1}(\mathbb{R}_{\geq 0}))}(p_1^{-1} F) \ , \\ &\simeq R\mathrm{pr}_* R\Gamma_{E^* \times (\mathrm{Re})^{-1}(\mathbb{R}_{\geq 0})}(R\gamma_* p_1^{-1} F) \ . \end{aligned}$$

Since $R\gamma_* p_1^{-1} F$ is $\mathbb{C}^\times$-conic with respect to the action of $\mathbb{C}^\times$ on $\mathbb{C}$, one easily deduces:

$$(10.3.31)\qquad F^\wedge \simeq \phi_t R\gamma_* p_1^{-1} F \ .$$

Similarly, using the second isomorphism in (10.3.30) we obtain:

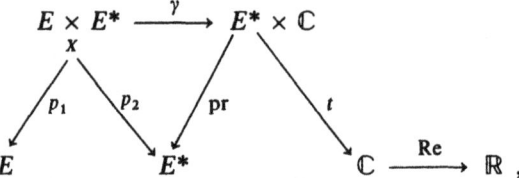

(10.3.32)
$$F^\wedge \simeq R\mathrm{pr}_{!}((R\gamma_! p_1^{-1}F)_{E^*\times(\mathrm{Re})^{-1}(\mathbb{R}_{\leq 0})})$$
$$\simeq \phi_t(R\gamma_! p_1^{-1}F) .$$

If F belongs to ${}^p\mathbf{D}^{\leq 0}_{w-\mathbb{C}-c}(E) \cap \mathbf{D}^b_{\mathbb{R}^+}(E)$ (resp. ${}^p\mathbf{D}^{\geq 0}_{w-\mathbb{C}-c}(E) \cap \mathbf{D}^b_{\mathbb{R}^+}(E)$), then $p_1^{-1}F$ belongs to ${}^p\mathbf{D}^{\leq n}_{w-\mathbb{C}-c}(E \times_X E^*)$ (resp. ${}^p\mathbf{D}^{\geq n}_{w-\mathbb{C}-c}(E \times_X E^*)$) by Proposition 10.3.3. Then we can apply Proposition 10.3.17 to the map γ and conclude by the t-exactness of ϕ_t (Corollary 10.3.13). □

Proposition 10.3.19. *Let Y be a closed submanifold of X with codimension n. Then $\nu_Y : \mathbf{D}^b_{w-\mathbb{C}-c}(X) \to \mathbf{D}^b_{w-\mathbb{C}-c}(T_Y X)$ and $\mu_Y[\operatorname{codim} Y] : \mathbf{D}^b_{w-\mathbb{C}-c}(X) \to \mathbf{D}^b_{w-\mathbb{C}-c}(T^*_Y X)$ are t-exact.*

Proof. We first describe a complex version of the normal deformation introduced in IV §1.

We can assume X is open in $\mathbb{C}^n \times \mathbb{C}^m$ and $Y = \{z' = 0\}$, where $z = (z', z'')$, $z' \in \mathbb{C}^n$, $z'' \in \mathbb{C}^m$. Let f be the map from $\mathbb{C} \times \mathbb{C}^n \times \mathbb{C}^m$ given by $(t, z', z'') \mapsto (tz', z'')$ and let $\tilde{X} = f^{-1}(X)$. We still denote by f the restriction of f to $\tilde{X}$.

Let $Z = t^{-1}(0) \subset \tilde{X}$. The map Tf gives a morphism $T_Z f : T_Z \tilde{X} \to T_Y X$. Identifying $T_Y X$ with $Y \times \mathbb{C}^n$ and $T_Z \tilde{X}$ with $Z \times \mathbb{C}$, $T_Z f$ is given by $(z', z'', \tau) \mapsto (z'', \tau z')$. Let $s : Z \to T_Z \tilde{X}$ be the section given by $(z', z'') \mapsto (z', z'', 1)$. Then $T_Z f$ identifies $s(Z)$ and $T_Y X$.

Now applying Proposition 4.2.5 to the smooth map $T_Z f|_{\dot{T}_Z \tilde{X}}$, we get the isomorphisms:

$$\nu_Y(F) \simeq (T_Z f \circ s)^{-1} \nu_Y(F)$$
$$\simeq s^{-1} (T_Z f)^{-1} \nu_Y(F)$$
$$\simeq s^{-1} \nu_Z(f^{-1}F)$$
$$\simeq \psi_t(f^{-1}F) .$$

Since $F \mapsto (f^{-1}F[1])|_{t \neq 0}$ and $\psi_t[-1]$ preserve the t-structure associated to the middle perversity, the result concerning ν_Y follows. Then we apply Proposition 10.3.18 to complete the proof. □

Corollary 10.3.20. (i) *Let $F \in \mathrm{Ob}({}^p\mathbf{D}^{\leq 0}_{w-\mathbb{C}-c}(X))$ and $G \in \mathrm{Ob}({}^p\mathbf{D}^{\geq 0}_{w-\mathbb{C}-c}(X))$. Then $\mu hom(F, G)[\dim X]$ belongs to ${}^p\mathbf{D}^{\geq 0}_{w-\mathbb{C}-c}(T^*X)$.*

(ii) *Assume the base ring A is a field. Then if F and G belong to ${}^p\mathbf{D}^0_{\mathbb{C}-c}(X)$, $\mu hom(F, G)[\dim X]$ belongs to ${}^p\mathbf{D}^0_{\mathbb{C}-c}(T^*X)$.*

Proof. (i) follows from Proposition 10.3.6 and the preceding proposition applied to the diagonal of $X \times X$.

(ii) follows from (i) and Proposition 8.4.14, since

$$\mathrm{D}_{T^*X}(\mu hom(F, G)[\dim X]) \simeq \mu hom(G, F) \otimes or_X[\dim X] . \quad \square$$

Exercises to Chapter X

Exercise X.1. Let $\mathbf{D}$ be a triangulated category with a t-structure. For $X \in \mathrm{Ob}(\mathbf{D})$, let $\mathrm{d}^n(X): \tau^{>n}X \to (\tau^{\leqslant n}X)[1]$ be the morphism d constructed in Proposition 10.1.4.

(i) Prove that, for $b \geqslant a$

$$\begin{array}{ccc} \tau^{>a}X & \xrightarrow{\mathrm{d}^a(X)} & (\tau^{\leqslant a}X)[1] \\ \downarrow & & \downarrow \\ \tau^{>b}X & \xrightarrow[\mathrm{d}^b(X)]{} & (\tau^{\leqslant b}X)[1] \end{array}$$

commutes.

(ii) Prove that:

$$\begin{array}{ccc} (\tau^{>n}X)[1] & \xrightarrow{\mathrm{d}^n(X)[1]} & (\tau^{\leqslant n}X)[1] \\ \wr\downarrow & & \downarrow\wr \\ \tau^{>n-1}(X[1]) & \xrightarrow[\mathrm{d}^{n-1}(X[1])]{} & \tau^{\leqslant n-1}(X[1]) \end{array}$$

anticommutes.

Exercise X.2. Let $(X, \mathcal{O}_X)$ be a Noetherian separated scheme such that $\mathcal{O}_{X,x}$ is a regular ring for any x, i.e. $\mathrm{gld}(\mathcal{O}_{X,x}) < \infty$. We assume X has equidimension d. Let $\mathbf{D}^b_{\mathrm{coh}}(\mathcal{O}_X)$ be the full subcategory of $\mathbf{D}^b(\mathcal{O}_X)$ consisting of objects with coherent cohomologies. Define the duality functor $\mathrm{D}(F) = R\mathcal{H}om_{\mathcal{O}_X}(F, \mathcal{O}_X)[d]$ from $\mathbf{D}^b_{\mathrm{coh}}(\mathcal{O}_X)$ to $\mathbf{D}^b_{\mathrm{coh}}(\mathcal{O}_X)$.

(i) (a) Prove that $\mathbf{D}^b(\mathrm{Coh}(\mathcal{O}_X)) \simeq \mathbf{D}^b_{\mathrm{coh}}(\mathcal{O}_X)$, where $\mathrm{Coh}(\mathcal{O}_X)$ is the abelian category of coherent $\mathcal{O}_X$-modules; (this holds for all Noetherian schemes X).

(b) Prove that $\mathrm{D} \circ \mathrm{D} \simeq \mathrm{id}$.

(ii) Let us define the full subcategories of $\mathbf{D}^b_{\mathrm{coh}}(\mathcal{O}_X)$ by

$$^p\mathbf{D}^{\leqslant 0}_{\mathrm{coh}}(\mathcal{O}_X) = \{F \in \mathbf{D}^b_{\mathrm{coh}}(\mathcal{O}_X);\ \dim \mathrm{supp}(\tau^{>p(k)}F) < k \text{ for any } k\}$$

$$^p\mathbf{D}^{\geqslant 0}_{\mathrm{coh}}(\mathcal{O}_X) =$$
$$\{F \in \mathbf{D}^b_{\mathrm{coh}}(\mathcal{O}_X);\ H^j_Z(F) = 0 \text{ for any closed subset } Z \text{ and } j < p(\dim Z)\}\ .$$

(Here p denotes a perversity.)

(a) Prove that D sends $^p\mathbf{D}^{\leqslant 0}_{\mathrm{coh}}(\mathcal{O}_X)$ and $^p\mathbf{D}^{\geqslant 0}_{\mathrm{coh}}(\mathcal{O}_X)$ into $^{p^*}\mathbf{D}^{\geqslant 0}_{\mathrm{coh}}(\mathcal{O}_X)$ and $^{p^*}\mathbf{D}^{\leqslant 0}_{\mathrm{coh}}(\mathcal{O}_X)$, respectively.

(b) Prove that $(^p\mathbf{D}^{\leqslant 0}_{\mathrm{coh}}(\mathcal{O}_X), {}^p\mathbf{D}^{\geqslant 0}_{\mathrm{coh}}(\mathcal{O}_X))$ is a t-structure.

(c) Prove that $U \mapsto {}^p\mathbf{D}^0_{\mathrm{coh}}(\mathcal{O}_U)$ is a stack (cf. Proposition 10.2.9).

(Hint: use a result of A. Grothendieck [5]: for any coherent ideal I and any $F \in \mathrm{Ob}(\mathbf{D}^b(\mathcal{O}_X))$ with quasi-coherent cohomology sheaves, setting $Z = \mathrm{supp}(\mathcal{O}_X/I)$, we have $H^n_Z(F) \xrightarrow{\sim} \varinjlim_k \mathcal{E}xt^n_{\mathcal{O}_X}(\mathcal{O}_X/I^k, F)$.)

Exercise X.3. Let A be a discrete valuation ring.

Prove that any t-structure on $\mathbf{D}^b(\mathfrak{Mod}^f(A))$ is either the natural one or its dual up to shift (in other words, setting $X = \mathrm{Spec}(A)$, it is $({}^p\mathbf{D}^{\leqslant 0}_{\mathrm{coh}}(\mathcal{O}_X), {}^p\mathbf{D}^{\geqslant 0}_{\mathrm{coh}}(\mathcal{O}_X))$ for some perversity p, (cf. Exercise X.2). (Hint: use Exercise I.18 and for a t-structure $(\mathbf{D}^{\leqslant 0}, \mathbf{D}^{\geqslant 0})$ on $\mathbf{D}^b(\mathfrak{Mod}^f(A))$, prove first that $A \in \mathrm{Ob}(\mathbf{D}^{\geqslant 0})$, $A \notin \mathrm{Ob}(\mathbf{D}^{\geqslant 1})$ imply $A \in \mathrm{Ob}(\mathbf{D}^{\leqslant 0})$.)

Exercise X.4. Let X be a real analytic manifold and let $F \in \mathrm{Ob}(\mathbf{D}^b_{w-\mathbb{R}-c}(X))$. Set $d = \dim \mathrm{supp}(F)$. Assuming $F \in \mathrm{Ob}({}^p\mathbf{D}^{\leqslant 0}_{w-\mathbb{R}-c}(X))$ (resp. ${}^p\mathbf{D}^{\geqslant 0}_{w-\mathbb{R}-c}(X)$) prove that $H^j(X; F) = 0$ for $j > p(d) + d$ (resp. $H^j_c(X; F) = 0$ for $j < p(d)$).

Exercise X.5. Let X be a real analytic manifold, U a subanalytic open subset, j the embedding $U \hookrightarrow X$ and let p be a perversity. For $F \in \mathrm{Ob}({}^p\mathbf{D}^0_{w-\mathbb{R}-c}(X))$, one defines ${}^pj_!j^{-1}F = {}^pH^0(j_!j^{-1}F)$ and ${}^pj_*j^{-1}F = {}^pH^0(Rj_*j^{-1}F)$, (cf. X §1). Prove that the following two conditions are equivalent:

(i) F is isomorphic to the image of ${}^pj_!j^{-1}F \to {}^pj_*j^{-1}F$ in ${}^p\mathbf{D}^0_{w-\mathbb{R}-c}(X)$,

(ii) F has neither non-zero sub-objects nor non-zero quotient objects supported by $X \setminus U$ in ${}^p\mathbf{D}^0_{w-\mathbb{R}-c}(X)$.

Exercise X.6. Let X be a complex manifold. Let $F \in \mathrm{Ob}(\mathbf{D}^b_{w-\mathbb{C}-c}(X))$. Prove that $SS(F) = \bigcup_j SS({}^pH^j(F))$.

Exercise X.7. Let X be a complex manifold and let $0 \to F' \to F \to F'' \to 0$ be an exact sequence in ${}^p\mathbf{D}^0_{w-\mathbb{C}-c}(X)$. Prove that $SS(F) = SS(F') \cup SS(F'')$.

Exercise X.8. We assume the base ring A is a field. Let X be a complex manifold of dimension n, F a perverse sheaf on X.

(i) Let Z be a closed subset of X. Prove that $\mathrm{H}^{-n}(F)|_Z$ satisfies "the principle of analytic continuation", that is, the support of any section of this sheaf on an open subset U of Z, is both open and closed in U.

(ii) Assume now Z is a closed subanalytic subset of X and let $x \in Z$ be a non-isolated point of Z. Prove that $(H^n_Z(F))_x = 0$. (Hint: use a duality argument.)

Exercise X.9. Let Y be a complex hypersurface of a complex manifold X. Prove that $M_Y[\dim X - 1]$ is a perverse sheaf for any A-module M.

Exercise X.10. Let $F \in \mathrm{Ob}(\mathbf{D}^b_{w-\mathbb{R}-c}(X))$, let Z be a locally closed subanalytic subset of X, and let $j \in \mathbb{Z}$. Prove that $H^j_Z(F) = 0$ for all $j < r$ if and only if $\dim(Z \cap \mathrm{cosupp}^j(F)) < j - r$ for any j.

Notes

At the origin of the theory of perverse sheaves, one finds on one hand the "intersection cohomology" of Goresky-MacPherson [1], a successful attempt to generalize Poincaré duality to singular spaces, and on the other hand the theory of holonomic $\mathscr{D}$-modules (and both theories are closely related, cf. Chapter XI). In fact, already in 1975, Kashiwara [3] showed that if $\mathscr{M}$ is a holonomic $\mathscr{D}$-module on a complex manifold X, then the complex $R\mathscr{H}om_{\mathscr{D}}(\mathscr{M}, \mathscr{O}_X)[\dim_{\mathbb{C}} X]$ is perverse, and he gave the following formulation of the "Riemann-Hilbert problem" (cf. Ramis [1, p.287]): to define a full abelian subcategory "holreg" of the abelian category of holonomic $\mathscr{D}$-modules, such that the functor $R\mathscr{H}om_{\mathscr{D}_X}(*, \mathscr{O}_X)$ induces an equivalence of categories from D^b(holreg) to $D^b_{\mathbb{C}-c}(X)$. Note that this problem was solved in the 80's (cf. Kashiwara [6] and Mebkhout [1]). Such a correspondence immediately implies that the category $D^b_{\mathbb{C}-c}(X)$ contains an abelian subcategory which corresponds to the abelian subcategory holreg of D^b(holreg): this is the category of perverse sheaves.

The theory of perverse sheaves is due to Gabber and Beilinson-Bernstein-Deligne [1], and most of the results of this chapter are well-known. Those concerning t-structures in §1 are extracted from Beilinson-Bernstein-Deligne (loc. cit.) as well as Theorem 10.2.8. The crucial result which asserts that perversity is preserved by the vanishing-cycle functor is due to Goresky-MacPherson [3] (after Beilinson-Bernstein-Deligne (loc. cit.) in the étale case). The microlocal interpretation of perversity was performed in Kashiwara-Schapira [3] when the base ring is $\mathbb{C}$. Finally let us also mention an interesting construction of perverse sheaves in MacPherson-Vilonen [1].

Chapter XI. Applications to $\mathcal{O}$-modules and $\mathcal{D}$-modules

Summary

By its definition, a complex manifold X is endowed with the sheaf of rings $\mathcal{O}_X$ of holomorphic functions. The structure of $\mathcal{O}_X$, and the theory of $\mathcal{O}_X$-modules, are now well-understood, and it is not our aim to explain this theory from the start. We shall content ourselves with a few basic facts concerning the algebraic structure of $\mathcal{O}_X$, its flabby dimension and the operations on $\mathcal{O}_X$. References are made to Banica-Stanasila [1], Cartan [2], Hörmander [1], Serre [1].

Next, we introduce the sheaf of rings $\mathcal{D}_X$ of finite order holomorphic differential operators on X. Here again the theory of $\mathcal{D}_X$-modules is now well-understood, and we shall be rather brief on this subject, having in mind to make understood the main notions, in particular that of characteristic variety, and to explain the operations on $\mathcal{D}_X$-modules. We shall also recall the classical Cauchy-Kowalewski theorem and its extension to $\mathcal{D}_X$-modules, and we shall derive the formula:

(11.0.1) $$SS(R\mathcal{H}om_{\mathcal{D}_X}(\mathcal{M}, \mathcal{O}_X)) \subset \mathrm{char}(\mathcal{M}) ,$$

where $\mathrm{char}(\mathcal{M})$ is the characteristic variety of the $\mathcal{D}_X$-module $\mathcal{M}$. (In fact, this inclusion is an equality, cf. §4.) As an application of (11.0.1) one immediately obtains, with the help of the results of VIII §5 the constructibility of the complex $R\mathcal{H}om_{\mathcal{D}_X}(\mathcal{M}, \mathcal{O}_X)$ when $\mathcal{M}$ is holonomic, and one also easily proves that this complex is perverse.

For a more detailed exposition of the theory of $\mathcal{D}_X$-modules, we refer to Björk [1], Kashiwara [5] and Schapira [2].

Then we study "microlocally" the sheaf $\mathcal{O}_X$. After having introduced the ring $\mathcal{E}_X^{\mathbb{R}}$ of microlocal operators we sketch the proof of an important theorem which asserts that one can locally "quantize" holomorphic contact transformations over $\mathcal{O}_X$. We end this chapter by introducing the sheaf $\mathcal{C}_M$ of Sato microfunctions on a real analytic manifold M. Using (11.0.1), and the results of Chapters V and VI, it is an easy exercise to recover many classical results of the theory of linear partial differential equations, in particular those concerning elliptic equations or the analytic wave front set, micro-hyperbolic systems and propagation of singularities.

As it should be clear, the aim of this chapter is not to give a complete or systematic treatment of the theory of analytic (micro-)differential equations, but rather to introduce the reader to it, and in particular to make him better understand the basic paper of Sato-Kawai-Kashiwara [1], under the light of the theory of micro-support of sheaves.

In this chapter all sheaves, unless otherwise specified, are sheaves of $\mathbb{C}$-vector spaces.

11.1. The sheaf $\mathcal{O}_X$

Let X be a complex manifold of complex dimension n, and let $\mathcal{O}_X$ be the sheaf of rings of holomorphic functions on X. First of all, we shall describe the relations between X and the underlying real analytic manifold $X^{\mathbb{R}}$.

One denotes by $\bar{X}$ the complex conjugate of X. This is a complex manifold such that $\bar{X}^{\mathbb{R}} = X^{\mathbb{R}}$, but the holomorphic functions on $\bar{X}$ are the anti-holomorphic functions on X. Hence we have the isomorphism of rings: $\mathcal{O}_X \xrightarrow{\sim} \mathcal{O}_{\bar{X}}$, $u \mapsto \bar{u}$ with $\overline{au} = \bar{a}\bar{u}$ for $a \in \mathbb{C}$ and $u \in \mathcal{O}_X$.

Identify $X^{\mathbb{R}}$ with the diagonal in $X \times \bar{X}$. Then $X \times \bar{X}$ is a complexification of $X^{\mathbb{R}}$. If $\mathcal{A}_{X^{\mathbb{R}}}$ denotes the sheaf of (complex valued) real analytic functions on $X^{\mathbb{R}}$, then $\mathcal{A}_{X^{\mathbb{R}}} = \mathcal{O}_{X \times \bar{X}}|_{X^{\mathbb{R}}}$.

Moreover

$$TX^{\mathbb{R}} \underset{\mathbb{R}}{\otimes} \mathbb{C} \simeq TX \underset{X}{\times} T\bar{X} .$$

The composite map

$$TX^{\mathbb{R}} \to TX^{\mathbb{R}} \underset{\mathbb{R}}{\otimes} \mathbb{C} \simeq TX \underset{X}{\times} T\bar{X} \to TX$$

defines the isomorphism of real analytic vector bundles over $X^{\mathbb{R}}$:

$$TX^{\mathbb{R}} \simeq (TX)^{\mathbb{R}} \tag{11.1.1}$$

and by duality the isomorphism:

$$T^*X^{\mathbb{R}} \xrightarrow{\sim} (T^*X)^{\mathbb{R}} . \tag{11.1.2}$$

For a real C^1-function ϕ on $X^{\mathbb{R}}$, this last isomorphism associates the vector $\partial\phi(x) \in T_x^*X$ to the vector $d\phi(x) \in T_x^*X^{\mathbb{R}}$, where $d\phi$ is the real differential and $\partial\phi$ its holomorphic component.

Let α_X (resp. $\alpha_{X^{\mathbb{R}}}$) be the complex (resp. real) canonical 1-form on T^*X (resp. $T^*X^{\mathbb{R}}$). Then $\alpha_{\bar{X}} = \overline{\alpha_X}$, and $\alpha_{X \times \bar{X}} = \alpha_X + \alpha_{\bar{X}}$. Thus:

$$\alpha_{X^{\mathbb{R}}} = 2\operatorname{Re}\alpha_X . \tag{11.1.3}$$

Let $z = (z_1, \ldots, z_n)$ be a system of holomorphic coordinates on X, $(z;\zeta)$ the associated coordinates on T^*X so that $\alpha_X = \sum_j \zeta_j dz_j$. If $z = x + \sqrt{-1}y$, $\zeta = \xi + \sqrt{-1}\eta$, then:

$$\alpha_{X^{\mathbb{R}}} = 2 \sum_j (\xi_j dx_j - \eta_j dy_j) \ . \tag{11.1.4}$$

It induces the Hamiltonian field $H_h^{\mathbb{R}}$, for a real valued function h on $T^*X^{\mathbb{R}}$.

From now on, if there is no risk of confusion, we shall identify $(T^*X)^{\mathbb{R}}$ and $T^*X^{\mathbb{R}}$, and even sometimes write T^*X or X instead of $T^*X^{\mathbb{R}}$ or $X^{\mathbb{R}}$.

Next, let us recall some basic notions related to that of coherency.

Let $(X, \mathscr{A}_X)$ be a ringed space. Unless otherwise specified, an $\mathscr{A}_X$-module means a left $\mathscr{A}_X$-module.

An $\mathscr{A}_X$-module $\mathscr{M}$ is finite free if it is isomorphic to $\mathscr{A}_X^N$ for some $N \in \mathbb{N}$. It is locally finite free if each $x \in X$ has an open neighborhood U such that $\mathscr{M}|_U$ is $\mathscr{A}_X|_U$-finite free.

An s-presentation of an $\mathscr{A}_X$-module $\mathscr{M}$ is an exact sequence of $\mathscr{A}_X$-modules:

$$\mathscr{M}_s \to \cdots \to \mathscr{M}_0 \to \mathscr{M} \to 0 \ . \tag{11.1.5}$$

One says this presentation is finite free or locally finite free etc., if the $\mathscr{M}_j$'s have the corresponding property.

A resolution of length m is an ∞-presentation with $\mathscr{M}_j = 0$ for $j > m$.

An $\mathscr{A}_X$-module $\mathscr{M}$ is called locally of finite type (resp. of finite presentation) if it admits locally a finite free 0-presentation (resp. 1-presentation), i.e. if locally on X, there exists an exact sequence $\mathscr{A}_X^N \to \mathscr{M} \to 0$ (resp. $\mathscr{A}_X^{N_1} \to \mathscr{A}_X^{N_0} \to \mathscr{M} \to 0$).

Definition 11.1.1. (i) *One says an $\mathscr{A}_X$-module $\mathscr{M}$ is coherent if $\mathscr{M}$ is locally of finite type and if, for any open set U, any sub-$\mathscr{A}_U$-module locally of finite type is locally of finite presentation.*

(ii) *One says $\mathscr{A}_X$ is coherent if $\mathscr{A}_X$ is coherent as a left $\mathscr{A}_X$-module.*

(iii) *One says $\mathscr{A}_X$ is Noetherian if it is coherent, for each $x \in X$ the stalk $\mathscr{A}_{X,x}$ is Noetherian and finally, for any open subset U of X, any increasing family of coherent submodules of a coherent $\mathscr{A}_X|_U$-module is locally stationary.*

Let $\mathfrak{Mod}_{\mathrm{coh}}(\mathscr{A}_X)$ denote the category of left coherent modules over a coherent ring $\mathscr{A}_X$. Then $\mathfrak{Mod}_{\mathrm{coh}}(\mathscr{A}_X)$ is an abelian category.

Let us come back to the situation where X is a complex manifold of complex dimension n.

Theorem 11.1.2. (i) *The sheaf of rings $\mathcal{O}_X$ is Noetherian (and in particular coherent).*

(ii) *Let $\mathscr{M}$ be a coherent $\mathcal{O}_X$-module. Then locally on X, $\mathscr{M}$ admits a finite free resolution of length n.*

The coherency of $\mathcal{O}_X$ is a result of Oka [1]. Property (ii) is known as the **Hilbert syzygy theorem**.

Theorem 11.1.3. *Let Z be a locally closed subset of X and let $x \in X$, $x \notin \operatorname{Int} Z$. Then:*

$$H_Z^j(\mathcal{O}_X)_x = 0 \qquad \text{for} \qquad j \notin [1, n] \ .$$

Of course, the vanishing of $H_Z^0(\mathcal{O}_X)_x$ is equivalent to the well-known *principle of analytic continuation.* The vanishing of $H_Z^j(\mathcal{O}_X)_x$ for $j > n$ is due to Malgrange [1] (cf. II §9.11).

Finally we shall discuss the operations on $\mathcal{O}_X$.

Let $(Y, \mathcal{O}_Y)$ be another complex manifold. There is a natural morphism of sheaves on $X \times Y$:

$$\mathcal{O}_X \boxtimes \mathcal{O}_Y \to \mathcal{O}_{X \times Y} \ . \tag{11.1.6}$$

Now assume to be given a holomorphic map $f: Y \to X$. There is a natural morphism of sheaves on Y:

$$f^{-1}\mathcal{O}_X \to \mathcal{O}_Y \ . \tag{11.1.7}$$

This morphism is nothing but the composition by f, which sends a holomorphic function ϕ on X to the holomorphic function $\phi \circ f$ on Y. The direct image is not so easy to describe.

Let $\mathcal{O}_X^{(p)}$ denote the sheaf of holomorphic p-forms on X, and set:

$$\Omega_X = \mathcal{O}_X^{(n)} \otimes or_X \tag{11.1.8}$$

where or_X denotes the orientation sheaf on $X^{\mathbb{R}}$, and $n = \dim_{\mathbb{C}} X$.

Of course the complex manifold X is orientable, and most of the time, one can forget or_X.

The sheaf Ω_X is an invertible $\mathcal{O}_X$-module, i.e. it is locally free of rank one on $\mathcal{O}_X$. If $\mathcal{L}$ is an invertible $\mathcal{O}_X$-module, one sets:

$$\mathcal{L}^{\otimes -1} = \mathcal{H}om_{\mathcal{O}_X}(\mathcal{L}, \mathcal{O}_X) \ . \tag{11.1.9}$$

Of course $\mathcal{L}^{\otimes -1} \otimes_{\mathcal{O}_X} \mathcal{L} \simeq \mathcal{O}_X$.

A volume element is a generator of Ω_X over $\mathcal{O}_X$. The existence of a volume element is only locally guaranteed.

Theorem 11.1.4. *There exists a natural morphism in* $\mathbf{D}^b(X)$:

$$Rf_!\Omega_Y[\dim_{\mathbb{C}} Y] \to \Omega_X[\dim_{\mathbb{C}} X] \ .$$

Moreover this morphism is functorial with respect to the composition of maps.

Recall (cf. II. §9) that this morphism may be obtained as follows.

Let $\mathcal{D}b_Y^{(p,q)}$ be the sheaf of distribution forms on $Y^{\mathbb{R}}$ of bidegree (p, q) with respect to Y and $\bar{Y}$. Set $m = \dim_{\mathbb{C}} Y$, $n = \dim_{\mathbb{C}} X$, $l = m - n$. There is an "integration" morphism

$$f_!\mathcal{D}b_{Y^{\mathbb{R}}}^{(p,q)} \otimes or_Y \to \mathcal{D}b_{X^{\mathbb{R}}}^{(p-l,q-l)} \otimes or_X$$

which gives the desired morphism by replacing Ω_Y and Ω_X by their Dolbeault resolutions with coefficients in $\mathscr{D}\mathscr{b}_Y^{(m,\cdot)}$ and $\mathscr{D}\mathscr{b}_X^{(n,\cdot)}$ respectively.

As an application of this theorem, let us construct the **fundamental class** associated to a submanifold Z of X of (complex) codimension d.

Let $\mathscr{I}$ be the defining ideal of Z in X, that is, the sheaf of ideals of $\mathcal{O}_X$ generated by the sections vanishing on Z. There is a natural morphism:

$$\mathscr{I}/\mathscr{I}^2 \to \mathcal{O}_X^{(1)} \otimes_{\mathcal{O}_X} \mathcal{O}_Z \tag{11.1.10}$$

given by $f \mapsto df$. From this morphism one gets:

$$\Lambda^d(\mathscr{I}/\mathscr{I}^2) \to \mathcal{O}_X^{(d)} \otimes_{\mathcal{O}_X} \mathcal{O}_Z \ . \tag{11.1.11}$$

But $\bigwedge^d(\mathscr{I}/\mathscr{I}^2) \simeq \mathcal{O}_Z^{(n-d)\otimes -1} \otimes_{\mathcal{O}_X} \mathcal{O}_X^{(n)}$. Therefore we get the morphism:

$$\mathcal{O}_Z \to \mathcal{O}_Z^{(n-d)} \otimes_{\mathcal{O}_X} \mathcal{O}_X^{(d)} \otimes \mathcal{O}_X^{(n)\otimes -1} \ . \tag{11.1.12}$$

On the other hand, Theorem 11.1.4 defines the morphism:

$$\Omega_Z \to H_Z^d(\Omega_X) \ .$$

Taking its tensor product with $\mathcal{O}_X^{(d)} \otimes \mathcal{O}_X^{(n)\otimes -1}$, and combining with (11.1.12), we get the morphism:

$$\mathcal{O}_Z \to H_Z^d(\mathcal{O}_X^{(d)}) \otimes or_{Z/X} \ . \tag{11.1.13}$$

Definition 11.1.5. *The image of the section* 1 *of* $\mathcal{O}_Z$ *by the morphism* (11.1.13) *is called the fundamental class of* Z *in* X *and denoted* δ_Z.

Example 11.1.6. Assume Z is defined by the equations $f_1 = \cdots = f_d = 0$ with $df_1 \wedge \cdots \wedge df_d \neq 0$ on Z. Then δ_Z is the class of $\dfrac{1}{(2\pi i)^d}\left(\dfrac{df_1}{f_1} \wedge \cdots \wedge \dfrac{df_d}{f_d}\right)$.

Here $1/f_1 \dots f_d$ is the image in $H_Z^d(X;\mathcal{O}_X)$ of the element of $H^{d-1}(X\backslash Z;\mathcal{O}_X)$ obtained by considering the open covering $\{f_j \neq 0\}_{j=1,\dots,d}$ of $X\backslash Z$ (cf. Exercise II.26).

11.2. $\mathscr{D}_X$-modules

First, let us recall a few basic facts on filtered rings and modules, (cf. Schapira [2]).

All rings are unitary (i.e. with unit 1), and unless otherwise specified, "module" means "left unitary module", (i.e. left module on which the action of 1 is the identity).

A **filtered ring** A indexed by $\mathbb{Z}$ is a ring (still denoted by A) endowed with a family of subgroups $\{A_k\}_{k\in\mathbb{Z}}$ such that:

$$(11.2.1) \qquad \begin{cases} A_k \subset A_{k+1} \textit{ for any } k, 1 \in A_0, A_k \cdot A_l \subset A_{k+l} \textit{ for any } k, l \textit{ and} \\ A = \bigcup_k A_k \ . \end{cases}$$

A **filtered module** M over A is an A-module M endowed with a family of subgroups $\{M_k\}_{k \in \mathbb{Z}}$ such that:

$$(11.2.2) \quad M_k \subset M_{k+1} \textit{ for any } k,\ A_l \cdot M_k \subset M_{k+l} \textit{ for any } k, l \textit{ and } M = \bigcup_k M_k \ .$$

Two filtrations $\{M_k\}_k$ and $\{M'_k\}_k$ on M are said equivalent if there exists an $r \in \mathbb{N}$ such that $M_{k-r} \subset M'_k \subset M_{k+r}$ for all $k \in \mathbb{Z}$. In particular a filtration on M is equivalent to its r-shifted filtration, denote $M_{[r]}$, defined by:

$$(M_{[r]})_k = M_{k+r} \qquad \textit{for} \qquad k \in \mathbb{Z} \ .$$

A morphism of filtered A-modules $\psi : M \to N$ is a morphism of A-modules such that $\psi(M_k) \subset N_k$ for all k.

Consider an exact sequence of A-modules, $0 \to L \to M \underset{\psi}{\to} N \to 0$, and assume M is endowed with a filtration. The induced filtration on L is defined by setting $L_k = L \cap M_k$. The image filtration on N is defined by setting $N_k = \psi(M_k)$.

A sequence of filtered A-modules $0 \to L \to M \to N \to 0$ is called **strictly exact** if it is an exact sequence of A-modules and the filtration on L (resp. N) is the induced (resp. image) filtration. One also defines the filtered direct sum $M \oplus M'$ of two filtered modules M and M' by setting:

$$(M \oplus M')_k = M_k \oplus M'_k \ .$$

Hence the category of filtered A-modules is additive. One shall take care that this category is not abelian: e.g. the filtered morphism $A \underset{\text{id}}{\longrightarrow} A_{[1]}$ is a monomorphism and an epimorphism but is not an isomorphism of filtered A-modules in general.

By definition a finite free filtered A-module is a filtered A-module isomorphic to a finite direct sum of modules of the type $A_{[r]}$.

A finite free s-presentation of a filtered A-module M is a strictly exact sequence of filtered A-modules:

$$(11.2.3) \qquad L_s \to \cdots \to L_0 \to M \to 0$$

where the L_j's are finite free.

A filtered module M is said to be of finite type if it admits a finite free 0-presentation. In such a case one says the filtration on M is **good**.

The filtered ring A is said to be **Noetherian** (or "filtered Noetherian") if any sub-object of a module of finite type in the category of filtered A-modules is of finite type. This is equivalent to saying that for any A-module M endowed with a good filtration and any submodule N of M endowed with a filtration such that $N \to M$ is a filtered morphism, this filtration is good. If the filtered ring A is Noetherian, and moreover for any $\alpha \in A_{-1}$, $1 - \alpha$ is invertible, then one says A is **Zariskian.**

This definition of filtered Noetherian rings slightly differs from Definition 1.1.2 of Chapter II of Schapira [2]. In fact, as pointed out by Van Oystaeyen the definition of Schapira (loc. cit.) is too weak to imply that gr(A) is Noetherian and in particular, Proposition 1.1.7 of this author is partly false. (This mistake has no implication for the rest of the book. To correct it, it is enough to replace the weak definition of filtered Noetherian ring by the definition we have just given above.)

One defines the **graded ring** gr(A) by:

$$\mathrm{gr}(A) = \bigoplus_k A_k/A_{k-1} \ . \tag{11.2.4}$$

Similarly if M is a filtered A-module one defines the (graded) gr(A)-module gr(M) by:

$$\mathrm{gr}(M) = \bigoplus_k M_k/M_{k-1} \ . \tag{11.2.5}$$

The **order** of an element $u \in M$ is the smallest $k \in \mathbb{Z} \cup \{-\infty\}$ such that $u \in M_k$, where $M_{-\infty} = 0$ by convention. It is denoted by ord(u).

One denotes by σ_k the projection $M_k \to \mathrm{gr}(M)$. If $\mathrm{ord}(u) = k$, one writes $\sigma(u)$ instead of $\sigma_k(u)$.

Proposition 11.2.1. *Assume* gr(A) *is Noetherian (as a graded ring) and* $A_k = 0$ *for* $k < 0$. *Then* A *is filtered Noetherian.*

The proof is an easy exercise.

Now assume gr(A) is commutative. One defines the **Poisson bracket** on gr(A) by setting for $\bar{a} \in A_k/A_{k-1}$, $\bar{b} \in A_l/A_{l-1}$:

$$\{\bar{a}, \bar{b}\} = \sigma_{k+l-1}([a, b]) \tag{11.2.6}$$

where a (resp. b) is any element of A_k (resp. A_l) with $\sigma_k(a) = \bar{a}$ (resp. $\sigma_l(b) = \bar{b}$), $[a, b] = ab - ba$.

Let M be an A-module of finite type. One can endow it with a good filtration by choosing a set of generators $(v_j)_{j=1}^N$ and setting $M_k = \sum_j A_k v_j$. Moreover two good filtrations on a module of finite type are clearly equivalent. Then one proves easily that the radical of the annihilator of gr(M) is a graded ideal of gr(A) which depends only on M, not on the good filtration. One denotes it by Icar(M). Thus:

$$\begin{cases} \bar{a} \in \mathrm{Icar}(M), \bar{a} \textit{ homogeneous of order } k \Leftrightarrow \exists l \in \mathbb{N}, \forall a \in A_{kl} \\ \textit{with } \sigma(a) = \bar{a}^k, aM_n \subset M_{n+kl-1} \ \forall n \ . \end{cases} \tag{11.2.7}$$

Assume A is filtered Noetherian, and consider an exact sequence:

$$0 \to L \to M \to N \to 0 \tag{11.2.8}$$

of A-modules of finite type. One may endow M with a good filtration, L (resp. N) with the induced (resp. image) filtration. Then (11.2.8) becomes an exact sequence of filtered modules, and the sequence:

$$0 \to \mathrm{gr}(L) \to \mathrm{gr}(M) \to \mathrm{gr}(N) \to 0 \tag{11.2.9}$$

is itself exact. This implies:

$$\mathrm{Icar}(M) = \mathrm{Icar}(L) \cap \mathrm{Icar}(N) \ . \tag{11.2.10}$$

Moreover we have the important result, due to Gabber [1].

Theorem 11.2.2. *Assume* $\mathrm{gr}(A)$ *is a Noetherian commutative* $\mathbb{Q}$*-algebra. Then for any* A*-module* M *of finite type,* $\mathrm{Icar}(M)$ *is involutive, i.e.:*

$$\{\mathrm{Icar}(M), \mathrm{Icar}(M)\} \subset \mathrm{Icar}(M) \ .$$

Now let X be a topological space. A sheaf of filtered rings $\mathcal{A}$ on X is a sheaf of rings $\mathcal{A}$ endowed with a family of sheaves of subgroups $\{\mathcal{A}_k\}_{k \in \mathbb{Z}}$ such that $\mathcal{A}_k \subset \mathcal{A}_{k+1}$, 1 is a section of $\mathcal{A}_0$, $\mathcal{A}_k \cdot \mathcal{A}_l \subset \mathcal{A}_{k+l}$ and $\mathcal{A}$ is the union of the $\mathcal{A}_k$'s. One defines similarly the notion of filtered $\mathcal{A}$-module, finite free filtered $\mathcal{A}$-module, etc.

A filtration on an $\mathcal{A}$-module $\mathcal{M}$ is said to be good if, locally on X, there exists a filtered exact sequence $\mathcal{L} \to \mathcal{M} \to 0$, where $\mathcal{L}$ is finite free.

Proposition 11.2.3. *Let* $\mathcal{A}$ *be a sheaf of filtered rings on* X*. Assume that* $\mathrm{gr}(\mathcal{A})$ *is coherent (as a graded ring) (resp. Noetherian) and for each* $x \in X$*, the filtered ring* $\mathcal{A}_{X,x}$ *is Zariskian. Then:*

(i) *$\mathcal{A}$ is coherent (resp. Noetherian) as a sheaf of rings.*
(ii) *Let $\mathcal{M}$ be an $\mathcal{A}$-module endowed with a good filtration. Then $\mathcal{M}$ is coherent as an $\mathcal{A}$-module if and only if $\mathrm{gr}(\mathcal{M})$ is coherent (as a graded module).*
(iii) *Assume $\mathcal{M}$ is coherent and endowed with a good filtration. Let $\mathcal{N}$ be a coherent sub-module of $\mathcal{M}$. Then the induced filtration on $\mathcal{N}$ is good.*

For the proof, cf. Björk [1] or Schapira [2].

Now assume X is a complex manifold of complex dimension n. We denote by $\mathcal{D}_X$ the **sheaf of rings of finite-order holomorphic differential operators.** This is the sub-algebra of $\mathcal{H}om_{\mathbb{C}}(\mathcal{O}_X, \mathcal{O}_X)$ generated by $\mathcal{O}_X$ and by the vector fields.

The filtration of $\mathcal{D}_X$ is defined recursively by setting:

$$\begin{cases} \mathcal{D}_X(m) = 0 & \text{for} \quad m < 0 \ , \\ \mathcal{D}_X(m) = \{P \in \mathcal{D}_X; [P, \mathcal{O}_X] \subset \mathcal{D}_X(m-1)\} \ . \end{cases} \tag{11.2.11}$$

In particular, $\mathcal{D}_X(0) = \mathcal{O}_X$.

The graded ring $\mathrm{gr}(\mathcal{D}_X)$ is naturally isomorphic to $\mathcal{O}_{[T^*X]}$, where $\mathcal{O}_{[T^*X]}$ denotes the sub-ring of $\pi_*(\mathcal{O}_{T^*X})$ consisting of sections which are polynomials on the fibers of the vector bundle $\pi : T^*X \to X$.

Let $(x) = (x_1, \dots, x_n)$ be a local system of holomorphic coordinates on X, $(x;\xi)$ the associated coordinates on T^*X. A differential operator P of order m is written:

$$P(x; D_x) = \sum_{|\alpha| \leq m} a_\alpha(x) D_x^\alpha \ , \tag{11.2.12}$$

where the a_α's are holomorphic functions, $\alpha = (\alpha_1, \dots, \alpha_n)$, $|\alpha| = \alpha_1 + \dots + \alpha_n$, $D_x^\alpha = D_1^{\alpha_1} \dots D_n^{\alpha_n}$, and $D_j = \dfrac{\partial}{\partial x_j}$ (also denoted D_{x_j}).

The **principal symbol** $\sigma_m(P)$ is the function on T^*X:

$$\sigma_m(P)(x;\xi) = \sum_{|\alpha| = m} a_\alpha(x) \xi^\alpha \ . \tag{11.2.13}$$

The principal symbol $\sigma_m(P)$ is intrinsically defined on T^*X. One can also consider the *total symbol* of P,

$$P(x, \xi) = \sum_{|\alpha| \leq m} a_\alpha(x) \xi^\alpha$$

but this function depends on the local coordinates $(x_1, \dots, x_n)$. The composition $P \circ Q$ of two differential operators is given (in a choice of local coordinates) by the **Leibniz formula**:

$$(P \circ Q)(x, \xi) = \sum_{\alpha \in \mathbb{N}^n} \frac{1}{\alpha!} \frac{\partial^\alpha}{\partial \xi^\alpha} P(x;\xi) \cdot \frac{\partial^\alpha}{\partial x^\alpha} Q(x;\xi) \ ,$$

where $\alpha! = \alpha_1! \dots \alpha_n!$ for $\alpha = (\alpha_1, \dots, \alpha_n) \in \mathbb{N}^n$.

Applying Proposition 11.2.3 and well-known results on $\mathscr{O}_{[T^*X]}$, we get:

Proposition 11.2.4. (i) *The ring $\mathscr{D}_X$ is right and left coherent and Noetherian.*

(ii) *Let $\mathscr{M}$ be a coherent $\mathscr{D}_X$-module endowed with a good filtration. Then $\mathrm{gr}(\mathscr{M})$ is coherent. Moreover if $\mathscr{N}$ is a coherent sub-module of $\mathscr{M}$, the induced filtration on $\mathscr{N}$ is good.*

Let $\mathscr{M}$ be a coherent $\mathscr{D}_X$-module. One can locally endow $\mathscr{M}$ with a good filtration and define the graded ideal $\mathrm{Icar}(\mathscr{M})$ of $\mathrm{gr}(\mathscr{D}_X)$. This is a coherent graded ideal which depends only on $\mathscr{M}$, and which, therefore, is globally defined. Its nullvariety in T^*X, that is, the set of common zeroes of all sections of this ideal, is called the **characteristic variety** of $\mathscr{M}$, and denoted $\mathrm{char}(\mathscr{M})$. By (11.2.10), we find that if $0 \to \mathscr{L} \to \mathscr{M} \to \mathscr{N} \to 0$ is an exact sequence of coherent $\mathscr{D}_X$-modules, then:

$$\mathrm{char}(\mathscr{M}) = \mathrm{char}(\mathscr{L}) \cup \mathrm{char}(\mathscr{N}) \ . \tag{11.2.14}$$

Assume $\mathscr{M}$ is endowed with a good filtration and set:

$$\widetilde{\mathrm{gr}}(\mathscr{M}) = \mathscr{O}_{T^*X} \underset{\pi^{-1}\mathscr{O}_{[T^*X]}}{\otimes} \pi^{-1}\,\mathrm{gr}(\mathscr{M}) \ .$$

The functor $\mathcal{O}_{T^*X} \otimes_{\pi^{-1}\mathcal{O}_{[T^*X]}} \pi^{-1}(\cdot)$ being exact on the category of $\mathcal{O}_{[T^*X]}$-modules, one has:

$$\text{char}(\mathcal{M}) = \text{supp}(\widetilde{\text{gr}}(\mathcal{M})) \ . \tag{11.2.15}$$

By Theorem 11.2.2 and Proposition 11.2.3, char($\mathcal{M}$) is a closed analytic subset of T^*X, involutive and conic for the action of $\mathbb{C}^\times$ (cf. Exercise VIII.8).

In particular $\dim_{\mathbb{C}}(\text{char}(\mathcal{M})) \geqslant n$. When $\dim_{\mathbb{C}}(\text{char}(\mathcal{M})) \leqslant n$, one says $\mathcal{M}$ is **holonomic**.

Example 11.2.5. Assume $\mathcal{M}$ has one generator u with the defining relations $P_j u = 0$, $j = 1, \ldots, N$. Let $\mathcal{I}$ denote the left ideal of $\mathcal{D}_X$ generated by the P_j's. Then $\mathcal{M} \simeq \mathcal{D}_X/\mathcal{I}$, and:

$$\text{char}(\mathcal{M}) = \{(x;\xi); \sigma(P)(x;\xi) = 0 \text{ for any } P \in \mathcal{I}\} \ .$$

Of course the set char($\mathcal{M}$) may be strictly smaller than the set $\{(x;\xi); \sigma(P_j)(x;\xi) = 0 \text{ for } j = 1,\ldots,N\}$. In fact, although the P_j's generate $\mathcal{I}$, it may happen that the $\sigma(P_j)$'s do not generate gr($\mathcal{I}$). However the ideal gr($\mathcal{I}$) is locally finitely generated.

As a corollary of the involutivity of char($\mathcal{M}$), one obtains:

Proposition 11.2.6. *Let $\mathcal{M}$ be a coherent $\mathcal{D}_X$-module. Then locally on X, $\mathcal{M}$ admits a finite free resolution of length n.*

In other words, locally on X, there exists a complex:

$$0 \longrightarrow \mathcal{D}_X^{N_n} \longrightarrow \cdots \longrightarrow \mathcal{D}_X^{N_1} \xrightarrow[P_0]{} \mathcal{D}_X^{N_0} \longrightarrow 0 \ , \tag{11.2.16}$$

which is exact except in degree 0, and $\mathcal{M} \simeq \mathcal{D}_X^{N_0}/\mathcal{D}_X^{N_1} P_0$.

Example 11.2.7. Let Z be a closed submanifold of complex codimension d in X.

The homomorphism $\mathcal{E}xt^d_{\mathcal{O}_X}(\mathcal{O}_Z, \mathcal{O}_X) \simeq H^d_Z(R\mathcal{H}om_{\mathcal{O}_X}(\mathcal{O}_Z, \mathcal{O}_X)) \to H^d_Z(\mathcal{O}_X)$ is injective, and one denotes by $\mathcal{B}_{Z|X}$ or else by $H^d_{[Z]}(\mathcal{O}_X)$ the $\mathcal{D}_X$-module generated by its image. If $(x_1, \ldots, x_n)$ is a system of local coordinates on X such that $Z = \{x_1 = \cdots = x_d = 0\}$, then $\mathcal{B}_{Z|X}$ is generated by the class u of $\dfrac{1}{x_1 \ldots x_d}$, and this generator satisfies the relations:

$$x_1 u = \cdots = x_d u = D_{d+1} u = \cdots = D_n u = 0 \ .$$

In other words:

$$\mathcal{B}_{Z|X} \simeq \mathcal{D}_X/(\mathcal{D}_X x_1 + \cdots + \mathcal{D}_X x_d + \mathcal{D}_X D_{d+1} + \cdots + \mathcal{D}_X D_n) \ .$$

Note that $\mathcal{B}_{Z|X}$ is coherent, and char($\mathcal{B}_{Z|X}$) $= T^*_Z X$. By considering the regular sequence $(x_1, \ldots, x_d, D_{d+1}, \ldots, D_n)$ in $\mathcal{D}_X$ and the associated Koszul complex (cf. e.g. Schapira [2 Appendix B.4]), one gets a free resolution of length n of $\mathcal{B}_{Z|X}$.

Let $\mathfrak{Mod}(\mathscr{D}_X)$ (resp. $\mathfrak{Mod}(\mathscr{D}_X^{op})$) denote the abelian category of left (resp. right) $\mathscr{D}_X$-modules.

The categories $\mathfrak{Mod}(\mathscr{D}_X)$ and $\mathfrak{Mod}(\mathscr{D}_X^{op})$ are equivalent. In fact, there exists a natural isomorphism:

(11.2.17) $$\mathcal{O}_X^{(n)} \simeq \mathscr{E}xt^n_{\mathscr{D}_X}(\mathcal{O}_X, \mathscr{D}_X) \ ,$$

which makes $\mathcal{O}_X^{(n)}$ a right coherent $\mathscr{D}_X$-module. If $\mathscr{M}$ is a left $\mathscr{D}_X$-module one endows $\mathcal{O}_X^{(n)} \otimes_{\mathcal{O}_X} \mathscr{M}$ with a structure of a right $\mathscr{D}_X$-module as follows.

For $v \in \mathcal{O}_X^{(n)}$, $u \in \mathscr{M}$ and a vector field θ, one sets:

$$(v \otimes u)\theta = (v\theta) \otimes u - v \otimes \theta u \ ,$$

and then one can extend this action to an action of $\mathscr{D}_X$ on $\mathcal{O}_X^{(n)} \otimes_{\mathcal{O}_X} \mathscr{M}$. One defines similarly the left action of $\mathscr{D}_X$ on $\mathscr{N} \otimes_{\mathcal{O}_X} \mathcal{O}_X^{(n)\otimes -1}$ for a right $\mathscr{D}_X$-module $\mathscr{N}$.

The functors $\mathscr{M} \mapsto \mathcal{O}_X^{(n)} \otimes_{\mathcal{O}_X} \mathscr{M}$ and $\mathscr{N} \mapsto \mathscr{N} \otimes_{\mathcal{O}_X} \mathcal{O}_X^{(n)\otimes -1}$ provide the desired equivalence.

Denote by $\mathbf{D}^b_{\mathrm{coh}}(\mathscr{D}_X)$ (resp. $\mathbf{D}^b_{\mathrm{coh}}(\mathscr{D}_X^{op})$) the full subcategory of the derived category $\mathbf{D}^b(\mathfrak{Mod}(\mathscr{D}_X))$ (resp. $\mathbf{D}^b(\mathfrak{Mod}(\mathscr{D}_X^{op}))$) consisting of objects with coherent cohomologies.

The category $\mathfrak{Mod}(\mathscr{D}_X)$ has enough injectives (by Proposition 2.4.3), and for each $x \in X$, $\mathrm{gld}(\mathscr{D}_{X,x}) \leqslant n$. (This is easily deduced from Proposition 11.2.6.) Therefore the functors $R\mathscr{H}om_{\mathscr{D}_X}(\cdot,\cdot)$ and $\cdot \otimes^L_{\mathscr{D}_X} \cdot$ are well defined on $\mathbf{D}^b(\mathscr{D}_X)^\circ \times \mathbf{D}^b(\mathscr{D}_X)$ and $\mathbf{D}^b(\mathscr{D}_X^{op}) \times \mathbf{D}^b(\mathscr{D}_X)$ with values in $\mathbf{D}^+(\mathbb{C}_X)$ and $\mathbf{D}^b(\mathbb{C}_X)$, respectively.

To end this section, we shall briefly recall the main operations on $\mathscr{D}_X$-modules.

There is an interesting involution on $\mathbf{D}^b_{\mathrm{coh}}(\mathscr{D}_X)$ (thus also on $\mathbf{D}^b_{\mathrm{coh}}(\mathscr{D}_X^{op})$) which is similar to the functor of duality (cf. III §4) in sheaf theory. One sets:

(11.2.18) $$\mathscr{K}_X = \mathscr{D}_X \underset{\mathcal{O}_X}{\otimes} \Omega_X^{\otimes -1}[\dim_{\mathbb{C}} X] \ .$$

Note that $\mathscr{K}_X$ is naturally endowed with two structures of a left $\mathscr{D}_X$-module. If $\mathscr{M} \in \mathbf{D}^b_{\mathrm{coh}}(\mathscr{D}_X)$ one defines:

(11.2.19) $$\mathscr{M}^* = R\mathscr{H}om_{\mathscr{D}_X}(\mathscr{M}, \mathscr{K}_X) \ .$$

This is clearly an object of $\mathbf{D}^b(\mathfrak{Mod}(\mathscr{D}_X))$, thanks to the structure of bimodule of $\mathscr{K}_X$ (cf. Exercise II.23), and one checks immediately that $\mathscr{M}^* \in \mathrm{Ob}(\mathbf{D}^b_{\mathrm{coh}}(\mathscr{D}_X))$. Moreover:

(11.2.20) $$\mathscr{M}^{**} \simeq \mathscr{M} \ .$$

Now, let Y be another manifold.

If $\mathscr{M}$ (resp. $\mathscr{N}$) is a left $\mathscr{D}_X$-module (resp. $\mathscr{D}_Y$-module) one defines a left $\mathscr{D}_{X\times Y}$-module by setting:

(11.2.21) $$\mathscr{M} \underline{\boxtimes} \mathscr{N} = \mathscr{D}_{X\times Y} \underset{\mathscr{D}_X \boxtimes \mathscr{D}_Y}{\otimes} (\mathscr{M} \boxtimes \mathscr{N}) \ .$$

It is easily verified that if $\mathcal{M}$ and $\mathcal{N}$ are coherent, then so is $\mathcal{M} \underline{\boxtimes} \mathcal{N}$ and

$$\text{(11.2.22)} \qquad \operatorname{char}(\mathcal{M} \underline{\boxtimes} \mathcal{N}) = \operatorname{char}(\mathcal{M}) \times \operatorname{char}(\mathcal{N}) \ .$$

Finally let $f: Y \to X$ be a holomorphic map.

One endows the sheaf $\mathcal{O}_Y \otimes_{f^{-1}\mathcal{O}_X} f^{-1}\mathcal{D}_X$ with its natural structure of a right $f^{-1}\mathcal{D}_X$-module. One also endows it with a structure of a left $\mathcal{D}_Y$-module as follows.

Let Θ_X (resp. Θ_Y) denote the sheaf of holomorphic vector fields on X (resp. on Y). If v is a section of Θ_Y, the differential $f' : TY \to Y \times_X TX$ applied to v defines a section $f'(v)$ of $\mathcal{O}_Y \otimes_{f^{-1}\mathcal{O}_X} f^{-1}\Theta_X$ which, locally on Y, may be written as a finite sum $\sum_j a_j \otimes w_j$. Consider a section of $\mathcal{O}_Y \otimes_{f^{-1}\mathcal{O}_X} f^{-1}\mathcal{D}_X$ of the type $a \otimes u$. One sets:

$$v(a \otimes u) = v(a) \otimes u + \sum_j aa_j \otimes w_j \circ u \ .$$

If one chooses local coordinates systems $(x_1, \ldots, x_n)$ on X, $(y_1, \ldots, y_m)$ on Y so that $f = (f_1, \ldots, f_n)$, then:

$$\text{(11.2.23)} \qquad D_{y_k}(a \otimes u) = \frac{\partial a}{\partial y_k} \otimes u + \sum_{j=1}^{n} a \frac{\partial f_j}{\partial x_k} \otimes D_{x_j} \circ u \ .$$

The left action of Θ_Y on $\mathcal{O}_Y \otimes_{f^{-1}\mathcal{O}_X} f^{-1}\mathcal{D}_X$ is then naturally extended to a left action of $\mathcal{D}_Y$.

Definition 11.2.8. *The sheaf $\mathcal{O}_Y \otimes_{f^{-1}\mathcal{O}_X} f^{-1}\mathcal{D}_X$ endowed with its structure of $(\mathcal{D}_Y, f^{-1}\mathcal{D}_X)$-bimodule is denoted by $\mathcal{D}_{Y\to X}$. Its canonical section $1 \otimes 1$ is denoted by $1_{Y\to X}$.*

The bimodule $\mathcal{D}_{Y\to X}$ defines two functors:

$$\mathcal{D}_{Y\to X} \overset{L}{\underset{f^{-1}\mathcal{D}_X}{\otimes}} f^{-1}(\cdot) : \mathsf{D}^b(\mathcal{D}_X) \to \mathsf{D}^b(\mathcal{D}_Y)$$

$$\cdot \overset{L}{\underset{\mathcal{D}_Y}{\otimes}} \mathcal{D}_{Y\to X} : \mathsf{D}^b(\mathcal{D}_Y^{op}) \to \mathsf{D}^b(f^{-1}\mathcal{D}_X^{op}) \ .$$

If $g: Z \to Y$ is another holomorphic map, there is a canonical isomorphism:

$$\text{(11.2.24)} \qquad \mathcal{D}_{Z\to Y} \overset{L}{\underset{g^{-1}\mathcal{D}_Y}{\otimes}} g^{-1}\mathcal{D}_{Y\to X} \simeq \mathcal{D}_{Z\to X} \ .$$

Example 11.2.9. (i) Let $x = (x', x'')$ be a system of coordinates on X and let $Y = \{x \in X; x' = 0\}$. Then as a $\mathcal{D}_X^{op}$-module, $\mathcal{D}_{Y\to X} \simeq \mathcal{D}_X/(x')\cdot\mathcal{D}_X$ where $(x')\cdot\mathcal{D}_X$ is the right ideal generated by the coordinates (x'). A section of this sheaf may be written as $P(x'', D_{x'}, D_{x''})$, (i.e.: a differential operator on X not depending on x').

(ii) Let $x = (x', x'')$ be a system of coordinates on Y and let $f : Y \to X$ be the projection $(x', x'') \mapsto (x'')$. Then $\mathscr{D}_{Y\to X} \simeq \mathscr{D}_Y/(D_{x'})\cdot \mathscr{D}_Y$ and a section of this sheaf may be written as $P(x, x'', D_{x''})$, (i.e.: a differential operator on Y not depending on $D_{x'}$).

Definition 11.2.10. (i) *Let $\mathscr{M}$ be a left $\mathscr{D}_X$-module (or more generally, $\mathscr{M} \in \mathrm{Ob}(\mathbf{D}^b(\mathscr{D}_X))$). One defines the inverse image of $\mathscr{M}$ by f, denoted $\underline{f}^{-1}\mathscr{M}$, by:*

$$\underline{f}^{-1}\mathscr{M} = \mathscr{D}_{Y\to X} \overset{L}{\underset{f^{-1}\mathscr{D}_X}{\otimes}} f^{-1}\mathscr{M} \in \mathrm{Ob}(\mathbf{D}^b(\mathscr{D}_Y)) .$$

(ii) *Let $\mathscr{N}$ be a right $\mathscr{D}_Y$-module (or more generally, $\mathscr{N} \in \mathrm{Ob}(\mathbf{D}^b(\mathscr{D}_Y^{op}))$). One defines the direct image (resp. the proper direct image) of $\mathscr{N}$ by f, denoted $\underline{f}_*\mathscr{N}$ (resp. $\underline{f}_!\mathscr{N}$) by:*

$$\underline{f}_*\mathscr{N} = Rf_*\left(\mathscr{N} \overset{L}{\underset{\mathscr{D}_Y}{\otimes}} \mathscr{D}_{Y\to X}\right) \in \mathrm{Ob}(\mathbf{D}^b(\mathscr{D}_X^{op}))$$

(resp.

$$\underline{f}_!\mathscr{N} = Rf_!\left(\mathscr{N} \overset{L}{\underset{\mathscr{D}_Y}{\otimes}} \mathscr{D}_{Y\to X}\right) \in \mathrm{Ob}(\mathbf{D}^b(\mathscr{D}_X^{op}))) .$$

Let us recall, without proofs, the main results concerning inverse images, due to Kashiwara [5]. (Concerning direct images, cf. Kashiwara [4], Houzel-Schapira [1], Schneiders [1] and Boutet de Monvel-Malgrange [1].)

We shall denote, as usual, by f_π and ${}^t f'$ the natural maps from $Y \times_X T^*X$ to T^*X and T^*Y, respectively.

Definition 11.2.11. *Let $\mathscr{M}$ be a left coherent $\mathscr{D}_X$-module. One says f is non-characteristic for $\mathscr{M}$ if:*

$$T_Y^*X \cap f_\pi^{-1}(\mathrm{char}(\mathscr{M})) \subset Y \times_X T_X^*X .$$

Proposition 11.2.12. *Assume f is non-characteristic for a coherent $\mathscr{D}_X$-module $\mathscr{M}$. Then:*

(i) $H^j(\underline{f}^{-1}\mathscr{M}) = 0$ *for* $j \neq 0$,
(ii) $H^0(\underline{f}^{-1}\mathscr{M})$ *(simply denoted by $\underline{f}^{-1}\mathscr{M}$) is a coherent $\mathscr{D}_Y$-module,*
(iii) $\mathrm{char}(\underline{f}^{-1}\mathscr{M}) = {}^t f' f_\pi^{-1}(\mathrm{char}(\mathscr{M}))$.

11.3. Holomorphic solutions of $\mathscr{D}_X$-modules

Let us first recall, without proof, the classical **Cauchy-Kowalewski theorem**, in its refined version due to Leray [3]. (cf. e.g. Schapira [2]).

Let $x = (x_1, \dots, x_n)$ be the holomorphic coordinates on $\mathbb{C}^n$, $(x;\xi)$ the associated coordinates on $T^*\mathbb{C}^n$. We endow $\mathbb{C}^n$ with its usual Hermitian structure ($\langle x, x'\rangle = \sum_j x_j \bar{x}'_j$) and denote by $B(x_o, \rho)$ the open ball with center x_o and radius ρ. We also use the notations $x = (x_1, x')$, and denote by $B'(x_o, \rho)$ the intersection of $B(x_o, \rho)$ with the hyperplane $\{x \in \mathbb{C}^n; x_1 = x_{o,1}\}$ where $x_o = (x_{o,1}, \dots, x_{o,n})$.

Let X be an open subset of $\mathbb{C}^n$, and let P be a holomorphic differential operator of order m defined on X. Then P is written as:

$$P = \sum_{|\alpha| \leq m} a_\alpha(x) D_x^\alpha \ . \tag{11.3.1}$$

One makes the hypothesis:

$$a_{(m,0,\dots,0)} \equiv 1 \ . \tag{11.3.2}$$

Let $x_o \in X$, and set:

$$Y_{x_o} = \{x \in X; x_1 = x_{o,1}\} \ .$$

For f a section of $\mathcal{O}_X$, define the m first traces of f on Y_{x_o}, a section of $(\mathcal{O}_{Y_{x_o}}^m)$, by:

$$\gamma_{x_o}(f) = (f|_{Y_{x_o}}, D_{x_1} f|_{Y_{x_o}}, \dots, D_{x_1}^{m-1} f|_{Y_{x_o}}) \ . \tag{11.3.3}$$

Consider the **Cauchy problem**:

$$\begin{cases} Pf = g \\ \gamma_{x_o}(f) = (h) \ . \end{cases} \tag{11.3.4}$$

Theorem 11.3.1. *Let $x \in X$. There exist $r > 0$, $\rho_o > 0$, $\delta > 0$ such that for any $\rho > 0$ with $\rho \leq \rho_o$, any $x_o \in X$ with $|x - x_o| \leq r$, any $g \in \mathcal{O}_X(B(x_o, \rho))$, any $(h) \in \mathcal{O}_{Y_{x_o}}^m(B'(x_o, \rho))$, the Cauchy problem* (11.3.4) *has a unique solution $f \in \mathcal{O}_X(B(x_o, \delta\rho))$.*

We shall first deduce a useful extension result, due to Zerner [1] (cf. Hörmander [3 Theorem 11.4.7]).

Proposition 11.3.2. *Let ϕ be a real C^1-function on X such that $\sigma(P)(x; \partial\phi(x)) \neq 0$ on X. Let $\Omega = \{x \in X; \phi(x) < 0\}$ and let $f \in \mathcal{O}_X(\Omega)$ be such that Pf extends holomorphically on a neighborhood of $x_o \in \partial\Omega$. Then f extends holomorphically in a neighborhood of x_o.*

Proof. If $d\phi(x_o) = 0$, P is invertible in a neighborhood of x_o and the conclusion follows. Assume $d\phi(x_o) \neq 0$ and choose a local coordinate system $(x_1, \dots, x_n)$ near x_o, such that $x_o = 0$, $\phi(x) = \operatorname{Re} x_1 - \psi(\operatorname{Im} x_1, x')$, with $d\psi(0) = 0$. We may assume P satisfies (10.3.1) and (10.3.2).

Set $x_\varepsilon = (-\varepsilon, 0, \dots, 0)$, and keep the notations of Theorem 11.3.1. Consider the Cauchy problem:

$$\begin{cases} Pf_\varepsilon = Pf \\ \gamma_{x_\varepsilon}(f_\varepsilon) = \gamma_{x_\varepsilon}(f) . \end{cases}$$

Then $f_\varepsilon = f$, and f_ε is holomorphic in a ball $B(x_\varepsilon, \delta R)$ where δ may be chosen independent of ε, and R satisfies:

$$-\varepsilon < \psi(0, x') \qquad \text{for} \qquad |x'| < R .$$

Since $d\psi(0) = 0$, $\psi(0, x') = o(|x'|)$ and $B(x_\varepsilon, \delta R)$ is a neighborhood of 0 for ε small enough. □

Theorem 11.3.3. *Let X be a complex manifold, $\mathscr{M}$ a coherent $\mathscr{D}_X$-module. Then:*

$$SS(R\mathscr{H}om_{\mathscr{D}_X}(\mathscr{M}, \mathscr{O}_X)) = \operatorname{char}(\mathscr{M}) .$$

Proof. We shall only prove here the inclusion $\cdot \subset \cdot$. See Remark 11.4.5 for the other inclusion.

(a) First assume $\mathscr{M} = \mathscr{D}_X/\mathscr{D}_X P$. Let $p \in T^*X$, with $\sigma(P)(p) \neq 0$. If $p \in T^*_X X$, P is an invertible function on a neighborhood of p, and $R\mathscr{H}om_{\mathscr{D}_X}(\mathscr{M}, \mathscr{O}_X) = 0$ on this neighborhood. Assume $p \notin T^*_X X$ and choose a local coordinate system such that $p = (x_\circ; \xi_\circ)$, with $\xi_\circ = (1, 0, \ldots, 0)$.

We shall apply Proposition 5.1.1. Set:

$$H_\varepsilon = \{x \in \mathbb{C}^n; \operatorname{Re}\langle x - x_\circ, \xi_\circ\rangle \geqslant -\varepsilon\} ,$$

$$L_\varepsilon = \{x \in \mathbb{C}^n; \operatorname{Re}\langle x - x_\circ, \xi_\circ\rangle = -\varepsilon\} ,$$

$$\gamma_\delta = \{x \in \mathbb{C}^n; \operatorname{Im} x_1 = 0, \operatorname{Re} x_1 \geqslant \delta |x'|\} .$$

We choose $0 < R \ll 1$, $\delta \gg 0$ such that:

$$(11.3.5) \qquad \sigma(P)(x; \xi) \neq 0 \qquad \text{for} \qquad |x - x_\circ| \leqslant R , \qquad \xi \in \gamma_\delta^\circ \setminus \{0\} .$$

Then we choose $0 < \varepsilon \ll 1$ such that $(x + \gamma_\delta) \cap L_\varepsilon \subset B(x_\circ, R)$ for $|x - x_\circ| \leqslant \frac{R}{2}$.

The object $R\mathscr{H}om_{\mathscr{D}_X}(\mathscr{M}, \mathscr{O}_X)$ may be represented by the complex:

$$0 \longrightarrow \mathscr{O}_X \xrightarrow[P]{} \mathscr{O}_X \longrightarrow 0 .$$

Moreover if K is convex and compact, a well-known result asserts that $H^i(K; \mathscr{O}_X) = 0$ for $i > 0$, that is, $R\Gamma(K; \mathscr{O}_X) \simeq \mathscr{O}_X(K)$ (cf. Hörmander [1]). To prove that $p \notin SS(R\mathscr{H}om_{\mathscr{D}_X}(\mathscr{M}, \mathscr{O}_X))$ it is thus enough to prove that the two complexes:

$$0 \longrightarrow \mathscr{O}_X((x + \gamma_\delta) \cap H_\varepsilon) \xrightarrow[P]{} \mathscr{O}_X((x + \gamma_\delta) \cap H_\varepsilon) \longrightarrow 0$$

and

$$0 \longrightarrow \mathcal{O}_X((x+\gamma_\delta)\cap L_\varepsilon) \xrightarrow[P]{} \mathcal{O}_X((x+\gamma_\delta)\cap L_\varepsilon) \longrightarrow 0$$

are quasi-isomorphic for $|x - x_\circ| \leqslant \frac{R}{2}$.

Since P is surjective on the space $\mathcal{O}_X((x+\gamma_\delta)\cap L_\varepsilon)$, by Theorem 11.3.1, it remains to show that:

(11.3.6)

$$\begin{cases} f \in \mathcal{O}_X((x+\gamma_\delta)\cap L_\varepsilon)\ , \\ Pf \in \operatorname{Im}(\mathcal{O}_X((x+\gamma_\delta)\cap H_\varepsilon) \to \mathcal{O}_X((x+\gamma_\delta)\cap L_\varepsilon)) \Rightarrow f \in \mathcal{O}_X((x+\gamma_\delta)\cap H_\varepsilon)\ . \end{cases}$$

But (11.3.6) follows easily from Proposition 11.3.2, in view of (11.3.5), by a similar argument to that of the proof of Proposition 5.1.1.

(b) Next, we treat the general case. Let $p \in T^*X$, $p \notin \operatorname{char}(\mathcal{M})$. We choose a system of generators $(u_1, \dots, u_N)$ of $\mathcal{M}$ in a neighborhood of $\pi(p)$. For each j, there exists an operator P_j such that $P_j u_j = 0$ and $\sigma(P_j)(p) \neq 0$. Set $\mathcal{L} = \bigoplus_{j=1}^N \mathcal{D}_X/\mathcal{D}_X P_j$, and define the $\mathcal{D}_X$-linear morphism $\psi : \mathcal{L} \to \mathcal{M}$ by setting $\psi(1 \bmod \mathcal{D}_X P_j) = u_j$. Let $\mathcal{K} = \operatorname{Ker}\psi$. We get the exact sequence of coherent $\mathcal{D}_X$-modules:

(11.3.7) $$0 \to \mathcal{K} \to \mathcal{L} \to \mathcal{M} \to 0\ .$$

Let U be a neighborhood of p such that $U \cap \operatorname{char}(\mathcal{M}) = \varnothing$, $U \cap \operatorname{char}(\mathcal{L}) = \varnothing$, (hence $U \cap \operatorname{char}(\mathcal{K}) = \varnothing$). Let ϕ be a real C^1-function on X and $x_\circ \in X$ such that $\phi(x_\circ) = 0$, $d\phi(x_\circ) \in U$.

Denote for short by $S_\phi(\cdot)$ the functor $(R\Gamma_{\{\phi \geqslant 0\}} R\mathcal{H}om_{\mathcal{D}_X}(\cdot, \mathcal{O}_X))_{x_\circ}$. By the first part of the proof, $S_\phi(\mathcal{L}) = 0$. Applying $S_\phi(\cdot)$ to the exact sequence (11.3.7) and taking the cohomology, we get the long exact sequence:

$$0 \to H^0(S_\phi(\mathcal{M})) \to H^0(S_\phi(\mathcal{L})) \to H^0(S_\phi(\mathcal{K})) \to H^1(S_\phi(\mathcal{M})) \to \cdots$$

Then $H^0(S_\phi(\mathcal{M})) = 0$. Since $\mathcal{K}$ satisfies the same hypotheses as $\mathcal{M}$, $H^0(S_\phi(\mathcal{K})) = 0$. Arguing by induction, we get $H^j(S_\phi(\mathcal{M})) = 0$ for all j, which completes the proof. □

Remark 11.3.4. The inclusion proved above will be the most useful in the applications.

Now we shall study some operations on holomorphic solutions of $\mathcal{D}$-modules. Let $\mathcal{M}$ (resp. $\mathcal{N}$) be a left $\mathcal{D}_X$-module (resp. $\mathcal{D}_Y$-module). There is a canonical morphism:

(11.3.8) $$R\mathcal{H}om_{\mathcal{D}_X}(\mathcal{M}, \mathcal{O}_X) \boxtimes R\mathcal{H}om_{\mathcal{D}_Y}(\mathcal{N}, \mathcal{O}_Y) \to R\mathcal{H}om_{\mathcal{D}_{X\times Y}}(\mathcal{M} \underline{\boxtimes} \mathcal{N}, \mathcal{O}_{X\times Y})\ .$$

Let $f : Y \to X$ be a holomorphic map. If $\mathscr{M}$ and $\mathscr{L}$ are two left $\mathscr{D}_X$-modules there is a natural morphism in $\mathbf{D}^b(\mathbb{C}_Y)$:

$$f^{-1}R\mathscr{H}om_{\mathscr{D}_X}(\mathscr{M}, \mathscr{L}) \to R\mathscr{H}om_{\mathscr{D}_Y}(\underline{f}^{-1}\mathscr{M}, \underline{f}^{-1}\mathscr{L}) \ . \tag{11.3.9}$$

This morphism is obtained as follows (cf. Exercise II.23). First replace $\mathscr{L}$ by a complex $\mathscr{L}^\cdot$ of injective $\mathscr{D}_X$-modules. Next choose a bounded resolution $\mathscr{P}^\cdot$ of $\mathscr{D}_{Y\to X}$ by $(\mathscr{D}_Y, f^{-1}\mathscr{D}_X)$-bimodules, flat over $f^{-1}\mathscr{D}_X$. Finally choose a $\mathscr{D}_Y$-injective resolution $\mathscr{I}^\cdot$ of $\mathscr{P}^\cdot \otimes_{f^{-1}\mathscr{D}_X} \mathscr{L}^\cdot$. Then we have the morphisms:

$$\begin{aligned} f^{-1}R\mathscr{H}om_{\mathscr{D}_X}(\mathscr{M}, \mathscr{L}) &\stackrel{\sim}{\to} f^{-1}\mathscr{H}om_{\mathscr{D}_X}(\mathscr{M}, \mathscr{L}^\cdot) \\ &\to \mathscr{H}om_{f^{-1}\mathscr{D}_X}(f^{-1}\mathscr{M}, f^{-1}\mathscr{L}^\cdot) \\ &\to \mathscr{H}om_{\mathscr{D}_Y}\left(\mathscr{P}^\cdot \underset{f^{-1}\mathscr{D}_X}{\otimes} f^{-1}\mathscr{M}, \mathscr{P}^\cdot \underset{f^{-1}\mathscr{D}_X}{\otimes} f^{-1}\mathscr{L}^\cdot\right) \\ &\to \mathscr{H}om_{\mathscr{D}_Y}\left(\mathscr{P}^\cdot \underset{f^{-1}\mathscr{D}_X}{\otimes} f^{-1}\mathscr{M}, \mathscr{I}^\cdot\right) \\ &\simeq R\mathscr{H}om_{\mathscr{D}_Y}(\underline{f}^{-1}\mathscr{M}, \underline{f}^{-1}\mathscr{L}) \ . \end{aligned}$$

It is possible to extend the Cauchy-Kowalewski Theorem to systems.

Theorem 11.3.5. *Assume f is non-characteristic for $\mathscr{M}$. Then the natural morphism*

$$f^{-1}R\mathscr{H}om_{\mathscr{D}_X}(\mathscr{M}, \mathscr{O}_X) \to R\mathscr{H}om_{\mathscr{D}_Y}(\underline{f}^{-1}\mathscr{M}, \mathscr{O}_Y)$$

is an isomorphism.

Proof. Consider the maps:

$$Y \underset{g}{\hookrightarrow} Y \times X \underset{p}{\longrightarrow} X$$

where g is the graph map $(y \mapsto (y, f(y)))$ and p the projection. In view of (11.2.24) it is enough to prove the result separately for p and for g. In case of p, we have to prove:

$$p^{-1}R\mathscr{H}om_{\mathscr{D}_X}(\mathscr{M}, \mathscr{O}_X) \simeq R\mathscr{H}om_{\mathscr{D}_{Y\times X}}(\mathscr{O}_Y \underline{\boxtimes} \mathscr{M}, \mathscr{O}_{Y\times X}) \ .$$

This can be easily reduced to the case $\mathscr{M} = \mathscr{D}_X$ and then this is nothing but the Poincaré lemma.

Hence one may assume from the beginning that f is an immersion. Arguing by induction on codim Y, we may even assume Y is a hypersurface. Then the same proof as for Theorem 11.3.3 reduces to the case where $\mathscr{M} = \mathscr{D}_X/\mathscr{D}_X P$.

We choose a system of local coordinates $(x_1, \ldots, x_n)$ with $Y = \{x_1 = 0\}$ and we may assume P satisfies (11.3.1), (11.3.2). Let us calculate the induced system

$\mathcal{D}_{Y\to X} \otimes_{\mathcal{D}_X} (\mathcal{D}_X/\mathcal{D}_X P) \simeq \mathcal{D}_X/(x_1\mathcal{D}_X + \mathcal{D}_X P)$. The **Weïerstrass Division Theorem** (for differential operators) asserts that for any $S \in \mathcal{D}_{X,x}$ there exists a unique pair $(Q, R) \in \mathcal{D}_{X,x}^2$ such that

$$\begin{cases} S = Q \circ P + R \ , \\ R = \sum_{j=0}^{m-1} R_j(x, D_{x'}) D_1^j \ . \end{cases}$$

Thus S may be uniquely written as:

$$S = Q \circ P + x_1 T + \sum_{j=0}^{m-1} \tilde{R}_j(x', D_{x'}) D_{x_1}^j \ .$$

It follows that $\mathcal{D}_{Y\to X} \otimes_{\mathcal{D}_X} (\mathcal{D}_X/\mathcal{D}_X P)$ is a free module of rank m over $\mathcal{D}_Y$, generated by $1_{Y\to X} \otimes u, \dots, 1_{Y\to X} \otimes D_{x_1}^{m-1} u$, where u is the generator $1 \bmod \mathcal{D}_X P$ of $\mathcal{D}_X/\mathcal{D}_X P$.

Then Theorem 11.3.4 is translated as:

$$\begin{cases} \mathrm{Ker}(\mathcal{O}_X \xrightarrow[P]{} \mathcal{O}_X)|_Y \xrightarrow[\gamma_Y]{\sim} \mathcal{O}_Y^m \ , \\ \mathrm{Coker}(\mathcal{O}_X \xrightarrow[P]{} \mathcal{O}_X)|_Y = 0 \ , \end{cases}$$

and this is a particular case of Theorem 10.3.1. □

Remark 11.3.6. There also exists a similar result, for direct images, due to Schneiders [1].

We shall end this section by proving that the complex of holomorphic solutions of a holonomic $\mathcal{D}_X$-module is perverse.

Theorem 11.3.7. *Assume $\mathcal{M}$ is a holonomic $\mathcal{D}_X$-module. Then $R\mathcal{H}om_{\mathcal{D}_X}(\mathcal{M}, \mathcal{O}_X)$ belongs to $\mathbf{D}^b_{\mathbb{C}-c}(X)$ (i.e.: is $\mathbb{C}$-constructible) and moreover $R\mathcal{H}om_{\mathcal{D}_X}(\mathcal{M}, \mathcal{O}_X)[\dim_{\mathbb{C}} X]$ is perverse.*

Proof. Set $F = R\mathcal{H}om_{\mathcal{D}_X}(\mathcal{M}, \mathcal{O}_X)[\dim_{\mathbb{C}} X]$.

(a) First we prove that F is $\mathbb{C}$-constructible.

By Theorem 11.3.3 and Theorem 8.5.5, F is weakly $\mathbb{C}$-constructible. Hence it remains to show that for any $x_\circ \in X, j \in \mathbb{Z}$, the vector space $H^j(F)_{x_\circ}$ is finite-dimensional.

Choose a local chart in a neighborhood of $x_\circ$ and denote by $B(x_\circ, \varepsilon)$ the open ball with center $x_\circ$ and radius ε. Set $F_\varepsilon = R\Gamma(B(x_\circ, \varepsilon); F)$. By Lemma 8.4.7 the natural morphisms $F_\varepsilon \to F_{\varepsilon'}$ are quasi-isomorphisms for $0 < \varepsilon' \leqslant \varepsilon \ll 1$. To calculate F_ε we choose (using Proposition 11.2.6) a finite free presentation of $\mathcal{M}$:

$$0 \longrightarrow \mathcal{D}_X^{N_n} \longrightarrow \cdots \xrightarrow[P_0]{} \mathcal{D}_X^{N_0} \longrightarrow \mathcal{M} \longrightarrow 0 \ .$$

Then F_ε is represented by the complex:

$$0 \longrightarrow \mathcal{O}_X^{N_0}(B(x_\circ, \varepsilon)) \xrightarrow[P_0]{} \cdots \longrightarrow \mathcal{O}_X^{N_n}(B(x_\circ, \varepsilon)) \longrightarrow 0$$

(since $H^j(B(x_\circ, \varepsilon); \mathcal{O}_X) = 0$ for $j \neq 0$). The space $\mathcal{O}_X(B(x_\circ, \varepsilon))$ is naturally endowed with a structure of a Fréchet space, and the restriction morphism $\mathcal{O}_X(B(x_\circ, \varepsilon)) \to \mathcal{O}_X(B(x_\circ, \varepsilon'))$ is compact if $0 < \varepsilon' < \varepsilon$. Therefore the result follows from the next lemma, whose proof is a direct application of a theorem of Schwartz [1].

Lemma 11.3.8. *Let $F^\cdot$ and $G^\cdot$ be two bounded complexes of Fréchet spaces and linear continuous maps and $u: F^\cdot \to G^\cdot$ a continuous linear morphism. Assume that for each j, the linear map $u^j: F^j \to G^j$ is compact and assume u is a quasi-isomorphism. Then for each $j \in \mathbb{Z}$, the spaces $H^j(F^\cdot)$ are finite dimensional.*

(b) Finally we shall prove that F is perverse. Let $n = \dim_{\mathbb{C}} X$. Let S be a complex submanifold of complex codimension d. Then $H^j_S(F) = 0$ for $j < d - n$, since $H^j_S(\mathcal{O}_X) = 0$ for $j < d$ (cf. (2.9.14)). Hence F belongs to ${}^p\mathbf{D}^{\geq 0}_{\mathbb{C}-c}(\mathbb{C}_X)$.

Now let $j \in \mathbb{N}$ be fixed, and set $Y = \operatorname{supp}(H^j(F))$.

Since F is $\mathbb{C}$-constructible, Y is a closed complex analytic set. We shall prove that $\dim_{\mathbb{C}} Y \leqslant -j$.

Since it is enough to prove this inequality at generic points of Y, we may assume by Proposition 8.2.10 that $\operatorname{char}(\mathcal{M}) \cap \pi^{-1}(Y) \subset T^*_Y X$. Let $x \in Y$ and choose a submanifold Z transversal to Y at x, with $\dim_{\mathbb{C}} Y + \dim_{\mathbb{C}} Z = n$. The embedding $f: Z \hookrightarrow X$ is non-characteristic for $\mathcal{M}$. Hence by Theorem 11.3.5, $F|_Z \simeq R\mathcal{H}om_{\mathcal{D}_Z}(\underline{f}^{-1}\mathcal{M}, \mathcal{O}_Z)[n]$.

Then $\underline{f}^{-1}\mathcal{M}$ admits a finite free resolution of length $\dim_{\mathbb{C}} Z$ by Proposition 11.2.6, and hence $H^k(F|_Z)_x = H^{k+n}(R\mathcal{H}om_{\mathcal{D}_Z}(\underline{f}^{-1}\mathcal{M}, \mathcal{O}_Z)) = 0$ for $k + n > \dim_{\mathbb{C}} Z$. Since $H^j(F|_Z)_x \neq 0$, we obtain $j \leqslant \dim_{\mathbb{C}} Z - n = -\dim_{\mathbb{C}} Y$, which is the desired result. □

11.4. Microlocal study of $\mathcal{O}_X$

Let X be a complex manifold of complex dimension n, S a complex submanifold of complex codimension d.

Proposition 11.4.1. *The complex $\mu_S(\mathcal{O}_X)$ is concentrated in degree d.*

Proof. The vanishing of $H^j(\mu_S(\mathcal{O}_X))$ for $j < d$ follows from assertion (2.9.14). The vanishing for $j > d$ is an easy exercise (cf. Sato-Kawai-Kashiwara [1]). □

Let Y be another complex manifold. We introduce the notation:

(11.4.1) $$\Omega_{X \times Y/Y} = \mathcal{O}_{X \times Y} \underset{q_1^{-1}\mathcal{O}_X}{\otimes} q_1^{-1}\Omega_X \ .$$

Definition 11.4.2. (i) *One sets:*

$$\mathscr{C}^{\mathbb{R}}_{S|X} = H^d(\mu_S(\mathcal{O}_X)) \ .$$

(ii) *Let $f: Y \to X$ be a morphism of complex manifolds. One sets:*

$$\mathscr{E}^{\mathbb{R}}_{Y\to X} = H^n(\mu_{\Delta_f}(\Omega_{X\times Y/Y}))\ ,$$

$$\mathscr{E}^{\mathbb{R}}_{X\leftarrow Y} = H^n(\mu_{\Delta_f}(\Omega_{X\times Y/X})^a)\ ,$$

where "a" is the antipodal map. When $f = \mathrm{id}_X$, one writes $\mathscr{E}^{\mathbb{R}}_X$ instead of $\mathscr{E}^{\mathbb{R}}_{X\to X}$.

The fundamental class of Δ_f in $X \times Y$ gives a section of $H^n_{\Delta_f}(\Omega_{X\times Y/Y})$, and it defines a global section of $\mathscr{E}^{\mathbb{R}}_{Y\to X}$. One denotes it by $1_{Y\to X}$ (in fact this definition agrees with Definition 11.2.8). When $f = \mathrm{id}_X$, one writes 1_X instead of $1_{X\to X}$. The sheaves $\mathscr{E}^{\mathbb{R}}_{Y\to X}$ and $\mathscr{E}^{\mathbb{R}}_{X\leftarrow Y}$ are supported by $T^*_{\Delta_f}(X \times Y) \simeq Y \times_X T^*X$.

We shall study the composition of the sheaves $\mathscr{E}^{\mathbb{R}}_{Y\to X}$.

If X, Y, Z are three manifolds, we denote by p_{ij} the projection from $T^*X \times T^*Y \times T^*Z$ to the (i,j)-component, and by p^a_{ij} the composition of p_{ij} and the antipodal map on the j-th component. We define similarly the projection p^a_{ij} defined on the product, $T^*W \times T^*X \times T^*Y \times T^*Z$, for four manifolds W, X, Y, Z.

Lemma 11.4.3. (i) *For $K_1 \in \mathrm{Ob}(\mathbf{D}^b(X \times Y))$, and $K_2 \in \mathrm{Ob}(\mathbf{D}^b(Y \times Z))$, there is a canonical morphism:*

$$Rp^a_{13!}(p^{a-1}_{12}\mu hom(K_1, \Omega_{X\times Y/Y}) \otimes p^{a-1}_{23}\mu hom(K_2, \Omega_{Y\times Z/Z}))$$

$$\to \mu hom(K_1 \circ K_2, \Omega_{X\times Z/Z})[-\dim_{\mathbb{C}} Y]\ .$$

(ii) *For $K_1 \in \mathrm{Ob}(\mathbf{D}^b(W \times X))$, $K_2 \in \mathrm{Ob}(\mathbf{D}^b(X \times Y))$ and $K_3 \in \mathrm{Ob}(\mathbf{D}^b(Y \times Z))$, the following diagram commutes:*

$$Rp^a_{14!}(p^{a-1}_{12}\mu hom(K_1, \Omega_{W\times X/X}) \otimes p^{a-1}_{23}\mu hom(K_2, \Omega_{X\times Y/Y}) \otimes p^{a-1}_{34}\mu hom(K_3, \Omega_{Y\times Z/Z}))$$

$$\swarrow \sim \qquad\qquad \sim \searrow$$

$$Rp^a_{13!}(p^{a-1}_{12}F_1 \otimes p^{a-1}_{23}\mu hom(K_3, \Omega_{Y\times Z/Z})) \qquad Rp^a_{13!}(p^{a-1}_{12}\mu hom(K_1, \Omega_{W\times X/X}) \otimes p^{a-1}_{23}F_2)$$

$$\downarrow \alpha \qquad\qquad\qquad\qquad \downarrow \beta$$

$$Rp^a_{13!}(p^{a-1}_{12}\mu hom(K_1 \circ K_2, \Omega_{W\times Y/Y})[-\dim X] \otimes p^{a-1}_{23}\mu hom(K_3, \Omega_{Y\times Z/Z})) \qquad Rp^a_{13!}(p^{a-1}_{12}\mu hom(K_1, \Omega_{W\times X/X}) \otimes p^{a-1}_{23}\mu hom(K_2 \circ K_3, \Omega_{X\times Z/Z})[-\dim X])$$

$$\searrow \qquad\qquad \swarrow$$

$$\mu hom(K_1 \circ K_2 \circ K_3, \Omega_{W\times Z/Z})[-\dim X - \dim Y]\ .$$

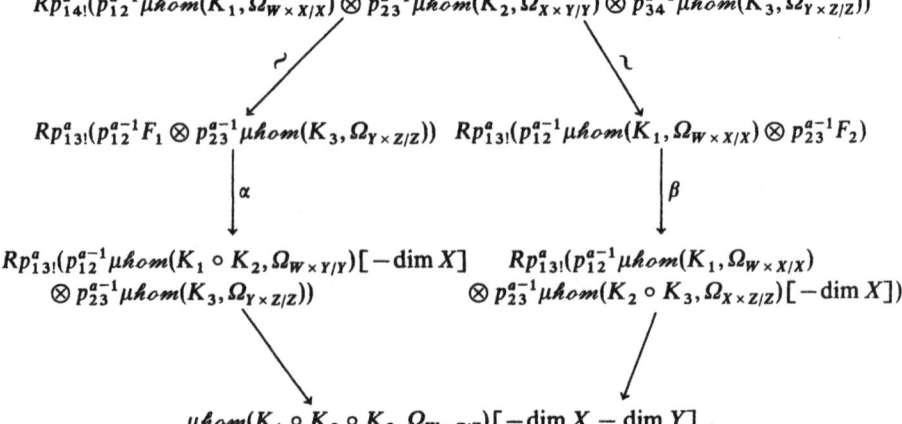

Here $F_1 = Rp^a_{13!}(p^{a-1}_{12}\mu hom(K_1, \Omega_{W\times X/X}) \otimes p^{a-1}_{23}\mu hom(K_2, \Omega_{X\times Y/Y})) \in \mathrm{Ob}(\mathbf{D}^b(T^*X \times T^*Y))$ and $F_2 = Rp^a_{13!}(p^{a-1}_{12}\mu hom(K_2, \Omega_{X\times Y/Y}) \otimes p^{a-1}_{23}\mu hom(K_3, \Omega_{Y\times Z/Z})) \in \mathrm{Ob}(\mathbf{D}^b(T^*Y \times T^*Z))$, and the morphisms α and β are the ones induced

by the morphisms given in (i):

$$F_1 \to \mu hom(K_1 \circ K_2, \Omega_{W\times Y/Y})[-\dim X] \qquad and$$

$$F_2 \to \mu hom(K_2 \circ K_3, \Omega_{X\times Z/Z})[-\dim Y] \ , \qquad respectively.$$

Proof. We can apply Proposition 4.4.11 to obtain the morphism:

$$Rp^a_{13!}(p^{a-1}_{12}\mu hom(K_1, \Omega_{X\times Y/Y}) \otimes p^{a-1}_{23}\mu hom(K_2, \Omega_{Y\times Z/Z}))$$
$$\to \mu hom(K_1 \circ K_2, \Omega_{X\times Y/Y} \circ \Omega_{Y\times Z/Z}) \ .$$

Combining with the morphisms (cf. Theorem 11.1.4):

$$\Omega_{X\times Y/Y} \circ \Omega_{Y\times Z/Z} \simeq Rp_{13!}(q^{-1}_{12}\Omega_{X\times Y/Y} \otimes q^{-1}_{23}\Omega_{Y\times Z/Z})$$
$$\to Rq_{13!}(\Omega_{X\times Y\times Z/Z})$$
$$\to \Omega_{X\times Z/Z}[-\dim_{\mathbb{C}} Y] \ ,$$

we get the result.

(ii) The proof is left to the reader. □

Let $g: Z \to Y$ and $f: Y \to X$ be morphisms of complex manifolds. We keep the same notations as for Proposition 4.4.9 (in particular, cf. diagram 4.4.13 where ${}^t f'$ is the map $Z \times_X T^*X \to Z \times_Y T^*Y$ and g_π the map $Z \times_X T^*X \to Y \times_X T^*X$).

Proposition 11.4.4. (i) *There are natural morphisms on* $Z \times_X T^*X$:

$${}^t f'^{-1}\mathscr{E}^{\mathbb{R}}_{Z\to Y} \otimes g^{-1}_\pi \mathscr{E}^{\mathbb{R}}_{Y\to X} \to \mathscr{E}^{\mathbb{R}}_{Z\to X} \ ,$$

$$g^{-1}_\pi \mathscr{E}^{\mathbb{R}}_{X\leftarrow Y} \otimes {}^t f'^{-1}\mathscr{E}^{\mathbb{R}}_{Y\leftarrow Z} \to \mathscr{E}^{\mathbb{R}}_{X\leftarrow Z} \ ,$$

and the image of $1_{Z\to Y} \otimes 1_{Y\to X}$ *by the first morphism is* $1_{Z\to X}$. *Moreover if* $h: W \to Z$ *is another morphism of complex manifolds, the composition of morphisms on* $p_3^{-1}\mathscr{E}^{\mathbb{R}}_{W\to Z} \otimes p_2^{-1}\mathscr{E}^{\mathbb{R}}_{Z\to Y} \otimes p_1^{-1}\mathscr{E}^{\mathbb{R}}_{Y\to X}$ *is associative, and similarly with the arrows reversed. Here* p_1, p_2 *and* p_3 *are the maps from* $W \times_X T^*X$ *to* $Y \times_X T^*X$, $Z \times_Y T^*Y$ *and* $W \times_Z T^*Z$ *respectively.*

(ii) *The sheaf* $\mathscr{E}^{\mathbb{R}}_X$ *is naturally endowed with a structure of a ring with unit* 1_X *and* $\mathscr{E}^{\mathbb{R}}_{Y\to X}$ *(resp.* $\mathscr{E}^{\mathbb{R}}_{X\leftarrow Y}$*) is a* $({}^t f'^{-1}\mathscr{E}^{\mathbb{R}}_Y, f^{-1}_\pi \mathscr{E}^{\mathbb{R}}_X)$*-bimodule (resp. an* $(f^{-1}_\pi \mathscr{E}^{\mathbb{R}}_X, {}^t f'^{-1}\mathscr{E}^{\mathbb{R}}_Y)$*-bimodule).*

(iii) *Let* $F \in \mathrm{Ob}(\mathbf{D}^b(\mathbb{C}_X))$ *and let* $j \in \mathbb{Z}$. *Then* $H^j(\mu hom(F, \mathcal{O}_X))$ *(resp.* $H^j(\mu hom(F, \Omega_X))$*) is naturally endowed with a structure of a left (resp. right)* $\mathscr{E}^{\mathbb{R}}_X$*-module.*

Proof. Using Lemma 11.4.3 and its notations, we get the morphism:

$$Rp^a_{13!}(p^{a-1}_{12}\mu hom(\mathbb{C}_{\Delta_f}, \Omega_{X\times Y/Y}) \times p^{a-1}_{23}\mu hom(\mathbb{C}_{\Delta_g}, \Omega_{Y\times Z/Z}))$$
$$\to \mu hom(\mathbb{C}_{\Delta_f} \circ \mathbb{C}_{\Delta_g}, \Omega_{X\times Z/Z})[-\dim_{\mathbb{C}} Y] \ .$$

Now $\mathbb{C}_{\Delta_f} \circ \mathbb{C}_{\Delta_g}$ is isomorphic to $\mathbb{C}_{\Delta_{f \circ g}}$ and its remains to take the cohomology of both sides. The associativity of the composition is left to the reader and the fact that $1_{Z \to Y} \otimes 1_{Y \to X}$ gives $1_{Z \to X}$ is a classical calculus on fundamental classes. The second morphism is defined similarly.

(ii) follows from (i).

(iii) The proof is similar to that of (i). □

Remark 11.4.5. The sheaf of rings $\mathscr{E}_X^{\mathbb{R}}$ over T^*X is called the ring of **microlocal operators** on X. Its restriction to the zero-section X of T^*X is denoted $\mathscr{D}_X^\infty$ and is called the **ring of infinite-order differential operators**. This last ring naturally contains the ring $\mathscr{D}_X$ of (finite-order) differential operators of Section 2. In fact the ring $\mathscr{E}_X^{\mathbb{R}}$ contains a subring, denoted $\mathscr{E}_X$ and called the ring of (finite-order) microdifferential operators, whose restriction to X is $\mathscr{D}_X$. It is not our intention to introduce the ring $\mathscr{E}_X$ in this book, and we refer to Sato-Kawai-Kashiwara [1], Björk [1] or Schapira [2] for its study. Let us only mention that the ring $\mathscr{E}_X$ is much easier to manipulate than $\mathscr{E}_X^{\mathbb{R}}$. In particular it is coherent and Noetherian.

Let $\mathscr{M}$ be a coherent $\mathscr{D}_X$-module. Using the sheaf $\mathscr{E}_X$ (and the fact that $\mathscr{E}_X^{\mathbb{R}}$ is faithfully flat over $\mathscr{E}_X$) one can show:

$$\text{char}(\mathscr{M}) = \text{supp}\left(\mathscr{E}_X^{\mathbb{R}} \underset{\pi^{-1}\mathscr{D}_X}{\otimes} \pi^{-1}\mathscr{M}\right) . \tag{11.4.2}$$

Since $\text{char}(\mathscr{M}) = \text{char}(\mathscr{M}^*)$ (cf. (10.2.19)), one gets:

$$\begin{aligned} \text{char}(\mathscr{M}) &= \text{supp}\left(\mathscr{E}_X^{\mathbb{R}} \underset{\pi^{-1}\mathscr{D}_X}{\otimes} \pi^{-1}\mathscr{M}^*\right) \\ &= \text{supp}(R\mathscr{H}om_{\pi^{-1}\mathscr{D}_X}(\pi^{-1}\mathscr{M}, \mathscr{E}_X^{\mathbb{R}})) \\ &= \text{supp}(\mu_\Delta R\mathscr{H}om_{\mathscr{D}_X}(\mathscr{M}, \mathscr{O}_{X \times X})) . \end{aligned} \tag{11.4.3}$$

Then one easily completes the proof of Theorem 11.3.3 (cf. Kashiwara-Schapira, [3, Theorem 10.1.1]).

Now we shall apply Theorem 7.5.11. Let X, Y, Z be three complex manifolds, $\Lambda_1 \subset T^*(X \times Y)$ and $\Lambda_2 \subset T^*(Y \times Z)$ two conic Lagrangian complex submanifolds, $\tilde{p}_1 = (p_X, p_Y^a) \in \Lambda_1$, $\tilde{p}_2 = (p_Y, p_Z^a) \in \Lambda_2$. We make the hypotheses:

$$p_2^a|_{\Lambda_1} : \Lambda_1 \to T^*Y \textit{ and } p_1|_{\Lambda_2} : \Lambda_2 \to T^*Y \textit{ are transversal} . \tag{11.4.4}$$

Proposition 11.4.6. *Let* $K_1 \in \text{Ob}(\mathbf{D}^b(X \times Y; (p_X, p_Y^a)))$ *and* $K_2 \in \text{Ob}(\mathbf{D}^b(Y \times Z; (p_Y, p_Z^a)))$ *with* $SS(K_i) \subset \Lambda_i$ *in a neighborhood of* $\tilde{p}_i$ $(i = 1,2)$. *Assume* (11.4.4) *and also*: K_i *is simple with shift zero along* Λ_i $(i = 1, 2)$. *Then*:

(a) $K_1 \circ_\mu K_2 [-\dim_{\mathbb{C}} Y]$ *is simple with shift zero along* $\Lambda_1 \circ \Lambda_2$,

(b) *there is a natural morphism of* $(p_1^{-1}\mathscr{E}_X^{\mathbb{R}}, p_3^{-1}\mathscr{E}_Z^{\mathbb{R}})$*-bimodule in a neighborhood of* (p_X, p_Z^a):

$$(11.4.5)\quad p^a_{13*}\left(p^{a-1}_{12}H^0(\mu hom(K_1,\Omega_{X\times Y/Y}))\underset{p_2^{-1}\mathscr{E}_Y^{\mathbb{R}}}{\otimes} p^{a-1}_{23}H^0(\mu hom(K_2,\Omega_{Y\times Z/Z}))\right)$$

$$\to H^0\left(\mu hom\left(K_1\underset{\mu}{\circ}K_2[-\dim_{\mathbb{C}}Y],\Omega_{X\times Z/Z}\right)\right),$$

where p^a_{ij} *is the composite of the projection* p_{ij} *and the antipodal map on the j-th space.*

Proof. (a) follows from Theorem 7.5.11 (cf. Exercise VII.5).

(b) By Proposition 7.3.1, we may assume that (K_1,K_2) satisfies (7.3.3) and (7.3.4), and $K_1\circ_\mu K_2=\text{“}\varinjlim_V\text{”}\,(K_1)_{X\times V}\circ K_2$, where V ranges over an open neighborhood system of $\pi(p_Y)$.

By Lemma 11.4.3, we have morphisms

$$Rp^a_{13!}(p^{a-1}_{12}\mu hom(K_1,\Omega_{X\times Y/Y})\otimes p^{a-1}_{23}\mu hom(K_2,\Omega_{Y\times Z/Z}))$$

$$\to \mu hom(K_1\circ K_2[-\dim_{\mathbb{C}}Y],\Omega_{X\times Z/Z})$$

$$\to \mu hom\left(K_1\underset{\mu}{\circ}K_2[-\dim_{\mathbb{C}}Y],\Omega_{X\times Z/Z}\right).$$

Now apply the same lemma, with X, Y, Y, Z and K_1, $\mathbb{C}_{\Delta_Y}$ and K_2. Then the two morphisms α and β from $Rp^a_{13!}(p^{a-1}_{12}\mu hom(K_1,\Omega_{X\times Y/Y})\otimes\mathscr{E}_Y^{\mathbb{R}}\otimes p^{a-1}_{23}(K_2,\Omega_{Y\times Z/Z}))$ to $\mu hom(K_1\circ_\mu K_2[-\dim_{\mathbb{C}}Y],\Omega_{X\times Z/Z})$ obtained by $\mu hom(K_1,\Omega_{X\times Y/Y})\otimes\mathscr{E}_Y^{\mathbb{R}}\to\mu hom(K_1,\Omega_{X\times Y/Y})$ and $\mathscr{E}_Y^{\mathbb{R}}\otimes\mu hom(K_2,\Omega_{Y\times Z/Z})\to\mu hom(K_2,\Omega_{Y\times Z/Z})$ are equal. Hence taking the cohomology groups, we obtain the result. □

With the help of Proposition 11.4.6, we are now ready to make operate complex contact transformations on the image of $\mathcal{O}_X$ in $\mathbf{D}^b(X;p)$.

Let Y be a complex manifold of the same dimension as X and let U_X and U_Y be open subsets of T^*X and T^*Y, respectively. Let $\Lambda\subset U_X\times U_Y^a$ be a closed complex Lagrangian submanifold and assume:

$$(11.4.7)\quad p_1|_\Lambda:\Lambda\to U_X\quad\text{and}\quad p_2^a|_\Lambda:\Lambda\to U_Y\qquad\text{are isomorphisms}.$$

Denote by χ the complex contact transformation associated to Λ, that is, $\chi=(p_1|_\Lambda)\circ(p_2^a|_\Lambda)^{-1}$. Let $K\in\mathrm{Ob}(\mathbf{D}^b(X\times Y))$ satisfying:

$$(11.4.8)\quad\begin{cases}K\text{ is cohomologically constructible},\\ (p_1^{-1}(U_X)\cup p_2^{a-1}(U_Y))\cap SS(K)\subset\Lambda,\\ K\text{ is simple with shift }0\text{ along }\Lambda.\end{cases}$$

Note that given Λ satisfying (10.4.7), there always locally exists K satisfying (10.4.8): this follows by the same argument as for the proof of Corollary 7.2.2.

By Theorem 7.2.1, the functor $\Phi_K : G \to Rq_{1!}(K \otimes q_2^{-1}G)$ from $\mathbf{D}^b(Y; U_Y)$ to $\mathbf{D}^b(X; U_X)$ is well-defined and is an equivalence of categories. Let $p = (p_X, p_Y^a) \in \Lambda$ and let:

$$s \in H^0(\mu hom(K, \Omega_{X \times Y/X}))_p \ . \tag{11.4.9}$$

By Theorem 6.1.2, s defines a morphism:

$$K \to \Omega_{X \times Y/X} \qquad \text{in} \qquad \mathbf{D}^b(X \times Y; p) \ .$$

We get the chain of morphisms in $\mathbf{D}^b(X; p_X)$:

$$\begin{aligned} \Phi_{K[n]}(\mathcal{O}_Y) &= Rq_{1!}(K \otimes q_2^{-1}\mathcal{O}_Y)[n] \\ &\to Rq_{1!}(\Omega_{X \times Y/X} \otimes q_2^{-1}\mathcal{O}_Y[n]) \\ &\to Rq_{1!}\Omega_{X \times Y/X}[n] \\ &\to \mathcal{O}_X \ . \end{aligned}$$

Hence, to such an s as in (11.4.9) we have associated a morphism:

$$\alpha(s) : \Phi_{K[n]}(\mathcal{O}_Y) \to \mathcal{O}_X \qquad \text{in} \qquad \mathbf{D}^b(X; p_X) \ . \tag{11.4.10}$$

Let us explain how to compose such morphisms.

Let Z be a third manifold of the same dimension, U_Z an open subset of T^*Z, $\Lambda' \subset U_Y \times U_Z^a$ a Lagrangian manifold, $K' \in \mathrm{Ob}(\mathbf{D}^b(Y \times Z))$ and assume that (U_Y, U_Z, Λ', K') satisfies the same hypothesis as (11.4.8). Let $p' = (p_Y, p_Z^a) \in \Lambda$ and let $s' \in H^0(\mu hom(K', \Omega_{Y \times Z/Y}))_{p'}$. Set $\Lambda'' = \Lambda \circ \Lambda'$, $K'' = K \circ K'[n]$. Then K'' is a simple sheaf with shift zero along Λ'' and

$$\Phi_{K[n]} \circ \Phi_{K'[n]} = \Phi_{K''[n]} \ .$$

We get the morphisms in $\mathbf{D}^b(X; p)$:

$$\Phi_{K''[n]}(\mathcal{O}_Z) \xrightarrow[\Phi_{K[n]}(\alpha(s'))]{} \Phi_{K[n]}(\mathcal{O}_Y) \xrightarrow[\alpha(s)]{} \mathcal{O}_X \ .$$

Proposition 11.4.7. *One has:*

$$\alpha(s) \circ \Phi_{K[n]}(\alpha(s')) = \alpha(s \circ s') \ ,$$

where $s \circ s'$ is the image of $s \otimes s'$ by the morphism of Proposition 11.4.6.

For the proof we refer to Kashiwara-Schapira [3, Lemma 11.1.4].

Corollary 11.4.8. *Let $p \in T^*X$. There is a natural morphism of rings:*

$$\mathcal{E}^{\mathbb{R}}_{X,p} \to \mathrm{Hom}_{\mathbf{D}^b(X;p)}(\mathcal{O}_X, \mathcal{O}_X) \ .$$

Proof. This is a particular case of Proposition 11.4.7 when $\Lambda = \Lambda' = T^*_\Delta(X \times X)$ and $K = K' = \mathbb{C}_\Delta[-n]$. □

Theorem 11.4.9. (i) *Assume* (11.4.7), (11.4.8) *and let* $p = (p_X, p_Y^a) \in \Lambda$. *Then there exists* $s \in H^0(\mu hom(K, \Omega_{X\times Y/X}))_p$ *such that*:

$$(11.4.11)\quad \begin{cases} \mathscr{E}^{\mathbb{R}}_{X,p_X} \ni P \mapsto Ps \quad \text{and} \quad \mathscr{E}^{\mathbb{R}}_{Y,p_Y} \ni Q \mapsto sQ \qquad \text{give isomorphisms} \\ \mathscr{E}^{\mathbb{R}}_{X,p_X} \xrightarrow{\sim} H^0(\mu hom(K, \Omega_{X\times Y/X}))_p \xleftarrow{\sim} \mathscr{E}^{\mathbb{R}}_{Y,p_Y} \ . \end{cases}$$

(ii) *For such an s, the morphism* $\alpha(s): \Phi_{K[n]}(\mathcal{O}_Y) \to \mathcal{O}_X$ *is an isomorphism in* $\mathbf{D}^b(X; p_X)$, *and* $\alpha(s)$ *is compatible with the action of* $\mathscr{E}^{\mathbb{R}}_{X,p_X}$ *on* $\mathcal{O}_X$ *in* $\mathbf{D}^b(X; p_X)$ *and* $\mathscr{E}^{\mathbb{R}}_{Y,p_Y}$ *on* $\mathcal{O}_Y$ *in* $\mathbf{D}^b(Y; p_Y)$.

Proof. (i) We shall admit here the existence of s satisfying (11.4.11). The construction of such an s is performed in Kashiwara [5] (cf. also Schapira [2, Chapter I, §5]). (In (loc. cit.), the construction is done in the frame of the rings $\mathscr{E}_X$ and $\mathscr{E}_Y$, but the arguments still hold with $\mathscr{E}^{\mathbb{R}}_X$ and $\mathscr{E}^{\mathbb{R}}_Y$.)

(ii) Let $K^* = r_* K$, where r is the canonical map $X \times Y \to Y \times X$ (cf. VII §2), and let $s' \in H^0(\mu hom(K^*, \Omega_{Y\times X/Y}))_{p'}$ with $p' = (p_Y, p_X^a)$, s' satisfying condition (11.4.11). Then $s'' = s \circ s'$ defines an automorphism of $\mathscr{E}^{\mathbb{R}}_{X,p_X}$. Since $K \circ K^* \simeq \mathbb{C}_\Delta$ (cf. the proof of Theorem 7.2.1), $\mu hom(K \circ K^*[-n], \Omega_{X\times X/X})$ is isomorphic to $\mathscr{E}^{\mathbb{R}}_X$ and $s \circ s'$ is invertible. By Proposition 11.4.6 this implies that $\alpha(s)$ has a right inverse. One proves similarly that $\alpha(s)$ has a left inverse. Hence $\alpha(s)$ is an isomorphism.

Finally, again by Proposition 11.4.7, we have if $P \in \mathscr{E}^{\mathbb{R}}_{X,p_X}$, $Q \in \mathscr{E}^{\mathbb{R}}_{Y,p_Y}$:

$$\alpha(s) \circ \Phi_{K[n]}(Q) = \alpha(s \circ Q) = \alpha(P \circ s) = P \circ \alpha(s) \ . \quad \square$$

Definition 11.4.10. *The isomorphism* $\alpha(s)$ *of Theorem* 11.4.9 *is called a "quantized contact transformation" above* χ.

Any contact transformation may be locally quantized, but the "quantization" is not unique, (there are many sections s satisfying (11.4.11)).

Corollary 11.4.11. *In the situation of Theorem* 11.4.9, *let s satisfying* (11.4.11). *Then for any* $G \in \mathrm{Ob}(\mathbf{D}^b(Y))$, $\alpha(s)$ *defines an isomorphism in a neighborhood of* $p_X : \chi_* \mu hom(G, \mathcal{O}_Y) \simeq \mu hom(\Phi_{K[n]}(G), \mathcal{O}_X)$.

Recall that χ is the contact transformation associated to Λ.

Proof. Apply Theorems 7.2.1 and 11.4.9. □

This Corollary has many applications, in particular when choosing $G = \mathbb{C}_N$ for a real submanifold N of Y, but we shall not develop it here (cf. Kashiwara-Schapira [3]).

Remark 11.4.12. For $F \in \mathrm{Ob}(\mathbf{D}^b(X))$, $H^j(\mu hom(F, \mathcal{O}_X))$ has the structure of an $\mathcal{E}_X^{\mathbb{R}}$-module as shown in Proposition 11.4.3. But we do not know if $\mu hom(F, \mathcal{O}_X)$ can be defined as an object of $\mathbf{D}(\mathcal{E}_X^{\mathbb{R}})$.

11.5. Microfunctions

Let M be a real analytic manifold of dimension n, X a complexification of M.

Definition 11.5.1. *One sets:*

$$\mathcal{C}_M = H^n(\mu_M(\mathcal{O}_X) \otimes or_{M/X}) \ ,$$

$$\mathcal{B}_M = \mathcal{C}_M|_M = H^n_M(\mathcal{O}_X) \otimes or_{M/X} \ .$$

One calls $\mathcal{C}_M$ *(resp.* $\mathcal{B}_M$*) the sheaf of Sato microfunctions (resp. hyperfunctions) on* M.

As usual, in the definition of $\mathcal{C}_M$ we wrote $or_{M/X}$ instead of $\pi^{-1}or_{M/X}$, for short.

Proposition 11.5.2. (i) *The complex* $\mu_M(\mathcal{O}_X)$ *is concentrated in degree* n.

(ii) *The sheaf* $\mathcal{C}_M|_{\dot{T}^*_M X}$ *is conically flabby* (*i.e. its direct image on* $\dot{T}^*_M X/\mathbb{R}^+$ *is flabby*).

(iii) *The sheaf* $\mathcal{B}_M$ *is flabby.*

(iv) *There is a natural exact sequence of sheaves on* M:

$$0 \to \mathcal{A}_M \to \mathcal{B}_M \to \dot{\pi}_* \mathcal{C}_M \to 0 \ . \tag{11.5.1}$$

Recall that $\mathcal{A}_M = \mathcal{O}_X|_M$ is the sheaf of real analytic functions on M.

Proof. (i) follows from assertion (2.9.14) and Theorem 11.1.3.

(ii) Using Corollary 11.4.11, one can show that $\mathcal{C}_M|_{\dot{T}^*_M X}$ is locally isomorphic to a sheaf $H^1(\mu_N(\mathcal{O}_X))$ where N is the smooth boundary of a strictly pseudo-convex open subset Ω of X. If γ denotes the map $\dot{T}^*_M X \to \dot{T}^*_M X/\mathbb{R}^+$, this implies that $\gamma_* \mathcal{C}_M$ is locally isomorphic to $H^1_{(X \setminus \Omega)}(\mathcal{O}_X|_{\bar{\Omega}})$, and this sheaf is flabby by (2.9.16).

(iii) Let Z be a locally closed subset of M. Then $R\Gamma_Z \mathcal{B}_M = R\Gamma_Z(\mathcal{O}_X) \otimes or_{M/X}[n]$ is concentrated in degree zero, by Theorem 11.1.3.

(iv) is a particular case of the distinguished triangle (4.3.1). □

Note that it follows from Corollary 11.4.11 that one can make locally operate real analytic contact transformations on the sheaf $\mathcal{C}_M$.

Denote by sp the isomorphism:

$$\mathrm{sp} : \mathcal{B}_M \xrightarrow{\sim} \pi_* \mathcal{C}_M \ . \tag{11.5.2}$$

Definition 11.5.3. *Let u be a hyperfunction on M. The support of* $\mathrm{sp}(u)$ *in* T_M^*X *is called the singular support of u and denoted* $\mathrm{SS}(u)$.

Hence $\mathrm{SS}(u)$ is a closed conic subset of T_M^*X and $\mathrm{SS}(u) \subset M \times_X T_X^*X$ if and only if u is real analytic.

Let us explain the notion of **boundary value** of a holomorphic function (cf. Schapira [3]). Let Ω be an open subset of X satisfying:

$$\text{(11.5.3)} \quad \begin{cases} \bar{\Omega} \supset M \text{ and the inclusion } j : \Omega \hookrightarrow X \text{ is homeomorphic to} \\ \text{the inclusion of an open convex subset into } \mathbb{R}^{2n}\text{, locally on } X \ . \end{cases}$$

By this hypothesis, $R\mathcal{H}om(\mathbb{C}_{\bar{\Omega}}, \mathbb{C}_X) \simeq \mathbb{C}_\Omega$. Applying the functor $R\mathcal{H}om(\cdot, \mathbb{C}_X)$ to the morphism $\mathbb{C}_{\bar{\Omega}} \to \mathbb{C}_M$ we get the morphism in $\mathbf{D}^b(X)$:

$$\text{(11.5.4):} \qquad \omega_{M/X} \to \mathbb{C}_\Omega \ .$$

(Recall that $\omega_{M/X} \simeq R\mathcal{H}om(\mathbb{C}_M, \mathbb{C}_X) \simeq or_{M/X}[-n]$.)

Applying the functor $\mu hom(\cdot, \mathcal{O}_X)$ to (11.5.4) we get a morphism that we still denote by b:

$$\text{(11.5.5)} \qquad \mathrm{b} : \mu hom(\mathbb{C}_\Omega, \mathcal{O}_X) \to \mu_M(\mathcal{O}_X) \otimes or_{M/X}[n]$$

and in particular:

$$\text{(11.5.6)} \qquad \mathrm{b} : j_* j^{-1}\mathcal{O}_X \to \mathcal{B}_M \ .$$

It is this last morphism which is usually called the "boundary value morphism".

If $f \in \Gamma(\Omega; \mathcal{O}_X)$, f defines an element of $H^0(T^*X; \mu hom(\mathbb{C}_\Omega, \mathcal{O}_X))$. Since $\mu hom(\mathbb{C}_\Omega, \mathcal{O}_X)$ is supported by $\mathrm{SS}(\mathbb{C}_\Omega)$, we get by (11.5.5):

$$\text{(11.5.7)} \qquad \mathrm{SS}(\mathrm{b}(f)) \subset T_M^*X \cap \mathrm{SS}(\mathbb{C}_\Omega) \ ,$$

(cf. Delort-Lebeau [1] for further developments).

Note that any hyperfunction may be obtained as a sum of boundary value of holomorphic functions. When one "translates" the flabbiness of the sheaf $\mathscr{C}_M$ in terms of boundary values, one obtain the famous *edge of the wedge theorem.* We refer to Martineau [2].

Let us now study the microfunction solutions of $\mathcal{D}_X$-modules. In fact most of the results we shall state hold in the more general frame of $\mathcal{E}_X$ or even $\mathcal{E}_X^{\mathbb{R}}$-modules (recall that $\mathscr{C}_M$ is an $\mathcal{E}_X^{\mathbb{R}}$-module by Proposition 11.4.4), but we prefer to restrict ourselves to $\mathcal{D}_X$-modules for the sake of simplicity.

Proposition 11.5.4. *Let $\mathcal{M}$ be a coherent $\mathcal{D}_X$-module. Then:*

$$\text{(i)} \qquad \mathrm{SS}(R\mathcal{H}om_{\mathcal{D}_X}(\mathcal{M}, \mathscr{C}_M)) \subset C_{T_M^*X}(\mathrm{char}(\mathcal{M})) \ .$$

In particular:

$$\text{(ii)} \qquad \mathrm{supp}(R\mathcal{H}om_{\mathcal{D}_X}(\mathcal{M}, \mathscr{C}_M)) \subset T_M^*X \cap \mathrm{char}(\mathcal{M}) \ .$$

Proof. Apply Theorem 6.4.1 to $F = R\mathcal{H}om_{\mathcal{D}_X}(\mathcal{M}, \mathcal{O}_X)$ and Theorem 11.3.3. □

If P is a differential operator we see by this result that P defines an isomorphism of the sheaf $\mathcal{C}_M$ on the set $T_M^*X \setminus \{\sigma(P)^{-1}(0)\}$, and that if u is a hyperfunction then $\mathrm{SS}(u) \subset \mathrm{SS}(Pu) \cup \{\sigma(P)^{-1}(0)\}$.

Definition 11.5.5. (i) *A coherent $\mathcal{D}_X$-module $\mathcal{M}$ is called elliptic if $\dot{T}_M^*X \cap \mathrm{char}(\mathcal{M}) = \varnothing$.*

(ii) *Let $\theta \in T_{T_M^*X}T^*X$. One says that θ is micro-hyperbolic for $\mathcal{M}$ if $\theta \notin C_{T_M^*X}(\mathrm{char}(\mathcal{M}))$.*

If $\theta \in T^*M$ (recall the embedding $T^*M \hookrightarrow T_{T_M^*X}T^*X$, VI §2), *one simply says that θ is hyperbolic for $\mathcal{M}$.*

By Proposition 11.5.4, if $\mathcal{M}$ is elliptic then the morphism:

$$R\mathcal{H}om_{\mathcal{D}_X}(\mathcal{M}, \mathcal{A}_M) \to R\mathcal{H}om_{\mathcal{D}_X}(\mathcal{M}, \mathcal{B}_M)$$

is an isomorphism (apply $R\mathcal{H}om_{\mathcal{D}_X}(\mathcal{M}, \cdot)$ to the exact sequence 11.5.1).

If θ is micro-hyperbolic then $\theta \notin \mathrm{SS}(R\mathcal{H}om_{\mathcal{D}_X}(\mathcal{M}, \mathcal{C}_M))$. If θ is hyperbolic, then $\theta \notin \mathrm{SS}(R\mathcal{H}om_{\mathcal{D}_X}(\mathcal{M}, \mathcal{B}_M))$.

A classical matter in the theory of partial differential equations is that of **propagation of singularities.**

Let U be an open subset of T^*X, f a holomorphic function on U. Assume that $\mathrm{Im}\, f$ vanishes on T_M^*X. Then for each $p \in U \cap T_M^*X$, the bicharacteristic curve b_p passing through p of the involutive submanifold $\{\mathrm{Im}\, f = 0\}$ of $(T^*X)^{\mathbb{R}}$ is contained in T_M^*X in a neighborhood of p (cf. the Appendix).

Proposition 11.5.6. *Let $\mathcal{M}$ be a coherent $\mathcal{D}_X$-module and let f be a holomorphic function on $U \subset T^*X$. Assume:*

$$\mathrm{Im}\, f|_{T_M^*X \cap U} = 0 \ , \qquad \mathrm{char}(\mathcal{M}) \cap U \subset \{f = 0\} \ .$$

Then for any bicharacteristic b of the manifold $\{\mathrm{Im}\, f = 0\}$, the sheaves $\mathcal{E}xt^j_{\mathcal{D}_X}(\mathcal{M}, \mathcal{C}_M)|_b$ are locally constant, ($j \in \mathbb{N}$).

Proof. By Proposition 11.5.4, $\mathrm{SS}(R\mathcal{H}om_{\mathcal{D}_X}(\mathcal{M}, \mathcal{C}_M))$ is contained in $C_{T_M^*X}(\{f^{-1}(0)\}) = \{\theta \in T^*T_M^*X; \langle \theta, H_f \rangle = 0\}$. Therefore it is enough to apply Proposition 5.4.5 (ii). □

Examples 11.5.7. (i) Let X be a complex manifold. On $X^{\mathbb{R}}$ the $\mathcal{D}_{X \times \bar{X}}$-module $\mathcal{D}_X \boxtimes \mathcal{O}_{\bar{X}}$ is elliptic. This is the **Cauchy-Riemann system**. Note that:

$$\mathcal{O}_X \simeq R\mathcal{H}om_{\mathcal{D}_{\bar{X}}}(\mathcal{O}_{\bar{X}}, \mathcal{B}_{X^{\mathbb{R}}}) \simeq R\mathcal{H}om_{\mathcal{D}_{X \times \bar{X}}}(\mathcal{D}_X \boxtimes \mathcal{O}_{\bar{X}}, \mathcal{B}_{X^{\mathbb{R}}}) \ . \tag{11.5.8}$$

(ii) Let $(z) = (z_1, \ldots, z_n)$ be a system of holomorphic coordinates on X, $(z;\zeta)$ the associated coordinates on T^*X, with $z = x + \sqrt{-1}y$, $\zeta = \xi + \sqrt{-1}\eta$, and let $M = \{z \in X; y = 0\}$ so that $T_M^*X = \{y = \xi = 0\}$.

One says an operator P is elliptic or hyperbolic, etc., if the associated system $\mathcal{M} = \mathcal{D}_X/\mathcal{D}_X P$ is elliptic or hyperbolic, etc. Then P is elliptic if $\sigma(P)(x; i\eta) \neq 0$ for $\eta \neq 0$. In particular the **Laplace operator** $\Delta = \sum_{j=1}^n D_j^2$ is elliptic.

(iii) Let $(z;\zeta)$ be as above and let $p = (x_\circ; i\eta_\circ) \in T_M^*X$, $\theta_\circ \in (T_{T_M^*X}T^*X)_p$. Then $\theta_\circ$ is micro-hyperbolic for P at p if:

$$\text{(11.5.9)} \quad \begin{cases} \sigma(P)((x; i\eta) + \varepsilon\theta) \neq 0 \\ \text{for} \quad |x - x_\circ| + |\eta - \eta_\circ| + |\theta - \theta_\circ| + \varepsilon \ll 1 \,, \qquad \varepsilon > 0 \,. \end{cases}$$

In fact, using the local version of the Bochner's tube theorem (cf. Komatsu [2]) it is enough to check (11.5.9) for $\theta = \theta_\circ$. The **wave operator,** $\square = D_1^2 - \sum_{j=2}^n D_j^2$ satisfies (11.5.9) with $\eta_\circ = 0$, $\theta = dx_1$.

(iv) Assume $\sigma(P)$ is real on T_M^*X. If u is a microfunction solution of the equation $Pu = 0$, supp(u) will be a union of integral curves of the Hamiltonian field $H^{\mathbb{R}}_{\operatorname{Im}\sigma(P)}$. Recall that for a real function h on T^*X, $H_h^{\mathbb{R}}$ is given by (cf. 11.1.4):

$$H_h^{\mathbb{R}} = \frac{1}{2}\sum_{j=1}^n \left(\frac{\partial h}{\partial \xi_j}\frac{\partial}{\partial x_j} - \frac{\partial h}{\partial x_j}\frac{\partial}{\partial \xi_j} + \frac{\partial h}{\partial y_j}\frac{\partial}{\partial \eta_j} - \frac{\partial h}{\partial \eta_j}\frac{\partial}{\partial y_j}\right) .$$

Therefore:

$$H^{\mathbb{R}}_{\operatorname{Im}\sigma(P)}|_{T_M^*X} = \frac{1}{2}\sum_{j=1}^n \left(\frac{\partial \operatorname{Re}\sigma(P)}{\partial x_j}\frac{\partial}{\partial \eta_j} - \frac{\partial \operatorname{Re}\sigma(P)}{\partial \eta_j}\frac{\partial}{\partial x_j}\right) .$$

We shall now describe the main operations on microfunctions. Let N be another real analytic manifold, Y a complexification of N.

The natural morphism (Proposition 4.3.6):

$$\mu_M(\mathcal{O}_X) \boxtimes \mu_N(\mathcal{O}_Y) \to \mu_{M\times N}(\mathcal{O}_X \boxtimes \mathcal{O}_Y)$$

combined with the morphism $\mathcal{O}_X \boxtimes \mathcal{O}_Y \to \mathcal{O}_{X\times Y}$, defines the morphism:

$$\text{(11.5.10)} \qquad \mathscr{C}_M \boxtimes \mathscr{C}_N \to \mathscr{C}_{M\times N} \,.$$

Now let $f: N \to M$ be a real analytic map and still denote by f a complexification $f: Y \to X$. Consider the maps (cf. IV §3):

$$\text{(11.5.11)} \qquad T_N^*Y \xleftarrow[{}^tf'_N]{} N \underset{M}{\times} T_M^*X \xrightarrow[f_{N\pi}]{} T_M^*X \,.$$

By Proposition 4.3.5, there is a natural morphism:

$$R^tf'_{N!}(\omega_{N/M} \otimes f_{N\pi}^{-1}\mu_M(\mathcal{O}_X)) \to \mu_N(f^{-1}\mathcal{O}_X \otimes \omega_{Y/X}) \,.$$

Combining with the morphism $f^{-1}\mathcal{O}_X \to \mathcal{O}_Y$ we get the morphism:

$$\text{(11.5.12)} \qquad R^tf'_{N!}f_{N\pi}^{-1}\mathscr{C}_M \to \mathscr{C}_N \,.$$

It means that if u is a microfunction defined on an open subset U of T_M^*X and if W is an open subset of T_N^*Y such that: ${}^tf'_N: f_{N\pi}^{-1}(\operatorname{supp}(u)) \cap ({}^tf'_N)^{-1}W \to W$ is proper, then one can define the **inverse image** f^*u of u by f, as a microfunction on W. In particular, if u is a hyperfunction on M and ${}^t f'_N$ is proper on $f_{N\pi}^{-1}(\mathrm{SS}(u))$, then one can define the hyperfunction f^*u and $\mathrm{SS}(f^*u)$ is contained in ${}^t f'_N f_{N\pi}^{-1}(\mathrm{SS}(u))$. When $f: N \to M$ is an immersion, one usually writes $u|_N$ instead of f^*u.

To define direct images, introduce the sheaf $\mathscr{V}_M = \mathscr{A}_M^{(n)} \otimes or_M$ of analytic densities on M (and similarly for $\mathscr{V}_N$). By Proposition 4.3.4 there is a natural morphism:

$$Rf_{N\pi}{}^t f_N'^{-1} \mu_N(\Omega_Y) \to \mu_M(Rf_! \Omega_Y) \ .$$

Combining with the integration morphism (Theorem 11.1.4) we get the morphism:

$$Rf_{N\pi!}{}^t f_N'^{-1}\left(\mathscr{C}_N \underset{\mathscr{A}_N}{\otimes} \mathscr{V}_N\right) \to \mathscr{C}_M \underset{\mathscr{A}_M}{\otimes} \mathscr{V}_M \ . \tag{11.5.13}$$

In particular, if v is a "hyperfunction density" on N, i.e. $v \in \Gamma(N; \mathscr{B}_N \otimes_{\mathscr{A}_N} \mathscr{V}_N)$ and f is proper on $\mathrm{supp}(v)$, then one can define the **direct image** $f_* v$ of v by f, a hyperfunction density on M and $\mathrm{SS}(f_* v)$ is contained in $f_{N\pi}{}^t f_N'^{-1}(\mathrm{SS}(v))$. Notice that if $f: N \to M$ is smooth, $N \times_M T_M^*X$ is a closed submanifold V of T_N^*Y and $\mathrm{SS}(f_* v)$ is contained into the image by f_π of $\mathrm{SS}(v) \cap V$. One often writes $\int_f v$ instead of $f_* v$.

One can generalize the preceding constructions by considering the operations on microfunction solutions of microdifferential systems. However, we shall restrict ourselves to $\mathscr{D}_X$-modules and we shall only consider inverse images.

Let $f: N \to M$ be as precedingly. Using the morphism of (11.3.9) and Proposition 4.3.5, we get the natural morphism:

$$R^t f'_{N!} f_{N\pi}^{-1} R\mathscr{H}om_{\mathscr{D}_X}(\mathscr{M}, \mathscr{C}_M) \to R\mathscr{H}om_{\mathscr{D}_Y}(\underline{f}^{-1}\mathscr{M}, \mathscr{C}_N) \ . \tag{11.5.14}$$

In particular assume f is non-characteristic for $\mathscr{M}$. Then $\underline{f}^{-1}\mathscr{M}$ is concentrated in degree zero, and moreover ${}^t f'_N$ is finite on $f_{N\pi}^{-1}(\mathrm{supp}(R\mathscr{H}om_{\mathscr{D}_X}(\mathscr{M}, \mathscr{C}_M)))$, by Proposition 11.5.4, since ${}^t f'$ is finite on $f_\pi^{-1}(\mathrm{char}(\mathscr{M}))$. Hence, in this case, (11.5.14) induces for each $j \in \mathbb{N}$, the morphism:

$${}^t f'_{N!} f_{N\pi}^{-1} \mathscr{E}xt^j_{\mathscr{D}_X}(\mathscr{M}, \mathscr{C}_M) \to \mathscr{E}xt^j_{\mathscr{D}_Y}(\underline{f}^{-1}\mathscr{M}, \mathscr{C}_N) \ . \tag{11.5.15}$$

In other words, the inverse image f^*u of a microfunction u solution of a system $\mathscr{M}$ is a microfunction solution of the system $\underline{f}^{-1}\mathscr{M}$.

Applying Theorem 6.7.1 to $F = R\mathscr{H}om_{\mathscr{D}_X}(\mathscr{M}, \mathscr{O}_X)$ and using Theorem 11.3.5, we can solve the Cauchy problem for microfunction solutions of micro-hyperbolic system:

Proposition 11.5.8. *Let V be an open subset of T_N^*Y. Assume:*

(i) *f is non-characteristic for $\mathscr{M}$ (i.e.: $T_Y^*X \cap f_\pi^{-1}(\mathrm{char}(\mathscr{M})) \subset Y \times_X T_X^*X$),*
(ii) *$f_{N\pi}|_{{}^t f_N'^{-1}(V)}: {}^t f_N'^{-1}(V) \to T_M^*X$ is non-characteristic for $C_{T_M^*X}(\mathrm{char}(\mathscr{M}))$,*
(iii) *${}^t f'^{-1}(V) \cap f_\pi^{-1}(\mathrm{char}(\mathscr{M})) \subset Y \times_X T_M^*X$.*

Then the morphism (11.5.14) *is an isomorphism on V.*

Example 11.5.9. Let $(z;\zeta)$ denote the coordinates on T^*X, as in Example 11.5.7 (ii) and set $\zeta = (\zeta_1, \zeta')$. Let N be the hypersurface $\{x_1 = 0\}$ of M, and let P be a differential operator of order m, hyperbolic in the directions $\pm dx_1$, that is, satisfying:

$$\sigma(P)(x; i\eta + t\theta) \neq 0 \qquad \text{for any } t \in \mathbb{R}\backslash\{0\} \ , \qquad \text{any } (x;\eta) \in \mathbb{R}^n \times \mathbb{R}^n$$

with $\theta = (1, 0, \ldots, 0)$.

Consider the Cauchy problem:

$$\text{(11.5.16)} \qquad \begin{cases} Pu = 0 \ , & \gamma(u) = (w) \\ \text{with} & \gamma(u) = (u|_N, \ldots, \partial^{m-1}u/\partial^{m-1}x_1|_N) \ . \end{cases}$$

Applying Proposition 11.5.8 we obtain that for any section $(w) \in \mathscr{B}_N^n$ there is a unique section $u \in \mathscr{B}_M|_N$ solution of (11.5.16). Moreover $P: \mathscr{B}_M|_N \to \mathscr{B}_M|_N$ is surjective.

Exercises to Chapter XI

Exercise XI.1. (i) Let $\mathscr{M}$ be a left coherent $\mathscr{D}_X$-module and let $\mathscr{M}^*$ be defined as in (11.2.19). Prove the isomorphism:

$$\text{(11.E.1)} \qquad \Omega_X \overset{L}{\underset{\mathscr{D}_X}{\otimes}} \mathscr{M}[-n] \simeq R\mathscr{H}om_{\mathscr{D}_X}(\mathscr{M}^*, \mathscr{O}_X)$$

where $n = \dim_{\mathbb{C}} X$.

(ii) One sets $\mathrm{DR}(\mathscr{M}) = \Omega_X \otimes^L_{\mathscr{D}_X} \mathscr{M}$. Prove that if $\mathscr{M}$ is holonomic, $\mathrm{DR}(\mathscr{M})$ is a perverse sheaf on X.

Exercise XI.2. (i) Prove that the morphism (11.3.8) is an isomorphism if one assumes $\mathscr{M}$ or $\mathscr{N}$ holonomic.

(ii) Similarly let $\mathscr{M}$ (resp. $\mathscr{N}$) be a coherent $\mathscr{D}_X$-module (resp. $\mathscr{D}_Y$-module). Assume $\mathscr{M}$ or $\mathscr{N}$ is holonomic. Prove the isomorphism:

$$\text{(11.E.2)} \qquad \left(\Omega_X \overset{L}{\underset{\mathscr{D}_X}{\otimes}} \mathscr{M}\right) \boxtimes \left(\Omega_Y \overset{L}{\underset{\mathscr{D}_Y}{\otimes}} \mathscr{N}\right) \simeq \Omega_{X\times Y} \overset{L}{\underset{\mathscr{D}_{X\times Y}}{\otimes}} (\mathscr{M} \underline{\boxtimes} \mathscr{N}) \ .$$

(Hint: use Theorem 11.3.7 and some functional analysis.)

Exercise XI.3. Let M be a real analytic manifold of dimension n, X a complexification of M. Let $\mathscr{M}$ be a holonomic $\mathscr{D}_X$-module. Prove that:

$$\text{(11.E.3)} \qquad \mathscr{E}xt^j_{\mathscr{D}_X}(\mathscr{M}, \mathscr{C}_M) = 0 \qquad \text{for} \quad j \geqslant n \ .$$

(Hint: use Exercise X.8.)

Exercise XI.4. Let M be a real analytic manifold, X a complexification of M, K a compact subset of M. Prove that one can naturally identify $\Gamma_K(M; \mathcal{B}_M \otimes_{\mathcal{A}_M} \mathcal{V}_M)$ to $(\mathcal{O}_X(K))'$, the dual space of the topological vector space $\mathcal{O}_X(K) = \Gamma(K; \mathcal{O}_X)$. (In particular, $\Gamma_K(M; \mathcal{B}_M)$ is naturally endowed with a topology of an FS-space.)

See Sato [1], Martineau [1], Schapira [1].

Exercise XI.5. Let M be a real analytic manifold, X a complexification of M and let $\mathcal{M}$ be a coherent $\mathcal{D}_X$-module. Assume:

(11.E.4) *globally on X, $\mathcal{M}$ admits a finite free resolution* (*as* (11.2.16)) ,

(11.E.5) *$\mathcal{M}$ is elliptic.*

Let $\Omega \subset\subset M$ be a relatively compact open subset with smooth boundary. Assume:

(11.E.6) *$\partial\Omega$ is hyperbolic for $\mathcal{M}$* (*i.e.*: $\forall \theta \in \dot{T}^*_{\partial\Omega}M$, $\theta \notin C_{T^*_MX}(\text{char}(\mathcal{M}))$) .

Prove that the spaces $H^j(R\Gamma(\Omega, R\mathcal{H}om_{\mathcal{D}_X}(\mathcal{M}, \mathcal{B}_M)))$ are finite-dimensional.

(Hint: let $\mathcal{M}^\cdot$ be a finite free resolution of $\mathcal{M}$. Prove the isomorphism $R\text{Hom}_{\mathcal{D}_X}(\mathcal{M}^\cdot, \mathcal{A}_M(\bar{\Omega})) \simeq R\text{Hom}_{\mathcal{D}_X}(\mathcal{M}^\cdot, \mathcal{B}_M(\Omega))$. Then use that $\mathcal{B}_M(\Omega) \simeq (\Gamma_{\partial\Omega}(M; \mathcal{B}_M) \to \Gamma_{\bar{\Omega}}(M; \mathcal{B}_M))[1]$ is a complex of FS-spaces, $\mathcal{A}_M(\bar{\Omega})$ is a DFS-space and conclude by functional analysis.)

See Schapira-Schneiders [1] or Schapira [4] for further developments.

Exercise XI.7. Let X be a complex manifold, $\mathcal{M}$ a coherent $\mathcal{D}_X$-module. Prove that if $\theta \in T^*X$ is non-characteristic for $\mathcal{M}$ (i.e. $\theta \notin \text{char}(\mathcal{M})$) then θ, considered as a vector of $T^*X^{\mathbb{R}}$, is hyperbolic for $\mathcal{M} \underline{\boxtimes} \mathcal{O}_{\bar{X}}$ (i.e.: $\theta \notin C_{T^*_{X^{\mathbb{R}}}(X \times \bar{X})}(\text{char}(\mathcal{M}) \times T^*_{\bar{X}}\bar{X})$).

Exercise XI.8. Let X be a complex manifold, $\mathcal{M}$ a coherent $\mathcal{D}_X$-module satisfying (11.E.4) and let $\Omega \subset\subset X$ be an open subset with smooth boundary. Assume $\dot{T}^*_{\partial\Omega}X \cap \text{char}(\mathcal{M}) = \emptyset$. Prove that the spaces $H^j(R\Gamma(\Omega; R\mathcal{H}om_{\mathcal{D}_X}(\mathcal{M}, \mathcal{O}_X)))$ are finite-dimensional.

(Hint: use Exercises XI.6 and XI.7.)

See Bony-Schapira [1] for a particular case.

Exercise XI.9. Let $f: Y \to X$ be a morphism of complex manifolds.

(i) Prove that there is a natural morphism in $\mathbf{D}^b(\mathcal{D}_X^{op})$:

$$Rf_!\left(\Omega_Y \overset{L}{\underset{\mathcal{D}_Y}{\otimes}} \mathcal{D}_{Y\to X}\right)[\dim_{\mathbb{C}} Y] \to \Omega_X[\dim_{\mathbb{C}} X] .$$

(This is a refinement of Theorem 11.1.4, cf. Schneiders [1].)

(ii) Let $\mathcal{N}$ be a right coherent $\mathcal{D}_Y$-module. Construct the morphisms:

(11.E.7) $$Rf_!(R\mathcal{H}om_{\mathcal{D}_Y}(\mathcal{N}, \Omega_Y[\dim_{\mathbb{C}} Y])) \to R\mathcal{H}om_{\mathcal{D}_X}(\underline{f}_*\mathcal{N}, \Omega_X[\dim_{\mathbb{C}} X]) ,$$

(11.E.8) $$Rf_*(R\mathcal{H}om_{\mathcal{D}_Y}(\mathcal{N}, \Omega_Y[\dim_{\mathbb{C}} Y])) \to R\mathcal{H}om_{\mathcal{D}_X}(\underline{f}_!\mathcal{N}, \Omega_X[\dim_{\mathbb{C}} X]) .$$

(iii) Let $f : N \to M$ be a morphism of real analytic manifolds, and still denote by f a complexification $f : Y \to X$ of f. Let $\mathcal{N}$ be a coherent right $\mathcal{D}_Y$-module. Construct the natural morphism:

$$Rf_{N\pi!}{}^t f_N'^{-1} R\mathcal{H}om_{\mathcal{D}_Y}\left(\mathcal{N}, \mathcal{C}_N \underset{\mathcal{A}_N}{\otimes} \mathcal{V}_N\right) \to R\mathcal{H}om_{\mathcal{D}_X}\left(\underline{f}_*\mathcal{N}, \mathcal{C}_M \underset{\mathcal{A}_M}{\otimes} \mathcal{V}_M\right) .$$

(iv) In the situation of (iii), assume f is proper over $N \cap \operatorname{supp}(\mathcal{N})$. Construct the natural morphism:

(11.E.9)
$$\Gamma\left(N; \mathcal{H}om_{\mathcal{D}_Y}\left(\mathcal{N}, \mathcal{B}_N \underset{\mathcal{A}_N}{\otimes} \mathcal{V}_N\right)\right) \to \Gamma\left(M; \mathcal{H}om_{\mathcal{D}_X}\left(H^0(\underline{f}_*\mathcal{N}), \mathcal{B}_M \underset{\mathcal{A}_M}{\otimes} \mathcal{V}_M\right)\right)$$

(i.e.: if v is a hyperfunction density on N solution of the system $\mathcal{N}$, then $\int_f v$ is solution of the system $H^0(\underline{f}_*\mathcal{N})$).

Exercise XI.10. Let X be a complex manifold, N a real analytic submanifold of codimension d satisfying:

(11.E.10) $$TN \underset{N}{+} \sqrt{-1}TN = N \underset{X}{\times} TX .$$

Let Y be a complexification of N and consider the commutative diagrams of maps, where δ is the diagonal embedding:

$$\begin{array}{ccc} Y & \overset{g}{\hookrightarrow} & X \times \overline{X} \\ \uparrow & & \uparrow \delta \\ N & \underset{i}{\hookrightarrow} & X \end{array} \qquad \begin{array}{ccc} Y & \overset{g}{\hookrightarrow} & X \times \overline{X} \\ & {}_{f}\searrow & \downarrow q_1 \\ & & X \end{array} .$$

Set $\mathcal{M} = \underline{g}^{-1}(\mathcal{D}_X \boxtimes \mathcal{O}_{\overline{X}})$.

(i) Prove that: $\overline{Y} \times_X T^*X = \operatorname{char}(\mathcal{M})$. (Note that f is smooth.)

(ii) Prove the isomorphism:

$$f^{-1}\mathcal{O}_X \simeq R\mathcal{H}om_{\mathcal{D}_Y}(\mathcal{M}, \mathcal{O}_Y) .$$

(iii) Prove the isomorphism:

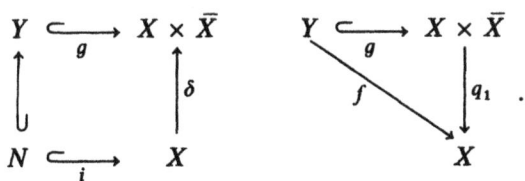

$$\mu_N(\mathcal{O}_X) \otimes or_{N/X}[d] \simeq R\mathcal{H}om_{\mathcal{D}_Y}(\mathcal{M}, \mathcal{C}_N) .$$

(Hint: use Corollary 6.7.4.)
See Kashiwara-Kawai [1].

Note: The $\mathcal{D}_Y$-module $\mathcal{M}$ is called the **induced Cauchy-Riemann system**.

Exercise XI.11. Let M be a real analytic manifold, N a submanifold, X a complexification of M, Y a complexification of N in X. Let $\mathcal{M}$ be a coherent $\mathcal{D}_X$-module such that Y is non-characteristic for $\mathcal{M}$.

(i) Prove the isomorphism on Y:

$$R\mathcal{H}om_{\mathcal{D}_Y}(\mathcal{M}_Y, \mathcal{O}_Y) \simeq R\mathcal{H}om_{\mathcal{D}_X}(\mathcal{M}, R\Gamma_Y(\mathcal{O}_X)) \otimes \omega_{Y/X}^{\otimes -1}$$

(use Corollary 5.4.11).

(ii) Prove the isomorphism on N:

$$R\mathcal{H}om_{\mathcal{D}_Y}(\mathcal{M}_Y, \mathcal{B}_N) \simeq R\mathcal{H}om_{\mathcal{D}_X}(\mathcal{M}, \Gamma_N(\mathcal{B}_M)) \otimes \omega_{N/M}^{\otimes -1} .$$

(iii) Let Ω be an open subset of M such that $\bar{\Omega} \supset N$ and the inclusion $\Omega \hookrightarrow M$ is locally homeomorphic to the inclusion of a convex open subset into $\mathbb{R}^n$. Construct the boundary value morphism:

$$\mathrm{b}: R\mathcal{H}om_{\mathcal{D}_X}(\mathcal{M}, \Gamma_\Omega(\mathcal{B}_M))|_N \to R\mathcal{H}om_{\mathcal{D}_Y}(\mathcal{M}_Y, \mathcal{B}_N) .$$

(Hint: use (11.5.4). See Schapira [3].)

Exercise XI.12. Let X be a complex manifold, U an open subset of T^*X, $p \in U$ and let $\mathcal{M}^\cdot$ be a bounded complex of finite free left $\mathcal{E}_X^{\mathbb{R}}$-modules on U:

$$\mathcal{M}^\cdot : 0 \to (\mathcal{E}_X^{\mathbb{R}})^{N_d} \longrightarrow \cdots \underset{P_0}{\longrightarrow} (\mathcal{E}_X^{\mathbb{R}})^{N_0} \longrightarrow 0 .$$

(The P_j's are matrices, acting on the right, of sections of $\mathcal{E}_X^{\mathbb{R}}$ on U.)

(i) Using Corollary 11.4.8, show that the complex:

$$0 \longrightarrow \mathcal{O}_X^{N_0} \underset{P_0}{\longrightarrow} \cdots \longrightarrow \mathcal{O}_X^{N_d} \longrightarrow 0$$

is well-defined in $\mathbf{D}^b(X;p)$. We denote it by $\mathrm{Sol}_p(\mathcal{M}^\cdot)$.

(ii) Prove that $SS(\mathrm{Sol}_p(\mathcal{M}^\cdot)) \subset \mathrm{supp}(\mathcal{M}^\cdot)$ in a neighborhood of p. (Recall that $\mathrm{supp}(\mathcal{M}^\cdot) = \bigcup_j \mathrm{supp}(H^j(\mathcal{M}^\cdot))$.)
See Kashiwara-Schapira [3, Chapter 11, §4].

Notes

It is beyond the scope of this book to discuss in detail the history of the theory of coherent $\mathcal{O}_X$-modules. Let us only emphasize that the names of Oka [1],

Cartan [2] and Serre [1] are attached to it, and refer to Hörmander [1] for a review.

Similarly, we shall not review the theory of $\mathscr{D}_X$-modules, and refer to the Notes in Schapira [2], simply recalling that this theory appeared in the 70's, with Kashiwara [1] and Bernstein [1, 2].

The classical Cauchy-Kowalewski theorem has been refined by Leray [3], who obtained Theorem 11.3.1. It was then extended to systems, in two different directions. One, due to Kashiwara [1], concerns the "Cauchy problem" (Theorem 11.3.5). The other one is related to "propagation": this is Theorem 11.3.3, due to Kashiwara-Schapira [2, 3] after partial results of Zerner [1], Bony-Schapira [1] and Kashiwara [5]. Theorem 11.3.7 was already obtained in Kashiwara [3]. It was the starting point of many important developments for which we refer to Björk [2].

The ring $\mathscr{E}_X$ of microlocal operators (Proposition 11.4.4), and the various modules naturally attached to it, were constructed by Sato-Kawai-Kashiwara [1] who developped Sato [2]'s idea of microlocalization and microfunctions. It is also Sato [1] who introduced in 1959 hyperfunctions as sums of boundary values of holomorphic functions and interpreted them in terms of local cohomology. The notion of microfunction now appears rather natural when studying these boundary values, and in particular when trying to understand the Martineau [2]'s "Edge of the Wedge theorem". It is an efficient tool in the study of linear partial differential equations, as shown in §5.

The theory of real quantized contact transformations goes back to Maslov [1], and has been systematically developped by Hörmander [2] in the C^∞-category, and by Sato-Kawai-Kashiwara [1] in the analytical case. These authors proved that the sheaves of microfunctions and microlocal operators are locally preserved by such transformations, which is a particular case of Corollary 11.4.11. Let us emphasize the fact that Theorem 11.4.9, due to Kashiwara-Schapira [3], is much stronger since it concerns the sheaf $\mathscr{O}_X$ itself, not its microlocalizations. Corollary 11.4.11 has nice applications to the vanishing of the cohomology of the microlocalization of $\mathscr{O}_X$ along real submanifolds, or of systems of microdifferential equations with simple characteristics. For the sake of brevity we have not included it here, and we refer to Kashiwara-Schapira [3, 4].

There is a wide literature dealing with analytic partial differential equations, and §5 is only a glance at this subject whose starting point is Proposition 11.5.4 (ii), due to Sato [2]. Typical features are Proposition 11.5.6 of Sato-Kawai-Kashiwara [1], which has now been considerably refined by many authors (cf. Tose [1] for a review) and hyperbolic equations, initiated by Bony-Schapira [2] (who treated the case of one operator), then developped by Kashiwara-Schapira [1] who obtained Propositions 11.5.4 (i) and 11.5.8. The same techniques are now used in the study of boundary value problems, which are not approached here (with the exception of formulas (11.5.5) and (11.5.7), due to Schapira [3]). Let us emphasize that the functor μhom allows us to define new sheaves of microfunctions and new wave front sets, well suited to boundary value problems (cf. Schapira (loc. cit.), Kataoka [1] and Uchida [1]).

To conclude, let us mention that when treating analytic singularities of distributions, most of people use the "Fourier-Bros-Iagolnitzer transformation" of Sjöstrand [1] (e.g. cf. Delort-Lebeau [1]). But it is now possible to extend the techniques of §4 and §5 to treat boundary values of holomorphic functions with moderate growth (i.e. distributions), by using the functor TH ("temperate homomorphisms") of Kashiwara [6], and its microlocalization, the functor $T\mu hom$, of Andronikof [1].

Appendix: Symplectic geometry

Summary

We collect here the basic tools of symplectic geometry which are used throughout the book. In §1 and §2 we discuss some basic notions concerning symplectic vector spaces and homogeneous symplectic manifolds. All results are well-known and more or less elementary. Hence we shall omit some proofs, and we refer to Arnold [2], Duistermaat [1], Abraham-Mardsen [1] and especially Hörmander [4 Chapter XXI].

In §3 we introduce the inertia index of a triplet of Lagrangian planes. Our presentation is close to that of Lion-Vergne [1] and for the reader's convenience, we give all proofs. We also collect some of its properties, that we need in Chapter 7, as exercises.

A.1. Symplectic vector spaces

In this section and the next one we shall treat symplectic spaces in the real case but the theory is the same in the complex case.

Let E be a real finite-dimensional vector space. A **symplectic form** σ on E is a non-degenerate alternate bilinear form on E. A vector space E endowed with a symplectic form σ is called a **symplectic vector space**. If E is symplectic, then $\dim E$ is even.

If (E_1, σ_1) and (E_2, σ_2) are two symplectic vector spaces, a linear map $u: E_1 \to E_2$ is called symplectic if $u^*\sigma_2 = \sigma_1$. If u is symplectic, then u is necessarily injective.

One denotes by $Sp(E)$ the group of symplectic automorphisms of a symplectic vector space E. It is a closed subgroup of the group $\mathrm{GL}(E)$ of linear automorphisms of E.

Example A.1.1. Let V be a real finite-dimensional vector space, and V^* its dual space. The space $E = V \oplus V^*$ is naturally endowed with a symplectic structure by setting for $(x;\xi)$ and $(x';\xi')$ in $V \oplus V^*$:

$$\sigma((x;\xi),(x';\xi')) = \langle x', \xi\rangle - \langle x, \xi'\rangle \ . \tag{A.1.1}$$

We call this form the natural symplectic form on E.

If $E = V \oplus V^*$ and u is a linear isomorphism of V, then the map $\begin{pmatrix} u & 0 \\ 0 & {}^t u^{-1} \end{pmatrix}$ is a symplectic automorphism of E.

Let (E, σ) be a symplectic vector space. Since σ is non-degenerate, it defines a linear isomorphism H from E^* to E by the formula:

$$\text{(A.1.2)} \qquad \langle \theta, v \rangle = \sigma(v, H(\theta)) \,, \qquad v \in E, \quad \theta \in E^* \,.$$

This isomorphism H is called the **Hamiltonian isomorphism**. If $\theta \in E^*$ one sometimes writes H_θ instead of $H(\theta)$ and calls H_θ the Hamiltonian vector of θ.

Since H is an isomorphism, the skew bilinear form on $E^*: (u, v) \mapsto \sigma(H_u, H_v)$ is a symplectic form on E^*. It is called the **Poisson bracket**, and denoted $\{u, v\}$. Thus:

$$\text{(A.1.3)} \qquad \{u, v\} = \sigma(H_u, H_v) = \langle v, H_u \rangle \,.$$

Let ρ be a linear subspace of E. One sets:

$$\text{(A.1.4)} \qquad \rho^\perp = \{x \in E; \sigma(x, \rho) = 0\} \,.$$

Then:

$$\rho^{\perp\perp} = \rho \,, \qquad (\rho_1 + \rho_2)^\perp = \rho_1^\perp \cap \rho_2^\perp \,, \qquad (\rho_1 \cap \rho_2)^\perp = \rho_1^\perp + \rho_2^\perp \,.$$

The space $\rho^\perp$ is called the orthogonal space to ρ.

Definition A.1.2. *A linear subspace ρ of E is called isotropic (resp. Lagrangian, resp. involutive) if $\rho \subset \rho^\perp$ (resp. $\rho = \rho^\perp$, resp. $\rho \supset \rho^\perp$).*

Some authors use "co-isotropic" instead of "involutive".

Note that if ρ is isotropic (resp. Lagrangian, resp. involutive) then $\dim \rho \leqslant n$ (resp. $= n$, resp. $\geqslant n$) where $n = \frac{1}{2} \dim E$.

A line (resp. a hyperplane) is always isotropic (resp. involutive). If $\dim \rho = n$ and if ρ is isotropic or else involutive, then ρ is Lagrangian.

Assume ρ is isotropic. Then the space $\rho^\perp/\rho$ is naturally endowed with a symplectic structure by setting $\sigma(\dot{x}, \dot{y}) = \sigma(x, y)$, where $\dot{x}$ (resp. $\dot{y}$) is the image of x (resp. y) in $\rho^\perp/\rho$. (We still denote by σ the symplectic form on $\rho^\perp/\rho$.)

If λ is a linear subspace of E one sets:

$$\text{(A.1.5)} \qquad \lambda^\rho = ((\lambda \cap \rho^\perp) + \rho)/\rho \,.$$

In particular, $E^\rho = \rho^\perp/\rho$. Then it is easily checked that:

$$\text{(A.1.6)} \qquad (\lambda^\perp)^\rho = (\lambda^\rho)^\perp \,.$$

In particular, if λ is Lagrangian in E, then λ^ρ is Lagrangian in E^ρ.

Conversely, denote by i the embedding $\rho^\perp \hookrightarrow E$ and by j the projection $\rho^\perp \to \rho^\perp/\rho$. Then if μ is a Lagrangian subspace of $\rho^\perp/\rho$, $ij^{-1}(\mu)$ is a Lagrangian subspace of E, containing ρ.

The symplectic space described in Example A.1.1 is not so special. In fact, we have:

Proposition A.1.3. *Let λ_0 be a Lagrangian subspace of E. Then there exists a Lagrangian subspace λ_1 of E such that $E = \lambda_0 \oplus \lambda_1$. Moreover for such a Lagrangian space λ_1, the map $u: \lambda_0 \oplus \lambda_0^* \to E$ defined by $u(x, y) = x - H(y)$ is a symplectic isomorphism. Here $H(y)$ is given by $\lambda_0^* \simeq (E/\lambda_1)^* \to E^* \underset{H}{\to} E$.*

Proof. (i) Let ρ be an isotropic space with $\rho \cap \lambda_0 = \{0\}$. If $\rho \neq \rho^\perp$, then $\rho^\perp \not\subset \lambda_0 + \rho$, otherwise $\rho \supset (\rho^\perp \cap \lambda_0)$, hence $\rho^\perp \cap \lambda_0 = \{0\}$, which contradicts $\dim \rho^\perp > n$. Choose $e \in \rho^\perp \backslash (\lambda_0 + \rho)$. Then $\rho + \mathbb{R}e$ is isotropic and $(\rho + \mathbb{R}e) \cap \lambda_0 = \{0\}$. Then we argue by induction on $\dim \rho$ to get λ_1.

(ii) Let x and x' belong to λ_0 and y and y' to λ_0^*. Then:

$$\sigma(x - H(y), x' - H(y')) = -\sigma(x, H(y')) + \sigma(x', H(y)) = -\langle y', x\rangle + \langle y, x'\rangle .$$

Hence the map u is symplectic. Since $\dim(\lambda_0 \oplus \lambda_0^*) = \dim E$, u is an isomorphism. □

Let (E, σ) be a symplectic vector space. We denote by E^a the space E endowed with the symplectic form $-\sigma$, i.e. $E^a = (E, -\sigma)$.

For two symplectic vector spaces (E_1, σ_1) and (E_2, σ_2), $E_1 \oplus E_2$ has also a structure of a symplectic space by $\sigma_1 \oplus \sigma_2$.

Let E_i, $(i = 1, 2, 3)$ be three symplectic vector spaces, and denote by p_{ij} the (i, j)-th projection defined on $E_1 \times E_2 \times E_3$ (e.g. p_{13} is the projection onto $E_1 \times E_3$).

Proposition A.1.4. *Let λ and μ be two Lagrangian subspaces of $E_1 \oplus E_2^a$ and $E_2 \oplus E_3^a$, respectively. Set $\lambda \circ \mu = p_{13}(p_{12}^{-1}\lambda \cap p_{23}^{-1}\mu)$. Then $\lambda \circ \mu$ is a Lagrangian subspace of $E_1 \oplus E_3^a$.*

Proof. The diagonal Δ of $E_2^a \oplus E_2$ is a Lagrangian subspace. Hence $\rho = \{0\} \times \Delta \times \{0\}$ is an isotropic subspace of $E_1 \oplus E_2^a \oplus E_2 \oplus E_3$. Since $E_1 \oplus E_3 = (E_1 \oplus E_2^a \oplus E_2 \oplus E_3)^\rho$ and $\lambda \circ \mu = (\lambda \oplus \mu)^\rho$, we get the result. □

Now we shall study symplectic bases. Let (E, σ) be a symplectic vector space, say of dimension $2n$.

A basis $(e_1, \dots, e_n; f_1, \dots, f_n)$ is called symplectic if denoting $(e_1^*, \dots, e_n^*; f_1^*, \dots, f_n^*)$ the dual basis on E^*, we have:

(A.1.7)
$$\sigma = \sum_{j=1}^{n} f_j^* \wedge e_j^* .$$

Of course, (A.1.7) is equivalent to:

$$\text{(A.1.8)} \qquad \begin{cases} \sigma(e_j, e_k) = \sigma(f_j, f_k) = 0 \ , \\ \sigma(e_j, f_k) = -\sigma(f_k, e_j) = -\delta_{j,k} \qquad 1 \leqslant j, k \leqslant n \ , \end{cases}$$

where δ_{jk} denotes the Kronecker symbol ($\delta_{jk} = 1$ if $j = k$, and is 0 otherwise).

For such a symplectic basis, the symplectic isomorphism H defined in (A.1.2) satisfies:

$$\text{(A.1.9)} \qquad H(e_j^*) = -f_j \ , \qquad H(f_j^*) = e_j \ , \qquad 1 \leqslant j \leqslant n \ .$$

Let J and K be two subsets of $\{1, \ldots, n\}$, and let $((e_j)_{j \in J}, (f_k)_{k \in K})$ be a linearly independent family satisfying the relations (A.1.8). Then it is easily proved that this family can be completed into a symplectic basis. In particular, if ρ is an isotropic (resp. Lagrangian, resp. involutive) subspace, then there exists a symplectic basis such that ρ is generated by $(e_1, \ldots, e_j)$ (resp. $(e_1, \ldots, e_n)$, resp. $(e_1, \ldots, e_n, f_1, \ldots, f_k)$) for some j and k.

Denote by $G(E, n)$ the Grassmannian manifold of n-dimensional linear subspaces of E (recall that $\dim E = 2n$). This is a compact manifold (cf. e.g. Griffith-Harris [1]). One denotes by $\Lambda(E)$ the subset of $G(E, n)$ consisting of Lagrangian subspaces. This is a closed (smooth) submanifold, called the **Lagrangian Grassmannian manifold**.

Let $\mu \in \Lambda(E)$. We set:

$$\text{(A.1.10)} \qquad \Lambda_\mu(E) = \{\lambda \in \Lambda(E); \lambda \cap \mu = \{0\}\}.$$

Assume E is endowed with a symplectic basis, and let $(x; \xi)$ denote the associated linear coordinates, (i.e. each $p \in E$ is written $p = \sum_{j=1}^n (x_j e_j + \xi_j f_j)$). For any $\lambda \in G(E, n)$, there exist $n \times n$ matrices A and B such that:

$$\text{(A.1.11)} \qquad \text{the matrix } (A, B) \text{ has rank } n \ , \qquad \lambda = \{(x; \xi); B\xi = Ax\} \ .$$

Then $\lambda \in \Lambda(E)$ if and only if $A^t B$ is symmetric.

Note that in this case $B\xi = Ax$ iff $x = {}^t Bz$, $\xi = {}^t Az$ for some $z \in \mathbb{R}^n$.

If μ is the Lagrangian subspace $\{x = 0\}$, then $\lambda \in \Lambda_\mu(E)$ if $\lambda = \{(x; \xi); \xi = Ax\}$ for some symmetric matrix A. Hence $\Lambda_\mu(E)$ is open and dense in $\Lambda(E)$ and isomorphic to $\mathbb{R}^{n(n+1)/2}$.

We shall sometimes say that a property "P" holds for **generic** λ (λ in $\Lambda(E)$) if there exists an open dense subset Ω of $\Lambda(E)$ such that the property "P" holds for $\lambda \in \Omega$.

We shall also encounter the following situation: λ is Lagrangian in E and contains a line ρ. We are looking for $\mu \in \Lambda(E)$, with $\mu \cap \lambda = \rho$. Consider the maps $i: \rho^\perp \to E$ and $j: \rho^\perp \to \rho^\perp/\rho$. Then it is enough to choose $\mu' \in \Lambda(E^\rho)$ with $\mu' \cap \lambda^\rho = 0$, and set $\mu = i(j^{-1}(\mu'))$. By abuse of language, we shall say that for generic μ with $\mu \supset \rho$, we have $\lambda \cap \mu = \rho$.

A.2. Homogeneous symplectic manifolds

All manifolds and morphisms of manifolds considered here will be real, of class C^∞ or real analytic. Unless otherwise specified, a real function on a manifold is supposed to be of class C^∞ or real analytic. However, most of the results we shall state still hold with suitable modifications for complex analytic manifolds.

Let X be a manifold. We denote by $\tau: TX \to X$ its tangent bundle and by $\pi: T^*X \to X$ its cotangent bundle. We denote by $\dot{T}X$ and $\dot{T}^*X$ the bundles TX and T^*X, with the zero-section removed, and we denote by $\dot{\tau}$ and $\dot{\pi}$ the maps τ and π restricted to $\dot{T}X$ and $\dot{T}^*X$, respectively. Let us recall that if M is a submanifold of X, the normal bundle T_MX and the conormal bundle T_M^*X to M in X are defined by the exact sequences of vector bundles on M:

$$\text{(A.2.1)} \qquad \begin{cases} 0 \to TM \to M \underset{X}{\times} TX \to T_MX \to 0 \ , \\ 0 \to T_M^*X \to M \underset{X}{\times} T^*X \to T^*M \to 0 \ . \end{cases}$$

Let $f: Y \to X$ be a morphism of manifolds. To f are associated the morphisms:

$$\text{(A.2.2)} \qquad \begin{cases} TY \xrightarrow[f']{} Y \underset{X}{\times} TX \xrightarrow[f_\tau]{} TX \ , \\ T^*Y \xleftarrow[{}^tf']{} Y \underset{X}{\times} T^*X \xrightarrow[f_\pi]{} T^*X \ . \end{cases}$$

In particular if one considers the projection $\pi: T^*X \to X$ we get the map ${}^t\pi'$: $T^*X \times_X T^*X \to T^*T^*X$.

If we restrict this map to the diagonal of $T^*X \times_X T^*X$, we get a map $T^*X \to T^*T^*X$, which is a section of the bundle $T^*T^*X \to T^*X$, that is, a differential form of degree 1 (one says: a 1-form). This 1-form on T^*X is called the **canonical 1-form** and denoted by α_X, or simply α if there is no risk of confusion.

Let $(x_1, \dots, x_n)$ be a system of local coordinates on X. Then the x_j's are real functions defined on some open subset U, satisfying $dx_1 \wedge \cdots \wedge dx_n \neq 0$ on U. At each $x \in U$, $(dx_1, \dots, dx_n)$ defines a basis of the vector space T_x^*X, and a vector $\xi \in T_x^*X$ is uniquely written as $\xi = \sum_{j=1}^n \xi_j dx_j$. The system $(x_1, \dots, x_n; \xi_1, \dots, \xi_n)$ is called the **coordinate system** on T^*X **associated** to the coordinate system $(x_1, \dots, x_n)$. It is easily checked that the canonical 1-form α_X is nothing but the form $\sum_{j=1}^n \xi_j dx_j$. Let $\sigma = d\alpha$. Hence $\sigma = \sum_{j=1}^n d\xi_j \wedge dx_j$ is a symplectic form on T^*X (i.e.: at each $p \in T^*X$, σ induces a symplectic structure on the vector space T_pT^*X). In other word the manifold T^*X is naturally endowed with a symplectic structure, by $d\alpha$. We can then extend to T^*X some of the notions introduced in §1.

A submanifold V of T^*X is called isotropic (resp. Lagrangian, resp. involutive) if at each $p \in V$, the tangent space T_pV has the corresponding property in T_pT^*X. If f is a real function defined on some open subset U of T^*X, the

Hamiltonian vector field H_f of f is the vector field on U, the image of df by the Hamiltonian isomorphism $H : T^*T^*X \simeq TT^*X$.

The **Poisson bracket** of two functions f and g is defined by:

$$\{f,g\} = H_f(g) = d\alpha(H_f, H_g) \ . \tag{A.2.3}$$

One checks the relations:

$$\begin{cases} \{f,g\} = -\{g,f\} \ , \\ \{f,hg\} = h\{f,g\} + g\{f,h\} \ , \\ \{\{f,g\},h\} + \{\{g,h\},f\} + \{\{h,f\},g\} = 0 \ . \end{cases} \tag{A.2.4}$$

In particular $[H_f, H_g] = H_{\{f,g\}}$ where $[u,v] = uv - vu$ is the commutator of the vector fields u and v.

If $(x_1,\ldots,x_n)$ is a system of local coordinates on X, $(x;\xi)$ the associated coordinates on T^*X and f is a real function on $U \subset T^*X$, we have:

$$H_f = \sum_{j=1}^{n} \left(\frac{\partial f}{\partial \xi_j}\frac{\partial}{\partial x_j} - \frac{\partial f}{\partial x_j}\frac{\partial}{\partial \xi_j} \right) . \tag{A.2.5}$$

A submanifold V of T^*X is involutive iff the Poisson bracket $\{f,g\}$ vanishes on V for any functions f and g which vanish on V. In fact the vector bundle $(TV)^\perp$ is generated by the vector fields H_f, with $f|_V = 0$. Thus $(TV)^\perp \subset TV$ is equivalent to $H_f(g) = 0$ for any f, g with $f|_V = 0$, $g|_V = 0$. Moreover by (A.2.4), we find that if V is involutive, the sub-bundle $(TV)^\perp$ of TV satisfies the Frobenius integrability conditions (i.e.: the sheaf of sections of $(TV)^\perp$ is closed under bracket $[\cdot,\cdot]$). By the Frobenius theorem (cf. Hörmander [4, Appendix C]), an involutive manifold V admits a foliation, and the leaves of this foliation are called the **bicharacteristic leaves** of V. Note that the dimension of the leaves is the codimension of V. In particular if V is Lagrangian, the leaves are open in V.

Example A.2.1. Let Z be a submanifold of X. The manifold $Z \times_X T^*X$ is involutive and the manifold $T^*_Z X$ is Lagrangian. Notice the extreme case where $Z = X$: we get the Lagrangian manifold $T^*_X X$, the zero-section of T^*X.

The 1-form α on T^*X induces a richer structure than merely that of a symplectic manifold, and we shall describe this **homogeneous symplectic structure** on T^*X.

Let $H(\alpha)$ be the image of α by the Hamiltonian isomorphism. If we have chosen coordinates $(x;\xi)$ as above, then:

$$\alpha = \sum_j \xi_j dx_j \ , \qquad H(\alpha) = -\sum_j \xi_j \frac{\partial}{\partial \xi_j} \ . \tag{A.2.6}$$

Thus $-H(\alpha)$ is just the radial vector field on the vector bundle T^*X (i.e. the infinitesimal generator of the action of $\mathbb{R}^+$ on T^*X). This vector field is also called the **Euler vector field**.

We say that a subset S of T^*X is **conic** (resp. **locally conic**) if it is invariant (resp. locally invariant) by the action of $\mathbb{R}^+$. Hence S is locally conic if its intersection with any orbit of $\mathbb{R}^+$ is open in this orbit. A function f defined on an open subset U of T^*X is said to be homogeneous if f satisfies the differential equation $H(\alpha)f = kf$ for some $k \in \mathbb{C}$. Note that a submanifold V is locally conic iff $H(\alpha)$ is tangent to V or equivalently iff V is locally defined by homogeneous equations.

A locally conic submanifold V is isotropic iff $\alpha|_V \equiv 0$, since $\langle \alpha, v \rangle = d\alpha(v, H(\alpha))$, for $v \in TT^*X$.

One says that a locally conic involutive submanifold V is **regular** if $\alpha|_V$ is everywhere different from zero. This is equivalent to the local existence of homogeneous functions $f_1, \ldots, f_r$ vanishing on V, with $r = \operatorname{codim} V$, such that:

$$\text{(A.2.7)} \qquad \begin{cases} \{f_i, f_j\} = 0 & \text{on } V \text{ for any } i,j \in \{1,\ldots,r\} \ , \\ df_1 \wedge \cdots \wedge df_r \wedge \alpha \neq 0 & \text{on } V \ . \end{cases}$$

This is again equivalent to saying that the Euler vector field is not tangent to any bicharacteristic leaves at any point.

Example A.2.2. Let Z be a submanifold of X. Then $Z \times_X T^*X$ is regular involutive outside of $T^*_Z X$.

Convention A.2.3. In this Appendix, unless otherwise specified, all submanifolds of T^*X are locally conic.

Let $\rho(p)$ denote the linear subspace of $T_p T^*X$ generated by the Euler vector field at p. If $p \in T^*_X X$, $\rho(p) = \{0\}$, otherwise $\rho(p)$ is a line.

If V is a (locally conic) submanifold, at each $p \in V$, $T_p V$ contains $\rho(p)$. Let $p \in T^*X$. One sets:

$$\text{(A.2.8)} \qquad \lambda_0(p) = T_p \pi^{-1}\pi(p) \ .$$

This is a Lagrangian linear subspace of $T_p T^*X$.

Let Λ be a Lagrangian submanifold. The corank of the projection $\pi|_\Lambda : \Lambda \to X$ is, by definition, the dimension of the space $T_p\Lambda \cap \lambda_0(p)$. On $\dot{T}^*X$, this corank is at least one since both $T_p\Lambda$ and $\lambda_0(p)$ contain $\rho(p)$. If this corank is constant, say d, then locally on Λ, $\pi(\Lambda)$ is a smooth submanifold M of X of codimension d, and $\Lambda = T^*_M X$. In particular if this corank is one at some p, then Λ is the conormal bundle to a hypersurface in a neighborhood of p.

Now let X and Y be two manifolds of the same dimension, U_X (resp. U_Y) an open subset of T^*X (resp. T^*Y). Let χ be a diffeomorphism from U_X onto U_Y. If $\chi^*(d\alpha_Y) = d\alpha_X$, one says that χ is a symplectic isomorphism. If moreover χ is homogeneous (i.e.: χ commutes with the action of $\mathbb{R}^+$), then $\chi^*(\alpha_Y) = \alpha_X$ and we shall say that χ is a **contact transformation**, although this is not really correct, since a contact structure is the structure obtained on the quotient space $\dot{T}^*X/\mathbb{R}^+$.

Let χ be a homogeneous diffeomorphism $U_X \xrightarrow{\sim} U_Y$, Λ_χ its graph in $U_X \times U_Y$. The inverse image $\chi^*(\beta)$ of a 1-form β on U_Y is characterized by the condition

$(\chi^*(\beta) - \beta)|_{\Lambda_\chi} \equiv 0$. Let Λ_χ^a denote the image of Λ_χ by the antipodal map on T^*Y. Then $\chi^*(\alpha_Y) = \alpha_X$ iff $(\alpha_X + \alpha_Y)|_{\Lambda_\chi^a} \equiv 0$, that is, iff Λ_χ^a is isotropic, hence iff it is Lagrangian. In other words, χ is a contact transformation iff Λ_χ^a is a (locally conic) Lagrangian submanifold of $T^*(X \times Y)$.

We call Λ_χ^a the Lagrangian manifold associated to the graph of the contact transformation χ.

Let (y) be a system of local coordinates on Y, and let $(y;\eta)$ denote the associated coordinates on T^*Y. Then a homogeneous map $\chi : U_X \to U_Y$ is defined by two sets of functions, f_j homogeneous of degree 0, g_k homogeneous of degree 1, $(1 \leqslant j, k \leqslant n)$, with $y_j = f_j$, $\eta_k = g_k$. The map χ is a contact transformation iff Λ_χ^a is involutive, that is, iff:

$$\{f_j, f_k\} = 0\ , \qquad \{g_j, g_k\} = 0\ , \qquad \{f_j, g_k\} = -\delta_{j,k}\ . \tag{A.2.9}$$

Example A.2.4. Let $(x;\xi)$ denote the coordinates on $T^*\mathbb{R}^n$, and let $\varphi(\xi)$ be a function homogeneous of degree one defined on some open subset U of $(\mathbb{R}^n)^*$ (e.g.: $\varphi(\xi) = (\sum_j \xi_j^2)^{1/2}$ on $\mathbb{R}^n \setminus \{0\}$). Then the map $\chi : (x;\xi) \mapsto (x + \varphi'(\xi); \xi)$ is a contact transformation.

The next result is a useful tool in order to construct contact transformations.

Proposition A.2.5. *Let V be a regular involutive submanifold of T^*X, $p \in V \cap \dot{T}^*X$. Let λ be a Lagrangian linear subspace of T_pT^*X such that $\rho(p) \subset \lambda \subset T_pV$. (Recall that $\rho(p)$ is the line generated by the Euler vector field.) Then there exists a Lagrangian manifold $\Lambda \subset T^*X$ such that $\Lambda \subset V$ and $T_p\Lambda = \lambda$.*

Proof. Let $n = \dim X$, $r = \operatorname{codim} V$. Let $(f_1, \dots, f_r)$ be a system of homogeneous functions vanishing on V and satisfying (A.2.7). If $r = n - 1$ we set $e = \alpha(p)$. Otherwise we choose $v \in \lambda^\perp \setminus (T_pV + \mathbb{R}H(\alpha))$ and set $e = H^{-1}(v)|_V$. By the classical theory of differential equations, we may find a function g on V such that:

$$\begin{cases} H(\alpha)|_V(g) = 0\ , & H_{f_j}|_V(g) = 0\ , \qquad (j \leqslant r) \\ dg(p) = e\ , & g(p) = 0\ , \end{cases}$$

because e is not tangent to the bicharacteristic leaf passing through p.

Set $V_1 = \{q \in V; g(q) = 0\}$. Then V_1 is a conic manifold which satisfies the required properties if $r = n - 1$ and otherwise V_1 is regular involutive. In this case we argue by induction on r. □

Let X, Y, Z be three manifolds. One denotes by p_1 and p_2 the first and second projection defined on $T^*X \times T^*Y$ or else on $T^*Y \times T^*Z$, and by p_{ij} the (i,j)-th projection defined on $T^*X \times T^*Y \times T^*Z$. We set $p_2^a = a \circ p_2$, where "a" is the antipodal map. Let $\Lambda_1 \subset T^*(X \times Y)$, $\Lambda_2 \subset T^*(Y \times Z)$ be two Lagrangian manifolds. Let $(p_X, p_Y^a) \in \Lambda_1$, $(p_Y, p_Z^a) \in \Lambda_2$, and assume:

$$\begin{cases} \textit{the maps } p_2^a|_{\Lambda_1} : \Lambda_1 \to T^*Y \textit{ and } p_1|_{\Lambda_2} : \Lambda_2 \to T^*Y \\ \textit{are transversal at } p_Y\ . \end{cases} \tag{A.2.10}$$

Then replacing Λ_1 and Λ_2 by $\Lambda_1 \cap U$ and $\Lambda_2 \cap V$, where U and V are sufficiently small open neighborhoods of (p_X, p_Y^a) and (p_Y, p_Z^a) respectively, the map p_{13} induces an isomorphism of $\Lambda_1 \times_{T^*Y} \Lambda_2$ with a Lagrangian manifold Λ of $T^*(X \times Z)$, and one sets:

$$\Lambda = \Lambda_1 \circ \Lambda_2 \ . \tag{A.2.11}$$

(Cf. Lemma 7.4.4 and Definition 7.4.5.)

Proposition A.2.6. *Let Λ_1 be a Lagrangian submanifold of $T^*(X \times Y)$, $(p_X, p_Y^a) \in \Lambda_1$ with $p_Y \notin T_Y^*Y$. Assume that the map $p_1|_{\Lambda_1} : \Lambda_1 \to T^*X$ is smooth. Then there exists a manifold Z of the same dimension as Y, and a Lagrangian manifold $\Lambda_2 \subset T^*(Y \times Z)$ defined in a neighborhood of (p_Y, p_Z^a), such that:*

(i) *$\Lambda_2 = T_S^*(Y \times Z)$, where S is a hypersurface of $Y \times Z$,*
(ii) *Λ_2 is associated to the graph of a contact transformation (i.e.: $p_1|_{\Lambda_2}$ and $p_2^a|_{\Lambda_2}$ are local isomorphisms),*
(iii) *$\Lambda_1 \circ \Lambda_2 = T_{S'}^*(X \times Z)$, where S' is a hypersurface of $X \times Z$.*

Proof. Set $E_X = T_{p_X}T^*X$, $E_Y = T_{p_Y}T^*Y$ and let E_Y^a denote the space E_Y endowed with the opposite symplectic structure. Then (p_1, p_2^a) defines a symplectic isomorphism $T_{(p_X, p_Y^a)}T^*(X \times Y) \simeq E_X \times E_Y^a$, and we shall identify a Lagrangian space in $T_{(p_X, p_Y^a)}T^*(X \times Y)$ and its image in $E_X \times E_Y^a$.

Let ρ_X denote the linear space generated by the Euler vector field at p_X in E_X and define similarly ρ_Y in E_Y, ρ_{XY} in $E_X \times E_Y^a$, ρ_{YY} in $E_Y \times E_Y^a$. Set:

$$\lambda_1 = T_{(p_X, p_Y^a)}\Lambda_1 \ , \qquad \lambda_{0X} = T_{p_X}\pi^{-1}\pi(p_X) \ , \qquad \lambda_{0Y} = T_{p_Y}\pi^{-1}\pi(p_Y) \ ,$$

and identify λ_1 and λ_{0X} to Lagrangian subspaces of $E_X \times E_Y^a$ and E_X, respectively.

By the hypothesis that $p_1|_{\lambda_1} : \lambda_1 \to E_X$ is surjective, we get that $p_2|_{\lambda_1} : \lambda_1 \to E_Y$ is injective (cf. Exercise A.4). Moreover $p_2^{-1}(\rho_Y) \cap \lambda_1 = \rho_{XY}$. Since $p_2(p_1^{-1}(\lambda_{0X}) \cap \lambda_1)$ is Lagrangian in E_Y (Proposition A.1.4), we have for a generic Lagrangian space $\lambda \subset E_Y$, with $\rho_Y \subset \lambda$:

$$\lambda \cap p_2(p_1^{-1}(\lambda_{0X}) \cap \lambda_1) = \rho_Y \ .$$

This implies $\rho_{XY} = p_2^{-1}(\lambda) \cap p_1^{-1}(\lambda_{0X}) \cap \lambda_1$, thus:

$$\begin{aligned} &\rho_{XY} = (\lambda_{0X} \times \lambda) \cap \lambda_1, \text{ for a generic Lagrangian space} \\ &\lambda \subset E_Y^a \text{ such that } \rho_Y \subset \lambda \ . \end{aligned} \tag{A.2.12}$$

Then for a generic Lagrangian space $\mu \subset E_Y \times E_Y^a$ with $\rho_{YY} \subset \mu$, we have:

$$p_1|_\mu : \mu \to E_Y \text{ and } p_2|_\mu : \mu \to E_Y^a \text{ are isomorphisms} \ , \tag{A.2.13}$$

$$(\lambda_{0Y} \times \lambda_{0Y}) \cap \mu = \rho_{YY} \ . \tag{A.2.14}$$

Since $\lambda = p_1(\mu \cap p_2^{-1}(\lambda_{0Y}))$ is generic we may assume further λ satisfies the condition (A.2.12). Then we have:

(A.2.15) $$(\lambda_1 \circ \mu) \cap (\lambda_{0X} \times \lambda_{0Y}) = \rho_{XY} .$$

Now take Y, p_Y and λ_{0Y} as Z, p_Z and λ_{0Z} respectively. By Proposition A.2.5 we may find a Lagrangian manifold $\Lambda_2 \subset T^*(Y \times Z)$ such that $T_{(p_Y, p_Z^a)}\Lambda_2 = \mu$. Then Λ_2 will satisfy all the required conditions, by (A.2.13), (A.2.14) and (A.2.15). □

Corollary A.2.7. *Let $\Lambda \subset \dot{T}^*X$ be a Lagrangian manifold, $p \in \Lambda$. Then there exists a contact transformation χ defined in a neighborhood of p such that $\chi(\Lambda)$ is the conormal bundle to a hypersurface, and moreover the Lagrangian manifold associated to the graph of χ is the conormal bundle to a hypersurface.*

Proof. This is nothing but Proposition A.2.5 when $X = \{\mathrm{pt}\}$. □

Corollary A.2.8. *Let $\Lambda \subset \dot{T}^*(X \times Y)$ be a Lagrangian manifold associated with a contact transformation (i.e.: $p_1|_\Lambda$ and $p_2^a|_\Lambda$ are local isomorphisms). Then, locally on Λ, there exists a manifold Z of the same dimension as X and two Lagrangian manifolds $\Lambda_1 \subset T^*(X \times Z)$, $\Lambda_2 \subset T^*(Z \times Y)$ such that Λ_1 and Λ_2 are associated with contact transformations, Λ_1 and Λ_2 are the conormal bundles to hypersurfaces of $X \times Z$ and $Z \times Y$ respectively, and $\Lambda = \Lambda_1 \circ \Lambda_2$.*

Proof. By Proposition A.2.6 there exists a contact transformation χ such that if Λ_2 is the Lagrangian manifold associated to the graph of χ, then Λ_2 and $\Lambda \circ \Lambda_2$ are the conormal bundles to hypersurfaces. Then $\Lambda = (\Lambda \circ \Lambda_2) \circ \Lambda_3$, where Λ_3 is the Lagrangian manifold associated to χ^{-1}. □

To end this section, let us recall the following well-known result.

Proposition A.2.9. *Let Λ be a conic submanifold of $\dot{T}^*X$, $p \in \Lambda$. Assume Λ is isotropic (resp. Lagrangian, resp. regular involutive). Then there exists a contact transformation χ defined in a neighborhood of p such that $\chi(p) = (0; dx_n) \in T^*\mathbb{R}^n$ and $\Lambda = \{(x;\xi); x = 0, \xi_1 = \cdots = \xi_r = 0\ (r < n)\}$ (resp. $\Lambda = \{(x;\xi); x = 0\}$, resp. $\Lambda = \{(x;\xi); \xi_1 = \cdots = \xi_p = 0, (p < n)\}$).*

The proof is left as an exercise.

A.3. Inertia index

Let (E, σ) be a real symplectic vector space of dimension $2n$ and let $\lambda_1, \lambda_2, \lambda_3$ be three Lagrangian subspaces. (In this section, unless otherwise specified, a subspace means a linear subspace.)

Definition A.3.1. *The inertia index of the triplet $(\lambda_1, \lambda_2, \lambda_3)$, denoted $\tau_E(\lambda_1, \lambda_2, \lambda_3)$* (or simply $\tau(\lambda_1, \lambda_2, \lambda_3)$) *is the signature of the quadratic form q defined on the $3n$*

dimensional vector space $\lambda_1 \oplus \lambda_2 \oplus \lambda_3$ *by*:

$$q(x_1, x_2, x_3) = \sigma(x_1, x_2) + \sigma(x_2, x_3) + \sigma(x_3, x_1) \ .$$

This index is also sometimes called the "Maslov index".

In a suitable basis of $\lambda_1 \oplus \lambda_2 \oplus \lambda_3$ one can represent q by a diagonal matrix whose diagonal entries consist of p_+-uples of $+1$, p_--uples of -1 and $3n - p_+ - p_-$-uples of 0. Then the **signature** of q, denoted $\operatorname{sgn}(q)$ is equal to $p_+ - p_-$.

The index τ has the following properties.

Theorem A.3.2. (i) $\tau(\lambda_1, \lambda_2, \lambda_3)$ *is alternating with respect to the permutations of the* λ_j's, *that is*:

$$\tau(\lambda_1, \lambda_2, \lambda_3) = -\tau(\lambda_2, \lambda_1, \lambda_3) = -\tau(\lambda_1, \lambda_3, \lambda_2) \ .$$

(ii) τ *satisfies the "cocycle condition": for any quadruplet* λ_1, λ_2, λ_3, λ_4 *of Lagrangian spaces, we have*:

$$\tau(\lambda_1, \lambda_2, \lambda_3) - \tau(\lambda_1, \lambda_2, \lambda_4) + \tau(\lambda_1, \lambda_3, \lambda_4) - \tau(\lambda_2, \lambda_3, \lambda_4) = 0 \ .$$

(iii) *When the Lagrangian subspaces* λ_1, λ_2, λ_3 *move continuously in the Lagrangian Grassmannian* $\Lambda(E)$ *in such a manner that* $\dim(\lambda_1 \cap \lambda_2)$, $\dim(\lambda_2 \cap \lambda_3)$ *and* $\dim(\lambda_3 \cap \lambda_1)$ *remain constant, then* $\tau(\lambda_1, \lambda_2, \lambda_3)$ *remains constant.*

(iv) $\tau(\lambda_1, \lambda_2, \lambda_3) \equiv n + \dim(\lambda_1 \cap \lambda_2) + \dim(\lambda_2 \cap \lambda_3) + \dim(\lambda_3 \cap \lambda_1) \bmod 2\mathbb{Z}$.

(v) *If* ρ *is an isotropic space contained in* $(\lambda_1 \cap \lambda_2) + (\lambda_2 \cap \lambda_3) + (\lambda_3 \cap \lambda_1)$ *then*:

$$\tau_E(\lambda_1, \lambda_2, \lambda_3) = \tau_{E^\rho}(\lambda_1^\rho, \lambda_2^\rho, \lambda_3^\rho) \ .$$

(vi) *Let* $E^a = (E, -\sigma)$ *be the vector space* E *endowed with the symplectic form* $-\sigma$. *Then*:

$$\tau_{E^a}(\lambda_1, \lambda_2, \lambda_3) = -\tau_E(\lambda_1, \lambda_2, \lambda_3) \ .$$

(vii) *Let* (E_1, σ_1) *and* (E_2, σ_2) *be two symplectic vector spaces, and let* λ_1, λ_2, λ_3 *(resp.* μ_1, μ_2, μ_3*) be a triplet of Lagrangian subspaces of* E_1 *(resp.* E_2*). Then*:

$$\tau_{E_1 \oplus E_2}(\lambda_1 \oplus \mu_2, \lambda_2 \oplus \mu_2, \lambda_3 \oplus \mu_3) = \tau_{E_1}(\lambda_1, \lambda_2, \lambda_3) + \tau_{E_2}(\mu_1, \mu_2, \mu_3) \ .$$

(Here $E_1 \oplus E_2$ *is endowed with the symplectic form* $\sigma_1 \oplus \sigma_2$.*)*

Proof. (i) is clear since the quadratic form $q(x_1, x_2, x_3)$ is alternating with respect to the permutations of the λ_j's.

(iii) Set, for $x = (x_1, x_2, x_3)$ and $y = (y_1, y_2, y_3)$ in $\lambda_1 \oplus \lambda_2 \oplus \lambda_3$:

$$B(x, y) = \sigma(x_1, y_2) + \sigma(x_2, y_3) + \sigma(x_3, y_1) + \sigma(y_1, x_2) + \sigma(y_2, x_3) + \sigma(y_3, x_1) \ .$$

The totally isotropic space I of q is by the definition the space:

(A.3.1) $$I = \{x \in \lambda_1 \oplus \lambda_2 \oplus \lambda_3; B(x, y) = 0 \text{ for any } y\} \ .$$

since $B(x, y) = \sigma(y_1, x_2 - x_3) + \sigma(y_2, x_3 - x_1) + \sigma(y_3, x_1 - x_2)$, we have:

$$I = \{(x_1, x_2, x_3) \in \lambda_1 \oplus \lambda_2 \oplus \lambda_3; x_2 - x_3 \in \lambda_1, x_3 - x_1 \in \lambda_2, x_1 - x_2 \in \lambda_3\} \ .$$

Set $y_1 = x_2 + x_3 - x_1$, $y_2 = x_3 + x_1 - x_2$, $y_3 = x_1 + x_2 - x_3$. Then:

$$y_1 \in \lambda_2 \cap \lambda_3 \ , \qquad y_2 \in \lambda_3 \cap \lambda_1 \ , \qquad y_3 \in \lambda_1 \cap \lambda_2 \ .$$

Moreover

$$2x_1 = y_2 + y_3 \ , \qquad 2x_2 = y_3 + y_1 \ , \qquad 2x_3 = y_1 + y_2 \ .$$

Therefore, by the linear transformation $(x_1, x_2, x_3) \mapsto (y_1, y_2, y_3)$, I is isomorphic to $(\lambda_1 \cap \lambda_2) \oplus (\lambda_2 \cap \lambda_3) \oplus (\lambda_3 \cap \lambda_1)$. Since the rank of q (denoted $\operatorname{rk}(q)$) is $3n - \dim I$, we get:

(A.3.2) $$\operatorname{rk}(q) = 3n - \dim(\lambda_1 \cap \lambda_2) - \dim(\lambda_2 \cap \lambda_3) - \dim(\lambda_3 \cap \lambda_1) \ .$$

With the hypotheses of (iii), we get that $\operatorname{rk}(q)$ is constant. Since q moves continuously when the λ_j's move continuously, this implies (iii).

(iv) We have:

(A.3.3) $$\operatorname{sgn}(q) \equiv \operatorname{rk}(q) \bmod 2\mathbb{Z} \ .$$

Hence (iv) follows from (A.3.2).

(ii) Assume for a while that λ_1 and λ_2 intersect transversally and denote by p_1 and p_2 the projections:

$$p_i : E = \lambda_1 \oplus \lambda_2 \to \lambda_i \qquad (i = 1, 2) \ .$$

For x and y in E we have:

$$\begin{aligned} \sigma(p_1(x), y) &= \sigma(p_1(x), p_1(y) + p_2(y)) \\ &= \sigma(p_1(x), p_2(y)) \\ &= \sigma(x, p_2(y)) \ . \end{aligned}$$

Lemma A.3.3. *Suppose λ_1 and λ_2 are transversal. Then $\tau(\lambda_1, \lambda_2, \lambda_3)$ is equal to the signature of the quadratic form q_3 on λ_3 defined by:*

$$q_3(x_3) = -\sigma(p_1(x_3), p_2(x_3)) = -\sigma(p_1(x_3), x_3) \ .$$

Proof of Lemma A.3.3. Let $x = (x_1, x_2, x_3) \in \lambda_1 \oplus \lambda_2 \oplus \lambda_3$. Then:

$$\begin{aligned} q(x) &= \sigma(x_1, x_2) + \sigma(x_2, x_3) + \sigma(x_3, x_1) \\ &= \sigma(x_1, x_2) - \sigma(p_1(x_3), x_2) - \sigma(x_1, p_2(x_3)) \\ &= \sigma(x_1 - p_1(x_3), x_2 - p_2(x_3)) - \sigma(p_1(x_3), p_2(x_3)) \ . \end{aligned}$$

By the linear transformation $(x_1, x_2, x_3) \mapsto (x_1 - p_1(x_3), x_2 - p_2(x_3), x_3)$, the quadratic form $(x_1, x_2, x_3) \mapsto \sigma(x_1 - p_1(x_3), x_2 - p_2(x_3))$ is equivalent to the quadratic form $(x_1, x_2, x_3) \mapsto \sigma(x_1, x_2)$. Hence its signature is zero, which proves the Lemma. □

Lemma A.3.4. *Let λ_j $(j = 1, 2, 3)$ and μ be four Lagrangian subspaces such that $\lambda_j \cap \mu = \{0\}$, $(j = 1, 2, 3)$. Then:*

(A.3.4) $$\tau(\lambda_1, \lambda_2, \lambda_3) = \tau(\lambda_1, \lambda_2, \mu) + \tau(\lambda_2, \lambda_3, \mu) + \tau(\lambda_3, \lambda_1, \mu) .$$

Proof of Lemma A.3.4. By Lemma (A.3.3), the right hand side of (A.3.4) is the signature of the quadratic form on $\lambda_1 \oplus \lambda_2 \oplus \lambda_3$:

$$q'(y_1, y_2, y_3) = \sigma(p_1(y_2), y_2) + \sigma(p_2(y_3), y_3) + \sigma(p_3(y_1), y_1) ,$$

where now p_j denote the projection:

$$p_j : \lambda_j \oplus \mu \to \lambda_j \qquad (j = 1, 2, 3) .$$

Consider the linear automorphism of $\lambda_1 \oplus \lambda_2 \oplus \lambda_3$ defined by:

$$x_1 = y_1 + p_1(y_2) , \qquad x_2 = y_2 + p_2(y_3) , \qquad x_3 = y_3 + p_3(y_1)$$

$$y_1 = (x_1 - p_1(x_2) + p_1(x_3))/2 , \qquad y_2 = (x_2 - p_2(x_3) + p_2(x_1))/2 ,$$

$$y_3 = (x_3 - p_3(x_1) + p_3(x_2))/2 .$$

An easy calculation shows:

(A.3.5) $$q(x_1, x_2, x_3) = q'(y_1, y_2, y_3) ,$$

which proves the lemma. □

End of the proof of Theorem A.3.2. Choose a Lagrangian subspace μ transversal to all λ_j's $(j = 1, 2, 3, 4)$ and apply (A.3.4). Then (ii) follows in view of (i).

(v) We shall decompose the proof in several steps.

(a) First assume $\rho \subset \lambda_1 \cap \lambda_2 \cap \lambda_3$. Then the quadratic form q on $\lambda_1 \oplus \lambda_2 \oplus \lambda_3$ is the pull-back of the corresponding quadratic form on $\lambda_1^\rho \oplus \lambda_2^\rho \oplus \lambda_3^\rho$, by the surjective map $\lambda_1 \oplus \lambda_2 \oplus \lambda_3 \to \lambda_1^\rho \oplus \lambda_2^\rho \oplus \lambda_3^\rho$. The assertion follows in that case.

(b) Now assume $\rho \subset \lambda_2 \cap \lambda_3$ and $\lambda_3^\rho = \lambda_1^\rho$. Consider the quadratic form q'' on $\lambda_1 \oplus \lambda_2 \oplus (\lambda_1 \cap \rho^\perp) \oplus \rho$ defined by:

$$q''(x_1, x_2, u, v) = \sigma(x_1, x_2) + \sigma(x_2, u + v) + \sigma(u + v, x_1) .$$

Then $q''(x_1, x_2, u, v) = \sigma(x_1, x_2) + \sigma(x_2, u) + \sigma(v, x_1) = \sigma(x_1 - u, x_2 - v)$. Hence the signature of q'' is zero. By the hypothesis, $\lambda_3 = (\lambda_1 \cap \rho^\perp) + \rho$. This implies that the signature of q on $\lambda_1 \oplus \lambda_2 \oplus \lambda_3$ is that of q''. Thus $\tau = 0$ in that case.

(c) Set $\tilde{\lambda}_i = (\lambda_i \cap \rho^\perp) + \rho = (\lambda_i + \rho) \cap \rho^\perp$, $i = 1, 2, 3$. We have:

(A.3.6) $$\lambda_1^{\tilde{\lambda}_1 \cap \lambda_2} = \tilde{\lambda}_1^{\tilde{\lambda}_1 \cap \lambda_2} .$$

In fact $\rho \subset \lambda_1 + \lambda_2 \cap \lambda_3 \subset \lambda_1 + \lambda_2 \cap \rho^\perp$ implies

$$\rho \subset \lambda_1 + (\lambda_1 + \rho) \cap \lambda_2 \cap \rho^\perp \subset \lambda_1 + \tilde{\lambda}_1 \cap \lambda_2 \ .$$

We obtain $\tilde{\lambda}_1 \subset \lambda_1 + (\tilde{\lambda}_1 \cap \lambda_2)$ thus $\tilde{\lambda}_1 \subset [\lambda_1 + (\tilde{\lambda}_1 \cap \lambda_2)] \cap (\tilde{\lambda}_1 + \lambda_2)$ which gives:

$$\tilde{\lambda}_1^{\tilde{\lambda}_1 \cap \lambda_2} = \lambda_1^{\tilde{\lambda}_1 \cap \lambda_2} \ . \tag{A.3.7}$$

Since both spaces in (A.3.7) are Lagrangian, (A.3.6) follows. Similarly we have:

$$\lambda_1^{\tilde{\lambda}_1 \cap \lambda_3} = \tilde{\lambda}_1^{\tilde{\lambda}_1 \cap \lambda_3} \ . \tag{A.3.8}$$

Therefore we get by (b):

$$\tau(\lambda_1, \tilde{\lambda}_1, \lambda_j) = 0 \qquad \text{for} \qquad j = 2, 3 \ . \tag{A.3.9}$$

Now we have:

$$\begin{aligned} \tau(\lambda_1, \lambda_2, \lambda_3) &= \tau(\lambda_1, \lambda_2, \tilde{\lambda}_1) + \tau(\lambda_2, \lambda_3, \tilde{\lambda}_1) + \tau(\lambda_3, \lambda_1, \tilde{\lambda}_1) \\ &= \tau(\tilde{\lambda}_1, \lambda_2, \lambda_3) \ . \end{aligned}$$

Repeating this argument, we obtain

$$\tau(\lambda_1, \lambda_2, \lambda_3) = \tau(\tilde{\lambda}_1, \tilde{\lambda}_2, \tilde{\lambda}_3)$$

and the term on the right-hand side eqoals $\tau(\lambda_1^\rho, \lambda_2^\rho, \lambda_3^\rho)$ by (a).

(vi) and (vii) are obvious. □

Remark A.3.5. The "cocycle condition", written in Theorem A.3.2 (ii), may be visualized by figure A.3.1.

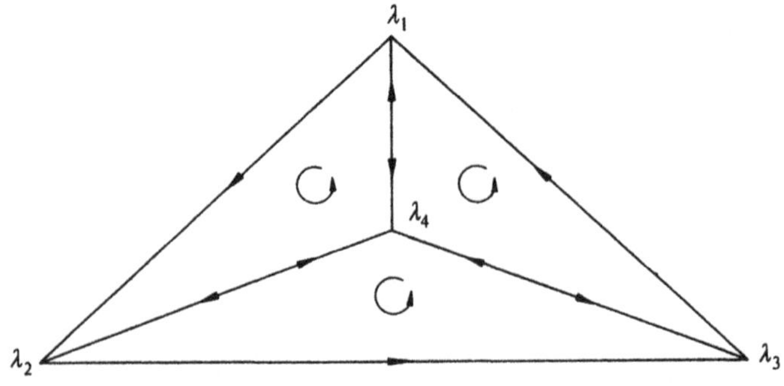

$$\tau(\lambda_1, \lambda_2, \lambda_3) = \tau(\lambda_1, \lambda_2, \lambda_4) + \tau(\lambda_2, \lambda_3, \lambda_4) + \tau(\lambda_3, \lambda_1, \lambda_4)$$

Fig. A.3.1

We shall calculate explicitly the Maslov index in a special case. Assume E endowed with a symplectic basis, and denote by $(x;\xi)$ the associated linear coordinates. Let A and B be two $(n \times n)$-matrices satisfying the hypothesis (A.1.11) with $A'B$ symmetric, and define:

$$\lambda_1 = \{x = 0\} , \qquad \lambda_2 = \{\xi = 0\} , \qquad \lambda_3 = \{(x;\xi); Ax = B\xi\} .$$

Proposition A.3.6. *One has:*

$$\tau(\lambda_1, \lambda_2, \lambda_3) = -\operatorname{sgn}(A'B) .$$

Proof. Denote by p_1 and p_2 the projections from $E = \lambda_1 \oplus \lambda_2$ to λ_1 and λ_2 respectively. By Lemma A.3.3, $-\tau(\lambda_1, \lambda_2, \lambda_3)$ is the signature of the quadratic form q_3 on λ_3, defined by:

$$\begin{aligned} q_3((x;\xi)) &= \sigma(p_1(x;\xi), p_2(x;\xi)) \\ &= \sigma(\xi, x) \\ &= \langle \xi, x \rangle . \end{aligned}$$

The map from $\mathbb{R}^n$ to λ_3 given by $z \mapsto ({}^tBz, {}^tAz)$ is a linear isomorphism. Hence $\langle \xi, x\rangle = \langle z, A'Bz\rangle$, and the result follows. □

Let $\lambda_1, \ldots, \lambda_N$ be Lagrangian subspaces with $N \geqslant 3$ and let μ be another Lagrangian subspace. By Theorem A.3.2 (ii) we have the identity:

$$\text{(A.3.10)} \quad \begin{cases} \tau(\lambda_1, \lambda_2, \lambda_3) + \tau(\lambda_1, \lambda_3, \lambda_4) + \cdots + \tau(\lambda_1, \lambda_{N-1}, \lambda_N) \\ \quad = \tau(\lambda_1, \lambda_2, \mu) + \tau(\lambda_2, \lambda_3, \mu) + \cdots + \tau(\lambda_{N-1}, \lambda_N, \mu) + \tau(\lambda_N, \lambda_1, \mu) . \end{cases}$$

This can be visualized by Figure A.3.2 (in which $N = 5$).

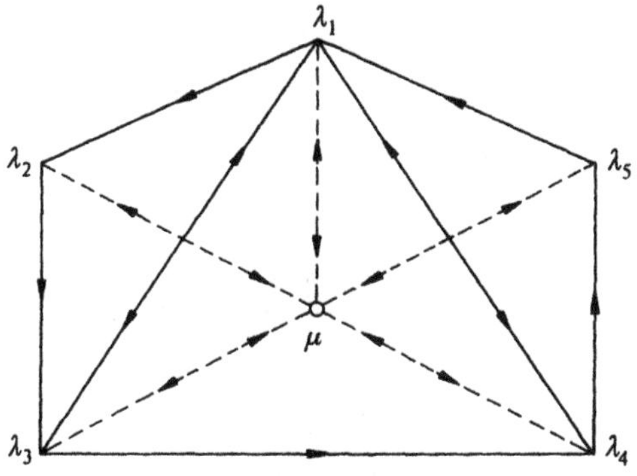

Fig. A.3.2

Definition A.3.7. *Let $\lambda_1, \dots, \lambda_N$ be Lagrangian subspaces, $N \geqslant 3$. One defines the index $\tau(\lambda_1, \dots, \lambda_N)$ as the left hand side of* (A.3.10).

Proposition A.3.8.
(i) $\tau(\lambda_1, \lambda_2, \dots, \lambda_N) = \tau(\lambda_2, \lambda_3, \dots, \lambda_N, \lambda_1) = -\tau(\lambda_N, \lambda_{N-1}, \dots, \lambda_1)$.
(ii) *Assume* $N \geqslant 4$ *and let* $j \in \{3, \dots, N-1\}$. *Then* $\tau(\lambda_1, \dots, \lambda_N) = \tau(\lambda_1, \dots, \lambda_j) + \tau(\lambda_1, \lambda_j, \lambda_{j+1}, \dots, \lambda_N)$.
(iii) *When the λ_j's move continuously in such a manner that* $\dim(\lambda_1 \cap \lambda_2)$, $\dim(\lambda_2 \cap \lambda_3)$, $\dots$, $\dim(\lambda_{N-1} \cap \lambda_N)$ *and* $\dim(\lambda_N \cap \lambda_1)$ *remain constant, then* $\tau(\lambda_1, \dots, \lambda_N)$ *remains constant.*
(iv) $\tau(\lambda_1, \dots, \lambda_N) \equiv nN + \dim(\lambda_1 \cap \lambda_2) + \cdots + \dim(\lambda_N \cap \lambda_1) \bmod 2\mathbb{Z}$.

Proof. (i) and (ii) are obvious by the definition and Theorem A.3.2. (iii) and (iv) follow immediately from Theorem A.3.2 and (A.3.10) when choosing the Lagrangian space μ transversal to all λ_j's. □

As before, denote by E^a the space E endowed with the symplectic form $-\sigma$, and denote by "a" the identity $E \to E^a$. If λ is a subspace of E, we denote by λ^a its image by a. Then, we have $\tau_{E^a}(\lambda_1^a, \lambda_2^a, \lambda_3^a) = -\tau_E(\lambda_1, \lambda_2, \lambda_3)$.

Proposition A.3.9. *Let $\lambda_1, \lambda_2, \mu_1, \mu_2$ be four Lagrangian subspaces of E, and denote by Δ the diagonal of $E^a \oplus E$ (which is Lagrangian). Then:*

$$\tau_{E^a \oplus E}(\lambda_1^a \oplus \lambda_2, \mu_1^a \oplus \mu_2, \Delta) = \tau_E(\lambda_1, \lambda_2, \mu_2, \mu_1) \ . \tag{A.3.11}$$

Proof. Denote by τ the left hand side of (A.3.11). Then $\tau = \tau_1 + \tau_2 + \tau_3$, with:

$$\tau_1 = \tau(\lambda_1^a \oplus \lambda_2, \mu_1^a \oplus \mu_2, \lambda_1^a \oplus \mu_2) \ ,$$

$$\tau_2 = \tau(\mu_1^a \oplus \mu_2, \Delta, \lambda_1^a \oplus \mu_2) \ ,$$

$$\tau_3 = \tau(\Delta, \lambda_1^a \oplus \lambda_2, \lambda_1^a \oplus \mu_2) \ .$$

Then:

$$\tau_1 = \tau(\lambda_1^a, \mu_1^a, \lambda_1^a) + \tau(\lambda_2, \mu_2, \mu_2) = 0 \ ,$$

$$\tau_2 = \tau(\mu_1^a, \mu_2^a, \lambda_1^a) \ ,$$

$$\tau_3 = \tau(\lambda_1, \lambda_2, \mu_2) \ .$$

(To calculuate τ_2 and τ_3 we apply Theorem A.3.2 (v) with $\rho = \{0\} \oplus \mu_2$ and $\rho = \lambda_1^a \oplus \{0\}$, respectively.) Therefore $\tau = \tau(\lambda_1, \lambda_2, \mu_2) - \tau(\mu_1, \mu_2, \lambda_1) = \tau(\lambda_1, \lambda_2, \mu_2, \mu_1)$. □

To end this section, let us describe the action of the symplectic group $Sp(E)$ on the space $\Lambda^3(E)$ of triplets of Lagrangian subspaces of E.

For $r = (r_0, r_1, r_2, r_3, d) \in \mathbb{N}^4 \times \mathbb{Z}$ we set:

$$\begin{aligned} W_r = \{(\lambda_1, \lambda_2, \lambda_3) \in \Lambda^3(E) \ ; \qquad & \dim(\lambda_1 \cap \lambda_2 \cap \lambda_3) = r_0 \ , \\ \dim(\lambda_1 \cap \lambda_2) = r_3 \ , \qquad & \dim(\lambda_2 \cap \lambda_3) = r_1 \ , \\ \dim(\lambda_3 \cap \lambda_1) = r_2 \ , \qquad & \tau(\lambda_1, \lambda_2, \lambda_3) = d\} \ . \end{aligned}$$

Then $Sp(E)$ acts naturally on $\Lambda^3(E)$, and the W_r's are invariant by $Sp(E)$. Consider the conditions:

$$\text{(A.3.12)} \quad \begin{cases} 0 \leqslant r_0 \leqslant r_1, r_2, r_3 \leqslant n \ , & r_1 + r_2 + r_3 \leqslant n + 2r_0 \ , \\ |d| \leqslant n + 2r_0 - (r_1 + r_2 + r_3) \ , & d \equiv n + r_1 + r_2 + r_3 \mod 2\mathbb{Z} \ . \end{cases}$$

It can be easily shown that (A.3.12) is a necessary and sufficient condition in order that W_r is non-empty. The action of $Sp(E)$ on $\Lambda^3(E)$ has only finitely many orbits, and these orbits are the W_r, where r satisfies (A.3.12). Since we do not use this result, we leave the proof as an exercise.

Exercises to the Appendix

Exercise A.1. Let $(x) = (x_1, \ldots, x_n)$ be a system of local coordinates on X, and $(x; \xi)$ the associated coordinates on T^*X. Let $\varphi = (\varphi_1, \ldots, \varphi_p) : X \to \mathbb{R}^p$ be a smooth map, and let $S = \{x; \varphi(x) = 0\}$, (hence $d\varphi_1 \wedge \cdots \wedge d\varphi_p \neq 0$ and S is smooth). Let $x_\circ \in S$, $p = (x_\circ; \sum_i a_i d\varphi_i(x_\circ)) \in T_S^*X$. Prove that:

$$T_S^*X = \left\{(x;\xi); \varphi(x) = 0, \xi_j = \sum_{i=1}^{p} \lambda_i \frac{\partial \varphi_i}{\partial x_j}, \lambda_i \in \mathbb{R}\right\} ,$$

$$\begin{aligned} T_p T_S^*X = \Bigg\{(x;\xi); & \sum_{j=1}^{n} \frac{\partial \varphi_i}{\partial x_j}(x_\circ) x_j = 0, \ i = 1, \ldots, p, \\ & \xi_j = \sum_{i=1}^{p} \frac{\partial \varphi_i}{\partial x_j}(x_\circ)\lambda_i + \sum_{k=1}^{n} \sum_{i=1}^{p} a_i \frac{\partial^2 \varphi_i(x_\circ)}{\partial x_k \partial x_j} x_k, \ \lambda_i \in \mathbb{R}\Bigg\} . \end{aligned}$$

Exercise A.2. Let Λ be a closed conic Lagrangian submanifold of T^*X. Prove that there exists a submanifold M of X such that $\Lambda = T_M^*X$.

Exercise A.3. Let X and Y be two manifolds of the same dimension, f a real function on $X \times Y$ with $df \neq 0$ on $S = \{f = 0\}$. Set $\Lambda = \dot{T}_S^*(X \times Y)$.

Prove that $p_1|_\Lambda$ and $p_2^a|_\Lambda$ are local isomorphisms (from Λ to T^*X and Λ to T^*Y, respectively) if and only if the determinant $\begin{pmatrix} 0 & d_y f \\ d_x f & d_{xy}^2 f \end{pmatrix}$ does not vanish on S.

Exercise A.4. Let (E_1, σ_1) and (E_2, σ_2) be two symplectic vector spaces and let λ be a Lagrangian linear subspace of $E_1 \oplus E_2$. Prove that $p_1|_\lambda$ is injective if and only if $p_2|_\lambda$ is surjective.

Exercise A.5. Let $f\colon Y \to X$ be a morphism of manifolds and let $p \in Y \times_X T^*X$.

(a) Let Λ_Y be a Lagrangian submanifold of T^*Y. Assume that ${}^tf'$ is clean with respect to Λ_Y at p_Y. Prove that for U a sufficiently small neighborhood of p, $f_\pi(U \cap {}^tf'^{-1}(\Lambda_Y))$ is a smooth Lagrangian manifold.

(b) Let Λ_X be a Lagrangian submanifold of T^*X. Assume that f_π is clean with respect to Λ_X at p_X. Prove that for U a sufficiently small neighborhood of p, ${}^tf'(U \cap f_\pi^{-1}(\Lambda_X))$ is a smooth Lagrangian manifold.

Exercise A.6. Let E_1 and E_2 be two symplectic vector spaces, ν a Lagrangian subspace of $E_1 \oplus E_2^a$ and let λ_i and μ_i be two Lagrangian subspaces of E_i $(i = 1, 2)$. Prove:

$$\begin{aligned}\tau_{E_1 \oplus E_2^a}(\lambda_1 \oplus \lambda_2^a, \mu_1 \oplus \mu_2^a, \nu) &= \tau_{E_1}(\lambda_1, \mu_1, \nu \circ \mu_2) - \tau_{E_2}(\lambda_2, \mu_2, \lambda_1 \circ \nu^a)\\ &= \tau_{E_1}(\lambda_1, \mu_1, \nu \circ \lambda_2) - \tau_{E_2}(\lambda_2, \mu_2, \mu_1 \circ \nu^a)\ .\end{aligned}$$

Exercise A.7. Let (E, σ) be a complex symplectic vector space, and endow the real underlying vector space $E^{\mathbb{R}}$ with the real symplectic form $2\operatorname{Re}\sigma$. Let λ_1, λ_2, λ_3 be three complex Lagrangian subspaces of E. Prove that $\tau_{E^{\mathbb{R}}}(\lambda_1, \lambda_2, \lambda_3) = 0$.

Exercise A.8. Let E_i $(i = 1, 2, 3, 4)$ be a symplectic vector space and let $\lambda_i \subset E_i \oplus E_{i+1}^a$ $(i = 1, 2, 3)$ be a Lagrangian plane.

Prove that $(\lambda_1 \circ \lambda_2) \circ \lambda_3 = \lambda_1 \circ (\lambda_2 \circ \lambda_3)$.

Exercise A.9. Let (E_i, μ_i) be a pair of a symplectic vector space E_i and a Lagrangian plane μ_i of E_i $(i = 1, 2, 3, 4)$.

Then for Lagrangian planes $\lambda_1 \subset E_1 \oplus E_2^a$ and $\lambda_2 \subset E_2 \oplus E_3^a$, we define (cf. (7.5.10)):

$$\tau(\lambda_1 : \lambda_2) = \tau_{E_2}(\mu_2, \lambda_2 \circ \mu_3, \mu_1 \circ \lambda_1^a)\ .$$

Now, let λ_i be a Lagrangian plane of $E_i \oplus E_{i+1}^a$ $(i = 1, 2, 3)$. Prove the equality:

$$\tau(\lambda_1 : \lambda_2 \circ \lambda_3) + \tau(\lambda_2 : \lambda_3) = \tau(\lambda_1 \circ \lambda_2 : \lambda_3) + \tau(\lambda_1 : \lambda_2)\ .$$

(Hint: Using Exercise A.6, prove that both sides are equal to:

$$\tau_{E_2 \oplus E_3^a}(\mu_2 \oplus \mu_3^a, \mu_1 \circ \lambda_1^a \oplus \lambda_3^a \circ \mu_4^a, \lambda_2).)$$

Exercise A.10. Let E_i be a symplectic vector space and λ_i and μ_i two Lagrangian subspaces of E_i $(i = 1, 2, 3)$. Let ν and ν' be Lagrangian subspaces of $E_1 \oplus E_2^a$ and $E_2 \oplus E_3^a$, respectively. Prove:

$$\tau_{E_1 \oplus E_2^a}(\lambda_1 \oplus \lambda_2^a, \nu, \mu_1 \oplus \mu_2^a) + \tau_{E_2 \oplus E_3^a}(\lambda_2 \oplus \lambda_3^a, \nu', \mu_2 \oplus \mu_3^a)$$
$$- \tau_{E_1 \oplus E_3^a}(\lambda_1 \oplus \lambda_3^a, \nu \circ \nu', \mu_1 \oplus \mu_3^a)$$
$$= \tau_{E_2}(\lambda_2, \nu' \circ \lambda_3, \lambda_1 \circ \nu^a) - \tau_{E_2}(\mu_2, \nu' \circ \mu_3, \mu_1 \circ \nu^a).$$

(Hint: use Exercise A.6.)

Notes

Symplectic and contact geometry are classical subjects which go back to Hamilton and Jacobi, and that we shall not review here. As pointed out at the beginning of this appendix, the results of §1 and §2 are well-known and may be found for example in Hörmander [4].

In 1965, in order to calculate asymptotic expansions "in a neighborhood of a caustic" (i.e. when the projection of a smooth Lagrangian manifold has not a constant rank), Maslov [1] (cf. also Keller [1]) introduced the index of a closed curve in a Lagrangian submanifold of a symplectic space. His theory was clarified and reformulated by Arnold [1], then by Hörmander [2] and Leray [4] who defined the index of three Lagrangian planes intersecting transversally, until Kashiwara (cf. Lion-Vergne [1]) defined the index τ in the general case by the simple method we have given here. Note that in Lion-Vergne (loc. cit.) the index τ is generalized to the local field case.

Bibliography

Abraham, R. and Mardsen, J.E. [1]: Foundation of Mechanics. Cummings Publ. (1978)

Andronikof, E. [1]: Microlocalisation tempérée des distributions et des fonctions holomorphes I, II. C.R. Acad. Sci. **303**, 347–350 (1986) and **304**, 511–514 (1987)

Arnold, V. [1]: On a characteristic class entering into conditions of quantization. Funct. Anal. Appl. **1**, 1–13 (1967)

— [2]: Mathematical methods of classical mechanics. Springer, New York Berlin Heidelberg (1978)

Banica, C. and Stanasila, O. [1]: Méthodes algébriques dans la théorie globale des espaces complexes. Vol. I–II. Gauthier-Villars-Bordas (1977)

Beilinson, A.A., Bernstein, J. and Deligne, P. [1]: Faisceaux pervers. Astérisque **100** (1982)

Bengel, G. and Schapira, P. [1]: Décomposition microlocale analytique des distributions. Ann. Inst. Fourier Grenoble **29**, 101–124 (1979)

Bernstein, J. [1]: Modules over a ring of differential operators. Study of fundamental solutions of equations with constant coefficients. Funct. Anal. Appl. **5**, 89–101 (1971)

— [2]: The analytic continuation of generalized functions with respect to a parameter. Funct. Anal. Appl. **6**, 272–285 (1972)

Berthelot, P., Breen, L. and Messing, W. [1]: Théorie de Dieudonné cristalline II. Lect. Notes Math. 930. Springer, Berlin Heidelberg New York (1982)

Bierstone, E. and Milman, P.D. [1]: Semi-analytic and subanalytic sets. Publ. Math. I.H.E.S. **67**, 5–42 (1988)

Björk, J-E. [1]: Rings of differential operators. North-Holland Math. Lib. (1979)

— [2]: Analytic *D*-Modules. Kluwer Publ. To appear (1991)

Bloom, T. and Herrera, M. [1]: De Rham cohomology of an analytic space. Invent. Math. **7**, 275–296 (1969)

Bony, J-M. and Schapira, P. [1]: Existence et prolongement des solutions holomorphes des équations aux dérivées partielles. Invent. Math. **17**, 95–105 (1972)

— [2]: Solutions hyperfonctions du problème de Cauchy. In: Hyperfunctions and pseudo-differential equations, Komatsu H. (Ed.), Proceedings Katata 1971. Lect. Notes Math. 287, 82–98. Springer, Berlin Heidelberg New York (1973)

— [3]: Propagation des singularités analytiques pour les solutions des équations aux dérivées partielles. Ann. Inst. Fourier Grenoble **26**, 81–140 (1976)

Borel, A. [1]: The Poincaré duality in generalized manifolds. Michigan Math. J. **4**, 227–239 (1957)

Borel, A. et al [1]: Intersection cohomology. Progress in Math. **50**, Birkhäuser, Boston (1984)

Borel, A. and Haefliger, A. [1]: La classe d'homologie fondamentale d'un espace analytique. Bull. Soc. Math. France **89**, 461–513 (1961)

Borel, A. and Moore, J.C. [1]: Homology theory for locally compact spaces. Michigan Math. J. **7**, 137–159 (1960)

Bott, R. [1]: Lectures on Morse theory, old and new. Bull. Amer. Math. Soc. **7**, 331–358 (1982)

Bourbaki, N. [1]: Algèbre, Chapitre 10. Eléments de Mathématiques. Masson (1980)

Boutet de Monvel, L. and Malgrange, B. [1]: Le théorème de l'indice relatif. Ann. Sc. Ec. Norm. Sup. **23**, 151–192 (1990)

Bredon, G.E. [1]: Sheaf theory. McGraw-Hill (1967)

Benner, A.V. and Shubin, M.A. [1]: Atiyah-Bott-Lefschetz theorem for manifolds with boundary. Funct. Anal. Appl. **15**, 286–287 (1981)
Brylinski, J-L. [1]: (Co-)homologie d'intersection et faisceaux pervers. Sém. Bourbaki 585 (1981–82)
— [2]: Transformations canoniques, dualité projective, théorie de Lefschetz. Astérisque 140/141, 3–134 (1986)
Brylinski, J-L., Malgrange, B. and Verdier, J-L. [1]: Transformée de Fourier géométrique I. C.R. Acad. Sci. **297**, 55–58 (1983)
Brylinski, J-L., Dubson, A. and Kashiwara, M. [1]: Formule de l'indice pour les modules holonomes et obstruction d'Euler locale. C.R. Acad. Sci. **293**, 573–576 (1981)
Cartan, H. [1]: Sém. Ec. Norm. Sup. (1950–51), Benjamin (1967).
— [2]: Sém. Ec. Norm. Sup. (1951–52, 1953–54, 1960–61), Benjamin (1967)
Cartan, H. and Chevalley, C. [1]: Sém. Ec. Norm. Sup. (1955–56), Benjamin (1967)
Cartan, H. and Eilenberg, S. [1]: Homological Algebra. Princeton Univ. Press (1956)
D'Agnolo, A. and Schapira, P. [1]: Un théorème d'image inverse pour les faisceaux. Application au problème de Cauchy. C.R. Acad. Sci. **311**, 23–26 (1990)
Deligne, P. [1]: Cohomologie à support propre. In [SGA 4], exposé XVII
— [2]: Le formalisme des cycles évanescents. In [SGA 7], exposé XIII
Delort, J-M. [1]: Deuxième microlocalisation simultanée et front d'onde de produits. Ann. Sc. Ec. Norm. Sup. **23**, 257–310 (1990)
Delort, J-M. and Lebeau, G. [1]: Microfonctions lagrangiennes. J. Math. Pures Appl. **67**, 39–84 (1988)
Denkowska, Z., Lojasiewicz, S. and Stasica, J. [1]: Certaines propriétés élémentaires des ensembles sous-analytiques. Bull. Acad. Polon. Sci. Math. **27**, 529–536 (1979)
Dubson, A. [1]: Formules pour l'indice des complexes constructibles et D-modules holonomes. C.R. Acad. Sci. **298**, 113–164 (1984)
Duistermaat, J.J. [1]: Fourier integral operators. Lect. Notes Courant Inst. New York, (1973)
Freyd, P. [1]: Abelian categories, an introduction to the theory of functors. Harper and Row, New York (1964)
Fulton, W. [1]: Intersection theory. Springer, Berlin Heidelberg New York Tokyo (1984)
Gabber, O. [1]: The integrability of the caracteristic variety. Amer. J. Math. **103**, 445–468 (1981)
Gabriel, P. and Zisman, M. [1]: Calculus of fractions and homotopy theory. Springer, Berlin Heidelberg New York (1967)
Gabrielov, A.M. [1]: Projections of semi-analytic sets. Funct. Anal. Appl. **2**, 282–291 (1968)
Gelfand, S. and Manin, Yu. [1]: Methods of homological algebra I: introduction to cohomology theory and derived categories. Springer Berlin Heidelberg New York Tokyo (1991)
Ginsburg, V. [1]: A theorem on the index of differential systems and the geometry of manifolds with singularities. Soviet Math. Dokl. **31**, 309–313 (1985)
— [2]: Characteristic cycles and vanishing cycles. Invent. Math. **84**, 327–402 (1986)
Golovin, V.D. [1]: Cohomology of analytical sheaves and duality theorem. Nauka, Moscow (1986) (in Russian)
Godement, R. [1]: Topologie algébrique et théorie des faisceaux. Hermann Paris (1958)
Goresky, M. and MacPherson, R. [1]: Intersection homology II. Invent. Math. **71**, 77–129 (1983)
— [2]: Stratified Morse theory. Springer, Berlin Heidelberg (1988)
— [3]: Morse theory and intersection homology theory. In: Analyse et topologie sur les espaces singuliers. Asterisque 101/102, 135–192 (1983)
Grauert, H. [1]: On Levi's problem and the embedding of real analytic manifolds. Ann. Math. **68**, 460–472 (1958)
Griffiths, P. and Harris, J. [1]: Principles of algebraic geometry. John Wiley & Sons (1978)
Grothendieck, A. [1]: Sur quelques points d'algèbre homologique. Tohuku Math. J. **9**, 119–221 (1957)
— [2]: Techniques de descente et théorème d'existence en géométrie algébrique II. Le théorème d'existence en théorie formelle des modules. Sém. Bourbaki 195 (1959–60)
— [3]: Eléments de géométrie algébrique III. Publ. Math. I.H.E.S. **11** (1961) and **17** (1963)

— [4]: Résidus et dualité. Pré-notes pour un "Séminaire Hartshorne", manuscrit (1963)
— [5]: Dix exposés sur la théorie des schémas. North-Holland, Amsterdam (1968)
Guillemin, V., Quillen, D. and Sternberg, S. [1]: The integrability of caracteristics. Comm. Pure Appl. Math. **23**, 39–77 (1970)
Guillemin, V. and Pollack, A. [1]: Differential topology. Prentice Hall (1974)
Guillemin, V. and Sternberg S. [1]: Symplectic techniques in physics. Cambridge Univ. Press (1984)
Hardt, R.M. [1]: Stratification of real analytic mappings and images. Invent. Math. **28**, 193–208 (1975)
— [2]: Triangulation of subanalytic sets and proper light subanalytic maps. Invent. Math. **38**, 207–217 (1977)
Hartshorne, R. [1]: Residues and duality. Lect. Notes Math. 20. Springer, Berlin Heidelberg New York (1966)
Henry, J-P., Merle, M. and Sabbah, C. [1]: Sur la condition de Thom stricte pour un morphisme analytique complexe. Ann. Sci. Ec. Norm. Sup. **17**, 227–268 (1984)
Herrera, M. [1]: Integration on a semi-analytic set. Bull. Soc. Math. France **94**, 141–180 (1966)
Hilton P.J. and Stammbach, U. [1]: A course in homological algebra. Springer, New York Berlin Heidelberg (1970)
Hironaka, H. [1]: Subanalytic sets. In: Number theory, algebraic geometry and commutative algebra (in honour to Y. Akizuki). Kinokuniya, Tokyo 453–493 (1973)
— [2]: Introduction to real analytic sets and real analytic maps. Istituto "L. Tonelli" Pisa (1973)
— [3]: Stratification and flatness. Real and complex geometry, Oslo 1976, Sythoff & Noordhoff 199–265 (1977)
Hörmander, L. [1]: An introduction to complex analysis in several variables. Van Nostrand, Princeton (1966)
— [2]: Fourier integral operators I. Acta Math. **127**, 79–183 (1971)
— [3]: The analysis of linear partial differential operators I. Springer, Berlin Heidelberg New York Tokyo (1983)
— [4]: The analysis of linear partial differential operators III-IV. Springer, Berlin Heidelberg New York Tokyo (1985)
Hotta, R. and Kashiwara, M. [1]: The invariant holonomic system on a semi-simple Lie algebra. Invent. Math. **75**, 327–358 (1984)
Houzel, C. and Schapira, P. [1]: Images directes des modules différentiels. C.R. Acad. Sci. **298**, 461–464 (1984)
Illusie, L. [1]: Deligne's *l*-adic Fourier transform. Proc. Symp. Pure Math. **46**, 151–163 (1987)
Iversen, B. [1]: Cohomology of sheaves. Springer, Berlin Heidelberg New York Tokyo (1987)
— [2]: Cauchy residues and de Rham homology. L'Enseig. Math. **35**, 1–17 (1989)
Kashiwara, M. [1]: Algebraic study of systems of partial differential equations. Thesis, Univ. Tokyo (1970)
— [2]: Index theorem for maximally overdetermined systems of linear differential equations. Proc. Japan Acad. **49**, 803–804 (1973)
— [3]: On the maximally overdetermined systems of linear differential equations I. Publ. R.I.M.S. Kyoto Univ. **10**, 563–579 (1975)
— [4]: *b*-functions and holonomic systems. Invent. Math. **38**, 33–53 (1976)
— [5]: Systems of microdifferential equations. Progress in Math. 34. Birkhäuser, Boston (1983)
— [6]: The Riemann-Hilbert problem for holonomic systems. Publ. R.I.M.S. Kyoto Univ. **20**, 319–365 (1984) (or: Faisceaux constructibles et systèmes holonomes d'équations aux dérivées partielles à points singuliers réguliers. Sém Eq. Dér. Part. Publ. Ec. Polyt. (1979/1980))
— [7]: Index theorem for constructible sheaves. In: Systèmes différentiels et singularités, A. Galligo, M. Maisonobe, Ph. Granger (Ed.). Astérisque 130, 193–209 (1985)
— [8]: Character, character cycle, fixed point theorem and group representation. Adv. Stud. Pure Math. **14**, 369–378 (1988)
Kashiwara, M. and Kawai, T. [1]: On the boundary value problems for elliptic systems of linear differential equations. Proc. Japan Acad. **48**, 712–715 (1971) and **49**, 164–168 (1972)

— [2]: Second microlocalisation and asymptotic expansions. In: Complex analysis, microlocal calculus and relativistic quantum theory, D. Iagolnitzer (Ed.). Lect. Notes Phys. 126, 21–76. Springer, Berlin Heidelberg New York (1980)

— [3]: On holonomic systems of microdifferential equations, III. Publ. RIMS, Kyoto Univ. **17**, 813–979 (1981)

Kashiwara, M. and Monteiro-Fernandes, T. [1]: Involutivité des variétés microcaractéristiques. Bull. Soc. Math. France **114**, 393–402 (1986)

Kashiwara, M. and Schapira, P. [1]: Micro-hyperbolic systems. Acta Math. **142**, 1–55 (1979)

— [2]: Micro-support des faisceaux. In: Journées complexes Nancy, Mai 1982 (Publ. Inst. Elie Cartan), or: C.R. Acad. Sci. **295**, 487–490 (1982)

— [3]: Microlocal study of sheaves. Astérisque 128 (1985)

— [4]: A vanishing theorem for a class of systems with simple characteristics. Invent. Math. **82**, 579–592 (1985)

Kataoka, K. [1]: Microlocal theory of boundary value problems I–II. J. Fac. Sci. Univ. Tokyo **27**, 355–399 (1980) and **28**, 31–56 (1981)

Katz, N. and Laumon, G. [1]: Transformation de Fourier et majoration de sommes d'exponentielles. Publ. Math. I.H.E.S. **62**, 146–202 (1985)

Keller, J.B. [1]: Corrected Bohr-Sommerfeld quantum conditions for nonseparable systems. Ann. Phys. **4**, 180–188 (1958)

Komatsu, H. [1]: Resolution by hyperfunctions of sheaves of solutions of differential equations with constant coefficients. Math. Ann. **176**, 77–86 (1968)

— [2]: A local version of the Bochner's tube theorem. J. Fac. Sci. Univ. Tokyo **19**, 201–214 (1972)

Kuo, J.C. [1]: The ratio test for analytic Whitney stratifications. In: Proc. Liverpool singularities symposium. Lect. Notes Math. 192. Springer, Berlin Heidelberg New York (1971)

Laumon, G. [1]: Transformée de Fourier, constante d'équations fonctionnelles et conjecture de Weil. Publ. Math. I.H.E.S. **65**, 131–210 (1987)

Lazzeri, F. [1]: Morse theory on singular spaces. Astérisque 7/8, 263–268 (1973)

Lebeau, G. [1]: Equations des ondes semi-linéaires II. Contrôle des singularités et caustiques non linéaires. Invent. Math. **95**, 277–323 (1988)

Leray, J. [1]: L'anneau d'homologie d'une représentation. C.R. Acad. Sci. **222**, 1367–1368. Structure de l'anneau d'homologie d'une représentation. Idem, 1419–1422 (1946)

— [2]: L'anneau spectral et l'anneau filtré d'homologie d'un espace localement compact et d'une application continue. J. Math. Pures Appl. **29**, 1–139 (1950)

— [3]: Problème de Cauchy I. Bull. Soc. Math. France **85**, 389–430 (1957)

— [4]: Lagrangian analysis. M.I.T. Press, Cambridge Mass. London (1981), or: Cours Collège de France (1976/77)

Lion, G. and Vergne, M. [1]: The Weil representation, Maslov index and theta series. Progress in Math. 6, Birkhäuser, Boston (1980)

Lojasiewicz, S. [1]: Ensembles semi-analytiques. Inst. Hautes Etudes Sci. Bures-sur-Yvette (1964)

— [2]: Triangulation of semi-analytic sets. Ann. Scuola Norm. Sup. Pisa **18**, 449–474 (1964)

MacLane, S. [1]: Homology theory. Academic Press (1963)

MacPherson, R. [1]: Chern classes for singular varieties. Ann. Math. **100**, 423–432 (1974)

MacPherson, R. and Vilonen, K. [1]: Elementary construction of perverse sheaves. Inv. Math. **84**, 403–435 (1986)

Malgrange, B. [1]: Faisceaux sur les variétés analytiques réelles. Bull. Soc. Math. France **85**, 231–237, (1957)

— [2]: Transformation de Fourier géométrique. Sém. Bourbaki, 692 (1987–88)

Martineau, A. [1]: Les hyperfonctions de M. Sato. Sém. Bourbaki 214 (1960–61)

— [2]: Théorème sur le prolongement analytique du type "Edge of the Wedge". Sem. Bourbaki 340 (1967/68)

Maslov, V.P. [1]: Theory of perturbations and asymptotic methods. Moscow Univ. Press (1965) (in Russian)

Mebkhout, Z. [1]: Une équivalence de catégories – Une autre équivalence de catégories. Comp. Math. **51**, 55–62 and 63–64 (1984)
Milnor, J.M. [1]: Morse theory. Ann. Math. Studies 51, Princeton Univ. Press (1963)
— [2]: Singular points of complex hypersurfaces. Ann. Math. Studies 61, Princeton Univ. Press (1968)
Mitchell, B. [1]: Theory of categories. Pure App. Math. 17, Academic Press (1965)
Northcott, D.G. [1]: An introduction to homological algebra. Cambridge Univ. Press (1960)
Oka, K. [1]: Sur les fonctions analytiques de plusieurs variables complexes. Iwanami Shoten, Tokyo (1961)
Pignoni, N. [1]: Density and stability of Morse functions on a stratified space. Ann. Sc. Norm. Sup. Pisa, 593–608 (1979)
Poly, J-B. [1]: Formule des résidus et intersection des chaines sous-analytiques. Thèse Univ. Poitiers (1974)
Ramis, J-P. [1]: Additif II à "variations sur le thème GAGA". Lect. Notes Math. 694. Springer, Berlin Heidelberg New York (1978)
de Rham, G. [1] Variétés différentiables. Hermann, Paris (1955) or: Differentiable manifolds. Springer, Berlin Heidelberg New York Tokyo (1984)
Remmert, R. and Stein, K. [1]: Über die wesentlichen Singularitäten analytischer Mengen. Math. Ann. **126**, 263–306 (1953)
Sabbah, C. [1]: Quelques remarques sur la géométrie des espaces conormaux. In: Systèmes différentiels et singularités, A. Galligo, M. Maisonobe, Ph. Granger (Ed.). Astérisque 130, 161–192 (1985)
Sato, M. [1]: Theory of hyperfunctions I–II. J. Fac. Sci. Univ. Tokyo **8**, 139–193 and 387–436 (1959–1960)
Sato, M. [2]: Hyperfunctions and partial differential equations. In: Proc. Int. Conf. on Functional Analysis and related topics, Tokyo Univ. Press, Tokyo, 91–94 (1969)
Sato, M., Kawai, T. and Kashiwara, M. [1]: Hyperfunctions and pseudo-differential equations. In: Hyperfunctions and Pseudo-differential equations, Komatsu H. (Ed.), Proceedings Katata 1971. Lect. Notes Math. 287, 265–529. Springer, Berlin Heidelberg New York (1973)
Schapira, P. [1]: Théorie des hyperfonctions. Lect. Notes Math. 126. Springer, Berlin Heidelberg New York (1970)
— [2]: Microdifferential systems in the complex domain. Springer, Berlin Heidelberg New York Tokyo (1985)
— [3]: Microfunctions for boundary values problems. In: Algebraic Analysis, 809–819 (Papers Dedicated to M. Sato). M. Kashiwara, T. Kawai (Ed.). Academic Press (1988)
— [4]: Sheaf theory for partial differential equations. Proc. Inter. Cong. Math. (1990) To appear
Schapira, P. and Tose, N. [1]: Morse inequalities for R-constructible sheaves. Adv. in Math. To appear
Schapira, P. and Schneiders, J-P. [1]: Paires elliptiques I. Finitude et dualité. C.R. Acad. Sci. **311**, 83–86 (1990) or: Finitude et classes caractéristiques pour les D-modules et les faisceaux constructibles. Sém. Eq. Dér. Part. Publ. Ec. Polyt. (1989/90)
Schneiders, J-P. [1]: Un théorème de dualité relative pour les modules différentiels. C.R. Acad. Sci. **303**, 235–238 (1986)
Schwartz, L. [1]: Homomorphismes et applications complètement continues. C.R. Acad. Sci. **236**, 2472–2473 (1953)
— [2]: Théorie des distributions. Hermann, Paris (1966)
Schwartz, M-H. [1]: Classes caractéristiques définies par une stratification d'une variété analytique complexe. C.R. Acad. Sci. **260**, 3262–3264 and 3535–3537 (1965)
[SHS]: Sém. Heidelberg-Strasbourg (1966–67). Dualité de Poincaré. Publ. I.R.M.A. 3, Strasbourg (1969)
[SGA 2]: Sém. géométrie algébrique (1962), by Grothendieck, A. Cohomologie locale des faisceaux cohérents et théorèmes de Lefschetz locaux et globaux. North-Holland, Amsterdam (1968)
[SGA 4]: Sém. géométrie algébrique (1963–64), by Artin, M., Grothendieck, A. and Verdier, J-L. Théorie des topos et cohomologie étale des schémas. Lect. Notes Math. 269, 270, 305. Springer, Berlin Heidelberg New York (1972–73)

[SGA 4½]: Sem. géométrie algébrique, by Deligne, P. Cohomologie étale. Lect. Notes Math. 569. Springer, Berlin Heidelberg New York (1977)

[SGA 5]: Sém. géométrie algébrique (1965–66) by Grothendieck, A. Cohomologie l-adique et fonctions L. Lect. Notes in Math. 589. Springer, Berlin Heidelberg New York (1977)

[SGA 6]: Sém. géométrie algébrique (1966–67) by Berthelot, P. Illusie, L. and Grothendieck, A. Théorie des intersections et théorème de Riemann-Roch. Lect. Notes Math. 225. Springer, Berlin Heidelberg New York (1971)

[SGA 7]: Sém. géométrie algébrique (1967–69). Groupes de monodromie en géométrie algébrique. Part I by Grothendieck, A. Lect. Notes Math. 288. Springer, Berlin Heidelberg New York, (1972). Part II by Deligne P. and Katz, N. Lect. Notes Math. 340. Springer, Berlin Heidelberg New York (1973)

Serre, J-P. [1]: Faisceaux algébriques cohérents. Ann. Math. **61**, 197–278 (1955)

Sjöstrand, J. [1]: Singularités analytiques microlocales. Astérisque 95 (1982)

Sullivan, D. [1]: Combinatorial invariants of analytic spaces. Proceedings of Liverpool singularities symposium I. Lect. Notes Math. 192, 165–168. Springer, Berlin Heidelberg New York (1971)

Tamm, M. [1]: Subanalytic sets in the calculus of variations. Acta Math. **146**, 167–199 (1981).

Thom, R. [1]: Ensembles et morphismes stratifiés. Bull. A.M.S. **75**, 240–284 (1969)

Teissier, B. [1]: Sur la triangulation des morphismes sous-analytiques. Publ. Math. I.H.E.S. 70 (1989)

Tose, N. [1]: Propagation theorem for sheaves and applications to microdifferential systems. J. Math. Pures Appl. **68**, 137–151 (1989)

Trotman, D. [1]: Comparing regularity conditions on stratifications. Proc. Symp. Pure Math. **40**, 575–585 (1983)

— [2]: Une version microlocale de la condition (w) de Verdier. Ann. Inst. Fourier Grenoble **39**, 825–829 (1989)

Uchida, M. [1]: Microlocal analysis of diffraction at the corner of an obstacle. To appear, or: Sém. Eq. Dér. Part. Publ. Ec. Polyt. (1989/90)

Verdier, J-L. [1]: Dualité dans les espaces localement compacts. Sém. Bourbaki 300 (1965–66)

— [2]: Catégories dérivées, état 0. In [S.G.A. 4½]

— [3]: Stratifications de Whitney et théorème de Bertini-Sard. Invent. Math. 36, 295–312 (1976)

— [4]: Classe d'homologie associée à un cycle. In Sém. Géom. anal. A. Douady, J-L. Verdier (Ed.). Astérisque 36–37, 101–151 (1976)

— [5]: Spécialisation de faisceaux et monodromie modérée. In: Analyse et topologie sur les espaces singuliers. Astérisque 101–102, 332–364 (1981)

Wells, R.O. Jr [1]: Differential analysis on complex manifolds. Springer, New York Berlin Heidelberg (1980)

Whitney, H. [1]: Local properties of analytic varieties. In: Differential and combinatorial topology (A symposium in honor of Marston Morse). Princeton Univ. Press, 205–244 (1965)

— [2]: Tangents to an analytic variety. Ann. Math. **81**, 496–549 (1965)

— [3]: Complex analytic varieties. Addison Wesley, Reading Mass (1972)

Zerner, M. [1]: Domaine d'holomorphie des fonctions vérifiant une équation aux dérivées partielles. C.R. Acad. Sci. **272**, 1646–1648 (1971)

List of notations and conventions

General notations

$\mathbb{N}$:	set of non-negative integers
$\mathbb{Z}$:	ring of integers
$\mathbb{Q}$:	field of rational numbers
$\mathbb{R}$:	field of real numbers
$\mathbb{C}$:	field of complex numbers
$\mathbb{R}^+$:	multiplicative group of positive real numbers
$\mathbb{R}^-$:	the set of negative real numbers
$\mathbb{R}_{\geqslant 0}$ (resp. $\mathbb{R}_{\leqslant 0}$):	$\{c \in \mathbb{R};\ c \geqslant 0,$ (resp. $c \leqslant 0)\}$
$\mathbb{C}^\times$:	multiplicative group of non-zero complex numbers
$\# A$:	number of elements of a finite set A
$A \setminus B$:	the complementary set to B in A
$\delta_{i,j}$:	Kronecker symbol, $\delta_{i,j} = 0$ for $i \neq j$ and $\delta_{i,j} = 1$ for $i = j$
$\bar{S}$:	the closure of a subset S
∂S:	$\bar{S} \setminus S$ cf. (9.2.4)
$X \times_S Y$:	fiber product over S cf. Notations 2.3.13
$\mathbb{R}^n$:	Euclidian n-space
$\varinjlim$:	inductive limit
$\varprojlim$:	projective limit
"$\varinjlim$":	ind-object cf. I §11
"$\varprojlim$":	pro-object cf. I §11
$\{x_n\}_{n \in I}$:	a sequence indexed by I; ($I = \mathbb{N}$ or $\mathbb{Z}$)
$x_n \underset{n}{\to} x$:	the sequence $\{x_n\}_n$ converges to x
$\{\mathrm{pt}\}$:	the set consisting of a single element
$\square$:	means the square is Cartesian

Manifolds

$X, Y, \ldots$: real or complex manifolds

δ or $\delta_X : X \hookrightarrow X \times X$: the diagonal embedding

$\tau : TX \to X$: the tangent vector bundle to X

$\pi : T^*X \to X$: the cotangent vector bundle to X

$T_M X$: the normal vector bundle to a submanifold M of X cf. A.2.1

T_M^*X: the conormal vector bundle to a submanifold M of X cf. A.2.1
$T_X^*X \simeq X$: the zero section, identified to X
E_γ: the vector space E endowed with the γ-topology cf. III §5
X_γ: the space X endowed with the γ-topology cf. III §5
φ_γ: the continuous map $X \to X_\gamma$ cf. III §5
$f: Y \to X$: a morphism of manifolds
$f|_N: N \to M$: the morphism induced by f, $(N \subset Y, M \subset X)$
Tf: $TY \xrightarrow[f']{} T \times_X TX \xrightarrow[f_\tau]{} TX$ cf. (4.1.8)
$T_N f$: $T_N Y \xrightarrow[f'_N]{} N \times_M T_M X \xrightarrow[f_{N\tau}]{} T_M X$ cf. (4.1.9)
$T^*Y \xleftarrow[{}^t f']{} Y \times_X T^*X \xrightarrow[f_\pi]{} T^*X$ cf. (4.3.2)
$T_N^*Y \xleftarrow[{}^t f'_N]{} N \times_M T_M^*X \xrightarrow[f_{N\pi}]{} T_M^*X$ cf. (4.3.3)
T_Y^*X: $\ker({}^t f': Y \times_X T^*X \to T^*Y)$ cf. (4.3.4)
a_X: the map $X \to \{pt\}$
$\bar{X}$: the complex conjugate manifold associated to a complex manifold X cf. X §1
$X^{\mathbb{R}}$: the real underlying manifold to a complex manifold X cf. X §1
$\dim Y/X$, $\operatorname{codim}_X Y$: the relative dimension and codimension cf. Notation 3.3.8
$\dim X$, $\dim_{\mathbb{R}} X$: dimension of X cf. II §9 and Notation 3.3.8 (at the exception of XI §3)
$\dim_{\mathbb{C}} X$: complex dimension of X cf. II §9

Vector bundles

$\tau: E \to Z$: a vector bundle over Z
$\pi: E^* \to Z$: the dual vector bundle
Z is identified to the zero-section of a vector bundle over Z
$\dot{E} = E \backslash Z$, $\dot{E}^* = E^* \backslash Z$
$\dot{\tau} = \tau|_{\dot{E}}$, $\dot{\pi} = \pi|_{\dot{E}^*}$
a: the antipodal map on E
S^a: the image of S by a, $S \subset E$
S°: the polar set to S cf. (3.7.6)
e: the Euler vector field cf. V §5
$A + B$: sum in a vector bundle cf. (5.4.5)
$S = \dot{E}/\mathbb{R}^+$: sphere bundle associated to the vector bundle E cf. (3.6.3)
$T^*(E/Z)$: the relative cotangent bundle cf. (5.5.3)

Normal cones

$C_M(S)$: normal cone of S along M cf. Definition 4.1.1
$C(S_1, S_2)$: normal cone of $S_1 \times S_2$ along the diagonal cf. Definition 4.1.1

$C_\mu(A,B)$, $f^{\#}(A,B)$, $f_\infty^{\#}(A,B)$, $f^{\#}(A)$, $f_\infty^{\#}(A)$, $A \hat{+} B$, $A \underset{\infty}{\hat{+}} B$: cf. Definition 6.2.3
$N(S) = TX \backslash C(X \backslash S, S)$: cf. Definition 5.3.6
$N^*(S) = N(S)^\circ$: cf. Definition 5.3.6
$\tilde{X}_M$: normal deformation of M in X cf. IV §1

Symplectic geometry

σ: symplectic form cf. A §1
α_X or α: canonical 1-form on T^*X cf. A §2
H: Hamiltonian isomorphism cf. A §1
H_f: Hamiltonian vector field cf. A §2
$\{\cdot,\cdot\}$: Poisson bracket cf. A §2
$\tau(\cdot,\cdot,\cdot)$: inertia index cf. A §3
E^ρ: $\rho^\perp/\rho$, ρ isotropic in E cf. A §1
$\lambda^\rho = (\lambda \cap \rho^\perp + \rho)/\rho$, λ Lagrangian, ρ isotropic cf. A §1
$\lambda_1 \circ \lambda_2$: composition of Lagrangian planes cf. A §1
$\Lambda_1 \circ \Lambda_2$: composition of Lagrangian manifolds cf. Definition 7.4.5
Λ_φ: Lagrangian manifold associated to φ cf. (7.5.1)
$\lambda_\circ(p) = T_p \pi^{-1}\pi(p)$ cf. (7.5.2)
$\lambda_\Lambda(p) = T_p\Lambda$
τ_φ: cf. (7.5.3)
$\tau(\lambda_1 : \lambda_2)$: cf. (7.5.10)

Algebra

A: a ring (all rings are unitary)
A-module: left A-module (all modules are unitary)
A^{op}: the opposite ring to A (an A^{op}-module is a right A-module)
$\mathscr{C}$: a category, $\mathrm{Ob}(\mathscr{C})$, $\mathrm{Hom}_{\mathscr{C}}(\cdot,\cdot)$ cf. I §1
$\mathscr{C}^\circ$: the opposite category cf. I §1
id: the identity morphism
$\mathscr{C}^\vee$: the category of functors from $\mathscr{C}^\circ$ to $\mathfrak{Set}$
$\mathscr{C}^\wedge$: $\mathscr{C}^{\circ\vee\circ}$
Ker, Coker, Im, Coim: cf. I §2
$\mathbf{C}^*(\mathscr{C})$: $* = \emptyset, +, -, b$: categories of complexes of $\mathscr{C}$ cf. I §3
$\mathbf{K}^*(\mathscr{C})$: $* = \emptyset, +, -, b$: cf. Definition 1.3.4
$\mathrm{Ht}(X,Y)$: the group of morphisms from X to Y homotopic to zero cf. I §3
$\mathrm{K}(\mathscr{C})$: Grothendieck group of a category $\mathscr{C}$ cf. Exercise I.27
$X[k]$: translated complex cf. Definition 1.3.2
$\tau^{\leqslant}, \tau^{\geqslant}$: truncation functors cf. (1.3.10), (1.3.11), X §1
$H^k(X)$: k-th cohomology object of a complex X cf. Definition 1.3.5

$M(f)$: mapping cone of f cf. I §4
$X \longrightarrow Y \longrightarrow Z \xrightarrow{+1}$: triangle cf. Notation 1.5.8
$\mathscr{C}_S$: localization of $\mathscr{C}$ by S cf. Definition 1.6.2
$\mathscr{C}/N$: localization of $\mathscr{C}$ by N cf. Notation 1.6.8
$\mathbf{D}^*(\mathscr{C})$, $* = \emptyset, +, -, b$: derived categories cf. Definition 1.7.1
RF: right derived functor of F cf. Definition 1.8.1
LF: left derived functor of F cf. I §8
$\mathbf{D}^*_{\mathscr{C}'}(\mathscr{C})$: the full subcategory of $\mathbf{D}^*(\mathscr{C})$ consisting of complexes whose cohomology objects belong to $\mathscr{C}'$ cf. I §7
$H_I(X)$, $H_{II}(X)$, $s(X)$: complexes associated to a double complex cf. I §9
$N \otimes_A M$ or $N \otimes M$: tensor product (over a ring A)
$\mathrm{Hom}_A(N, M)$ or $\mathrm{Hom}(N, M)$: group of homomorphisms (over a ring A)
$\mathrm{Tor}_n^A(N, M) = H^{-n}(N \otimes_A^L M)$ cf. Example 1.10.12
$\mathrm{Ext}_A^n(N, M) = H^n(R\mathrm{Hom}_A(N, M))$
M-L : Mittag-Leffler cf. I §12
$\oplus$: direct sum cf. I §2
$X \times_Z Y$: product over Z cf. Exercise I.6
$X \oplus_Z Y$: direct sum over Z cf. Exercise I.6
$\mathrm{Ext}^j(X, Y) = \mathrm{Hom}_{\mathbf{D}(\mathscr{C})}(X, Y[j])$ cf. Exercise I.17
$\mathrm{hd}(\mathscr{C})$: homological dimension of $\mathscr{C}$ cf. Exercise I.17
$\mathrm{gld}(A)$: global homological dimension of A cf. Exercise I.28
$\mathrm{wgld}(A)$: weak global homological dimension of A cf. Exercise I.29
tr: trace cf. Exercise I.32
χ: Euler-Poincaré index cf. Exercise I.32
$\mathrm{b}_j(V) = \dim H^j(V)$ cf. (5.4.17)
$\mathrm{b}_j^*(V) = (-1)^j \sum_{k \leqslant j} (-1)^k \mathrm{b}_k(V)$ cf. Exercise I.34 and (5.4.18)

Sheaves

$F, G, H, \ldots$: sheaves, $\mathscr{R}$ a sheaf of rings on a space X
$\mathscr{H}om_{\mathscr{R}}(F, G)$ or $\mathscr{H}om(F, G)$: sheaf of $\mathscr{R}$-homomorphisms of F in G cf. Definition 2.2.7
$\mathrm{Hom}(F, G) = \Gamma(X; \mathscr{H}om(F, G))$
$\mathscr{R}^{\mathrm{op}}$: the opposite ring
$F \otimes_{\mathscr{R}} G$ or $F \otimes G$: tensor product sheaf of F and G (over $\mathscr{R}$) cf. Definition 2.2.8
$F|_Z$: inverse image of F on Z
$F_x = F|_{\{x\}}$: the stalk of F at x
$s|_Z$, s_x: the restriction of a section s to Z and the germ of s at x
$\mathrm{supp}(s)$: the support of a section s
$\Gamma(X; F)$: global section of F on X
$\Gamma(Z; F) = \Gamma(Z; F|_Z)$
$f^{-1}F$: inverse image of a sheaf F cf. Definition 2.3.1
f_*F: direct image of a sheaf F cf. Definition 2.3.1

$f_! F$: direct image with proper supports cf. (2.5.1)
$f^{\#}$: see Definition 2.7.4
F_Z: sheaf on X such that $F_Z|_Z = F|_Z, F_Z|_{X \setminus Z} = 0$, Z locally closed cf. II §3
$\Gamma_Z(F)$: subsheaf of F consisting of sections supported by Z cf. II §3
$M_X = a_X^{-1} M$: constant sheaf on X with stalk M (M: an A-module)
$M_Z = (M_X)_Z$, $Z \subset X$, Z locally closed
$F \boxtimes_S G$: external tensor product (a sheaf on $X \times_S Y$) cf. Notation 2.3.12
$\Gamma_c(X; F) = a_{X!} F$: global section with compact supports cf. (2.5.2)
$Rf_*, R\Gamma_Z, R\Gamma(X, \cdot), Rf_!, \otimes^L, \boxtimes^L, R\mathcal{H}om$: derived functors of the preceding ones cf. II §6
$H_Z^j(F), H_Z^j(X; F), H_c^j(X; F), \mathcal{E}xt_{\mathcal{R}}^j(F, G), \mathrm{Ext}_{\mathcal{R}}^j(F, G), \mathcal{T}or_j^{\mathcal{R}}(F, G)$: cf. Notation 2.6.8
$\mathrm{wgld}(\mathcal{R})$: cf. Definition 2.6.2
$\mathrm{supp}(F) =$ closure of $\bigcup_j \mathrm{supp}\, H^j(F)$ cf. (2.6.34) and II §2
$\mathcal{C}^\cdot(\mathcal{U}; F)$: Čech complex associated to a family of open subsets cf. II §8
$f^!$: right adjoint of $Rf_!$ cf. III §1
$\mathrm{D}_X F = R\mathcal{H}om(F, \omega_X)$ cf. Definition 3.1.16
$\mathrm{D}'_X F = R\mathcal{H}om(F, A_X)$ cf. Definition 3.1.16
$\int_X$: the morphism $H_c^n(X; or_X) \to A$ cf. (3.3.15)
$F^\wedge$: Fourier-Sato transform cf. Definition 3.7.8
$F^\vee$: inverse Fourier-Sato transform cf. Definition 3.7.8
Φ_K, Ψ_K: functors associated to the kernel K cf. Definitions 3.6.1 and 7.1.3
$K_1 \circ K_2$: composition of kernels, cf. (3.6.2) and Proposition 7.1.2
$K_1 \circ_\mu K_2$: microlocal composition of kernels cf. Definition 7.3.2
$\nu_M(F)$: specialisation of F along M cf. Definition 4.2.2
$\mu_M(F)$: microlocalization of F along M cf. Definition 4.3.1
$f_\mu^{-1}, f_\mu^!, f_*^\mu, f_!^\mu$: microlocal operations cf. VI §1
$\mu hom(G \to F), \mu hom(F \leftarrow G), \mu hom(F, G)$: microlocalisation functors cf. Definition 4.4.1
$\mathrm{SS}(F)$: micro-support of F cf. V §1
$\mathbf{D}^*(X) = \mathbf{D}^*(A_X), (* = \varnothing, +, b, -)$: derived category of the category of sheaves of A-modules cf. II §6
$N(X, Y; \Omega_X, \Omega_Y)$: category of kernels cf. Definition 7.1.1
$\mathbf{N}(X, Y; p_X, p_Y)$: category of kernels cf. Definition 7.3.7
ϕ_f: vanishing-cycle functor cf. VIII §6
Ψ_f: nearby-cycle functor cf. VIII §6
${}^pH^k$: perverse cohomology cf. X §2

Special sheaves

or_X: orientation sheaf cf. Definition 3.3.3
$or_{Y/X}$: relative orientation sheaf, cf. (3.3.3)
ω_X: dualizing complex cf. Definition 3.1.16
$\omega_{Y/X}$: relative dualizing complex cf. Definition 3.1.16

$\mathscr{D}\mathscr{b}_M$: sheaf of distributions cf. II §9.6
$\mathscr{B}_M$: sheaf of hyperfunctions cf. II §9.6 and Definition 10.5.1
$\mathscr{V}_M$: sheaf of densities on a manifold cf. II §9.5
$\mathscr{C}_M$: sheaf of microfunctions cf. Definition 11.5.1

Sheaves on complex manifolds

$\mathcal{O}_X$: sheaf of holomorphic functions on X
$\mathcal{O}_X^{(p)}$: sheaf of holomorphic p-forms
Ω_X: $\mathcal{O}_X^{(n)} \otimes or_X$ $(n = \dim_{\mathbb{C}} X)$
$\mathscr{D}_X$: sheaf of rings of holomorphic differential operators of finite order
$\mathscr{D}_{Y \to X}$: bimodule of differential operators from Y to X cf. Definition 11.2.8
$\mathrm{Icar}(\mathscr{M})$: cf. (11.2.7)
$\mathrm{char}(\mathscr{M})$: characteristic variety of a $\mathscr{D}_X$-module $\mathscr{M}$ cf. (11.2.14)
$\mathscr{K}_X = \mathscr{D}_X \otimes_{\mathcal{O}_X} \Omega_X^{\otimes -1}[\dim_{\mathbb{C}} X]$ cf. (11.2.18)
$\mathscr{N}^* = \mathrm{R}\mathscr{H}om_{\mathscr{D}_X}(\mathscr{N}, \mathscr{K}_X)$ cf. (11.2.19)
$\boxtimes$,: external tensor product in the category of $\mathscr{D}_X$-modules cf. (11.2.21)
$\underline{f}^{-1}\mathscr{M}$: inverse image in the category of $\mathscr{D}_X$-modules cf. Definition 11.2.10
$\underline{f}_*\mathscr{N}$, $\underline{f}_!\mathscr{N}$: direct image in the category of $\mathscr{D}_X$-modules cf. Definition 11.2.10
$\mathscr{E}_X^{\mathbb{R}}$, $\mathscr{E}_{Y \to X}^{\mathbb{R}}$, $\mathscr{E}_{X \leftarrow Y}^{\mathbb{R}}$: ring and bimodules of holomorphic microlocal operators cf. Definition 11.4.2 and Proposition 11.4.3
$\mathscr{C}_{S|X}^{\mathbb{R}}$: sheaf of microfunctions on a complex submanifold S cf. Definition 11.4.2

Categories

$\mathfrak{Set}$: category of sets
$\mathfrak{Ab}$: category of abelian groups
$\mathfrak{Mod}(A)$: abelian category of left A-modules
$\mathfrak{Mod}^f(A)$: category of finitely generated left A-modules
$\mathfrak{Mod}(\mathscr{R})$: category of sheaves of $\mathscr{R}$-modules ($\mathscr{R}$: a sheaf of rings on X)
$\mathfrak{Mod}(A_X)$: category of sheaves of A-modules on X
$\mathfrak{Sh}(X) = \mathfrak{Mod}(\mathbb{Z}_X)$: category of sheaves of abelian groups on X
$\mathbf{D}(A) = \mathbf{D}(\mathfrak{Mod}(A))$ cf. Notations 1.7.14 and 2.6.1
$\mathbf{D}(X) = \mathbf{D}(A_X) = \mathbf{D}(\mathfrak{Mod}(A_X))$ cf. Notations 2.6.11
$\mathbf{D}^b(X;\Omega)$: localization of $\mathbf{D}^b(X)$ on $\Omega \subset T^*X$ cf. VI §1
$\mathbf{D}^b(X;p) = \mathbf{D}^b(X;\{p\})$: localization of $\mathbf{D}^b(X)$ at p cf. VI §1
$\mathbf{D}^+_{\mathbb{R}^+}(E)$: subcategory of $\mathbf{D}^+(E)$ consisting of conic objects cf. Definition 3.7.1
$\mathfrak{Cons}(\mathrm{S})$, w-$\mathfrak{Cons}(\mathrm{S})$: categories of constructible and weakly constructible sheaves on S cf. VIII §1

$\mathbf{D}^b_{\mathsf{S}-c}(\mathsf{S})$ and $\mathbf{D}^b_{w-\mathsf{S}-c}(\mathsf{S})$: subcategories of $\mathbf{D}^b(|\mathsf{S}|)$ consisting of objects with constructible and weakly constructible cohomologies cf. VIII §1

$\mathbb{R}$-$\mathfrak{Cons}(X)$ and w-$\mathbb{R}$-$\mathfrak{Cons}(X)$: categories of $\mathbb{R}$-constructible and weakly $\mathbb{R}$-constructible sheaves on X cf. Definition 8.4.3

$\mathbf{D}^b_{\mathbb{R}-c}(X)$ and $\mathbf{D}^b_{w-\mathbb{R}-c}(X)$: subcategories of $\mathbf{D}^b(X)$ consisting of objects with $\mathbb{R}$-constructible and weakly $\mathbb{R}$-constructible cohomology cf. VIII §4

$\mathbf{D}^b_{\mathbb{C}-c}(X)$ and $\mathbf{D}^b_{w-\mathbb{C}-c}(X)$: cf. VIII §5

$\mathrm{K}_{\mathbb{R}-c}(X)$: Grothendieck group of $\mathbf{D}^b_{\mathbb{R}-c}(X)$ cf. IX §7

${}^p\mathbf{D}^{\leq 0}_{w-\mathbb{R}-c}(X)$, ${}^p\mathbf{D}^{\geq 0}_{w-\mathbb{R}-c}(X)$, etc.: t-structure associated to a perversity p cf. X §2 and §3

${}^\mu\mathbf{D}^{\leq 0}_{w-\mathbb{R}-c}(X)$, ${}^\mu\mathbf{D}^{\geq 0}_{w-\mathbb{R}-c}(X)$, etc.: t-structure defined microlocaly cf. X §3

Cycles, traces and constructible functions

$\chi(F)(x)$, $\chi_c(F)(x)$, $\chi(X;F)$, $\chi_c(X;F)$: Euler-Poincaré indices cf. IX §1

tr_X: trace morphism cf. (9.1.3)

$\int_X$: $H^0_c(X,\omega_X)\to k$ cf. III §3 and IX §1

$\mathscr{CS}_p(F)$: sheaf of subanalytic p-chains with values in F cf. Definition 9.2.1

$\mathscr{CS}_p^X = \mathscr{CS}_p(A_X) = \mathscr{CS}_p$

$\partial_p:\mathscr{CS}_p\to\mathscr{CS}_{p-1}$: boundary operator cf. (9.2.10)

$\mathscr{ZS}_p^X$: sheaf of subanalytic p-cycles cf. Definition 9.2.5

$C_1\cap C_2$: intersection of two cycles cf. Definition 9.2.12

$\#(C_1\cap C_2)$: intersection number of two cycles cf. Definition 9.2.12

$\mathscr{L}_X$: sheaf of Lagrangian cycles cf. Definition 9.3.1

f^*, f_*: inverse and direct images of Lagrangian cycles cf. Definition 9.3.3

$\boxtimes$: external product of cycles cf. (9.3.2)

$[T^*_Y X]$: the Lagrangian cycle associated to $Y\hookrightarrow X$ cf. Example 9.3.4

$[\sigma_0]$: cycle associated to the zero section cf. Definition 9.3.5

Index

Grundlehren der mathematischen Wissenschaften

A Series of Comprehensive Studies in Mathematics

A Selection

190. Faith: Algebra: Rings, Modules, and Categories I
191. Faith: Algebra II, Ring Theory
192. Mal'cev: Algebraic Systems
193. Pólya/Szegö: Problems and Theorems in Analysis I
194. Igusa: Theta Functions
195. Berberian: Baer*-Rings
196. Arthreya/Ney: Branching Processes
197. Benz: Vorlesungen über Geometrie der Algebren
198. Gaal: Linear Analysis and Representation Theory
199. Nitsche: Vorlesungen über Minimalflächen
200. Dold: Lectures on Algebraic Topology
201. Beck: Continuous Flows in the Plane
202. Schmetterer: Introduction to Mathematical Statistics
203. Schoeneberg: Elliptic Modular Functions
204. Popov: Hyperstability of Control Systems
205. Nikol'skiĭ: Approximation of Functions of Several Variables and Imbedding Theorems
206. André: Homologie des Algébres Commutatives
207. Donoghue: Monotone Matrix Functions and Analytic Continuation
208. Lacey: The Isometric Theory of Classical Banach Spaces
209. Ringel: Map Color Theorem
210. Gihman/Skorohod: The Theory of Stochastic Processes I
211. Comfort/Negrepontis: The Theory of Ultrafilters
212. Switzer: Algebraic Topology – Homotopy and Homology
215. Schaefer: Banach Lattices and Positive Operators
217. Stenström: Rings of Quotients
218. Gihman/Skorohod: The Theory of Stochastic Processes II
219. Duvant/Lions: Inequalities in Mechanics and Physics
220. Kirillov: Elements of the Theory of Representations
221. Mumford: Algebraic Geometry I: Complex Projective Varieties
222. Lang: Introduction to Modular Forms
223. Bergh/Löfström: Interpolation Spaces. An Introduction
224. Gilbarg/Trudinger: Elliptic Partial Differential Equations of Second order
225. Schütte: Proof Theory
226. Karoubi: K-Theory. An Introduction
227. Grauert/Remmert: Theorie der Steinschen Räume
228. Segal/Kunze: Integrals and Operators
229. Hasse: Number Theory
230. Klingenberg: Lectures on Closed Geodesics
231. Lang: Elliptic Curves: Diophantine Analysis
232. Gihman/Skorohod: The Theory of Stochastic Processes III
233. Stroock/Varadhan: Multidimensional Diffusion Processes
234. Aigner: Combinatorial Theory
235. Dynkin/Yushkevich: Controlled Markov Processes
236. Grauert/Remmert: Theory of Stein Spaces
237. Köthe: Topological Vector Spaces II
238. Graham/McGehee: Essays in Commutative Harmonic Analysis
239. Elliott: Probabilistic Number Theory I
240. Eliott: Probabilistic Number Theory II

241. Rudin: Function Theory in the Unit Ball of C^n
242. Huppert/Blackburn: Finite Groups II
243. Huppert/Blackburn: Finite Groups III
244. Kubert/Lang: Modular Units
245. Cornfeld/Fomin/Sinai: Ergodic Theory
246. Naimark/Štern: Theory of Group Representations
247. Suzuki: Group Theory I
248. Suzuki: Group Theory II
249. Chung: Lectures from Markov Processes to Brownian Motion
250. Arnold: Geometrical Methods in the Theory of Ordinary Differential Equations
251. Chow/Hale: Methods of Bifurcation Theory
252. Aubin: Nonlinear Analysis on Manifolds. Monge-Ampère Equations
253. Dwork: Lectures on p-adic Differential Equations
254. Freitag: Siegelsche Modulfunktionen
255. Lang: Complex Multiplication
256. Hörmander: The Analysis of Linear Partial Differential Operators I
257. Hörmander: The Analysis of Linear Partial Differential Operators II
258. Smoller: Shock Waves and Reaction-Diffusion Equations
259. Duren: Univalent Functions
260. Freidlin/Wentzell: Random Perturbations of Dynamical Systems
261. Bosch/Güntzer/Remmert: Non Archimedian Analysis – A Systematic Approach to Rigid Analytic Geometry
262. Doob: Classical Potential Theory and Its Probabilistic Counterpart
263. Krasnosel'skiĭ/Zabreĭko: Geometrical Methods of Nonlinear Analysis
264. Aubin/Cellina: Differential Inclusions
265. Grauert/Remmert: Coherent Analytic Sheaves
266. de Rham: Differentiable Manifolds
267. Arbarello/Cornalba/Griffiths/Harris: Geometry of Algebraic Curves, Vol. I
268. Arbarello/Cornalba/Griffiths/Harris: Geometry of Algebraic Curves, Vol. II
269. Schapira: Microdifferential Systems in the Complex Domain
270. Scharlau: Quadratic and Hermitian Forms
271. Ellis: Entropy, Large Deviations, and Statistical Mechanics
272. Elliott: Arithmetic Functions and Integer Products
273. Nikol'skiĭ: Treatise on the Shift Operator
274. Hörmander: The Analysis of Linear Partial Differential Operators III
275. Hörmander: The Analysis of Linear Partial Differential Operators IV
276. Liggett: Interacting Particle Systems
277. Fulton/Lang: Riemann-Roch Algebra
278. Barr/Wells: Toposes, Triples and Theories
279. Bishop/Bridges: Constructive Analysis
280. Neukirch: Class Field Theory
281. Chandrasekharan: Elliptic Functions
282. Lelong/Gruman: Entire Functions of Several Complex Variables
283. Kodaira: Complex Manifolds and Deformation of Complex Structures
284. Finn: Equilibrium Capillary Surfaces
285. Burago/Zalgaller: Geometric Inequalities
286. Andrianov: Quadratic Forms and Hecke Operators
287. Maskit: Kleinian Groups
288. Jacod/Shiryaev: Limit Theorems for Stochastic Processes
289. Manin: Gauge Field Theory and Complex Geometry
290. Conway/Sloane: Sphere Packings, Lattices and Groups
291. Hahn/O'Meara: The Classical Groups and K-Theory
292. Kashiwara/Schapira: Sheaves on Manifolds
293. Revuz/Yor: Continuous Martingales and Brownian Motion
294. Knus: Quadratic and Hermitian Forms over Rings

www.ingramcontent.com/pod-product-compliance
Lightning Source LLC
Chambersburg PA
CBHW072017011225
36163CB00003B/192
* 9 7 8 3 5 4 0 5 1 8 6 1 7 *